More information about this subseries at http://www.springer.com/series/7407

Lecture Notes in Computer Science

Founding Editors

Gerhard Goos
Karlsruhe Institute of Technology, Karlsruhe, Germany

Juris Hartmanis
Cornell University, Ithaca, NY, USA

Editorial Board Members

Elisa Bertino
Purdue University, West Lafayette, IN, USA

Wen Gao
Peking University, Beijing, China

Bernhard Steffen
TU Dortmund University, Dortmund, Germany

Gerhard Woeginger
RWTH Aachen, Aachen, Germany

Moti Yung
Columbia University, New York, NY, USA

Igor Farkaš · Paolo Masulli ·
Sebastian Otte · Stefan Wermter (Eds.)

Artificial Neural Networks and Machine Learning – ICANN 2021

30th International Conference on Artificial Neural Networks
Bratislava, Slovakia, September 14–17, 2021
Proceedings, Part V

Springer

Editors
Igor Farkaš (ID)
Comenius University in Bratislava
Bratislava, Slovakia

Paolo Masulli (ID)
iMotions A/S
Copenhagen, Denmark

Sebastian Otte (ID)
University of Tübingen
Tübingen, Baden-Württemberg, Germany

Stefan Wermter (ID)
Universität Hamburg
Hamburg, Germany

ISSN 0302-9743   ISSN 1611-3349   (electronic)
Lecture Notes in Computer Science
ISBN 978-3-030-86382-1   ISBN 978-3-030-86383-8   (eBook)
https://doi.org/10.1007/978-3-030-86383-8

LNCS Sublibrary: SL1 – Theoretical Computer Science and General Issues

This Springer imprint is published by the registered company Springer Nature Switzerland AG
The registered company address is: Gewerbestrasse 11, 6330 Cham, Switzerland

Preface

Research on artificial neural networks has progressed over decades, in recent years being fueled especially by deep learning that has proven, albeit data-greedy, efficient in solving various, mostly supervised, tasks. Applications of artificial neural networks, especially related to artificial intelligence, affect our lives, providing new horizons. Examples range from autonomous car driving, virtual assistants, and decision support systems to healthcare data analytics, financial forecasting, and smart devices in our homes, just to name a few. These developments, however, also provide challenges, which were not imaginable previously, e.g., verification of raw data, explaining the contents of neural networks, and adversarial machine learning.

The International Conference on Artificial Neural Networks (ICANN) is the annual flagship conference of the European Neural Network Society (ENNS). Last year, due to the COVID-19 pandemic, we decided not to hold the conference but to prepare the ICANN proceedings in written form. This year, due to the still unresolved pandemic, the Organizing Committee, together with the Executive Committee of ENNS decided to organize ICANN 2021 online, since we felt the urge to allow research presentations and live discussions, following the now available alternatives of online conference organization. So for the first time, ENNS and the Organizing Committee prepared ICANN as an online event with all its challenges and sometimes unforeseeable events!

Following a long-standing successful collaboration, the proceedings of ICANN are published as volumes within the Lecture Notes in Computer Science Springer series. The response to this year's call for papers resulted, unexpectedly, in a record number of 557 article submissions (a 46% rise compared to previous year), of which almost all were full papers. The paper selection and review process that followed was decided during the online meeting of the Bratislava organizing team and the ENNS Executive Committee. The 40 Program Committee (PC) members agreed to check the submissions for the formal requirements and 64 papers were excluded from the subsequent reviews. The majority of the PC members have doctoral degrees (80%) and 75% of them are also professors. We also took advantage of filled-in online questionnaires providing the reviewers' areas of expertise. The reviewers were assigned one to four papers, and the papers with undecided scores also received reports from PC members which helped in making a final decision.

In total, 265 articles were accepted for the proceedings and the authors were requested to submit final versions. The acceptance rate was hence about 47% when calculated from all initial submissions. A list of PC members and reviewers who agreed to publish their names is included in the proceedings. With these procedures we tried to keep the quality of the proceedings high while still having a critical mass of contributions reflecting the progress of the field. Overall we hope that these proceedings will contribute to the dissemination of new results by the neural network community during these challenging times and we hope that we can have a physical ICANN in 2022.

Finally, we very much thank the Program Committee and the reviewers for their invaluable work.

September 2021

Igor Farkaš
Paolo Masulli
Sebastian Otte
Stefan Wermter

Organization

Organizing Committee

Cabessa Jérémie	Université Paris 2 Panthéon-Assas, France
Kerzel Matthias	University of Hamburg, Germany
Lintas Alessandra	University of Lausanne, Switzerland
Malinovská Kristína	Comenius University in Bratislava, Slovakia
Masulli Paolo	iMotions A/S, Copenhagen, Denmark
Otte Sebastian	University of Tübingen, Germany
Wedeman Roseli	Universidade do Estado do Rio de Janeiro, Brazil

Program Committee Chairs

Igor Farkaš	Comenius University in Bratislava, Slovakia
Paolo Masulli	iMotions A/S, Denmark
Sebastian Otte	University of Tübingen, Germany
Stefan Wermter	University of Hamburg, Germany

Program Committee

Andrejková Gabriela	Pavol Jozef Šafárik University in Košice, Slovakia
Atencia Miguel	Universidad de Malaga, Spain
Bodapati Jyostna Devi	Indian Institute of Technology, Madras, India
Bougie Nicolas	Sokendai/National Institute of Informatics, Japan
Boža Vladimír	Comenius University in Bratislava, Slovakia
Cabessa Jérémie	Université Paris 2 Panthéon-Assas, France
Di Nuovo Alessandro	Sheffield Hallam University, UK
Duch Włodzisław	Nicolaus Copernicus University, Poland
Eppe Manfred	Universität Hamburg, Germany
Fang Yuchun	Shanghai University, China
Garcke Jochen	Universität Bonn, Germany
Gregor Michal	University of Žilina, Slovakia
Guckert Michael	Technische Hochschule Mittelhessen, Germany
Guillén Alberto	University of Granada, Spain
Heinrich Stefan	University of Tokyo, Japan
Hinaut Xavier	Inria, France
Humaidan Dania	University of Tübingen, Germany
Jolivet Renaud	University of Geneva, Switzerland
Koprinkova-Hristova Petia	Bulgarian Academy of Sciences, Bulgaria
Lintas Alessandra	University of Lausanne, Switzerland
Lü Shuai	Jilin University, China
Micheli Alessio	Università di Pisa, Italy

Oravec Miloš	Slovak University of Technology in Bratislava, Slovakia
Otte Sebastian	University of Tübingen, Germany
Peltonen Jaakko	Tampere University, Finland
Piuri Vincenzo	University of Milan, Italy
Pons Rivero Antonio Javier	Universitat Politècnica de Catalunya, Barcelona, Spain
Schmidt Jochen	TH Rosenheim, Germany
Schockaert Cedric	Paul Wurth S.A., Luxembourg
Schwenker Friedhelm	University of Ulm, Germany
Takáč Martin	Comenius University in Bratislava, Slovakia
Tartaglione Enzo	Università degli Studi di Torino, Italy
Tetko Igor	Helmholtz Zentrum München, Germany
Triesch Jochen	Frankfurt Institute for Advanced Studies, Germany
Vavrečka Michal	Czech Technical University in Prague, Czech Republic
Verma Sagar	CentraleSupélec, Université Paris-Saclay, France
Vigário Ricardo	Nova School of Science and Technology, Portugal
Wedemann Roseli	Universidade do Estado do Rio de Janeiro, Brazil
Wennekers Thomas	Plymouth University, UK

Reviewers

Abawi Fares	University of Hamburg, Germany
Aganian Dustin	Technical University Ilmenau, Germany
Ahrens Kyra	University of Hamburg, Germany
Alexandre Frederic	Inria Bordeaux, France
Alexandre Luís	University of Beira Interior, Portugal
Ali Hazrat	Umeå University, Sweden
Alkhamaiseh Koloud	Western Michigan University, USA
Amaba Takafumi	Fukuoka University, Japan
Ambita Ara Abigail	University of the Philippines Diliman, Philippines
Ameur Hanen	University of Sfax, Tunisia
Amigo Galán Glauco A.	Baylor University, USA
An Shuqi	Chongqing University, China
Aouiti Chaouki	Université de Carthage, Tunisia
Arany Adam	Katholieke Universiteit Leuven, Belgium
Arnold Joshua	University of Queensland, Australia
Artelt André	Bielefeld University, Germany
Auge Daniel	Technical University of Munich, Germany
Bac Le Hoai	University of Science, Vietnam
Bacaicoa-Barber Daniel	University Carlos III of Madrid, Spain
Bai Xinyi	National University of Defense Technology, China
Banka Asif	Islamic University of Science & Technology, India
Basalla Marcus	University of Liechtenstein, Liechtenstein
Basterrech Sebastian	Technical University of Ostrava, Czech Republic
Bauckhage Christian	Fraunhofer IAIS, Germany
Bayer Markus	Technical University of Darmstadt, Germany

Bečková Iveta	Comenius University in Bratislava, Slovakia
Benalcázar Marco	Escuela Politécnica Nacional, Ecuador
Bennis Achraf	Institut de Recherche en Informatique de Toulouse, France
Berlemont Samuel	Orange Labs, Grenoble, France
Bermeitinger Bernhard	Universität St. Gallen, Switzerland
Bhoi Suman	National University of Singapore, Singapore
Biesner David	Fraunhofer IAIS, Germany
Bilbrey Jenna	Pacific Northwest National Lab, USA
Blasingame Zander	Clarkson University, USA
Bochkarev Vladimir	Kazan Federal University, Russia
Bohte Sander	Universiteit van Amsterdam, The Netherlands
Bouchachia Abdelhamid	Bournemouth University, UK
Bourguin Grégory	Université du Littoral Côte d'Opale, France
Breckon Toby	Durham University, UK
Buhl Fred	University of Florida, USA
Butz Martin V.	University of Tübingen, Germany
Caillon Paul	Université de Lorraine, Nancy, France
Camacho Hugo C. E.	Universidad Autónoma de Tamaulipas, Mexico
Camurri Antonio	Università di Genova, Italy
Cao Hexin	OneConnect Financial Technology, China
Cao Tianyang	Peking University, China
Cao Zhijie	Shanghai Jiao Tong University, China
Carneiro Hugo	Universität Hamburg, Germany
Chadha Gavneet Singh	South Westphalia University of Applied Sciences, Germany
Chakraborty Saikat	C. V. Raman Global University, India
Chang Hao-Yuan	University of California, Los Angeles, USA
Chang Haodong	University of Technology Sydney, Australia
Chen Cheng	Tsinghua University, China
Chen Haopeng	Shanghai Jiao Tong University, China
Chen Junliang	Shenzhen University, China
Chen Tianyu	Northwest Normal University, China
Chen Wenjie	Communication University of China, China
Cheng Zhanglin	Chinese Academy of Sciences, China
Chenu Alexandre	Sorbonne Université, France
Choi Heeyoul	Handong Global University, South Korea
Christa Sharon	RV Institute of Technology and Management, India
Cîtea Ingrid	Bitdefender Central, Romania
Colliri Tiago	Universidade de São Paulo, Brazil
Cong Cong	Chinese Academy of Sciences, China
Coroiu Adriana Mihaela	Babes-Bolyai University, Romania
Cortez Paulo	University of Minho, Portugal
Cuayáhuitl Heriberto	University of Lincoln, UK
Cui Xiaohui	Wuhan University, China
Cutsuridis Vassilis	University of Lincoln, UK

Cvejoski Kostadin	Fraunhofer IAIS, Germany
D'Souza Meenakshi	International Institute of Information Technology, Bangalore, India
Dai Feifei	Chinese Academy of Sciences, China
Dai Peilun	Boston University, USA
Dai Ruiqi	INSA Lyon, France
Dang Kai	Nankai University, China
Dang Xuan	Tsinghua University, China
Dash Tirtharaj	Birla Institute of Technology and Science, Pilani, India
Davalas Charalampos	Harokopio University of Athens, Greece
De Brouwer Edward	Katholieke Universiteit Leuven, Belgium
Deng Minghua	Peking University, China
Devamane Shridhar	KLE Institute of Technology, Hubballi, India
Di Caterina Gaetano	University of Strathclyde, UK
Di Sarli Daniele	Università di Pisa, Italy
Ding Juncheng	University of North Texas, USA
Ding Zhaoyun	National University of Defense Technology, China
Dold Dominik	Siemens, Munich, Germany
Dong Zihao	Jinan University, China
Du Songlin	Southeast University, China
Edwards Joshua	University of North Carolina Wilmington, USA
Eguchi Shu	Fukuoka University, Japan
Eisenbach Markus	Ilmenau University of Technology, Germany
Erlhagen Wolfram	University of Minho, Portugal
Fang Tiyu	University of Jinan, China
Feldager Cilie	Technical University of Denmark, Denmark
Ferianc Martin	University College London, UK
Ferreira Flora	University of Minho, Portugal
Fevens Thomas	Concordia University, Canada
Friedjungová Magda	Czech Technical University in Prague, Czech Republic
Fu Xianghua	Shenzhen University, China
Fuhl Wolfgang	Universität Tübingen, Germany
Gamage Vihanga	Technological University Dublin, Ireland
Ganguly Udayan	Indian Institute of Technology, Bombay, India
Gao Ruijun	Tianjin University, China
Gao Yapeng	University of Tübingen, Germany
Gao Yue	Beijing University of Posts and Telecommunications, China
Gao Zikai	National University of Defense Technology, China
Gault Richard	Queen's University Belfast, UK
Ge Liang	Chongqing University, China
Geissler Dominik	Relayr GmbH, Munich, Germany
Gepperth Alexander	ENSTA ParisTech, France
Gerum Christoph	University of Tübingen, Germany
Giancaterino Claudio G.	Catholic University of Milan, Italy
Giese Martin	University Clinic Tübingen, Germany

Gikunda Patrick	Dedan Kimathi University of Technology, Kenya
Goel Anmol	Guru Gobind Singh Indraprastha University, India
Göpfert Christina	Bielefeld University, Germany
Göpfert Jan Philip	Bielefeld University, Germany
Goyal Nidhi	Indraprastha Institute of Information Technology, India
Grangetto Marco	Università di Torino, Italy
Grüning Philipp	University of Lübeck, Germany
Gu Xiaoyan	Chinese Academy of Sciences, Beijing, China
Guo Hongcun	China Three Gorges University, China
Guo Ling	Northwest University, China
Guo Qing	Nanyang Technological University, Singapore
Guo Song	Xi'an University of Architecture and Technology, China
Gupta Sohan	Global Institute of Technology, Jaipur, India
Hakenes Simon	Ruhr-Universität Bochum, Germany
Han Fuchang	Central South University, China
Han Yi	University of Melbourne, Australia
Hansen Lars Kai	Technical University of Denmark, Denmark
Haque Ayaan	Saratoga High School, USA
Hassen Alan Kai	Leiden University, The Netherlands
Hauberg Søren	Technical University of Denmark, Denmark
He Tieke	Nanjing University, China
He Wei	Nanyang Technological University, Singapore
He Ziwen	Chinese Academy of Sciences, China
Heese Raoul	Fraunhofer ITWM, Germany
Herman Pawel	KTH Royal Institute of Technology, Sweden
Holas Juraj	Comenius University in Bratislava, Slovakia
Horio Yoshihiko	Tohoku University, Japan
Hou Hongxu	Inner Mongolia University, China
Hu Ming-Fei	China University of Petroleum, China
Hu Ting	Hasso Plattner Institute, Germany
Hu Wenxin	East China Normal University, China
Hu Yanqing	Sichuan University, China
Huang Chenglong	National University of Defense Technology, China
Huang Chengqiang	Huawei Technology, Ltd., China
Huang Jun	Chinese Academy of Sciences, Shanghai, China
Huang Ruoran	Chinese Academy of Sciences, China
Huang Wuliang	Chinese Academy of Sciences, Beijing, China
Huang Zhongzhan	Tsinghua University, China
Iannella Nicolangelo	University of Oslo, Norway
Ienco Dino	INRAE, France
Illium Steffen	Ludwig-Maximilians-Universität München, Germany
Iyer Naresh	GE Research, USA
Jalalvand Azarakhsh	Ghent University, Belgium
Japa Sai Sharath	Southern Illinois University, USA
Javaid Muhammad Usama	Eura Nova, Belgium

Jia Qiaomei	Northwest University, China
Jia Xiaoning	Inner Mongolia University, China
Jin Peiquan	University of Science and Technology of China, China
Jirak Doreen	Istituto Italiano di Tecnologia, Italy
Jodelet Quentin	Tokyo Institute of Technology, Japan
Kai Tang	Toshiba, China
Karam Ralph	Université Franche-Comté, France
Karlbauer Matthias	University of Tübingen, Germany
Kaufhold Marc-André	Technical University of Darmstadt, Germany
Kerzel Matthias	University of Hamburg, Germany
Keurulainen Antti	Bitville Oy, Finland
Kitamura Takuya	National Institute of Technology, Japan
Kocur Viktor	Comenius University in Bratislava, Slovakia
Koike Atsushi	National Institute of Technology, Japan
Kotropoulos Constantine	Aristotle University of Thessaloniki, Greece
Kovalenko Alexander	Czech Technical University, Czech Republic
Krzyzak Adam	Concordia University, Canada
Kurikawa Tomoki	Kansai Medical University, Japan
Kurpiewski Evan	University of North Carolina Wilmington, USA
Kurt Mehmet Necip	Columbia University, USA
Kushwaha Sumit	Kamla Nehru Institute of Technology, India
Lai Zhiping	Fudan University, China
Lang Jana	Hertie Institute for Clinical Brain Research, Germany
Le Hieu	Boston University, USA
Le Ngoc	Hanoi University of Science and Technology, Vietnam
Le Thanh	University of Science, Hochiminh City, Vietnam
Lee Jinho	Yonsei University, South Korea
Lefebvre Grégoire	Orange Labs, France
Lehmann Daniel	University of Greifswald, Germany
Lei Fang	University of Lincoln, UK
Léonardon Mathieu	IMT Atlantique, France
Lewandowski Arnaud	Université du Littoral Côte d'Opale, Calais, France
Li Caiyuan	Shanghai Jiao Tong University, China
Li Chuang	Xi'an Jiaotong University, China
Li Ming-Fan	Ping An Life Insurance of China, Ltd., China
Li Qing	The Hong Kong Polytechnic University, China
Li Tao	Peking University, China
Li Xinyi	Southwest University, China
Li Xiumei	Hangzhou Normal University, China
Li Yanqi	University of Jinan, China
Li Yuan	Defence Innovation Institute, China
Li Zhixin	Guangxi Normal University, China
Lian Yahong	Dalian University of Technology, China
Liang Nie	Southwest University of Science and Technology, China
Liang Qi	Chinese Academy of Sciences, Beijing, China

Liang Senwei	Purdue University, USA
Liang Yuxin	Northwest University, China
Lim Nengli	Singapore University of Technology and Design, Singapore
Liu Gongshen	Shanghai Jiao Tong University, China
Liu Haolin	Chinese Academy of Sciences, China
Liu Jian-Wei	China University of Petroleum, China
Liu Juan	Wuhan University, China
Liu Junxiu	Guangxi Normal University, China
Liu Qi	Chongqing University, China
Liu Shuang	Huazhong University of Science and Technology, China
Liu Shuting	University of Shanghai for Science and Technology, China
Liu Weifeng	China University of Petroleum, China
Liu Yan	University of Shanghai for Science and Technology, China
Liu Yang	Fudan University, China
Liu Yi-Ling	Imperial College London, UK
Liu Zhu	University of Electronic Science and Technology of China, China
Long Zi	Shenzhen Technology University, China
Lopes Vasco	Universidade da Beira Interior, Portugal
Lu Siwei	Guangdong University of Technology, China
Lu Weizeng	Shenzhen University, China
Lukyanova Olga	Russian Academy of Sciences, Russia
Luo Lei	Kansas State University, USA
Luo Xiao	Peking University, China
Luo Yihao	Huazhong University of Science and Technology, China
Ma Chao	Wuhan University, China
Ma Zeyu	Harbin Institute of Technology, China
Malialis Kleanthis	University of Cyprus, Cyprus
Manoonpong Poramate	Vidyasirimedhi Institute of Science and Technology, Thailand
Martinez Rego David	Data Spartan Ltd., UK
Matsumura Tadayuki	Hitachi, Ltd., Tokyo, Japan
Mekki Asma	Université de Sfax, Tunisia
Merkel Cory	Rochester Institute of Technology, USA
Mirus Florian	Intel Labs, Germany
Mizuno Hideyuki	Suwa University of Science, Japan
Moh Teng-Sheng	San Jose State University, USA
Mohammed Elmahdi K.	Kasdi Merbah university, Algeria
Monshi Maram	University of Sydney, Australia
Moreno Felipe	Universidad Católica San Pablo, Peru
Morra Lia	Politecnico di Torino, Italy

Morzy Mikołaj	Poznań University of Technology, Poland
Mouček Roman	University of West Bohemia, Czech Republic
Moukafih Youness	International University of Rabat, Morocco
Mouysset Sandrine	University of Toulouse, France
Müller Robert	Ludwig-Maximilians-Universität München, Germany
Mutschler Maximus	University of Tübingen, Germany
Najari Naji	Orange Labs, France
Nanda Abhilasha	Vellore Institute of Technology, India
Nguyen Thi Nguyet Que	Technological University Dublin, Ireland
Nikitin Oleg	Russian Academy of Sciences, Russia
Njah Hasna	University of Sfax, Tunisia
Nyabuga Douglas	Donghua University, China
Obafemi-Ajayi Tayo	Missouri State University, USA
Ojha Varun	University of Reading, UK
Oldenhof Martijn	Katholieke Universiteit Leuven, Belgium
Oneto Luca	Università di Genova, Italy
Oota Subba Reddy	Inria, Bordeaux, France
Oprea Mihaela	Petroleum-Gas University of Ploiesti, Romania
Osorio John	Barcelona Supercomputing Center, Spain
Ouni Achref	Institut Pascal UCA, France
Pan Yongping	Sun Yat-sen University, China
Park Hyeyoung	Kyungpook National University, South Korea
Pateux Stéphane	Orange Labs, France
Pecháč Matej	Comenius University in Bratislava, Slovakia
Pecyna Leszek	University of Liverpool, UK
Peng Xuyang	China University of Petroleum, China
Pham Viet	Toshiba, Japan
Pietroń Marcin	AGH University of Science and Technology, Poland
Pócoš Štefan	Comenius University in Bratislava, Slovakia
Posocco Nicolas	Eura Nova, Belgium
Prasojo Radityo Eko	Universitas Indonesia, Indonesia
Preuss Mike	Universiteit Leiden, The Netherlands
Qiao Peng	National University of Defense Technology, China
Qiu Shoumeng	Shanghai Institute of Microsystem and Information Technology, China
Quan Hongyan	East China Normal University, China
Rafiee Laya	Concordia University, Canada
Rangarajan Anand	University of Florida, USA
Ravichandran Naresh Balaji	KTH Royal Institute of Technology, Sweden
Renzulli Riccardo	University of Turin, Italy
Richter Mats	Universität Osnabrück, Germany
Robine Jan	Heinrich Heine University Düsseldorf, Germany
Rocha Gil	University of Porto, Portugal
Rodriguez-Sanchez Antonio	Universität Innsbruck, Austria
Rosipal Roman	Slovak Academy of Sciences, Slovakia

Rusiecki Andrzej	Wroclaw University of Science and Technology, Poland
Salomon Michel	Université Bourgogne Franche-Comté, France
Sarishvili Alex	Fraunhofer ITWM, Germany
Sasi Swapna	Birla Institute of Technology and Science, India
Sataer Yikemaiti	Southeast University, China
Schaaf Nina	Fraunhofer IPA, Germany
Schak Monika	University of Applied Sciences, Fulda, Germany
Schilling Malte	Bielefeld University, Germany
Schmid Kyrill	Ludwig-Maximilians-Universität München, Germany
Schneider Johannes	University of Liechtenstein, Liechtenstein
Schwab Malgorzata	University of Colorado at Denver, USA
Sedlmeier Andreas	Ludwig-Maximilians-Universität München, Germany
Sendera Marcin	Jagiellonian University, Poland
Shahriyar Rifat	Bangladesh University of Engineering and Technology, Bangladesh
Shang Cheng	Fudan University, China
Shao Jie	University of Electronic Science and Technology of China, China
Shao Yang	Hitachi Ltd., Japan
Shehu Amarda	George Mason University, USA
Shen Linlin	Shenzhen University, China
Shenfield Alex	Sheffield Hallam University, UK
Shi Ying	Chongqing University, China
Shrestha Roman	Intelligent Voice Ltd., UK
Sifa Rafet	Fraunhofer IAIS, Germany
Sinha Aman	CNRS and University of Lorraine, France
Soltani Zarrin Pouya	Institute for High Performance Microelectronics, Germany
Song Xiaozhuang	Southern University of Science and Technology, China
Song Yuheng	Shanghai Jiao Tong University, China
Song Ziyue	Shanghai Jiao Tong University, China
Sowinski-Mydlarz Viktor	London Metropolitan University, UK
Steiner Peter	Technische Universität Dresden, Germany
Stettler Michael	University of Tübingen, Germany
Stoean Ruxandra	University of Craiova, Romania
Su Di	Beijing Institute of Technology, China
Suarez Oscar J.	Instituto Politécnico Nacional, México
Sublime Jérémie	Institut supérieur d'électronique de Paris, France
Sudharsan Bharath	National University of Ireland, Galway, Ireland
Sugawara Toshiharu	Waseda University, Japan
Sui Yongduo	University of Science and Technology of China, China
Sui Zhentao	Soochow University, China
Swiderska-Chadaj Zaneta	Warsaw University of Technology, Poland
Szandała Tomasz	Wroclaw University of Science and Technology, Poland

Šejnová Gabriela	Czech Technical University in Prague, Czech Republic
Tang Chenwei	Sichuan University, China
Tang Jialiang	Southwest University of Science and Technology, China
Taubert Nick	University Clinic Tübingen, Germany
Tek Faik Boray	Isik University, Turkey
Tessier Hugo	Stellantis, France
Tian Zhihong	Guangzhou University, China
Tianze Zhou	Beijing Institute of Technology, China
Tihon Simon	Eura Nova, Belgium
Tingwen Liu	Chinese Academy of Sciences, China
Tong Hao	Southern University of Science and Technology, China
Torres-Moreno Juan-Manuel	Université d'Avignon, France
Towobola Oluyemisi Folake	Obafemi Awolowo University, Nigeria
Trinh Anh Duong	Technological University Dublin, Ireland
Tuna Matúš	Comenius University in Bratislava, Slovakia
Uelwer Tobias	Heinrich Heine University Düsseldorf, Germany
Van Rullen Rufin	CNRS, Toulouse, France
Varlamis Iraklis	Harokopio University of Athens, Greece
Vašata Daniel	Czech Technical University in Prague, Czech Republic
Vásconez Juan	Escuela Politécnica Nacional, Ecuador
Vatai Emil	RIKEN, Japan
Viéville Thierry	Inria, Antibes, France
Wagner Stefan	Heinrich Heine University Düsseldorf, Germany
Wan Kejia	Defence Innovation Institute, China
Wang Huiling	Tampere University, Finland
Wang Jiaan	Soochow University, China
Wang Jinling	Ulster University, UK
Wang Junli	Tongji University, China
Wang Qian	Durham University, UK
Wang Xing	Ningxia University, China
Wang Yongguang	Beihang University, China
Wang Ziming	Shanghai Jiao Tong University, China
Wanigasekara Chathura	University of Auckland, New Zealand
Watson Patrick	Minerva KGI, USA
Wei Baole	Chinese Academy of Sciences, China
Wei Feng	York University, Canada
Wenninger Marc	Rosenheim Technical University of Applied Sciences, Germany
Wieczorek Tadeusz	Silesian University of Technology, Poland
Wiles Janet	University of Queensland, Australia
Windheuser Christoph	ThoughtWorks Inc., Germany
Wolter Moritz	Rheinische Friedrich-Wilhelms-Universität Bonn, Germany

Wu Ancheng	Pingan Insurance, China
Wu Dayan	Chinese Academy of Sciences, China
Wu Jingzheng	Chinese Academy of Sciences, China
Wu Nier	Inner Mongolia University, China
Wu Song	Southwest University, China
Xie Yuanlun	University of Electronic Science and Technology of China, China
Xu Dongsheng	National University of Defense Technology, China
Xu Jianhua	Nanjing Normal University, China
Xu Peng	Technical University of Munich, Germany
Yaguchi Takaharu	Kobe University, Japan
Yamamoto Hideaki	Tohoku University, Japan
Yang Gang	Renmin University of China, China
Yang Haizhao	Purdue University, USA
Yang Jing	Guangxi Normal University, China
Yang Jing	Hefei University of Technology, China
Yang Liu	Tianjin University, China
Yang Sidi	Concordia University, Canada
Yang Sun	Soochow University, China
Yang Wanli	Harbin Institute of Technology, China
Yang XiaoChen	Tianjin University of Technology, China
Yang Xuan	Shenzhen University, China
Yang Zhao	Leiden University, The Netherlands
Yang Zhengfeng	East China Normal University, China
Yang Zhiguang	Chinese Academy of Sciences, China
Yao Zhenjie	Chinese Academy of Sciences, China
Ye Kai	Wuhan University, China
Yin Bojian	Centrum Wiskunde & Informatica, The Netherlands
Yu James	Southern University of Science and Technology, China
Yu Wenxin	Southwest University of Science and Technology, China
Yu Yipeng	Tencent, China
Yu Yue	BNU-HKBU United International College, China
Yuan Limengzi	Tianjin University, China
Yuchen Ge	Hefei University of Technology, China
Yuhang Guo	Peking University, China
Yury Tsoy	Solidware, South Korea
Zeng Jia	Jilin University, China
Zeng Jiayuan	University of Shanghai for Science and Technology, China
Zhang Dongyang	University of Electronic Science and Technology of China, China
Zhang Jiacheng	Beijing University of Posts and Telecommunications, China
Zhang Jie	Nanjing University, China
Zhang Kai	Chinese Academy of Sciences, China

Zhang Kaifeng	Independent Researcher, China
Zhang Kun	Chinese Academy of Sciences, China
Zhang Luning	China University of Petroleum, China
Zhang Panpan	Chinese Academy of Sciences, China
Zhang Peng	Chinese Academy of Sciences, China
Zhang Wenbin	Carnegie Mellon University, USA
Zhang Xiang	National University of Defense Technology, China
Zhang Xuewen	Southwest University of Science and Technology, China
Zhang Yicheng	University of Lincoln, UK
Zhang Yingjie	Hunan University, China
Zhang Yunchen	University of Electronic Science and Technology of China, China
Zhang Zhiqiang	Southwest University of Science and Technology, China
Zhao Liang	University of São Paulo, Brazil
Zhao Liang	Dalian University of Technology, China
Zhao Qingchao	Harbin Engineering University, China
Zhao Ying	University of Shanghai for Science and Technology, China
Zhao Yuekai	National University of Defense Technology, China
Zheng Yuchen	Kyushu University, Japan
Zhong Junpei	Plymouth University, UK
Zhou Shiyang	Defense Innovation Institute, China
Zhou Xiaomao	Harbin Engineering University, China
Zhou Yucan	Chinese Academy of Sciences, China
Zhu Haijiang	Beijing University of Chemical Technology, China
Zhu Mengting	National University of Defense Technology, China
Zhu Shaolin	Zhengzhou University of Light Industry, China
Zhu Shuying	The University of Hong Kong, China
Zugarini Andrea	University of Florence, Italy

Contents – Part V

Spiking Neural Networks

Text Understanding I

Text Understanding II

Transfer and Meta Learning

Video Processing

Representation Learning

SageDy: A Novel Sampling and Aggregating Based Representation Learning Approach for Dynamic Networks

Jiaming Wu[1], Meng Liu[1], Jiangting Fan[1], Yong Liu[1(✉)], and Meng Han[2(✉)]

[1] Heilongjiang University, Harbin, China
liuyong123456@hlju.edu.cn
[2] Kennesaw State University, Georgia, USA
menghan@kennesaw.edu

Abstract. Served as an important role in numerous machine learning models, network representation learning, also known as graph embedding, aims to represent large-scale networks by mapping nodes into a low-dimensional space. However, transitional approaches mainly focus on the learning on static graphs instead of dynamic situation. Consider the broad existence of dynamic network in real world, this paper proposes a novel framework SageDy (sampling and aggregating on dynamic networks) to address the challenge in dynamic network for representation learning. In SageDy, we first propose a sampling method and a novel aggregator function to achieve high-quality representation; Then we developed two influence factors to measure the time interval influence and information upheaval influence. Finally, a temporal attention network is well introduced to model temporal information. The extensive experiments on four real-world network datasets demonstrate that SageDy could well fit the demand of dynamic network representation and significantly outperform other state-of-the-art methods.

Keywords: Dynamic network · Network representation learning · Sampling · Aggregator function · Influence factor

1 Introduction

Representation learning, which aims to represent the real-world networks by mapping nodes into a low-dimensional space, has received widespread attention from both academia and industry. By utilizing node representations, researchers could address many downstream tasks, such as link prediction, node clustering, and node classification, etc. In practice, lots of networks in many real-world are highly dynamic, such as the user connection in social network when new friend relationship are established. Effective network representations learning

---

J. Wu and M. Liu— Equally contributed and should be considered co-first authors

© Springer Nature Switzerland AG 2021
I. Farkaš et al. (Eds.): ICANN 2021, LNCS 12895, pp. 3–15, 2021.
https://doi.org/10.1007/978-3-030-86383-8_1

approach for dynamic networks could benefit the research community a lot on accurately obtaining node representations for many downstream tasks. However, majority of the current research on network representation learning are still focusing on the static network [3, 8, 10] without the consideration of the network dynamics. Besides, network representation learning on dynamic network is much more challenging than the same task on static network due to the unstable network topology structure and the properties on each of the time snapshot in dynamic network. The effective network representation learning algorithm on a dynamic network have to well address the structural information of dynamic network at a specific point-in-time, but also the evolution of the network.

For now, dynamic networks representation approaches have been developed in the temporal networks and static snapshots two main research categories. Dividing a dynamic network into static snapshots with different timestamps is an intuitive and convenient way and collecting dynamic network data in the form of static snapshots are more straightforward than in the form of temporal networks. Temporal networks only consider the interactive process between a pair of nodes ignoring more information on structure. Therefore, majority of researchers conducted research on network representation learning for dynamic networks in the form of static snapshots. However, these existed methods only consider the first-order neighbor interaction of a node but not the higher-order neighbor interaction [15]. Besides, the performance of current studies are still lack of good performance in capturing network structure due to the big amount of data processed. The main challenges in this area are (1) how to effectively capture the high-order information of network nodes; (2) how to better capture the network structure information; and (3) how to capture the dependence of the network structure on time.

In this paper, an novel representation learning framework named SageDy[1] is proposed in the form of static snapshots to address the above challenges. SageDy composes two modules called structural module and temporal module, respectively. In structural module, SageDy could capture the structural information of the network by using our proposed sampling method and a new aggregation function. The sampling method adopts node priority strategy, which defines a weight for each node. In temporal module, we introduce two influence factors to measure the influence of time interval and the influence of information upheaval. In addition, based on the idea of attention, we develop a temporal attention network to capture the evolution process of node representations and thus to obtain the entire network structure evolution.

It is worth to mention that SageDy allows a flexible model to utilize the structural module and temporal module by using graph neural network and recurrent neural network to get multiple model variants. Therefore, the different components we proposed in SageDy could be utilized with other approaches separately as the variants of our model, although the integrally SageDy outperforms all other models combination in our comprehensive evaluation. We test SageDy on the multiple downstream tasks on four dynamic network datasets. Compared

---

[1] The source code and datasets can be obtained from https://github.com/wjm199717/SageDy.

with multiple state-of-the-art baselines and different model variants. We demonstrate that our proposed method SageDy significantly outperform other baselines. Our contributions in this paper are summarized as follows:

(1) We propose a new sampling and aggregation . The sampling and aggregator function could significantly improve the efficiency on the learning process.

(2) To address the challenges on dynamic network measurement, we introduce two influence factors to measure time influence and information upheaval.

(3) We adopt a temporal attention network to model temporal information on the dynamic network, which could capture all the information not only includes the structure during the time internal but also the network evolution.

(4) Comprehensive experiences are conducted with real life dynamic networks to evaluate the performance of our proposed methods and the advantages of our components in SageDy as a plugin methods.

2   Related Work

Traditional network representation learning is generally accompanied by dimensionality reduction techniques, such as matrix decomposition [1], or constructing an affinity graph to embed the graph into a low-dimensional space [11].

With the increasing use of deep learning, researchers began to combine deep learning with network representation learning. Perozzi et al. [8] used random walk to generate a sequence of nodes, and then generated the node representation. Hamilton [4] proposed a trainable neighbor aggregation function through extended graph convolution. But none of these methods can be applied to dynamic networks. Whether it is a traditional method or a deep learning method, they were initially applied to static network representation learning. However, the networks involved in lives are constantly changing. How to capture the information in the dynamic network will become particularly important, which will help us in the next practical operation.

In dynamic networks, many methods have been proposed in recent years [6,7]. Most methods use temporal smoothness regularization to ensure the stability of continuous time embedding. In the early dynamic network research, Trivedi et al. [12] applied the neural network method to the field of knowledge graphs. He used recurrent neural networks to model time information. Recently, Zhou et al. [14] took the triadic closure [5] as guidance, and greatly improved the dynamic network representation learning. However, these methods mainly study the evolution process of connecting edges between nodes. They often only consider the first-order proximity between nodes, ignoring the high-order proximity.

3   Problem Definition

We consider the problem of dynamic network representation learning. A dynamic network G is defined as a series of snapshots $G = \{G^1, G^2, ..., G^T\}$, where T is the number of time steps. Each static snapshot is a weighted undirected graph $G^t = (V^t, E^t)$ with node set V^t and edge set E^t at time t. We aim to learn a low-dimensional embedding vector e_v^t for each node v in the network. The vector representation e_v^t can capture the temporal change in topological structure around v.

4   Method

In this section, we will describe the structure of our proposed SageDy. The structure of SageDy consists of two modules, the one is composed of the sampling and the aggergator function. The other one is composed of a temporal attention network. The general structure of SageDy is shown in Fig. 1.

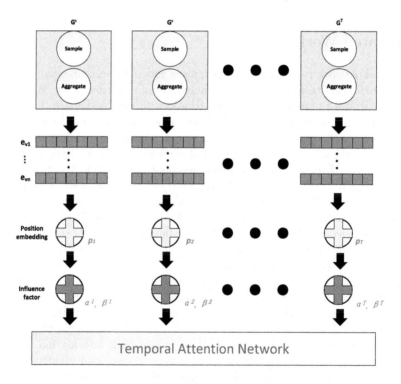

Fig. 1. The architecture of SageDy

4.1   Structure Module

The structure module is composed of sampling and aggeregation. In the sampling and aggeregation, we use node priority sampling strategy to sample neighbors of each node, and then aggeregate feature information from sampled neighbors. The sampling and aggregation are shown in Algorithm 1.

We adopt node priority sampling based on node weight in the sampling process. We first introduce a new definition of node weight. Let v be a source node, and u be a neighbor of v. The weight of neighbor u is defined in Eq. (1).

$$W_u = E_{(v,u)} * W_{(v,u)} - ||h_v^0 - h_u^0||_2 \qquad (1)$$

Where $E_{(v,u)}$ is the number of edges between v and u, $W_{(v,u)}$ is the weight between v and u, which is normalized to between 0 and 1, and h_v^0 and h_u^0 is initial embedding of v and u respectively.

After we sample N neighbors $N(v)$ of source node v, we propose a new aggergator function to aggregate feature information from sampled neighbors $N(v)$, which is defined in Eq. (2).

$$h_v^k = \sigma(W_1 \cdot MEAN(CONCAT(W_2 \cdot h_v^{k-1}, W_3 \cdot H_{N(v)}^k)) + b) \qquad (2)$$

Where $h_v^k \in R^{F' \times 1}$ denotes the embedding of node v at k^{th} iteration, $h_v^{k-1} \in R^{F \times 1}$ denotes the embedding of node v at $(k-1)^{th}$ iteration, and $H_{N(v)}^k \in R^{F \times N}$ denotes the embedding matrix of sampled neighbors of node v. Here N is the number of sampled neighbors. $W_1 \in R^{F' \times F}$, $W_2 \in R^{F \times F}$ and $W_3 \in R^{F \times F}$ are learnable parameter matrix, where F is the dimension size of node embedding at $(k-1)^{th}$ iteration and F' is the dimension size of node embedding at k^{th} iteration. W_2 is used to capture the features of source node, W_3 is used to capture the features of neighbor nodes, and W_1 is used to project aggregation results into a new space. $CONCAT$ denotes a splice along a dimension and $MEAN$ represents taking the average along a dimension. σ is a nonlinear activation function and b denotes a bias.

Algorithm 1: structural module embedding generation

Input   : Graph $G^t(V^t, E^t)$; Input features $\{X_v \in R^D, \forall v \in V\}$; Number of iterations K; Weight matrices W; Non-linearity σ;

Output: embeddings $e_v^t \in R^{F'}$ for all $v \in V^t$

1 $h_v^0 = X_v, \forall v \in V^t$;
2 **for** $v \in V^t$ **do**
3   use node priority sampling to sample N neighbors $N(v)$ of v;
4 **end**
5 **for** $k = 1...K$ **do**
6   **for** $v \in V^t$ **do**
7          $h_v^k = \sigma(W_1 \cdot MEAN(CONCAT(W_2 \cdot h_v^{k-1}, W_3 \cdot H_{N(v)}^k)) + b)$
8   **end**
9      $h_v^k = h_v^k / ||h_v^k||_2, \forall v \in V^t$;
10 **end**
11 $e_v^t = h_v^K, \forall v \in V^t$;

4.2   Temporal Module

Position Embedding. As shown in Fig. 1, before putting node embeddings into the temporal module, we need to perform three operations. First of all, we embed the position information on time step t into node embeddings. The position embedding is defined in Eq. (3).

$$e_v^t = e_v^t + p_t \tag{3}$$

where $p_t \in R^{F'}$ denotes a position embedding at time step t that is a learnable parameter.

Influence Factor of Interval. After position embedding, we introduce an influence factor of interval α_{t_i} at time step t_i. Intuitively, the influence of α_{t_i} decays over time. The closer to the current time, the greater the influence of α_{t_i}. The influence of α_{t_i} on the current time should be measured by the time interval. Therefore, the influence factor α_{t_i} is defined in Eq. (4).

$$\alpha_{t_i} = exp(-\delta_{t_i}|t_c - t_i|) \tag{4}$$

Where δ_{t_i} is a time dependent learnable parameter, t_c represents the current time, and t_i denotes the historical time.

As shown in Eq. (4), the influence factor α_{t_i} of interval is between 0 and 1, which helps us extract some past information from node embeddings at time step t_i. The closer to the current time, the more information is extracted. In Eq. (4), we make full use of the property of the exponential function. For example, if the current time is t_n, due to the nature of the exponential function, then there is a huge difference in the amount of information captured at time t_{n-1} and time t_{n-2}, which emphasizes the importance of time in a short time interval from the current time. At the same time, the amount of information captured at $t_{n-s}(s >> 0)$ and t_{n-s+1} are almost the same and close to 0, which emphasizes that the influence of historical behaviors far away from the current time on the current behaviors can be ignored.

Influence Factor of Upheaval. Next, we analyze the magnitude of the information change in the network. In real life, when a person's information has undergone tremendous change, the information at that time is very important. For example, a scholar initially studied the field of data mining, he and other scholars in the field of data mining have published many papers together. As time goes by, the scholar's research field converted to computer vision, then he will have most likely collaborate with scholars in computer vision. Therefore, we propose the second influence factor β to capture this important information upheaval. The main influence of β is on the node representation at the time when the information changes drastically, which is defined in Eq. (5) and Eq. (6).

$$t_i = \arg\max_{t_h}(||e_v^{t_h+1} - e_v^{t_h}||_2) \tag{5}$$

$$\beta_{t_i} = exp(-\lambda_{t_i}|t_c - t_i|) + 1 \tag{6}$$

Where t_i represents the time when the information changes drastically, $e_v^{t_h}$ denotes the node representation of a history time. As shown in Eq. (6), the influence factor β_{t_i} of upheaval is between 1 and 2, which will be used to amplify the effect of embeddings at upheaval time. Furthermore, the closer to the current time, the greater influence factor β_{t_i} of upheaval. Please note that the value of β_{t_i} is larger than 0 at upheaval time, and the value of β_{t_i} is 0 at other time.

After obtaining the influence factor α_{t_i} of interval and the influence factor β_{t_i} of upheaval, we update the node embeddings $e_v^{t_i}$ at time t_i, which is shown in Eq. (7).

$$e_v^{t_i} = (e_v^{t_i} * \alpha_{t_i} + e_v^{t_i} * \beta_{t_i})/2 \tag{7}$$

In this way, the node embeddings $e_v^{t_i}$ at time t_i is affected by both time interval and drastic change.

Temporal Attention Network. To better model temporal information in a dynamic network, we introduce a temporal attention network. The input of this temporal attention network is a sequence of representations of nodes at different times. We define the input representations of node v as $\{e_v^1, e_v^2, ..., e_v^T\}$, $e_v^t \in R^{F'}$, where T is the number of time steps, F' is the dimension size of input representations and e_v^t denotes the input representation of node v at time t. The output representations of node v is defined as a new representation sequence $z_v = \{z_v^1, z_v^2, ..., z_v^T\}$, $z_v^t \in R^{D'}$, where D' is the dimension size of output representations and z_v^t denotes the output representation of node v at time t. We then pack the input sequence of representations across time. We denote it as $E_v \in R^{T \times F'}$. The output sequence of representations across time is also packed together and it is denoted as $Z_v \in R^{T \times D'}$.

To obtain the output representation of node v at time t, we use the scaled dot-product form of attention where the queries, keys, and values are set as the input node representations. The queries, keys, and values are first transformed to a different space by using projection matrices $W_q \in R^{F' \times D'}$, $W_k \in R^{F' \times D'}$ and $W_v \in R^{F' \times D'}$ respectively. The temporal attention network is defined in Eq. (8), Eq. (9) and Eq. (10).

$$a_v^{ij} = (((E_v \cdot W_q) \cdot (E_v \cdot W_k)^T)_{ij}/\sqrt{D'}) + M_{ij} \tag{8}$$

$$\Gamma_v^{ij} = exp(a_v^{ij})/\sum_{k=1}^{T} exp(a_v^{ik}) \tag{9}$$

$$Z_v = \Gamma_v \cdot (X_v \cdot W_v) \tag{10}$$

where $\Gamma_v \in R^{T \times T}$ is the attention weight matrix obtained by the multiplicative attention function. $M \in R^{T \times T}$ is a mask matrix with each entry $M_{ij} \in \{-\infty, 0\}$. When $M_{ij} = -\infty$, the softmax function results will be a zero attention weight, $\Gamma_v^{ij} = 0$, which switches off the attention from time i to j. To encode the temporal order, we define M as follows.

$$M_{ij} = \begin{cases} -\infty, & i \leq j \\ 0, & i > j \end{cases} \tag{11}$$

4.3   Model Optimization

We use a binary cross-entropy objective function at different time t to ensure the learned representations capture both structure and temporal information.

$$L_v = \sum_{t=1}^{T}(\sum_{u \in N_{walk(v)}^t} -log(\sigma(<e_u^t, e_v^t>)) - w_n \cdot \sum_{u' \in P_n^t(v)} log(1 - \sigma(<e_{u'}^t, e_v^t>)))$$

(12)

Where σ is the sigmoid function, $\langle , \rangle$ is the inner product operation. $N_{walk}^t(v)$ is the set of nodes that co-occur with v on fixed-length random walks at time t, $P_n^t(v)$ is a negative sampling distribution of node v at time t, and w_n denotes negative sampling ratio which is a tunable hyper-parameter to balance the positive and negative samples.

Table 1. The statistics of the datasets

Datasets	Enron	UCI	Yelp	ML-10M
Nodes	143	1,809	6,569	20,537
Edges	2,347	16,822	95,361	43,760
Time steps	16	13	13	16

5   Experiment

5.1   Datasets

Our datasets are divided into two types, communication network and rating network respectively. In our datasets, the number of edges between nodes may be arbitrary. We list the statistical information of the data in Table 1.

Communication Network: We use two publicly available communication network datasets. They are the Enron dataset and UCI dataset. In the Enron dataset, the connected edges between nodes represent the email interactions between core employees in the company. In the UCI dataset, the connected edges between nodes represent messages sent between users on the social network platform.

Rating Network: We use two binary rating networks from Yelp and Movie-LensD. The nodes on Yelp can be divided into two categories: user and business. ML-10M is an version of MovieLensD in which the tagging behavior of Movie-Lenss users on movies are recorded. The nodes on ML-10M can be divided into two categories: user and movies. There is an edge between a user and a movie, which means that the user likes the movie.

5.2   Baseline Methods

SDNE [13]: The method constructs an optimization function by preserving the first-order and second-order proximity between nodes to capture the structure on the network.

Table 2. Experiment results on link prediction

Metric	Method	Enron	UCI	ML-1OM	Yelp
Macro-AUC	SDNE	0.623225	0.518755	0.530690	0.527683
	GraphSAGE	0.531635	0.661733	0.709014	0.509783
	DySAT	0.602393	0.672845	0.828921	**0.718437**
	DynGEM	0.697223	0.732145	0.819603	0.659402
	GCN+TAN	0.691549	0.725744	0.803032	0.702221
	GCN+LSTM	0.607339	0.751909	0.802099	0.690479
	GAT+LSTM	0.722315	0.699811	0.801046	0.694516
	SageDy	**0.736003**	**0.783821**	**0.848656**	0.671986
Micro-AUC	SDNE	0.604167	0.508215	0.542353	0.522205
	GraphSAGE	0.531636	0.661744	0.708837	0.509791
	DySAT	0.678333	0.647853	0.814275	**0.717064**
	DynGEM	0.602393	0.722645	0.736912	0.660214
	GCN+TAN	0.679149	0.720422	0.806577	0.707012
	GCN+LSTM	0.607253	0.741941	0.803037	0.715342
	GAT+LSTM	0.712829	0.709533	0.812273	0.690144
	SageDy	**0.739441**	**0.780881**	**0.839849**	0.656802

GraphSAGE [4]: This method applies the translation invariance idea of CNN. GraphSAGE by sampling and aggregating to integrate the information from neighbor nodes. It can capture higher-order information between nodes.

DySAT [9]: This method uses two graph attention networks to capture the structural information and temporal information on the network respectively.

DyGEM [2]: This method was designed based on recent advance in deep autoencoders for graph embeddings.

"Convolution Neural Network + Recurrence Neural Network": To further verify the efficiency of SageDy, we integrate different graph neural networks with different recurrent neural networks to construct several baseline methods, e.g., GCN+LSTM, GAT+LSTM and GCN+TAN. TAN denotes the temporal attention network proposed in this paper.

5.3 Variants

"SageDy-" denotes the variants of SageDy. We use "SageDy-L" to denote the SageDy method that use a recurrent neural network LSTM instead of the temporal attention network in this paper. "SageDy-NI" denotes the SageDy method without the influence factors. "SageDy-NS" denotes the SageDy method without our proposed sampling method. In order to highlight the contribution of our proposed aggregator function, "SageDy-Mean" and "SageDy-Meanpooling" denote the SageDy methods where our aggergator function is replaced with mean and menpooling aggergator function in GraphSAGE [4] respectively. Note that our aggregator function, mean and meanpooling aggregator function are all related to mean operation.

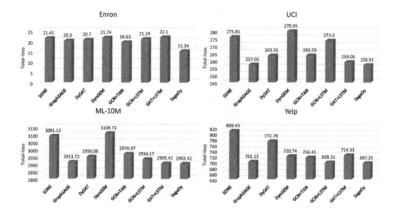

Fig. 2. Experiment results on clustering task.

5.4   Link Prediction Results

Given two nodes x and y, we define the Hadamard product between their representations as a feature, which is indicative for an edge between x and y. On each dataset, we randomly hold out 80% of the total edges as training set and the remaining as testing set. A logistic regression classifier is trained on training set, and then to the prediction is made on testing set. We use ROC-AUC as metric for link prediction, the experimental results are shown in Table 2.

The experimental results show that our method has the best performance on Enron, UCI and ML-10M datasets. We also notice that SageDy performs worse in Yelp dataset. By examining the data carefully, we find that there is only one edge between any pair of nodes on Yelp, which makes SageDy unable to effectively capture the interaction change between nodes. In the baseline methods, DySAT and DynGEM perform better. This is because that they are dynamic network representation methods that can capture temporal change. SDNE and GraphSAGE are static network representation methods that can only capture static structure ignoring temporal change, which makes them poor in link prediction. The baselines based on "convolution neural network + recurrence neural network" can capture temporal change on networks, and thus they are obviously better than static representation methods SDNE and GraphSAGE.

5.5 Clustering Results

In this subsection, we apply all the methods to learn node embeddings on all datasets, then employ the t-SNE method to project node embeddings into a 2-dimensional space, and finally use K-means to cluster points in the 2-dimensional space. Here, we use the clustering total-loss as metric to evaluate the performance of various methods. The clustering total-loss is defined as the sum of the distances between nodes and the centroid in the cluster. The smaller the total-loss, the better the performance. As shown in Fig. 2, the total-loss of SageDy is the lowest on all datasets, which illustrates that SageDy significantly outperforms all baselines.

Table 3. Ablation study on link prediction

Metric	Method	Enron	UCI	ML-1OM	Yelp
Macro-AUC	SageDy	**0.736003**	**0.783821**	**0.848656**	**0.671986**
	SageDy-L	0.650691	0.669977	0.800024	0.641626
	SageDy-NI	0.707049	0.778554	0.817408	0.633107
	SageDy-NS	0.662471	0.775216	0.799707	0.666730
	SageDy-Mean	0.726207	0.722507	0.826060	0.624810
	SageDy-Meanpooling	0.544671	0.694610	0.719099	0.541770
Micro-AUC	SageDy	**0.739441**	**0.780881**	**0.839849**	**0.676802**
	SageDy-L	0.650567	0.666499	0.811908	0.647780
	SageDy-NI	0.729085	0.685718	0.825345	0.634087
	SageDy-NS	0.723417	0.710602	0.810842	0.668621
	SageDy-Mean	0.726944	0.710187	0.785178	0.605958
	SageDy-Meanpooling	0.539029	0.698626	0.733885	0.550320

5.6 Ablation Study

In this subsection, we construct several variants of SageDy to investigate the effect of each component on the performance of SageDy. We use "SageDy-" to represent the variants of SageDy. Our experimental results are shown in Table 3 and Fig. 3. The experimental results show that our method SageDy is superior to its variants on both link prediction and node clustering, which illustrate that our sampling method, aggregator function, influence factor and temporal attention network all play an important role in SageDy.

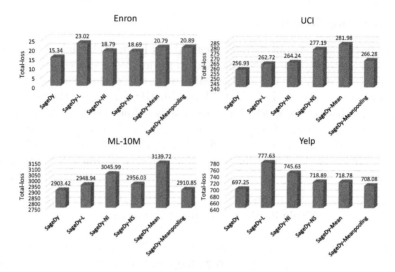

Fig. 3. Ablation study on clustering task.

6   Conclusion

In this paper, we propose a novel dynamic network representation learning method SageDy composed of structural module and temporal module. We model the structural information and temporal information respectively. Experimental results show that SageDy has a significant performance gain over several state-of-the-art baselines. In the future, we will add attributes into edges to further improve the performance of SageDy.

Acknowledgment. This work was supported by the National Natural Science Foundation of China (No. 61972135), the Natural Science Foundation of Heilongjiang Province in China (No. LH2020F043), the Innovation Talents Project of Science and Technology Bureau of Harbin in China (No. 2017RAQXJ094), the Postgraduate Innovative Scientific Research Project of Heilongjiang University in China (No. YJSCX2021-198HLJU), and Project Intelligentization and Digitization for Airline Revolution (No. 2018R02008).

References

1. Ahmed, A., Shervashidze, N., Narayanamurthy, S., Josifovski, V., Smola, J.A.: Distributed large-scale natural graph factorization. In: WWW, pp. 37–48 (2013)
2. Goyal, P., Kamra, N., He, X., Liu, Y.: Dyngem: Deep embedding method for dynamic graphs. arXiv: Social and Information Networks (2018)
3. Grover, A., Leskovec, J.: node2vec: scalable feature learning for networks. In: KDD, pp. 855–864 (2016)
4. Hamilton, L.W., Ying, R., Leskovec, J.: Inductive representation learning on large graphs. In: NIPS, pp. 1024–1034 (2017)

5. Kossinets, G., Watts, J.D.: Empirical analysis of an evolving social network. Science **311**(5757), 88–90 (2006)
6. Kumar, S., Zhang, X., Leskovec, J.: Predicting dynamic embedding trajectory in temporal interaction networks, pp. 1269–1278 (2019)
7. Lee, B.J., Nguyen, H.G., Rossi, A.R., Ahmed, K.N., Koh, E., Kim, S.: Temporal network representation learning. arXiv: Learning (2019)
8. Perozzi, B., Al-Rfou', R., Skiena, S.: Deepwalk: online learning of social representations. In: KDD, pp. 701–710 (2014)
9. Sankar, A., Wu, Y., Gou, L., Zhang, W., Yang, H.: Dynamic graph representation learning via self-attention networks. In: ICLR (2019)
10. Tang, J., Qu, M., Wang, M., Zhang, M., Yan, J., Mei, Q.: Line: large-scale information network embedding. In: WWW (2015)
11. Tenenbaum, B.J., Silva, D.V., Langford, J.: A global geometric framework for nonlinear dimensionality reduction. Science **290**(5500), 2319–2323 (2000)
12. Trivedi, R., Dai, H., Wang, Y., Song, L.: Know-evolve: deep temporal reasoning for dynamic knowledge graphs. In: ICML, pp. 3462–3471 (2017)
13. WANG, D., Cui, P., Zhu, W.: Structural deep network embedding. In: KDD (2016)
14. Zhou, L.K., Yang, Y., Ren, X., Wu, F., Zhuang, Y.: Dynamic network embedding by modeling triadic closure process. In: AAAI (2018)
15. Zuo, Y., Liu, G., Lin, H., Guo, J., Hu, X., Wu, J.: Embedding temporal network via neighborhood formation. In: KDD, pp. 2857–2866 (2018)

CuRL: Coupled Representation Learning of Cards and Merchants to Detect Transaction Frauds

Maitrey Gramopadhye, Shreyansh Singh[✉], Kushagra Agarwal,
Nitish Srivasatava, Alok Mani Singh, Siddhartha Asthana, and Ankur Arora

AI Garage, Mastercard, Gurugram, India
shreyansh.singh@mastercard.com

Abstract. Payment networks like Mastercard or Visa process billions of transactions every year. A significant number of these transactions are fraudulent that cause huge losses to financial institutions. Conventional fraud detection methods fail to capture higher-order interactions between payment entities i.e., cards and merchants, which could be crucial to detect out-of-pattern, possibly fraudulent transactions. Several works have focused on capturing these interactions by representing the transaction data either as a bipartite graph or homogeneous graph projections of the payment entities. In a homogeneous graph, higher-order cross-interactions between the entities are lost and hence the representations learned are sub-optimal. In a bipartite graph, the sequences generated through random walk are stochastic, computationally expensive to generate, and sometimes drift away to include uncorrelated nodes. Moreover, scaling graph-learning algorithms and using them for real-time fraud scoring is an open challenge.

In this paper, we propose CuRL and tCuRL, coupled representation learning methods that can effectively capture the higher-order interactions in a bipartite graph of payment entities. Instead of relying on random walks, proposed methods generate coupled session-based interaction pairs of entities which are then fed as input to the skip-gram model to learn entity representations. The model learns the representations for both entities simultaneously and in the same embedding space, which helps to capture their cross-interactions effectively. Furthermore, considering the session constrained neighborhood structure of an entity makes the pair generation process efficient. This paper demonstrates that the proposed methods run faster than many state-of-the-art representation learning algorithms and produce embeddings that outperform other relevant baselines on fraud classification task.

Keywords: Fraud detection · Payment network · Representation learning · Skip-gram

---

M. Gramopadhye and S. Singh—Equal contribution.

I. Farkaš et al. (Eds.): ICANN 2021, LNCS 12895, pp. 16–29, 2021.
https://doi.org/10.1007/978-3-030-86383-8_2

1   Introduction

Detecting fraudulent transactions has been an ever-present problem in the payment industry with an average number of fraudulent transactions attempted per merchant increasing at a rate of 34% per annum [25] and total worldwide fraud losses projected to reach $35.67 billion in 2023 [28].

Conventional fraud detection solutions have moved from manually coded rules to either traditional classification models or anomaly detection-based models [5,22,26]. The use of deep learning models has further improved the performance as they better capture the complex interactions between transaction features. Given the sequential nature of transactions, LSTM and other seq2seq models have previously been used for fraud detection. However, most of these methods [1,9] treat each transaction independently and fail to capture any interaction information between the payment entities, which can be useful in detecting any out of pattern transaction.

Recent works have focused on capturing the interaction between payment entities by leveraging graph-based methods for learning card and merchant representations which serve as useful features for downstream supervised tasks. However, methods that consider the transaction network as a graph of homogeneous entities [12,18,24,27], capture merchant-merchant relation or card-card relation but miss out on cross-entity interaction information. Whereas, most methods that consider the heterogeneous nature of the transaction graph rely on generating sequences through random walks [4,7] which take into account the connections of the entities but without the temporal information. Furthermore, these random walks are stochastic and sometimes drift away to include uncorrelated nodes thus generating sub-optimal representations. Moreover, these graph-learning algorithms are computationally expensive, and scaling them for real-time fraud detection is a big challenge.

In this paper, we propose CuRL, a scalable method of learning card and merchant representations from a bipartite transaction graph. We explain how our method effectively captures the cross-interactions between cards and merchants along with their homogeneous interactions. Capturing these interactions allows our model to retain more information in the entity embeddings, which in turn function as richer features for downstream fraud detection. We adapt word2vec's skip-gram [19] as our embedding generation model and sample the neighborhood of a node in the transaction graph to form coupled pairs (Card-Card, Merchant-Merchant, Card-Merchant) to train the skip-gram. Additionally, as new transactions come in, the model has to be periodically retrained. To facilitate this, our model focuses on being lightweight and samples only the immediate neighborhood of a node. As an extension to CuRL, we also propose tCuRL, a session-based sampling method that further reduces the size of the neighborhood sampled for each node by removing uncorrelated neighbors. Thus, further reducing the training time for the embedding generation model and improving the quality of the representations.

We empirically compare our methods with other embedding generation methods and show that embeddings generated by CuRL and tCuRL, when included as features for fraud detection, beat all baseline methods. Moreover, since most of the graph-based models are computationally expensive on large graphs, we also introduce the embedding generation time as a metric while benchmarking.

The organization of the paper is as follows. In Sect. 2, we cover the related work done in the field of fraud detection and embedding generation, particularly in the payment industry domain. Section 3 talks about the synthetic data generation and data preparation process. Section 4 describes the proposed approach for card and merchant embedding generation. Section 5 has the details for the conducted experiments and Sect. 6 shows the comparison of our method against the baselines. Section 7 concludes our work.

2   Related Work

Fraud detection has been an active area of research for a long time. We present here an overview of the fraud detection techniques and methods that have been applied in detecting different financial frauds. Bayes method [2], SVM [30], decision trees and neural networks [26] have been applied extensively for solving fraud detection problems. The general problem with above-mentioned models is that they usually rely on hand-crafted features and manual feature engineering to capture transaction information. Deep learning methods like [20,31] use autoencoders and denoising autoencoders to transform transaction features to a lower dimension to classify the transaction as fraud. [16] tried to solve the fraud detection problem as a transaction sequence classification technique and hence used an LSTM model, but it failed to also capture the payment entity interactions.

To capture these interactions, several works have considered the transaction network as a bipartite graph with cards and merchants as nodes. [14] was a graph-based method for detecting fraudulent behavior in social media networks. [8,15] model fraud detection as graph-based anomaly detection. Other methods involve creating node embeddings using the network structure. Earlier approaches for node embedding generation involved using hand-crafted features based on network properties [10,13]. [21,29] propose methods to analyze the transaction graph and extract network features manually to use along with intrinsic transaction features relevant for fraud detection task.

Recent advancements in representational learning for natural language processing opened new ways of feature learning for discrete objects such as words. In particular, the Skip-gram model [19] aims to learn continuous feature representations for words by optimizing a neighborhood preserving likelihood objective. Inspired by the Skip-gram model, recent research established an analogy for networks by representing a network as a "document" [24,27]. The same way as a document is an ordered sequence of words, one could sample sequences of nodes from the underlying network and turn a network into an ordered sequence of nodes. However, there are many possible sampling strategies for nodes, resulting in different learned feature representations. A popular theme we noticed in earlier works was that of using random walks to sample the neighbors of a node. node2vec [12] performs a biased random walk to obtain the neighborhood while DeepWalk [24] deploys a truncated random walk for social network embedding, and metapath2vec [7] uses meta-path-based random walks. Unlike Deepwalk, metapath2vec defines metapaths along which we want the walker to move. Pin2Vec [18], originally created for bipartite graphs, forms homogeneous graphs for each entity and trains separate skip-gram models for them. Deep-Trax [3] learns embeddings for cards and merchants from time-constrained random walks taken on the transaction graph. HitFraud [4] forms a heterogeneous information network from the transactions and analyzes it to form graph-based features by generating meta-paths on the graph. A limitation of such models is that they often tend to drift away to include uncorrelated nodes while creating the sequences. In practice, we also noticed that sampling a neighborhood using random walks is a computationally expensive process. These observations are evident in our results section (Sect. 6) as well. BigGraph [17] is another algorithm for learning node embeddings for large graphs, with up to billions of nodes and trillions of edges. It learns node embeddings through knowledge graph implementation. BiNE, short for Bipartite Network Embedding [11], generates sequences that preserve the long-tail distribution of nodes in a bipartite graph. It however uses a biased random walk generator to generate node neighborhood for subsequent sequence generation.

3 Dataset

3.1 Description of Data

Pursuant to internal controls to protect data, confidentiality, and privacy, we don't use real transaction data for this paper. As this is an exploratory research, we have used synthetically generated transaction data for our experiments, created by applying SMOTE-NC [6] on real transaction data. Data created is synthetic and cannot be traced back to any original transaction. Synthetic data was generated for binary classes - fraud and non-fraud. It consists of 537k transactions, of which $1,100$ transactions are fraudulent (0.2% of transactions). There are $125,019$ unique cards and 220 unique merchants in the dataset. Our dataset mimics the skewed distribution of cards found in transaction data. Several cards have just 1–2 transactions whereas a few cards have many transactions.

From the dataset, we only use the card number and merchant ID to generate embeddings for cards and merchants. Note that the card numbers and merchant IDs used are from the synthetic dataset and not the real data. The same dataset is used for all the experiments and the embeddings generated are then used to train the model for the downstream task of fraud detection.

3.2 SMOTE-NC

SMOTE-NC (Synthetic Minority Oversampling Technique) [6] is a data augmentation technique for generating synthetic data. It is a variant of SMOTE and is used when the data to be generated has continuous as well as non-continuous features. Since we need to generate synthetic data for both the classes, we train SMOTE-NC on two different datasets, one for each class.

Table 1. Distribution of transaction amount in real data with 30-day and 14-day average

Statistic	txn_amt	30d_avg	14d_avg
count	538,125	538,125	538,125
mean	46.31	0.86	0.82
std	150.45	0.27	0.34
min	0.01	0.00	0.00
25%	8.97	0.88	0.83
50%	16.98	1.00	1.00
75%	43.96	1.00	1.00
max	44,418.73	1.00	1.00

Table 2. Distribution of transaction amount in synthetic data with 30-day and 14-day average

Statistic	txn_amt	30d_avg	14d_avg
count	538,125	538,125	538,125
mean	46.28	0.87	0.83
std	144.30	0.24	0.30
min	0.01	0.00	0.00
25%	9.08	0.87	0.82
50%	17.02	0.98	0.99
75%	43.95	1.00	1.00
max	36,661.17	1.00	1.00

3.3 Distribution Similarity

Distribution similarity between the synthetic data and real data will ensure that a model performing well on synthetic data, will perform well on the real data as well. Due to privacy concerns, we cannot compare the two datasets at the individual data point level but in Tables 1 and 2 we show the comparisons on an aggregate level. Furthermore, we also use the Evaluation Framework of SDV [23] to calculate the quality of the synthetic data. The Chi-squared test on the two datasets gives a value of 0.998 and the Kolmogorov–Smirnov test gives a value of 0.885. These metrics show that the two datasets are very similar.

4   Methodology

4.1   CuRL

The CuRL framework, visually explained in the system architecture Fig. 1, generates Card and Merchant embedding by a novel strategy of inducing Card-Card, Merchant-Merchant, and Merchant-Card pairs from bipartite transaction graph. These pairs of C-M, M-M and C-C are then fed to word2vec based skip-gram model to learn entities representations in the same embedding space.

To create M-M, C-M and C-C pairs, joint neighborhood (N_i) is calculated collectively for card (C_i) and Merchant (M_i), for each transaction (T_i). As shown in Eq. 2, N_i is defined as the union of the individual neighborhoods of C_i (N_i^C) and M_i (N_i^M). From Eq. 1, N_i^C consists of all merchants that C_i transacted with, in an analysis window w, immediately before and after T_i, i.e. w merchants the C_i transacted with each before and after T_i. Similarly, N_i^M consists of the cards which transacted with M_i immediately before and after T_i, within the analysis window. Also, relationship between C_i and M_i is also included in N_i. After calculating the joint neighborhood N_i, all nodes in N_i are paired with both C_i and M_i to train the skip-gram model. We consider M_k to be the position of M_i in its own neighborhood. Similarly for C_k and C_i.

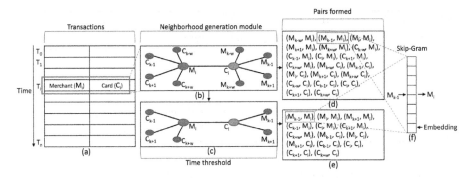

Fig. 1. CuRL/tCuRL embedding generation process: (a) For each transaction T_i, the pair of card C_i and merchant M_i, is selected (b) The neighborhood generation module constructs the joint neighborhood of the card and merchant (c) The time threshold used in tCuRL, helps in removing uncorrelated interactions by limiting the card and merchant neighborhood to a dynamic time window (d,e) $card - card$, $card - merchant$, and $merchant - merchant$ pairs are formed from the collected neighborhood (f) The skip-gram model is trained on these coupled pairs to predict M_i and C_i from their neighborhood to learn cross-entity as well as homogenous relationships. The model weights are then used as embeddings for the cards and merchants

$$N_i^C = \bigcup_{-w \leq j \leq w} M_{k+j} \quad and \quad N_i^M = \bigcup_{-w \leq j \leq w} C_{k+j} \tag{1}$$

$$N_i = N_i^C \bigcup N_i^M \tag{2}$$

And the pairs formed for transaction T_i are,

$$\{(C,x); \ \forall x \in N_i\} \bigcup \{(M,x); \ \forall x \in N_i\} \tag{3}$$

As shown in Eq. 3, CuRL includes both 1st and 2nd order neighbors to form $card - card$, $card - merchant$, $merchant - merchant$ entity pairs which helps in capturing historic interaction pattern of entities involved.

4.2 tCuRL

Optimizing the analysis window to derive card and merchant neighborhoods is essential in building a robust solution. Hence, session-based CuRL is introduced in this paper named tCuRL which decides on analysis window (henceforth called session) based on card's current spend frequency. Session constrained card and merchant neighborhood for a transaction would ensure that only relevant interactions are captured between entities and the performance of the model can be improved. To explain this further, let us again consider N_i^C in Eq. (1) for card C. There is a possibility that difference between the time of the transactions between C and M_k and between C and M_{k+1} is large. Keeping both

---

Algorithm 1: The CuRL Framework

Result: Vector representations of Payment Entities

for *Transaction T_i; $\forall 0 \leq i \leq n$* **do**

    $M_i = T_i(Merchant)$;

    $C_i = T_i(Card)$;

    $N_i^M = neighborhood(M_i)$;

    #*The cards M_i transacts with, immediately before and after T_i;*

    $N_i^C = neighborhood(C_i)$;

    #*The merchants C_i transacts with, immediately before and after T_i;*

    $N_i^M = Time\text{-}threshold(N_i^M)$;

    $N_i^C = Time\text{-}threshold(N_i^C)$;

    # *All cards/merchants that transacted with M_i/C_i before or after the time-threshold from T_i are removed (this step is skipped in CuRL);*

    $N_i = N_i^M \bigcup N_i^C$;

    $P_i^M = \{(M_i,x) \ \forall \ x \in N_i\}$;

    $P_i^C = \{(C_i,x) \ \forall \ x \in N_i\}$;

    $P_i = P_i^M \bigcup P_i^C$;

end

$Pairs = \bigcup_{1 \leq i \leq n} P_i$;

$Embeddings = skipgram(Pairs)$;

#*The skip-gram model is trained to learn the entity embeddings;*

---

these merchants in the same card neighborhood would add noise and force the model to learn spurious interactions. To incorporate this tCuRL uses a dynamic analysis window, which ensures that in such a case, M_k and M_{k+1} would never come together in the same neighborhood. While calculating the dynamic analysis window we don't consider the merchants because, in the transaction network, merchants usually have a much higher degree than cards and also have a more uniform timestamp difference distribution. We found that a static window size works best for calculating merchant neighborhoods. Algorithm 1 provides the pseudo-code for CuRL and tCuRL.

5   Experiments

In this paper, we benchmark CuRL and tCuRL with five baseline algorithms of entity representation learning and evaluate the performance of the embeddings in fraud detection task on the metrics - AUCPR and F_1 score. The paper also benchmarks models on running time to create entity embeddings.

5.1   Baseline Models

The size of embeddings is kept the same i.e., 128 in all the baseline models for fair benchmarking. After we create the embeddings for the cards and merchants, we use an internal classification tool to classify the transactions into fraud and non-fraud. The features fed into the fraud classification task are - card embedding and merchant embedding.

node2vec - The paper uses the StellarGraph implementation[1] of node2vec [12] in the experiments. The maximum length of random walk is set to 50 with 10 random walks per node. The skip-gram model uses a window size of 20, the min count is set to 0 and the number of epochs is set to 50. The *return hyperparameter* p is taken as 0.5 and the *in-out hyperparameter* q is set to 2.

metapath2vec - The paper again uses the StellarGraph implementation(See footnote 1) of metapath2vec [7]. For the experiments, we consider the metapaths as *card-merchant-card* and *merchant-card-merchant* and define the maximum length of the random walk to be 50 with 2 random walks per root node. The word2vec based skip-gram model is used with a window size of 20, the min count set to 0 and the number of epochs is 50.

BigGraph - Although the dataset considered for the experiments is relatively small, BigGraph [17] is taken into consideration since transaction networks involve billions of transactions. For the experiments, we use the default parameters as used by the authors in the official code repository[2] of the paper. The learning rate is set to 0.001 and the model is run for 30 epochs.

LINE - For learning the node representations using LINE we use the implementation provided by the authors[3]. The negative sample size is set to 5 and 0.025 is taken as the starting learning rate. The order-1 and order-2 embeddings are concatenated as recommended by the authors.

BiNE - We used the official implementation provided by the authors[4]. The default parameters are used i.e., 32 maximum walks per vertex, learning rate of 0.01, walk-stopping probability of 0.15, and window size of 5.

Pin2Vec - This is our own implementation of Pin2Vec [18] as we could not find any implementation by the authors. We trained the skip-gram models for 50 epochs to generate embeddings. We use a window size of 5 to pick neighbors from the sequences generated, to train the skip-gram models. We also use a timed approach with Pin2Vec and we name it **tPin2Vec**. tPin2Vec uses a variable window size for picking neighbors from sequences to train the skip-gram model. The calculation of the variable window sizes is similar to what we described in Sect. 4.2 for tCuRL. For experimentation purposes, tPin2Vec uses the same parameters as Pin2Vec.

5.2   Proposed Models

CuRL generates pairs of cards and merchants which are used to train a single skip-gram model which learns the embeddings for cards and merchants. To get the best possible results we trained the skip-gram model for 30 epochs (ep) to

---

[1] https://github.com/stellargraph/stellargraph.

[2] https://github.com/facebookresearch/PyTorch-BigGraph.

[3] https://github.com/tangjianpku/LINE.

[4] https://github.com/clhchtcjj/BiNE.

generate embeddings of length 128 (len). We use a window size of 15 (ws) to pick entity neighbors from the transaction graph. The min count parameter ($minc$) of the skip-gram model was set to 3.

tCuRL incorporates a time threshold to have a variable window size for forming pairs to train the skip-gram model. The cards are grouped into bins based on the number of transactions they are making. Further, for each bin, a time threshold is decided for the time between two transactions in a session, in a way such that 95% of the consecutive transactions have a time difference lesser than the threshold. We tune this percentage as a hyperparameter, $perc\_thresh$. The other parameters have the same values as CuRL.

6   Results

All the models were trained on a machine with a 2.6 GHz 6-Core Intel Core i7 processor with 16 GB of RAM. The runtimes are given in Table 4 while Table 3 gives the fraud detection metrics for the different algorithms. Section 6.1 compares our models' performance with the baseline models. In Sect. 6.2, we show the effect of the various parameters of the **tCuRL** model on its performance.

Table 3. Fraud classification results contrasting CuRL and tCuRL with the baseline methods on relevant metrics

Algorithm	AUCPR	F_1 score	Precision	Recall
metapath2vec	0.156	0.212	0.298	0.113
node2vec	0.159	0.230	0.267	0.158
BigGraph	0.771	0.833	0.890	0.770
LINE	0.777	0.823	0.865	**0.792**
BiNE	0.776	0.826	0.865	0.790
Pin2Vec	0.764	0.820	0.890	0.760
tPin2Vec	0.782	0.817	0.870	0.770
CuRL	0.797	0.829	0.878	0.785
tCuRL	**0.847**	**0.846**	**0.920**	0.783

Table 4. Comparing embedding generation time of CuRL and tCURL with the baseline methods

Algorithm	Time
metapath2vec	42 min
node2vec	77 min
BigGraph	3 min
LINE	50 s
BiNE	20 h
Pin2Vec	50 s
tPin2Vec	2.33 min
CuRL	3 min
tCuRL	3.13 min

6.1   Fraud Detection

We found that **tCuRL** outperforms all the baselines in terms of the AUCPR score. In general for imbalanced datasets like ours, AUCPR gives a good measure of performance. For the sake of comparison, we also present the results on other metrics as well. Methods that use simple random walks to generate sequences like metapath2vec and node2vec perform quite poorly. BiNE also employs random walks but preserves entity relationships and is able to achieve better performance.

However, these methods require a lot of time to generate embeddings. BigGraph, Pin2Vec, and tPin2Vec are efficient in generating embeddings but they fail to capture the cross-interactions between cards and merchants. The embedding generation process using LINE is also efficient but it learns the embeddings for first and second order neighbors separately rather than learning a single embedding. From Table 3, we see that CuRL and tCuRL outperform all the baseline models and also have an efficient embedding generation process, as shown in Table 4.

Furthermore, to show the quality of our generated embeddings, we use t-SNE to plot the transaction embeddings from tCuRL in a two-dimensional space. The transaction embeddings are the concatenation of the card and merchant embeddings involved in the transaction. It can be seen in Fig. 2 that most of the fraudulent transactions separate out from the non-fraudulent transactions, which makes it easier for the classification model to identify the frauds.

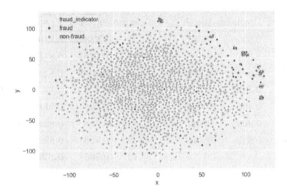

Fig. 2. t-SNE plot of the transaction embeddings from tCuRL (concatenation of card and merchant embeddings) mapped to a 2-D space.

6.2 Parameter Sensitivity

CuRL and tCuRL involve many parameters and this section examines how changing them affects the performance. Since CuRL and tCuRL share all parameters except the percentage threshold and Table 3 shows that tCuRL outperforms CuRL, we limit our parameter sensitivity experiments to tCuRL.

Table 5 shows that a smaller embedding size in the case of our model gives a better AUCPR. However, we keep an embedding size of 128 as transaction datasets are much larger in practice and a larger embedding size can store more information. This also helps to maintain uniformity among experiments. Similarly, Table 6 shows that increasing the window size improves the AUCPR for the model. This can be because increasing the window size while generating pairs allows the model to learn from a larger entity neighborhood. From Table 7, we see that the AUCPR of the model is highest at the middle min count values. A low min count can lead to a lot of noise during training. Whereas, a large min

Table 5. Effect of varying embedding length on performance metrics

len	AUCPR	F$_1$ score	Precision	Recall
16	**0.789**	0.821	0.874	0.774
32	0.786	0.816	0.874	0.774
64	0.778	0.816	0.865	0.772
128	0.784	0.819	0.864	**0.778**
256	0.766	**0.824**	**0.894**	0.764
512	0.780	0.823	0.876	0.776

Table 6. Effect of varying window size for neighborhood calculation

ws	AUCPR	F$_1$ score	Precision	Recall
2	0.773	0.819	0.865	0.778
3	0.761	0.813	0.867	0.765
5	**0.784**	0.819	0.864	0.778
7	0.778	0.833	**0.909**	0.769
10	0.782	0.819	0.863	**0.779**
12	0.780	**0.828**	0.904	0.764
15	**0.784**	0.821	0.883	0.767

Table 7. Effect of varying minimum count of cards for skip-gram training

minc	AUCPR	F$_1$ score	Precision	Recall
1	0.784	0.819	0.864	0.778
2	0.775	0.811	0.861	0.766
3	**0.789**	0.825	0.879	0.777
4	0.783	0.828	**0.891**	0.773
5	0.774	0.811	0.859	0.768
6	0.788	0.823	0.883	0.771
8	0.766	0.822	0.863	**0.785**
10	0.765	**0.829**	0.885	0.780

Table 8. Effect of varying number of epochs for skip-gram training

ep	AUCPR	F$_1$ score	Precision	Recall
5	0.765	0.816	0.862	0.774
10	0.767	0.816	0.864	0.773
15	0.781	0.823	0.873	0.778
20	0.783	0.819	0.864	0.778
25	0.784	0.819	0.864	0.778
30	**0.787**	0.823	0.876	0.776
35	0.782	0.820	0.869	0.776
40	0.778	**0.824**	0.867	**0.783**
45	0.778	0.819	**0.893**	0.756
50	0.786	0.819	0.876	0.769

Table 9. Varying the percentage of transactions that have time intervals below the threshold

perc_ thresh	AUCPR	F$_1$ score	Precision	Recall
60	0.780	0.816	0.864	0.774
65	0.778	0.823	0.947	0.801
70	0.787	0.820	0.872	0.774
75	0.790	0.819	0.864	0.778
80	0.784	0.819	0.864	0.778
85	0.790	0.834	**0.905**	0.774
90	0.802	**0.836**	0.868	**0.805**
95	**0.810**	0.833	0.879	0.792
100	0.797	0.829	0.878	0.785

count is also not acceptable as we could be missing out on important entity relationships. It can be seen from Table 8 that the AUCPR of the model drops when the number of epochs is too large, which could be because of the model overfitting on the training data. The results in Table 9 show that a higher AUCPR score is obtained on increasing the percentage threshold. Note that CuRL is the case when the percentage threshold is 100% and time between transactions is not considered. Also, it can be observed that reducing the percentage threshold to just 95% from 100% effectively removes the noise and improves the AUCPR considerably. Reducing it further leads to loss of information.

7   Conclusion

In this paper, we have proposed a method to generate embeddings for cards and merchants from transaction data. Our method introduced a novel technique for creating pairs of cards and merchants, that effectively captured their cross interactions, before applying the skip-gram model to generate the embeddings. We also introduced a dynamic session-based approach to reduce the noise in the embedding generation process by limiting the entity neighborhoods created while generating pairs. We also discussed our model's scalability and efficiency, while processing large number of transactions, and have compared the results with the baselines.

Future work directions could include testing the performance of our generated embeddings for other financial downstream tasks. Additional card and merchant features (geographical data, spend information) can also be incorporated in the embedding generation process. Researchers can also look into a more extensive hyperparameter testing for the experiments. Furthermore, efficient dynamic updation of embeddings is also something that can be looked into.

References

1. Akhilomen, J.: Data mining application for cyber credit-card fraud detection system. In: Perner, P. (ed.) ICDM 2013. LNCS (LNAI), vol. 7987, pp. 218–228. Springer, Heidelberg (2013). https://doi.org/10.1007/978-3-642-39736-3_17
2. Bahnsen, A.C., Stojanovic, A., Aouada, D., Ottersten, B.: Improving credit card fraud detection with calibrated probabilities. In: SDM (2014)
3. Bruss, C., et al.: DeepTrax: embedding graphs of financial transactions. In: 2019 18th IEEE International Conference on Machine Learning and Applications (2019)
4. Cao, B., Mao, M., Viidu, S., Yu, P.S.: HitFraud: a broad learning approach for collective fraud detection in heterogeneous information networks. In: 2017 IEEE International Conference on Data Mining (ICDM), pp. 769–774 (2017)
5. Chandola, V., et al.: Anomaly detection: a survey. ACM Comput. Surv. **41**, 1–58 (2009)
6. Chawla, N.V., Bowyer, K.W., Hall, L.O., Kegelmeyer, W.P.: Smote: synthetic minority over-sampling technique. J. Artif. Intell. Res. **16**, 321–357 (2002)
7. Dong, Y., Chawla, N.V., Swami, A.: metapath2vec: scalable representation learning for heterogeneous networks. In: KDD 2017, pp. 135–144. ACM (2017)

8. Eberle, W., Holder, L.: Mining for insider threats in business transactions and processes. In: 2009 IEEE Symposium on Computational Intelligence and Data Mining, pp. 163–170 (2009)

9. El hlouli, F.Z., Riffi, J., et al.: Credit card fraud detection based on multilayer perceptron and extreme learning machine architectures. In: International Conference on Intelligent Systems and Computer Vision (2020)

10. Gallagher, B., Eliassi-Rad, T.: Leveraging label-independent features for classification in sparsely labeled networks: an empirical study. In: Giles, L., Smith, M., Yen, J., Zhang, H. (eds.) SNAKDD 2008. LNCS, vol. 5498, pp. 1–19. Springer, Heidelberg (2010). https://doi.org/10.1007/978-3-642-14929-0_1

11. Gao, M., Chen, L., He, X., Zhou, A.: BiNE: bipartite network embedding. In: SIGIR 2018. Association for Computing Machinery (2018)

12. Grover, A., Leskovec, J.: Node2vec: scalable feature learning for networks. In: Proceedings of the 22nd ACM SIGKDD International Conference on Knowledge Discovery and Data Mining, KDD 2016. Association for Computing Machinery (2016)

13. Henderson, K., et al.: It's who you know: graph mining using recursive structural features. In: KDD (2011)

14. Hooi, B., Song, H.A., Beutel, A., Shah, N., Shin, K., Faloutsos, C.: Fraudar: bounding graph fraud in the face of camouflage. In: Proceedings of the 22nd ACM SIGKDD International Conference on Knowledge Discovery and Data Mining (2016)

15. Huang, D., Mu, D., Yang, L., Cai, X.: Codetect: financial fraud detection with anomaly feature detection. IEEE Access 6, 19161–19174 (2018)

16. Jurgovsky, J., Granitzer, M., et al.: Sequence classification for credit-card fraud detection. Expert Syst. Appl. 100, 234–245 (2018)

17. Lerer, A., Wu, L., et al.: PyTorch-BigGraph: a large-scale graph embedding system. In: Proceedings of the 2nd SysML Conference, Palo Alto, CA, USA (2019)

18. Liu, D.C., et al.: Related pins at pinterest: the evolution of a real-world recommender system. In: Proceedings of the 26th International Conference on World Wide Web Companion (2017)

19. Mikolov, T., Chen, K., Corrado, G.S., Dean, J.: Efficient estimation of word representations in vector space (2013)

20. Misra, S., Thakur, S., Ghosh, M., Saha, S.K.: An autoencoder based model for detecting fraudulent credit card transaction. Procedia Comput. Sci. 167, 254–262 (2020). International Conference on Computational Intelligence and Data Science

21. Molloy, I., et al.: Graph analytics for real-time scoring of cross-channel transactional fraud. In: Grossklags, J., Preneel, B. (eds.) FC 2016. LNCS, vol. 9603, pp. 22–40. Springer, Heidelberg (2017). https://doi.org/10.1007/978-3-662-54970-4_2

22. Moschini, G., Houssou, R., Bovay, J., Robert-Nicoud, S.: Anomaly and fraud detection in credit card transactions using the ARIMA model (2020)

23. Patki, N., Wedge, R., Veeramachaneni, K.: The synthetic data vault. In: 2016 IEEE International Conference on Data Science and Advanced Analytics (DSAA) (2016)

24. Perozzi, B., Al-Rfou, R., Skiena, S.: DeepWalk: online learning of social representations. In: Proceedings of the 20th ACM SIGKDD International Conference on Knowledge Discovery and Data Mining (2014)

25. Salim Hasham, R.H., Wavra, R.: Combating payments fraud and enhancing customer experience (2018). https://mck.co/2Qi4ead

26. Shen, A., et al.: Application of classification models on credit card fraud detection. In: International Conference on Service Systems and Service Management (2007)

27. Tang, J., Qu, M., Wang, M., Zhang, M., Yan, J., Mei, Q.: Line: large-scale information network embedding. In: Proceedings of the 24th International Conference on World Wide Web (2015)
28. The Nilson Report, Issue 1164: Card Fraud Worldwide 2010–2027 - Card Fraud Losses Reach $27.85 billion (2019). https://bit.ly/3uZ1v4D
29. Van Vlasselaer, V., et al.: APATE: a novel approach for automated credit card transaction fraud detection using network-based extensions. Decis. Support Syst. **75**, 38–48 (2015)
30. Zheng, E.H., Zou, C., Sun, J., Chen, L., Li, P.: SVM-based cost-sensitive classification algorithm with error cost and class-dependent reject cost. In: 2010 Second International Conference on Machine Learning and Computing, pp. 233–236 (2010)
31. Zou, J., Zhang, J., Jiang, P.: Credit card fraud detection using autoencoder neural network (2019)

Revisiting Loss Functions for Person Re-identification

Dustin Aganian$^{(\boxtimes)}$, Markus Eisenbach, Joachim Wagner, Daniel Seichter,
and Horst-Michael Gross

Ilmenau University of Technology, 98693 Ilmenau, Germany
dustin.aganian@tu-ilmenau.de

Abstract. Appearance-based person re-identification is very challenging, i.a. due to changing illumination, image distortion, and differences
in viewpoint. Therefore, it is crucial to learn an expressive feature embedding that compensates for changing environmental conditions. There are
many loss functions available to achieve this goal. However, it is hard to
judge which one is the best. In related work, the experiments are only performed on the same datasets, but the use of different setups and different
training techniques compromises the comparability. Therefore, we compare the most widely used and most promising loss functions under identical conditions on three different setups. We provide insights into why some
of the loss functions work better than others and what additional benefits they provide. We further propose sequential training as an additional
training trick that improves the performance of most loss functions. In our
conclusion, we provide guidance for future usage an d research regarding
loss functions for appearance-based person re-identification. Source code
is available (Source code: https://www.tu-ilmenau.de/neurob/data-sets-
code/re-id-loss/).

Keywords: Person re-identification · Deep learning · Representation
learning · Loss functions · Softmax loss · Triplet hard loss · Ring loss ·
Center loss · Additive angular margin loss · Circle loss

1 Introduction and Related Work

For appearance-based person re-identification (ReID), there are two strategies
for learning a feature embedding using deep convolutional networks (see Fig. 1):
First, the training can be formulated as a classification problem, where each
person in the training set represents a separate class. By fitting the model this
way, the penultimate layer forms a meaningful feature embedding that can subsequently be used for ReID. Second, the feature embedding can be learned directly
using tuples of person images including a match and a mismatch. The objective
is to create feature vectors being more similar to each other for matching pairs
than for mismatches. In both strategies, the choice of a suitable loss function is
crucial.

This work has received funding from the Carl Zeiss Foundation as part of the project
E4SM under grant agreement no. P2017-01-005.

I. Farkaš et al. (Eds.): ICANN 2021, LNCS 12895, pp. 30–42, 2021.
https://doi.org/10.1007/978-3-030-86383-8_3

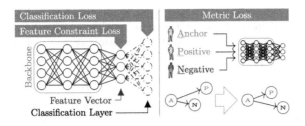

Fig. 1. Basic concepts for learning a feature embedding by applying different loss functions. Left: Formulation as classification problem. Right: Triplet-based training.

Figure 2 shows applicable loss functions, categorized into three types. The state of the art for ReID most often uses a combination of up to three loss functions simultaneously for training. Usually, one loss function from each category shown in Fig. 2 is selected. A popular baseline [10] composes softmax loss [1], triplet loss [11], and center loss [16]. Therefore, recent work mainly builds on these loss functions, despite there being more advanced loss functions available in each of the categories.

The single influence of these loss functions is rarely evaluated. In [6], the performance of softmax loss [1], a.k.a. ID loss, and its extensions multiplicative angular margin loss (MAML, A-Softmax, SphereFace) [8,9], additive cosine margin loss (ACML, CosFace) [13,15], and additive angular margin loss (AAML, ArcFace) [2] is compared. However, this comparison is done using a very weak baseline (see Table 1). There is also related work that evaluates the individual influence of loss functions for ReID, e.g., center loss [16] was used in [7] and ring

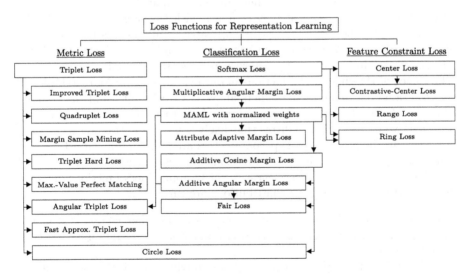

Fig. 2. Systematization of loss functions. Arrows show which loss functions improve previous ones. In our analysis, we incorporate loss functions of all three categories.

Table 1. Results of related work using a ResNet50 backbone on the Market-1501 dataset compared to our results using the same backbone with modern training techniques [10] and an identical setup for each loss function. A "—" signifies that the loss function was not evaluated on Market-1501 in the paper.

Loss	ReID	Paper		Ours	
		Rank 1	mAP	Rank 1	mAP
Center Loss	[7]	—	—	94.4	85.3
Ring Loss	[5]	—	—	94.7	86.1
Triplet Hard Loss	[4]	84.9	69.1	92.4	82.7
Softmax Loss, a.k.a. ID loss	[6]	71.0	46.3	93.1	82.0
MAML (with $\|\mathbf{w}_j\|_2 = 1$),	[6]	78.4	56.0		
a.k.a. A-Softmax (SphereFace)	[3]	94.4	83.6	95.2	87.9
	[12]	92.4	83.8		
ACML (CosFace)	[6]	77.8	56.2	95.4	87.7
AAML (ArcFace)	[6]	79.2	57.3	95.5	88.1
Circle Loss	[12]	94.2	84.9	95.5	88.4

loss [19] was used in [5], or introduced new loss functions, like circle loss [12], triplet hard loss [4], or other triplet loss extensions as in Fig. 2. All these papers share the same problem: They either compare with a weak baseline (mainly softmax loss, triplet loss, or MAML) or do not compare with other loss functions at all but use the loss functions in another context. Other loss functions (Attribute Adaptive Margin Loss, Viewpoint-aware Loss) need additional label information (attributes, camera IDs, etc.) and, thus, are not comparable. A second problem can be seen in Table 1: Even if they compare with the same baseline (MAML with $\|\mathbf{w}_j\|_2 = 1$), the results are not comparable at all. This is mainly due to different setups that include only subsets of new training techniques introduced in recent years that significantly improve results [10]. Table 1 also shows the performance of the single loss functions in our experiments on Market-1501 [18] using the same backbone as the papers above – a ResNet50 – and the training techniques of [10]. Our results are consistently better than the results reported in the respective papers. This confirms that likely some of the training techniques are missing in these papers or hyperparameters are not carefully tuned.

In order to find out, which loss functions of each category are most useful for ReID, we evaluate the effectiveness of solely applied loss functions on three different setups. Therefore, our contributions are three-fold:

1. Our evaluation is the first attempt towards a more fair comparison of the most promising and most often used loss functions for ReID under identical conditions on strong setups.
2. Due to different setups, we gain insights into why some of the loss functions work better than others and what additional benefits they provide.
3. We propose sequential training as an additional training trick that improves the performance of most loss functions.

2   Loss Functions

In our comparison, we include the typical baselines for the three categories of loss functions – triplet (hard) loss as metric loss, softmax loss as classification loss, and center loss as feature constraint loss – as well as additive angular margin loss as the most promising classification loss without adaptive margin and without side information based on benchmark results in ReID and face identification, ring loss as the most promising feature constraint loss based on benchmark results in face identification, and circle loss as the most promising loss with weighted similarities based on how far they are from the optimization goal. Circle loss can be formulated as metric loss or classification loss. We report the classification loss results since, in our experiments, this version clearly outperforms the metric loss version.

2.1   Metric Loss

Most metric loss functions train a feature vector $\mathbf{x}$ directly with triplets of an anchor $\mathbf{x}^a$, a positive $\mathbf{x}^p$, and a negative $\mathbf{x}^n$ [4,11]. As shown in Fig. 1, the objective is to make the distances of matches $d(\mathbf{x}^a, \mathbf{x}^p)$ much smaller than the distances of mismatches $d(\mathbf{x}^a, \mathbf{x}^n)$. The metric loss functions differ in how the triplets are compiled and how deviations from the objective are penalized.

Triplet Hard Loss (THL) [4] builds triplets of P random classes and K images per class for each mini-batch. Then, for each person i and anchor a, the hardest positive $\hat{p}_i^a$ and hardest negative $\hat{n}_i^a$ triplets of this batch that violate the objective the most are selected. Thus, the loss L_{THL} is calculated as in Eq. 1 with either a hard [11] or a soft margin [4].

$$L_{\text{THL}} = \frac{1}{PK} \sum_{i=1}^{P} \sum_{a=1}^{K} f(\hat{p}_i^a, \hat{n}_i^a) \tag{1}$$

$$f_{Hard}(\hat{p}_i^a, \hat{n}_i^a) = \max\left(\hat{p}_i^a + m - \hat{n}_i^a, 0\right) \qquad \hat{p}_i^a = \max_{p=1...K} \|\mathbf{x}_i^a - \mathbf{x}_i^p\|_2$$

$$f_{Soft}(\hat{p}_i^a, \hat{n}_i^a) = \ln\left(1 + e^{\hat{p}_i^a - \hat{n}_i^a}\right) \qquad \hat{n}_i^a = \min_{\substack{j=1...P \\ n=1...K \\ j \neq i}} \|\mathbf{x}_i^a - \mathbf{x}_j^n\|_2$$

2.2   Classification Loss

When using a classification loss, a classification layer with each person in the training set as a separate class is added after the feature vector $\mathbf{x}$ during training (see Fig. 1). The general equation for a mini-batch of size N and K classes, with y_i being the class label of the i-th mini-batch sample, is as follows:

$$L_{\text{Class}} = \frac{1}{N} \sum_{i=1}^{N} - \log\left(\frac{e^{f_p(\mathbf{x}_i, \mathbf{w}_{y_i})}}{e^{f_p(\mathbf{x}_i, \mathbf{w}_{y_i})} + \sum_{\substack{k=1...K \\ k \neq y_i}} e^{f_n(\mathbf{x}_i, \mathbf{w}_k)}}\right) \tag{2}$$

Softmax Loss (SL) [1], a.k.a. ID loss, applies a cross entropy loss on a softmax output layer. Therefore, f_p and f_n in Eq. 2 are defined as:

$$f_p^{\mathrm{SL}}(\mathbf{x}_i, \mathbf{w}_{y_i}) = \mathbf{x}_i \cdot \mathbf{w}_{y_i} + b_{y_i} \qquad f_n^{\mathrm{SL}}(\mathbf{x}_i, \mathbf{w}_k) = \mathbf{x}_i \cdot \mathbf{w}_k + b_k \qquad (3)$$

Additive Angular Margin Loss (AAML) [2] is an extension of SL. It normalizes the feature ($\|\mathbf{x}_i\|_2 = \gamma$) and weight vectors ($\|\mathbf{w}_j\|_2 = 1$), removes the bias ($b_j = 0$), and adds a decision margin m regarding the angle θ_{x_i,w_j} in order to increase the inter-class distance and decrease the inner-class variance. Therefore, f_p and f_n in Eq. 2 are defined as:

$$f_p^{\mathrm{AAML}}(\mathbf{x}_i, \mathbf{w}_{y_i}) = \gamma \cos(\theta_{\mathbf{x}_i,\mathbf{w}_{y_i}} + m) \qquad (4)$$

$$f_n^{\mathrm{AAML}}(\mathbf{x}_i, \mathbf{w}_k) = \gamma \cos(\theta_{\mathbf{x}_i,\mathbf{w}_k})$$

$$\text{with } \cos(\theta_{\mathbf{x}_i,\mathbf{w}_j}) = s_{\mathbf{x}_i,\mathbf{w}_j} = \frac{\mathbf{x}_i \cdot \mathbf{w}_j}{\|\mathbf{x}_i\|_2 \cdot \|\mathbf{w}_j\|_2}$$

Circle Loss (CirL) [12] implements a circular optimization goal. Therefore, f_p and f_n in Eq. 2 are defined as:

$$f_p^{\mathrm{CirL}}(\mathbf{x}_i, \mathbf{w}_{y_i}) = \gamma \cdot \max(-(s_{\mathbf{x}_i,\mathbf{w}_{y_i}} - 1) + m, 0) \cdot (s_{\mathbf{x}_i,\mathbf{w}_{y_i}} - 1 + m) \qquad (5)$$

$$f_n^{\mathrm{CirL}}(\mathbf{x}_i, \mathbf{w}_k) = \gamma \cdot \max(s_{\mathbf{x}_i,\mathbf{w}_k} + m, 0) \cdot (s_{\mathbf{x}_i,\mathbf{w}_k} - m)$$

2.3 Feature Constraint Loss

Loss functions of this type restrict the feature vector in a classification setting. Usually, it is scaled by a weighting factor λ and added to the classification loss.

Center Loss (CenL) [16] forces the model to learn feature vectors with low distances to their respective class centers $\mathbf{c}_{y_i}$. Therefore, the inner-class distance is reduced. For a mini-batch of size N the loss L_{CenL} is defined as:

$$L_{\mathrm{CenL}} = \frac{\lambda}{2} \sum_{i=1}^{N} \|\mathbf{x}_i - \mathbf{c}_{y_i}\|_2^2 \qquad (6)$$

Ring Loss (RL) [19] forces the model to keep the feature vector on a hypersphere with a radius R, which leads to more robust feature vectors. The loss L_{RL} with the trainable parameter R is defined as:

$$L_{\mathrm{RL}} = \frac{\lambda}{2N} \sum_{i=1}^{N} (\|\mathbf{x}_i\|_2 - R)^2 \qquad (7)$$

3 Experiments

In our experiments, we compare the performance of the six loss functions described above under identical conditions on three strong setups. In addition, we analyze the costs of a setup that follows best practices in real-world applications, the complexity of training, the influence of different initializations in a sequential training setting, and the generalization abilities.

3.1 Setup

In order to achieve comparability with most state-of-the-art ReID papers, we use a ResNet50 as backbone, pre-trained on ImageNet unless otherwise stated (PyTorch weights as in state of the art). For optimization, we either used SGD with momentum of 0.9 or Adam with standard parameters. The extensive hyperparameter search for each loss function included the learning rate, batch size, weight decay, batch composition, as well as scaling/weighting factors and the margin if applicable. All experiments were conducted using TensorFlow 2.

For assessing the ReID performance, we report the mean Average Precision (mAP) and the rank-1 accuracy (rank 1) in percentage values. We applied the single query protocol of [18] by running the evaluation code of [4]. The similarities of the feature vectors were calculated with cosine similarity in favor of the Euclidean distance since results are better regardless of the loss function used. For metric losses, it does not make a significant difference to the results, but for all other loss functions, the results always improve by a few percentage points.

Baseline Setups

In our experiments, we compare the models trained with the different loss functions on three baseline setups: The strong baseline as described in [10], the Multi Granular Network (MGN) architecture as described in [14], and a simple baseline, that we derived from [10]. In all setups, we use ResNet50 as architectural backbone. We trained and tested on the Market-1501 [18] dataset.[1]

Modifications for Simple Baseline: With the simple baseline, we follow two strategies: First, we intend to leave some room for improvements by the loss functions and, therefore, deliberately abandon some of the training techniques described in [10] that are not essential. Namely, these are label smoothing, a change of the last stride in the backbone, random erasing augmentation, and a batch normalization neck.

Second, we want this training setup to follow best practices in real-world applications. These include the use of validation data to avoid optimizing on the test data, and the use of side data to learn a better generalization. We used DukeMTMC-reID and CUHK03-NP as side data. The validation set is obtained by splitting off 10% of the training data. For this, we randomly drew from the 904 persons with the most samples. This resulted in 36,823 training samples incorporating 2,220 different persons for our simple baseline setup. We further employ an additional fully connected layer after the last ResNet block. This can reduce the size of the feature vector, which is particularly relevant for some applications with real-time and data-storage constraints. In the following, we simply refer to this setup as baseline.

---

[1] We also evaluated the performance on DukeMTMC-reID, and CUHK03-NP, but since the results are very similar and due to space restrictions, we decided in favor of reporting only the Market-1501 results in this paper.

3.2   Costs of a Setup for Real-World Applications

While using these best practices in our baseline setup brings us closer to a real-world application, this choice comes with a cost that diminishes the test results on the typical publicly available datasets in ReID. Therefore, we only use these techniques for comparison on our first baseline. For the strong baseline and MGN architecture, we follow the typical protocol from the state of the art in order to ensure our results are comparable.

In the following, these costs are described numerically using an example. In this example, we have trained the baseline with THL and SL. Our experiment reveals that if we add the training data of DukeMTMC-reID and CUHK03-NP as side data to the training and not just use the Market-1501 training data, the test result on the Market-1501 test data worsens by about 2 percentage points (p.p.) for the mAP and by about 1 p.p. for rank 1. If we split off validation data from the training data, then the mAP worsens by another 3 p.p. and rank 1 by about 2 p.p. These declines were expected since when using side data, the network no longer overspecializes for the test data on the dataset and when using validation, the training data is somewhat reduced and the hyperparameters are no longer optimized on the test data. Furthermore, the use of the additional fully connected layer also results in a decrease in performance by slightly under 2 p.p. for mAP and 1 p.p. for rank 1. However, it is not possible to avoid this if smaller feature vectors are needed on the target system. Likewise, as also stated in [17], this shows that an additional fully connected layer, which is still often used in current works (e.g. in MGN [14]), does not necessarily have to be advantageous.

3.3   Complexity of Training

All loss functions achieve bad results with randomly initialized weights. Therefore, transferring ImageNet weights is crucial for representation learning in ReID. In the following, we report challenges we have faced during training:

Easy to Train: We did not face any difficulties in learning a feature embedding for ReID using *SL*. Similarly, it is easy to train with a constraint loss (*RL, CenL*). We found that a good choice of the loss weighting factor λ is important. The larger the feature vector is, the smaller λ has to be chosen since, at the beginning of the training, the ℓ_2 norm of a large feature vector is huge, resulting in a huge constraint loss. Otherwise, the constraint loss would dominate the overall loss, which is composed of SL and the constraint loss. For *THL*, in our experiments, it was crucial to apply some kind of learning rate warm-up. Except for that, there is no difficulty.

Hard to Train: Contrarily to SL, while it is easy and straightforward to implement the SL extensions *AAML* and *CirL*, these advanced loss functions are quite difficult to handle. When initializing with ImageNet weights, the trained models performed poorly if training converged at all. In order to overcome the problems of having a margin right from the beginning of the training, we first pre-trained with SL on ReID data, transferred the weights, and then started training with the SL extension. We refer to this approach as sequential training. Figure 5 gives evidence that pre-training on ReID data simplifies the problem to be solved (see Sect. 3.5).

3.4   Experimental Comparison

For each loss function included in our analysis, we performed an extensive hyper-parameter search to ensure a fair comparison. Figure 3 shows the individual results[2] using the best hyperparameter combinations for all tested network architectures. The best models were determined based on the largest mAP, as the mAP is closer to a realistic use case, as not only the simplest example from many gallery matches is assessed in the evaluation, as it is the case for rank 1.

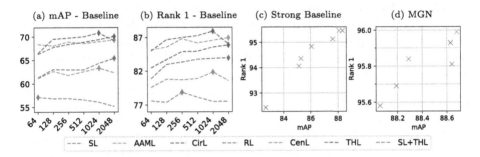

Fig. 3. (a, b) mAP and rank 1 baseline results for different feature vector sizes trained with the best hyperparameters for each of the loss functions. (c) mAP and rank 1 results for our strong baseline and (d) the MGN architecture for each of the loss functions. All results are reported on the Market-1501 [18] test set.

Baseline: We evaluated six different feature vector sizes for each of the loss functions. Figure 3 (a) and (b) show the results for each loss function. The best result is highlighted by a diamond. Based on mAP, CirL performs best, followed by THL and AAML. Next up are the feature constraint losses RL and CenL, which were both used together with SL, whereas SL alone scores significantly worse. When comparing the results over different feature vector sizes, it becomes apparent that the largest feature vector (2048) does not always provide the best result, e.g., for CirL, the best result is achieved with 1024, SL works best with 64. Often, in more complex architectures, such as the local-feature-focused MGN, additional fully connected layers are added in different heads after the last ResNet block. Our results show that depending on the loss function that is applied on such a head with an additional fully connected layer, the feature vector size might be adjusted to achieve better results.

---

[2] The standard deviation for 8 training runs of the best loss function in each case is $\sigma_{mAP} = 0.226\%, \sigma_{r1} = 0.418\%$ for the baseline setup, $\sigma_{mAP} = 0.079\%, \sigma_{r1} = 0.114\%$ for the strong baseline, and $\sigma_{mAP} = 0.046\%, \sigma_{r1} = 0.098\%$ for MGN.

Strong Baseline, MGN: To get a better comparison of the loss functions, we also tested them on two appropriate state-of-the-art architectures, the strong baseline of [10] and the MGN architecture [14]. The results of these experiments can be found on Fig. 3 (c) and (d). Here, we also trained SL together with THL (SL + THL), as this is a common approach. The best loss functions on these two architectures are AAML and CirL, followed by SL+THL. For the strong baseline, RL, CenL, and SL follow next, and THL scored the worst. On the MGN architecture, all loss functions, with the exception of THL,[3] produce results in a very close range of values. In this case, SL + THL corresponds to the original MGN setup proposed in [14]. For THL, it was applied on all heads, and for all other loss functions, the THL heads were removed and SL was extended with a constraint loss or replaced with a classification loss.

Findings: In general, all experiments have shown that AAML and CirL achieve the best results and far surpass the still often used SL. The comparable performance of AAML and CirL also shows the necessity to always compare with strong reference approaches on a strong setup when introducing a new loss function, since the postulated superiority of CirL for ReID in [12] is only due to a weaker reference approach.

It can be seen that all the loss functions perform differently depending on the architecture. For the baseline, which corresponds to a typical architecture a metric loss is applied to, THL is on a par with AAML and CirL. On the strong baseline, however, THL is among the weakest loss functions. Mainly this is due to training techniques in the strong baseline and MGN that help the classification loss functions to learn features better suited for cosine similarity, but do not benefit (Euclidean-distance-based) metric loss functions equally strong.

Furthermore, RL beats CenL on the strong baseline. However for MGN it is the other way around. This shows that the architecture plays a decisive role when choosing a loss function. It also shows that when developing new loss functions, a comparison with existing ones should always be done on various setups.

3.5   Generalization Ability and Sequential Training

Two aspects will be examined in this section. First, the generalization abilities of the loss functions are investigated by using validation data. Second, the sequential training of loss functions is examined in more detail. These investigations are performed on our baseline setup, since it follows best practices as described in Sect. 3.1 and Sect. 3.2. Especially generalization abilities are a quality that is in great demand in real-world applications and is often disregarded in the state of the art with the typical approach of hyperspecialization of training methods on public datasets. Fig. 4 shows the test and validation results of the best models from Fig. 3 (a) and (b). Furthermore, the results from our sequential training experiments are shown. The models with ImageNet in the index were initialized with weights from an ImageNet training and the models with SL in the index

---

[3] Results for THL (mAP: 79.38% and rank 1: 90.80%) are omitted in Fig. 3(d) for visualization purposes.

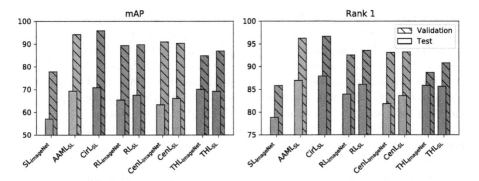

Fig. 4. Comparison of validation and test results. The validation split is taken from the training set of Market-1501, DukeMTMC-reID, and CUHK03-NP. The test data are from Market-1501.

were initialized with weights from a previous ImageNet-initialized training with SL on ReID data.

Generalization: First, we examine the difference between validation and test results, and thus the generalization abilities of the loss functions. As mentioned in Sect. 3.1, we split the training data into a training and a validation set. Images in the training and validation set are disjoint but comprise the same persons. In contrast, the test set contains different persons. Therefore, on the validation set, we measure the model's ability to distinguish known persons, while on the test set, we measure the generalization ability to unknown data. CirL and AAML achieve the best validation results by a wide margin. The constraint losses, THL, and lastly SL follow in descending order. When comparing the mAP test results, THL is about as good as AAML and CirL. This shows that THL is less likely to memorize the classes in the training data and more likely to achieve good generalization ability. This is a reasonable explanation why THL is used in addition to SL (SL + THL) in current approaches, since according to these results, the additional use of a metric loss in addition to a classification loss induces the network to develop better generalization capabilities. This is also an important conclusion for practical applications where the generalization abilities of a network should be maximized.

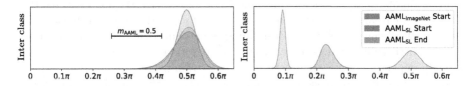

Fig. 5. Different initializations influence the distributions of inner-class angles $\theta_{x_i,w_{y_i}}$ and inter-class angles θ_{x_i,w_j} (training data).

Sequential Training: In order to train with AAML and CirL successfully, we had to initialize the models with weights from a previous training with SL on the ReID data. This approach is not described in the state of the art, but this is the easiest way to achieve similar results to the state of the art when training with these classification loss functions that introduce a margin. For a closer look at why this sequential training works so well, Fig. 5 shows the inner-class and inter-class angles when training with AAML[4]. As can be seen, in a typical initialization with ImageNet weights (gray), the inner- and inter-class angles are distributed in the same range of values and, thus, very large errors occur in the beginning of the training. If weights from a pre-training with SL on ReID data (purple) are used for initialization, then already at the beginning many inner-class angles are by m smaller than the inter-class angles, so the problem becomes simpler and the training easier to handle. Thus, right from the start the loss functions mainly focus on hard positives and hard negatives that violate the margin constraint.

Since this has led to many benefits for AAML and CirL, we further examined whether other loss functions that complicate the optimization target also benefit from such an approach. The results of these sequential training experiments are shown in Fig. 4. As it turns out, RL and CenL benefit from this approach. The optimizer has an easier problem to solve right at the beginning of the training. Thus, the training simplifies, which results in a better generalization. This can be seen in Fig. 4 when comparing the validation and test results with ($\mathrm{RL_{SL}}$, $\mathrm{CenL_{SL}}$) and without sequential training ($\mathrm{RL_{ImageNet}}$, $\mathrm{CenL_{ImageNet}}$). While validation results are similar, test results are better with sequential training.

On the other hand, the THL results worsen with sequential training. Most likely, many class-specific aspects of the training data are learned in the pre-training with SL. Thus, there is no way to learn an equally good generalization as in the less biased initialization with ImageNet weights. This is confirmed by the larger gap between validation and test results for $\mathrm{THL_{SL}}$ in comparison to $\mathrm{THL_{ImageNet}}$.

4   Conclusion

Our analysis is the first step towards a fairer comparison of loss functions for ReID. We compared the most widely used and most promising loss functions on three different setups. We confirmed the superiority of the softmax loss extension additive angular margin loss (AAML) and circle loss (CirL). We also analyzed the effect of sequential training, which is a necessity for AAML and CirL to perform well but also benefits constraint loss functions by improving the generalization ability. Furthermore, we observed that the performance of the loss functions strongly depends on the architecture and the training techniques used, which needs to be taken into consideration for loss function selection. Thus, a decisive ranking of all loss functions is not feasible as it would change depending on

---

[4] The angles of AAML are shown instead of those of CirL, because here the margin of AAML to be learned can be better visualized.

the training setup. Modern training techniques seem to benefit classification and constraint loss functions but do not improve the performance of (Euclidean-distance-based) metric loss functions, like triplet hard loss (THL), equally strong. However, THL tends to generalize better than classification and constraint loss functions. Therefore, it is understandable why metric loss functions are still in use as a complement to classification loss functions. Furthermore, we examined what costs are to be expected on a public test dataset when typical best practices are used for a real-world application.

As a consequence, for future work, we suggest to benchmark new loss functions on different setups, including a setup following best practices for real-world applications, and to compare against strong reference approaches. Based on our findings, we propose to always try sequential training instead of immediately starting training with ImageNet weights, as this is beneficial in most cases and allows stronger constraints to be enforced by loss functions. We also suggest to finally replace the softmax loss with a more advanced classification loss function.

References

1. Bridle, J.S.: Training stochastic model recognition algorithms as networks can lead to maximum mutual information estimation of parameters. In: NIPS (1990)
2. Deng, J., et al.: ArcFace: additive angular margin loss for deep face recognition. In: CVPR (2019)
3. Fan, X., et al.: SphereReID: deep hypersphere manifold embedding for person re-identification. In: JVCIR (2019)
4. Hermans, A., et al.: In defense of the triplet loss for person re-identification. arXiv (2017)
5. Jia, J., et al.: Frustratingly easy person re-identification: generalizing person re-id in practice. In: BMVC (2019)
6. Jie, S., et al.: A new discriminative feature learning for person re-identification using additive angular margin softmax loss. In: UCET (2019)
7. Jin, H., et al.: Deep person re-identification with improved embedding and efficient training. In: IJCB (2017)
8. Liu, W., et al.: Large-margin softmax loss for convolutional neural networks. In: ICML (2016)
9. Liu, W., et al.: SphereFace: deep hypersphere embedding for face recognition. In: CVPR (2017)
10. Luo, H., et al.: Bag of tricks and a strong baseline for deep person re-identification. In: CVPRW (2019)
11. Schroff, F., et al.: FaceNet: a unified embedding for face recognition and clustering. In: CVPR (2015)
12. Sun, Y., et al.: Circle loss: a unified perspective of pair similarity optimization. In: CVPR (2020)
13. Wang, F., et al.: Additive margin softmax for face verification. SPL **25**(7), 926–930 (2018)
14. Wang, G., et al.: Learning discriminative features with multiple granularities for person re-identification. In: ICM (2018)
15. Wang, H., et al.: CosFace: large margin cosine loss for deep face recognition. In: CVPR (2018)

16. Wen, Y., Zhang, K., Li, Z., Qiao, Yu.: A discriminative feature learning approach for deep face recognition. In: Leibe, B., Matas, J., Sebe, N., Welling, M. (eds.) ECCV 2016. LNCS, vol. 9911, pp. 499–515. Springer, Cham (2016). https://doi.org/10.1007/978-3-319-46478-7_31
17. Xiong, F., et al.: Good practices on building effective CNN baseline model for person re-identification. In: ICGIP (2019)
18. Zheng, L., et al.: Scalable person re-identification: a benchmark. In: ICCV (2015)
19. Zheng, Y., et al.: Ring loss: convex feature normalization for face recognition. In: CVPR (2018)

Statistical Characteristics of Deep Representations: An Empirical Investigation

Daeyoung Choi[ID], Kyungeun Lee[ID], Duhun Hwang[ID], and Wonjong Rhee[(✉)][ID]

Department of Intelligence and Information, Seoul National University, Seoul, Korea
{choid,ruddms0415,yelobean,wrhee}@snu.ac.kr

Abstract. In this study, the effects of eight representation regularization methods are investigated, including two newly developed rank regularizers (RR). The investigation shows that the statistical characteristics of representations such as correlation, sparsity, and rank can be manipulated as intended, during training. Furthermore, it is possible to improve the baseline performance simply by trying all the representation regularizers and fine-tuning the strength of their effects. In contrast to performance improvement, no consistent relationship between performance and statistical characteristics was observable. The results indicate that manipulation of statistical characteristics can be helpful for improving performance, but only indirectly through its influence on learning dynamics or its tuning effects.

Keywords: Deep learning · Representation learning · Regularization

1 Introduction

A learned representation can affect the performance of deep networks; the distributed and deep natures of the representation are the essential elements for the success of deep learning. Owing to the depth, deep networks have a greater expressiveness compared to other machine learning algorithms [11], or shallow networks [8,15]. In addition to the distributed and deep natures that have been intensively studied, a hidden layer's *representation characteristics* are considered to be important as well. Nonetheless, a relatively limited number of studies have been conducted on this matter. The goal of the present study is to better understand representation characteristics.

In the past several years, dropout [16] and batch normalization (BN) [12] have become essential regularization options. Additionally, manipulating representation characteristics has become increasingly popular for the improvement of performance [2,3,9]. The regularization methods often lead to improved performance, but rigorous explanation has been missing. Instead, it has been implied or conjectured that the manipulation of the representation can lead to improved

---

D. Choi and K. Lee—Authors contributed equally.

© Springer Nature Switzerland AG 2021
I. Farkaš et al. (Eds.): ICANN 2021, LNCS 12895, pp. 43–55, 2021.
https://doi.org/10.1007/978-3-030-86383-8_4

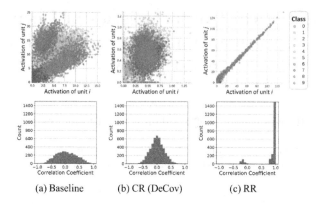

Fig. 1. Two representative units' activation scatter plots (upper plots) and histograms of the correlation coefficient distribution (lower plots) for the MNIST dataset. For a 6-layer MLP with 100 units for each layer, the fifth layer's activation vectors, calculated using 10,000 test samples, were used to select two neurons randomly and to generate the plots. (a) The baseline model shows a moderate correlation. (b) CR shows very low correlation. (c) RR shows very high correlation. Despite exhibiting totally different representation characteristics, the performances of the three models are comparable.

performance because of known and relevant concepts in machine learning (typically, a reduced generalization gap is quoted as the supporting empirical evidence). For instance, reduced co-adaptation (that is closely related to the correlation of the representation) has been put forth as a possible reason for the good performance of dropout; sparser or less correlated representations have been argued as better representations because the number of true underlying features must be limited. As a more general result on regularizing representation, [14] recently showed that aiming for certain activation characteristics ('disentangled representations' in their work, in contrast to 'representation characteristics' in our work) is not as universally meaningful as is often assumed in unsupervised learning.

In our study, statistical characteristics of deep representations are investigated for common supervised learning tasks. As a basic framework for the study, an extensive set of representation regularizers are considered, including a baseline model (no regularization). A total of eight regularization options are investigated to examine six characteristics of deep representations. Among the eight, rank regularizer (RR) and class-wise rank regularizer (cw-RR) were newly designed and tested in this study because of their association with important representation characteristics, such as correlation and rank. The regularizers aim to decrease the rank of representation (increase the correlation of representation) by reducing the stable rank of each mini-batch activation from the all-class samples (RR) or the same-class samples (cw-RR).

Some examples of the representations found with the regularizers are shown in Fig. 1. In the figure, correlation characteristics vary largely, depending on which regularizers are used. RR shows a strong correlation and has a performance comparable to CR as shown in the lower plot of Fig. 1. The comparable performance of RR under strong correlation is precisely the opposite of what has been conjectured for DeCov [3].

As we will show in Sect. 4, representation characteristics can be manipulated as intended, by applying a variety of regularizers. Additionally, all these regularizers, as a set, can be a useful tool for improving performance. However, the problem is that there is (perhaps unsurprisingly) no distinct pattern that can be used to assess which regularizer (representation characteristic) is likely to be helpful for a given task. All that can be concluded is that some regularizers would be helpful for any given task; but it is not possible to find which ones would qualify to be helpful for the same. In this paper, we do not claim that deep learning practitioners should not attempt to change representation characteristics owing to this problem. Instead, we empirically show that representation regularization can be a useful option for improving performance at the cost of tedious tuning. Despite of inconspicuous relationship between representation characteristics and performance, representation regularization can work as proxies that affect the learning dynamics of deep network training and thus indirectly improve the performance.

2   Related Works

Popular Regularization Methods. Many distinct regularization methods have been developed for deep learning. The most traditional methods are L1 and L2 weight regularizations (L1W, L2W). Dropout [16] and batch normalization (BN) [12] have shown large performance improvements in the context of many interesting tasks. With the extended definition of regularization in [10], many other methods such as data augmentation, adversarial training, and multi-task learning can be considered as regularization methods too. In this work, however, we limit our focus to the traditional, dropout, BN, and representation regularizations.

Representation Regularization Methods. A representation regularizer explicitly aims to modify a statistical property of the activation vectors, typically by using a penalty. One of the earliest representation regularizers is the L1 representation regularizer [9], and it applies an L1 penalty to the activation vector instead of the weight vectors. It encourages sparsity in the representation, and it is called L1R in this work. Similarly, [3] suggested DeCov that utilizes a penalizing loss function to reduce activation covariance among hidden units. [2] considered the extension to class-wise regularization and provide four representation regularizers: CR (Covariance regularizer), cw-CR (class-wise covariance regularizer), VR (variance regularizer), and cw-VR (class-wise variance regularizer). To note, CR is equivalent to DeCov.

Role of Explicit Regularizations. [17] showed that explicit regularizations such as L2W and dropout are not directly responsible for reducing or controlling the generalization error. Rather, they argue that performance improvement can be because of a tuning effect. [1] investigated the impact of explicit regularization on the memorization speed and generalization.

Table 1. Symbols and expressions of representation characteristics.

Characteristic	Symbol	Expression
AMPLITUDE	$|\bar{z}|$	$\mathbb{E}_i[|\,\mathbf{z}_{l,i}\,|]$
COVARIANCE	$\bar{c}$	$\mathbb{E}_{i\neq j}[|c_{i,j}|]$, where $c_{i,j} \triangleq \{\mathbf{C}_l\}_{i,j} = \mathbb{E}[(\mathbf{z}_{l,i}-\mu_{z_{l,i}})(\mathbf{z}_{l,j}-\mu_{z_{l,j}})]$
CORRELATION	$\bar{\rho}$	$\mathbb{E}_{i\neq j}[|\rho_{i,j}|]$, where $\rho_{i,j} \triangleq \{\mathbf{C}_l\}_{i,j}/\sigma_{z_{l,i}}\sigma_{z_{l,j}} = \mathbb{E}[(\mathbf{z}_{l,i}-\mu_{z_{l,i}})(\mathbf{z}_{l,j}-\mu_{z_{l,j}})]/\sigma_{z_{l,i}}\sigma_{z_{l,j}}$
SPARSITY	P_s	$\mathbb{E}_{i,n}[\mathbb{1}(z_{l,i}^n)]$, where $\mathbb{1}$ is an indicator function whose output is 1 only when $z_{l,i}^n = 0$
DEAD UNIT	P_d	$\mathbb{E}_i[\mathbb{1}(z_{l,i})]$, where $\mathbb{1}$ is an indicator function whose output is 1 only when $z_{l,i}^n = 0$ for all $n = 1,..,N$
RANK	r	$rank(\mathbf{C}_l)$; numerical evaluations are approximated as the stable rank $\|\mathbf{C}_l\|_F^2/\|\mathbf{C}_l\|_2^2$

Table 2. Penalty loss functions of representation regularizers.

Symbol	Penalty loss function	Description of regularization term
Ω_{CR}	$= \sum_{i\neq j}(c_{i,j})^2$	*Covariance* of representations calculated from all-class samples
$\Omega_{cw\text{-}CR}$	$= \sum_k \sum_{i\neq j}(c_{i,j}^{(k)})^2$	*Covariance* of representations calculated from **the same class samples**
Ω_{VR}	$= \sum_i v_i$	*Variance* of representations calculated from all-class samples
$\Omega_{cw\text{-}VR}$	$= \sum_k \sum_i v_{z_{l,i}}^{(k)}$	*Variance* of representations calculated from **the same class samples**
Ω_{L1R}	$= \sum_n \sum_i |z_{l,i}^n|$	*Absolute amplitude* of representations calculated from all-class samples
Ω_{RR}	$= \|\mathbf{Z}_l\|_F^2/\|\mathbf{Z}_l\|_2^2$	*Stable rank* of representations calculated from all-class samples
$\Omega_{cw\text{-}RR}$	$= \sum_k \left(\|\mathbf{Z}_l^{(k)}\|_F^2/\|\mathbf{Z}_l^{(k)}\|_2^2\right)$	*Stable rank* of representations calculated from **the same class samples**

3   Representation Characteristics and Regularizers

In this section, we elucidate a few mathematical formulations of representation characteristics and introduce representation regularizers.

3.1   Representation Characteristics

Consider a neural network $\mathcal{N}_\mathcal{A}$ whose architecture $\mathcal{A}$ is fixed and the weights for the lth layer are given by $\{\mathbf{W}_l\}$ and $\{\mathbf{b}_l\}$ after training. We write $\mathcal{N}_\mathcal{A} = (\mathbf{W}, \mathbf{b})$ to denote a network and $\mathbf{y}$ or $\mathcal{N}_\mathcal{A}(\mathbf{x})$ to refer to its deterministic output for a given input $\mathbf{x}$. The index l is omitted when the meaning is obvious. The activation vector of the lth layer for the given input $\mathbf{x}$ is denoted as $\mathbf{z}_l(\mathbf{x})$ or simply $\mathbf{z}_l$, and the ith element of $\mathbf{z}_l$ is denoted as $z_{l,i}$. The mean, variance, and standard deviation of $z_{l,i}$ over $p(\mathbf{x})$ are defined as $\mu_{z_{l,i}}$, $v_{z_{l,i}}$, and $\sigma_{z_{l,i}}$, respectively. We define class-wise statistics that are calculated using only samples of class k, out of a total of K labels in the mini-batch. Class-wise mean, covariance, and variance are defined as follows. $\mu_{z_{l,i}}^{(k)} = \mathbb{E}_{n \in S_k}[z_{l,i}^n]$. $c_{i,j}^{(k)} = \mathbb{E}_{n \in S_k}[(z_{l,i}^n - \mu_{z_{l,i}}^{(k)})(z_{l,j}^n - \mu_{z_{l,j}}^{(k)})]$. $v_{z_{l,i}}^{(k)} = c_{i,i}^{(k)}$. Here, S_k is the set that contains the indices of the samples with the class label k. Note that superscripts with and without parenthesis indicate class label and sample index, respectively.

The basic representation characteristics can be summarized as in Table 1. Since the true distribution of the data is not accessible, the numerical results are evaluated using the empirical distribution of the test dataset. $\mathbf{z}_l^n$ corresponds to the activation vector for the nth test data example, $\mathbf{x}^n$. For instance, $\mathbf{C}_l$ is calculated as the covariance matrix of N activation vectors $\{\mathbf{z}_l^1, ..., \mathbf{z}_l^N\}$ where $\mathbf{z}_l^n$ corresponds to the activation vector for the nth test data example, $\mathbf{x}^n$. Rank can be calculated by examining $\mathbf{C}_l$, but often there are small eigenvalues that hinder a proper assessment of the rank. Therefore, *stable rank* is evaluated instead.

3.2   Representation Regularizers

In this study, mainly eight options are considered: the baseline model (no regularizer) and seven models of regularizers (CR, cw-CR, VR, cw-VR, L1R, RR, cw-RR). We call the seven regularizers 'representation regularizers' because they are used to reduce generalization error by placing a penalty on the neurons' activations. Even though dropout and BN do not explicitly target to modify representation characteristics, they were also studied together because they certainly affect the representation characteristics. Regularization terms are added to the original cost function as penalty regularizers. The total cost function $\tilde{J}$ can be denoted as $\tilde{J} = J + \lambda \Omega(\mathbf{z})$, where λ is the loss weight ($\lambda \in [0, \infty)$). Each regularizer targets a different representation characteristic. For example, CR and VR reduce covariance and variance of the activations calculated from all-class samples, respectively. L1R decreases the absolute amplitude of activations calculated from all-class samples to make the activations sparser. Regularizers with the prefix 'cw-' are the class-wise counterparts of all-class regularizers. All the loss functions are summarized in Table 2.

Rank Regularizer. RR is designed to encourage a lower rank of representations and is used while training the network. Since the usual definition of rank can be very sensitive to small singular values, we use *stable rank* of the activation matrix $\mathbf{Z} = [\mathbf{z}_l^1, ..., \mathbf{z}_l^B]^T$ as a surrogate. Note that B instead of N activation vectors are used for each mini-batch. The stable rank of $\mathbf{Z}$ is defined as

$$\Omega_{RR} = \frac{\|\mathbf{Z}\|_F^2}{\|\mathbf{Z}\|_2^2} = \frac{\sum_i s_i^2}{\max_i s_i^2}, \tag{1}$$

where $\|\mathbf{Z}\|_F$ is the Frobenius norm, $\|\mathbf{Z}\|_2$ is the spectral norm, and $\{s_i\}$ are the singular values of $\mathbf{Z}$. From $\frac{\sum_i s_i^2}{\max_i s_i^2}$, it can be clearly seen that stable rank is upper-bounded by the rank that counts strictly positive singular values. As the spectral norm is based on singular value decomposition, calculating the derivative of the stable rank for every mini-batch is a computationally heavy operation. To reduce the computational burden, we apply an approximation using a special case of Hölder's inequality.

$$\Omega_{RR} = \frac{\|\mathbf{Z}\|_F^2}{\|\mathbf{Z}\|_2^2} = \frac{\text{trace}(\mathbf{Z}^T \mathbf{Z})}{\|\mathbf{Z}\|_2^2} \tag{2}$$

$$\geq \frac{\text{trace}(\mathbf{Z}^T \mathbf{Z})}{\|\mathbf{Z}\|_1 \|\mathbf{Z}\|_\infty} \tag{3}$$

$$= \frac{\sum_{i,n}(z_i^n)^2}{(\max_i \sum_{n=1}^{B} |z_i^n|)(\max_n \sum_{i=1}^{M} |z_i^n|)}, \tag{4}$$

where M is the number of units for the lth layer. The inequality $\|\mathbf{Z}\|_2 \leq \sqrt{\|\mathbf{Z}\|_1 \|\mathbf{Z}\|_\infty}$ is used, where $\|\mathbf{Z}\|_1$ is the maximum absolute column-wise sum of the matrix $\mathbf{Z}$ (sum of all activation values of unit i) and $\|\mathbf{Z}\|_\infty$ is the maximum absolute row-wise sum of the matrix $\mathbf{Z}$ (sum of all activation values of sample n). The extension of RR to cw-RR is straightforward.

4   Experiments

In this section, it is empirically shown that the regularization affects the statistical characteristics of deep representations. The relationship between performance and the representation characteristics is also examined. Finally, performance results on a variety of tasks and analysis are presented.

4.1   Experimental Settings

As examples of simple networks, we used a 6-layer MLP for the MNIST dataset, and a CNN with four convolutional layers and one fully-connected layer for the CIFAR-10/100 dataset. (In this paper, we call them 'MLP' and 'CNN' respectively, for convenience.) As examples of sophisticated networks, VGG-16 on the CIFAR-10/100, ResNet-18/50 on the ImageNet/Tiny-ImageNet datasets were used. For ResNet, a single fully-connected layer was added following the last average pooling layer. Validation performance was evaluated with different loss weights {0.001, 0.01, 0.1, 1, 10, 100, 1000}, and the one with the best validation performance for each regularizer and condition was chosen for testing. For ResNet, pre-trained models were fine-tuned with the regularizers. Each training trial was repeated five times. Unless mentioned; the mean and standard deviation of the five trials are reported. The mini-batch size was set to 100 for MLP (MNIST) and CNN (CIFAR-10), 128 for VGG-16 (CIFAR-10) and ResNet-50 (Tiny-ImageNet), and 256 for ResNet-18 (ImageNet). For CIFAR-100 that has 100 classes, mini-batch size of 500 was used to calculate meaningful class-wise statistics. Experiments with class-wise regularizers were not performed for ImageNet and Tiny-Imagenet datasets to avoid inefficient training of large mini-batch size.

4.2 Experimental Results

Effect of Regularization on Representation Characteristics. The representation characteristics were visually and quantitatively investigated, as shown in Fig. 2 and Table 3. In Fig. 2, it can be observed that the representation characteristics exhibit large variations depending on the choice of the regularizer. In particular, dropout shows a strong pair-wise correlation, as shown in the lower plot of Fig. 2(b). This is precisely the opposite of what has been believed for dropout. We discuss this phenomenon in Sect. 5. One can also notice that the scatter plot of BN results from the normalization of neuron activations. Even though not shown, the visualization of the CIFAR-10/100, the Tiny-ImageNet, and the ImageNet datasets showed similar patterns as in Fig. 2. (the patterns were less distinct for the class-wise regularizers).

Our quantitative result confirms the visualization. Each characteristic was obtained by applying the largest loss weight possible while maintaining comparable performance with the baseline model. The result confirms that the statistical characteristics targeted by each regularizer are manipulated as expected (**bold**)

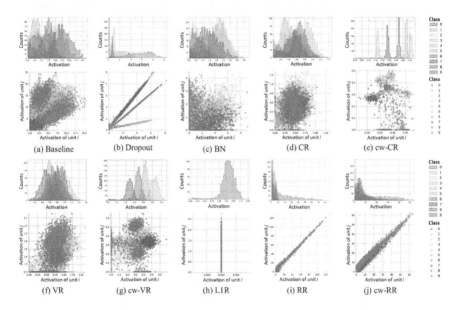

Fig. 2. Activation histogram of a single unit (upper plots) and the activation scatter plots of two randomly chosen units (lower plots) for a 6-layer MLP trained with the MNIST dataset. The plots were produced in the same way as in Fig. 1. (**upper**) The baseline has a large class-wise variance and inter-class overlaps; BN and CR show similar properties. Dropout looks completely different where activation values are more spread out. cw-CR and cw-VR show well-separated activation distributions because they are regularized class-wise. L1R increases the sparsity of representation. (**lower**) CR, RR, and cw-RR show completely different patterns. cw-CR and cw-VR show low correlation per class because they are regularized class-wise.

Table 3. Statistical characteristics of representations. The characteristics of MLP were generated in the same way as in Fig. 1. For ResNet, one fully-connected layer was added next to the last average pooling layer and regularizers were applied on it. One can observe that the characteristics are modified, as initially predicted (**bold**).

Data & Net	Reg.	Accuracy (%)	Amplitude	Covariance	Correlation	Sparsity	Dead unit	Rank
MNIST MLP	Baseline	97.15	4.93	2.08	0.27	0.34	0.13	2.41
	CR	97.50	0.50	**0.01**	**0.19**	0.40	0.03	7.12
	L1R	97.65	1.29	0.03	0.40	**0.97**	**0.39**	5.94
	RR	97.19	7.23	226.20	0.90	0.43	0.18	**1.00**
Tiny- ImageNet ResNet-50	Baseline	78.56	1.06	0.155	0.08	0.436	0.00	6.51
	CR	78.14	0.26	**0.007**	**0.04**	0.585	0.00	26.09
	L1R	78.32	0.22	0.016	0.05	**0.780**	0.00	5.36
	RR	77.99	1.59	0.204	0.12	0.155	0.00	**1.46**
ImageNet ResNet-18	Baseline	70.34	0.90	0.049	0.062	0.010	0.00	6.46
	CR	68.76	0.52	**0.005**	**0.051**	0.000	0.00	22.46
	L1R	69.51	0.83	0.067	0.078	**0.010**	0.00	2.40
	RR	69.75	0.92	20.448	0.968	0.012	0.00	**1.00**

in Table 3. In particular, RR regularizes the stable rank, and thus works as intended. RR (highly correlated representations) shows comparable performance to CR (decorrelated representations). This result is somewhat counter-intuitive to the conventional wisdom.

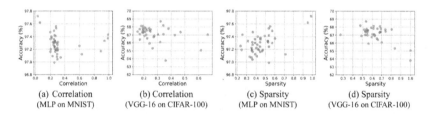

(a) Correlation (MLP on MNIST)   (b) Correlation (VGG-16 on CIFAR-100)   (c) Sparsity (MLP on MNIST)   (d) Sparsity (VGG-16 on CIFAR-100)

Fig. 3. Relationship between the representation characteristics and the performance on the MNIST (MLP) and the CIFAR-100 (VGG-16). Each blue point indicates a single pair of characteristic and performance, from the corresponding model that utilizes specific regularizer and loss weight. The red triangle indicates the baseline model. (Color figure online)

Relationship Between Representation Characteristics and Performance. To examine the relationship between representation characteristics and performance more precisely, the scatter plots of CORRELATION, SPARSITY, and performance were drawn in Fig. 3. Each circle corresponds to one characteristic and performance pair from a specific choice of model (regularizer and loss weight), and the red triangle is that of the baseline model. Only the points with comparable results to the baseline were drawn, and each model was trained and tested only once. One can observe that neither correlation nor sparsity has a clear relationship with performance. We analyze this phenomenon in the next subsection.

So far, the results have been shown when the regularizers are applied to the last fully-connected (FC) layer where the representation can be considered as the most processed feature set. We now investigate how differently regularizers behave when applied to different layers. In Fig. 4, we apply CR (top), L1R (middle), and RR (bottom) to each layer of the 6-layer MLP. The result confirms that the regularizers perform better than the baseline (red dotted line) when applied on layers 4 and 5. Conversely, when regularizers are applied to lower layers, the performance declines, as loss weight increases even though corresponding characteristics (blue lines) can be controlled (the blue dotted lines are each layer's characteristic of the baseline model). We conjecture that this is because low-level features that should flow to the upper layers with a rich level of information can be negatively affected by putting constraints on the activations.

Performance Improvement by Representation Regularization. We investigate if regularizers can indeed improve the performance for a given task condition. For instance, a regularizer might be effective when the task has a small number of data examples and another regularizer might be effective when the task has a large number of classes. The following task conditions were chosen

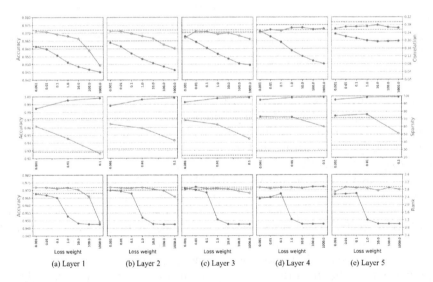

(a) Layer 1   (b) Layer 2   (c) Layer 3   (d) Layer 4   (e) Layer 5

Fig. 4. Layer dependence of regularizations. Each plot was generated with the MLP on the MNIST in the same manner, as shown in Fig. 1. Regularizers were applied to all the layers. The top, middle, and bottom rows correspond to results of CR, L1R, and RR; the red and blue dotted lines indicate the baseline model's performance and the characteristics of each of its layers, respectively. Note that some models of L1R are excluded because they cannot be trained with loss weights that are greater than 0.1. (Color figure online)

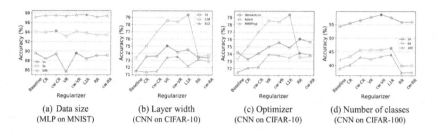

(a) Data size   (b) Layer width   (c) Optimizer   (d) Number of classes
(MLP on MNIST)   (CNN on CIFAR-10)   (CNN on CIFAR-10)   (CNN on CIFAR-100)

Fig. 5. Results of the 'task condition' experiment. Each color indicates different task conditions such as data size, number of units, choice of optimizer, and number of classes. The result shows that the seven regularizers often outperform the baseline; however, the best performing regularizer cannot be specified for the given task condition.

Table 4. Test accuracy (%) of MLP, VGG-16, and ResNet-50 models on the MNIST, the CIFAR-10/100, and the Tiny-ImageNet datasets. For Tiny-ImageNet, we did not experiment with the class-wise regularizers because their mini-batch size is required to be much larger than the number of classes, which leads to inefficient training.

Regularizer	MLP on MNIST	VGG-16 on CIFAR-10	VGG-16 on CIFAR-100	ResNet-50 on Tiny-ImageNet
Baseline	97.15 ± 0.11	92.26 ± 0.14	67.11 ± 0.44	78.53 ± 0.09
CR	97.50 ± 0.05	92.39 ± 0.14	67.07 ± 1.20	78.41 ± 0.08
cw-CR	97.51 ± 0.10	92.31 ± 0.16	67.54 ± 0.22	–
VR	97.35 ± 0.11	92.40 ± 0.22	67.38 ± 0.45	77.84 ± 0.18
cw-VR	97.58 ± 0.06	92.46 ± 0.27	$\mathbf{67.63 \pm 0.32}$	–
L1R	$\mathbf{97.65 \pm 0.08}$	92.46 ± 0.10	65.56 ± 0.31	78.54 ± 0.13
RR(ours)	97.19 ± 0.10	92.21 ± 0.12	67.37 ± 0.29	$\mathbf{78.57 \pm 0.13}$
cw-RR(ours)	97.43 ± 0.08	$\mathbf{92.56 \pm 0.08}$	67.45 ± 0.60	–

for the experiment: a learning task with 1k, 5k, or 50k data size, 32, 128, or 512 layer width, a specific dataset, a small number of classes, or a specific optimizer. We performed experiments on the MNIST and the CIFAR-10/100 datasets using the eight regularization setups and the four task conditions. All the regularizers are applied to the last FC layer only.

We first investigated simple MLP (MNIST) and CNN (CIFAR-10,100) models. The results in Fig. 5 indicate that the regularizers are generally beneficial for improving the performance. Even though no single representation characteristic consistently outperforms the rest, it is possible to improve performance by using the regularizers as a set and by choosing the best performing regularizer for the given task. On the other hand, we have performed an extensive experiment in addition to the results shown in Fig. 5 but we were not able to observe any meaningful relationship between a type of task condition and a type of regularizer.

We have also investigated more sophisticated networks of VGG-16 and ResNet-50 as shown in Table 4. Even for the sophisticated networks, we were able to affect representation characteristics using the regularizers and achieve a mild performance improvement. While the performance improvements are clearly observable, again it was impossible to identify a meaningful relationship.

4.3   Analysis

In this subsection, we provide possible reasons for the above experimental results.

Equivalent Networks. Infinitely many global optima exist for deep neural networks [7]. By re-arranging the hidden units or by properly scaling the incoming and outgoing weights of ReLU networks, one can easily construct equivalent networks with different representation characteristics but with exactly same outputs [6]. Therefore, statistical characteristics such as correlation and covariance can be easily altered without affecting performance, simply by choosing one of the equivalent networks. The easiness of constructing equivalent networks clearly indicates that at least some of the statistical characteristics of representation do not need to have a certain property (e.g. low correlation) when the best performance is achieved.

Tuning Effect of Representation Regularization. [17] have shown that explicit and implicit regularization can affect the performance, but possibly only at the tuning level. Nonetheless, regularization like weight decay is often considered to be an important part of training as in the training of ResNet. For representation regularizers, our study shows that their effect is most likely at the tuning level, too, although CR, cw-CR, and cw-VR could be explained by feature independence or compactness. Because of the neural network's large capacity, even VR can be useful for performance improvement despite its counter-intuitive design. Typically, previous works on regularizers have considered only a small number of regularizers in each work. By evaluating only a small number of regularizers over a small number of task conditions, it can be easy to identify a possibly meaningful relationship between a regularizer and a task condition. When many regularizers and many task conditions are evaluated as in our work, however, it becomes apparent that a strong relationship is difficult to observe. We conclude that it is risky and likely to be incorrect to imply or conjecture that an intuitive manipulation of representation can lead to improved performance.

5   Discussion and Future Work

Representation Characteristics of Dropout. Dropout was conceived as a way of preventing a large coadaptation [16], and DeCov was conceived partly based on the observation that covariance is reduced with dropout [3]. For our experiments, however, it turned out that covariance is reduced simply because of the scaling down effect of the activation vectors, and the normalized metric, i.e., correlation, actually increased. When dropout is used, a neural network is trained to perform well even when randomly chosen units are dropped. To achieve this, it is essential to have the task-relevant information spread over multiple units (such that the surviving units can still carry the information), and thus the average pair-wise mutual information must stay positive. In fact, it perfectly makes sense to expect an increased amount of information spread when dropout is used. On the other hand, correlation is a metric of linear relationship. Therefore, the increase in correlation indicates that the spread information's linear relationship becomes stronger with dropout. Unfortunately, we could not devise a way to analyze if much more complicated relationships among the units are weakened by dropout. Such a result would confirm reduced coadaptation in the form of complicated relationships.

Learning Dynamics. In the linear least square case, a model converges along the eigenvectors of the covariance matrix at a rate depending on the magnitude of their corresponding eigenvalues [13]. Therefore, representation regularization affects not only representation characteristics but also learning dynamics. [5] proposed a method to reparameterize the weights of the neural network by implicitly whitening each layer's activations. The method improves the learning dynamics owing to reparameterization; thus, the networks can be trained more efficiently. Also, [4] proved various properties of learning dynamics in deep nonlinear neural networks by studying the case of binary classification under strong assumptions such as linear separability of the data.

While significant progress has been made in recent years, the learning dynamics of deep network remain largely an open question. Together with our experiment results, it can be concluded that representation characteristics, performance, and learning dynamics are all interwoven together. While we negatively concluded on the causal and direct effect of representation characteristics on the performance, causal effects via learning dynamics are still possible. Representation characteristics can certainly affect learning dynamics, and learning dynamics might affect the performance in a causal and explainable way. In this case, representation characteristics would indirectly affect the performance. An empirical study of all three elements remains as possible future work.

References

1. Arpit, D., et al.: A closer look at memorization in deep networks. arXiv preprint arXiv:1706.05394 (2017)
2. Choi, D., Rhee, W.: Utilizing class information for DNN representation shaping. In: Thirty-Third AAAI Conference on Artificial Intelligence (2019)
3. Cogswell, M., Ahmed, F., Girshick, R., Zitnick, L., Batra, D.: Reducing overfitting in deep networks by decorrelating representations. arXiv preprint arXiv:1511.06068 (2015)
4. Combes, R.T.D., Pezeshki, M., Shabanian, S., Courville, A., Bengio, Y.: On the learning dynamics of deep neural networks (2018)
5. Desjardins, G., Simonyan, K., Pascanu, R., et al.: Natural neural networks. In: Advances in Neural Information Processing Systems, pp. 2071–2079 (2015)
6. Dinh, L., Pascanu, R., Bengio, S., Bengio, Y.: Sharp minima can generalize for deep nets. In: Proceedings of the 34th International Conference on Machine Learning, vol. 70, pp. 1019–1028 (2017). JMLR.org

7. Du, S.S., Lee, J.D., Li, H., Wang, L., Zhai, X.: Gradient descent finds global minima of deep neural networks. arXiv preprint arXiv:1811.03804 (2018)
8. Eldan, R., Shamir, O.: The power of depth for feedforward neural networks. In: Conference on Learning Theory, pp. 907–940 (2016)
9. Glorot, X., Bordes, A., Bengio, Y.: Deep sparse rectifier neural networks. In: Proceedings of the Fourteenth International Conference on Artificial Intelligence and Statistics, pp. 315–323 (2011)
10. Goodfellow, I., Bengio, Y., Courville, A.: Deep Learning. MIT press, Cambridge (2016)
11. Hinton, G.E., et al.: Learning distributed representations of concepts. In: Proceedings of the Eighth Annual Conference of the Cognitive Science Society, vol. 1, p. 12, Amherst, MA (1986)
12. Ioffe, S., Szegedy, C.: Batch normalization: accelerating deep network training by reducing internal covariate shift. In: International Conference on Machine Learning, pp. 448–456 (2015)
13. LeCun, Y., Kanter, I., Solla, S.A.: Second order properties of error surfaces: learning time and generalization. In: Advances in Neural Information Processing Systems, pp. 918–924 (1991)
14. Locatello, F., Bauer, S., Lucic, M., Gelly, S., Schölkopf, B., Bachem, O.: Challenging common assumptions in the unsupervised learning of disentangled representations. In: International Conference on Machine Learning (2019)
15. Montufar, G.F., Pascanu, R., Cho, K., Bengio, Y.: On the number of linear regions of deep neural networks. In: Advances in Neural Information Processing Systems, pp. 2924–2932 (2014)
16. Srivastava, N., Hinton, G.E., Krizhevsky, A., Sutskever, I., Salakhutdinov, R.: Dropout: a simple way to prevent neural networks from overfitting. J. Mach. Learn. Res. **15**(1), 1929–1958 (2014)
17. Zhang, C., Bengio, S., Hardt, M., Recht, B., Vinyals, O.: Understanding deep learning requires rethinking generalization. arXiv preprint arXiv:1611.03530 (2016)

Reservoir Computing

Unsupervised Pretraining of Echo State Networks for Onset Detection

Peter Steiner[1(✉)] ⓘ, Azarakhsh Jalalvand[2,3] ⓘ, and Peter Birkholz[1] ⓘ

[1] Institute for Acoustics and Speech Communication, Technische Universität Dresden, Dresden, Germany
{peter.steiner,peter.birkholz}@tu-dresden.de
[2] IDLab, Ghent University–imec, Ghent, Belgium
azarakhsh.jalalvand@ugent.be
[3] Mechanical and Aerospace Engineering Department, Princeton University, Princeton, USA

Abstract. Note onset detection – the detection of the beginning of new note events – is a fundamental task for music analysis that can help to improve Automatic Music Transcription (AMT). The method for onset detection always follows a similar outline: An audio signal is transformed into an Onset Detection Function (ODF), which should have rather low values (i.e. close to zero) for most of the time, and pronounced peaks at onset times, which can then be extracted by applying peak picking algorithms on the ODF. Currently, Recurrent Neural Networks (RNNs) and Convolutional Neural Networks (CNNs) define the state of the art. In this paper, we build upon previous work about onset detection using Echo State Networks (ESNs) that have achieved comparable results to CNNs. We show that unsupervised pre-training of the ESN leads to similar results whilst reducing the model complexity.

Keywords: Echo State Networks · Clustering · Note onset detection

1 Introduction

Music analysis is a relatively complex task that is typically split in various subtasks, such as note onset detection and multipitch tracking. Musical onsets are the starting points of new note events. Detecting note onsets is the first step for further music analysis tasks, such as Beat Tracking [4] or Automatic Music Transcription [7].

In general, musical onset detection can be treated as a three-step approach [5] including the following steps (1) feature Extraction, (2) reduction – computation of an Onset Detection Function ODF, and (3) peak picking.

Starting from a raw audio signal, mostly spectral features and their derivatives (Spectral Flux) over time are extracted. Since onsets are accompanied by the increase and change of energy, they can be visually recognized in this spectral representation by edges. Thus, the first step is responsible to simplify the onset detection.

© Springer Nature Switzerland AG 2021
I. Farkaš et al. (Eds.): ICANN 2021, LNCS 12895, pp. 59–70, 2021.
https://doi.org/10.1007/978-3-030-86383-8_5

The next step (reduction) computes a one-dimensional Onset Detection Function (ODF) from the typically high-dimensional feature vector sequence. In the ODF, onsets are characterized by strong peaks. There are various ways to compute an ODF from the feature vector sequence, with and without machine learning techniques. In [16], we have provided an extensive overview about reduction techniques.

Finally, peak picking algorithms are applied on the ODF to extract the actual onset times. In [6,12,15], neural networks such as bidirectional LSTM networks, Convolutional Neural Networks (CNNs) and Echo State Networks (ESNs) have all achieved state-of-the-art results in learning the ODF from feature vectors. The methods achieved F-Measures of 0.873, 0.903 and 0.886 on the Böck dataset [2], respectively.

Recently, in [14], we have shown that ESNs strongly benefit from a cluster-based input weight initialization, leading to equivalent or superior results compared to a baseline ESN whilst needing significantly less reservoir neurons, which consequently reduces the required amount of free parameters and training time.

In this paper, we built upon our results from [14,15] and investigated whether the unsupervised pre-training improves the performance of ESNs for note onset detection while lowering the model complexity. As the Böck dataset [2] also contains audio files that are not labeled and thus excluded from the standard cross validation, we furthermore investigated whether the pre-training benefits from additional data.

The remainder of this paper is structured as follows. Section 2 introduces the outline of onset detection with the basic ESN and the pre-trained KM-ESN. The experimental setup including the dataset is described in Sect. 3. In Sect. 4, we discuss the impact of the pre-training on the onset detection results and compare our result to the state of the art. Finally, we summarize our conclusion and future work in Sect. 5.

2   Onset Detection with Echo State Networks

Echo State Networks (ESNs) [8] belong to the class of Recurrent Neural Networks (RNNs) together with e.g. Long-Short-Term Memory Cells (LSTMs). However, in contrast to LSTMs, only very small parts of the weights in ESNs are trained using linear regression. Due to the recurrent connections inside the reservoir, which is the core element of an ESN, information from previous inputs is retained for a certain amount of time. This so called memory can be tuned by the hyper-parameters of the reservoir and subsequently act as a short- or long-term memory. As the number of neurons inside the reservoir is typically higher than the input features, the reservoir transforms the input space into a high-dimensional feature space, in which onsets can be separated with hyper-planes from non-onsets. Thus, the Onset Detection Function ODF is a multi-linear function of the reservoir states.

2.1   Feature Extraction

In this paper, we used our findings from [15] and divided the audio signal $s[k]$ (sampling frequency 44.1 kHz) in frames of the lengths $\{1024, 2048, 4096\}$ samples starting at the same time index and a frame rate 100 Hz. Each frame was windowed using a Hann window, and afterwards magnitude spectra were computed. As in [6], we applied a triangular filterbank with a logarithmic frequency spacing and 12 filters per octave to the magnitude spectra. We stacked the output of all three filterbanks, applied the $\log_{10}$ to the magnitude plus 1, and finally added the "Super-Flux" features [3] to obtain the final feature set with N^{in} features.

2.2   Echo State Network

The main outline of a basic ESN is depicted in Fig. 1a. It mainly consists of the following components:

- The input weight matrix $\mathbf{W}^{\text{in}}$ passes the N^{in} features to the reservoir with N^{res} neurons.
- The reservoir weight matrix $\mathbf{W}^{\text{res}}$ interconnects the neurons inside the reservoir.
- The output weight $\mathbf{W}^{\text{out}}$ connects the neurons to the output node.

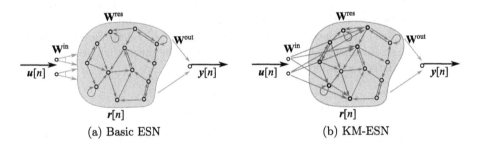

(a) Basic ESN   (b) KM-ESN

Fig. 1. Main outline of ESNs. In basic ESN, the input weights are randomly initialized. In KM-ESN, the input weights are the normed centroids of a K-Means algorithm applied to the training data. As there can be more neurons than centroids, not every neuron necessarily receives input information in case of the KM-ESN.

Typically, $\mathbf{W}^{\text{in}}$ and $\mathbf{W}^{\text{res}}$ are initialized from a random uniform distribution between ± 1 and standard normal distribution, respectively. Furthermore, they are typically initialized sparsely, which means that each neuron inside the reservoir receives a very small fraction of $K^{\text{in}} = 10$ input features and previous outputs of $K^{\text{rec}} = 10$ reservoir neurons. As $\mathbf{W}^{\text{res}}$ needs to fulfill the *Echo State Property* [8], $\mathbf{W}^{\text{res}}$ is normalized to its maximum absolute eigenvalue. The hyperparameters α_{u} (input scaling) and ρ (spectral radius) are factors to multiply $\mathbf{W}^{\text{in}}$ and $\mathbf{W}^{\text{res}}$ with, respectively.

The key difference between ESNs and typical RNN architectures is that $\mathbf{W}^{\text{in}}$ and $\mathbf{W}^{\text{res}}$ are initialized randomly, with no more optimization during the training. Only the output weights $\mathbf{W}^{\text{out}}$ are trained using linear regression.

With $\mathbf{r}[n]$ representing the reservoir state, Eqs. (1) and (2) describe the behaviour of ESN.

$$\mathbf{r}[n] = (1 - \lambda)\mathbf{r}[n-1] + \lambda \tanh\left(\mathbf{W}^{\text{in}}\mathbf{u}[n] + \mathbf{W}^{\text{res}}\mathbf{r}[n-1] + \mathbf{w}^{\text{bi}}\right) \qquad (1)$$

$$y[n] = \mathbf{W}^{\text{out}}\mathbf{r}[n] \qquad (2)$$

Equation (1) is a leaky integration of the reservoir states $\mathbf{r}[n]$ with $\lambda \in (0, 1]$ being the leakage. Every neuron in the reservoir has a constant bias input from the bias weight vector $\mathbf{w}^{\text{bi}}$, which is initialized and fixed from a uniform distribution between ± 1 and then scaled with the factor α_{bi}. Equation (2) shows that the one-dimensional output $y[n]$ is actually a linear combination of a given reservoir state $\mathbf{r}[n]$ (i.e. $\mathbf{W}^{\text{out}}$ is a row vector).

For training, all reservoir states are collected in the reservoir state collection matrix $\mathbf{R}$. To add the intercept term for linear regression, every reservoir state $\mathbf{r}[n]$ is expanded by a constant of 1. The desired outputs $d[n]$, which are 0 for non-onsets, 1 for onsets and 0.5 for frames around onsets, are collected into the desired output collection vector $\mathbf{D}$. Afterwards, $\mathbf{W}^{\text{out}}$ is obtained using ridge regression (Eq. (3)), to prevent overfitting to the training data. The regularization parameter $\epsilon = 0.01$ penalizes large values in $\mathbf{W}^{\text{out}}$, and $\mathbf{I}$ is the identity matrix. The size of the output weight vector $1 \times (N^{\text{res}} + 1)$ determines the total number of free parameters to be trained in ESNs. The output $y[n]$ corresponds to the onset detection function ODF.

$$\mathbf{W}^{\text{out}} = \left(\mathbf{R}\mathbf{R}^{\text{T}} + \epsilon \mathbf{I}\right)^{-1}\left(\mathbf{D}\mathbf{R}^{\text{T}}\right) \qquad (3)$$

The basic ESN described so far, can be extended in several directions. Here, we focus on the K-Means-based initialized ESN (KM-ESN) and on bidirectional architectures.

KM-ESN: Although randomly initialized ESNs have achieved state-of-the-art results in various directions, several publications, such as [10, 11] argue that there should exist better approaches that incorporate more prior or biologically plausible knowledge. Consequently, in [14], we have proposed the KM-ESN, which utilizes normalized centroids obtained from a K-Means algorithm applied to all feature vectors of the training data as input weights instead of randomly initialized input weights. The motivation behind the KM-ESN is that

1. using the K-Means-based pre-training tunes the input weights in a way that few reservoir states respond strongly to an input vector instead of many reservoir states responding with similar strengths.

2. a purely random weight initialization leads not to the best results [13],
3. passing feature vectors into the reservoir via $\mathbf{W}^{\text{in}}$ using Eq. (1) is closely related to computing the cosine similarity between the feature vectors and all weights of one neuron,
4. unsupervised pre-training allows to train the ESN with additional unlabeled data.

The outline of the KM-ESN is visualized in Fig. 1b. The key difference between the basic ESN and the KM-ESN are the pre-trained input weights. As the number of centroids (K) is task-dependent and needs to be chosen carefully, we theoretically allow that only a part of the reservoir neurons receives input information. However, as the dataset in this paper is relatively complex, we kept $K = N^{\text{res}}$ in our experiments.

Bidirectional Reservoirs: In the case of bidirectional reservoirs [9], the input is fed through the ESN normally and reversed in time. The reservoir states of both directions are collected and concatenated. As the output is computed using the concatenated reservoir states, bidirectional ESNs have the double number of free parameters in W^{out}. For example, the number of parameters for a reservoir with 500 neurons in the bidirectional case is 500 in the forward and 500 in the backward path.

2.3 Peak Picking

In [16], we have discussed that the output of an ESN after linear regression indicates an onset or non-onset, and would be zero for a non-onset and one for an onset, ideally. We used exactly the same peak picking algorithm as in our previous study. The algorithm itself is a simple threshold-based peak picking algorithm originally proposed in [12]. At first, the ODF was smoothed using a Hamming window with 5 samples. Next, local maxima greater than a tunable threshold δ were detected. The locations of the resulting peaks were considered to be onsets.

3 Experimental Setup

3.1 Dataset

For training and evaluation of the ESN models, we used the publicly available dataset[1] introduced by Böck in [2], which consists of all important types of onsets and various musical genres. Because the dataset was already used for several evaluations of algorithms for onset detection, we could directly compare the results of our cross-validation with state-of-the-art algorithms.

The dataset consists of 321 audio files together with onset annotations and 84 audio files without annotations that can be ignored according to Sebastian

---

[1] Download links: https://github.com/CPJKU/onset_db/.

Böck. All audio files are sampled at 44.1 kHz and the total duration of all labelled audio files is 102 min. In total, there are 27 700 onset labels included. The labelled dataset is already split into eight folds for an 8-fold cross validation. We used six folds to train the ESN and one subset as a validation set to tune the hyper-parameters. Afterwards, we rotated the folds and repeated the optimization for another set of training and validation folds. This procedure was repeated eight times until every subset has been used for validation exactly one time. The hyper-parameters for the final evaluation were obtained by determining the lowest mean loss across all eight validation losses.

After fixing the hyper-parameters, the final model was trained using seven folds and tested on the last unseen fold. Again, this was repeated for eight times until each fold has been used for testing exactly one time. The results we report later are the mean values across the eight repetitions.

As we investigated the impact of utilizing additional data for the unsuper-vised weight initialization, we add the 84 audio files without onset annotations to the training dataset of the K-Means algorithm. We did not split the unlabeled files in eight subsets for cross validation but used all of them for every fold.

3.2 Measurements

As in [15,16], we compared our results with the state-of-the-art algorithms and report different measures using the Madmom library [1] with the same settings as used in [12]. The detected onset times were compared to the reference onset times. If an onset was detected in a time-window of ± 25 ms around a reference, it was considered as a true positive (TP). If no onset was detected in the window around a reference, it was considered as a false negative (FN). If any onset was detected outside the window, it was a false positive (FP). With these notations, the measurements Precision P (4), Recall R (5), F-Measure F (6) are defined.

$$P = \frac{\mathrm{TP}}{\mathrm{TP} + \mathrm{FP}} \tag{4}$$

$$R = \frac{\mathrm{TP}}{\mathrm{TP} + \mathrm{FN}} \tag{5}$$

$$F = 2 \cdot \frac{P \cdot R}{P + R} \tag{6}$$

In this work, F served as the objective function to determine the peak picking threshold δ.

The K-Means algorithm aims to partition N feature vectors into K sets $S_1, S_2, \ldots, S_K$ and thereby minimizes the within-cluster sum of squares (SSE) in Eq. (7)

$$\text{SSE} = \sum_{k=1}^{K} \sum_{\mathbf{u}[n] \in S_k} \|\mathbf{u}[n] - \mu_k\|^2 \,, \tag{7}$$

where μ_k is the k-th centroid. With the SSE, we determined a suitable reservoir size for hyper-parameter tuning of the ESN.

3.3 Implementation and Optimization Strategy

The algorithm was developed in Python 3 and will be available in our Github repository[2]. The optimization process was conducted using a sequence of grid and line searches as in [15].

The optimization workflow to fix the hyper-parameters of the KM-ESN was slightly different: Before the reservoir design, we computed the average SSE across all eight folds for different K, which is presented in Fig. 2. A strong decay could be observed until $K = 200$. Afterwards, SSE continued to decrease with a much slower slope.

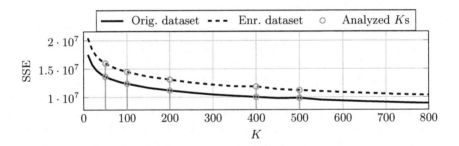

Fig. 2. Summed Squared Error (SSE) as a function of number of clusters (K) for applying K-Means algorithm on Böck music dataset.

Comparing the SSE of the original and the enriched dataset, no strong differences could be observed in the global behaviour. The SSE of the enriched dataset was always larger, because there were more training examples for the K-Means algorithm and the SSE is not normalized to the number of training examples.

We sampled $K \in \{50, 100, 200, 400, 500\}$ and then optimized the hyper-parameters using the workflow from [15]. As we observed that the determined hyper-parameters of the KM-ESN with 200 neurons were similar to larger reservoirs, we did not optimize the hyper-parameters for larger reservoirs with more

---

[2] https://github.com/TUD-STKS/PyRCN.

than 500 neurons. For the experiments in Sect. 4, we fixed the hyper-parameters of the KM-ESN with 500 neurons and only increased the reservoir size as usual. The same observation also occurred in case of the KM-ESN with the enriched dataset. Table 1 shows the determined hyper-parameters for the basic ESN and for the KM-ESN without and with the enriched dataset.

Table 1. List of all hyper-parameters to be tuned. The values show the search range, the step size and the final values for all investigated architectures.

Hyper-parameter	Range	Step size	Final values		
			basic ESN	KM-ESN	KM-ESN (augm.)
Input scaling α_u	$[0.1, 1.5]$	0.1	0.4	0.1	0.1
Spectral radius ρ	$[0, 1.0]$	0.1	0.3	0.6	0.8
Bias scaling α_{bi}	$[0, 1.0]$	0.1	0.2	0.6	0.6
Leakage λ	$[0.1, 1.0]$	0.1	1.0	1.0	1.0
Threshold δ	$[0.2, 0.5]$	0.02	0.42	0.42	0.42

4   Results and Discussion

4.1   Basic ESN Vs. KM-ESN

In Fig. 3, results of different ESN architectures are summarized. In general, larger reservoirs performed better than small reservoirs. Especially the F-Measure of the basic ESN increased from 0.73 to 0.85. The KM-ESN always outperformed the basic ESN, especially in case of smaller reservoirs. The only exception occurred for the smallest model with 50 neurons, where the basic ESN performed slightly better. A KM-ESN with only 1000 neurons reaches almost the same performance as a basic ESN with 5000 neurons. This shows that the pre-training of the input weights influenced the detection results. In case of reservoir sizes up to 10 000, the performances of the basic ESN and KM-ESN reached a maximum of $F \approx 0.85$ and $F \approx 0.86$, respectively.

In [14], where we have introduced the KM-ESN, an outcome for speech was that we only needed a very limited number of 300 clusters for speech analysis. This is contrary to the results presented here, where even $K = 10\,000$ still improved the results. In contrast to speech, music is relatively complex: Several instruments can play simultaneously, and each instrument can play one or more notes at the same time. For example, a typical piano has 88 notes and the pianist can play up to ten notes simultaneously. This leads to a lot of possible note combinations. Therefore in case of polyphonic music, it would be logical to consider each centroid as the representative of a combination of notes, rather

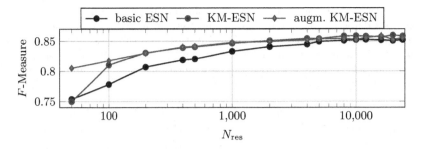

Fig. 3. F-Measures for the unidirectional basic ESN and for the unidirectional KM-ESN with and without the enriched dataset. The KM-ESN always outperformed the basic ESN. The enriched dataset only helped the smallest KM-ESN to improve the results.

than an isolated note. Consequently, a large number of centroids can be expected for music-related tasks.

A potential advantage of the KM-ESN is that it can be pre-trained with basically any meaningful data. NFrom Fig. 3 it can be observed that mostly the smallest KM-ESN benefited from the enriched dataset. If the reservoir size is increased, the KM-ESN with and without the enrichment strategy achieved similar results, and for very large reservoirs, the performance of the KM-ESN with the enriched dataset even slightly decreased. This means that the K-Means algorithm with many centroids found subclusters in the training dataset which were not representative for the test dataset.

4.2   Final Results

In Table 2, the final results of this work are compared to the current state-of-the-art algorithms, namely the bidirectional LSTM [6], the best ESN results from [15], and the CNN [12]. The proposed model clearly outperformed the large bidirectional ESN increasing the F-Measure from 0.84 to 0.861 and 0.866 for the unidirectional KM-ESN with 20 000 neurons and for the bidirectional KM-ESN with 10 000 neurons, respectively. Obviously, with a causal model, we have almost reached the performance of the non-causal bidirectional LSTM [6] with less parameters N_{params} to be trained. Comparing the results of single- and multi-layer ESN in [15], one would expect that stacking KM-ESNs can minimize or close the gap between the performance of CNN and ESN with much fewer parameters. As the KM-ESN closely follows the basic ESN, it can be stacked as usual, e.g. as in [15].

Figure 4 shows the time signal, spectrogram and ODF with target and predicted onsets for an example of the test dataset. In all ODFs, we can see that potential onsets are characterized by prominent peaks. In case of the KM-ESN, the peaks are more prominent and fewer peaks in non-onset positions occurred. Thus, the KM-ESN is able to outperform the basic ESN.

Table 2. Performance of large reservoirs evaluated using the 8-fold cross validation on the Böck dataset. All the reference models were evaluated on the same dataset by the authors. With the utilized "Super-Flux" features and the basic ESN, we were able to outperform the reference single-layer ESN model from [15]. With the bidirectional KM-ESN models without and with the enrichment strategy, we almost reached the performance of the bidirectional LSTM [6].

Architecture	Precision	Recall	F-Measure	N_{params}	Mode	Features
Basic ESN 12000u	0.877	0.859	0.854	12 001	uni	Super-Flux
Basic ESN 16000b	0.892	0.826	0.858	32 001	bi	Super-Flux
KM-ESN 20000u	0.875	0.846	0.861	20 001	uni	Super-Flux
KM-ESN 10000b	0.889	0.844	0.866	20 001	bi	Super-Flux
KM-ESN 16000u (augm.)	0.870	0.850	0.860	16 001	uni	Super-Flux
KM-ESN 8000b (augm.)	0.879	0.855	0.866	16 001	bi	Super-Flux
Bidirectional LSTM [6]	0.892	0.855	0.873	20 225	bi	Super-Flux
ESN 1L-24000b [15]	0.881	0.804	0.840	48 001	bi	Spectral Flux
ESN 2L-28000b-5000b [15]	0.920	0.855	0.886	66 002	bi	Spectral Flux
CNN [12]	0.917	0.889	0.903	289 406	–	Super-Flux

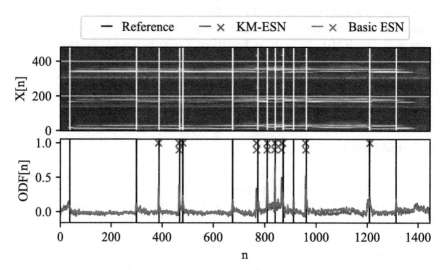

Fig. 4. Features, reference onsets and ODF for a basic ESN and a KM-ESN of a clarinet excerpt from the test set. The onsets were often relatively soft and difficult to recognize. Both, the basic ESN and the KM-ESN have several false negative onsets. However, as the peaks in case of the KM-ESN are more prominent and thus the onsets around $n = \{400, 500\}$ could be detected.

5   Conclusion and Outlook

We have successfully extended our approach [15] for onset detection by utilizing the novel KM-ESN [14] with input weights pretrained using the unsupervised

K-Means algorithm. With this model, we have removed randomness from the basic ESN and, especially for small reservoir sizes, we have achieved promising results. Specifically, a KM-ESN of only 1000 neurons reached almost the same performance as a bidirectional basic ESN model with 5000. Data enrichment seemed to be not very useful as there was no improvement when increasing the amount of training data for the K-Means algorithm.

In the future, we will extend this approach in various directions: We have shown that a large number of centroids was required for this seemingly simple task of onset detection. This might be due to the complexity of music signals, and is contrary to speech-related tasks, in which we utilized a very low number of centroids. If we can predict or at least robustly estimate the number of required clusters from a given unlabeled dataset task-dependently, we could speed up the reservoir design by sampling fewer K for hyperparameter optimization. Furthermore, it easily allows to switch to sparse KM-ESNs as soon as the reservoir size is getting larger than the estimated K.

With the existing KM-ESN model, we can do transfer learning towards other music-related tasks. As the combination of multipitch tracking and onset detection has been shown to be beneficial for note-based automatic music transcription [7], we can use exactly the same pre-trained set of input weights for onset and for multipitch models. Only the hyper-parameters need to be tuned and the outputs need to be trained task-dependently.

Acknowledgements. The parameter optimizations were performed on a Bull Cluster at the Center for Information Services and High Performance Computing (ZIH) at TU Dresden. The research is also partially funded by the Special Research Fund of Ghent University (BOF19/PDO/134).

References

1. Böck, S., Korzeniowski, F., Schlüter, J., Krebs, F., Widmer, G.: Madmom: a new Python audio and music signal processing library. In: Proceedings of the 24th ACM International Conference on Multimedia, pp. 1174–1178. ACM (2016)
2. Böck, S., Krebs, F., Schedl, M.: Evaluating the online capabilities of onset detection methods. In: Proceedings of the 13th International Society for Music Information Retrieval Conference, ISMIR 2012, Mosteiro S. Bento Da Vitória, Porto, Portugal, 8–12 October 2012, pp. 49–54 (2012). http://ismir2012.ismir.net/event/papers/049-ismir-2012.pdf
3. Böck, S., Widmer, G.: Maximum filter vibrato suppression for onset detection. In: Proceedings of the 16th International Conference on Digital Audio Effects (DAFx), Maynooth, Ireland, September 2013, vol. 7 (2013)
4. Böck, S., Schedl, M.: Enhanced beat tracking with context-aware neural networks. In: Proceedings of the International Conference on Digital Audio Effects (DAFx-2011), pp. 135–139, Montreal, Quebec, Canada, 19–23 September 2011. https://www.dafx.de/paper-archive/2011/Papers/31_e.pdf
5. Dixon, S.: Onset detection revisited. In: Proceedings of the International Conference on Digital Audio Effects (DAFx-06), pp. 133–137, Montreal, Quebec, Canada, 18–20 September 2006. http://www.dafx.ca/proceedings/papers/p_133.pdf

6. Eyben, F., Böck, S., Schuller, B.W., Graves, A.: Universal onset detection with bidirectional long short-term memory neural networks. In: Proceedings of the 11th International Society for Music Information Retrieval Conference, ISMIR 2010, Utrecht, Netherlands, 9–13 August 2010, pp. 589–594 (2010). http://ismir2010.ismir.net/proceedings/ismir2010-101.pdf
7. Hawthorne, C., et al.: Onsets and frames: dual-objective piano transcription (2017)
8. Jaeger, H.: The "echo state" approach to analysing and training recurrent neural networks. Technical report GMD Report 148, German National Research Center for Information Technology (2001). http://www.faculty.iu-bremen.de/hjaeger/pubs/EchoStatesTechRep.pdf
9. Jalalvand, A., Demuynck, K., Neve, W.D., Martens, J.P.: On the application of reservoir computing networks for noisy image recognition. Neurocomputing **277**, 237–248 (2018)
10. Lukoševičius, M., Jaeger, H., Schrauwen, B.: Reservoir computing trends. KI - Künstliche Intelligenz **26**(4), 365–371 (2012). https://doi.org/10.1007/s13218-012-0204-5
11. Scardapane, S., Wang, D.: Randomness in neural networks: an overview. WIREs Data Mining Knowl. Discov. **7**(2), e1200 (2017)
12. Schlüter, J., Böck, S.: Improved musical onset detection with Convolutional Neural Networks. In: 2014 IEEE International Conference on Acoustics, Speech and Signal Processing (ICASSP), pp. 6979–6983, May 2014. https://doi.org/10.1109/ICASSP.2014.6854953
13. Schrauwen, B., Verstraeten, D., van Campenhout, J.: An overview of reservoir computing: theory, applications and implementations. In: Proceedings of the 15th European Symposium on Artificial Neural Networks, pp. 471–482 (2007)
14. Steiner, P., Jalalvand, A., Birkholz, P.: Cluster-based input weight initialization for echo state networks. IEEE Transactions Neural Networks and Learning Systems (2021, submitted)
15. Steiner, P., Jalalvand, A., Stone, S., Birkholz, P.: Feature engineering and stacked echo state networks for musical onset detection. In: 2020 25th International Conference on Pattern Recognition (ICPR), pp. 9537–9544 (2020)
16. Steiner, P., Stone, S., Birkholz, P.: Note onset detection using echo state networks. In: Böck, R., Siegert, I., Wendemuth, A. (eds.) Studientexte zur Sprachkommunikation: Elektronische Sprachsignalverarbeitung 2020, pp. 157–164. TUDpress, Dresden (2020)

Canary Song Decoder: Transduction and Implicit Segmentation with ESNs and LTSMs

Nathan Trouvain[1,2,3] and Xavier Hinaut[1,2,3(✉)]

[1] INRIA Bordeaux Sud-Ouest, Talence, France
xavier.hinaut@inria.fr
[2] LaBRI, Bordeaux INP, CNRS, UMR 5800, Bordeaux, France
[3] Institut des Maladies Neurodégénératives, Université de Bordeaux, CNRS, UMR 5293, Bordeaux, France

Abstract. Domestic canaries produce complex vocal patterns embedded in various levels of abstraction. Studying such temporal organization is of particular relevance to understand how animal brains represent and process vocal inputs such as language. However, this requires a large amount of annotated data. We propose a fast and easy-to-train transducer model based on RNN architectures to automate parts of the annotation process. This is similar to a speech recognition task. We demonstrate that RNN architectures can be efficiently applied on spectral features (MFCC) to annotate songs at time frame level and at phrase level. We achieved around 95% accuracy at frame level on particularly complex canary songs, and ESNs achieved around 5% of word error rate (WER) at phrase level. Moreover, we are able to build this model using only around 13 to 20 min of annotated songs. Training time takes only 35 s using 2 h and 40 min of data for the ESN, allowing to quickly run experiments without the need of powerful hardware.

Keywords: Birdsong · Echo State Networks · Long Short Terms Memory · RNN · Audio classification · MFCC

1 Introduction

Birdsongs are a common resource to study sensorimotor learning of complex sequences of gestures. Many songbirds species, like the Bengalese finch or the canary, organize their songs on top of small stereotypical units, called *notes* or *syllables*. In the case of canaries, songs are composed of around 10 to 30 different syllables classes [12]. Individuals in a canary population might share syllables to some extent, due to the learning process of these vocalizations involving imitation from tutors, but each individual's repertoire is a unique combination of syllable types [19]. Canaries chain syllables of a same class together in repetitive patterns to form *phrases* (Fig. 1), that are then chained to form songs. Phrase length (i.e. the number of repetitions of a syllable within a phrase) is uncorrelated to syllable type, but might play a role in song syntax [13]. Markowitz *et al.* [13]

© Springer Nature Switzerland AG 2021
I. Farkaš et al. (Eds.): ICANN 2021, LNCS 12895, pp. 71–82, 2021.
https://doi.org/10.1007/978-3-030-86383-8_6

also demonstrated the existence of a complex temporal structure in canary songs, where the order of phrases can be approximated by a high order Markov chain. This makes canary songs of a particular relevance to understand the mechanisms underlying the sequential organization of gestures in animal brains.

However, investigating the origin of the sequential organization of bird songs requires large amount of data. Studies like [13] should be more numerous in order to decipher general syntactic rules in canary songs, but this is limited by the available datasets of annotated song recordings. The annotation process is a task consisting in (a) segmenting the songs, by identifying temporal patterns onset and offset in time, and (b) classifying and labelling these temporal patterns, similarly to what is done for human speech recognition datasets. This process is done by hand, and is as time consuming as it is error prone.

We propose to use Recurrent Neural Networks (RNN) to process spectral features of canary songs to fulfill both the labelling and segmentation tasks. These RNNs can be trained on small datasets of only a dozen of canary songs, allowing to automatize most of the annotation process. They operate at phrase level, segmenting canary songs into sequences of repetitive patterns, which is supposed to be sufficient for analysis like [13]. We compare simple neural architectures like a single Long Short Term Memory (LSTM) layer or Echo States Networks (ESN) [8]. We chose these methods to demonstrate that lightweight algorithms, in comparison to more intensive deep learning methods, can be successfully used on difficult tasks similar to speech recognition, while being less time and energy consuming. Importantly, the limited number of parameters that need to be learned enables one to apply them on limited amount of data while not overfitting on it; this is particularly interesting because it limits the amount of data necessary to be hand-labelled beforehand to train the model.

1.1   Related Work

Automatic methods for song annotation of various bird species have been developed in the past years, to try leveraging the large audio recordings available. These methods make use of various techniques, like Dynamic Time Warping (DTW) [1,10,17], Hidden Markov Models (HMM) [3], Support Vector Machines (SVM) [16], unsupervised clustering [15] or combinations of Convolutional Neural Networks (CNN) and HMM [11]. While they can usually be applied on birdsongs where vocal patterns can be easily identified and segmented like Bengalese finch songs, these methods are known to fail on birdsongs with more complex temporal patterns like canary songs, where segmentation of songs into sequences of phrases or syllables is usually done by hand with only partial automation possible using thresholding techniques. Ongoing work described in Cohen *et al.* [4] solves the segmentation problem by using machine learning techniques to try to extract the position of syllables on spectrograms of the songs while classifying them using CNNs and LSTMs. We propose a similar approach, using simpler neural networks to avoid overfitting and to allow fast training and deployment of models on hardware with limited capacity. However, unlike [4], we focused on segmenting songs at phrases (i.e. consecutive repetitions of syllables) level and not at single syllables level. Phrase annotations are indeed sufficient to perform analysis like [13].

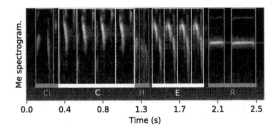

Fig. 1. Labeled canary song excerpt. Canary songs are made of a sequence of phrases, which are themselves made of repetitions of syllables. Each box delimits a *syllable*. The sequence of phrases is here *Ci-C-H-E-R* (sequence of labels on the figure). Sound is represented as Mel-spectrogram (i.e. spectrogram with Mel scale).

2   Methods

2.1   Song Transduction Task

Our task lies at the level of phrase annotation: each phrase is attributed a label corresponding to the syllable type that composed it, and each phrase label is delimited in time (expressed in seconds relative to the beginning of the audio sequence). An annotation is the combination of these absolute timestamps and of a syllable label. Figure 1 gives an example of an annotated portion of song, with groups of syllables (gray and white boxes) identified with arbitrary letters.

The task of canary song annotation can then be broken down into two *sequence-to-sequence* subtasks:

- an **audio-to-frame** transduction task – each time frame (each column of pixel in Fig. 1) representing the preprocessed audio signal must be annotated with a corresponding phrase label.
- an **audio-to-phrases** transduction and segmentation task – the song must be transcribed as a sequence of tokens representing the phrase labels in the same fashion as the available dataset, by grouping in time the annotations found during the audio-to-frame subtask. In Fig. 1, the sequence of phrases to reconstruct is *Ci-C-H-E-R*.

To evaluate the fulfillment of these subtasks, we use two different metrics. The audio-to-frame task is evaluated using the frame accuracy (ACC) [2]. The ACC score is computed by assessing the correctness of the predicted labels for each time frame of data representing the song. Additionally, we also compute a global macro averaged *F1*-measure over all the annotated songs to take into account the dataset imbalance (i.e. the *F1*-measure is computed for all labels and then their unweighted mean is computed).

The audio-to-phrases task is evaluated using the word error rate (WER) measure at the phrase level (i.e. a "word" in this context is considered to be a phrase). The WER normalises the number of editions, deletions and substitutions necessary to align perfectly a predicted sequence with the correspondent expected one.

It is thus defined as the Levenshtein distance between the predicted and expected sequences divided by the exact number of phrases in the expected sequence. All scores in Sect. 3 are presented as mean ± standard deviation, figures included (plain areas of color represent standard deviation boundaries, curves represent mean).

We propose to train two RNN architectures on this task: an ESN (Sect. 2.4) and an LSTM (Sect. 2.5).

2.2 Available Data

We use a corpus of 459 songs recorded from a single canary, for a total song duration of approximately 3 h 20 min. The songs were annotated by one human experimenter, and checked by other experimenters when the syllable category was unclear. They were then corrected by a second human experimenter assisted by early versions of the models presented in Sect. 2.4. The corrected version of the dataset contains 27 different types of phrases, each identified by a unique arbitrary label. We finally added three other classes of phrases. A *cri* class (*call* in French) is used to annotate all canary vocalizations that are not part of a sound, and are simple calls. A *SIL* (silence) class is used to annotate all segments where the bird is not singing. A *TRASH* class is used to annotate phrases that are impossible to classify clearly.

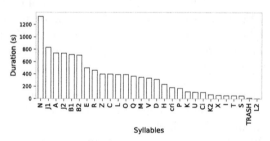

Fig. 2. Cumulated duration of the different types of phrases in the dataset.

The phrases distribution is highly unbalanced inside the dataset, as presented in Fig. 2. As we want the model trained to be as good as possible at recognizing all types of syllables, we split the dataset in order to ensure that all types of syllables appear at least once in the training dataset. Once this condition is filled and the training set contains representative songs, we keep splitting the dataset until the training set contains 368 songs (around 2 hours 39 min), and the testing set contains 91 songs (around 41 min). The dataset is available on Zenodo (zenodo.org) [7].

2.3 Data Preprocessing

Speech recognition tasks generally need to convert raw audio signals to a format having a good trade-of between accuracy and number of features. We chose Mel-Frequency Cepstral Coefficients (MFCC) as a representation for canary songs:

they allow to extract frequency features of the audio signal in a biologically meaningful way, while compressing the spectral representation down to a few coefficients. We mainly based our preprocessing steps on human audio standards (making some adjustments based on the frequency bands of canary vocalizations),
assuming that we only need to extract the information that could have been perceived by human annotators. Indeed, canary syllables often form a clear identifi-
able pattern on a spectrogram, because it does not include complex harmonics.

MFCC computations where performed using the *Librosa* [14] Python library.
Song spectrograms are first extracted using Short Time Fourier Transform every
11 ms (often called *frame stride*) and computed on overlapping windows of 23 ms
(often called *window width*)[1], using a Hanning window to reduce edge effects.
Then, we set the frequency range of a 128 filters Mel filterbank to [500 Hz;
8 kHz], as canaries vocal patterns occur below 8 kHz and as the [0 Hz; 500 Hz]
bandwidth represents mostly noise. Mel filters are appliyed on spectrograms
through a dot product, and a final Discrete Cosine Transform (DCT) creates
cepstral representations. We extracted 13 cepstral coefficients as usually fed to
MFCC-based speech recognition models. We also computed the first (Δ) and
second (Δ^2) derivatives of the MFCC signal, in order to provide the models
with gestures dynamics. We therefore feed all models with a total of 39 features
per timeframe. No normalization is applied to the MFCC representations, except
a cosine liftering as described in [9], with a factor of 40, which helps to linearize
the variance of the coefficients. We did not apply any other normalization to
avoid any loss in representativeness in the extracted features.

2.4   Presentation of the Echo State Network Model

We used ESNs with leaky integrator sigmoid neurons as transducers of the songs
audio signals. ESNs can be described as randomized RNNs using simple learning
rules to update the parameters of a *readout* layer of neurons. Appart from this
readout layer of neurons, no other parameters are learned. All the other parameters are randomly chosen to build the *reservoir* and the input layer, and are
kept fixed during the whole life cycle of the network. The reservoir is the main
component of ESNs. It is a randomly, sparsely connected pool of neuronal units
in charge of unfolding the temporal dynamics of the input data in a high dimensional space. The reservoir is randomly, sparsely connected to the sequential
input stream of data through connections defined in the input layer. Activities
(also known as *states*) of neuronal units in the reservoir are described following
the Eq. 1:

$$x[t] = (1 - \alpha)x[t-1] + \alpha \tanh(\mathbf{W^{in}} \cdot u[t] + \mathbf{W} \cdot x[t-1]) \qquad (1)$$

where $x[t]$ is the vector storing the activities of the N neurons at time t, and $u[t]$
is the current input vector at time t. The parameter α is the leaking rate (LR),
controlling the time constant of the ESN, and set to 9×10^{-2}. $\mathbf{W} \in \mathbb{R}^{N \times N}$

---

[1] *frame stride* and *window width* are respectively called *hop_length* and *win_length* in
librosa.

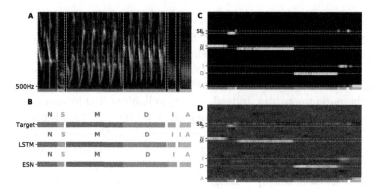

Fig. 3. An example predicted sequence of ESN and LSTM models on a song bout. **A** Mel-scale spectrogram of the song bout used as example. **B** Target annotation sequence (blank represent silence *SIL* class) and predicted annotation sequence from LSTM and ESN. Colored lines represents segment of consecutive frames sharing the same annotation label, indicated on top. Frame accuracy can be conceptualized as measuring the lines alignment in length and color between the targets and the predictions. WER only measures the validity of the colors (and thus labels) sequence (e.g. *N-S-M-D-I-A* in the figure). **C** LSTM chromagram-like output activations for all frame representing the song. The y axis represents phrases classes. The x axis represents time in frames. Each frame is labeled following the argmax of the activations for this frame. **D** Same as **C** with the ESN. Because output activation of ESN is linear, the outputs values tends to be noisier than the outputs of the LSTM, that are normalized using a *softmax* activation function.

stores connections weights of reservoir neurons, with $N = 1000$. All weights are randomly initialized from a uniform distribution using the method described by [5] with a spectral radius of 0.7 and a connection density of 20% (i.e. non-zero weights). $\mathbf{W^{in}} \in \mathbb{R}^{I \times N}$ stores connections weights going from the I-dimensional inputs to the N-dimensional reservoir. These weights are randomly chosen in $\{-1, 1\}$, with a connection density of 20%, and are then scaled by a constant factor called input scaling (IS). Because the input data is composed of three different sets of features (MFCCs, Δ, Δ^2) with different distributions, we chose to apply three different ISs to the corresponding connections weights. We therefore define the MFCC input scaling (ISS), the Δ input scaling (ISD), and the Δ^2 input scaling (ISD2), respectively set to 10^{-3}, 5×10^{-3} and 5×10^{-3}. Finally, we use a linear regression with $L2$ regularization (also called Tikhonov regression or *ridge*), with a regularization coefficient of 10^{-4}, to use the same as the LSTM model described in Sect. 2.5, to learn the $\mathbf{W^{out}} \in \mathbb{R}^{N \times U}$ matrix. The latter stores the readout connections between the N-dimensional reservoir and the U-dimensional output space. In our case, U is set to 30, the number of phrase labels. The outputs $\hat{y}[t]$ of the model at time frame t are then computed following $\hat{y}[t] = \mathbf{W^{out}} \cdot x[t]$, given the reservoir state $x[t]$. The predicted class index $c[t]$ in the repertoire is then defined as the index of the maximum value in the prediction vector.

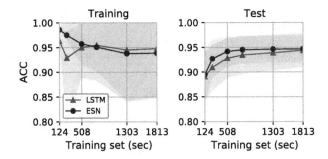

Fig. 4. Frame accuracy of models trained on subsets of songs of increasing size, on the training set (left) and on the testing set (right). Thirty songs represents around 780 s of song, 50 songs represents around 1303 s of song, 70 songs represents around 1813 s of song.

All hyperparameters presented were optimized during a random search made on a subset of 100 songs including all syllable types.

2.5   Presentation of the LSTM Model

We used a single LSTM (with forget gate) as originally presented in [6]. We set the number of units inside the LSTM to 72, in order to make the comparison fair with ESNs in term of total number of trainable parameters, which is around 30, 000. After adding a fully connected layer of units with softmax activation function on top of the LSTM to outputs class belonging probabilities, the model has a total of 34, 446 parameters. A $L2$ regularization is applied to the weights of the fully connected layer, with a regularization coefficient of 10^{-4}. We then trained the LSTM using Adam gradient descent algorithm, with a learning rate of 10^{-3}. Loss was computed using a log cross-entropy measure. Training was automatically stopped using the validation accuracy value, after a performance stagnation or decrease of 20 consecutive epochs. On average, LSTMs were trained on 181 ± 36 (std) epochs before achieving their best measured performance on validation data. The validation accuracy value is either computed on a validation fold of the training dataset during 5-folds cross-validation or on the test dataset during training with the whole training dataset.

3   Results

3.1   Performance of Transduction

Performance of the models on the audio-to-frame task was evaluated using a 5-fold cross-validation on the training dataset defined in Sect. 2.2. We trained 30 random initializations of ESNs for each experiment, against 5 random initializations of LSTMs. This imbalance was motivated by the important training time of LSTMs, and by the fact that LSTMs models are expected to converge

Table 1. Average scores obtained with a 5-fold cross-validation over all training songs and several models instances.

Model	Average frame accuracy (ACC)	Median frame accuracy	F1 (macro avg.)
LSTM	0.931 ± 0.104	0.951	0.865
ESN	0.935 ± 0.09	0.952	0.877

Table 2. Average WER obtained after training on the full training set.

Model	ESN	LSTM
WER (%)	5.3 ± 7.2	27.4 ± 16.2

toward similar solutions independently of their initializations, as fully trained neural networks, while ESNs rely on a randomized layer of neurons kept fixed during training.

The frame accuracy (ACC) was computed for all songs, and then averaged over the whole corpus, the folds and the random instances of the models. There is no significative difference in performance between the models medians (Wilcoxon rank-sum test over all accuracy measures, for all songs and all folds, $W = 10016240$, $p = 0.28$), and they both achieve an accuracy rate slightly higher than 93% in mean and 95% in median. We completed ACC with an F1-measure, which is significantly lower than ACC (around 0.86). This F1-measure is computed using a macro-averaging of precision and recall over all the classes of phrases, i.e. all classes are given equal importance in the computation, unlike with the accuracy score. This metric therefore gives insights on how well the models truly perform for all syllables types, without taking into account the imbalance of the classes in the dataset. All the metrics values are given in Table 1. An example of models outputs can be found in Fig. 3.

Finally, all models instances were trained on the whole training set and evaluated on the test set defined in Sect. 2.2. ESNs achieve an accuracy rate of $94.4\% \pm 2.7\%$, while LSTMs reach a lower accuracy rate of $93.9\% \pm 10.0\%$. ESNs accuracy median (95.4%) is however significantly lower than LSTMs accuracy median (95.9%). (Wilcoxon rank-sum test over all accuracy measures on test set, $W = 534682.5$, $p = 2.92 \times 10^{-8}$).

Comparison with [4] can not be done fairly, as our method operate at phrase level and not at syllable level, and as we did not use the same dataset. We discuss the possibility of extending this work in Discussion. However, our method shows performance at frame level belonging to the same range of performance exhibited in [4] (between 92%–98% accuracy)[2].

---

[2] As the work in [4] seems to not having been reviewed yet, we will not make further comparisons to avoid any mistakes than could originate from unverified results.

3.2 Performance of Models on Reduced Dataset

The aim of canary songs annotation automation is to save as much time and resources as possible for the experimenters working with these songs. Although our models are supervised, the amount of annotated data required for the models to produce acceptable results can be low enough to still significantly save time. We tried to assess the minimum amount of data necessary to reach the performances described in Sect. 3.1.

Figure 4 shows the evolution of the ACC score on the test set given the number of songs in the training dataset, using 5 instances of LSTMs and 30 instances of ESNs on random subsets of data. ESNs appear to be less sensitive to reduction of the training dataset than LSTMs, and reach back their peak performance being trained on 30 to 50 songs, which represents approximately 780 ± 55 to 1303 ± 87 seconds of song (13 min to 21 min 43 s). Low overfitting is observed with this training set sizes, as the test and train ACC are comparable. LSTMs, on the other hand, reach back their peak performance defined in Sect. 3.1 with a training set containing at least 70 songs, which represents approximately 1813 ± 87 s (around 30 min of song).

3.3 Sequence Extraction

In order to perform the audio-to-phrases task, we reconstructed phrases from the models predictions made at frame level for the audio-to-frame task. To do so, we simply annotated consecutive frames sharing the same annotation with a single label. Segmentation of songs is therefore implicitly performed by the RNN: we consider that phrases onset and offset are delimited by uninterrupted sequences of frames with the same label. Figure 3 shows that this method allows to accurately segment the songs without any post processing, because ESNs and LSTMs shift their outputs activations in time following the onset and offset of the phrases. Figure 5 shows that ESNs significantly outperform LSTMs on the audio-to-phrases task, by achieving a $5.3\% \pm 7.2\%$ WER on the testing set, being

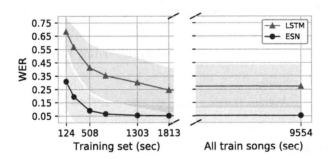

Fig. 5. WER of models trained on subsets of songs of increasing size (left) and WER of models trained on the whole training set (right). Song time is equivalent to the one explained in Fig. 4.

trained on the whole training set. LSTMs achieve a $27.4\% \pm 16.2\%$ WER. This difference is explained by a higher instability of LSTMs predictions onsets and offsets, creating small noisy occurrences of phrases where they are not expected, as visible in Fig. 3B. WER scores are summarized in Table 2. While these errors are insignificant at frame level, because of the important imbalance in duration between phrase classes, they can seriously hinder the WER, as it is not weighted by phrases durations. Finally, ESNs are able to reach their peak performance using a subset of 30 to 50 songs, the same number they require in Sect. 3.2 to reach their peak ACC. This allows us to recommend to use ESNs to annotate songs and to train them on at least around 20 min of song to obtain acceptable results.

3.4   Time Constraint

All trainings where performed on an *Intel Core* i7-9850H with 12 cores operating at a frequency of 2.60 GHz, on a computer equiped with 31.1 GiB of RAM. The training time considered takes into account the tool used to build the models – *ReservoirPy* [18] for the ESNs, *TensorFlow* and *Keras* for the LSTMs – and the policy used to stop the training, when applicable. Training times were averaged over the 10-folds cross-validation. Average training time for ESNs is significantly shorter than average training time for LSTMs, as ESNs only require one epoch of training. LSTMs on the other hand were trained using an early stopping policy based on validation accuracy with a patience of 20 epochs, and only reach top performance after around 180 epochs of training. These measures are nevertheless only given as very broad indications, as the comparison is not completely fair: ESNs were able to benefit from a parallelized training policy over the 12 cores of the processor, while the LSTMs were only partially benefiting from it. Also, most deep learning models such as LSTMs are nowadays trained on specialized hardware like GPUs. In our case, training on GPU makes the training longer, as our LSTMs are quite small, and as the flow of data between CPU and GPU adds a significant overhead. However, regarding the very short training time of ESNs, we can safely hypothesized that such performances are out of reach for an LSTM, even with further optimizations (Table 3).

Table 3. Average training time for the two models during 10-fold cross-validation.

Model	LSTM	ESN
Average training time (s)	2930 ± 222	$\mathbf{35 \pm 1}$

4   Discussion

Canary is one of the most complex singer that can be found among song birds. Its songs display sophisticated syntactic rules, as demonstrated in [13], which make it an appropriate candidate for deeper analysis of the emergence and learning of sequential organization of gestures. However, these analysis heavily rely on the quantity of available data to untangle the temporal complexity of the songs. These data are usually hand-annotated, and the full annotation process has to be repeated for each individual, because each canary produces specific syllables.

In this context, we proposed an efficient yet lightweight solution using RNNs to help with the annotation process, by shortening significantly the amount of time necessary to annotate large amount of data. Our method achieves a low error rate at phrase level (around 5% when using ESN) which should be acceptable for syntactic analysis. We applied this method on a canary that was producing particularly complex syllables in order to build robust classifiers applicable to any domestic canary songs. A more extensive study should be conducted soon, using other publicly available datasets. We also make ours public, to contribute to any field of research requiring annotated canary songs. Additionally, a more extensive study would also allows to perform a fair comparison with [4] by performing the annotation task at syllable level, and to assess the quality of the annotated sequence of phrases by comparing their temporal structure with the results of [13]. In any case, we confidently make the hypothesis that our method allows faster computation that the deep learning solution presented in [4], at least when using an ESN, while reaching similar performance. Indeed, the full training of an ESN only take around 40 seconds using 2 h 39 min of data.

Furthermore, our method is rooted in speech recognition area by using MFCCs as representation of the audio signals, in order to extract the relevant spectral information and reduce its dimensionality. While this approach was criticized in [11], a previous study successfully using CNN and HMM to decode Bengalese finch songs, we empirically show that MFCCs seem to remain a good approximation of spectral information, at least for canary vocalizations. We also demonstrated that RNNs like ESNs can outperform more complicated and widely used techniques like LSTMs, within a fair comparison setup, on tasks deemed as difficult and closely related to speech recognition.

We finally make the hypothesis that our method could be successfully applied to other species, song birds, mammals, or even insects. Our method has the advantage of being easy to experiment with, as the training time for ESNs is very short compared to other deep learning algorithms, allowing for extensive optimization and statistical robustness, and does not require powerful hardware to run successfully.

Acknowledgment. We would like to thank Catherine Del Negro, Aurore Cazala and Juliette Giraudon for the recording and transcription of the canary data. We also thank Inria for the ADT grant Scikit-ESN.

References

1. Anderson, S.E., Dave, A.S., Margoliash, D.: Template-based automatic recognition of birdsong syllables from continuous recordings. J. Acoust. Soc. Am. **100**(2), 1209–1219 (1996)

2. Bay, M., Ehmann, A.F., Downie, J.S.: Evaluation of multiple-F0 estimation and tracking systems. In: ISMIR (2009)
3. Chu, W., Blumstein, D.T.: Noise robust bird song detection using syllable pattern-based hidden Markov models. In: 2011 IEEE ICASSP, pp. 345–348 (2011)
4. Cohen, Y., Nicholson, D., Gardner, T.J.: TweetyNet: a neural network that enables high-throughput, automated annotation of birdsong. bioRxiv p. 2020.08.28.272088 (2020)
5. Gallicchio, C., Micheli, A., Pedrelli, L.: Fast spectral radius initialization for recurrent neural networks. In: Oneto, L., Navarin, N., Sperduti, A., Anguita, D. (eds.) INNSBDDL 2019. PINNS, vol. 1, pp. 380–390. Springer, Cham (2020). https://doi.org/10.1007/978-3-030-16841-4_39
6. Gers, F.A., Schmidhuber, J., Cummins, F.: Learning to forget: continual prediction with LSTM. In: ICANN 1999, vol. 2, pp. 850–855 (1999)
7. Giraudon, J., Trouvain, N., Cazala, A., Del Negro, C., Hinaut, X.: Labeled songs of domestic canary M1–2016-spring (Serinus canaria), May 2021
8. Jaeger, H.: The "echo state" approach to analysing and training recurrent neural networks-with an erratum note'. German National Research Center for Information Technology GMD Technical Report 148 (2001)
9. Juang, B.H., Rabiner, L., Wilpon, J.: On the use of bandpass liftering in speech recognition. IEEE Trans. Acoust. Speech Signal Process. **35**(7), 947–954 (1987)
10. Kaewtip, K., Alwan, A., O'Reilly, C., Taylor, C.E.: A robust automatic birdsong phrase classification: a template-based approach. J. Acoust. Soc. Am. **140**(5), 3691–3701 (2016)
11. Koumura, T., Okanoya, K.: Automatic recognition of element classes and boundaries in the birdsong with variable sequences. PLOS ONE **11**(7), e0159188 (2016)
12. Leitner, S., Catchpole, C.K.: Syllable repertoire and the size of the song control system in captive canaries (Serinus canaria). J. Neurobiol. **60**(1), 21–27 (2004)
13. Markowitz, J.E., Ivie, E., Kligler, L., Gardner, T.J.: Long-range order in canary song. PLOS Comput. Biol. **9**(5), e1003052 (2013)
14. McFee, B., et al.: Librosa/librosa: 0.8.0. Zenodo (2020)
15. Sainburg, T., Thielk, M., Gentner, T.Q.: Latent space visualization, characterization, and generation of diverse vocal communication signals. bioRxiv, p. 870311 (2020)
16. Tachibana, R.O., Oosugi, N., Okanoya, K.: Semi-automatic classification of birdsong elements using a linear support vector machine. PLOS ONE **9**(3), e92584 (2014)
17. Tan, L.N., Alwan, A., Kossan, G., Cody, M.L., Taylor, C.E.: Dynamic time warping and sparse representation classification for birdsong phrase classification using limited training data. J. Acoust. Soc. Am. **137**(3), 1069–1080 (2015)
18. Trouvain, N., Pedrelli, L., Dinh, T.T., Hinaut, X.: *ReservoirPy*: an efficient and user-friendly library to design echo state networks. In: Farkaš, I., Masulli, P., Wermter, S. (eds.) ICANN 2020. LNCS, vol. 12397, pp. 494–505. Springer, Cham (2020). https://doi.org/10.1007/978-3-030-61616-8_40
19. Waser, M.S., Marler, P.: Song learning in canaries. J. Comp. Physiol. Psychol. **91**(1), 1–7 (1977)

Which Hype for My New Task? Hints and Random Search for Echo State Networks Hyperparameters

Xavier Hinaut[1,2,3(✉)] 🆔 and Nathan Trouvain[1,2,3] 🆔

[1] INRIA Bordeaux Sud-Ouest, Talence, France
xavier.hinaut@inria.fr
[2] LaBRI, Bordeaux INP, CNRS, UMR 5800, Talence, France
[3] Institut des Maladies Neurodégénératives, Université de Bordeaux, CNRS, UMR 5293, Bordeaux, France

Abstract. In learning systems, hyperparameters are parameters that are not learned but need to be set a priori. In Reservoir Computing, there are several parameters that needs to be set a priori depending on the task. Newcomers to Reservoir Computing cannot have a good intuition on which hyperparameters to tune and how to tune them. For instance, beginners often explore the reservoir sparsity, but in practice this parameter is not of high influence on performance for ESNs. Most importantly, many authors keep doing suboptimal hyperparameter searches: using grid search as a tool to explore more than two hyperparameters, while restraining the spectral radius to be below unity. In this short paper, we give some suggestions, intuitions, and give a general method to find robust hyperparameters while understanding their influence on performance. We also provide a graphical interface (included in *ReservoirPy*) in order to make this hyperparameter search more intuitive. Finally, we discuss some potential refinements of the proposed method.

Keywords: Reservoir computing · Echo state networks · Hyperparameters · Random search · Grid search · Effective spectral radius

1 Introduction

1.1 Disclaimer

This short paper aims to give some keys to reservoir newcomers to know how to set hyperparameters. It does not aim to be comprehensive like [12] or to discuss studies on hyperparameter search, but rather to give a few useful practical hints to always keep in mind. The main originality lies in the general method proposed, supplemented with ReservoirPy graphics, applied on two common tasks.

© Springer Nature Switzerland AG 2021
I. Farkaš et al. (Eds.): ICANN 2021, LNCS 12895, pp. 83–97, 2021.
https://doi.org/10.1007/978-3-030-86383-8_7

1.2   You Have a New Task

Suppose you have already some practical experience with Reservoir Computing (RC) [13], in particular with Echo State Networks [8]. For instance, you read the nice guide of Lukoševičius [12] and you want to apply RC on a new task. How do you choose your HyperParameters (HPs) for that new task (for which you have no assumption of which HP values would work)?

If you only want good results and do not want to know how the HPs influence the performance of your task, you can use an optimizer (e.g. a Bayesian [17] or evolutionary one [1]) to set your HPs once and that's it. The problem is that if you want to know if you have found robust HPs with your optimized search, you will run the optimizer a few times more and will probably end up with several sets of HP values that are all different. Unfortunately, making an average of values for each HP is probably not a good idea given the interdependencies between HPs (see Subsect. 2.6). In the end, you would probably want to know a little more on the robustness of the HPs you found. Moreover, knowing which HPs are important will help you to adapt them if needed. For example if at some point you obtain more diverse data (e.g. on a classification task) and you would like to increase the reservoir size to gain performance. You will need to perform again the optimization but this time with more data and an increased size of reservoir: this will cost you much more time to optimize.

Conversely, you want to find good HPs for your task, and this time you are willing to know more about the robustness of the HP values found and understand which HPs are the most important for your task. Thus, you will probably do a grid search because that's what other people usually do, and because it provides you nicely discretized maps showing gradients of performance depending on your HPs. The problem is, by doing this, you will probably miss some of the best performances you could expect. Moreover, it will cost you a lot of computational time, much more than what you would need with an optimizer: because by exploring v values for each of your p HPs, you will need to perform p^v evaluations[1].

Thus, what should you do to disentangle this seemingly impossible trade-off? This is what we will argue in this paper. First, we discuss theoretical aspects of reservoir computing and interpret their impact on practical purposes while explaining why grid search is not a good idea. Then, we provide a general practical method for hyperparameter exploration along with graphical tools. Afterwards, we show how this method applies to two tasks of time series prediction. Finally, we discuss some potential refinements of the proposed method.

2   From Theory to Practice

2.1   At the Edge of Chaos

Like other dynamical systems (e.g. cellular automata [9]), reservoirs have rich and "interesting" dynamics at the edge of chaos, or *edge of stability* [2]. For most

---

[1] We call evaluations the training of instances of reservoirs for sets of HP values.

tasks, you want to have reservoir dynamics with contractive property in order not to have diverging dynamics that could destroy the behavior of your outputs. In other words, contractive property means that the reservoir will forget its current state and inputs after some time. While in chaotic regime small differences in states or inputs are amplified dramatically. Legenstein et al. [10] have proposed an interesting computational predictive measure to find these edge of stability dynamics in the HP space. The measure is based on the idea that, in order to generalize well, a reservoir needs a mixture of two properties: (1) the ability to represent similar inputs in a similar way, and (2) the ability to separate inputs. Taking into account the fact that dynamics with contractive properties enable property (1), and chaotic dynamics enable property (2), what we need for a good generalization ability is to have both (1) and (2) at the same time, namely, at the intersection of contractive and chaotic dynamics: at the edge of stability. As they show in their computational study, their predictive generalization measure often overlaps with regions of the HP space with edge of stability dynamics.

That being said, remember that there is no such optimal reservoir dynamics for all tasks. Indeed, the desired dynamics depend heavily on your task. Thus, such computational predictive measures are probably useful only for some class of tasks.

2.2   Why You Shouldn't Care About the *Echo State Property*

In his seminal paper, Jaeger defines the Echo State Property (ESP) [8]. To simplify shortly, ESP states that a spectral radius (SR) following $SR < 1$ is a theoretical condition to have a contracting system in the absence of inputs. One should remember that originally this theoretical condition is interpolated from linear systems. However, we usually do not use linear reservoirs, as most people use hyperbolic tangent (*tanh*) activation function. Thus, even if refinements of the ESP were made, this condition should not be a *constraining* rule but only a rough guide, because reservoir dynamics depend on the inputs fed; it was shown that for some tasks optimal SR could be superior to 1.

Knowing that ESP is mainly a theoretical guide to set the SR is not enough. If you use reservoir with leaky neurons (i.e. with a *leak rate* inferior to one), then you should also know what the *effective spectral radius* is: it is what you should care about instead of the SR itself. Jaeger et al. [3] introduced the notion of *effective spectral radius* for reservoirs with leaky-integrator neurons. The idea is roughly the following: the leak rate has a kind of "cooling" effect on the dynamics, thus for a constant spectral radius if one decreases the leak rate, the dynamics will "slow down": the same happens for the *effective spectral radius*. Consequently, the more your decrease the leak rate, the more you can increase the spectral radius, while still having the ESP.

2.3   The Myth of Grid Search

Often, if people do not tune HPs by hand, they use grid search, and we have done it for too long too. Grid search means that you explore each HP by changing its values in a predefined grid, such as for each possible value of one HP you will have tested all possible values of the other HPs.

Bergsta et al. [6] explained why grid search is suboptimal compared to random search. In Fig. 1 we can see a schematic explanation of why random search performs better. As Bergsta et al. [6] pointed out: "Grid search allocate too many trials to the exploration of dimensions that do not matter and suffer from poor coverage in dimensions that are important." Thus, you should prefer random exploration over grid search. One advantage of grid search is to be able to guide intuition by visualizing a map of the performance depending on the HPs values. As we will show in Sect. 3, similar maps could be obtained with random search. Moreover, random search is particularly adapted to reservoir because of the little computation cost of the training compared to other methods based on back-propagation. A good view of the hyperspace can be obtained at low cost.

In fact, grid search wastes evaluations to explore unimportant HPs. Whereas, random search do not waste evaluations because for all HP no value is explored twice (i.e. no need to evaluate the same value several time for one HP in order to make a grid to explore other HPs). With grid search, by exploring v values for each of p HPs, one needs to perform p^v evaluations. While for random search, the number of values explored v is equal to the total number of evaluations.

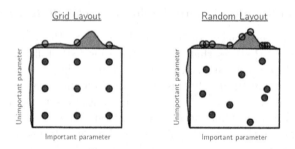

Fig. 1. Random search samples more efficiently than grid search. Image from [6].

2.4   Pieces of Advice and Hints on HPs Choice

Like in other computing paradigms, all HPs do not have the same influence on the performance of your task. The most important HPs, like Lukoševičius suggests [12] are the *spectral radius, input scaling, leaking rate (or leak rate), number of units in the reservoir* and *feedback scaling* (in case you have feedback from the readout units to the reservoir). If you do not want to explore too many HPs you should care only about them. For instance, newcomers often consider the sparsity of the reservoir to be important, but it is not, in particular if you generate the weights with a Gaussian distribution, many weights will be close to zero in any case. A rule of thumb that we use for sparsity is to keep 20% of non-zero connections in input and reservoir weights matrices, whatever the size of your reservoir is. For most HPs, in particular the ones recommended by

Lukoševičius, one should not use a uniform exploration, but rather a logarithmic exploration. This means that you will be likely to explore low values as much as high values of your searching ranges (with the same sampling at a given order of magnitude). For your first HPs search on a new task, we recommend to use to explore HPs with three levels of magnitude between the smallest value and the highest value (e.g. 1 and 1000).

2.5 A General Method with Random Search

We consider a simple ESN with leaky integrator units, without feedback weights and with readout weights regularization. The main steps of the proposed method are (including two preliminary steps):

- Pre1: Take the minimal reservoir size N you would like to use (you can change N after the HP search)
- Pre2: Fix one of the following HP: IS (Input Scaling), SR (Spectral radius) or LR (Leak Rate) (to avoid conflicting interdependencies)
- A: 1st random search (large range): explore all HP within a wide range of values.
- B1: 2nd random search (narrow range): remove unimportant HPs and narrow the ranges of HP values to the $x\%$ best values (e.g. 10% best values is a good choice when doing more than 100 evaluations.).
- B2: Choose median or best HP values (but the ridge) depending on *loss vs. HP* cloud shape
- C: (optional) Check robustness of previously fixed HPs (in *2nd* step) and adapt if needed
- D1: (optional) Explore the trade-off between loss and computation time when varying the reservoir size N
- D2: Find best regularization parameter (ridge) for chosen N
- E: Evaluate test set on chosen HPs for 30 reservoir instances

2.6 *ReservoirPy* HP Interdependency Plot

The proposed method uses HP interdependency plots in order to have a visual evaluation of the loss obtained as a function of all HP explored along with the interactions between HPs. Implementation details are provided here[2].

3 An Illustrated Example

We will make the exploration for two well known tasks of chaotic time series prediction: Mackey-Glass [14] and Lorenz [11] time series prediction. Both time series where generated using *ReservoirPy* library, and details about their characteristics can be found in the documentation.

---

[2] https://hal.inria.fr/hal-03203318/document.

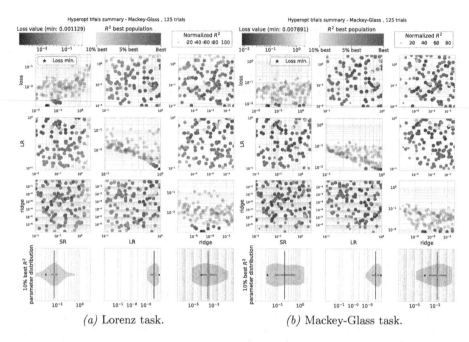

(a) Lorenz task.   *(b)* Mackey-Glass task.

Fig. 2. (a) Small search dependence plot for Lorenz task. Two different patterns of interaction between error (or loss) and parameter value can be observed: the U-shaped clouds (top-left scatter plot and bottom right scatter plot) and the skewed-linear clouds (middle scatter plot). (b) Small search dependence plot for Mackey-Glass task. Here, if the dependence to the leak rate seems to be the same as in the Lorenz task, we can see that the SR has little influence on the error value, unlike in the Lorenz task, and is therefore task dependent.

For the other parameters that need to be set we do as follows. We set the *washout* (i.e. number of timesteps ignored at the beginning) to the size of the reservoir N. After N timesteps, it is indeed expected that all neural nodes in the reservoir graph have received signals from inputs and surrounding units. The random seeds used to initialize the 5 instances of the model used during cross validation are kept fixed for all evaluations. We finally fixed the reservoir size N to 100, to shorten the computation time during the searches.

We chose 5 parameters to explore: *spectral radius* (SR), *leak rate* (LR), *connectivity probability of W* (W proba), *connectivity probability of* W_{in} (W_{in} proba), and *ridge* regularization parameter. We included known "unimportant" HPs like the connectivity probability of matrices in order to illustrate how to deal with these kinds of parameter throughout the method we propose. We also chose to fix at least one of the more important HPs, to reduce the complexity of the search: IS will be kept constant and equal to 1 during this first step.

If we would have liked to do a grid search on 5 HPs and limit our search to about 100 evaluations, we could have taken 3 values per HP for a total of $5^3 = 125$ evaluations. For grid search exploring only 3 values is very small, that is why exploring more than 2 HPs with grid search is too computationally expensive. Whereas for random search, 125 evaluations is enough to have a first idea of the good HP ranges. Taking 25 evaluations per HP explored is a good rule of thumb.

3.1   1st Random Search

For both tasks, we used a "task agnostic" range of HPs to be explored (with the type of distribution chosen *a priori*, specified in parentheses): SR $= [10^{-2}, 10^1]$ (log-uniform); LR $= [10^{-3}, 1]$ (log-uniform); W proba $= [10^{-2}, 1]$ (uniform); W_{in} proba $= [10^{-2}, 1]$ (uniform); ridge $= [10^{-3}, 10^{-9}]$ (log-uniform). IS is kept fixed and equal to 1. A minimum error of 1.2×10^{-3} is obtained at this stage of the evaluation for the Lorenz task, and 7.8×10^{-3} for the Mackey-Glass task[3].

3.2   2nd Random Search

We can then use this information to redefine the range of possible values of parameters. One has to include the $x\%$ best results inside this range. In other words, the new range must include previous $x\%$ best values, with an error margin set depending on the trend observed in the influence of the parameter over performance, e.g. if it seems that expanding a bit the range towards positive side of a parameter axis would allow to explore possibly interesting values, we recommend expanding it. With 125 evaluations, $x = 10$ seems to be a good compromise. Also, one can fix parameters that seem to have no particular influence on the model performance (e.g. sparsity of W and W_{in}[4]). We can therefore fix these parameters to a specific value for the rest of the exploration. In both tasks, we chose to fix W proba and W_{in} to the median value of the top $x = 10\%$ trials.

Lorenz Task. For Lorenz task, we redefined the range of HPs as follows: SR $= [10^{-2}, 3]$ (log-uniform); LR $= [10^{-1}, 1]$ (log-uniform); W proba $= 0.6$ (fixed); W_{in} proba $= 0.5$ (fixed); ridge $= [10^{-7}, 1]$ (log-uniform).

Mackey-Glass Task. For Mackey-Glass task, we redefined the range of HPs as follows: SR $= [10^{-2}, 3]$ (log-uniform); LR $= [10^{-1}, 1]$ (log-uniform); W proba $= 0.6$ (fixed); W_{in} proba $= 0.5$ (fixed); ridge $= [10^{-6}, 10]$ (log-uniform). Figure 2 summarizes the results of this second random search for both tasks.

3.3   Best Values Obtained

During this second stage of search, the minimum error reached for Lorenz task lowered to 1.1×10^{-3}, while the minimum error reached for Mackey-Glass remained close to the value obtained during the large search.

---

[3] A figure summarizing the results of this first random search on the Lorenz time series prediction task is available here: https://hal.inria.fr/hal-03203318/document.

[4] W proba and W_{in} proba have little influence on the error distribution (diagonal plots) and no linear dependence with the other parameters can be seen on figure available here: https://hal.inria.fr/hal-03203318/document.

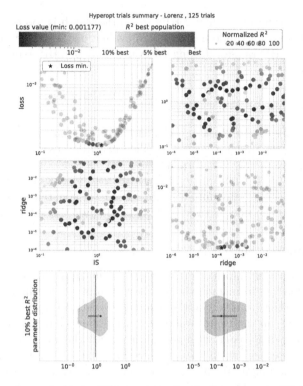

Fig. 3. Restricted search on the IS parameter, for the Lorenz task. With a value of 1, we can see that the IS yields the best results and reaches the minimum of the U-shaped error cloud that can be seen in the top-left plot.

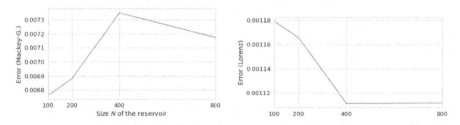

Fig. 4. Error obtained with the best ridge for sizes of reservoir in $\{100, 200, 400, 800\}$, on the Mackey-Glass task (left) and Lorenz task (right). Note that variations in error are really small for both tasks.

Looking at the graphs, we select values given the shape of the estimated distribution of the error metrics. Two patterns generally arise: the U-shape clouds and the skewed clouds. For the parameters provoking U-shaped clouds of error, we set the value of the parameter to the median of the $x = 10\%$ best values, e.g. $SR = 7.10^{-2}$ for the Lorenz prediction task, as observed in Fig. 2. For the HPs clearly skewed with a linear trend (still on a logarithmic scale), we take the best parameter found, e.g. $LR = 0.98$, in the case of the Lorenz prediction task. We also double checked this values on the coupled parameters plots around the diagonal of plots, to make sure that no picked value would poorly interact with another parameter. We can also see in Fig. 2 that the spectral radius seems to play a not so important role in the Mackey-Glass task. The error distribution regarding to this parameter is quite flat, with unclear dependence compared to the Lorenz task. This shows that the SR can be a task dependent parameter, and be robust against variations for a wide range of values.

Lorenz Task. For Lorenz task, using the method described above, we finally chose the following parameters: SR = 7.10^{-2} (fixed); LR = 0.96 (fixed); W proba = 0.6 (fixed); W_{in} proba = 0.5 (fixed); ridge = $[10^{-6}, 10^{-1}]$ (log-uniform).

Mackey-Glass Task. For Mackey-Glass task, we finally chose the following parameters: SR = 0.12 (log-uniform); LR = 0.97 (log-uniform); W proba = 0.6 (fixed); W_{in} proba = 0.5 (fixed); ridge = $[10^{-6}, 10]$ (log-uniform).

The ridge was not fixed until the end of the process. We only kept narrowing its explored range. Regularization parameters should not be fixed until all other parameters are found: they are meant to optimize the model regarding to its complexity, in our case regarding to the number of neuronal units of the reservoir, the parameter N. We assess the impact of this parameter on the performance in Sect. 3.5.

3.4 Checking the Input Scaling

Before evaluating the model using the parameters found, we recommend assessing the interdependence of the chosen parameters and the parameters kept constant during evaluation. In our case, the input scaling (IS) was fixed to 1, but as it controls the amplitude of the signals fed to the reservoir, it can have great influence on the model performance. We performed a simple check consisting in defining a range of one or two order of magnitude around the fixed value of the IS, and running a random search on this range. The ridge will also be kept variable, as it has not been fixed yet. Results of this evaluation for the Lorenz task can be found in Fig. 3 and on Fig. 2 for the Mackey-Glass task. Although it appears that a maximum of performance can be reached with a fixed value of 1 in the case of the Lorenz task, with this value placed just below the minimum of the error, we can see that the distribution of error is skewed towards the left of this value in the case of the Mackey-Glass task. We therefore redefine the IS value for Mackey-Glass to 0.3.

3.5 Searching for the Best Ridge

Finally, we propose to explore the dependence of the ridge on the reservoir size N. We explored the ridge in a range of $[10^{-5}, 10^{-1}]$, and used four different reservoir sizes N = {100, 200, 400, 800}. By selecting the most adapted ridge for each size, i.e. the ridge corresponding to the minimum error produced with each reservoir size, we can see the relation existing between the error, the ridge and the reservoir size in Fig. 4. Using this information, one can then select the pair {ridge, N} that suits the most their needs and resources in terms of performance and computational power. Assuming that we have enough computational power to use a 800 neurons reservoir at least, we select the final values for ridge and N:

Lorenz Task. Because a size of 400 seems to already give good performance as shown in Fig. 4, we will keep this value instead of the maximum of 800. We finally chose: $N = 400$; ridge $= 2 \times 10^{-4}$.

Mackey-Glass Task. In the case of Mackey-Glass, we obtained best results with a size of 100, as shown in Fig. 4. Even if counter-intuitive, this result is probably due to the simplicity of the task performed, or to a low dependency between the reservoir size and the chosen parameters. We finally chose: $N = 100$; ridge $= 5 \times 10^{-4}$.

To conclude this experimentation, we evaluated the selected HPs on a test data set, with 30 different random initializations. The test data set is defined as the 2000 time steps following the 6000 time steps used during cross validation. This 6000 time steps are used as training steps during the evaluation. The results obtained are presented in Table 1:

Table 1. Results averaged over 30 random instances on the test set with the chosen HPs.

Task	NRMSE	R^2
Lorenz	4.5×10^{-4} ($\pm 1 \times 10^{-4}$)	0.99999 ($\pm 2 \times 10^{-6}$)
Mackey-G.	9.78×10^{-3} ($\pm 4.56 \times 10^{-3}$)	0.99805 ($\pm 2 \times 10^{-3}$)

Our final results outperformed the previously reached error values in both tasks. However, the difference in performance between previously reached values and final values remains relatively low, and could be attributed to chance. A better evaluation could be done using more folds during cross validation, and more random initializations to avoid any bias from both the data and the initialization. Nevertheless, our method helped finding ranges of HPs able to produce satisfying models in terms of performance, and without relying on too many assumptions and hypothesis, as the ones generally made about SR value for instance. Moreover, we evaluated the model regarding the influence of almost all possible parameters, ensuring robustness of the results.

4   Discussion

We first discussed theoretical aspects of reservoir computing and which impact they should have on practical purposes. Hopefully, we convinced the reader not to perform grid search for more than two hyperparameters (HPs), given the high number of evaluations that such suboptimal method would require[5]. Then, we provided a general practical method for HP exploration. Finally, we showed how this method applies to the Lorenz and Mackey-Glass chaotic time series prediction tasks, and provided several figures illustrating the influence of HPs on performance. To our knowledge, neither grid search nor other optimizer provide such an amount of information and intuitions for reservoir computing HP exploration. In summary, we proposed a method that is better than a compromise between the graphical intuitions that grid search would provide, and the optimum performance that an optimizer provides. Our method probably uses more evaluations than most efficient optimizers (even if evolutionary methods are generally demanding in terms of number of evaluations). Given the precision of the performance obtained throughout the paper, the number of evaluations used could be reduced.

Our study illustrates that the importance of HPs depends on the task. Indeed, we showed in Fig. 2 that the spectral radius is more robust to changes (i.e. it offers a wider range of eligible values) for the Mackey-Glass than for the Lorenz task. Moreover, the interdependencies between HPs are visible, for instance on Fig. 2 for *SR vs. ridge* subplot we can see a red-orange diagonal: if the SR is changed, the ridge has to be adapted, and vice versa[6]. In future work, we would like to integrate these interdependencies within the HP search. For example by defining a HP Y as a linear function of a HP X. Searching the best values for Y would thus be redefined as finding the best values of a and b in the following equation: $Y = a.X + b$.[7] One can argue that this adds more HPs, but one can assume that finding such interdependencies in such a way is less demanding in number of evaluations needed.

The tasks we have chosen are famous tasks for reservoir benchmarking, however they are rather simple to solve, given that a 100-unit reservoir already solves the task well. In future work, we will apply the proposed method to more complex and more computationally demanding tasks which would not enable hundreds of evaluations to be performed. Indeed, in one of our recent study [7] we used parts of the method proposed here. We applied it to a more difficult task (i.e. language processing) and among other things, in supplementary material of [7] we showed the effect of the regularization on a *loss vs. reservoir size* plot. Given the point-clouds drawn, we could see the best loss given the reservoir size both with optimal regularization and without any regularization at the same time.

---

[5] With grid search, by exploring v values for each of p HPs, one needs to perform p^v evaluations. While for random search, v is equal to the total number of evaluations.

[6] For interdependency for *IS vs. ridge* see https://hal.inria.fr/hal-03203318/document.

[7] Of course as we plot many variables with log scales, the equation would often look like $log(Y) = a.log(X) + b$.

The method we proposed in this study is probably applicable to other machine learning paradigms than RC. Thus reducing the time and computational resources needed for HP search: this is a step towards more environment-friendly methods. In one of our recent studies [16], we explored a new way of viewing the influence of HPs. With UMAP visualizations (Uniform Manifold Approximation and Projection [15]) we explored the influence of HPs on the internal dynamics of the reservoir. For instance, increasing the spectral radius or the leak rate is fragmenting the UMAP representations of the internal states.

Finally, we would like to enjoin authors to think about simple rules while writing their next paper on Reservoir Computing: (1) please provide all the HPs you have used with the details of how you found them, in order to enable others to have a chance to reproduce your work; (2) demonstrate that the HPs you have found or chosen are robust again perturbations or slight modifications (like in the supplementary material of [7]), instead of just picking the results from a black-box optimization search.

5   Supplementary Material

5.1   Implementation details

We used the *ReservoirPy* library [5], which has been interfaced with *hyperopt* [4], to make the following experiments and figures in order to dive into the exploration of HPs. We illustrated our proposed method with two different tasks of chaotic time series prediction. These tasks consist in predicting the value of the series at time step $t+1$ given its value at time step t. To asses the performance of our models on these tasks, we performed cross validation. Each fold is composed of a train and a validation data set. The folds are defined as continuous slices of the original series, with an overlap in time, e.g. the validation set of the first fold would be the training set of the second, and the two sets would be two adjacent sequences of data in time in the series. For the last fold of the time series, the train set is defined as the last available slice of data, while the train set of the first fold is used as validation set.

For all tasks, we used a 3-fold cross validation measure on a time series composed of 6000 time steps, i.e. each fold is composed of 4000 time steps, with a train set and a validation set of 2000 time steps each. We used two metrics to perform this measure: the Normalized Root Mean Squared Error, defined in Eq. 1, and the R^2 correlation coefficient, defined in Eq. 2:

$$\text{NRMSE}(y, \hat{y}) = \frac{\sqrt{\frac{1}{N} \sum_{t=0}^{N-1} (y_t - \hat{y}_t)^2}}{\max y - \min y} \tag{1}$$

$$R^2(y, \hat{y}) = 1 - \frac{\sum_{t=0}^{N-1} (y_t - \hat{y}_t)^2}{\sum_{t=0}^{N-1} (y_t - \bar{y})^2} \tag{2}$$

where y is a time series defined over N time steps, $\hat{y}$ is the estimated time series predicted by the model, and $\bar{y}$ is the average value of the time series y over time. NRMSE was used as an error measure, which we expect to reach a value near 0, while R^2 was used as a score, which we expect to reach 1, its maximum possible value. All measures were made by averaging this two metrics across all folds, with 5 different initializations of the models for each fold (Figs. 5 and 6).

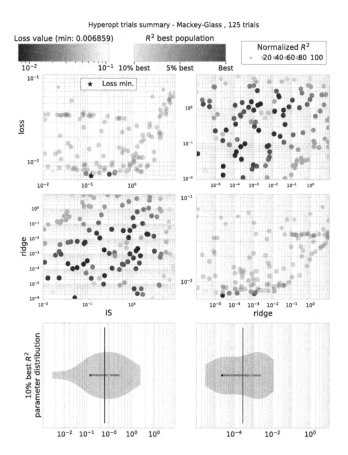

Fig. 5. Restricted search on the IS parameter, for the Mackey-Glass task. The fixed value of 1 defined at the beginning of the search for IS might not be the optimal value given all the other parameters chosen. In the case of the Mackey-Glass task, we can clearly see in the top-left and bottom-left plots that better results are achieved with lower values, with the top 10% trials distribution being placed around a median of 0.3.

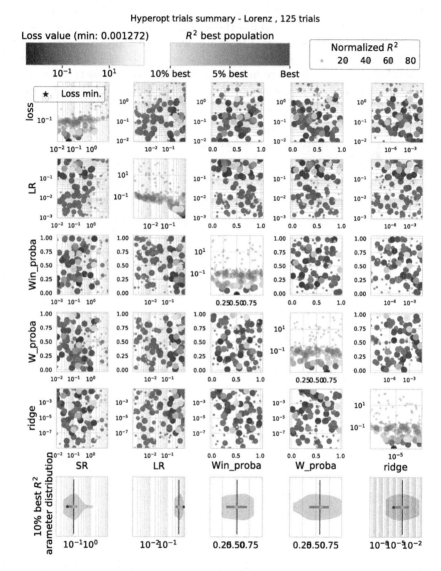

Fig. 6. Large search dependence plot for Lorenz task, with 125 trials. Diagonal of the plot matrix displays interactions between all parameters explored and the error value, also referred to as loss value. Top 10% trials in terms of score (here, R^2) are represented using colors from yellow (top 10) to red (top 1) in all plots. Other plots on the matrix display the interactions of all possible couple of parameters. In these plots, the value of the error is represented using shades of purple and blue, on a logarithmic scale, while score is represented using different circle sizes. Circle size is normalized regarding to the score values. Because R^2 can take values between $-\infty$ and 1, the smallest dots represents negative values. Finally, the bottom row of plots display the parameter distribution of the top 10% of trials, in terms of score, and for each parameter.

References

1. Ferreira, A., et al.: An approach to reservoir computing design and training. Expert Syst. Appl. **40**(10), 4172–4182 (2013)
2. Schrauwen, B., et al.: An overview of reservoir computing: theory, applications and implementations. In: Proceedings of ESANN, pp. 471–482 (2007)
3. Jaeger, H., et al.: Optimization and applications of echo state networks with leaky-integrator neurons. Neural Netw. **20**(3), 335–352 (2007)
4. Bergstra, J., et al.: Hyperopt: a python library for optimizing the hyperparameters of machine learning algorithms. In: Proceedings SciPy, pp. 13–20. Citeseer (2013)
5. Trouvain, N., Pedrelli, L., Dinh, T.T., Hinaut, X.: *ReservoirPy*: an efficient and user-friendly library to design echo state networks. In: Farkaš, I., Masulli, P., Wermter, S. (eds.) ICANN 2020. LNCS, vol. 12397, pp. 494–505. Springer, Cham (2020). https://doi.org/10.1007/978-3-030-61616-8_40
6. Bergstra, J., Bengio, Y.: Random search for hyper-parameter optimization. J. Mach. Learn. Res. **13**, 281–305 (2012)
7. Hinaut, X.: Which input abstraction is better for a robot syntax acquisition model? phonemes, words or grammatical constructions? In: ICDL-EpiRob (2018)
8. Jaeger, H.: The "echo state" approach to analysing and training recurrent neural networks. GNRCIT GMD Technical report, Bonn, Germany. 148, 34 (2001)
9. Langton, C.G.: Computation at the edge of chaos: phase transitions and emergent computation. Physica D **42**(1–3), 12–37 (1990)
10. Legenstein, R., Maass, W.: Edge of chaos and prediction of computational performance for neural circuit models. Neural Netw. **20**(3), 323–334 (2007)
11. Lorenz, E.N.: Deterministic nonperiodic flow. J. Atmo. Sci. **20**(2), 130–141 (1963)
12. Lukoševičius, M.: A practical guide to applying echo state networks. In: Montavon, G., Orr, G.B., Müller, K.-R. (eds.) Neural Networks: Tricks of the Trade. LNCS, vol. 7700, pp. 659–686. Springer, Heidelberg (2012). https://doi.org/10.1007/978-3-642-35289-8_36
13. Lukoševičius, M., Jaeger, H.: Reservoir computing approaches to recurrent neural network training. Comput. Sci. Rev. **3**(3), 127–149 (2009)
14. Mackey, M.C., Glass, L.: Oscillation and chaos in physiological control systems. Science **197**(4300), 287–289 (1977)
15. McInnes, L., Healy, J., Saul, N., Großberger, L.: UMAP: uniform manifold approximation and projection. J. Open Source Softw. **3**(29), 861 (2018)
16. Variengien, A., Hinaut, X.: A journey in ESN and LSTM visualisations on a language task. arXiv preprint arXiv:2012.01748 (2020)
17. Yperman, J., Becker, T.: Bayesian optimization of hyper-parameters in reservoir computing. arXiv preprint arXiv:1611.05193 (2016)

Semi- and Unsupervised Learning

A New Nearest Neighbor Median Shift Clustering for Binary Data

Gael Beck[1,2], Mustapha Lebbah[1(✉)], Hanene Azzag[1],
and Tarn Duong[1]

[1] Computer Science Laboratory of Paris North (LIPN, CNRS UMR 7030),
Sorbonne Paris Nord University, 93430 Villetaneuse, France
{beck,mustpha.lebbah,hanane.azzag,duong}@lipn.univ.paris13.fr
[2] HephIA SAS, 30, rue de Gramont, 75002 Paris, France
Gael@hephia.com

Abstract. We describe in this paper the theory and practice behind a new modal clustering method for binary data. Our approach (BinNNMS) is based on the nearest neighbor median shift. The median shift is an extension of the well-known mean shift, which was designed for continuous data, to handle binary data. We demonstrate that BinNNMS can discover accurately the location of clusters in binary data with theoretical and experimental analyses.

Keywords: Density gradient ascent · Hamming distance · Mean shift

1 Introduction

The goal of clustering (unsupervised learning) is to assign cluster membership to unlabeled candidate points where the number and location of these clusters are unknown. Clustering is an important step in the exploratory phase of data analysis, and it becomes more difficult when applied to binary or mixed data. Binary data occupy a special place in many application fields: behavioral and social research, survey analysis, document clustering, and inference on binary images.

Clusters are formed usually from a process that minimizes the dissimilarities inside the clusters and to maximizes the dissimilarities between clusters. A popular clustering algorithm for binary data is the k-modes [8], and it is similar to the k-means clustering [14] wherein the modes are used instead of the means for the prototypes of the clusters. Other clustering algorithms have been developed using a matching dissimilarity measure for categorical points instead of Euclidean distance [12], and a frequency-based method to update modes in the clustering process [10].

In this paper, we focus on the mean shift clustering [5,6], which is another generalization of the k-means clustering. Mean shift clustering belongs to the class of modal clustering methods where the arbitrarily shaped clusters are defined in terms of the basins of attraction to the local modes of the data density, created by the density gradient ascent paths. In the traditional characterization of the mean

© Springer Nature Switzerland AG 2021
I. Farkaš et al. (Eds.): ICANN 2021, LNCS 12895, pp. 101–112, 2021.
https://doi.org/10.1007/978-3-030-86383-8_8

shift, these gradient ascent paths are computed from successive iterations of the mean of the nearest neighbors of the current prototype. Due to its reliance on mean computations, it is not suited to be directly applied to binary data. Our contribution is the presentation of a modified mean shift clustering which is adapted to binary data. It is titled Nearest Neighbor Median Shift clustering for binary data (BinNNMS). The main novelty is the that the cluster prototypes are updated via iterations on the majority vote of their nearest neighbors. We demonstrate that this majority vote corresponds to the median of the nearest neighbors with respect to the Hamming distance [7].

The rest of the paper is organized as follows: Sect. 2 introduces the traditional mean-shift algorithm for continuous data, Sect. 3 presents our new median shift clustering procedure for binary data BinNNMS, and Sect. 4 describes the results of the BinNNMS compared to the k-modes clustering.

2 Nearest Neighbor Mean Shift Clustering for Continuous Data

The mean shift clustering proceeds in an indirect manner based on local gradients of the data density, and without imposing an ellipsoidal shape to clusters or that the number of clusters be known, as is the case for k-means clustering. For a candidate point $\boldsymbol{x}$, the mean shift method generates a sequence of points $\{\boldsymbol{x}_0, \boldsymbol{x}_1, \dots\}, \boldsymbol{x}_j \in \mathbb{R}^d, j = 1, 2, \dots$, which follows the gradient density ascent. The theoretical mean shift recurrence relation is

$$\boldsymbol{x}_{j+1} = \boldsymbol{x}_j + \frac{\mathbf{A}\mathsf{D}f(\boldsymbol{x}_j)}{f(\boldsymbol{x}_j)} \tag{1}$$

for a given positive-definite matrix $\mathbf{A}$, for $j \geq 1$ and $\boldsymbol{x}_0 = \boldsymbol{x}$. The output from Eq. (1) is the sequence $\{\boldsymbol{x}_j\}_{j \geq 0}$ which follows the density gradient ascent $\mathsf{D}f$ to a local mode of the density function f.

To derive the formula for the nearest neighbor mean shift for a random sample $\boldsymbol{X}_1, \dots, \boldsymbol{X}_n$ drawn from a common density f, we replace the density f and density gradient $\mathsf{D}f$ by their nearest neighbor estimates

$$\hat{f}_{\mathrm{NN}}(\boldsymbol{x}; k) = n^{-1}\delta_{(k)}(\boldsymbol{x})^{-d}\sum_{i=1}^{n}\frac{K((\boldsymbol{x} - \boldsymbol{X}_i)}{\delta_{(k)}(\boldsymbol{x}))}$$

$$\mathsf{D}\hat{f}_{\mathrm{NN}}(\boldsymbol{x}; k) = n^{-1}\delta_{(k)}(\boldsymbol{x})^{-d-1}\sum_{i=1}^{n}\frac{\mathsf{D}K((\boldsymbol{x} - \boldsymbol{X}_i)}{\delta_{(k)}(\boldsymbol{x}))} \tag{2}$$

where K is a kernel function and $\delta_{(k)}(\boldsymbol{x})$ as the k-th nearest neighbor distance to $\boldsymbol{x}$, i.e. $\delta_{(k)}(\boldsymbol{x})$ is the k-th order statistic of the Euclidean distances $\|\boldsymbol{x} - \boldsymbol{X}_1\|, \dots, \|\boldsymbol{x} - \boldsymbol{X}_n\|$. These nearest neighbor estimators were introduced by [13] and elaborated by [5,6] for the mean shift.

These authors established that the beta family kernels are computationally efficient for estimating f and $\mathsf{D}f$ for continuous data. The uniform kernel is the most widely known member of this beta family, and it is defined as $K(\boldsymbol{x}) = v_0^{-1} \mathbf{1}\{\boldsymbol{x} \in B_d(\mathbf{0}, 1)\}$ where $B_d(\boldsymbol{x}, r)$ is the d-dimensional hyper-ball centered at $\boldsymbol{x}$ with radius r and v_0 is the hyper-volume of the unit d-dimensional hyper-ball $B_d(\mathbf{0}, 1)$. With this family of kernels, and the choice $\mathbf{A} = (d+2)^{-1} \delta_{(k)}(\boldsymbol{x}) \mathbf{I}_d$, the nearest neighbor mean shift becomes

$$\boldsymbol{x}_{j+1} = k^{-1} \sum_{\boldsymbol{X}_i \in \mathrm{NN}_k(\boldsymbol{x}_j)} \boldsymbol{X}_i \tag{3}$$

where $\mathrm{NN}_k(\boldsymbol{x})$ is the set of the k nearest neighbors of $\boldsymbol{x}$. For the derivation of Eq. (3), see [4,6]. This nearest neighbor mean shift has a simple interpretation since in the mean shift recurrence relation, the next iterate $\boldsymbol{x}_{j+1}$ is the sample mean of the k nearest neighbors of the current iterate $\boldsymbol{x}_j$. On the other hand, as these iterations calculate the sample mean, the mean shift is not directly applicable to binary data.

3 Nearest Neighbor Median Shift Clustering for Binary Data

A categorical feature, which has a finite (usually small) number of possible values, can be represented by a binary vector, i.e. a vector which is composed solely of zeroes and ones. These categorical features can either ordinal (which have an implicit order) or can be nominal (which no order exists). Table 1 presents the two main types of the coding for a categorical feature into a binary vector, additive and disjunctive, for an example of 3-class categorical feature.

Table 1. Additive and disjunctive coding for a 3-class categorical feature.

Class	Additive coding	Disjunctive coding
1	1 0 0	1 0 0
2	1 1 0	0 1 0
3	1 1 1	0 0 1

The usual Euclidean distance is not adapted to measuring the dissimilarities between binary vectors. A popular alternative is the Hamming distance $\mathcal{H}$ [11]. The Hamming distance between two binary vectors $\boldsymbol{x}_1 = (x_{11}, \ldots, x_{1d})$ and $\boldsymbol{x}_2 = (x_{21}, \ldots, x_{2d})$, $\boldsymbol{x}_j \in \{0, 1\}^d, j \in 1, 2$, is defined as:

$$\mathcal{H}(\boldsymbol{x}_1, \boldsymbol{x}_2) = \sum_{j=1}^{d} |x_{1j} - x_{2j}|$$
$$= d - (\boldsymbol{x}_1 - \boldsymbol{x}_2)^\top (\boldsymbol{x}_1 - \boldsymbol{x}_2). \tag{4}$$

Equation (4) measures the number of mismatches between the two vectors $\boldsymbol{x}_1$ and $\boldsymbol{x}_2$: as the inner product $(\boldsymbol{x}_1 - \boldsymbol{x}_2)^\top (\boldsymbol{x}_1 - \boldsymbol{x}_2)$ counts the number of elements which agree in both $\boldsymbol{x}_1$ and $\boldsymbol{x}_2$, then $d - (\boldsymbol{x}_1 - \boldsymbol{x}_2)^\top (\boldsymbol{x}_1 - \boldsymbol{x}_2)$ counts the number of disagreements.

The Hamming distance is the basis from which we define the median center of a set of observations $\mathcal{X} = \{\boldsymbol{X}_1, \ldots, \boldsymbol{X}_n\}, \boldsymbol{X}_i \in \{0,1\}^d, i = 1, \ldots, n$. Importantly the median center of the set of binary vectors, as a measure of the centrality of the values, remains a binary vector, unlike the mean vector which can take on intermediate values. The median center of $\mathcal{X}$ is a point $\boldsymbol{w} = (w_1, \ldots, w_d)$ which minimizes the inertia of $\mathcal{X}$, i.e.

$$\boldsymbol{w} = \operatorname*{argmin}_{\boldsymbol{x} \in \{0,1\}^d} \mathcal{I}(\boldsymbol{x}) \qquad (5)$$

where

$$\mathcal{I}(\boldsymbol{x}) = \sum_{i=1}^n \pi_i \mathcal{H}(\boldsymbol{X}_i, \boldsymbol{x}) = \sum_{i=1}^n \sum_{j=1}^d \pi_i \mathcal{I}(x_j),$$

and π_i are the weights and $\mathcal{I}(x_j) = |X_{ij} - x_j|$.

Each component w_j of $\boldsymbol{w}$ minimizes $\mathcal{I}(x_j)$. In the case where all the weights are set to 1, $\pi_i = 1, i = 1, \ldots, n$, the w_j can be easily computed since it is the most common value in the observations of the j-th feature. This is denoted as $\operatorname{maj}(\mathcal{X})$, the component-wise majority vote winner among the data points. Hence the median center is the majority vote, $\boldsymbol{w} = \operatorname{maj}(\mathcal{X})$.

If we minimize the cost function in Eq. (5) using the dynamic clusters [3] then this leads to the k-modes clustering. Like the k-means algorithm, the k-modes operates in two steps: (a) an assignment step which assigns each candidate point $\boldsymbol{x}$ to the nearest cluster with respect to the Hamming distance, and (b) an optimization step which computes the median center as the majority vote. These two steps are executed iteratively until the value of $\mathcal{I}(\boldsymbol{x})$ converges.

Now we show how the median center can be utilized to define a new modal clustering for binary data based on the mean shift paradigm. In Sect. 2, the beta family kernels were used in the mean shift for continuous data. The most commonly used smoothing kernel, introduced by [1], for binary data is the Aitchison and Aitken kernel:

$$K_\lambda(\boldsymbol{x}) = \lambda^{d - \boldsymbol{x}^\top \boldsymbol{x}} (1 - \lambda)^{\boldsymbol{x}^\top \boldsymbol{x}}, \quad \boldsymbol{x} \in \{0,1\}^d.$$

Observe that the exponent for λ is the Hamming distance of $\boldsymbol{x}$. The tuning parameter $\frac{1}{2} \le \lambda \le 1$ controls the spread of the probability mass around the origin $\boldsymbol{0}$:

- For $\lambda = 1/2$, then $K_{1/2}(\boldsymbol{x}) = (1/2)^d$, which assigns a constant probability to all points $\boldsymbol{x}$, regardless of its distance from $\boldsymbol{0}$.
- For $\lambda = 1$, $K_1(\boldsymbol{x}) = \mathbf{1}\{\boldsymbol{x} = \boldsymbol{0}\}$, which assigns all the probability mass to $\boldsymbol{0}$.
- For intermediate values of λ, we have intermediate assignment of between point and uniform probability mass.

Using K_λ, the corresponding kernel density estimate is

$$\tilde{f}(\boldsymbol{x}; \lambda) = n^{-1} \sum_{i=1}^{n} \lambda^{[d-(\boldsymbol{x}-\boldsymbol{X}_i)^\top(\boldsymbol{x}-\boldsymbol{X}_i)]}(1-\lambda)^{[(\boldsymbol{x}-\boldsymbol{X}_i)^\top(\boldsymbol{x}-\boldsymbol{X}_i)]}. \qquad (6)$$

Since the gradient of the kernel K_λ is $\mathsf{D}K_\lambda(\boldsymbol{x}) = 2\boldsymbol{x}\log((1-\lambda)/\lambda)K_\lambda(\boldsymbol{x})$, the density gradient estimate is

$$\mathsf{D}\tilde{f}(\boldsymbol{x}; \lambda) = 2\log(\lambda/(1-\lambda))n^{-1}\left[\sum_{i=1}^{n} \boldsymbol{X}_i K_\lambda(\boldsymbol{x}-\boldsymbol{X}_i) - \boldsymbol{x}\sum_{i=1}^{n} K_\lambda(\boldsymbol{x}-\boldsymbol{X}_i)\right]. \qquad (7)$$

To progress in our development of a nearest neighbor median shift for binary data, we focus on the point mass kernel $K_1(\boldsymbol{x}) = \mathbf{1}\{\boldsymbol{x} = \boldsymbol{0}\}$. In order that ensure that it is amenable for the median shift, we modify K_1 with two main changes:

1. K_1 is multiplied by the indicator function $\mathbf{1}\{\boldsymbol{x} \in B_d(\boldsymbol{0}, 1)\}$
2. the indicator function $\mathbf{1}\{\boldsymbol{x} = \boldsymbol{0}\}$, which places the point mass at the center $\boldsymbol{0}$, is replaced an indicator that places it on $\mathrm{maj}(B_d(\boldsymbol{0}, 1))$, where $\mathrm{maj}(B_d(\boldsymbol{0}, 1))$ is the majority vote winner/median center of the data points $\boldsymbol{X}_1, \ldots, \boldsymbol{X}_n$ inside of $B_d(\boldsymbol{0}, 1)$.

This second modification results in an asymmetric kernel as the point mass is no longer always placed in the centre of the unit ball. This modified, asymmetric kernel L is

$$L(\boldsymbol{x}) = \mathbf{1}\{\boldsymbol{x} = \mathrm{maj}(B_d(\boldsymbol{0}, 1))\}\mathbf{1}\{\boldsymbol{x} \in B_d(\boldsymbol{0}, 1)\}.$$

Since L is not directly differentiable, we define its derivative indirectly via $\mathsf{D}K_1$ and the convention that $\log(\lambda/(1-\lambda)) = 1$ for $\lambda = 1$. As $\mathsf{D}K_\lambda(\boldsymbol{x})|_{\lambda=1} = 2\boldsymbol{x}K_1(\boldsymbol{x})$ then analogously we define $\mathsf{D}L(\boldsymbol{x}) = 2\boldsymbol{x}L(\boldsymbol{x})$. To obtain the corresponding estimators, we substitute $L, \mathsf{D}L$ for $K, \mathsf{D}K$ in $\tilde{f}, \mathsf{D}\tilde{f}$ in Eqs. (6)–(7) to obtain $\hat{f}, \mathsf{D}\hat{f}$:

$$\hat{f}(\boldsymbol{x}; k) = n^{-1}\delta_{(k)}(\boldsymbol{x})^{-d}\sum_{i=1}^{n} L((\boldsymbol{x}-\boldsymbol{X}_i)/\delta_{(k)}(\boldsymbol{x}))$$

$$\mathsf{D}\hat{f}(\boldsymbol{x}; k) = 2\delta_{(k)}(\boldsymbol{x})^{-d-1}n^{-1}\left[\sum_{i=1}^{n} \boldsymbol{X}_i L((\boldsymbol{x}-\boldsymbol{X}_i)/\delta_{(k)}(\boldsymbol{x})) - \boldsymbol{x}\sum_{i=1}^{n} L((\boldsymbol{x}-\boldsymbol{X}_i)/\delta_{(k)}(\boldsymbol{x}))\right]. \qquad (8)$$

To obtain a nearest neighbor mean shift recurrence relation for binary data, we substitute $\hat{f}, \mathsf{D}\hat{f}$ for $f, \mathsf{D}f$ is Eq. (1). For these estimators, the appropriate choice of $\mathbf{A} = \frac{1}{2}\delta_{(k)}(\boldsymbol{x})\mathbf{I}_d$. Then we have

$$\begin{aligned}
\boldsymbol{x}_{j+1} &= \boldsymbol{x}_j + \frac{\delta_{(k)}(\boldsymbol{x})}{2}\frac{\mathsf{D}\hat{f}(\boldsymbol{x}_j; k)}{\hat{f}(\boldsymbol{x}_j; k)} \\
&= \frac{\sum_{i=1}^{n} \boldsymbol{X}_i L((\boldsymbol{x}_j - \boldsymbol{X}_i)/\delta_{(k)}(\boldsymbol{x}_j))}{\sum_{i=1}^{n} L((\boldsymbol{x}_j - \boldsymbol{X}_i)/\delta_{(k)}(\boldsymbol{x}_j))}.
\end{aligned}$$

We can simplify this ratio if we observe that the scaled kernel is

$$L((\boldsymbol{x} - \boldsymbol{X}_i)/\delta_{(k)}(\boldsymbol{x})) = \mathbf{1}\{\boldsymbol{X}_i \mathrm{maj}(B_d(\boldsymbol{x}, \delta_{(k)}(\boldsymbol{x})))\} \cdot \mathbf{1}\{\boldsymbol{X}_i \in B_d(\boldsymbol{x}, \delta_{(k)}(\boldsymbol{x}))\};$$

and that $B_d(\boldsymbol{x}, \delta_{(k)}(\boldsymbol{x}))$ comprises the k nearest neighbors of $\boldsymbol{x}$, then $\mathbf{1}\{\boldsymbol{X}_i \in B_d(\boldsymbol{x}, \delta_{(k)}(\boldsymbol{x}))\} = \mathbf{1}\{\boldsymbol{X}_i \in \mathrm{NN}_k(\boldsymbol{x})\}$. If m is the number of nearest neighbors of $\boldsymbol{x}_j$ which coincide with the majority vote, then

$$
\begin{aligned}
\boldsymbol{x}_{j+1} &= \frac{\sum_{\boldsymbol{X}_i \in \mathrm{NN}_k(\boldsymbol{x}_j)} \boldsymbol{X}_i \mathbf{1}\{\boldsymbol{X}_i = \mathrm{maj}(\mathrm{NN}_k(\boldsymbol{x}_j))\}}{\sum_{\boldsymbol{X}_i \in \mathrm{NN}_k(\boldsymbol{x}_j)} \mathbf{1}\{\boldsymbol{X}_i = \mathrm{maj}(\mathrm{NN}_k(\boldsymbol{x}_j))\}} \\
&= \frac{m \cdot \mathrm{maj}(\mathrm{NN}_k(\boldsymbol{x}_j))}{m} \\
&= \mathrm{maj}(\mathrm{NN}_k(\boldsymbol{x}_j)).
\end{aligned}
\tag{9}
$$

Therefore in the median shift recurrence relation in Eq. (9), the next iterate $\boldsymbol{x}_{j+1}$ is the median center of the k nearest neighbors of the current iterate $\boldsymbol{x}_j$. Thus, once the binary gradient ascent has terminated, the converged point can be decoded using Table 1), allowing for its unambiguous symbolic interpretation. The gradient ascent paths towards the local modes produced by Eq. (9) form the basis of Algorithm 1, our nearest neighbor median shift clustering for binary data method (BinNNMS).

The inputs to BinNNMS are the data sample $\boldsymbol{X}_1, \ldots, \boldsymbol{X}_n$ and the candidate points $\boldsymbol{x}_1, \ldots, \boldsymbol{x}_m$ which we wish to cluster (these can be the same as $\boldsymbol{X}_1, \ldots, \boldsymbol{X}_n$, but this is not required); and the tuning parameters: the number of nearest neighbors k_1 used in BGA task, the maximum number of iterations $j_{\max}$, and the tolerance under which two cluster centres are considered form a single cluster ε. The output are the cluster labels of the candidate points $\{\mathrm{c}(\boldsymbol{x}_1), \ldots, \mathrm{c}(\boldsymbol{x}_m)\}$.

The aim of the ε-proximity cluster labeling step is to gather all points which are under a threshold ε. In order to apply this method we have to build the Hamming similarity matrix which has a $O(n^2)$ time complexity. We initialize the process by taking first point and cluster with it all point whose distance is less than ε. Thus we apply this iterative exploration process by adding the nearest neighbors. Once the first cluster is generated, we take another point from the reduced similarity matrix and repeat the process, until all points are assigned a cluster label. A notable problem still remains with the choice of main tuning parameter ε: we set it to be the average of distance from each point to their k_2 nearest neighbors.

Algorithm 1. BinNNMS – Nearest neighbor median shift clustering for binary data

---

Input: $\{X_1, \ldots, X_n\}, \{x_1, \ldots, x_m\}, k_1, k_2, j_{\max}$
Output: $\{c(x_1), \ldots, c(x_m)\}$
/* **BGA task**: compute binary gradient ascent paths */
1: **for** $\ell := 1$ to m **do**
2:      $j := 0;\ x_{\ell,0} := x_\ell;$
3:      $x_{\ell,1} := \text{maj}(\text{NN}_{k_1}(x_{\ell,0}));$
4:      **while** $j < j_{\max}$ **do**
5:          $j := j + 1;$
6:          $x_{\ell,j+1} := \text{maj}(\text{NN}_{k_1}(x_{\ell,j}));$
7:      $x_\ell^* := x_{\ell,j};$
    /* ε-**proximity cluster labeling task**: create clusters by merging near final iterates*/
8: **for** $\ell_1, \ell_2 := 1$ to m **do**
9:      **if** $\mathcal{H}(x_{\ell_1}^*, x_{\ell_2}^*) \leq \varepsilon(k_2)$ **then** $c(x_{\ell_1}^*) := c(x_{\ell_2}^*);$

---

4   Numerical Experiments

In this section, we present an experimental comparison of the BinNNMS to the k-modes clustering (as outlined in Sect. 3). Table 2 lists the details of the dataset obtained from the UCI Machine learning repository [2].

- The Zoo data set contains $n = 101$ animals described with 16 categorical features: 15 of the variables are binary and one is numeric with 6 possible values. Each animal is labelled 1 to 7 according to its class. Using disjunctive coding for the categorical variable with 6 possible values, the data set consists of a 101×21 binary data matrix.
- The Digits data concerns a dataset consisting of the handwritten numerals ("0"–"9") extracted from a collection of Dutch utility maps. There are 200 samples of each digit so there is a total of $n = 2000$ samples. As each sample is a 15×16 binary pixel image, the dataset consisted of a 2000×240 binary data matrix.
- The Spect dataset describes the cardiac diagnoses from Single Proton Emission Computed Tomography (SPECT) images. Each patient is classified into two categories: normal and abnormal; there are $n = 267$ samples which are described by 22 binary features.
- The Soybean data is about 19 classes, but only the first 15 have been justified as it appears that the last four classes are not well-defined. There are 35 categorical attributes, with both nominal and ordinal features.
- The Car dataset contains examples with the structural information of the vehicle is removed. Each instance is classified into 4 classes. This database is highly unbalanced since the distribution of the classes is $(70.02\%, 22.22\%, 3.99\%, 3.76\%)$.

Table 2. Overview of experimental datasets.

Dataset	size (n)	#features (d)	#classes (M)
Zoo	101	26	7
Digits	2000	240	10
Spect	267	22	2
Soybean	307	97	18
Car	1728	15	4

4.1   Comparison of the k-Modes and the BinNNMS Clustering

To evaluate the clustering quality, we compare the known cluster labels in Table 2 to the estimated cluster labels from BinNNMS and k-modes. For comparability, the k-modes clustering is also based on the binary median center from Eq. (5). Values of the Adjusted Rand Index (ARAND) [9] and the normalized mutual information (NMI) [15] close to one indicate highly matched cluster labels, and values close to zero for the NMI/less than zero for the ARAND) indicate mismatched cluster labels. Our scala codes to reproduce all results are available at https://github.com/Clustering4Ever/Clustering4Ever.

Table 3 reports the results in terms of the NMI and ARAND after 10 runs of the BinNNMS and k-modes. Unlike BinNNMS, the k-modes clustering requires an a priori number of clusters k, then we set k to be whichever value between the target number of classes from Table 2, or to be the number of clusters obtained from the BinNNMS clustering gives the highest clustering accuracy. The BinNNMS, apart from the Car dataset, outperforms the k-modes algorithm on Zoo, Digits, Spect, and Soybean datasets. Upon further investigation for the Car dataset, recall that the distribution of the cluster labels is highly unbalanced which leads the BinNNMS giving a single class (i.e. no clustering). These unbalanced clusters also translate into low values of the NMI and ARAND for the k-modes clustering.

4.2   Comparison of the Quantization Errors for the BinNNMS

An important and widely used measure of resolution, the quantization error, is computed based on Hamming distances between the data points and the cluster prototypes:

$$Error = \frac{1}{n} \sum_{m=1}^{M} \sum_{x_j \in \mathcal{C}_m} \mathcal{H}(x_j, w_m) \tag{10}$$

where $\{\mathcal{C}_1, \ldots, \mathcal{C}_M\}$ is the set of M clusters, x is a point assigned to cluster $\mathcal{C}_m$, and w_m is the prototype.median center of cluster $\mathcal{C}_m$.

The right hand column in Fig. 1 shows the evolution of the quantization errors for the BinNNMS with different values of k_1 with respect to the target cluster prototypes. As the quantization errors decrease this implies that the data points

Table 3. Comparison of clustering quality indices (NMI and ARAND) for k-modes and BinNNMS. The bold value indicates the most accurate clustering for the dataset.

	NMI		
Dataset	k-modes	k	BinNNMS
Digits	0.360 ± 0.011	40	$\mathbf{0.880 \pm 0.000}$
Zoo	0.789 ± 0.023	8	$\mathbf{0.945 \pm 0.000}$
Soybean	0.556 ± 0.000	40	$\mathbf{0.743 \pm 0.000}$
Spect	0.135 ± 0.000	47	$\mathbf{0.145 \pm 0.000}$
Car	$\mathbf{0.039 \pm 0.019}$	4	Single class
	ARAND		
Dataset	k-modes	k	BinNNMS
Digits	0.166 ± 0.021	40	$\mathbf{0.876 \pm 0.000}$
Zoo	0.675 ± 0.032	8	$\mathbf{0.904 \pm 0.000}$
Soybean	0.178 ± 0.000	40	$\mathbf{0.331 \pm 0.000}$
Spect	$\mathbf{-0.009 \pm 0.055}$	2	-0.019 ± 0.000
Car	0.016 ± 0.039	4	Single class

converge toward their cluster prototypes, and that the decreasing intra-cluster distance further facilitates the clustering process. Thus at the end of the training phase, the data points converge towards to their local mode. In comparison with the ARAND scores in Table 3, the magnitude of the decrease in the quantization errors is inversely proportional to the cluster quality indices. That is, the largest decrease for the Digits dataset implies that BinNNMS clustering achieves here the highest ARAND score.

If we run the labeling phase during the BGA phase for a fixed k_1 then we compute the intermediate prototypes $\boldsymbol{w}_m$ of the clusters $\mathcal{C}_m$ during the binary gradient ascent BGA task. Since BinNNMS provides clusters as the basins of attraction to the local median created by the binary gradient ascent paths, the left column of Fig. 1 shows the quantization error with respect to the intermediate median centers/prototypes. In this case we compute at each iteration 7 modes for Zoo dataset, 10 modes for the Digits, 18 modes for Soybean and 2 modes for Spect datasets using ground truth. These quantization errors decrease to an asymptote for all datasets as the iteration number increases.

Visual Comparison of k-modes and BinNNMS on the Digit Dataset:

to obtain a visual representations of binary digit dataset we have the possibility to transform the binary vector into a binary image where each pixel represent one dimension. We present here prototypes of ground truth and clustering results. Figure 2 show the cluster prototypes provided by k-modes and BinNNMS, displayed as 15×16 binary pixel images. For the k-modes image, the cluster prototype for the "4" digit has been incorrectly associated with the "9" cluster. On the other hand, the BinNNMS image correctly identifies all ten digits from "0" to "9".

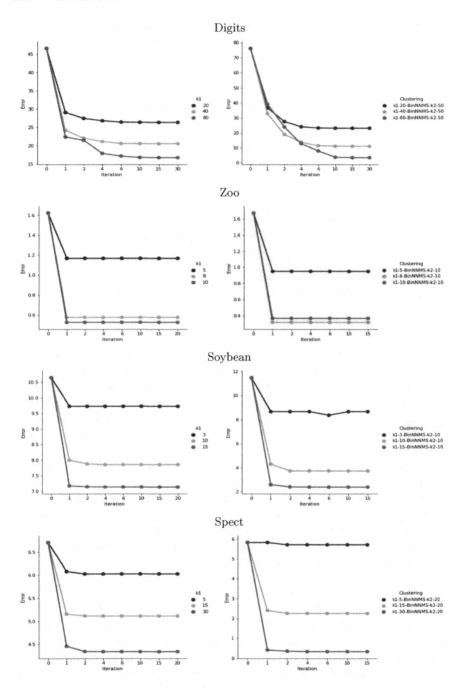

Fig. 1. Evolution of quantization errors as a function of the k_1 and k_2 tuning parameters in BinNNMS for the Digits, Zoo, Soybean and Spect datasets. Left. Quantization errors between the data points and the target prototypes. Right. Quantization errors between the data points and the intermediate median centers in the BGA task and the cluster prototypes.

k-modes   BinNNMS

Fig. 2. Comparison of the k-modes and BinNNMS clustered images for the Digits dataset.

5 Conclusion

In this paper, we have proposed a new and efficient modal clustering method for binary data. We introduced a mathematical analysis of the nearest neighbor estimators for binary data. This was then combined with the Aitchison and Aitken kernel in order to generalize the traditional mean shift clustering to the median shift clustering for binary data (BinNNMS). Experimental evaluation for a number of experimental datasets demonstrated that the BinNNMS outperformed the k-modes clustering in terms of visual criteria, as well as quantitative clustering quality criteria such as the adjusted Rand index, the normalized mutual information and the quantization error. In the future we envisage to make our algorithm as automatic as possible by optimizing the choice of the tuning parameters, and to implement a scalable version for Big Data by using approximate nearest neighbor searches.

References

1. Aitchison, J., Aitken, C.G.G.: Multivariate binary discrimination by the kernel method. Biometrika **63**, 413–420 (1976)
2. Dheeru, D., Karra Taniskidou, E.: UCI machine learning repository (2017). http://archive.ics.uci.edu/ml
3. Diday, E., Simon, J.C.: Clustering Analysis, pp. 47–94. Springer, Berlin (1976). https://doi.org/10.1007/978-3-642-96303-2_3
4. Duong, T., Beck, G., Azzag, H., Lebbah, M.: Nearest neighbour estimators of density derivatives, with application to mean shift clustering. Pattern Recogn. Lett. **80**, 224–230 (2016). https://doi.org/10.1016/j.patrec.2016.06.021
5. Fukunaga, K., Hostetler, L.: Optimization of k-nearest-neighbor density estimates. IEEE Trans. Inform. Theory **19**, 320–326 (1973)
6. Fukunaga, K., Hostetler, L.: The estimation of the gradient of a density function, with applications in pattern recognition. IEEE T. Inform. Theory **21**, 32–40 (1975). https://doi.org/10.1109/TIT.1975.1055330
7. Hamming, R.W.: Error detecting and error correcting codes. Bell Syst. Tech. J. **29**, 147–160 (1950). https://doi.org/10.1002/j.1538-7305.1950.tb00463.x

8. Huang, Z.: Clustering large data sets with mixed numeric and categorical values. In: The First Pacific-Asia Conference on Knowledge Discovery and Data Mining, pp. 21–34 (1997)
9. Hubert, L., Arabie, P.: Comparing partitions. J. Classif. **2**, 193–218 (1985)
10. Lebbah, M., Badran, F., Thiria, S.: Topological map for binary data. In: ESANN 2000, 8th European Symposium on Artificial Neural Networks, Bruges, Belgium, 26–28 April 2000, Proceedings, pp. 267–272 (2000)
11. Leisch, F., Weingessel, A., Dimitriadou, E.: Competitive learning for binary valued data. In: Niklasson, L., Bodén, M., Ziemke, T. (eds.) ICANN 1998. PNC, pp. 779–784. Springer, London (1998). https://doi.org/10.1007/978-1-4471-1599-1_120
12. Li, T.: A unified view on clustering binary data. Mach. Learn. **62**, 199–215 (2006). https://doi.org/10.1007/s10994-005-5316-9
13. Loftsgaarden, D.O., Quesenberry, C.P.: A nonparametric estimate of a multivariate density function. Ann. Math. Statist. **36**, 1049–1051 (1965). https://doi.org/10.1214/aoms/1177700079
14. MacQueen, J.: Some methods for classification and analysis of multivariate observations. In: Proceedings of the Fifth Berkeley Symposium on Mathematical Statistics and Probability, Volume 1: Statistics, pp. 281–297. University of California Press, Berkeley, USA (1967). https://projecteuclid.org/euclid.bsmsp/1200512992
15. Strehl, A., Ghosh, J.: Cluster ensembles - a knowledge reuse framework for combining multiple partitions. J. Mach. Learn. Res. **3**, 583–617 (2002)

Self-supervised Multi-view Clustering for Unsupervised Image Segmentation

Tiyu Fang[1] , Zhen Liang[1], Xiuli Shao[2], Zihao Dong[1(✉)], and Jinping Li[1(✉)]

[1] School of Information Science and Engineering, University of Jinan,
Jinan 250022, China
{ise_dongzh,ise_lijp}@ujn.edu.cn
[2] College of Computer Science, Nankai University, Tianjin 300350, China

Abstract. At present, the main idea of CNN-based unsupervised image segmentation is clustering a single image in the framework of CNNs. However, the single image clustering is very difficult to obtain enough supervision information for network learning. For solving this problem, we propose a Self-supervised Multi-view Clustering (SMC) structure for unsupervised image segmentation to mine additional supervised information. Based on the observation that the predicted pixel-level labels and the input images have the same spatial features, the multi-view images acquired by data augmentation are clustered to obtain the multi-view results and the proposed SMC uses the differences among these results to learn self-supervised information. Moreover, a Hybrid Self-supervised (HS) loss is proposed to make full use of the self-supervised information for further improving the prediction accuracy and the convergence speed. Extensive experiments in BSD500 and PASCAL VOC 2012 datasets demonstrate the superiority of our proposed approach.

Keywords: Unsupervised image segmentation · Convolutional neural network · Self-supervised Multi-view Clustering

1 Introduction

Image segmentation is a basic problem in the computer vision field, which can be seen as a pixel-level classification problem. In the supervised image segmentation, a large number of images and corresponding labels are used to train classifiers for completing the classification of pixels. Unlike supervised image segmentation, due to the lack of training data and ground truth, unsupervised image segmentation methods usually segment a single image into an arbitrary number (≥ 2) of meaningful regions, which is obviously a more complex segmentation task. To address the challenge, this paper studies the unsupervised image segmentation problem without any prior knowledge.

Unsupervised image segmentation can be seen as a clustering process since there are no labels as supervising clue. After the features of the pixels are extracted, the categories of the pixels can be obtained by feature clustering. In

© Springer Nature Switzerland AG 2021
I. Farkaš et al. (Eds.): ICANN 2021, LNCS 12895, pp. 113–125, 2021.
https://doi.org/10.1007/978-3-030-86383-8_9

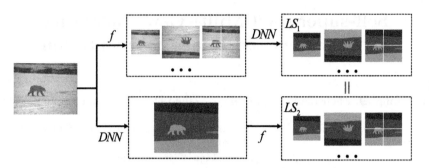

Fig. 1. The basic principles of the proposed method.

the traditional methods, k-means clustering and the graph-based model [7] are used for unsupervised image segmentation, but these methods only extract features manually, which makes it difficult to segment images accurately. Recently, some works [11,12] began to focus on the application of CNNs in unsupervised image segmentation. Their main idea is to apply gradient descent for the single image clustering. However, these works only obtain extremely limited supervised information, which results in the slow convergence speed and low segmentation accuracy. Therefore, it is vital to learn sufficient supervised information without priori knowledge. Inspired by the data augmentation in supervised learning, as shown in Fig. 1, the augmentation method f is used to obtain a multi-view image set from a single image, and the multi-view label set LS_1 can be predicted by the deep neural networks (DNN). Meantime, a single label can also employ the same f to get the corresponding multi-view label set LS_2. Due to the same spatial consistency between input images and output pixel-level labels, LS_1 should be consistent with LS_2 after the network converges sufficiently. However, in the process of network convergence, the gap between LS_1 and LS_2 always exists and we can learn strong self-supervision information by constraining the implicit gap.

Driven by this, in this paper, we propose a Self-supervised Multi-view Clustering (SMC) approach for unsupervised image segmentation. Firstly, this method makes full use of image augmentation operations such as scale, angle and position. Then, the spatial consistency of input images and output labels is learned to provide self-supervised information during the network iteration. Moreover, for fully employing self-supervised information, we design a Hybrid Self-supervised (HS) loss function to optimize the network in terms of label error correction and supervision information enhancement. Sufficient experiments prove that the proposed method can achieve favorable segmentation for images.

The main contributions of this paper are as follows:

1. We propose a Self-supervised Multi-view Clustering (SMC) method to obtain self-supervised information for unsupervised image segmentation through the spatial consistency of input and output in the network.

2. Based on proposed method, a Hybrid Self-supervised (HS) loss function is designed to improve the efficiency of self-supervised information from the label correction and supervision information enhancement.
3. The experimental results on PASCAL VOC2012 and BSD500 datasets prove that the proposed method can achieve better performance than recent methods.

2   Related Work

With the development of deep learning, supervised image segmentation [2,13,15, 23] has made a great breakthrough. In this section, we mainly focus on related works, including unsupervised image segmentation and self-supervised learning.

Unsupervised Image Segmentation: Unsupervised segmentation models mainly use image itself for segmentation. In traditional methods, k-means clustering is often applied for unsupervised segmentation based on the similarity between data and K cluster centroids. In addition, a graph-based segmentation method [7] is used to obtain the segmentation results by greedy clustering in the graph representation of image. However, these methods heavily rely on various prior knowledge. Recently, a few learning-based methods have been proposed [4,11,12,14,21]. For instance, MSLRR [14] realizes unsupervised segmentation by superpixel affinity matrix. Because superpixels are used in the method, the segmentation boundaries of image are only be fixed on the superpixels. W-net [21] obtains the segmentation results by minimizing the normalized cut between the reconstructed image and the encoded image, but the way relies on a lot of data. In [4], Croitoru et al. propose a dual student teacher system for unsupervised segmentation, whereas this method only conducts foreground and background segmentation. Asako et al. propose an unsupervised segmentation method based on feature clustering [11,12], which can realize image segmentation in the basic CNNs. However, the method still has great speed and accuracy limitations because of the lack of supervision information.

Self-supervised Learning: The core idea of self-supervised learning is using pretext tasks to mine supervision information from data. At present, the pretext tasks of self-supervised learning can be summarized into three types: context based, temporal based and contrastive based. The context-based approaches mainly excavate the supervision information through jigsaw [5], inpainting [17], color prediction [22], image transformation [8] and so on. The temporal-based approaches mainly use the relationship between adjacent frames in the video [16], multi-view data [18] and other constraints for self-supervised learning. The contrastive-based approaches are to obtain the supervision information by using the gap of samples [3,9,20], whose strong supervision performance is particularly concerned. Similar to this way, in this paper, we use the spatial consistency of the input image and the output label in the network, and propose a SMC method to achieve accurate image segmentation.

3   Method

The overall structure of the method is shown in Fig. 2. First, we illustrate the motivation of our work, and then we introduce the proposed SMC method. Finally, the loss function of the method is discussed.

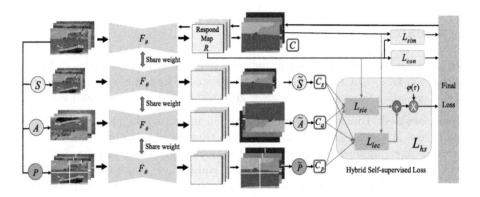

Fig. 2. Overall structure of the method.

3.1   Motivation

Image segmentation is essentially a pixel-level classification and the main solution for it is as follows. For simplicity, we use k to represent arbitrary pixel position in the image. Let $I = \{I_1, ..., I_n\}$ represent input image, where n is the number of pixels in image. First, the feature extraction function $E_{\theta_1} : I \longrightarrow V$ is applied to extract the feature set V, where V is equivalent to $\{v_1, ..., v_n\}$, $v_k \in R^p$ represents the p-dimensional feature vector for arbitrary pixel and θ_1 denotes the parameter set in E_{θ_1}. Then, the class discriminant function $J_{\theta_2} : V \longrightarrow C$ is used to get categories of pixels $C = \{c_1, ..., c_n\}$, where $c_k \in (1, ..., q)$ means that the pixel can be divided into q classes and θ_2 denotes the parameter set in J_{θ_2}. For input image I, when θ_1 and θ_2 are known, we can get C through the above equation. On the contrary, when C is known, we can use the relationship between C and I to train θ_1 and θ_2, which can be as a standard supervised classification. However, in the current unsupervised manner, C is unknown when training θ_1 and θ_2. For putting this in practice, we need to solve two sub problems: predicting C with fixed θ_1 and θ_2, and training θ_1 and θ_2 with fixed C.

For solving the two sub problems, the CNN-based structure is adopted. We expand F_θ in Fig. 2 to get the structure, as shown in Fig. 3. F_θ consists of E_{θ_1} and J_{θ_2}, where $\theta = \theta_1 \bigcup \theta_2$. In the forward propagation, θ_1 and θ_2 are fixed. We optimize C through this process. In the back propagation, C is fixed. We can optimize θ_1 and θ_2 by gradient descent. When the network is sufficiently convergent, the features of each pixel can be extracted by M convolution components. Then we can get respond map $R = \{r_1, ..., r_n\}$ by one-dimensional convolution

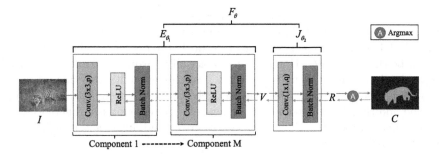

Fig. 3. The network structure of F_θ.

of q channels and batch norm, where $r_k = \{r_k^1, ..., r_k^q\}$ denotes the value set of each pixel in q channels. Finally, the segmentation results C are obtained by $c_k = argmax(r_k^1, ..., r_k^q)$. The above steps are essentially clustering in CNNs. However, in the process of clustering, it is extremely difficult to obtain sufficient supervision information for optimizing θ_1 and θ_2, which greatly hinders the convergence of network. For solving the above problem, the SMC is proposed, which will be described in the next section.

3.2 Self-supervised Multi-view Clustering

In supervised learning, data augmentation is often used to increase the diversity of input samples and labels. Inspired by this, we think that the way can still be achieved in unsupervised learning. Only different from supervised learning, unsupervised learning can augment input samples from multiple views. Because our method mainly uses the spatial consistency of input images and output labels, the basic spatial transformations including scale transformation S, angle transformation A and position transformation P are used to augment images in the method. After obtaining multi-view images, how to combine them with the CNN-based unsupervised segmentation method has become a problem that must be solved. In F_θ, two-dimensional convolutions are used for image feature extraction and classification. Therefore, the output through the network always keeps the spatial characteristics of the input. Based on this characteristic, it can be concluded that the input and output of the ideal convergent network should have the following equivariant laws:

$$
\begin{cases}
C = F_\theta(I) \\
C_s = \widetilde{S}(F_\theta(S(I))) \\
C_a = \widetilde{A}(F_\theta(A(I))) \\
C_p = \widetilde{P}(F_\theta(P(I))) \\
C = C_s = C_a = C_p,
\end{cases}
\tag{1}
$$

where C_s, C_a, C_p represent the processed labels and $\widetilde{S}$, $\widetilde{A}$, $\widetilde{P}$ represent the inverse operations of S, A and P respectively, which aims at restoring the obtained labels. In the ideal convergent network, the processed labels C_s, C_a, C_p should be the same as C. However, in the actual network iteration, the implicit constraint is ignored. To make up for it, we cluster multi-view images in the network to get multi-view labels respectively, the implicit constraint is used to provide additional self-supervision for network by reducing the differences and errors among these labels. The reduction of errors and differences is mainly realized by Hybrid Self-supervised (HS) loss, which will be detailed in Sect. 3.3.

3.3   Loss Function

By the proposed method, we get the multi-view labels, which include C_s, C_a and C_p. For mining more supervised information from these labels, a Hybrid Self-supervised (HS) loss is designed for network optimization, as shown below:

$$L_{hs} = \varphi(\tau)(L_{sie} + \eta L_{lec}), \qquad (2)$$

where η is a constant. Here, the relationship between the above labels and response map R is considered. Although C_s, C_a and C_p are obtained through restoration, these labels still contain a lot of supervision information, which can be extracted by comparing R. Therefore, in our method, the cross entropy loss of these labels and R is used to form L_{sie} for enhancing supervision:

$$L_{sie} = -\sum_{k=1}^{n}\sum_{i=1}^{q}(\varepsilon(i - c_{s,k})\ln r_k^i + \varepsilon(i - c_{a,k})\ln r_k^i$$
$$+ \varepsilon(i - c_{p,k})\ln r_k^i), \qquad (3)$$

where

$$\varepsilon(t) = \begin{cases} 1 & \text{if } t = 0 \\ 0 & \text{otherwise.} \end{cases} \qquad (4)$$

In addition, the differences between C_s, C_a, C_p and C are also fully considered. It can be seen that C is obtained by clustering, which is a pseudo label. Therefore, in the network iteration, there is a certain error in C, which greatly affects the accuracy and convergence speed of network. For correcting this error, L1 norm between C_s, C_a, C_p and C is minimized to make the network converge faster and more accurately. The specific formula is defined as follows:

$$L_{lec} = \sum_{k=1}^{n} \|C_k - C_{s,k}\|_1 + \|C_k - C_{a,k}\|_1$$
$$+ \|C_k - C_{p,k}\|_1. \qquad (5)$$

By minimizing the above two functions, the network can achieve faster convergence. However, due to excessive supervision information, the entire network learning process is rough and easy to cause simple schemes: changing the categories in the image into one category. For refining the network learning process, $\varphi(\tau)$ is introduced into the loss function:

$$\varphi(\tau) = \begin{cases} \dfrac{e^{-a\frac{\tau^2-\tau+2}{2}}}{1+e^{-\lambda}} & \lambda \le \lambda_{limit} \\[4mm] \dfrac{e^{-a\frac{\tau^2-\tau+2}{2}}}{1+e^{\lambda}} & \lambda > \lambda_{limit}, \end{cases} \tag{6}$$

where λ represents the number of network iterations, a and λ_{limit} are constants. $\tau \in \{1, 2, 3, 4\}$ is the number of labels with the same number of categories in C_s, C_a, C_p and C. For example, when C_s, C_a, C_p and C have the same number of categories, τ is 4. When C_s, C_a, C_p and C have the different number of categories, τ is 1.

In addition to the HS loss, the constraints of feature similarity and spatial continuity are considered to form the final loss. The cross entropy loss between C and R is calculated to restrain feature similarity:

$$L_{sim} = \sum_{k=1}^{n} \sum_{i=1}^{q} -\varepsilon (i - c_k) \ln r_k^i. \tag{7}$$

For introducing the spatial continuity into the network, in a similar manner to [19], the following function is used to constrain spatial continuity:

$$L_{con} = \sum_{a=1}^{W-1} \sum_{\beta=1}^{H-1} \|r_{\alpha+1,\beta} - r_{\alpha,\beta}\|_1 + \|r_{\alpha,\beta+1} - r_{\alpha,\beta}\|_1, \tag{8}$$

where W and H represent the width and height of input image, $r_{\alpha,\beta}$ represents the feature value at (α, β) in R. In summary, the final loss of the proposed method is defined as following:

$$L = L_{hs} + L_{sim} + \mu L_{con}, \tag{9}$$

where μ represents the weight for the constraint on spatial continuity. The specific process of the whole method is shown in the Algorithm 1.

Algorithm 1. Self-supervised Multi-view Clustering

Input: $I = \{I_1, ..., I_n\}$ $I_k \in R^3$

Output: $C = \{c_1, ..., c_n\}$ $c_k \in (1, ..., q)$

1: $\{\theta = \theta_1 \bigcup \theta_1\} \longleftarrow Init()$

2: **for** $\lambda = 1 : T$ **do**

3:      $V \longleftarrow E_{\theta_1}(I), R \longleftarrow J_{\theta_2}(V), C = argmax(R)$

4:      $C_s \leftarrow \widetilde{S}\left(F_\theta(S(I))\right)$

5:      $C_a \leftarrow \widetilde{A}\left(F_\theta(A(I))\right)$

6:      $C_p \leftarrow \widetilde{P}\left(F_\theta(P(I))\right)$

7:      $L \longleftarrow L_{hs} + L_{sim} + \mu L_{con}$

8:      $\theta \longleftarrow Update(L)$

9: **end for**

---

4   Experiments

4.1   Implementation Details

We evaluate the proposed method on the validation set of PASCAL VOC 2012 dataset [6] and the testing set of BSD500 dataset [1]. In all experiments, M is set to 3, and $p = q = 100$. For loss function, $\mu = 5$, $a = 0.07$, $\eta = 0.05$ and $\lambda_{limit} = 100$. Through S, the image is reduced by 0.5 times. Using A, the image is rotated by 180°. It should be noted that we use clipping for position transformation P. By P, the image is divided into 2×2 image regions. In this paper, all the results of the experiments are evaluated by the mean intersection over union (mIOU). Here, mIOU is calculated as the mean IOU of each segment in the ground truth (GT) and the predicted segment that has the largest IOU with the GT segment.

Table 1. Comparison of mIOU for unsupervised segmentation on the PASCAL VOC 2012 and BSD500 datasets.

Methods	VOC2012	BSD500 all	BSD500 fine	BSD500 coarse	Mean
KC, $K = 2$	0.3166	0.1223	0.0865	0.1972	0.1807
KC, $K = 17$	0.2383	0.2404	0.2208	0.2648	0.2411
GS [7] $\sigma = 100$	0.2682	0.3135	**0.2951**	0.3255	0.3006
GS [7] $\sigma = 500$	0.3647	0.2768	0.2238	0.3659	0.3078
IIC [10], $K = 2$	0.2729	0.0896	0.0537	0.1733	0.1474
IIC [10], $K = 20$	0.2005	0.1724	0.1513	0.2071	0.1828
Baseline/SP [11]	0.3082	0.2261	0.1690	0.3239	0.2568
Baseline/DC [12]	0.3520	0.3050	0.2592	0.3739	0.3225
Ours	**0.3833**	**0.3156**	0.2662	**0.3913**	**0.3391**

4.2   Effect of Self-supervised Multi-view Clustering

For evaluating the accuracy of SMC, we have conducted sufficient comparative experiments on PASCAL VOC 2012 and BSD500. The experimental results are shown in Table 1. Referring to the experimental setup in [12], k-means clustering (KC), Invariant Information Clustering (IIC) [10] and graph-based segmentation method (GS) [7] are chosen the comparative methods. We adjust the parameters K, σ of these methods to obtain the best two results for the experimental comparison. In addition, the recent CNN-based unsupervised segmentation methods based on superpixels (SP) [11] and differentiable clustering (DC) [12] are used as the baselines. It should be noted that there are several types of ground truth in BSD500 dataset, we evaluate the dataset from three aspects, which can be defined as "all" using all ground truth files, "fine" using a single ground truth file per image that contains the largest number of segments and "coarse" using the ground truth file that contains the smallest number of segments. It can be seen

Fig. 4. Visual comparison of our method and baselines. (a) Input images. (b) Ground truth. (c) SP [11]. (d) DC [12]. (e) Our method.

from the experimental results that the proposed method has a greater accuracy improvement than the baseline methods, and has a better accuracy advantage on most datasets. The visual comparison between SMC and baselines is shown in Fig. 4. Moreover, for exploring the convergence speed of SMC, we also study its segmentation accuracy under different iterations. The two baselines mentioned above adopt the same optimization strategy, so we choose the better baseline based on differential clustering [12] for experimental comparison. The results are shown in Fig. 5. From the figure, we can see that under the same number of iterations, the SMC has a greater advantage than baseline in accuracy.

4.3 Ablation Experiments

Here, a series of ablation experiments are designed for experimental verification. First of all, for proving the effectiveness of the proposed loss function, we conduct ablation experiments on each component of the loss function. The specific results are shown in Table 2. It can be seen from the experimental results that although L_{sie} and L_{lec} can improve the final accuracy slightly, when they are combined to form L_{hs}, the final accuracy can be greatly improved.

For evaluating the effect of different image augmentation operations for SMC, we conduct ablation studies on the three operations of S, A and P in BSD500 all dataset, The results are shown in Table 3. From the results, we can see that the proposed method can effectively improve the segmentation results by adding three operations. And among these operations, scale transformation S plays a relatively important role in the segmentation accuracy gain. When the three operations are used at the same time, the proposed method

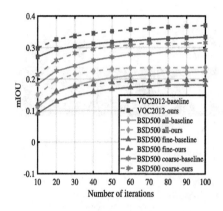

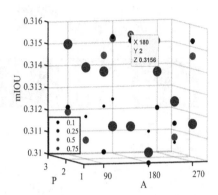

Fig. 5. Convergence curve of segmentation accuracy.

Fig. 6. The parameters ablation results of our method with S, A, P.

Table 2. Ablation results about each component of loss function.

L_{sim}	L_{con}	L_{sie}	L_{lec}	VOC2012	BSD500 all	BSD500 fine	BSD500 coarse
√				0.3358	0.3007	0.2619	0.3506
√	√			0.3520	0.3050	0.2592	0.3739
√	√	√		0.3697	0.3062	0.2610	0.3749
√	√		√	0.3776	0.3083	0.2621	0.3793
√	√	√	√	**0.3833**	**0.3156**	**0.2662**	**0.3913**

can get the best segmentation effect. In addition, when all three operations are applied to our method, we also perform ablation experiments on the parameters of these operations in BSD500 all dataset. For S, the scale range of image reduction is $\{0.1, 0.25, 0.5, 0.75\}$. For A, the angle of image rotation is selected from $\{90°, 180°, 270°\}$. And for P, the image is clipped into 3 specifications as $\{1 \times 1, 2 \times 2, 3 \times 3\}$. Through different parameter combinations, we have obtained segmentation results under different image augmentation specifications. For facilitating the display of experimental results, we use the transformation of A as the X axis, the transformation of P as the Y axis, the different scales of S as the types of nodes, and mIOU as the Z axis to establish a three-dimensional coordinate system, which is shown in the Fig. 6. From the results, it can be seen that when the reduction ratio is 0.5, the angle of rotation is 180° and the clipping specification is 2×2, the segmentation accuracy of SMC is better than others.

In order to evaluate the sensitivity of the parameters in the proposed method, the values of λ_{limit} and η are analyzed in BSD500 all. The experimental results are shown in the Table 4. From the experimental results, we can know when λ_{limit} is 100 and η is 0.05, the accuracy of the proposed method is better than others.

Table 3. Ablation results about different image augmentation operations of SMC.

S		$\checkmark$			$\checkmark$	$\checkmark$		$\checkmark$
A			$\checkmark$		$\checkmark$		$\checkmark$	$\checkmark$
P				$\checkmark$		$\checkmark$	$\checkmark$	$\checkmark$
mIOU	0.3050	0.3081	0.3067	0.3053	0.3121	0.3101	0.3078	**0.3156**

Table 4. Ablation results of λ_{limit} and η.

λ_{limit}	10	50	100	150	200
mIOU	0.3067	0.3100	**0.3156**	0.3101	0.3080

η	0.01	0.05	0.1	0.5	1
mIOU	0.3131	**0.3156**	0.3133	0.3124	0.3120

5 Conclusion

In this paper, we propose a Self-supervised Multi-view Clustering (SMC) method to learn self-supervised information from multi-view images. Based on the observation that the predicted pixel-level labels and the input images have the same spatial features, the multi-view labels can be predicted by clustering the multi-view images and the self-supervised information is mined out by using the differences among these labels. Meantime, for making the best of self-supervised information, a Hybrid Self-supervised (HS) loss function is proposed to further improve the convergence speed and prediction accuracy of our method. In the future work, more effective multi-view clustering schemes can be further explored to improve the unsupervised image segmentation.

References

1. Arbelaez, P., Maire, M., Fowlkes, C., Malik, J.: Contour detection and hierarchical image segmentation. IEEE Trans. Pattern Anal. Mach. Intell. **33**(5), 898–916 (2010)
2. Chen, L.-C., Zhu, Y., Papandreou, G., Schroff, F., Adam, H.: Encoder-decoder with atrous separable convolution for semantic image segmentation. In: Ferrari, V., Hebert, M., Sminchisescu, C., Weiss, Y. (eds.) ECCV 2018, Part VII. LNCS, vol. 11211, pp. 833–851. Springer, Cham (2018). https://doi.org/10.1007/978-3-030-01234-2_49
3. Chen, T., Kornblith, S., Norouzi, M., Hinton, G.: A simple framework for contrastive learning of visual representations. arXiv preprint arXiv:2002.05709 (2020)
4. Croitoru, I., Bogolin, S.V., Leordeanu, M.: Unsupervised learning of foreground object segmentation. Int. J. Comput. Vis. **127**(9), 1279–1302 (2019)

5. Doersch, C., Gupta, A., Efros, A.A.: Unsupervised visual representation learning by context prediction. In: Proceedings of the IEEE International Conference on Computer Vision, pp. 1422–1430. IEEE (2015)
6. Everingham, M., Eslami, S.A., Van Gool, L., Williams, C.K., Winn, J., Zisserman, A.: The pascal visual object classes challenge: a retrospective. Int. J. Comput. Vis. **111**(1), 98–136 (2015)
7. Felzenszwalb, P.F., Huttenlocher, D.P.: Efficient graph-based image segmentation. Int. J. Comput. Vis. **59**(2), 167–181 (2004)
8. Gidaris, S., Singh, P., Komodakis, N.: Unsupervised representation learning by predicting image rotations. arXiv preprint arXiv:1803.07728 (2018)
9. He, K., Fan, H., Wu, Y., Xie, S., Girshick, R.: Momentum contrast for unsupervised visual representation learning. In: Proceedings of the IEEE Conference on Computer Vision and Pattern Recognition, pp. 9729–9738. IEEE (2020)
10. Ji, X., Henriques, J.F., Vedaldi, A.: Invariant information clustering for unsupervised image classification and segmentation. In: Proceedings of the IEEE International Conference on Computer Vision, pp. 9865–9874. IEEE (2019)
11. Kanezaki, A.: Unsupervised image segmentation by backpropagation. In: 2018 IEEE International Conference on Acoustics, Speech and Signal Processing, pp. 1543–1547. IEEE (2018)
12. Kim, W., Kanezaki, A., Tanaka, M.: Unsupervised learning of image segmentation based on differentiable feature clustering. IEEE Trans. Image Process. **29**, 8055–8068 (2020)
13. Kolesnikov, A., Lampert, C.H.: Seed, expand and constrain: three principles for weakly-supervised image segmentation. In: Leibe, B., Matas, J., Sebe, N., Welling, M. (eds.) ECCV 2016, Part IV. LNCS, vol. 9908, pp. 695–711. Springer, Cham (2016). https://doi.org/10.1007/978-3-319-46493-0_42
14. Liu, X., Xu, Q., Ma, J., Jin, H., Zhang, Y.: MSLRR: a unified multiscale low-rank representation for image segmentation. IEEE Trans. Image Process. **23**(5), 2159–2167 (2014)
15. Long, J., Shelhamer, E., Darrell, T.: Fully convolutional networks for semantic segmentation. In: Proceedings of the IEEE Conference on Computer Vision and Pattern Recognition, pp. 3431–3440. IEEE (2015)
16. Misra, I., Zitnick, C.L., Hebert, M.: Shuffle and learn: unsupervised learning using temporal order verification. In: Leibe, B., Matas, J., Sebe, N., Welling, M. (eds.) ECCV 2016, Part I. LNCS, vol. 9905, pp. 527–544. Springer, Cham (2016). https://doi.org/10.1007/978-3-319-46448-0_32
17. Pathak, D., Krahenbuhl, P., Donahue, J., Darrell, T., Efros, A.A.: Context encoders: feature learning by inpainting. In: Proceedings of the IEEE Conference on Computer Vision and Pattern Recognition, pp. 2536–2544. IEEE (2016)
18. Sermanet, P., et al.: Time-contrastive networks: self-supervised learning from video. In: 2018 IEEE International Conference on Robotics and Automation, pp. 1134–1141. IEEE (2018)
19. Shibata, T., Tanaka, M., Okutomi, M.: Misalignment-robust joint filter for cross-modal image pairs. In: Proceedings of the IEEE International Conference on Computer Vision, pp. 3295–3304. IEEE (2017)
20. Wang, Y., Zhang, J., Kan, M., Shan, S., Chen, X.: Self-supervised equivariant attention mechanism for weakly supervised semantic segmentation. In: Proceedings of the IEEE Conference on Computer Vision and Pattern Recognition, pp. 12275–12284. IEEE (2020)
21. Xia, X., Kulis, B.: W-net: a deep model for fully unsupervised image segmentation. arXiv preprint arXiv:1711.08506 (2017)

22. Zhang, R., Isola, P., Efros, A.A.: Colorful image colorization. In: Leibe, B., Matas, J., Sebe, N., Welling, M. (eds.) ECCV 2016, Part III. LNCS, vol. 9907, pp. 649–666. Springer, Cham (2016). https://doi.org/10.1007/978-3-319-46487-9_40
23. Zhou, B., Khosla, A., Lapedriza, A., Oliva, A., Torralba, A.: Learning deep features for discriminative localization. In: Proceedings of the IEEE Conference on Computer Vision and Pattern Recognition, pp. 2921–2929. IEEE (2016)

Evaluate Pseudo Labeling and CNN for Multi-variate Time Series Classification in Low-Data Regimes

Dino Ienco[1(✉)], Davi Pereira-Santos[2], and André C. P. L. F. de Carvalho[2]

[1] INRAE, UMR TETIS, Univ. Montpellier, Montpellier, France
dino.ienco@inrae.fr
[2] ICMC, Sao Carlos, Brazil
{davips,andre}@icmc.usp.br

Abstract. Nowadays, huge amount of data are being produced by a large and diverse family of sensors (e.g., remote sensors, biochemical sensors, wearable devices). These sensors typically measure multiple variables over time, resulting in data streams that can be profitably organized as multivariate time-series. In practical scenarios, the speed at which such information is collected often makes the data labeling a difficult task. This results in a low-data regime scenario where only a small set of labeled samples is available and standard supervised learning algorithms cannot be employed. To cope with the task of multi-variate time series classification in low-data regime scenarios, here, we propose a framework that combines convolutional neural networks (CNNs) with self-training (pseudo labeling) in a transductive setting (test data are already available at training time). Our framework, named $ResNet^{IPL}$, wraps a CNN based classifier into an iterative procedure that, at each step, enlarges the training set with new samples and their associated pseudo labels. An experimental evaluation on several benchmarks, coming from different domains, has demonstrated the value of the proposed approach and, more generally, the ability of the deep learning approaches to effectively deal with scenarios characterized by low-data regimes.

1 Introduction

A vast amount of information is generated by a widespread and diverse family of sensors like remote sensors, biochemical sensors and wearable devices. They typically measure multiple variables over time, resulting in data streams that can be profitably organized as multivariate time-series. Due to the ubiquitous nature of multivariate time-series, conceiving classification methods especially tailored for such kind of data is crucial [15]. In a more realistic but challenging scenario, only a limited set of samples, among the available data, is associated with label information resulting in a low-data regime scenario that requires effective semi-supervised learning methods [2].

© Springer Nature Switzerland AG 2021
I. Farkaš et al. (Eds.): ICANN 2021, LNCS 12895, pp. 126–137, 2021.
https://doi.org/10.1007/978-3-030-86383-8_10

Regarding the semi-supervised classification of time series data, [9] introduces an approach that firstly uses hierarchical clustering to group labeled and unlabeled data, then propagates label information inside each cluster and, finally, employs a one nearest neighbors (1NN) classifier with dynamic time warping (DTW) to perform classification. A similar approach is proposed in [5] where clustering and a 1NN classifier are combined to perform classification of univariate time series when only a few labeled data are available. [18] recently introduces a multivariate time series classification based on neural attentional prototype network to train the feature representation based on their distance to class prototypes considering low-data regimes. The proposed deep learning based method works considering both fully supervised and semi-supervised settings.

Another family of semi-supervised methods for time series classification is based on self-training (or self-labeling) approaches, whose goal is to enlarge the original labeled set selecting the unlabeled samples with the most confident predictions [14]. In conjunction with the self-training framework, the 1NN classifier is typically used as the base learner, as it has been effective for time series classification tasks [2].

Recently, [2] evaluated several machine learning based approaches to deal with the semi-supervised classification of univariate time series. The approach proposed by the authors is to couple standard classifiers with pseudo labeling and self-training strategies. The results underline that, among all considered classifiers, 1NN still exhibits superior performance as base learner. Unfortunately, the study is limited to univariate time series while multivariate time-series are becoming more and more predominant nowadays. Additionally, it totally ignores the recent advent of deep learning (DL) approaches in the time series community [4,18]. Indeed, despite the recent findings reported in [4] where a DL strategy (residual-based convolutional neural networks) exhibits superior performance in standard fully supervised classification tasks, no discussion is reported about the appropriateness of such approaches when only few labeled time-series data are available to learn a classification model. This fact indicates that, considering the classification of time series data in a low-data regime scenario, the use of deep learning approaches is still an under explored field of research.

We propose, here, a framework that combines deep learning with pseudo labeling in a *transductive setting*, i.e., when test data are already available during model training. Our approach is motivated by the lack of studies that explore the application of deep learning to low-data regime scenarios, with a particular emphasis on multivariate time series classification tasks. The proposed framework, referred to as $ResNet^{IPL}$ (ResNet with Incremental Pseudo Labeling), wraps the deep learning based classifier into an iterative procedure that, at each step, enlarges the training set with new samples and their associated pseudo labels. The sample selection stage leverages the classifier prediction on unlabeled data and chooses those samples that minimize the relative entropy associated to the model output distribution. To assess the performance of the proposed framework as well as its generality, we conduct an extensive experimental evaluation on several multivariate time-series benchmarks coming from different domains.

The rest of the manuscript is organized as follows: the $ResNet^{IPL}$ framework is introduced in Sect. 2, experimental settings as well as experimental evaluation are described in Sect. 3 while Sect. 4 draws conclusions and possible follow-ups.

2   Methodology

In this section, we describe our proposed incremental pseudo labeling procedure for the classification of multi-variate time series data considering a low-data regime scenario.

The general procedure is depicted in Algorithm 1. Due to the fact that we are considering a transductive scenario, test samples are available at training time. The procedure takes as input the set of training samples (X_{train}) with the associated labels (Y_{train}), the test samples X_{test}, the number of iterations of the incremental procedure (T), and the number of samples per class added at each iteration to the training set (k). The output of the procedure is a multi-variate time series classification model that is trained with both original and pseudo labeled samples. At the beginning, the classification model ($Classifier$) is initialized and then trained on the original labeled samples X_{train}, Y_{train} (Line 1–2). Then, the incremental process starts (Line 4–13). At each iteration, the classification model is applied on the current test data (X_{test}) and the class distribution for each test sample (derived by the softmax layer of the classification model) is obtained (Line 5). The class distribution, the unlabeled data X_{test} and the k parameter are employed by the sample selection procedure. This procedure extracts, for each class, k reliable examples according to the class distribution previously outputted by the classification model. The set of selected samples and their associated pseudo labels are referred to as X_{sel} and $pseudolLabel$, respectively (Line 6). Successively, the training and testing sets are updated according to the results of the sample selection procedure (Line 7–9). Finally, the classification model is initialized again and trained on the new set of training samples that combines both original and pseudo label information (Line 10–11).

Regarding the general procedure depicted in Algorithm 1, two points must be defined in order to deploy such strategy: firstly, the choice of the classification model and, secondly, the implementation of the sample selection procedure.

Concerning the classification model, we base our choice on the findings reported in a recent literature survey [4]. Among several deep learning architectures for time series classification, the Residual Network $ResNet$ model proposed in [16] exhibits superior behavior. Due to this fact, we choose such architecture as classification model in our study. The network is composed of three residual blocks followed by a GAP (Global Average Pooling) layer and a final softmax classifier whose number of neurons is equal to the number of classes in a dataset. Each residual block is first composed of three convolutions whose output is added to the residual block's input and then fed to the next layer. The number of filters for all convolutions is fixed to 64, with the ReLU activation function that is

---

Algorithm 1. Incremental Pseudo Labeling procedure

Require: $X_{train}, Y_{train}, X_{test}, T, k$.
Ensure: $Classifier$.
1:   $Classifier \leftarrow$ initModel()
2:   $Classifier \leftarrow$ TrainModel(Classifier, X_{train}, Y_{train})
3:   $i \leftarrow 0$
4: **while** i < T **do**
5:      classDistrib $\leftarrow$ Classify($classifier, X_{test}$)
6:      X_{sel}, pseudoLabel $\leftarrow$ SampleSelection(X_{test}, classDistrib, k)
7:      $X_{train} \leftarrow X_{train} \cup X_{sel}$
8:      $Y_{train} \leftarrow Y_{train} \cup pseudoLabel$
9:      $X_{test} \leftarrow X_{test} - X_{sel}$
10:     $Classifier \leftarrow$ initModel()
11:     $Classifier \leftarrow$ TrainModel(Classifier, X_{train}, Y_{train})
12:     $i \leftarrow i + 1$
13: **end while**
14: **return** $Classifier$

---

preceded by a batch normalization operation. In each residual block, the filter's length is set to 8, 5 and 3 respectively for the first, second and third convolution.

The second point involves the definition of a sample selection strategy. Such a strategy is mainly based on the analysis of the class distribution output by the classification model. More in detail, for each sample x_t we exploit the class distribution $pd(x_t)$. $pd(x_t)$ is the probability distribution over all possible classes that corresponds to the softmax output of the classification model regarding the sample x_t. Our strategy selects unlabeled samples on which the classifier has the highest confidence. To this purpose, we consider as surrogate of the confidence measure the entropy over the classifier output probability distribution. The entropy measure has already demonstrated its quality in pseudo labeling strategies to select valuable samples in the context of image analysis and semantic segmentation [10]. In our case, we adopt a normalized version of the entropy measure defined as follows:

$$H(x_t) = -\frac{\sum_{c \in C} pd_c(x_t) \times \log(pd_c(x_t))}{\log(|C|)} \tag{1}$$

where C is the set of possible classes and $pd_c(x_t)$ is the probability of sample x_t to belong to class $c \in C$. Samples with low entropy values correspond to time series on which the classifier has high confidence in its prediction. The *SampleSelection*() procedure is summarized in Algorithm 2.

The procedure takes as input the set of test samples (X_{test}), the class distribution obtained by the classification model $ClassDistrib$ and the parameter k corresponding to the number of per-class samples to select. The output of the procedure is the set of the selected samples (X_{sel}) with their associated pseudo labels ($pseudoLabel$).

We can note that the set of selected samples (X_{sel}) with the associated pseudo labels ($pseudoLabel$) is obtained class by class (Line 3–8). For each class $c \in C$, we select the samples that the classifier judges to belong to that class (Line 4). Successively, the selected samples (X_c) are ranked in ascending order w.r.t. the entropy measure defined in Eq. 1. Finally, the top K samples ($K = \{x_i \mid x_i \in$

$X_c, 1 \leq i < k\}$) are added to the final set along with their corresponding pseudo labels. $[c]^k$ indicates a vector where the class value c is repeated k times.

Algorithm 2. SampleSelection(X_{test}, ClassDistrib, k)

Require: X_{test}, ClassDistrib, k.
Ensure: X_{sel}, pseudoLabel.
  1: $X_{sel} \leftarrow \emptyset$
  2: $pseudoLabel \leftarrow \emptyset$
  3: **for all** $c \in C$ **do**
  4:     $X_c \leftarrow \{x | x \in X_{test}, [\underset{v \in C}{\operatorname{argmax}} \, ClassDistrib_v(x)] = c\}$
  5:     rank X_c in ascending order considering the entropy measure defined in Equation 1
  6:     $K \leftarrow \{x_i \mid x_i \in X_c, 1 \leq i < k\}$
  7:     $X_{sel} \leftarrow X_{sel} \cup K$
  8:     $pseudoLabel \leftarrow pseudoLabel \cup [c]^k$
  9: **end for**
10: **return** X_{sel}, $pseudoLabel$

---

3   Experimental Evaluation

In this section we assess the behavior of our framework considering five real world multivariate time series benchmarks. To evaluate the performance of our proposal, we compare $ResNet^{IPL}$ with several competing and baseline approaches.

3.1   Competitors and Method Ablations

For the comparative study, we consider the following competitors:

- A one nearest neighbors classifier (1NN) coupled with the DTW measure [3]. 1NN is a well recognized and widely adopted classifier in the time series classification domain [2,7]. We name such competitor $1NN_{DTW}$.
- A graph-based semi-supervised learning approach since we are considering a transductive scenario. Among the different available methods, we adopted the CAMLP (Confidence-Aware Modulated Label Propagation) approach [17] as a representative one. Since CAMLP requires the construction of a K-nearest-neighbors graph to perform its propagation process, we chose to set K equals to 20, according to the study proposed in [11], and construct the K-nearest-neighbors graph leveraging, also in this case, the DTW similarity measure. We name this competitor as $GBSSL_{DTW}$.
- The $ResNet$ approach proposed in [4] without the incremental pseudo labeling strategy. This competitor can be seen as an ablation of the proposed framework. We name this baseline as $ResNet$.
- The recent TapNet approach [18] which introduces a multivariate time series classification with attentional prototypical deep neural network. Due to the transductive setting considered in our work, we adopt the semi-supervised version that exploits unlabelled data during training.

For the $ResNet$ [1] and TapNet models [2], we use their available implementations.

Furthermore, we couple $1NN_{DTW}$ and $GBSSL_{DTW}$ with the proposed incremental pseudo labeling strategy. These additional competitors are referred to as $1NN_{DTW}^{IPL}$ and $GBSSL_{DTW}^{IPL}$, respectively.

$ResNet$ is implemented via the Tensorflow 2 python library [3] while the implementation of the remain competitors is based on TSLEAN [13] and SCIKIT-learn [1] python libraries.

3.2 Data and Experimental Settings

The evaluation has been carried out by performing experiments on five benchmarks coming from disparate application domains and characterized by contrasted features in terms of number of samples, number of attributes (dimensions) and time length. All benchmarks, except $Dordogne$ – which was obtained contacting the authors of [6], are available online.

Table 1. Benchmarks characteristics

Dataset	# Samples	# Dims	Min/Max length	Avg. length	# Classes
Dordogne	9 919	6	23/23	23	7
GTZAN	600	33	128/128	128	6
HAR	10 299	9	128/128	128	6
JapVowel	640	12	7/29	15	9
SpeechCom	23 682	40	14/32	31	10

The characteristics of the five benchmarks are reported in Table 1. For each benchmark, we consider different amount of per-class labeled samples. The amount of per-class labeled samples ranges in the set {10, 15, 20, 25}. This means that, for instance, considering the value 10, ten samples per class are randomly chosen and used to compose the initial training set, and the rest of them is considered as the test set. For all the methods that involve the incremental pseudo labeling procedure, 10 samples per class are moved, at each round (according to the strategy specified in Sect. 2) from the test set to the training set and associated to the pseudo labels estimated by the specific learning algorithm. Classification performances are assessed by F-Measure metric [12] considering the original test set. F-Measure is chosen as metric due to its ability to take into account possible class imbalance scenarios.

The obtained results are averaged over five different runs for each given method and benchmark, due to the non deterministic nature of the sample selection. Finally, the average value is reported.

---

[1] https://github.com/hfawaz/dl-4-tsc.

[2] https://github.com/xuczhang/tapnet.

[3] Code will be available upon acceptance.

3.3   Quantitative Results

Tables 2, 3, 4, 5, 6 depict the performance results, in terms of F-Measure (average and standard deviation), of the different competing approaches varying the amount of available label data. We can observe that $ResNet$ always outperforms the competing approaches ($1NN$ and $GBSSL$) considering all the five benchmarks as well as all the training size. Regarding non deep learning approaches ($1NN$ and $GBSSL$), we can note that no approach systematically outperforms the other. Despite $GBSSL$ is not largely adopted by the time series classification community, it exhibits comparable behavior w.r.t. the commonly adopted $1NN$ method. In addition, coupling such competitors with the incremental pseudo labeling framework always ameliorate the method performances no matter the training size. Regarding the $1NN$ approach behavior, this is in line with the experimental findings reported in [2] for univariate time series.

Table 2. F-Measure results over the Dordogne benchmark varying the amount of labelled examples per class. We report average and standard deviation. Bold and underlined text indicate best and second-best results, respectively.

	10	15	20	25
$1NN_{DTW}$	14.25 ± 1.58	14.99 ± 3.18	16.49 ± 2.43	16.05 ± 1.67
$GBSSL_{DTW}$	60.03 ± 1.58	61.82 ± 2.20	63.00 ± 2.30	64.24 ± 1.40
$1NN_{DTW}^{IPL}$	58.32 ± 2.07	60.81 ± 2.61	61.02 ± 1.76	61.71 ± 1.54
$GBSSL_{DTW}^{IPL}$	62.57 ± 2.50	63.71 ± 2.99	64.53 ± 2.46	65.24 ± 1.91
TapNet	60.97 ± 1.97	63.78 ± 2.52	65.33 ± 2.17	67.20 ± 1.17
$ResNet$	$\mathbf{64.70} \pm 2.46$	$\underline{66.47} \pm 2.65$	$\mathbf{68.94} \pm 1.99$	$\mathbf{70.35} \pm 1.46$
$ResNet^{IPL}$	$\underline{63.15} \pm 2.31$	$\mathbf{67.11} \pm 2.78$	$\underline{68.69} \pm 2.64$	$\underline{70.02} \pm 1.35$

Table 3. F-Measure results over the GTZAN benchmark varying the amount of labelled examples per class. We report average and standard deviation. Bold and underlined text indicate best and second-best results, respectively.

	10	15	20	25
$1NN_{DTW}$	16.59 ± 2.09	17.02 ± 1.66	15.47 ± 2.09	16.26 ± 1.59
$GBSSL_{DTW}$	47.77 ± 1.11	49.84 ± 2.26	50.48 ± 2.32	51.96 ± 1.72
$1NN_{DTW}^{IPL}$	62.86 ± 2.24	65.33 ± 2.10	67.23 ± 3.05	68.05 ± 2.68
$GBSSL_{DTW}^{IPL}$	63.58 ± 1.81	65.74 ± 2.15	67.05 ± 0.55	69.30 ± 1.24
TapNet	44.98 ± 1.08	45.83 ± 1.75	47.85 ± 1.09	48.40 ± 1.59
$ResNet$	$\underline{64.90} \pm 1.48$	$\underline{69.18} \pm 3.06$	$\underline{71.59} \pm 3.72$	$\underline{72.90} \pm 2.74$
$ResNet^{IPL}$	$\mathbf{67.25} \pm 1.98$	$\mathbf{72.83} \pm 1.63$	$\mathbf{74.18} \pm 0.96$	$\mathbf{76.65} \pm 2.20$

Regarding the proposed framework, we observe that incremental pseudo labeling achieves better performances than its counterpart without pseudo labeling on three benchmarks ($GTZAN$, $JapWovel$ and $SpeechCom$) while on the

Table 4. F-Measure results over the HAR varying the amount of labelled examples per class. We report average and standard deviation. Bold and underlined text indicate best and second-best results, respectively.

	10	15	20	25
$1NN_{DTW}$	16.82 ± 2.29	16.92 ± 3.11	15.59 ± 2.40	17.25 ± 3.16
$GBSSL_{DTW}$	60.29 ± 4.01	65.43 ± 3.64	69.13 ± 1.98	70.48 ± 1.44
$1NN_{DTW}^{IPL}$	78.90 ± 2.12	81.34 ± 2.11	83.07 ± 1.79	84.45 ± 1.39
$GBSSL_{DTW}^{IPL}$	60.61 ± 3.86	65.94 ± 3.10	69.67 ± 1.61	70.93 ± 0.96
TapNet	65.41 ± 2.24	67.49 ± 2.15	69.53 ± 1.93	70.41 ± 1.78
ResNet	<u>87.87</u> ± 2.71	<u>89.66</u> ± 2.51	**90.96** ± 0.71	**90.26** ± 0.90
$ResNet^{IPL}$	**88.28** ± 2.13	**90.15** ± 1.23	<u>89.10</u> ± 2.64	<u>88.98</u> ± 2.14

Table 5. F-Measure results over the JapVowel benchmark varying the amount of labelled examples per class. We report average and standard deviation. Bold and underlined text indicate best and second-best results, respectively.

	10	15	20	25
$1NN_{DTW}$	6.88 ± 1.57	8.68 ± 3.37	7.73 ± 1.65	7.60 ± 2.65
$GBSSL_{DTW}$	88.07 ± 1.14	89.14 ± 0.58	89.92 ± 0.78	89.90 ± 0.62
$1NN_{DTW}^{IPL}$	92.01 ± 0.78	92.65 ± 0.78	93.14 ± 0.89	93.76 ± 0.69
$GBSSL_{DTW}^{IPL}$	90.39 ± 0.87	91.08 ± 0.84	92.04 ± 0.51	93.15 ± 0.90
TapNet	76.88 ± 1.89	82.49 ± 1.64	84.83 ± 1.63	86.89 ± 1.29
ResNet	<u>94.08</u> ± 0.46	<u>96.28</u> ± 0.85	<u>97.52</u> ± 0.26	<u>97.33</u> ± 1.18
$ResNet^{IPL}$	**97.25** ± 0.61	**97.40** ± 0.60	**98.03** ± 0.49	**98.23** ± 0.61

Table 6. F-Measure results over the SpeechCommand benchmark varying the amount of labelled examples per class. We report average and standard deviation. Bold and underlined text indicate best and second-best results, respectively.

	10	15	20	25
$1NN_{DTW}$	8.27 ± 0.79	8.67 ± 0.83	9.02 ± 0.70	9.54 ± 0.85
$GBSSL_{DTW}$	11.82 ± 0.38	12.81 ± 0.35	13.54 ± 0.32	14.00 ± 0.29
$1NN_{DTW}^{IPL}$	28.85 ± 0.93	29.81 ± 1.25	29.73 ± 0.54	30.30 ± 0.57
$GBSSL_{DTW}^{IPL}$	14.42 ± 0.54	14.95 ± 0.30	15.39 ± 0.36	15.62 ± 0.31
TapNet	12.86 ± 0.62	13.52 ± 0.73	13.70 ± 0.46	14.28 ± 0.40
ResNet	<u>46.29</u> ± 3.66	<u>58.51</u> ± 3.05	<u>67.96</u> ± 1.42	<u>73.59</u> ± 0.57
$ResNet^{IPL}$	**59.29** ± 4.42	**69.36** ± 1.46	**74.91** ± 1.52	**78.37** ± 0.67

rest of the datasets (*Dordogne* and *HAR*) the behaviours are comparable. In addition, we can observe that $ResNet^{IPL}$ systematically outperforms the recent TapNet approach over all the considered benchmark no matter the amount of labelled samples we consider as initial training set. Interestingly, we can note that $ResNet$ and $ResNet^{IPL}$ always take advantage when the quantity of labeled samples increases compared to all the other competing approaches. This phenomena is clearly evident considering the *SpeechCom* benchmark (Table 6). Here, we can see that *ResNet* and $ResNet^{IPL}$ generally ameliorate their behavior when more labeled samples are available.

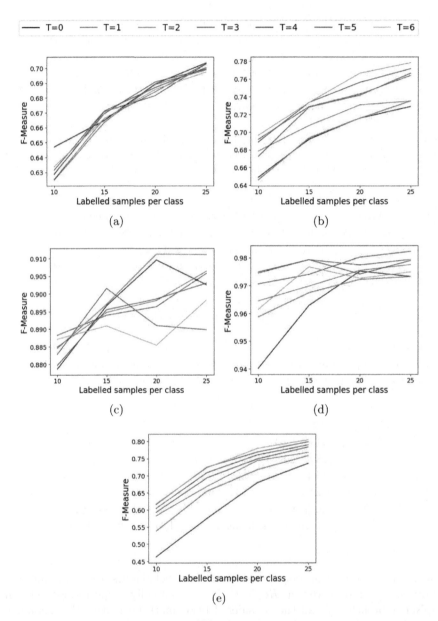

Fig. 1. Sensitivity analysis of $ResNet^{IPL}$ w.r.t. the T parameter, varying the amount of available training data, on the considered benchmarks: (a) Dordogne (b) GTZAN (c) HAR (d) JapVowel and (e) SpeechCom.

Figure 1 reports the sensitivity analysis of $ResNet^{IPL}$ regarding the T parameter (the number of iterations of the incremental pseudo labeling process). More in detail, we vary the T parameter in the interval $[0, 6]$, where $T = 0$ corresponds to the behaviour of $ResNet$. Consistently with the results reported in Tables 2, 3, 4, 5, 6, we can observe two different types of behaviors. Regarding $GTZAN$, $JapWovel$ and $SpeechCom$ (resp. Figure 1b, Fig. 1d and Fig. 1e), we can note that increasing the number of iterations (T) results, in general, in higher value of F-Measure. This is particularly evident for small training data (with a number of labeled samples per class equal or lesser than 15). A different behavior is exhibited by the $Dordogne$ and HAR datasets where the parameter T of the incremental pseudo labeling procedure does not really influence the obtained performances in terms of F-Measure.

To sum up the obtained findings, we highlight that, also when extreme low-data regime scenarios are considered, deep learning approaches still exhibit high performances for the classification of multi-variate varying-length time series when compared to standard methods; and, the incremental pseudo labeling strategy clearly ameliorates the results, in terms of F-Measure, considering three benchmarks over five, while on the remaining test cases the performance are comparable w.r.t. the ablation variant that does not involve pseudo labeling.

3.4 Visual Inspection

Figure 2 depicts the visualization of the embeddings obtained considering the model trained on the original training set (Fig. 2a) and, successively, by $ResNet^{IPL}$ for the values 2, 4 and 6 (Fig. 2b, Fig. 2c, and Fig. 2d, respectively) of the T parameter (the number of iterations in the iterative pseudo labeling process) for the $JapWovel$ benchmark.

The original training set involves 10 labeled samples per class. The embeddings are obtained considering the output of the last convolutional layer. We visualize 30 samples per class coming from the test data by means of the two dimensional projection supplied by the T-SNE method [8]. We underline that the same set of test samples is considered over the four different cases. Each colour represents a different class. We can clearly observe that as the T parameter increases, the cluster structure associated to the underlying data distribution emerges. While the visualisation related to the embeddings obtained by the model trained on the original training set (Fig. 2a) exhibits evident confusion among most of the classes, we can note that, the incremental pseudo labeling procedure allows to reduce confusions and to recover a clear cluster structure. More in detail, we can see that, when a value of $T = 4$ (Fig. 2c) is considered, the majority of confusions disappear.

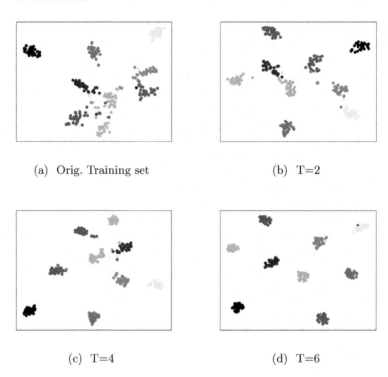

(a) Orig. Training set   (b) T=2

(c) T=4   (d) T=6

Fig. 2. T-SNE Feature visualization of the same 30 per-class test samples belonging to the *Jap Wovel* dataset considering the representation learnt from (a) the original (5 labelled samples per class) training data (b) the proposed framework with T = 2 (c) the proposed framework with T = 4 and (d) the proposed framework with T = 6. The original label set involves 10 labeled samples per class.

4   Conclusion

In this paper, we proposed a framework that combines CNN and self-training to deal with multi-variate time series classification considering low-data regime scenarios. The proposed framework works in a transductive fashion and it leverages the entropy associated to the classifier prediction to select new samples to enlarge the training set. The evaluation on real-world benchmarks has demonstrated the effectiveness of $ResNet^{IPL}$ w.r.t. recent classification framework and, more generally, the value of deep learning-based strategies to deal with low-data regime scenarios in the context of multi-variate time series classification.

References

1. Buitinck, L., et al.: API design for machine learning software: experiences from the scikit-learn project. In: ECML PKDD Workshop: Languages for Data Mining and Machine Learning, pp. 108–122 (2013)

2. Castellanos, M.G., Bergmeir, C., Triguero, I., Rodríguez, Y., Benítez, J.M.: Self-labeling techniques for semi-supervised time series classification: an empirical study. Knowl. Inf. Syst. **55**(2), 493–528 (2018)
3. Chen, Y., Hu, B., Keogh, E.J., Batista, G.E.A.P.A.: DTW-D: time series semi-supervised learning from a single example. In: KDD, pp. 383–391. ACM (2013)
4. Fawaz, H.I., Forestier, G., Weber, J., Idoumghar, L., Muller, P.: Deep learning for time series classification: a review. Data Min. Knowl. Discov. **33**(4), 917–963 (2019)
5. Frank, J., Mannor, S., Pineau, J., Precup, D.: Time series analysis using geometric template matching. IEEE Trans. Pattern Anal. Mach. Intell. **35**(3), 740–754 (2013)
6. Gbodjo, Y.J.E., Ienco, D., Leroux, L.: Toward spatio-spectral analysis of sentinel-2 time series data for land cover mapping. IEEE Geosci. Remote Sens. Lett. **17**(2), 307–311 (2020)
7. Geler, Z., Kurbalija, V., Radovanovic, M., Ivanovic, M.: Comparison of different weighting schemes for the KNN classifier on time-series data. Knowl. Inf. Syst. **48**(2), 331–378 (2016)
8. van der Maaten, L., Hinton, G.: Visualizing Data Using t-SNE. J. Mach. Learn. Res. **9**, 2579–2605 (2008)
9. Marussy, K., Buza, K.: SUCCESS: a new approach for semi-supervised classification of time-series. In: Rutkowski, L., Korytkowski, M., Scherer, R., Tadeusiewicz, R., Zadeh, L.A., Zurada, J.M. (eds.) ICAISC 2013. LNCS (LNAI), vol. 7894, pp. 437–447. Springer, Heidelberg (2013). https://doi.org/10.1007/978-3-642-38658-9_39
10. Saporta, A., Vu, T., Cord, M., Pérez, P.: ESL: entropy-guided self-supervised learning for domain adaptation in semantic segmentation. CoRR abs/2006.08658 (2020)
11. de Sousa, C.A.R., Rezende, S.O., Batista, G.E.A.P.A.: Influence of graph construction on semi-supervised learning. In: ECML/PKDD, vol. 8190, pp. 160–175 (2013)
12. Tan, P.N., Steinbach, M., Kumar, V.: Introduction to Data Mining, 1st edn. Addison-Wesley Longman Publishing Co. Inc., Boston (2005)
13. Tavenard, R., et al.: Tslearn, a machine learning toolkit for time series data. J. Mach. Learn. Res. **21**(118), 1–6 (2020)
14. Triguero, I., García, S., Herrera, F.: Self-labeled techniques for semi-supervised learning: taxonomy, software and empirical study. Knowl. Inf. Syst. **42**(2), 245–284 (2015)
15. Wang, X., Mueen, A., Ding, H., Trajcevski, G., Scheuermann, P., Keogh, E.J.: Experimental comparison of representation methods and distance measures for time series data. Data Min. Knowl. Discov. **26**(2), 275–309 (2013)
16. Wang, Z., Yan, W., Oates, T.: Time series classification from scratch with deep neural networks: a strong baseline. In: IJCNN, pp. 1578–1585. IEEE (2017)
17. Yamaguchi, Y., Faloutsos, C., Kitagawa, H.: CAMLP: confidence-aware modulated label propagation. In: SDM, pp. 513–521. SIAM (2016)
18. Zhang, X., Gao, Y., Lin, J., Lu, C.: TapNet: multivariate time series classification with attentional prototypical network. In: AAAI, pp. 6845–6852 (2020)

Deep Variational Autoencoder with Shallow Parallel Path for Top-N Recommendation (VASP)

Vojtěch Vančura$^{(\boxtimes)}$ (ID) and Pavel Kordík (ID)

Faculty of Information Technology, Czech Technical University in Prague,
Prague, Czech Republic
{vancurv,pavel.kordik}@fit.cvut.cz
https://fit.cvut.cz/en/

Abstract. The recently introduced Embarrasingelly Shallow Autoencoder (EASE) algorithm presents a simple and elegant way to solve the top-N recommendation task. In this paper, we introduce Neural EASE to further improve the performance of this algorithm by incorporating techniques for training modern neural networks. Also, there is a growing interest in the recsys community to utilize variational autoencoders (VAE) for this task. We introduce Focal Loss Variational AutoEncoder (FLVAE), benefiting from multiple non-linear layers without an information bottleneck while not overfitting towards the identity. We show how to learn FLVAE in parallel with Neural EASE and achieve state-of-the-art performance on the MovieLens 20M dataset and competitive results on the Netflix Prize dataset.

Keywords: Recommender systems · Variational autoencoders

1 Introduction

With the increasing amount of information on the Web, Recommender Systems (RS) are an important way to overcome infobesity. On the other hand, companies like NetFlix, Youtube, Amazon, or Google are making significant revenues from recommendations[1]. Thus, RS are gaining more attention over the past two decades.

As online companies grow, RS have to scale to millions of active users and millions of items. Speed of training and recall are also increasingly important as available content often changes dynamically, and RS need to react in real-time.

Proper evaluation of RS is also an increasingly important topic as offline evaluation is often a biased predictor of the online performance [17]. For offline

---

[1] According to [25] 80% movies on NetFlix, 60% videos on Youtube are watched based on recommendations; recommendations are responsible of 35% sale revenues on Amazon [22].

© Springer Nature Switzerland AG 2021
I. Farkaš et al. (Eds.): ICANN 2021, LNCS 12895, pp. 138–149, 2021.
https://doi.org/10.1007/978-3-030-86383-8_11

evaluation, the recsys community shifted towards Top-N approaches [9] as evaluating the performance based on root mean squared error on top of a predicted rating matrix can be very misleading.

In the Top-N recommendation scenario [3], RS is recommending N most relevant items for every user. This is the typical case in various domains, including media, news, or e-commerce.

Various approaches have been proposed for solving the Top-N recommendation task, including collaborative filtering with matrix factorization to give an example. Recently, sparse-data autoencoders [10,13,16,18,19,23] gained much attention and were providing state-of-the-art results in solving this task. We examined various proposed models, including denoising autoencoders, variational autoencoders, and a shallow autoencoder called EASE [20], which despite being a simple linear model, is providing competitive and explainable results while addressing the biggest problem with sparse autoencoders: overfitting towards identity.

Inspired by [2], our motivation was to build a RS model, that is as elegant and explainable as EASE while leveraging the potential of deep autoencoders to model complex nonlinear patterns in the data.

In order to do this, we had to overcome several issues, most importantly, the overfitting towards identity.

Traditionally, overfitting towards identity is addressed by using dropout in the input layer [10,13,19]. However, this approach is not effective enough and is not enabling the usage of deep architectures with millions of parameters.

In this work, we propose three major contributions to address the issues mentioned above. We propose:

- the usage of focal loss for training autoencoders for Top-N recommendation.
- a simple yet effective data augmentation technique to prevent Top-N recommending autoencoders from overfitting towards identity.
- a joint-learning technique based on the Hadamard product for training different combinations of various models.

As a demonstration, we build the VASP, a Variational Autoencoder with a Shallow parallel Path. VASP combines deep Variational Autoencoder and a neural variant of shallow EASE jointly trained together to model both linear and non-linear patterns in the data. VASP was able to achieve state of the art performance on the MovieLens 20M dataset and competitive results on the Netflix Prize dataset.

2   Related Work

Matrix Factorization (MF) has been the first choice model for many years since the team "BellKor's Pragmatic Chaos" won the Netflix Prize [12].

In 2016 Sedhain et al. proposed AutoRec [18], an autoencoder-based model for collaborative filtering with explicit ratings that outperforms all current baseline models. After emergence of variational autoencoders (VAE), the collaborative filtering model MultVAE was proposed in 2018 by Liang et al. [13] This model uses multinomial log-likelihood for data distribution.

Several techniques for improving the MultVAE were proposed recently. H. Steck proposed EASE [20], the Embarrasingelly Shallow Autoencoder with no hidden layers as opposed to deep architectures. This approach was able to beat state-of-the-art models when introduced. RecVAE [19] uses a separate regularization term in the form of the Kullback-Leibler divergence between the actual parameter distribution and the distribution in previous training step preventing instability during training. H+VAMP [10] implements a variational autoencoder with a variational mixture of posteriors prior (Vamp Prior) with the goal to learn better latent representations of user-items interactions.

During the evolution of recommender systems, many simple, shallow (linear, wide), and complex deep architectures have been proposed. Cheng et al. proposed a combination of those two approaches into a single framework called Wide & Deep Learning [2] and introduced a technique called joint training. The authors also point out the distinction between joint training and ensembling. We are inspired by this work, but our approach is quite different. We do not process item attributes, just the interactions, therefore deep path is not design to encode items but to find nonlinear interaction patterns. Also, voting scheme is different.

In [4], the authors propose ensembling of pre-trained recommender models by variational autoencoder. However, joint-learning of such a model seems to be problematic from the perspective of scaling and practical usability.

Many other deep learning techniques originally developed for computer vision or natural language processing were later successfully used in other fields such as recommender systems. Residual networks [7] are a good example. Another example is using the approach from [8] for dense layers in RecVAE [19].

We follow this trend by adopting Focal Loss (FL) from [14] for recommendation systems. This novel approach is used for imbalanced classes in object detection addressing the imbalance between the background class and other classes. That means that the loss is higher for examples in the training set that are difficult to classify. This perfectly fits the situation in collaborative filtering, where some items are more popular than others. It is more difficult to recommend niche items as they do not have many interactions. Higher loss for these items push recommender system to focus more on cold start and niche items.

Another essential idea while training an autoencoder on sparse data is to prevent overfitting towards learning the identity function between the input and the output layer of the autoencoder [21]. In [24], the authors proposed the Split-Brain Autoencoder, which prevents learning identity by splitting input image into two separate channels: grayscale channel X_1 and color channels X_2. Learning is then performed in a separate way by training two networks, F_1 to perform automatic colorization by learning X_2 by showing X_1 and F_2 to make a grayscale prediction by learning X_2 by showing X_1. On the other hand, our approach uses only one neural network with automated data augmentation as a preprocessing step.

3   Our Approach

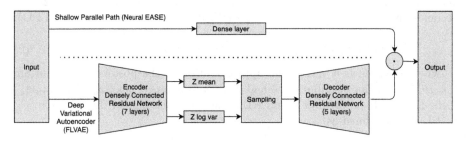

Fig. 1. VASP architecture

Following notation from [13], we index users as $u \in \{1, ..., U\}$, items as $i \in \{1, ..., I\}$, and user-item interaction matrix $X \in \mathbb{N}^{U \times I}$. Lowercase $\mathbf{x}_u = [x_{u1}, x_{u2}, ..., x_{uI}]^T \in \mathbb{N}^I$ denotes the interaction history and $\hat{\mathbf{x}}_u = [\hat{x}_{u1}, \hat{x}_{u2}, ..., \hat{x}_{uI}]^T \in \mathbb{N}$ predicted ratings of the user u.

3.1   Neural EASE (NEASE)

Following [20], the EASE model can be described as:

$$\hat{\mathbf{x}}_u = W \cdot \mathbf{x}_u, \tag{1}$$

where $W \in \mathbb{R}^{|I| \times |I|}$ is the weight matrix. The diagonal elements of W are constrained to zero to prevent learning the identity function between the input and output. In [20] the authors proposed using the square loss between the data $\mathbf{x}_u$ and the predicted scores $\hat{\mathbf{x}}_u$ because this training objective has a closed-form solution. The authors also suggest that using more complex loss functions may lead to better prediction accuracy with higher computational costs.

To enable running the model in parallel to a deep autoencoder, we interpret the EASE model (1) as a single-layer perceptron without bias nodes and with forced zeros on the diagonal, which can be trained with any suitable loss function using backpropagation.

Our experiments with several different loss functions are described in Sect. 6.

3.2   MultVAE with Focal Loss (FLVAE)

Consistently with any other variational autoencoder [11], the MultVae model's generative process starts by sampling a k-dimensional latent representation $\mathbf{z}_u$ from a standard Gaussian prior [13]. Then, under an assumption that interaction history $\mathbf{x}_u$ has been drawn from a multinomial distribution, a neural network $f_\theta(\cdot)$ is used to produce a probability distribution $\pi(\mathbf{z}_u)$ over I items:

$$\mathbf{z}_u \sim N(0, \mathbf{I}_k),$$
$$\pi(\mathbf{z}_u) \propto exp\{f_\theta(\mathbf{z}_u)\},$$
$$\mathbf{x}_u \sim Mult(\mathbf{z}_u, \pi(\mathbf{z}_u))$$

The variational autoencoder then aims to maximize the average marginal likelihood $p(\mathbf{z}_u|\mathbf{x}_u) = \int p(\mathbf{x}_u|\mathbf{z}_u)p(\mathbf{z}_u)dz$. Since $f_\theta(\cdot)$ is a neural network, $p(\mathbf{z}_u|\mathbf{x}_u)$ becomes intractable and it is approximated with evidence lower bound (ELBO):

$$\log p(\mathbf{x}_u; \theta) \geq \mathbb{E}_q[\log p(\mathbf{x}_u|\mathbf{z}_u) - KL(q(\mathbf{z}_u|\mathbf{x}_u)||p(\mathbf{z}_u))] \tag{2}$$

where $q(\mathbf{x}_u; \phi)$ is a variational approximation of the posterior distribution, $p(\mathbf{x}_u; \theta)$ is the prior distribution, ϕ and θ are parameters of $p(\mathbf{x}_u|\mathbf{z}_u)$, $\log p(\mathbf{x}_u|\mathbf{z}_u)$ is the log-likelihood for user u and KL is the Kullback-Leibler divergence.

FL [14] is defined as

$$FL(p_t) = -\alpha_t(1 - p_t)^\gamma log(p_t), \tag{3}$$

where p_t is:

$$p_t = \begin{cases} \hat{x}_{ui} & if \ x_{ui} = 1 \\ 1 - \hat{x}_{ui} & otherwise \end{cases} \tag{4}$$

and α_t, γ act as hyperparameters.

Since maximising log-likelihood is the same as minimising cross-entropy and focal loss can be understood as a form of weighted cross-entropy, ELBO (2) can be easily rewritten:

$$\log p(\mathbf{x}_u; \theta) \geq \mathbb{E}_q[\alpha_t(1 - p(\mathbf{x}_u|\mathbf{z}_u)^\gamma \log p(\mathbf{x}_u|\mathbf{z}_u) - \\ -KL(q(\mathbf{z}_u|\mathbf{x}_u)||p(\mathbf{z}_u))] \tag{5}$$

3.3 VASP

Recommender model m can be expressed as a function $m(\cdot) : \mathbf{x}_u \to \hat{\mathbf{x}}_u$. If m uses a sigmoid function on the output, it's obvious that $\hat{x}_{uI} \in \ <0,1>$. We propose joint-learning with Hadamard product [15], denoted $\odot$ for combining any $n \in \mathbb{N}$ of recommender models m_n as

$$m_n(\mathbf{x}_u) = \bigodot_{j=1}^n m_j = m_1(\mathbf{x}_u) \odot m_2(\mathbf{x}_u) \odot ... \odot m_n(\mathbf{x}_u) \tag{6}$$

since $m_n(\mathbf{x}_u) = \hat{\mathbf{x}}_{nu}$ and $\mathbf{x}_u$ is the same for all models in combination, (6) can be directly rewritten as

$$m_n(\mathbf{x}_u) = \hat{\mathbf{x}}_{nu} = \bigodot_{j=1}^n \hat{x}_{ju} \tag{7}$$

while

$$\hat{x}_{nu} \in \langle 0, 1 \rangle. \tag{8}$$

While in Wide & Deep [2], networks are combined with the summation (logical OR), in VASP we use the Hadamard product (logical AND), meaning that both networks have to agree. In [2], activation of a single network is sufficient for positive output.

Our proposed VASP architecture (Fig. 1.) uses a combination of two models, NEASE and FLVAE ensembled by element-wise multiplication (7):

$$m_{VASP}(\mathbf{x}_u) = m_{FLVAE}(\mathbf{x}_u) \odot m_{EASE}(\mathbf{x}_u)$$
$$= \hat{\mathbf{x}}_{FLVAEu} \odot \hat{\mathbf{x}}_{NEASEu} \qquad (9)$$

To satisfy condition (8), we add sigmoid function to (1):

$$\hat{\mathbf{x}}_{NEASEu} = \sigma(W \cdot \mathbf{x}_u) \qquad (10)$$

Since m_{FLVAE} and m_{EASE} are both fully differentiable, m_{VASP} is also fully differentiable and backpropagation can be used to optimize the m_{VASP}.

3.4 Data Augmentation to Prevent Learning Identity

Inspired by [24], we prevent learning identity by splitting the input interactions x_u before every training epoch randomly into two parts, $\mathbf{x}_{Au}$ and $\mathbf{x}_{Bu}$ so:

$$x_{Aui} = \begin{cases} 0 & if\ x_{ui} = 0 \\ 1 - x_{Bui} & otherwise \end{cases}$$
$$x_{Bui} = \begin{cases} 0 & if\ x_{ui} = 0 \\ 1 - x_{Aui} & otherwise \end{cases} \qquad (11)$$

while

$$\sum_{i=1}^{I} x_{Aui} \approx \sum_{i=1}^{I} x_{Bui} \qquad (12)$$

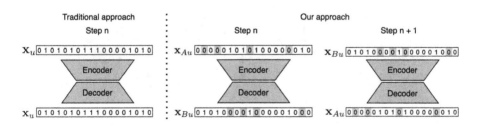

Fig. 2. Data augmentation to prevent overfitting towards identity

The autoencoder is learning x_{Bui} by showing x_{Aui} in one training step and then x_{Aui} by showing x_{Bui} in another. Thus autoencoder is still seeing all the data, but does not see the identity and cannot learn it (as demonstrated by Fig. 2).

4 Interpreting VASP

In order to analyze the effect of shallow and deep components of the proposed ensemble model, we have implemented a workflow to produce a visualization of movie embeddings learned by individual models and the ensemble.

The principle was the following. We performed a sensitivity analysis of models by putting one-hot vectors to the input and generating output probabilities (reconstructions). These probabilities were then transformed to distances by linear scaling and then projected into a t-SNE plot. See Fig. 3 for further details.

5 Experimental Setup

To verify our assumptions by experiments, we implemented three models: neural variant of EASE, deep Variational Autoencoder, and then VASP: joint learning model consisting of NEASE and deep Variational Autoencoder ensembled by the Hadamard product as described in Sect. 3.3. We trained those models on two datasets: MovieLens20M and Netflix prize dataset, and compared results over various baselines, including current state-of-the-art models.

5.1 Datasets

MovieLens20M [6]. Dataset of 27000 movies rated by 138,000 users generating 20 million ratings in total. We preprocessed the dataset according to [13]: Since this dataset contains explicit ratings, we converted the data to implicit interactions by considering a valid interaction only a rating of four or higher[2]. Only users with five or more interactions remain in the dataset after preprocessing. We randomly choose 10000 users as a test set and train our models on the rest.

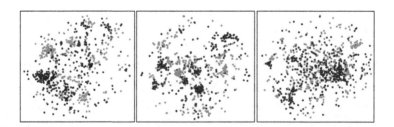

Fig. 3. Explaining VASP on MovieLens20M Dataset: Output of the joint model (left) was linearly decomposed to EASE component (middle) and FLVAE component (right) to demonstrate that EASE is learning more apparent linear dependencies and FLVAE the non-linear ones. Red = Horrors, Blue = Children movies, Green = Western movies and Yellow = Noir. (Color figure online)

---

[2] Note that we use implicit interactions because it is prevalent case for practical recommendation tasks.

Netflix Prize Dataset [1]. Dataset from Netflix prize - over 100 million ratings from 480000 randomly-chosen, anonymous Netflix customers over 17000 movie titles. We converted explicit ratings to implicit interactions by the same method as we used on the MovieLens20M. We randomly choose 40000 users as a test set and train our models on the rest.

5.2   Metrics

We evaluate our models in the same way as in [13]. First, we sample 80% of the test user's interactions as input for the model, and then we measure $Recall@k$ and $NDCG@k$ for predicted interactions against the remaining 20% of the user's interactions.

$NDCG$ for TOP-k recommended items, denoted $NDCG@k$ is defined as

$$NDCG@k = \frac{DCG@k}{IDCG@k} = \frac{\sum_{i=1}^{k} \frac{2^{rel_i}-1}{log_2(i+1)}}{\sum_{i=1}^{|R_k|} \frac{2^{rel_i}-1}{log_2(i+1)}} \tag{13}$$

where rel_i is relevance of the recommendation at position i and R_k is the list of those 20% interactions acting as "true" user interactions.

$Recall$ for TOP-k recommended items, denoted $Recall@k$ is defined as

$$Recall@k = \frac{|\hat{R}_k \cap R_k|}{|R_k|} \tag{14}$$

where $\hat{R}_k$ is the set of top-k items recommended by evaluated model.

5.3   Baselines

We chose several autoencoder-based models for a top-n recommendation, including MultDAE and MultVAE from [13], EASE from [20] and current state-of-the-art models, RecVAE [19] and H+VAMP [10] as baselines for performance evaluation.

5.4   Implemented Models

Neural EASE was implemented as a dense layer with forced zeros on diagonal by kernel constraint. We evaluate three different loss functions: mean squared error, cosine proximity loss and focal loss. Since EASE has forced zeros on the diagonal in the parameters matrix to prevent learning identity, no data augmentation was used.

Variational Autoencoder we implemented FLVAE with a densely connected residual network both in encoder and decoder. Sigmoid activation was used on the output.

Data augmentation described in Sect. 3.4 was used to prevent autoencoder in learning identity.

Default values of $\alpha_t = 0.25$ and $\gamma = 2.0$ in (3) was used.

VASP is build by connecting neural EASE and FLVAE by the Hadamard product as described in Sect. 3.3. Hyperparameters was the same as for plain FLVAE. We evaluated three variants: pre-trained EASE and FLVAE joined together as ensemble, jointly training form the start and alternating approach where FLVAE and EASE was training in every step separatedly. We prevent our model to learn identity by using data augmentation described in Sect. 3.4.

5.5 Hyperparameters

We used 2048 units for latent space and 4096 units for hidden layers. We used seven densely connected residual hidden layers for the encoder and five layers of the same architecture in the decoder. We trained our model for 50 epochs with a learning rate of 0.00005 and batches of 1024 samples. Then, we lower the learning rate to 0.00001 and trained for another 20 epochs. Then we performed finetuning with a learning rate of 0.000001 for another 20 epochs.

All models were implemented in Tensorflow [5] and the source code with notes for reproducing the results is publicly available on our GitHub page[3].

6   Results and Discussion

Neural EASE: We evaluated three different variants of the model based on the loss function used - mean squared error, cosine loss, and focal loss (see Table 1). We found that using cosine proximity loss leads to better performance of the model as authors of [20] expected.

FLVAE: Authors of MultVAE reject deeper architectures by stating that "going deeper does not improve performance." We investigated the matter and believed that overfitting towards identity was to blame. We address this issue by adopting data augmentation described in the Sect. 3.4. This approach successfully allowed us to build much bigger models than in [13,19] or [10] where 200 units for latent space and 600 in hidden layers were used.

Table 1. Results with different loss functions for the EASE model on MovieLens20M dataset.

Loss function used	NDCG@100	Recall@20	Recall@50
MSE	0.425	0.393	0.523
Cosine proximity	**0.431**	**0.403**	**0.532**
Focal loss	0.377	0.343	0.426

VASP: We evaluated three methods of training models connected by the Hadamard product (see Table 2).

---

[3] https://github.com/zombak79/vasp.

Table 2. Results with different training approach for the VASP model on Movie-Lens20M dataset.

Training approach	NDCG@100	Recall@20	Recall@50
Pretrained ensemble	0.442	0.414	0.545
Alternating training	0.436	0.401	0.543
Joint learning	**0.448**	**0.414**	**0.552**

First, we connected pre-trained models and evaluated them as an ensemble. Then we initialized the joint model and trained it from scratch. Lastly, we experimented with the alternating approach, where in one step was frozen weights of FLVAE, and in the next step, we weights of the EASE model were frozen instead. However, this approach did not perform better than joint-learning from the start.

Finally, we have compared our base models NEASE, FLVAE and jointy learned ensemble VASP to state of the art approaches (see Table 3). Significantly best performing models are in bold. Our VASP outperformed other models and achieved the state-of-the-art for MovieLens 20M dataset. It also performed quite well for the Netflix dataset (second highest ranking model).

In our future experiments, we will carefully analyse the H+Vamp Gated model that performed better on Netflix. We will try to put it in the ensemble with the Neural EASE or use the idea of Variational Mixture of Posteriors to improve performance of our deep FLVAE model.

Table 3. Results

	MovieLens 20M			Netflix Prize Dataset		
	NDCG@100	Recall@20	Recall@50	NDCG@100	Recall@20	Recall@50
Mult-DAE	0.419	0.387	0.524	0.380	0.344	0.438
Mult-VAE	0.426	0.395	0.537	0.386	0.351	0.444
EASE	0.420	0.391	0.521	0.393	0.362	0.445
RecVAE	0.442	**0.414**	**0.553**	0.394	0.361	0.452
H+Vamp Gated	0.445	**0.413**	0.551	**0.409**	**0.376**	**0.463**
Neural EASE	0.431	0.403	0.532	0.395	0.363	0.447
FLVAE	0.445	0.409	0.547	0.398	0.363	0.450
VASP	**0.448**	**0.414**	**0.552**	0.406	0.372	0.457

7   Conclusion

We proved EASE to be a compelling Top-N recommendation model that can still match current state-of-the-art baselines.

We proposed a data augmentation method to prevent overfitting to identity and experimentally proved that using this method leads to better performance of autoencoders used for top-n recommendation.

We proposed a novel joint-learning technique for training multiple models together. Using that we constructed VASP, Variational Autoencoder with parallel Shalow Path and experimentally proved, that variational autoencoder connected with parallel simple shallow linear model can match current sophisticated state-of-the-art models and even outperform them in some cases.

Acknowledgements. Our research has been supported by the Grant Agency of the Czech Technical University in Prague (SGS20/213/OHK3/3T/18), the Czech Science Foundation (GAČR 18-18080S), Recombee and VUSTE-APIS.

References

1. Bennett, J., Lanning, S.: The Netflix Prize. In: KDD Cup and Workshop (2007)
2. Cheng, H.-T., et al.: Wide & deep learning for recommender systems. In: RecSys 2017 - Proceedings of the 11th ACM Conference on Recommender Systems, pp. 396–397, June 2016
3. Cremonesi, P., Koren, Y., Turrin, R.: Performance of recommender algorithms on top-N recommendation tasks. In: RecSys 2010 - Proceedings of the 4th ACM Conference on Recommender Systems (2010)
4. Drif, A., Zerrad, H.E., Cherifi, H.: Ensvae: ensemble variational autoencoders for recommendations. IEEE Access **8**, 188335–188351 (2020)
5. Abadi, M., et al.: TensorFlow: large-scale machine learning on heterogeneous systems (2015). Software available from tensorflow.org
6. Maxwell Harper, F., Konstan, J.A.: The MovieLens datasets: history and context. ACM Trans. Interact. Intell. Syst. **5**(4) (2015)
7. He, K., Zhang, X., Ren, S., Sun, J.: Deep residual learning for image recognition. In: Proceedings of the IEEE Computer Society Conference on Computer Vision and Pattern Recognition, pp. 770–778, December 2016
8. Huang, G., Liu, Z., Van Der Maaten, L., Weinberger, K.Q.: Densely connected convolutional networks. In: Proceedings - 30th IEEE Conference on Computer Vision and Pattern Recognition, CVPR 2017, pp. 2261–2269, January 2017
9. Karypis, G.: Evaluation of item-based top-n recommendation algorithms. In: Proceedings of the Tenth International Conference on Information and Knowledge Management, pp. 247–254 (2001)
10. Kim, D., Suh, B.: Enhancing VAEs for collaborative filtering: flexible priors & gating mechanisms. In: RecSys 2019–13th ACM Conference on Recommender Systems (2019)
11. Kingma, D.P., Welling, M.: Auto-encoding variational Bayes. In: 2nd International Conference on Learning Representations, ICLR 2014 - Conference Track Proceedings (2014)
12. Koren, Y., Bell, R., Volinsky, C.: Matrix factorization techniques for recommender systems. Computer (2009)
13. Liang, D., Krishnan, R.G., Hoffman, M.D. Jebara, T.: Variational autoencoders for collaborative filtering. In: The Web Conference 2018 - Proceedings of the World Wide Web Conference, WWW 2018 (2018)

14. Lin, T.Y., Goyal, P., Girshick, R., He, K., Dollar, P.: Focal Loss for Dense Object Detection. IEEE Trans. Pattern Anal. Mach. Intell. **42**(2), 318–327 (2020)
15. Million, E.: The hadamard product. Creative commons (2007)
16. Ng, A., et al.: Sparse autoencoder. CS294A Lecture note **72**(2011), 1–19 (2011)
17. Rehorek, T., Kordik, P., et al.: Comparing offline and online evaluation results of recommender systems. In: In Proceedings of the REVEAL workshop at RecSyS conference (RecSyS 2018) (2018)
18. Sedhain, S., Menon, A.K., Sanner, S., Xie, L.: AutoRec. In: Proceedings of the 24th International Conference on World Wide Web - WWW 2015 Companion, pp. 111–112. ACM Press, New York (2015)
19. Shenbin, I., Alekseev, A., Tutubalina, E., Malykh, V., Nikolenko, S.I.: RecVAE: a new variational autoencoder for top-n recommendations with implicit feedback. In: WSDM 2020 - Proceedings of the 13th International Conference on Web Search and Data Mining, pp. 528–536. Association for Computing Machinery Inc., January 2020
20. Steck, H.: Embarrassingly shallow autoencoders for sparse data. In: The Web Conference 2019 - Proceedings of the World Wide Web Conference, WWW 2019 (2019)
21. Steck, H.: Autoencoders that don't overfit towards the identity. In: Advances in Neural Information Processing Systems, vol. 33 (2020)
22. Symeonidis, P., Zioupos, A.: Matrix and Tensor Factorization Techniques for Recommender Systems (2016)
23. Vincent, P., Larochelle, H., Bengio, Y., Manzagol, P.-A.: Extracting and composing robust features with denoising autoencoders. In: Proceedings of the 25th International Conference on Machine Learning, pp. 1096–1103 (2008)
24. Zhang, R., Isola, P., Efros, A.A.: Split-brain autoencoders: unsupervised learning by cross-channel prediction. In: Proceedings of the IEEE Conference on Computer Vision and Pattern Recognition, pp. 1058–1067 (2017)
25. Zhang, S., Yao, L., Sun, A., Tay, Y.: Deep learning based recommender system: a survey and new perspectives. ACM Comput. Surv. **52**(1) (2019)

Short Text Clustering with a Deep Multi-embedded Self-supervised Model

Kai Zhang[1,2(✉)], Zheng Lian[1,2], Jiangmeng Li[1,2], Haichang Li[1], and Xiaohui Hu[1]

[1] Institute of Software Chinese Academy of Sciences, Beijing 100190, China
{zhangkai2020c,lianzheng2017,jiangmeng2019,haichang,hxh}@iscas.ac.cn
[2] University of Chinese Academy of Sciences, Beijing 100049, China

Abstract. Short text clustering is challenging in the field of Natural Language Processing (NLP) since it is hard to learn the discriminative representations with limited information. In this paper, fused multi-embedded features are employed to enhance the representations of short texts. Then, a denoising autoencoder with an attention layer is adopted to extract low-dimensional features from the multi-embeddings against the disturbance of noisy texts. Furthermore, we propose a novel distribution estimation with jointly utilizing soft cluster assignment and the prior target distribution transition to better fine-tune the encoder. Combining the above work, we propose a deep multi-embedded self-supervised model(DMESSM) for short text clustering. We compare our DMESSM with the state-of-the-art methods in head-to-head comparisons on benchmark datasets, which indicates that our method outperforms them.

Keywords: Short text clustering · Autoencoder · Self-supervised clustering · Attention · Distribution estimation

1 Introduction

In recent years, with the rapid development of social media, e.g. Twitter, short texts are generated in large volumes which provide research resources for data mining tasks such as topic discovery, association analysis and intelligent recommendation. Short text clustering(STC) is the foundation of the above tasks and aims to group semantically similar short texts without supervision.

There are currently many technologies for STC. Traditionally, short texts are first represented as vectors by Term Frequency-Inverse Document Frequency (TF-IDF) or bag-of-words(BOW) models. Then a clustering algorithm is applied to divide them into different groups. Because of the short length of texts, the aforementioned vector representations are sparse, which may lead to poor clustering performance. In response to this problem, expanding the texts with external resources can enrich the representations. Another way is to use some topic models such as BTM [2] and GSDMM [20]. Recently, neural networks are widely used in STC and perform well. Using neural network embedding, the high-

---

The original version of this chapter was revised: The affiliation of three co-authors has been corrected. The correction to this chapter is available at
https://doi.org/10.1007/978-3-030-86383-8_55

I. Farkaš et al. (Eds.): ICANN 2021, LNCS 12895, pp. 150–161, 2021.
https://doi.org/10.1007/978-3-030-86383-8_12

dimensional data are embedded into a lower-dimensional space, which can reduce sparsity. Moreover, embedding and clustering can be jointly optimized.

We focus on the STC based on neural networks. Current word embedding methods in clustering are generally static such as word2vec [11] and Glove [13], which do not capture the semantics of words in their current context. However, pre-trained models with the attention mechanism can dynamically learn the embedding based on context. As for joint deep embedding and clustering with a neural network, autoencoder is commonly used whose anti-interference ability and feature extraction ability are limited. Besides, we find that the target distribution in DEC [18] may change the affiliations of the samples much compared with the soft assignment from the network.

In response to the above problems, this paper proposes a deep multi-embedded self-supervised model, referred as DMESSM. The main contributions are as follows: (a) Static embedding and dynamic embedding are fused to form muti-embeddings, which can express short texts better. (b) We use a denoising autoencoder with an attention block to learn features. Thus the model can reduce the interference of noisy texts and output the important low-dimensional features for following clustering. (c) We propose a new target distribution. Besides enhancing the clustering, it can preserve the order of soft assignment more than before, which brings better clustering performance. (d) We conduct experiments on four short text datasets. The experimental results demonstrate that our DMESSM is better than the state-of-the-art models in terms of accuracy.

2 Related Work

There are two major directions in STC now, which are using topic models to cluster directly and applying cluster algorithms on text representation. Several methods have been proposed in the literature. For topic models, Biterm topic modeling (BTM) learns topics from word co-occurrence patterns (i.e., biterms). GSDMM was proposed to extend dirichlet multinomial mixture model with collapsed gibbs sampling. For text representation and cluster algorithms, the term frequency-inverse document frequency (TF-IDF) or word embeddings [11,13] can express short texts. And an external knowledge resource called BabelNet [12] can be used to add more features. K-means [9] is a common clustering method.

In addition, deep learning can be used in text representation and clustering. Xu et al. (2015) [19] uses word2vec vector to represent the original text, and encode the vector with CNN, whose fitting target is the pre-train binary code B from word2vec vector based on the keyword features with a locality-preserving constraint, then applies k-means on the CNN encoder values. Xie et al. (2016) [18] uses autoencoder to reduce the word embedding's dimensionality, and proposed a joint optimization frame of deep embedding and clustering. Hadifar et al. (2019) [5] uses Smooth Inverse Frequency(SIF) embeddings [1] in order to improve the word embeddings and make clustering more efficient. Rakib et al. (2020) [15] proposes the iterative classification to enhance the clustering.

During optimizing the deep clustering model, minimizing the distance of clustering probability distribution and an auxiliary target distribution is an effective way [18]. Matten and Hinton (2008) [8] propose the Student's t-distribution. The

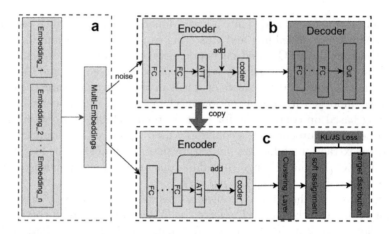

Fig. 1. An overview of DMESSM structure. (a) Combine many different embeddings into a multi-embeddings to express short texts. (b) Pretrain a denoising autoencoder. (c) Self-supervised clustering.

Kullback–Leibler divergence [7], is a measure of how one probability distribution is different from a second, reference probability distribution. Based on KL divergence , the Jensen–Shannon divergence [10] can also measure the similarity between two probability distributions.

The methods of embeddings have further developed with the emergence of pre-trained models such as BERT [4]. Before, static embeddings can not get the semantics of a word in the current context and the word order information. Using the pre-trained model based on transformers can get dynamic embeddings to improve the above things. SentenceBERT [16] is a Siamese bert-networks whose task is sentence pairs' text semantic similarity, which can generate sentence embedding. Poerner et al. (2020) [14] ensembles diverse pre-trained sentence encoders into sentence meta-embeddings.

3   Methods

We use three parts to achieve short text clustering which are multi-dimension embeddings, autoencoder and iterative clustering. Below, each part will be explained detailly how it works. The described setup is illustrated in Fig. 1.

3.1   Multi-Embeddings

Short texts can be embedded in different ways to obtain features that have different information. So we combine different embeddings to get the final multi-embeddings, which have more plentiful information. We assume that the different embedding method are E_1, E_2, ..., E_n, and the final embedding is M. Then we need to determine suitable E and a way to combine them into M.

Word2vec is a common method of word embeddings, which are functions that map word types to vectors. Then, Smooth Inverse Frequency(SIF) is a better way to get the sentence embeddings. First, the weight of each word is calculated as $\frac{a}{a+p(w)}$ with a being a hyperparameter and $p(w)$ being the empirical word frequency in the text corpus. Second, computing the first principal component of all the resulting vectors and removing it from the weighted embeddings to get SIF embeddings. It can be used simply as a static method.

Pre-trained models using transformer-networks and appropriate pre-training tasks can obtain dynamic embeddings. Besides, adding location information in the model training will make the word vector express order information. When using this way, we feed the current short text into the pre-trained model to get proper embeddings.

After obtaining source embeddings, we combine them to get a multi-dimension embedding. We use naive methods by just concatenating [21] or averaging [3] the source embeddings. Each embedding can be normalized to minimize the impact of the dimensional differences during the combination.

$$M^{conc}(s) = \begin{bmatrix} E_1(s) \\ E_2(s) \\ \dots \\ E_n(s) \end{bmatrix} \tag{1}$$

$$M^{avg}(s) = \sum_i \frac{E_i(s)}{N} \tag{2}$$

Singular Value Decomposition (SVD) [21] is a good way to compactify concatenated the source embeddings. Let $\mathbf{X}^{conc} \in \mathbb{R}^{|S| \times \sum_j d_j}$ with

$$\mathbf{x}_n^{conc} = M^{conc}(s_n) - \mathbb{E}_{s \in S}[M^{conc}(s)] \tag{3}$$

Let $\mathbf{USV}^T \approx \mathbf{X}^{conc}$ be the d-truncated SVD. The SVD embedding is:

$$M^{svd}(s) = \mathbf{V}^T(M^{conc}(s) - \mathbb{E}_{s' \in S}[M^{conc}(s')]) \tag{4}$$

3.2   Denoising Stacked Autoencoder

After multi-embeddings, we have a M to represent the short texts whose dimension is always high. Autoencoder [6] can convert high-dimensional data to low-dimensional codes by training a multilayer neural network with a small central layer to reconstruct high-dimensional input networks.

We use a denoising stacked autoencoder [17] with an attention layer in middle. The denoising autoencoder adds noise to the input data based on the Autoencoder, so that the learned encoder has strong robustness, thereby enhancing the generalization ability of the model. The attention mechanism is a powerful way to extract features that can improve important features and suppress unimportant features. We give a shallow model as an example in Fig. 2.

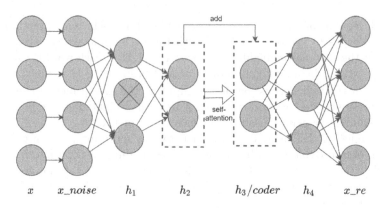

Fig. 2. Denoising stacked autoencoder with an attention layer

Given a multi-embeddings $x \in \mathbb{R}^{d \times n}$, where d is the dimensionality of the text features and n means the number of input data batch size. we add White Gaussian Noise(WGN) to get x_noise as the network input. The encoder which includes a feedforward neural network and a self-attention layer maps x_noise to a low-dimensional representation *coder*. Dropout is applied in the encoder to prevent overfitting. Then the decoder reconstructs an input x_re. We use Leaky ReLU as the activation function for all layers. Following [18] and [5], we choose the Mean Squared Error (MSE) as the loss function between x and x_re.

3.3 Self-supervised Iterative Clustering

After the first two parts, we get a relatively low-dimensional vector to express short text. We add a clustering layer whose parameters are the cluster centroids to the encoder from the trained autoencoder model. Then we do iterative clustering to get final results. Our model is self-training which has 3 steps.

First, we compute a soft assignment between the embedded points and cluster centroids. Following van der Matten and Hinton (2008) [8], we use the Student's t-distribution Q with a single degree of freedom as a kernel to measure the similarity between embedded points z_i and centroids μ_j:

$$q_{ij} = \frac{(1 + \|z_i - \mu_j\|^2)^{-1}}{\sum_j (1 + \|z_i - \mu_j\|^2)^{-1}} \tag{5}$$

where $z_i = f(x)$ corresponds to x after multi-embedded, q_{ij} can be regarded as the probability of assigning sample i to cluster j.

Second, design an auxiliary target distribution. The target distribution is critical to the final performance. Xie et al. proposed a target distribution that can put more emphasis on data points assigned with high confidence and normalize the loss contribution of each centroid. The target distribution is

$$p_{ij} = \frac{q_{ij}^2/f_j}{\sum_{j'} q_{ij'}^2/f_{j'}} \tag{6}$$

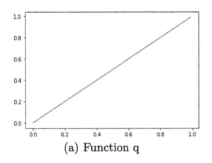
(a) Function q

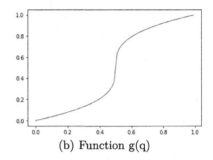
(b) Function g(q)

Fig. 3. Adjusting function

where $f_j = \sum_i q_{ij}$. This design has a problem that the value of probabilities may change the order in target distribution, causing the change of affiliation of clusters. In other words, an embedded point z' which belongs to cluster u_m in initial distribution Q would belong to cluster $u_n (m \neq n)$ in target distribution P. Assume that $q_{im} > q_{in}$, then $p_{im}/p_{in} = (\frac{q_{im}}{q_{in}})^2 \frac{f_n}{f_m}$, if there are many samples belonging to cluster n, f_n may be much larger than f_m, leading to $p_{im} > p_{in}$. Intuitively, increasing the value of $\frac{q_{im}}{q_{in}}$ appropriately can keep the order. Our work proposes that the target distribution's primary goal is putting more emphasis on data points assigned with high confidence and the order of probabilities should remain. Then we report an adjust function g. Specifically, it should have the characteristics: (1) the value ranges from 0 to 1. (2) The value in the middle of the function changes a lot and remains at the both ends, which is a s-shaped function. It can preserve the low probabilities lower, the high probabilities higher. (3) Because of $\sum_j q_{ij} = 1$, every value of q_{ij} is usually small. Thus, for low values of q_{ij}, if $q_{im} > q_{in}$, we make $g(q_{im}) > g(q_{in})$ to keep the order. For example, when $q_{11} = 0.05$ and $q_{21} = 0.1$, $\frac{g(q_{21})}{g(q_{11})} > \frac{q_{21}}{q_{11}}$. There are many functions that can achieve the goal, e.g. the power function. In this paper, we choose the following function and show it in Fig. 3:

$$g(q) = \frac{\sqrt[3]{2q-1}+1}{2} \tag{7}$$

The final target distribution is:

$$p_{ij} = \frac{g^2(q_{ij})/f_j}{\sum_{j'} g^2(q_{ij'})/f_{j'}} \tag{8}$$

The effectiveness of the adjusting function is detailed in Sect. 4.5.

Third, construct a loss function between two distributions. Xie et al. (2016) and Hadifar et al. (2019) both use the KL-divergence as a training objective, i.e. the training loss L is defined as:

$$KL(P\|Q) = \sum_i \sum_j p_{ij} \log \frac{p_{ij}}{q_{ij}} \tag{9}$$

Based on KL divergence, the JS divergence is symmetric, and its value ranges from 0 to 1, which is:

$$JS(P\|Q) = \frac{1}{2}KL(P\|\frac{P+Q}{2}) + \frac{1}{2}KL(Q\|\frac{P+Q}{2}) \tag{10}$$

Our work both uses KL divergence and JS divergence as loss functions to do clustering. Generally, JS divergence performs better.

4   Experiments

4.1   Dataset

We evaluate our model on some datasets for short text clustering:

StackOverflow. Published in Kaggle.com, this dataset have 3370528 samples from question and answer site stackoverflow. Same as Xu et al. (2017), we use 20000 question titles of 20 categories as our test dataset.

SearchSnippets. This dataset include short texts of 8 different topics from Web search snippets.

Tweet89. It consists of 2472 tweets grouped into 89 clusters which are from 2011 and 2012 microblog tracks published by Text REtrieval Conference.

20ngnewsshort. 20ngnews collected about 20,000 newsgroup documents, evenly divided into 20 topics. We choose the texts whose words are less than 30, constituting a short text dataset 20ngnewsshort which has 2029 texts.

4.2   Experimental Setups

For word embeddings, we use pre-trained word2vec vector from Xu et al. (2015) and Google. Then the hypermeter α in SIF is set as 0.1. We also use a pre-trained sentence-bert model to get dynamic sentence vector.

We set encoder's hidden layer size to d:500:500:2000:20 for Stackoverflow, d:500:500:200:200:20 for SearchSnippets, d:500:500:200:200 for Tweet89 and d:500:2000:20 for 20ngnewsshort, where d is the short text embedding dimension. We train the autoencoder for 15 epochs, with the batch size of 64, learning rate of 0.01 and momentum value of 0.9.

The number of clusters is consistent with the actual number in the data set to compare the different methods easily. The choice of centroid will affect the effect of clustering, so we restarted k-means++ 100 times to select the best centroids. During iterative clustering, we train the model by using the dataset 8 times.

4.3   Evaluation Metrics

To measure the performance of our method, we select two widely used metrics, the accuracy(ACC) and the normalized mutual information(NMI) [5,19].

Table 1. Clustering results for 4 short datasets.

Method	Stackoverflow		SearchSnippets		Tweet89		20ngnews	
	ACC	NMI	ACC	NMI	ACC	NMI	ACC	NMI
TF	13.5 ± 2.2	7.8 ± 2.5	24.7 ± 2.2	9.0 ± 2.3	52.9 ± 1.3	70.3 ± 1.7	13.1 ± 2.1	7.4 ± 1.2
TF-IDF	20.3 ± 4.0	15.6 ± 4.7	33.8 ± 3.9	21.4 ± 4.4	53.1 ± 2.3	76.2 ± 3.4	20.7 ± 2.4	18.8 ± 1.6
Word2vec	38.1 ± 2.4	36.5 ± 1.5	67.5 ± 0.1	51.5 ± 0.1	48.9 ± 0.9	77.2 ± 1.5	28.1 ± 0.2	28.4 ± 0.8
SIF	48.5 ± 1.3	45.8 ± 1.6	66.8 ± 0.2	50.6 ± 0.1	49.1 ± 1.2	76.8 ± 0.7	29.1 ± 0.6	30.2 ± 0.7
SBERT	63.2 ± 2.5	60.5 ± 2.2	67.2 ± 0.5	48.5 ± 0.5	51.8 ± 0.7	80.1 ± 1.0	31.3 ± 1.1	31.5 ± 0.6
STCC	51.1 ± 2.9	49.0 ± 1.5	77.0 ± 4.1	62.9 ± 1.7	–	–	–	–
SIF-Auto	59.8 ± 1.9	54.8 ± 1.0	77.1 ± 1.1	56.7 ± 1.0	54.5 ± 3.3	74.6 ± 3.2	28.2 ± 1.8	28.6 ± 1.3
DMESSM	**79.9 ± 0.3**	**70.7 ± 0.2**	**83.3 ± 0.2**	**65.0 ± 0.2**	**77.3 ± 2.2**	**85.8 ± 2.5**	**38.5 ± 0.6**	**37.7 ± 0.4**

Accuracy is defined as:

$$ACC = \frac{\sum_{i=1}^{n} \delta(y_i = map(c_i))}{n} \tag{11}$$

where n is the number of the text, $\delta(p = q)$ is the indicator function that equals one if $p = q$ and equals zero otherwise, c_i is the clustering label of text x_i, y_i is the true group label of text x_i, $map(c_i)$ is a function that maps each cluster label c_i to the equivalent label from the text data by Hungarian algorithm.

Normalized mutual information is defined as:

$$NMI(T, C) = \frac{MI(T, C)}{\sqrt{H(T)H(C)}} \tag{12}$$

where T is the true label sets, C is the predicted assignments, $MI(T, C)$ is the mutual information between T and C, H is the entropy and $\sqrt{H(T)H(C)}$ is used for normalizing the mutual information to be $[0, 1]$.

4.4   Results

Table 1 shows the evaluation results for both the baseline and our model DMESSM[1] on several datasets. We use the experiment results from Xu [19] and Hadifar [5] for Stackoverflow and SearchSnippets. As for the other datasets, we use the code[2] for SIF-Auto and achieve the simple programs to get other metrics.

We select the common feature engineering methods such as TF and TF-IDF. Besides, some sentence-embeddings including average word2vec, SIF and SBERT are used to represent texts. Neural networks and self-training can lead to improved cluster quality, which are shown in recent methods STCC and SIFAuto. Our model has achieved the best results on datasets of different sizes and categories, showing its superiority. For Stackoverflow and SearchSnippets which are widely used, our method both obtains the highest ACC of 79.9%/83.3% and highest NMI of 70.7%/65.0%. Compared with the basic self-training model SIF-Auto, DMESSM

---

[1] Our code is available at https://github.com/zkharryhhhh/DMESSM.
[2] https://github.com/hadifar/stc_clustering.

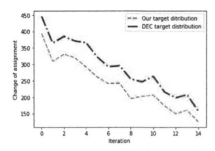 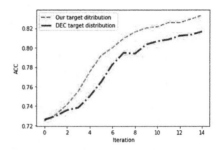

Fig. 4. Performance on change of assignment and accuracy of two distribution.

Table 2. Results by using different embedding methods directly.

Method	Stackoverflow		SearchSnippets		Tweet89		20ngnews	
	ACC	NMI	ACC	NMI	ACC	NMI	ACC	NMI
single:SIF	48.5 ± 1.3	45.8 ± 1.6	66.8 ± 0.2	50.6 ± 0.1	49.1 ± 1.2	76.8 ± 0.7	29.1 ± 0.6	30.2 ± 0.7
single:SBERT	63.2 ± 2.5	60.5 ± 2.2	67.2 ± 0.5	48.5 ± 0.5	51.8 ± 0.7	80.1 ± 1.0	31.3 ± 1.1	31.5 ± 0.6
multi:conc	67.1 ± 0.8	63.0 ± 0.6	**69.1 ± 0.2**	**50.8 ± 0.3**	54.7 ± 1.1	82.0 ± 0.3	**32.3 ± 1.0**	32.5 ± 1.3
multi:avg	67.0 ± 0.5	**63.2 ± 0.5**	68.6 ± 0.5	49.8 ± 0.2	54.4 ± 0.9	82.1 ± 0.5	32.2 ± 0.5	31.9 ± 0.7
multi:svd	**67.2 ± 0.6**	**63.2 ± 0.3**	69.0 ± 0.2	50.0 ± 0.1	**55.9 ± 1.9**	**82.5 ± 0.6**	32.2 ± 0.8	32.3 ± 0.5

achieves large improvement by 20.1%/16.1% and 6.2%/8.3%(ACC/NMI). For the small-scale dataset with many categories Tweet89, our model performs really well, increasing ACC by more than 20 points and NMI more than 10 points. Though the clustering for 20ngnewsshort is generally not good, DMESSM gets the best score in ACC of 38.5% and NMI of 37.7%.

4.5 Effectiveness of Our New Target Distribution

Take the dataset SearchSnippets as an example. We use the change of the assignment and clustering accuracy to compare our target distribution and that of DEC [18] in iterative clustering. The results in Fig. 4 show that our new distribution reduces the change of assignment and increases the accuracy.

4.6 Ablation Study

We conduct further ablation experiments to dissect our model and analysis the effect of the following structures.

Multi-embedding: we choose the two single embedding ways which perform better in clustering, SIF and SBERT. Then we use concatenating, averaging and SVD to get the multi-embeddings, which can significantly improve the clustering. To compare with the SIF embedding in [5], we use the same SIF representation. For SBERT, the values of the vectors are preprocessed and almost range from −1 to 1. And we directly use it without normalization. Table 2 shows that

Table 3. Results on 4 datasets that we use the same multi-embeddings but different network structure to do clustering without the self-training steps. DAE means denoising autoencoder and ATT means an attention layer.

Method	Stackoverflow		SearchSnippets		Tweet89		20ngnews	
	ACC	NMI	ACC	NMI	ACC	NMI	ACC	NMI
DAE	71.1 ± 1.4	61.3 ± 1.8	69.3 ± 2.1	48.5 ± 1.4	55.6 ± 1.1	**82.2 ± 0.8**	31.7 ± 0.4	31.2 ± 0.8
DAE+ATT	**73.8 ± 0.8**	**62.5 ± 1.2**	**72.3 ± 2.5**	**51.0 ± 1.3**	**56.7 ± 0.8**	82.0 ± 0.4	**32.2 ± 0.7**	**35.7 ± 0.3**

Table 4. Results with different target distribution and loss function. TD means the previous target distribution, adjustTD means our target distribution.

Method	Stackoverflow		SearchSnippets		Tweet89		20ngnews	
	ACC	NMI	ACC	NMI	ACC	NMI	ACC	NMI
TD+KL	79.1 ± 0.4	70.5 ± 0.3	78.2 ± 1.1	59.5 ± 0.4	71.3 ± 2.3	83.7 ± 0.8	36.0 ± 0.2	36.2 ± 0.3
adjustTD+KL	**79.9 ± 0.3**	**70.7 ± 0.2**	80.4 ± 0.5	61.1 ± 0.6	73.7 ± 1.5	84.9 ± 0.5	**38.5 ± 0.6**	**37.7 ± 0.4**
adjustTD+JS	79.2 ± 0.3	70.3 ± 0.2	**83.3 ± 0.2**	**65.0 ± 0.2**	**77.3 ± 2.2**	**85.8 ± 2.5**	36.2 ± 1.2	37.1 ± 0.4

almost all the multi-embeddings outperform their source embedding. The multi-embeddings can improve the clustering results by 1% to 4% compared with the best single source embedding.

Network Structure: while autoencoder is a common network to extract the features, we try to add an attention layer. For every dataset, we first train a denoising autoencoder to get the learned representations that are directly used for clustering. Then we use the same experimental setups but adding an attention layer. Table 3 shows that our method performs better except for NMI in Tweet89, improving the metrics by around 2%. More specifically, with the attention layer, ACCs in 4 datasets are improved by 2.7%/3.0%/1.1%/0.5% and NMIs in Stackoverflow/SearchSnippets/20ngnews are improved by 1.2%/2.5%/4.5%.

Clustering with New Distribution: using the same pre-trained model, we compare the target distribution in DEC and our adjusted target distribution in every dataset. As the results shown in Table 4, our target distribution is more effective than before. JS divergence and KL divergence are available and the former can make the clustering results better in most of the datasets.

4.7   Discussion

Our method significantly improves the effect of using a self-supervised model to do short text clustering and three stages in our method have been shown better than before. But there are also some disadvantages. The pre-trained model can seriously influence the final clustering results and the model trained by the AE or DAE is not very stable every time. Sometimes the clustering results using the features from the pre-trained model are even worse than that using the multi-embeddings directly. So we may change the network layers to get a relatively better pre-trained model. We also do the experiments using the same network layers such as d:500:500:2000:20, but the pre-trained model is not good on small-scale datasets. In addition, our clustering stage can always improve the clustering. Another thing to note is the chosen of loss function for denoising autoencoder. Mean squared error is a common choice in regression problems and is widely used in autoencoder. Besides, we randomly choose a dataset and conduct experiments with different loss functions, including mean squared error, mean absolute error, binary cross-entropy and mean squared logarithmic error. The denoising autoencoder with MSE can get the best representation for clustering.

5   Conclusion

In this paper, we proposed a method of multi-embeddings, added an attention mechanism into a dinoising autoencoder and improved the target distribution of short text clustering based on a joint deep embedded clustering frame. Our model DMESSM starts from an unsupervised method using SIF embeddings and sentence-bert embeddings, then does iterative clustering by using a denoising autoencoder and a clustering layer. The experimental study shows that our model can reach the most advanced level on multiple datasets.

Acknowledgments. This work was supported by the National Key Research and Development Program of China under Grant 2019YFB1405100, and the National Natural Science Foundation of China under Grants 61802380 and 62076232.

References

1. Arora, S., Liang, Y., Ma, T.: A simple but tough-to-beat baseline for sentence embeddings. In: 5th International Conference on Learning Representations, ICLR 2017 (2017)
2. Cheng, X., Yan, X., Lan, Y., Guo, J.: BTM: topic modeling over short texts. IEEE Trans. Knowl. Data Eng. **26**(12), 2928–2941 (2014)
3. Coates, J., Bollegala, D.: Frustratingly easy meta-embedding-computing meta-embeddings by averaging source word embeddings. arXiv preprint arXiv:1804.05262 (2018)
4. Devlin, J., Chang, M.W., Lee, K., Toutanova, K.: Bert: pre-training of deep bidirectional transformers for language understanding. arXiv preprint arXiv:1810.04805 (2018)
5. Hadifar, A., Sterckx, L., Demeester, T., Develder, C.: A self-training approach for short text clustering. In: Proceedings of the 4th Workshop on Representation Learning for NLP (RepL4NLP-2019), pp. 194–199 (2019)
6. Hinton, G.E., Salakhutdinov, R.R.: Reducing the dimensionality of data with neural networks. Science **313**(5786), 504–507 (2006)
7. Kullback, S., Leibler, R.A.: On information and sufficiency. Ann. Math. Stat. **22**(1), 79–86 (1951)
8. Van der Maaten, L., Hinton, G.: Visualizing data using t-SNE. J. Mach. Learn. Res. **9**(11) (2008)
9. MacQueen, J., et al.: Some methods for classification and analysis of multivariate observations. In: Proceedings of the Fifth Berkeley Symposium on Mathematical Statistics and Probability, Oakland, CA, USA, vol. 1, pp. 281–297 (1967)

10. Manning, C., Schutze, H.: Foundations of statistical natural language processing. MIT Press, Cambridge (1999)
11. Mikolov, T., Sutskever, I., Chen, K., Corrado, G., Dean, J.: Distributed representations of words and phrases and their compositionality. arXiv preprint arXiv:1310.4546 (2013)
12. Navigli, R., Ponzetto, S.P.: BabelNet: the automatic construction, evaluation and application of a wide-coverage multilingual semantic network. Artif. Intell. **193**, 217–250 (2012)
13. Pennington, J., Socher, R., Manning, C.D.: GloVe: global vectors for word representation. In: Proceedings of the 2014 Conference on Empirical Methods in Natural Language Processing (EMNLP), pp. 1532–1543 (2014)
14. Poerner, N., Waltinger, U., Schütze, H.: Sentence meta-embeddings for unsupervised semantic textual similarity. In: Proceedings of the 58th Annual Meeting of the Association for Computational Linguistics, pp. 7027–7034. ACL (2020)
15. Rakib, M.R.H., Zeh, N., Jankowska, M., Milios, E.: Enhancement of short text clustering by iterative classification. In: Métais, E., Meziane, F., Horacek, H., Cimiano, P. (eds.) NLDB 2020. LNCS, vol. 12089, pp. 105–117. Springer, Cham (2020). https://doi.org/10.1007/978-3-030-51310-8_10
16. Reimers, N., et al.: Sentence-BERT: sentence embeddings using Siamese BERT-networks. In: Proceedings of the 2019 Conference on Empirical Methods in Natural Language Processing. Association for Computational Linguistics (2019)
17. Vincent, P., Larochelle, H., Lajoie, I., Bengio, Y., Manzagol, P.A., Bottou, L.: Stacked denoising autoencoders: Learning useful representations in a deep network with a local denoising criterion. J. Mach. Learn. Res. **11**(12) (2010)
18. Xie, J., Girshick, R., Farhadi, A.: Unsupervised deep embedding for clustering analysis. In: International Conference on Machine Learning, pp. 478–487. PMLR (2016)
19. Xu, J., et al.: Short text clustering via convolutional neural networks. In: Proceedings of the 1st Workshop on Vector Space Modeling for Natural Language Processing, pp. 62–69 (2015)
20. Yin, J., Wang, J.: A Dirichlet multinomial mixture model-based approach for short text clustering. In: Proceedings of the 20th ACM SIGKDD International Conference on Knowledge Discovery and Data Mining, pp. 233–242 (2014)
21. Yin, W., Schütze, H.: Learning word meta-embeddings. In: Proceedings of the 54th Annual Meeting of the Association for Computational Linguistics (Volume 1: Long Papers), pp. 1351–1360 (2016)

Brain-Like Approaches to Unsupervised Learning of Hidden Representations - A Comparative Study

Naresh Balaji Ravichandran[1]([✉]), Anders Lansner[1,2]([✉]) [iD],
and Pawel Herman[1]([✉]) [iD]

[1] Computational Brain Science Lab, KTH Royal Institute of Technology,
Stockholm, Sweden
{nbrav,ala,paherman}@kth.se
[2] Department of Mathematics, Stockholm University, Stockholm, Sweden

Abstract. Unsupervised learning of hidden representations has been one of the most vibrant research directions in machine learning in recent years. In this work we study the brain-like Bayesian Confidence Propagating Neural Network (BCPNN) model, recently extended to extract sparse distributed high-dimensional representations. The usefulness and class-dependent separability of the hidden representations when trained on MNIST and Fashion-MNIST datasets is studied using an external linear classifier and compared with other unsupervised learning methods that include restricted Boltzmann machines and autoencoders.

Keywords: Neural networks · Bio-inspired · Hebbian learning · Unsupervised learning · Structural plasticity

1 Introduction

Artificial neural networks have made remarkable progress in supervised pattern recognition in recent years. In particular, deep neural networks have dominated the field largely due to their capability to discover hierarchies of hidden data representations. However, most deep learning methods rely extensively on supervised learning from labeled samples for extracting and tuning data representations. Given the abundance of unlabeled data there is an urgent demand for unsupervised or semi-supervised approaches to learning of hidden representations [1]. Although early concepts of greedy layer-wise pretraining allow for exploiting unlabeled data, ultimately the application of deep pre-trained networks to pattern recognition problems rests on label-dependent end-to-end weight fine tuning

Funding for the work is received from the Swedish e-Science Research Centre (SeRC), European Commission H2020 program, Grant Agreement No. 800999 (SAGE2), and Grant Agreement No. 801039 (EPiGRAM-HS). The simulations were performed on resources provided by Swedish National Infrastructure for Computing (SNIC) at the PDC Center for High Performance Computing, KTH Royal Institute of Technology.

I. Farkaš et al. (Eds.): ICANN 2021, LNCS 12895, pp. 162–173, 2021.
https://doi.org/10.1007/978-3-030-86383-8_13

[5]. At the same time, we observe a surge of interest in more brain plausible networks for unsupervised and semi-supervised learning problems that build on some fundamental principles of neural information processing in the brain [7,8]. Most importantly, these brain-like computing approaches rely on local learning rules and label-independent biologically compatible mechanisms to build data representations whereas deep learning methods predominantly make use of error back-propagation (backprop) for learning the weights. Although backprop as a learning algorithm is highly efficient for finding good representations from data, there are several issues that make it an unlikely candidate model for synaptic plasticity in the brain. The most apparent issue is that the synaptic strength between two biological neurons is expected to comply with Hebb's postulate, i.e. to depend only on the available local information provided by the activities of pre- and postsynaptic neurons. This is violated in backprop since synaptic weight updates need gradient signals to be communicated from distant output layers. Please refer to [14] for a detailed review of possible biologically plausible implementations of and alternatives to backprop.

In this work we utilize the MNIST and Fashion-MNIST datasets to compare two classical representation learning networks, the autoencoder (AE) and the restricted Boltzmann machine (RBM), with two brain-like approaches to unsupervised learning of hidden representations, i.e. the recently proposed model by Krotov and Hopfield (KH) [9], and the BCPNN model [19]. In particular, we qualitatively examine the extracted hidden representations and quantify their class-dependent separability using a simple linear classifier on top of all the networks under investigation. This classification step is not part of the learning strategy, and we use it merely to evaluate the resulting representations.

Special emphasis is on the feedforward BCPNN model with a single hidden layer, which frames the update and learning steps of the neural network as probabilistic computations. BCPNN has previously been used in abstract models of associative memory [11,21], action selection [2], and in application to data mining [18]. Spiking versions of BCPNN with biologically detailed Hebbian synaptic plasticity have also been developed to model different forms of cortical associative memory [6,15,23]. BCPNN architecture comprises many modules, referred to as hypercolumns (HCs), that in turn comprise a set of functional minicolumns (MCs) competing in a soft-winner-take-all manner. The abstract view of a HC in this cortical-like network is that it represents some attribute, e.g. edge orientation, in a discrete coded manner. A minicolumn conceptually represents one discrete value (a realization of the given attribute) and, as a biological parallel, it accounts for a local subnetwork of around a hundred recurrently connected neurons with similar receptive field properties [16]. Such an architecture was initially generalized from the primary visual cortex, but today has more support also from later experimental work and has been featured in spiking computational models of cortex [10,20].

Finally, in this work we highlight an additional mechanism called structural plasticity, introduced recently to the BCPNN framework [19], which enables self-organization and unsupervised learning of hidden representations. Structural plasticity learns a set of sparse connections while simultaneously learning the weights of the connections. This is in line with structural plasticity found in the brain, where there is continuous formation and removal of synapses in an activity-dependent manner [3].

2   Related Works

A popular unsupervised learning approach is to train a hidden layer to reproduce the input data as, for example, in AE and RBM. The AE and RBM networks trained with a single hidden layer are relevant here since learning weights of the input-to-hidden-layer connections relies on local gradients, and the representations can be stacked on top of each other to extract hierarchical features. However, stacked autoencoders and deep belief nets (stacked RBMs) have typically been used for pre-training procedures followed by end-to-end supervised fine-tuning (using backprop) [5]. The recently proposed KH model [9] addresses the problem of learning solely with local gradients by learning hidden representations only using an unsupervised method. In this network, the input-to-hidden connections are trained and additional (non-plastic) lateral inhibition provides competition within the hidden layer. For evaluating the representation, the weights are fixed, and a linear classifier trained with labels is used for the final classification. Our approach shares some common features with the KH model, e.g. learning hidden representations solely by unsupervised methods, and evaluating the representations by a separate classifier ([8] provides an extensive review of methods with similar goals).

All the aforementioned models employ either competition within the hidden layer (KH), or feedback connections from hidden to input (RBM and AE). The BCPNN uses only the feedforward connections, along with an implicit competition via a local softmax operation, the neural implementation of which would be lateral inhibition within a hypercolumn.

It is also observed that, for unsupervised learning, having sparse connectivity in the feedforward connections performs better than full connectivity [8]. Even networks employing supervised learning, like convolutional neural networks (CNNs), force a fixed spatial filter to obtain this sparse connectivity. The BCPNN model takes an alternative adaptive approach by using structural plasticity to obtain a sparse connectivity.

3   Bayesian Confidence Propagation Neural Network

Here we describe the BCPNN network architecture and update rules [11, 19, 21]. The feedforward BCPNN architecture contains two layers, referred to as the input layer and hidden layer (Fig. 1A). A layer consists of a set of HCs, each of which represents a discrete random variable X_i (upper case). Each HC, in turn, is composed of a set of MCs representing a particular value x_i (lower case) of X_i. The probability of X_i is then a multinomial distribution, defined as $p(X_i = x_i)$, such that $\sum_{x_i} p(X_i = x_i) = 1$. In the neural network, the activity of the MC is interpreted as $p(X_i = x_i)$, and the activities of all the MCs inside a HC sum to one.

Since the network is a probabilistic graphical model (see Fig. 1B), we can compute the posterior of a target HC in the hidden layer conditioned on all the source HCs in the input layer. We will use x's and y's to refer to the HCs in the input and hidden layers respectively. Computing the exact posterior $p(Y_j|X_1, ..., X_N)$ over the target HC is intractable, since it scales exponentially with the number of units. The naive Bayes assumption $p(X_1, .., X_N|Y_j) = \prod_{i=1}^{N} p(X_i|Y_j)$ allows us to write the posterior as follows:

$$p(Y_j|X_1, ..., X_N) = \frac{p(Y_j) \prod_{i=1}^{N} p(X_i|Y_j)}{p(X_1, ..., X_N)} \propto p(Y_j) \prod_{i=1}^{N} p(X_i|Y_j) \qquad (1)$$

When the network is driven by input data $\{X_1, .., X_N\} = \{x_1^D, .., x_N^D\}$, we can write the posterior probabilities of a target MC in terms of the source MCs as:

$$p(y_j|x_1^D, ..., x_N^D) \propto p(y_j) \prod_{i=1}^{N} p(x_i^D|y_j) = p(y_j) \prod_{i=1}^{N} \prod_{x_i} p(x_i|y_j)^{\mathbb{I}(x_i = x_i^D)} \qquad (2)$$

where $\mathbb{I}(\cdot)$ is the indicator function that equals 1 if its argument is true, and zero otherwise. We have written the posterior of the target MC as a function of all the source MCs (all x_i's). The log posterior can be written as:

$$\log p(y_j|x_1^D, ..., x_N^D) \propto \log p(y_j) + \sum_{i=1}^{N} \sum_{x_i} \mathbb{I}(x_i = x_i^D) \log p(x_i|y_j) \qquad (3)$$

Since the posterior is linear in the indicator function of data sample, $\mathbb{I}(x_i = x_i^D)$ can be approximated by its expected value defined as $\pi(x_i) = p(x_i = x_i^D)$. Except for $\pi(x_i)$, all the terms in the posterior are functions of the marginals $p(y_j)$ and $p(x_i, y_j)$. We define the terms bias $b(y_j) = \log p(y_j)$ and weight $w(x_i, y_j) = \log \frac{p(x_i, y_j)}{p(x_i)p(y_j)}$ in analogy with artificial neural networks. Note that in the weight term, we have added an additional $p(x_i)$ factor in the denominator to be consistent with previous BCPNN models [21,23], but this will not affect the computation since all terms independent of y_j will absorbed in the activity normalization.

The inference step to calculate the posterior probabilities of the target MCs conditioned on the input sample is given by the activity update equations:

$$h(y_j) = b(y_j) + \sum_{i=1}^{N} \sum_{x_i} \pi(x_i) w(x_i, y_j) \qquad (4)$$

$$\pi(y_j) = \frac{\exp h(y_j)}{\sum_k \exp h(y_k)} \qquad (5)$$

where $h(y_j)$ is the total input received by each target MC from which the posterior probability $\pi(y_j) = p(y_j|x_1^D, ..., x_N^D)$ is recovered by softmax normalization of all MCs within the HC.

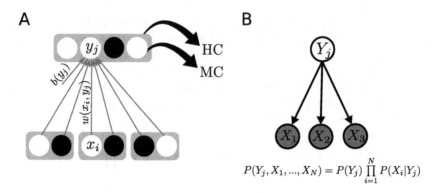

Fig. 1. Schematic of the BCPNN architecture with three input HCs and one hidden HC. **A.** The neural network model where the input HCs are binary and the hidden HC is multinomial (gray boxes). The MCs within the HC are the discrete values of the hidden variable (open and shaded circles inside the box). **B.** The equivalent probabilistic graphical model illustrating the generative process with each random variable (circle) representing a HC. The input HCs are observable (shaded circle) and the hidden HC is latent (open circle). The naive Bayes assumption renders the likelihood of generating inputs factorial.

As each data sample is presented, the learning step incrementally updates the marginal probabilities, weights, and biases as follows:

$$\tau_p \frac{dp(x_i)}{dt} = \pi(x_i) - p(x_i) \tag{6}$$

$$\tau_p \frac{dp(x_i, y_j)}{dt} = \pi(x_i)\,\pi(y_j) - p(x_i, y_j) \tag{7}$$

$$\tau_p \frac{dp(y_j)}{dt} = \pi(y_j) - p(y_j) \tag{8}$$

$$b(y_j) = \log\,p(y_j) \tag{9}$$

$$w(x_i, y_j) = \log \frac{p(x_i, y_j)}{p(x_i)\,p(y_j)} \tag{10}$$

where the parameter τ_p is the learning time constant. The sets of Eqs. 4–5 and Eqs. 6–10 define the activity and learning update equations of the BCPNN architecture respectively. For the network with multiple input HCs and one hidden HC (as in Fig. 1A), the computation is equivalent to a mixture model with each hidden MC representing one mixture component. For learning distributed representations from data, we can train multiple hidden HCs using the same principles. The network for unsupervised representation learning requires, in addition to the above computation, structural plasticity for learning a sparse set of connections from the input to hidden layer, which we discuss in detail in the next section.

3.1   Structural Plasticity

Structural plasticity builds a sparse set of connections from the input to hidden layer by iteratively improving from randomly initialized connections. We first define connections in terms of input HC to hidden HC, that is, when an input HC has an active connection to a hidden HC, all MCs within the input HC are connected to all MCs within the hidden HC. The connections are formulated as a connectivity matrix M, where $M_{ij} = 1$ when the connection from the ith input HC to jth hidden HC is active, or $M_{ij} = 0$ if silent[1] (see Fig. 2). Each M_{ij} is initialized as active stochastically with probability p_{ih}, with p_{ih} being the hyperparameter that controls the connection density. Once initialized, the total number of active incoming connections to each hidden HC is fixed whereas the outgoing connections from a source HC can be changed. For each connection, we compute the mutual information $I(X_i, Y_j)$ between the ith input HC and jth hidden HC and normalize by the number of active outgoing connections from the input HC to compute the score $\tilde{I}(X_i, Y_j)$:

$$\tilde{I}(X_i, Y_j) = \frac{I(X_i, Y_j)}{1 + \sum_k M_{ik}} \qquad (11)$$

Since the total number of active incoming connections is fixed, each hidden HC greedily maximizes the sum of score $\tilde{I}(X_i, Y_j)$ by silencing the active connection with the lowest score (change M_{ij} from 1 to 0) and activating the silent connection with the highest score (change M_{ij} from 0 to 1), provided that latter's score is higher than the former. We call this operation a flip and use a hyperparameter n_{flips} to set the number of flips made per training epoch.

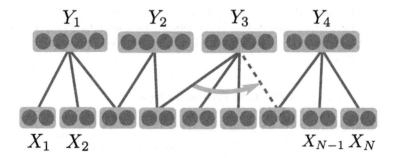

Fig. 2. Structural plasticity. The input layer contains N binary HCs, and the hidden layer contains four HCs (gray boxes). The existence of a connection between an input HC and hidden HC is shown as a blue line, i.e., $M_{ij} = 1$. Structural plasticity involves each hidden HCs silencing an active connection and activating another silent connection. The red arrow shows one such flip operation for the third hidden HC. (Color figure online)

---

[1] In analogy with biological synapses that can be non-existing, silent, or active, we adopt the term 'silent' for inactive connections.

4   Experiments

Here we first describe the experimental setup for the BCPNN and three other related models for unsupervised learning as described in Sect. 2. Next, we study qualitatively the hidden representations by examining the receptive fields formed by unsupervised learning. Finally, we provide quantitative evaluation of the representations learnt by the four models using class-dependent classification with a linear classifier.

4.1   Data

We ran the experiments on the MNIST [12] and Fashion-MNIST [24] datasets. Both datasets contain 60,000 training and 10,000 test samples of 28×28 greyscale images. The images were flattened to 784 dimensions and the pixel intensities were normalized to the range $[0, 1]$. The images acted as the input layer for the models and the labels were not used for unsupervised learning.

4.2   Models

We considered four network architectures: BCPNN (c.f. Sect. 3), AE, RBM and, KH. Each model had one hidden layer with 3000 hidden units.

BCPNN. The BCPNN network had an input layer with 784 HCs and 2 MCs per HC (pixel intensity was interpreted as probability of a binary variable) and a hidden layer with $n_{HC} = 30$ and $n_{MC} = 100$. The learning time constant was set as $\tau_p = 60$ to roughly match the training time for one epoch (for details, see [19]). The entire list of parameters and their values are listed in Table 1. The simulations were performed on code parallelized using MPI on a cluster of 2.3 GHz Xeon E5 processors and the training process took approximately fifteen minutes per run.

Table 1. BCPNN model parameters

Symbol	Value	Description
n_{epoch}	5	Number of epochs of unsupervised learning
n_{HC}	30	Number of HCs in hidden layer
n_{MC}	100	Number of MCs per HC in hidden layer
Δt	0.01	Time-step
τ_p	60	Learning time-constant
p_{ih}	8%	Connectivity from input to hidden layer
n_{flips}	16	Number of flips per epoch for structural plasticity
μ	10	Mean of poisson distribution for initializing MCs

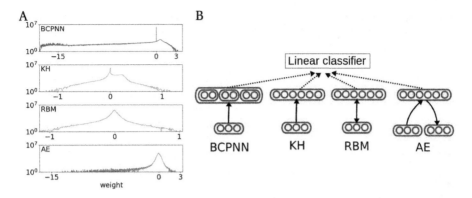

Fig. 3. A. Histogram of weights from the input layer to hidden layer. The horizontal axis has the minimum to maximum value of the weights as the range, and the vertical axis is in log scale. **B**. Schematic of the four unsupervised learning models under comparison and the supervised classifier. The dotted lines imply we use the representations of the hidden layer as input for the classifier.

KH. The KH network was reproduced from the original work using the code provided by Krotov and Hopfield [9]. We kept all the parameters as originally described, except for having 3000 hidden units instead of 2000, to be consistent in the comparison with other models.

RBM. For the RBM network, we used sigmoidal units for both input and hidden layer. The weights were trained using the Contrastive Divergence algorithm with one iteration of Gibbs sampling (CD-1). The learning rate α was set as 0.01 and the training was done in minibatches of 256 samples for 300 epochs.

AE. For the AE network, we used sigmoidal units for both hidden layer and reconstruction layer and two sets of weights, one for encoding from input to hidden layer and another for decoding from hidden to reconstruction layer. The weights were trained using the Adam optimizer and L2 reconstruction loss with an additional L1 sparsity loss on the hidden layer. The sparsity loss coefficient was determined by maximizing the accuracy of a held-out validation set of 10000 samples and set as $\lambda = $ 1e-7 for MNIST and $\lambda = $ 1e-5 for Fashion-MNIST. The training was in minibatches of 256 samples for 300 epochs.

4.3   Receptive Field Comparison

As can be observed in Fig. 3A, the distribution of weight values when trained on trained on MNIST considerably differs across the networks. It appears that the range of values for BCPNN corresponds to that reported for AE, whereas for KH and RBM, weights lie in a far narrower interval centered around 0. Importantly,

BCPNN has by far the highest proportion of zero weights (90%), which renders the connectivity truly sparse.

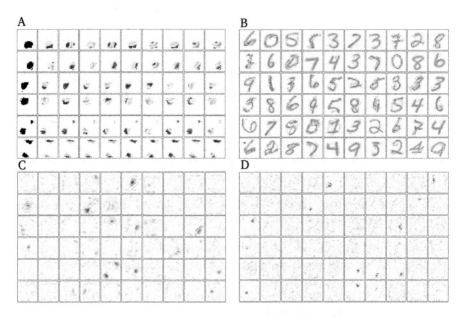

Fig. 4. Receptive fields of different unsupervised learning methods. For each model, the positive and negative values are normalized, such that blue, white, and red represent the lowest, zero, and highest value of weights. **A. BCPNN**: Each row corresponds to a randomly chosen HC and the constituent MCs of BCPNN. First column shows the receptive field of HC (black means $M_{ij} = 1$). The remaining columns show the receptive field of nine randomly chosen MCs out of 100 MCs within the HC. **B. KH, C. RBM, D. AE**: Receptive fields of 60 randomly chosen hidden units out of 3000. (Color figure online)

In Fig. 4, we visualize the receptive fields of the four unsupervised learning networks trained on the MNIST dataset. Firstly, it is straightforward to see that the receptive fields of all the networks differ significantly. The RBM (Fig. 4C) and AE (Fig. 4D) have receptive fields that are highly localized and span the input space, a characteristic of distributed representations. The KH model (Fig. 4B) has receptive fields that resemble the entire image, showing both positive and negative values over the image, as a result of Hebbian and anti-Hebbian learning [9]. Generally, local representations like mixture models and competitive learning, as opposed to distributed representations, tend to have receptive fields that resemble prototypical samples. With this distinction in mind, the receptive fields in the BCPNN should be closely examined (Fig. 4A). The receptive fields of HCs (first column) are localized and span the input space, much like a distributed representation. Within each HC however, the MCs have receptive fields (each row) resembling prototypical samples, like diverse sets of lines and strokes. This suggests that the BCPNN representations are "hybrid", with the higher-level HCs coding a distributed representation, and the lower level MCs coding a local representation.

4.4   Classification Performance

For all the four models of unsupervised learning, we employed the same linear classifier for predicting the labels (see Fig. 3B). This allowed us to consistently evaluate the representations learned by the different models. The linear classifier considers the hidden layer representations as the input and the labels as the output. The output layer consists of softmax units for the 10 labels. The classifier's weights were trained by stochastic gradient descent with the Adam optimizer using cross-entropy loss function. The training procedure used minibatches of 256 samples and a total of 300 training epochs.

Table 2. Accuracy comparison for MNIST and Fashion-MNIST datasets

Model	Hyperparameters	MNIST Train %	MNIST Test %	Fashion-MNIST Train %	Fashion-MNIST Test %
BCPNN	See Table 1	99.59 ± 0.01	98.31 ± 0.02	88.71 ± 0.05	86.31 ± 0.09
KH	See [9]	98.75 ± 0.01	97.39 ± 0.06[a]	87.49 ± 0.03	85.10 ± 0.05
RBM	$\alpha = 0.01$	98.92 ± 0.04	97.67 ± 0.10	88.13 ± 0.06	86.06 ± 0.12
AE	$\lambda = \{$1e-7,1e-5$\}$	100.0 ± 0.00	97.78 ± 0.09	88.52 ± 0.02	86.12 ± 0.08

[a] This is lower than the test accuracy of 98.54% reported by Krotov and Hopfield [9] who used a non-linear classifier with exponentiated ReLU activation and a non-linear loss function. Here we used instead a simpler linear classifier with softmax activation and cross-entropy loss.

The results of the classification are shown in Table 2. All the results presented here are the mean and standard deviation of the classification accuracy over ten random runs. We performed three independent comparisons of BCPNN with KH, RBM, and AE using the Kruskal-Wallis test to evaluate statistical significance. On both MNIST and Fashion-MNIST, BCPNN outperforms KH, RBM, and AE (all with $p < 0.0001$).

For the BCPNN model, we set $n_{HC} = 30$ and $n_{MC} = 100$ in order to match the size of the hidden layer with the other models. However, BCPNN also scales well with the size of the hidden layer. From manual inspection, we found the best performance is at $n_{HC} = 200$ and $n_{MC} = 100$, where the test accuracy for MNIST is 98.58 ± 0.05 and Fashion-MNIST is 88.87 ± 0.08.

5   Discussion

We have evaluated four different network models that can perform unsupervised representation learning using biologically plausible local learning rules. We made our assessment relying on the assumption that the usefulness of representations is reflected in their class separability, which can be quantified by classification performance, similar to recent unsupervised learning methods [17]. Learning representations without supervised fine-tuning is a harder task compared to similar networks with end-to-end backprop, since the information about the samples' corresponding labels cannot be utilized. Additionally, our unsupervised learning method relies on local Hebbian-like learning without any global optimisation criterion as in backprop learning, which makes the task even harder. Yet, we show that the investigated BCPNN model performs well, comparable to the 98.5% accuracy of multi-layer perceptron networks on MNIST with one hidden layer trained using end-to-end backprop [13].

We also showed that the recently proposed BCPNN model performs competitively against other unsupervised learning models. The modular structure of the BCPNN layer led to "hybrid" representations that differ from the well-known distributed and local representations. In contrast to the minibatch method of other unsupervised learning methods, learning in BCPNN was chosen to remain incremental using dynamical equations, since such learning is biologically feasible and useful in many autonomous engineering solutions. Despite the slow convergence properties of an incremental approach, BCPNN required only 5 epochs of unsupervised training, in comparison to 300 epochs for AE and RBM, and 1000 epochs for KH. The incremental learning, along with modular architecture, sparse connectivity, and scalability of BCPNN is currently also taken advantage of in dedicated VLSI design [22].

One important difference between current deep learning architectures and the brain concerns the abundance of recurrent connections in the latter. Deep learning architectures rely predominantly on feedforward connectivity. A typical cortical area receives only around 10% of synapses from lower order structures, e.g. thalamus, and the rest from other cortical areas [4]. These feedback and recurrent cortical connections are likely involved in associative memory, constraint-satisfaction e.g. for figure-ground segmentation, top-down modulation and selective attention [4]. Incorporating these important aspects of cortical computation can play a key role in improving our machine learning models and approaches.

It is important to note that the unsupervised learning models discussed in this work are proof-of-concept designs and not meant to directly model some specific biological system or structure. Yet, they may shed some light on the hierarchical functional organization of e.g. sensory processing streams in the brain. Further work will focus on extending our study to multi-layer architectures.

References

1. Bengio, Y., Courville, A., Vincent, P.: Representation learning: a review and new perspectives. IEEE Trans. Pattern Anal. Mach. Intell. **35**(8), 1798–1828 (2013)
2. Berthet, P., Hellgren Kotaleski, J., Lansner, A.: Action selection performance of a reconfigurable basal ganglia inspired model with Hebbian-Bayesian Go-NoGo connectivity. Front. Behav. Neurosci. **6**, 65 (2012)

3. Butz, M., Wörgötter, F., van Ooyen, A.: Activity-dependent structural plasticity. Brain Res. Rev. **60**(2), 287–305 (2009)
4. Douglas, R.J., Martin, K.A.: Recurrent neuronal circuits in the neocortex. Curr. Biol. **17**(13), R496–R500 (2007)
5. Erhan, D., Bengio, Y., Courville, A., Manzagol, P.A., Vincent, P., Bengio, S.: Why does unsupervised pre-training help deep discriminant learning? (2009)
6. Fiebig, F., Lansner, A.: A spiking working memory model based on Hebbian short-term potentiation. J. Neurosci. **37**(1), 83–96 (2017)
7. Hassabis, D., Kumaran, D., Summerfield, C., Botvinick, M.: Neuroscience-inspired artificial intelligence. Neuron **95**(2), 245–258 (2017)
8. Illing, B., Gerstner, W., Brea, J.: Biologically plausible deep learning–but how far can we go with shallow networks? Neural Netw. **118**, 90–101 (2019)
9. Krotov, D., Hopfield, J.J.: Unsupervised learning by competing hidden units. Proc. Natl. Acad. Sci. **116**(16), 7723–7731 (2019)
10. Lansner, A.: Associative memory models: from the cell-assembly theory to bio-physically detailed cortex simulations. Trends Neurosci. **32**(3), 178–186 (2009)
11. Lansner, A., Benjaminsson, S., Johansson, C.: From ANN to biomimetic information processing. In: Biologically Inspired Signal Processing for Chemical Sensing, pp. 33–43 (2009)
12. LeCun, Y.: The MNIST database of handwritten digits (1998). http://yann.lecun.com/exdb/mnist/
13. LeCun, Y., Bottou, L., Bengio, Y., Haffner, P.: Gradient-based learning applied to document recognition. Proc. IEEE **86**(11), 2278–2324 (1998)
14. Lillicrap, T.P., Santoro, A., Marris, L., Akerman, C.J., Hinton, G.: Backpropagation and the brain. Nat. Rev. Neurosci., 1–12 (2020)
15. Lundqvist, M., Herman, P., Lansner, A.: Theta and gamma power increases and alpha/beta power decreases with memory load in an attractor network model. J. Cogn. Neurosci. **23**(10), 3008–3020 (2011)
16. Mountcastle, V.B.: The columnar organization of the neocortex. Brain J. Neurol. **120**(4), 701–722 (1997)
17. Oord, A.V.D., Li, Y., Vinyals, O.: Representation learning with contrastive predictive coding. arXiv preprint arXiv:1807.03748 (2018)
18. Orre, R., Lansner, A., Bate, A., Lindquist, M.: Bayesian neural networks with confidence estimations applied to data mining. Comput. Stat. Data Anal. **34**(4), 473–493 (2000)
19. Ravichandran, N.B., Lansner, A., Herman, P.: Learning representations in Bayesian confidence propagation neural networks. In: International Joint Conference on Neural Networks (IJCNN) (2020)
20. Rockland, K.S.: Five points on columns. Front. Neuroanat. **4**, 22 (2010)
21. Sandberg, A., Lansner, A., Petersson, K.M., Ekeberg, O.: A Bayesian attractor network with incremental learning. Netw. Comput. Neural Syst. **13**(2), 179–194 (2002)
22. Stathis, D., et al.: eBrainii: a 3 kw realtime custom 3D DRAM integrated ASIC implementation of a biologically plausible model of a human scale cortex. J. Sig. Process. Syst., 1–21 (2020)
23. Tully, P.J., Hennig, M.H., Lansner, A.: Synaptic and nonsynaptic plasticity approximating probabilistic inference. Frontiers Synaptic Neurosci. **6**, 8 (2014)
24. Xiao, H., Rasul, K., Vollgraf, R.: Fashion-MNIST: a novel image dataset for benchmarking machine learning algorithms. arXiv preprint arXiv:1708.07747 (2017)

Spiking Neural Networks

A Subthreshold Spiking Neuron Circuit Based on the Izhikevich Model

Shigeo Sato[1]([envelope]) [iD], Satoshi Moriya[1] [iD], Yuka Kanke[1], Hideaki Yamamoto[1] [iD],
Yoshihiko Horio[1] [iD], Yasushi Yuminaka[2], and Jordi Madrenas[3] [iD]

[1] Research Institute of Electrical Communication, Tohoku University,
2-1-1 Katahira, Aoba-ku, Sendai, Miyagi 980-8577, Japan
shigeo@riec.tohoku.ac.ip
[2] Graduate School of Science and Technology, Gunma University, 1-5-1 Tenjin-cho,
Kiryu, Gunma 376-8515, Japan
[3] Department of Electronics Engineering, Universitat Politècnica de Catalunya,
Jordi Girona 1-3, 08034 Catalunya, Barcelona, Spain

Abstract. Low-power neuromorphic hardware is indispensable for edge
computing. In this study, we report the simulation results of a spiking neuron circuit. The circuit based on the Izhikevich neuron model is
designed to reproduce various types of spikes and is optimized for low-
voltage operation. Simulation results indicate that the proposed circuit
successfully operates in the subthreshold region and can be utilized for
reservoir computing.

Keywords: Subthreshold operation · Izhikevich model · Low-power
consumption

1 Introduction

Practical implementation of neuromorphic hardware is an important issue given
the increasing use of artificial intelligence (AI) in a society with IoT. In particular, dedicated application-specific hardware is indispensable for edge AI computing to minimise power consumption. Generally, efficient neuromorphic hardware
is obtained via analog implementation because necessary functions in a neural
network, such as a multiply-accumulate operation and thresholding function, can
be easily built with fewer transistors.

An approach to low-power operation involves utilising a spiking neuron similar to that in the biological brain. The use of a complex spiking neuron model

This study was supported in part by the Cooperative Research Project Program of the
Research Institute of Electrical Communication, Tohoku University; JSPS KAKENHI
(Grant Nos, 18H03325 and 20H00596); JST PRESTO (Grant Number JPMJPR18MB);
and JST CREST (Grant Number JPMJCR19K3), Japan.

© Springer Nature Switzerland AG 2021
I. Farkaš et al. (Eds.): ICANN 2021, LNCS 12895, pp. 177–181, 2021.
https://doi.org/10.1007/978-3-030-86383-8_14

as opposed to a simple integrate-and-fire model exhibits advantages in exploiting the computational power of neural networks. For example, complex neuron
dynamics is favourable for mapping input signals to high-dimensional neural
activities similar to that in reservoir computing [2,3].

Among various neuron models, the Izhikevich model [1] has received significant attention because of its ability to reproduce various neural activities. It can
be widely applied to engineering problems because it can be described by simple
differential equations. Among the MOS circuits based on the Izhikevich model,
the circuit proposed by Wijekoon and Dudek in 2008 [7] exhibits the simplest
structure and is significantly practical, although it operates in the strong inversion region at 3.3 V. Subthreshold operation of a circuit is indispensable to reduce
power consumption to meet the requirement of edge computing. Therefore, in
this study, we propose a subthreshold CMOS spiking neuron circuit based on
the Izhikevich model.

2   Izhikevich Neuron Model

The Izhikevich neuron model [1] can reproduce spiking and bursting behavior of
known types of cortical neurons with less computational cost. The dynamics is
described by the following equations:

$$\dot{v} = 0.04v^2 + 5v + 140 - u + I, \tag{1}$$

$$\dot{u} = a(bv - u), \tag{2}$$

$$\text{if } v \geq 30\,\text{mV, then} \begin{cases} v \leftarrow c \\ u \leftarrow u + d \end{cases}, \tag{3}$$

where v, u, and I denote the membrane potential of a neuron, recovery variable,
and synapse current, respectively.

3   Circuit Design

A few studies on MOS circuits are based on the Izhikevich neuron model [4,5,7].
Among the studies, the circuit proposed by Wijekoon and Dudek [7] is potentially
the simplest one and is composed of 14 MOS transistors. In particular, MOS
transistors of the circuit operate in the strong inversion region. They successfully
reproduced various spikes, such as RS, IB, CH, FS, and LTS, similar to the
original model.

Although the circuit is well optimized, some discrepancies between the original circuit and the Izhikevich model exist. Furthermore, it is significantly difficult to control parameters to reproduce specific pulses when the supply voltage
becomes low. Thus we have modified the circuit so that it fits for low voltage
operation [6]. Though it successfully operates at 1.0 V supply, the power consumption ($\sim$20 μW) is still large. To reduce power consumption so that the

circuit meets the requirements of edge computing, the operation in the weak inversion region is indispensable. Therefore, we improve the neuron circuit to operate at 0.3 V, as depicted in Fig. 1. The circuit parameters were carefully optimized so that the MOS transistors operated properly in the weak inversion region. In addition, the output of the comparator was amplified using CMOS inverters in series and used for a faster reset operation expressed in (3). This was because we avoided slow reset operation, which could cause discrepancy in the dynamical behaviour.

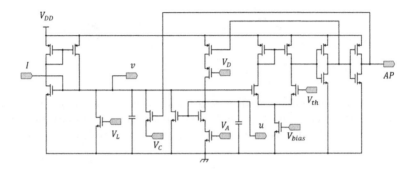

Fig. 1. Circuit diagram of the proposed circuit

4   Simulation Results

We performed SPICE simulations of the proposed neuron circuit before hardware implementation using the TSMC 65 nm technology. The supply voltage was set to 0.3 V. Figure 2 exhibits the examples of the reproduced spikes, where AP, U, and V denote the action potential, recovery variable, and membrane potential, respectively. AP was delivered to other neurons for the transmission of firing information. Given DC-current input $I = 3$ nA, various spikes of V were generated. We confirmed the reproduction of spike types of cortical neurons including RS, FS, IB, and CH. The external biases were $V_A = 0.1$ V, $V_C = 0.05$ V, and $V_D = 0.1$ V for RS; $V_A = 0.3$ V, $V_C = 0.05$ V, and $V_D = 0.3$ V for FS; $V_A = 0.1$ V, $V_C = 0.1$ V, and $V_D = 0.1$ V for IB; $V_A = 0.2$ V, $V_C = 0.11$ V, and $V_D = 0.1$ V for CH. $V_{th} = 0.2$ V, $V_{bias} = 0.2$ V, and $V_L = 0.1$ V are common for all spike types. The spike interval of V modulates based on U and the nature of the model. The power consumption depends on spike type and ranges from 3.2 nW to 7.6 nW.

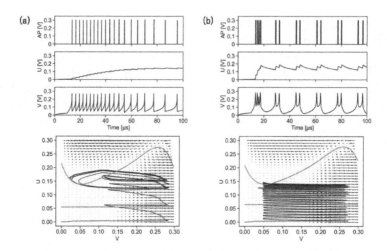

Fig. 2. Typical circuit transient responses to steps of DC current input $I = 3$ nA. Panels show the voltages at each terminal: action potential AP, recovery variable U, membrane potential V, and locus in the (V, U) plane. (a) regular spiking and (b) chattering.

5   Conclusion

For reduction in power consumption, we proposed a spiking neuron circuit operating in the weak inversion region. SPICE simulations confirmed that the power consumption was less than 10 nW. The circuit is applicable to various applications. Its complex dynamics originated from the model can improve the performance in reservoir computing.

References

1. Izhikevich, E.M.: Simple model of spiking neurons. IEEE Trans. Neural Netw. **14**(6), 1569–1572 (2003)
2. Jaeger, H.: The "echo state" approach to analysing and training recurrent neural networks. GMD Report 148, GMD - German National Research Institute for Computer Science (2001). http://www.faculty.jacobs-university.de/hjaeger/pubs/EchoStatesTechRep.pdf
3. Maass, W., Natschläger, T., Markram, H.: Real-time computing without stable states: a new framework for neural computation based on perturbations. Neural Comput. **14**(11), 2531–2560 (2002)
4. Mizoguchi, N., Nagamatsu, Y., Aihara, K., Kohno, T.: A two-variable silicon neuron circuit based on the Izhikevich model. Artif. Life Robot. **16**(3), 383–388 (2011)
5. Rangan, V., Ghosh, A., Aparin, V., Cauwenberghs, G.: A subthreshold a VLSI implementation of the Izhikevich simple neuron model. In: 2010 Annual International Conference of the IEEE Engineering in Medicine and Biology, pp. 4164–4167, August 2010

6. Tamura, Y., Moriya, S., Kato, T., Sakuraba, M., Horio, Y., Sato, S.: An Izhikevich model neuron MOS circuit for low voltage operation. In: Tetko, I.V., Kurková, V., Karpov, P., Theis, F. (eds.) ICANN 2019. LNCS, vol. 11727, pp. 718–723. Springer, Cham (2019). https://doi.org/10.1007/978-3-030-30487-4_55
7. Wijekoon, J.H., Dudek, P.: Compact silicon neuron circuit with spiking and bursting behaviour. Neural Net. **21**(2), 524–534 (2008). Advances in Neural Networks Research: IJCNN 2007

SiamSNN: Siamese Spiking Neural Networks for Energy-Efficient Object Tracking

Yihao Luo[1], Min Xu[1], Caihong Yuan[1,2], Xiang Cao[1], Liangqi Zhang[1], Yan Xu[1], Tianjiang Wang[1], and Qi Feng[1(✉)]

[1] Huazhong University of Science and Technology, Wuhan 430074, China
fengqi@hust.edu.cn
[2] Henan University, Kaifeng 475004, China

Abstract. Recently spiking neural networks (SNNs), the third-generation of neural networks has shown remarkable capabilities of energy-efficient computing, which is a promising alternative for deep neural networks (DNNs) with high energy consumption. SNNs have reached competitive results compared to DNNs in relatively simple tasks and small datasets such as image classification and MNIST/CIFAR, while few studies on more challenging vision tasks on complex datasets. In this paper, we focus on extending deep SNNs to object tracking, a more advanced vision task with embedded applications and energy-saving requirements, and present a spike-based Siamese network called SiamSNN. Specifically, we propose an optimized hybrid similarity estimation method to exploit temporal information in the SNNs, and introduce a novel two-status coding scheme to optimize the temporal distribution of output spike trains for further improvements. SiamSNN is the first deep SNN tracker that achieves short latency and low precision loss on the visual object tracking benchmarks OTB2013/2015, VOT2016/2018, and GOT-10k. Moreover, SiamSNN achieves notably low energy consumption and real-time on Neuromorphic chip TrueNorth.

Keywords: Spiking neural networks · Energy-efficient · Temporal information · Object tracking

1 Introduction

Nowadays Deep Neural Networks (DNNs) have shown remarkable performance in various scenarios [7,15,33]. However, it is hard to employ DNNs on embedded systems such as mobile devices due to the high computation cost and heavy energy consumption. Spiking neural networks (SNNs) are regarded as the third

Supported in part by the National Natural Science Foundation of China under Grant 61572214, 62006070, and Seed Foundation of Huazhong University of Science and Technology (2020kfyXGYJ114).

I. Farkaš et al. (Eds.): ICANN 2021, LNCS 12895, pp. 182–194, 2021.
https://doi.org/10.1007/978-3-030-86383-8_15

generation artificial neural networks, which are biologically-inspired computational models and have realized ultra-low power consumption on neuromorphic hardware (such as TrueNorth [18]). But they still lag behind DNNs in terms of accuracy [26]. One major reason is that the training algorithms can only train shallow SNNs [20]. To avoid this issue, an alternative way is converting trained DNNs to SNNs [4,6,22], which aims to convert DNNs to SNNs by transferring the well-trained models with weight rescaling and normalization methods.

Although SNNs have shown excellent energy-efficient potential, they have been limited to relatively simple tasks and small datasets (e.g., image classification on MNIST and CIFAR) [30]. Most critically, the insufficiency of spike-based modules for various functions outside classification limits their applications [10], it is necessary to enrich them for greater prevalence. Besides, SNNs of remarkable performance through conversion usually requires long latency [22,25], which causes the inability to reach real-time requirements. Exploiting the sparse and temporal dynamics of spike-based information is significant to achieve real-time algorithms [21].

Following the successes of the DNNs to SNNs conversion methods on image classification and object detection tasks [10], it is desirable to extend those results to solve visual object tracking, a more advanced vision task with embedded applications and energy-saving requirements. The state-of-the-art tracking models [12,27,31] require high-performance GPUs, which involve a drastic increase in computation and power requirements that tends to be intractable for embedded platforms [1]. With the aid of energy-efficient computing in SNNs, implementing a spike-based tracker is significant to realize energy-efficient tracking algorithms on embedded systems, which also extends the limitations of related theories and applications of SNNs. The problems of implementing it can be summarized as follows: 1) There is no effective similarity estimation function in SNN fashion for deep tracking. 2) The existed methods usually require long latency while real-time is a significant metric to tracking tasks.

In this paper, we focus on extending deep SNNs to object tracking with DNN-to-SNN conversion methods. To address the first issues, we introduce an optimized hybrid similarity estimation method in SNN fashion. The proposed method evaluates the similarity between feature maps of the exemplar image and candidate images in the form of spiking. And inspired by the neural phenomenon of long-term depression (LTD) [3], we design a two-status coding scheme to optimize the temporal distribution of output spike trains, which utilizes temporal information for shortening the latency to reach real-time and mitigating the accuracy degradation. Eventually, we present a spike-based Siamese network for object tracking called SiamSNN based on SiamFC [2]. To the best of our knowledge, SiamSNN is the first deep SNN for object tracking that achieves short latency and low precision loss of the original SiamFC on the tracking benchmarks VOT2016 [17], VOT2018 [11], and GOT-10k [8].

2   Related Work and Background

2.1   Conversion Methods of DNN-to-SNN

Converting trained DNNs to SNNs is an effective approach to obtain deep SNNs with relatively high accuracy. Cao *et al.* [4] convert DNNs to SNNs and achieve excellent performance with a comprehensive conversion scheme. Then Rueckauer *et al.* [22] introduce spike max-pooling and batch normalization (BN) for SNN. Different from DNNs, SNNs use event-driven binary spike trains in a certain period rather than a single continuous value between neurons. The widely used integrate-and-fire (IF) neuron model integrates post-synaptic potential (PSP) into the membrane potential V_{mem}. In other words, the spiking neuron i accumulates the input z into the membrane potential $V_{\text{mem},i}^l$ in the lth layer at each time step, which is described as

$$V_{\text{mem},i}^l(t) = V_{\text{mem},i}^l(t-1) + z_i^l(t) - V_{\text{th}}(t)\Theta_i^l(t), \tag{1}$$

where $\Theta_i^l(t)$ is a spike in the ith neuron in the lth layer, $V_{\text{th}}(t)$ is a certain threshold, and $z_i^l(t)$ is the input of the ith neuron in the lth layer, which is described as

$$z_i^l(t) = \sum_j w_{ji}^l V_{\text{th}}(t)\Theta_j^{l-1}(t) + b_i^l, \tag{2}$$

where w^l is the weight and b^l is the bias in the lth layer. Each neuron will generate a spike only if its membrane potential exceeds a certain threshold $V_{\text{th}}(t)$. The process of spike generation can be generalized as

$$\Theta_i^l(t) = U\left(V_{\text{mem},i}^l(t-1) + z_i^l(t) - V_{\text{th}}(t)\right), \tag{3}$$

where $U(\cdot)$ is a unit step function. Once the spike is generated, the membrane potential is reset to the resting potential. As shown in Eq. 1, we adopt the method of resetting by subtraction rather than resetting to zero, which increases firing rate and reduces the loss of information. The firing rate is defined as

$$\text{firing rate } = \frac{N}{T}, \tag{4}$$

where N is the total number of spikes during a given period T. The maximum firing rate will be 100% since the neuron generates a spike at each time step.

To prevent the neuron from over-activation and under-activation, Diehl *et al.* [6] suggest a data-based method to enable a sufficient firing rate and achieve nearly lossless accuracy on MNIST. Sengupta *et al.* [25] introduce a novel spike-normalization method for generating an SNN with deeper architecture and experiment on complex standard vision dataset ImageNet [23]. The conventional normalized weights w^l and biases b^l are calculated by

$$\tilde{w}^l = w^l \frac{\lambda^{l-1}}{\lambda^l} \quad \text{and} \quad \tilde{b}^l = \frac{b^l}{\lambda^l}, \tag{5}$$

where w are the weights of the original DNN, b are the biases, and λ is the 99.9th percentile of the maximum activation [22] in the layer l. And Spiking-YOLO [10] proposes a channel normalization for SNN object detection.

As for spike coding, the method of representing information with spike trains, rate and temporal coding are the most frequently used coding schemes in SNNs. Rate coding is based on the spike firing rate, which is used as the signal intensity, counted by the number of spikes that occurred during a period. Thus, it has been widely used in converted SNNs [4,22] for classification by comparing the firing rates of each category. However, rate coding requires a higher latency for more accurate information. And temporal coding uses timing information to mitigate long latency. Park et $al.$ [20] introduce a burst coding scheme inspired by neuroscience research and investigates a hybrid coding scheme to exploit different features of different layers in SNNs. Kim et $al.$ [9] propose weighted spikes by the phase function to make the most of the temporal information. The phase function of it is given by

$$V_{\text{th}}(t) = 2^{-(1+\ \text{mod}\ (t-1,K))} \tag{6}$$

where K is the period of the phase. Phase coding needs only K time steps to represent K-bit data (8-bit images usually) whereas rate coding would require 2^K time steps to represent the same data.

2.2 Object Tracking

Object tracking aims to estimate the position of an arbitrary target in a video sequence while its location only initializes in the first frame by a bounding box. Two main branches of recent trackers are based on correlation filter (CF) [5] and DNNs [13]. CF trackers train regressors in the Fourier domain and update the weights of filters to do online tracking. Motivated by CF, trackers based on Siamese networks with similarity comparison strategy have drawn great attention due to their high performance. SiamFC [2] introduces the correlation layer rather than fully connected layers to evaluate the regional feature similarity between two frames. It learns a function $f(z, x)$ to compare an exemplar image z to candidate images x and computes the similarity at all sub-windows on a dense grid in a single evaluation, which is defined as

$$f(z, x) = \varphi(z) * \varphi(x) + b \cdot \mathbb{1}, \tag{7}$$

where φ is the CNN to perform feature extraction and $b \cdot \mathbb{1}$ denotes a bias equated in every location. In the subsequent work, SiamRPN [13], SiamRPN++ [12], Siam R-CNN [27], SiamFC++ [31] obtain the state-of-the-art tracking performance. All of them are based on the Siamese architecture in SiamFC. Therefore, we select SiamFC as our base model to construct a Siamese Spiking Neural Network for energy-efficient object tracking.

For the aspect of energy and memory consumption, Liu et $al.$ [14] construct small, fast yet accurate trackers by a teacher-students knowledge distillation model. To exploit the low-powered nature of SNNs, Yang et $al.$ [32] propose

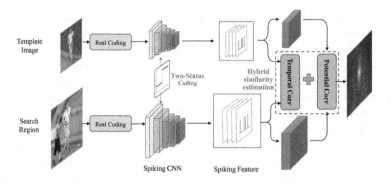

Fig. 1. Framework of SiamSNN. This framework consists of a converted Siamese spiking convolutional neural network, two-status coding scheme, and an optimized hybrid similarity estimation method (potential and temporal correlation).

a hybrid paradigm of ANN and SNN for efficient high-speed object tracking, called Dashnet. However, Dashnet is not an absolute deep SNN, the problem of insufficient computing power still exists while applying to resource-constrained systems. Luo *et al.* [16] show the possibility to construct a deep SNN for object tracking (the early version of SiamSNN) on some simple videos. But they lack effective methods and convincing performance on the benchmark. We propose two approaches to obtain a stronger spiking tracker in this work.

3   Proposed Methods

3.1   Model Overview

We detail the SiamSNN, the converted SiamFC in Fig. 1. SiamSNN consists of a converted Siamese spiking convolutional neural network, two-status coding scheme, and an optimized hybrid similarity estimation method (potential and temporal correlation). The Spiking CNN is consistent with the CNN model architecture in SiamFC [2], which can be obtained by Eqs. 1, 2, 3, 5. The hybrid similarity estimation method calculates the response map with the potential and temporal correlation of spike features. The two-status coding scheme optimizes the temporal distribution of output spike trains to improve the performance.

Similar to spiking softmax layer [22], Eq. 7 can be converted into a spiking form:

$$M_P(z, x) = \left(\sum_{t=1}^{T} \varphi(z(t))\right) * \left(\sum_{t=1}^{T} \varphi(x(t))\right), \tag{8}$$

where M_P is the potential response map, $z(t)$ and $x(t)$ are the input encoded spike trains of exemplar image and candidate image, φ is the Spiking CNN, T is the latency. We will refer to this simple method as potential similarity estimation (PSE) in the rest of the paper.

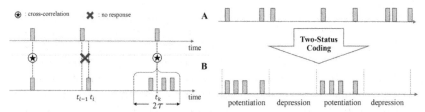

(a) Temporal correlation between spike train A and B.

(b) The distribution of spikes after two-status coding.

Fig. 2. Illustrations of the hybrid spiking similarity estimation and two-status coding.

3.2 Hybrid Spiking Similarity Estimation

PSE does not consider the temporal information of spike trains, which is the main difference between DNNs and SNNs and significant to information transmission [20]. We can also match the temporal similarity between spiking features of exemplar and candidate images as

$$M_T(z, x) = \sum_{t=1}^{T} \left(\varphi(z(t)) * \varphi(x(t)) \right), \tag{9}$$

which is similar to Eq. 8, M_T is the temporal response map. It calculates the correlation at each time step. We will refer to it as temporal similarity estimation (TSE) in the rest of the paper.

Intuitively, rely on the synchronization of spikes at each time step, TSE causes strict temporal consistency of correlation. For instance in Fig. 2(a), if spike train A fires at t_{i-1}, and spike train B fires at t_i, their temporal correlation equals 0 during (t_{i-2}, t_{i+1}), but they are highly similar measured by spike-distance. Motivated by PySpike [19], we optimize TSE and propose a response period with time weights during correlation operation. It is defined as:

$$M_{\Delta\tau}(z, x) = \sum_{t=1}^{T} \sum_{m=t-\tau}^{t+\tau} \left(\frac{\varphi(z(m)) * \varphi(x(t))}{2|t - m| + 1} \right), \tag{10}$$

where τ is the response period threshold. We set $\tau = 1$ by experiments, and the impacts of τ are analyzed in Fig. 3(b) of Sect. 4.2. As shown in Fig. 2(a), Eq. 10 makes spike trains A response with B during $[t_k - \tau, t_k + \tau)$. The response value will attenuate gradually when the time steps far from t_k. A large period has few contributions to the response value and also increases computation consumption. Thus, τ is usually set to a small number.

Considering PSE performs lossless on portions of sequences in OTB-2013, we calculate the final response map as follows:

$$f_{spike}(z, x) = \frac{1}{T^2} M_P(z, x) + \frac{1}{2T \cdot 2\tau} M_{\Delta\tau}(z, x) + b \cdot \mathbb{1}, \tag{11}$$

where the weights of M_P and $M_{\Delta\tau}$ aim to normalize them to the same magnitude, $b \cdot \mathbb{1}$ denotes the same bias in SiamFC. Noted that in SiamFC, $f(z,x)$ compares an exemplar image z to a candidate image x in the current frame of the same size. It generates a high value if the two images describe the same object and low otherwise. And $f_{spike}(z,x)$ is the SNN fashion of this function, not strictly calculates the similarity between two spike trains (e.g., the similarity equals 1 but response equals 0 between two zero spike trains). It is why choosing T instead of spike numbers as the normalization parameter to get the response value. The proposed similarity estimation method makes it possible to convert the state-of-the-art Siamese trackers into deep SNNs. In the rest of the paper, we will refer to hybrid spiking similarity estimation as HSE.

3.3   Two-Status Coding Scheme

Although Eq. 11 enhances the temporal correlation between two spike trains, the temporal distributions of spikes are random and disorganized, which makes a large portion of spike values underutilized. For shortening latency and reducing energy consumption, we expect that the output spike of each time step will contribute to the final response value.

Phase coding (Eq. 6) indicates that assigning periodicity to spike trains enhances their ability of information transmission, which brings efficient performance in image classification. Therefore, we draw on the experience of periodic spiking feature when estimating the similarity. According to [3], repetitive electrical activity can induce a persistent potentiation or depression of synaptic efficacy in various parts of the nervous system (LTP and LTD). Motivated by this neural phenomenon, we propose two-status coding scheme. It represents the voltage threshold of potentiation status and depression status. We define the function as follows:

$$V_{\text{th}}(t) = \begin{cases} \alpha \text{ if mod } (t/p, 2) = 1 \\ \beta \text{ otherwise ,} \end{cases} \tag{12}$$

where $\alpha \to +\infty$ to prevent neurons from generating spikes during depression status and β is often smaller than the normalized voltage threshold 1 to excite neurons spiking during potentiation status, p is a constant that controls the period of neuron state change, t is the current time step. We set $p = 5$ by observation of experiments and $\beta = 0.5$ by experiences in Sect. 4.2.

Our two-status coding makes neurons fire with the same value in potentiation status and accumulates membrane potential in depression status. It constrains equivalent spikes in a fixed periodic distribution, increase the density of spikes in potentiation status to enhance response value. Figure 2(b) shows the distribution of spikes after two-status coding from the original rate coding. The proposed method can also save energy by avoiding neurons generating spikes which are useless for HSE. Moreover, spiking neurons in the next layer are not consuming energy during depression status due to zero input.

Park et $al.$ [20] propose a hybrid coding scheme by the motivation that neurons occasionally use different neural coding schemes depending on their roles and locations in the brain. Inspired by this idea, we use two-status coding scheme to optimize the temporal distribution of output spikes, and real coding for the input layer due to its fast and accurate features.

4   Experiments

4.1   Datasets and Metrics

We evaluate our methods on OTB-2013 [29], OTB-2015 [28], VOT2016 [17], VOT2018 [11], and GOT-10k [8]. The simulation and implementations are based on TensorFlow. To assess and verify the accuracy of the proposed methods, we use the average overlap ratio as the basic metric. As for energy consumption, we average the number of frames in a random video.

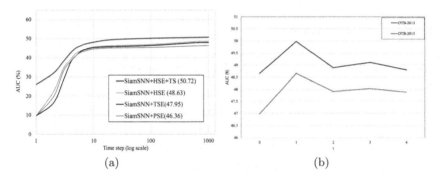

(a)   (b)

Fig. 3. (a). Results of SiamSNN on different configurations evaluated by latency and the AUC of success plots on OTB-2015. (b). Experimental results for different values of response period threshold on OTB-2013 and OTB-2015.

4.2   Ablation Experiments and Hyperparameters Settings

We conduct ablation experiments to analyze the effects of the proposed methods and determine the hyperparameters on OTB-2013 [29] and OTB-2015 [28].

To analyze the importance of each proposed component, we report the overall ablation studies in Fig. 3(a). We gradually add potential similarity estimation (PSE), potential similarity estimation (PSE), hybrid similarity estimation (HSE), and two-status coding (TS) on SiamSNN converted from SiamFC. As shown in Fig. 3(a), PSE is the baseline method and the reported scores are tested during 1000 time steps. The AUC score of PSE is 46.36. TSE improves tracking performance from 46.36 to 47.95. HSE brings 0.68 points higher AUC than TSE. Finally, TS obtains a 50.72 AUC score and reduces the latency. This result validates the effectiveness of our proposed methods. SiamSNN+HSE+TS achieves outstanding performance improvement than other configurations.

Hybrid Spiking Similarity Estimation. Figure 3(b) compares the results for several small values of response period threshold τ on OTB-2013 and OTB-2015. It can be seen that far-distance spikes have poor correlation and increasing the value of τ will bring more computational complexity. Thus, τ is set to 1.

Two-Status Coding. As mentioned before in Sect. 4.2, we set $\alpha \rightarrow +\infty$ ($\alpha =$ float('Inf') in python code) to prevent neurons from generating spikes during depression status. Refer to the exponential change of voltage threshold in burst coding [20], we set β to 0.5 to excite neurons spiking during potentiation status. It is inappropriate to set β to smaller values (e.g., 0.25 and 0.125), which will alter the balanced normalization in Eq. 2,5 and cause worse results. The latency requirements are presented in Fig. 3(a), we can see that SiamSNN converges rapidly at the beginning and then rises slowly. So in the two-status coding, we choose the state period p as 5 to improve the performance in shorter latency. To reach the maximum AUC, SiamSNN+HSE+TS requires approximately 20 time steps.

4.3   Experimental Results

Table 1. Experimental results with other state-of-the-art SNN methods on OTB2013 and OTB2015.

Similarity estimation	Coding	OTB2013	OTB2015	Latency
PSE	rate	48.03	46.25	600
PSE	Phase [9]	48.65	47.14	120
PSE	burst [20]	48.75	47.45	100
PSE	**two-status**	50.41	48.77	**20**
HSE	**two-status**	**51.15**	**49.31**	**20**

Comparison with Other Spike Coding Methods. To the best of our knowledge, SiamSNN is the first deep SNN model for object tracking. So we compare our coding method with phase [9] and burst coding [20] on OTB2013 and OTB2015. All experiments are performed with HSE due to lacking existed similarity estimation method for SNN tracking. As shown in Table 1, two-status coding outperforms phase and burst coding, especially on the latency.

Rate coding suffers from an intrinsic flaw about long latency [9]. Phase and burst coding achieve efficient information transmission, which shortens the latency. Our two-status coding optimizes the temporal distribution of output spike trains. It enhances the utilization of time steps for similarity estimation, which improves both accuracy and speed.

Tracking Results. Conversion methods of DNN-to-SNN will degrade the accuracy [10]. For further evaluation, we implement experiments on VOT-2016 [17], VOT2018 [11], and GOT-10k [8]. As shown in Table 2, compared to SiamFC on VOT benchmarks, the accuracy of SiamSNN drops 3.2% and 1.8% while the robustness becomes poor, and the degradations of EAO are 2.3% and 0.7%. As for GOT-10k, the AO and $SR_{0.5}$ score of SiamSNN degrades 3.4% and 2.6%. This is the first work that reports the comparable performance of SNNs for object tracking on VOT2016, VOT2018, and GOT-10k datasets.

Table 2. Experimental results of performance degradation on VOT2016, VOT2018 and GOT-10k.

Tracker	VOT-16			VOT-18			GOT-10k	
	A	R	EAO	A	R	EAO	AO	$SR_{0.5}$
SiamFC	0.529	0.49	0.233	0.478	0.69	0.183	0.348	0.353
SiamSNN	0.497	0.63	0.210	0.460	0.86	0.176	0.314	0.327

Energy Consumption and Latency Evaluation. To investigate the energy-efficient effect of SiamSNN, we evaluate energy consumption of our methods on TrueNorth [18] compared to SiamFC [2], SiamRPN++ [12], DSTfc [14] and ECO [5]. SiamRPN++ has deeper CNNs and stronger tracking performance that requires additional floating-point operations per second (FLOPS). DSTfc is a small, fast yet accurate tracker by teacher-students knowledge distillation. ECO is a state-of-art correlation-filter based tracker on CPU.

Table 3. Comparison of energy consumption of SiamSNN to other trackers.

Methods	FLOPs/SOPS	Power (W)	Time (ms)	Energy (J)
SiamFC	5.44E+09	250	11.63	2.91
SiamRPN++	1.42E+10	250	28.57	7.14
DSTfc	–	250	4.35	1.10
ECO	–	120	452.49	54.3
SiamSNN	**3.94E+08**	9.85E-04	20	**1.97E-05**

Refer to the reports of these works and [8,24], we summarize their hardware and speed information as follows. SiamFC(3 s) is run on a single NVIDIA GeForce GTX Titan X (250 W) and reaches 86 FPS (11.63 ms per frame). SiamRPN++ can achieve 35 FPS (28.57 ms per frame) on NVIDIA Titan Xp GPU (250 W). And DSTfc is run at 230 FPS (4.35 ms per frame) on Nvidia GTX 1080ti GPU (250 W). ECO is run on an Intel(R) Xeon(R) 2.0 GHz CPU (120 W) and achieves 2.21 FPS (452.49 ms per frame). And TrueNorth measures computation by synaptic operations per second (SOPS) and can deliver 400 billion SOPS per Watt, while FLOPs in modern supercomputers. And the time step is nominally 1 ms, set by a global 1 kHz clock [18]. We count the operations with the formula in [22].

The calculation results of processing one frame are presented in Table 3. The energy consumption of SiamSNN on TrueNorth is extremely lower than other trackers on GPU or CPU. Although GPUs are far more advanced computing technology, it is hard to employ DNNs on embedded systems through them. Neuromorphic chips with higher energy and computational efficiency have a promising development and application.

5   Conclusion

In this paper, we propose SiamSNN, the first deep SNN model for object tracking that reaches competitive performance to DNNs with short latency and low precision loss on OTB2013/2015, VOT2016/2018, and GOT-10k. Consequently, we propose an optimized hybrid spiking similarity estimation method and a two-status coding scheme for taking full advantage of temporal information in spike trains, which achieves real-time on TrueNorth. We aim at studying the spiking representation of SiamRPN and SiamRPN++ in the subsequent research. We believe that our methods can be applied in more spiking Siamese networks for energy-efficient tracking or other similarity estimation problems.

References

1. Basu, A., et al.: Low-power, adaptive neuromorphic systems: recent progress and future directions. IEEE J. Emerg. Sel. Topics Circ. Syst. 8(1), 6–27 (2018)
2. Bertinetto, L., Valmadre, J., Henriques, J.F., Vedaldi, A., Torr, P.H.S.: Fully-convolutional Siamese networks for object tracking. In: Hua, G., Jégou, H. (eds.) ECCV 2016. LNCS, vol. 9914, pp. 850–865. Springer, Cham (2016). https://doi.org/10.1007/978-3-319-48881-3_56
3. Bi, G.Q., Poo, M.M.: Synaptic modifications in cultured hippocampal neurons: dependence on spike timing, synaptic strength, and postsynaptic cell type. J. Neurosci. 18(24), 10464–10472 (1998)
4. Cao, Y., Chen, Y., Khosla, D.: Spiking deep convolutional neural networks for energy-efficient object recognition. Int. J. Comput. Vis. 113(1), 54–66 (2015)
5. Danelljan, M., Bhat, G., Shahbaz Khan, F., Felsberg, M.: Eco: efficient convolution operators for tracking. In: CVPR, pp. 6638–6646 (2017)
6. Diehl, P.U., Neil, D., Binas, J., Cook, M., Liu, S.C., Pfeiffer, M.: Fast-classifying, high-accuracy spiking deep networks through weight and threshold balancing. In: IJCNN, pp. 1–8. IEEE (2015)
7. Guo, J., Yuan, C., Zhao, Z., Feng, P., Luo, Y., Wang, T.: Object detector with enriched global context information. Multimed. Tools Appl.79(39), 29551–29571 (2020)
8. Huang, L., Zhao, X., Huang, K.: Got-10k: a large high-diversity benchmark for generic object tracking in the wild. IEEE Trans. Pattern Anal. Mach. Intell. 43, 1562–1577 (2019)
9. Kim, J., Kim, H., Huh, S., Lee, J., Choi, K.: Deep neural networks with weighted spikes. Neurocomputing 311, 373–386 (2018)
10. Kim, S., Park, S., Na, B., Yoon, S.: Spiking-yolo: Spiking neural network for energy-efficient object detection. In: AAAI (2020)
11. Matej, K., et al.: The sixth visual object tracking VOT2018 challenge results. In: ECCV 2018 Workshops, pp. 3–53 (2018)

12. Li, B., Wu, W., Wang, Q., Zhang, F., Xing, J., Yan, J.: SiamRPN++: evolution of Siamese visual tracking with very deep networks. In: CVPR, pp. 4282–4291 (2019)
13. Li, B., Yan, J., Wu, W., Zhu, Z., Hu, X.: High performance visual tracking with Siamese region proposal network. In: CVPR, pp. 8971–8980 (2018)
14. Liu, Y., Dong, X., Wang, W., Shen, J.: Teacher-students knowledge distillation for Siamese trackers. arXiv preprint arXiv:1907.10586 (2019)
15. Luo, Y., et al.: CE-FPN: enhancing channel information for object detection. arXiv preprint arXiv:2103.10643 (2021)
16. Luo, Y., et al.: A spiking neural network architecture for object tracking. In: Zhao, Y., Barnes, N., Chen, B., Westermann, R., Kong, X., Lin, C. (eds.) ICIG 2019. LNCS, vol. 11901, pp. 118–132. Springer, Cham (2019). https://doi.org/10.1007/978-3-030-34120-6_10
17. Kristan, M., et al.: The visual object tracking VOT2016 challenge results. In: Hua, G., Jégou, H. (eds.) ECCV 2016. LNCS, vol. 9914, pp. 777–823. Springer, Cham (2016). https://doi.org/10.1007/978-3-319-48881-3_54
18. Merolla, P.A., Arthur, J.V., Alvarez-Icaza, R., Cassidy, A.S., et al.: A million spiking-neuron integrated circuit with a scalable communication network and interface. Science 345(6197), 668–673 (2014)
19. Mulansky, M., Kreuz, T.: Pyspike-a python library for analyzing spike train synchrony. SoftwareX 5, 183–189 (2016)
20. Park, S., Kim, S., Choe, H., Yoon, S.: Fast and efficient information transmission with burst spikes in deep spiking neural networks. In: Design Automation Conference, p. 53. ACM (2019)
21. Roy, K., Jaiswal, A., Panda, P.: Towards spike-based machine intelligence with neuromorphic computing. Nature 575(7784), 607–617 (2019)
22. Rueckauer, B., Lungu, I.A., Hu, Y., Pfeiffer, M., Liu, S.C.: Conversion of continuous-valued deep networks to efficient event-driven networks for image classification. Front. Neurosci. 11, 682 (2017)
23. Russakovsky, O., Deng, J., Su, H., Krause, J., Satheesh, S., Ma, S., Huang, Z., Karpathy, A., Khosla, A., Bernstein, M., et al.: Imagenet large scale visual recognition challenge. Int. J. Comput. Vis. 115(3), 211–252 (2015)
24. Schuchart, J., Hackenberg, D., Schöne, R., Ilsche, T., Nagappan, R., Patterson, M.K.: The shift from processor power consumption to performance variations: fundamental implications at scale. Comput. Sci. Res. Dev. 31(4), 197–205 (2016). https://doi.org/10.1007/s00450-016-0327-2
25. Sengupta, A., Ye, Y., Wang, R., Liu, C., Roy, K.: Going deeper in spiking neural networks: VGG and residual architectures. Front. Neurosci. 13, 95 (2019)
26. Tavanaei, A., Ghodrati, M., Kheradpisheh, S.R., Masquelier, T., Maida, A.: Deep learning in spiking neural networks. Neural Netw. 111, 47–63 (2018)
27. Voigtlaender, P., Luiten, J., Torr, P.H.S., Leibe, B.: Siam R-CNN: visual tracking by re-detection. In: CVPR (2020)
28. Wu, Y., Lim, J., Yang, M.-H.: Object tracking benchmark. IEEE Trans. Pattern Anal. Mach. Intell. 37(9), 1834–1848 (2015)
29. Wu, Y., Lim, J., Yang, M.H.: Online object tracking: a benchmark. In: CVPR, pp. 2411–2418 (2013)
30. Wu, Y., Deng, L., Li, G., Zhu, J., Xie, Y., Shi, L.: Direct training for spiking neural networks: faster, larger, better. AAAI 33, 1311–1318 (2019)
31. Xu, Y., Wang, Z., Li, Z., Yuan, Y., Yu, G.: SiamFC++: towards robust and accurate visual tracking with target estimation guidelines. In: AAAI (2020)

32. Yang, Z., Wu, Y., Wang, G., Yang, Y., et al.: DashNet: a hybrid artificial and spiking neural network for high-speed object tracking. arXiv preprint arXiv:1909.12942 (2019)
33. Yuan, C., et al.: Learning deep embedding with mini-cluster loss for person re-identification. Multimed. Tools Appl. **78**(15), 21145–21166 (2019). https://doi.org/10.1007/s11042-019-7446-2

The Principle of Weight Divergence Facilitation for Unsupervised Pattern Recognition in Spiking Neural Networks

Oleg Nikitin$^{(\boxtimes)}$ ⓘ, Olga Lukyanova ⓘ, and Alex Kunin ⓘ

Computing Center, Far Eastern Branch of the Russian Academy of Sciences,
Khabarovsk 680000, Russia
{oleg,alex}@kanju.tech, olukyanova@ccfebras.ru

Abstract. Parallels between the signal processing tasks and biological neurons lead to an understanding of the principles of self-organized optimization of input signal recognition. In the present paper, we discuss such similarities among biological and technical systems. We propose adding the well-known STDP synaptic plasticity rule to direct the weight modification towards the state associated with the maximal difference between background noise and correlated signals. We use the principle of physically constrained weight growth as a basis for such weights' modification control. It is proposed that the existence and production of bio-chemical 'substances' needed for plasticity development restrict a biological synaptic straight modification. In this paper, the information about the noise-to-signal ratio controls such a substances' production and storage and drives the neuron's synaptic pressures towards the state with the best signal-to-noise ratio. We consider several experiments with different input signal regimes to understand the functioning of the proposed approach.

Keywords: Neural homeostasis · Spike-timing-dependent plasticity · Synaptic scaling · Adaptive control · Bio-inspired cognitive architectures · Neural networks · Neural network architectures

1 Introduction

Signal processing is a well-known field of computer science. Since the invention of radio signal transmission, there was a task of optimal signal recovery from noise. Correlators represent the mathematical approach to solve the noise filtering task. They perform a simple mathematical operation to estimate the correlation coefficient among all input sources and may be implemented as special hardware. The fundamental goal of the correlator is to find the correlated patterns in noise and to improve the signal-to-noise ratio [1]. The properties of noise have to be determined to recover the signal. Wiener pioneered in this area in [11] and used

---

The computing resources of the Shared Facility Center "Data Center of FEB RAS" (Khabarovsk) were used to carry out calculations.

© Springer Nature Switzerland AG 2021
I. Farkaš et al. (Eds.): ICANN 2021, LNCS 12895, pp. 195–206, 2021.
https://doi.org/10.1007/978-3-030-86383-8_16

Volterra expansion for the recovery in the recurrent circuit. These two sides of signal processing remind of the structure of the spiking model of a neural cell [4] with plastic Hebbian synapses [3]. Indeed, brain information processing might be viewed as a signal processing task. During the signal receiving, the brain aims to make sense of the surrounding environment. Some of the signals represent valuable objects in the form of repetitive patterns. Brain neurons should detect correlations in the input information to recognize such repetitions. For input object detection, neurons have to increase the divergence between reactions to the noise and the regularly appearing objects.

In the present article, we investigate spiking neurons with spike-timing-dependent plasticity (STDP) as active signal recovering devices, seeking correlations in the present input. We propose the modification of the STDP rule [3,10] incorporating plasticity substance pool restriction [7]. This modification sets the goal for the neuron to increase the weight divergence between random and correlated input and keep the firing rate of the neuron constrained. The presented model of a neural cell lets neural networks improve the signal-to-noise ratio of a whole network in a layered manner.

2 Signal Processing and Biological Groundings for Weight Divergence Optimization Approach

It is remarkable how much in common do modern approaches to signal processing and brain simulation have. However, the connection is often neglected, and a significant legacy of signal processing research seems to us under-looked in theoretical neuroscience. It may include understanding the brain as the optimal filtering device to improve the signal-to-noise ratio to function adaptively in an ever-changing world.

A single neuron or small neural assembly task is very similar to some radar stations. The passive radar station constantly scans the surrounding area in search of repetitive patterns, so as the neuron searching for patterns in input spike combinations. For passive signal receivers, it is essential to find correlated, and repetitive unknown patterns in the presence of incoherent noise [5]. The optimal detector for such a task should incorporate a set of correlation detectors and a nonlinear filter to recognize noise and signal parameters (see Fig. 1).

The most distinctive parts of the filter presented above are synchronized correlators for different signal reception and a recurrent filter to estimate noise and signal parameters. The basic structure of the correlator was proposed by Fano [1]. The basic idea of correlator lies in adaptive filtration and recurrent reverberation of input. Thus, the goal of the correlator is similar to the Hebb postulate. Indeed, the modern formulation of the STDP principle [3,10] raises from the correlation approach to the synaptic weight plasticity in biological neurons.

The filtering approach for noise parameters estimation was first proposed in [11]. However, it is notable that in this paper, Wiener used an approach that later became known as Volterra series expansion and [4] also used Volterra expansion

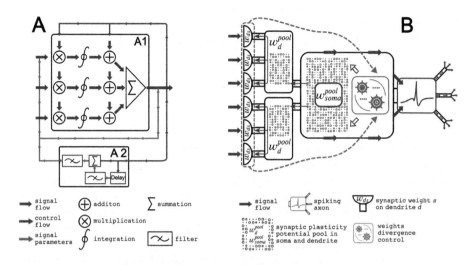

Fig. 1. The principal schematic of the optimal signal receiver and unsupervised learning neuron. A – the optimal signal receiver, A1 – the correlation receiver, A2 – the noise and signal parameters estimation filter, B – the spiking neuron with weight divergence maximization control.

for the reduction of the Hodgkin-Huxley model to a more simplified one. Thus, in principle, it shows that cell spiking behavior may, in theory, play the role of noise detector and active coordinator of synaptic plasticity as correlation detector filters.

In our previous work, we have introduced the model of a neuron with STDP synapses and a restricted plasticity pool, used for pro-active control of firing rate [7]. Here, we extend this approach to maximize the noise-to-signal discrimination efficiency of a spiking neuron in an unsupervised reception task. We propose that the goal of many neural cells might be to find and enhance the difference between the correlated repetitive input patterns and background noise. It is consistent with a biologically plausible model [8] and with earlier proposed research [6] and also, with optical hardware-based implementations of well-known neural networks [9].

Below, we propose the weight divergence maximization principle as guiding for the restriction of Hebbian plasticity. In this approach, like in the optimal filter proposed above, there is a set of correlators, performing a cross-correlation between different input signals by STDP plasticity restricted by the plasticity potential pool. This pool comprises an abstract substance responsible for synaptic growth and controlled by central cellular processes to improve the correlated signals selectivity. Furthermore, we suppose that some plasticity substance values are spent to let weights grow, restricting the weight amplification. In previous research, we have demonstrated that such an approach allows heterosynaptic plasticity and synaptic competition and amplifies the signal-to-noise ratio of Hebbian plasticity while keeping the firing frequencies of neurons under precise control. Below, we describe the modification of the described approach to amplify the correlated signal transduction in an unsupervised manner.

3 A Weight Divergence Maximization Approach to Hebbian Plasticity

The modification of membrane structures realizes synaptic plasticity in biological neurons at the expense of the secretion of proteins. In the present paper, we will call these weight modification structures the plasticity pool. It consists of abstract substances spending on weight modification. In [7] authors provide the detailed explanation and formulation of dependencies and equations. Here, we will give a brief overview of the basic principle of weight correction restriction. In this paper, we consider the Izhikevich [2] spiking neuron model composed of three dendritic compartments. Dendrites have synaptic inputs with a set of weights w_{ds}. All signals x_{ds} are weighted and summed and transferred to the dynamical excitability model. We suppose that a weight correction substance is secreted in the central compartment of the neuron. It is stored in the soma and called w_{soma}^{pool}. The weight plasticity substance is spent from w_{soma}^{pool} to refill the plasticity storage pools w_d^{pool} in d dendrites.

The plasticity pool w_d^{pool} is then used to provide the resources for synaptic weights amplification. The basic dynamics of weights modification in the paper is equivalent to asymmetric STDP rule [10]. The initial weight correction Δw_{ds}^{stdp} is calculated according to the standard STDP rule. We use restriction coefficient of the synaptic weight growth k_d^{wp} to determine the final update $\Delta w_{ds}^{stdp^{**}}$ of the synaptic weight of the synapse s on the dendrite d:

$$\Delta w_{ds}^{stdp^{**}} = \begin{cases} \Delta w_{ds}^{stdp}, & \text{if } \Delta w_{ds}^{stdp} < 0, \\ k_d^{wp} \cdot \Delta w_{ds}^{stdp}, & \text{if } \Delta w_{ds}^{stdp} > 0, \end{cases} \tag{1}$$

where k_d^{wp} is the restriction coefficient of the growth of synaptic weights depending on the deviation from the optimum frequency.

$$k_d^{wp} \sim \frac{w_d^{pool}}{\sum\limits_{s} \Delta w_{ds}^{stdp}}, \text{for all } \Delta w_{ds}^{stdp} > 0, \tag{2}$$

According to Eq. 2, weight growth restriction depends on the control of the w_d^{pool} and, consequently, of the w_d^{pool} as the initial source of plasticity restricting substance. The neuron is involved in signal separation from noise. We may use the w_{soma}^{pool} level control for the facilitation of correlated input signals separation from noise. In this regard, we calculate w_{soma}^{pool} for the next step depending on the maximum estimated standard deviation $\sigma(w_{ds})$ of synaptic weights.

$$\sigma(w_{ds}) = \sqrt{\frac{1}{d \cdot s - 1} \sum_d \sum_s (w_{ds} - \bar{w})^2}, \tag{3}$$

where w_{ds} – the current synaptic weight of the synapse s on the dendrite d in time t, $\bar{w}$ – the mean value of the w_{ds} in time t.

We propose an algorithm to maximize the synaptic weight divergence based on w_{soma}^{pool} allocation (see Algorithm 1). This approach can be called **weight divergence maximization (WDM)**.

Algorithm 1: The weight divergence maximization algorithm (WDM) based on w_{soma}^{pool} allocation by the standard deviation of synaptic weights.

Require: Δw_{ds}^{stdp}
1: Calculate w_{ds}^{min} (if $w_d^{pool} = 0$) and w_{ds}^{max} (if $w_d^{pool} = w_d^{res}$)
2: Calculate standard deviations: $\sigma(w_{ds}^{min})$ and $\sigma(w_{ds}^{max})$
3: Compare $\sigma(w_{ds}^{min})$ and $\sigma(w_{ds}^{max})$:
4: **if** $\sigma(w_{ds}^{min}) > \sigma(w_{ds}^{max})$ **then**
5:    Set $w_{soma}^{pool} = 0$
6: **else if** $\sigma(w_{ds}^{min}) < \sigma(w_{ds}^{max})$ **then**
7:    Set $w_{soma}^{pool} = \sum_d \sum_s \Delta w_{ds}^{stdp}$ for all $\Delta w_{ds}^{stdp} > 0$
8: **else if** $\sigma(w_{ds}^{min}) = \sigma(w_{ds}^{max})$ **then**
9:    Remain w_{soma}^{pool} the same as in the end of the previous step $t - 1$
10: **end if**
11: **return** w_{soma}^{pool}

Here, w_d^{res} is the maximum amplitude of growth of the sum of weights on the dendrite d in t previous steps. We calculate the theoretically possible minimum w_{ds}^{min} (if w_d^{pool} is equal to zero) and maximum w_{ds}^{max} (if w_d^{pool} is maximum possible) values of the weights. Next, we determine the standard deviation from the resulting values ($\sigma(w_{ds}^{min})$ and $\sigma(w_{ds}^{max})$) and compare them. Based on this comparison, we set the w_{soma}^{pool}, which should lead to maximization of the weights divergence for the current step.

3.1 Firing Rate-Controlling Approaches to WDM

The basic WDM algorithm should maximize the difference between informative and random inputs. However, it set no goal for the neuron firing rate, and it will fluctuate according to input signal parameters. The approach above seems to be plausible but not flexible enough. We cannot set it to some determined goal point. At the same time, for some tasks such as switching between different network oscillation frequencies, it is crucial to keep the target frequency. That is why it is necessary to drive the current firing rate θ_{real} to some set point as target firing rate θ_{target}. We calculate the difference ($\Delta\theta$) between θ_{target}

and θ_{real} to achieve the selected goal. If the firing rate is too low, we satisfy all the demand for plasticity modification ($\sum_d \sum_s \Delta w_{ds}^{stdp}$) and essentially act for the next step as general STDP. In the case when the firing rate θ_{real} is too high w_{soma}^{pool} is set to zero. Actions in both deviation cases are done with some probability $p(\Delta\theta)$, proportional to the current deviation. We call this approach a **firing rate optimization (FRO)** of the WDM (see Algorithm 2).

Algorithm 2: The weight divergence maximization algorithm with firing rate-controlling.

Require: θ_{real}, θ_{target}
1: Perform all the steps of the Algorithm 1
2: Calculate $\Delta\theta = \theta_{real} - \theta_{target}$
3: **if** $\Delta\theta > 0$ **then**
4:     Calculate the probability $p(\Delta\theta) = \beta^+ \cdot \Delta\theta$
5:     With $p(\Delta\theta)$ probability, set $w_{soma}^{pool} = 0$
6: **else if** $\Delta\theta < 0$ **then**
7:     Calculate the probability $p(\Delta\theta) = -\beta^- \cdot \Delta\theta$
8:     With $p(\Delta\theta)$ probability, set $w_{soma}^{pool} = \sum_d \sum_s \Delta w_{ds}^{stdp}$ for all $\Delta w_{ds}^{stdp} > 0$
9: **end if**
10: **return** w_{soma}^{pool}

A firing rate minimization represents the special case. Indeed, for the biological brain, it is crucial to minimize the energy spent on spiking. Hence, it is most apparent to minimize the firing rate along with STD maximization. We call it a **firing rate minimization (FRM)** of the WDM approach. In this case, the θ_{target} is set to zero, meaning that the frequency will always be minimal, while weights' standard deviation (SD) will be maximized. We introduce the amplification coefficients β^-/β^+ to control the sensitivity to minimization/maximization of firing rate and apply them to $p(\Delta\theta)$. In our experiments below, the β^+ and β^- equal to one.

4   Different Noise Filtering Tasks Examination

In order to test the possibilities of using the methods described above, we propose to carry out tests with various combinations of noise and input signals. For each synapse, the signal in the experiments is a time series of binary data, where the numbers of "1" appear at random with a specific predetermined frequency. Since the data is binary, we calculate the closeness of the correlation for correlated signals between different synapses using the Matthews correlation coefficient. Further, we compare the proposed models with the model of the basic STDP and **the model of backward calculation of plasticity potential reserve demand (PPD)** proposed in [7]. The neuronal plasticity regulation model changes the potential synaptic weight growth by controlling the plasticity potential pool to stabilize the artificial neuron's firing rate dynamically.

Experiments are carried out on the neuron with three dendrites, each of which has six synapses. We run 100 tests with 2400 steps. The results for the figures are averaged.

4.1   Detection of Signals with Different Correlation Coefficients

In the following experiment, the nature of the signal incoming to the neuron varied from noise to a correlated signal. We send correlated input to the chosen synapses during specific time ranges. The signal frequency for all synapses is 0.2 throughout the experiment. The correlation coefficient for chosen synapses from 100 to 300 steps and from 1300 to 1500 steps is 0.5, from 500 to 700 steps and from 1700 to 1900 steps is 0.7, and from 900 to 1100 steps and from 2100 to 2300 steps is 0.9. The rest of the synapses during these periods receive noise with a frequency of 0.2.

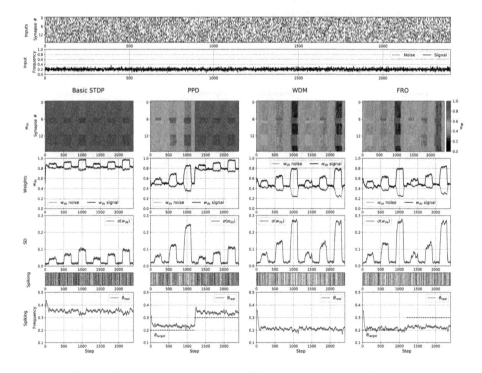

Fig. 2. Detection of signals with different correlation coefficients.

In Fig. 2, you can see that all models, including STDP, are capable of filtering noise by extracting correlated signals. The higher the correlation coefficient between the incoming signals, the better the correlated signal differs from the noise: with a low correlation coefficient, the SD value is significantly lower than with a high correlation coefficient, and vice versa. It can be seen that the basic STDP copes with this task worse than other models and also adheres to a high current firing rate in comparison with other models due to an increase in synapse weights. Increasing the target firing rate by 0.1 for the PPD model degrades its results significantly, making them close to the baseline PPD. It is due to the minimization of the w_{soma}^{pool} deficit, which removes the restriction on the growth of synapse weights. The same increase in the target firing rate for the FRO model results in a slight degradation in quality and a slight increase in the firing rate. The WDM model independently goes into a state with a current firing rate close to the frequency of the incoming signals. We can see that all the weight-restricted approaches (PPD, WDM, FRO) improve the signal-to-noise ratio relative to the standard STDP plasticity rule.

4.2   Detection of a Correlated Signal from Low-Frequency Noise

In the second experiment, we input the low-frequency noise and change the signal parameters for the chosen synapses.

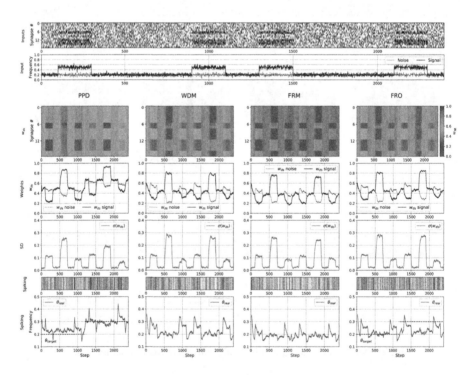

Fig. 3. Detection of the correlated signal from low-frequency noise with low target firing rate. Here, the input frequency is constant, and the correlation between input signals varies over time.

So, we apply the correlated signals to 2 synapses of one dendrite and 4 synapses of another. A signal of different frequencies arrives at these synapses with or without correlation in specified ranges. All other synapses always receive the noise.

We set the ranges for chosen synapses as follows: from 100 to 300 steps and from 1300 to 1500 steps, we input the noise with a frequency of 0.5, from 500 to 700 steps, and from 1700 to 1900 steps, synapses receive a correlated signal with a frequency of 0.2, from 900 to 1100 steps and from 2100 to 2300 steps a correlated signal with a frequency of 0.5 is received (high-frequency correlated input). The rest of the synapses during these periods receive noise with a frequency of 0.2. In all remaining step ranges, all synapses of the neuron receive noise with a frequency of 0.2.

The plot of synapse weights in Fig. 3 shows that all models filter out high-frequency noise and high-frequency correlated signals entering the neuron simultaneously with low-frequency noise.

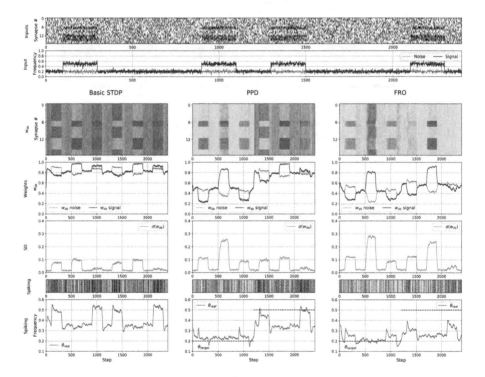

Fig. 4. Detection of a correlated signal from low-frequency noise with high target firing rate.

At the same time, the plots of the SD show that the FRM model manages to achieve slightly higher SD values on the ranges with high-frequency correlated input. Also, in these areas, this model shows the lowest firing rate. All weight divergence models operate at approximately the same firing rate during the periods of low-frequency correlated input. At the same time, the model aimed at maximizing weight divergence reaches the highest SD values. Setting the target firing rate to 0.3 worsens the test results for the PPD model but almost does not change the results of the FRO model.

Figure 4 shows the results of modeling the same task, but for the basic STDP model, for PPD and FRO with a target firing rate equal to 0.5. Again, we see that only basic STDP and PPD with a high target firing rate can filter high-frequency correlated signals from low-frequency noise. In the case of PPD, we achieve it through a significant reduction in the w_{soma}^{pool} deficit, which brings this model closer to the basic STDP. At the same time, PPD effectively separates high-frequency noise from low-frequency noise. In the case of FRO, when the target firing rate increase, the model still effectively separates the low-frequency correlated signal from the low-frequency noise but continues to filter out the high-frequency correlated signal, albeit to a lesser extent than with a low target firing rate.

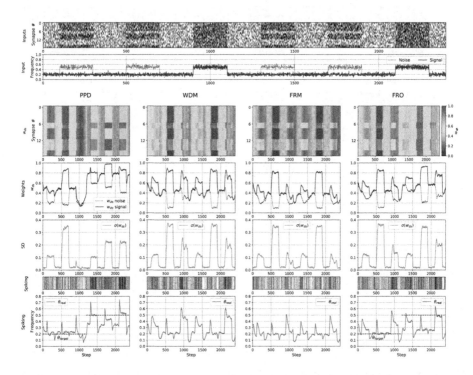

Fig. 5. Detection of a correlated signal from high-frequency noise.

4.3   Detection of Correlated Signal from High-Frequency Noise

In this experiment, we assess the change in signal parameters for chosen synapses using high-frequency noise. Thus, we apply the correlated signals to 2 synapses of one dendrite and 4 synapses of another. A signal of different frequencies arrives at these synapses with or without correlation in specified ranges. All other synapses always receive the noise. These parameters are the same as in the first experiment. In the difference with the last part, we present the functioning of models in the presence of more intense uncorrelated high-frequency noise. The ranges for the chosen synapses in this task are as follows: from 100 to 300 steps and from 1300 to 1500 steps, we input the noise with a frequency of 0.2, from 500 to 700 steps, and from 1700 to 1900 steps, synapses receive a correlated signal with a frequency of 0.2, from 900 to 1100 steps and from 2100 to 2300 steps we input a correlated signal with a frequency of 0.5. The rest of the synapses during these periods receive noise with a frequency of 0.5. In all remaining step ranges, all synapses of the neuron receive noise with a frequency of 0.2.

Figure 5 shows that in areas where low-frequency and low-frequency noise are combined, the PPD model reduces the weights at synapses with high-frequency noise, filtering them out. At the same time, the WDM model increases the weights of synapses with low-frequency noise. The FRM and FRO models simultaneously increase the weights of synapses with low-frequency noise and filter out synapses with high-frequency noise. Thus, all four models achieve approximately the same SD value in these areas. It is also unaffected by setting a high spike rate target. During the periods of low-frequency correlated input, all four models achieve high SD values. Setting a high target firing rate for the PPD model degrades the simulation results significantly. However, for the FRO model, the SD indicator decreases insignificantly. In the area with a high-frequency correlated input and a low target firing rate, the PPD model equally reduces the weights of all neuron synapses, filtering out the input signals for them. With the nature of the inputs, the PPD model can successfully filters high-frequency noise only when setting high target values of the firing rate. The FRM model also copes with this task worse than the WDM model, reducing its ability to filter out high-frequency noise from high-frequency correlated signals due to its tendency to reduce the firing rate after their sharp increase. Although it reaches various SD values, the FRO model successfully copes with this task both at high and low target values of the firing rate.

Thus, the models discussed above have both advantages and disadvantages, depending on the specific task. For example, the WDM model shows itself better than other models in detecting a low-frequency correlated signal, but it increases the firing rate when working with high-frequency signals. On the other hand, the FRM model allows one to restrain the growth of the firing rate, but this ability degrades the performance when high-frequency signals with high-frequency noise arrive. At the same time, the FRO model allows one to optimize the firing rate without significant losses in the quality of work. However, its practical application involves the setting of a target firing rate.

5   Discussion and Perspectives

In the present paper, we have drawn the connections between signal reception approaches and the biological basement of neural activities. We have proposed an unsupervised approach to noise filtering and signal reception based on Hebbian plasticity. This approach is biologically plausible and based on physical restrictions of the synaptic modifications of real biological neurons. Examining the performance of restricted STDP in the presence of different signals and noise intensities showed that the physically bounding of STDP plasticity and goal-directed control of the synaptic growth led to significant improvement of the signal-to-noise ratio of a single neuron as a signal receiver. The plasticity rules in the paper are unsupervised and local, as well as the standard STDP rule. The rule showed in the present paper was studied for one stimulated neuron and needs to be studied in networks, including networks consisting of excitatory and inhibitory neurons. The frequency control model (FRO) will stabilize the overall frequencies in the whole neural net, allow switching of neural assemblies in different oscillation regimes, and allow replay of different memorized patterns inside the minimal neural network circuit.

References

1. Fano, R.M.: On the signal-to-noise ratio in correlation detectors. Technical report, 186, Research Laboratory of Electronics, M.I.T. (1951)
2. Izhikevich, E.M.: Simple model of spiking neurons. IEEE Trans. Neural Netw. **14**(6), 1569–1572 (2003)
3. Kempter, R., Gerstner, W., van Hemmen, L.: Hebbian learning and spiking neurons. Phys. Rev. E **59**(4), 4498 (1999). https://doi.org/10.1103/PhysRevE.59.4498
4. Kistler, W.M., Gerstner, W., van Hemmen, L.: Reduction of the Hodgkin-Huxley equations to a single-variable threshold model. Neural Comput. **9**(5), 1015–1045 (1997). https://doi.org/10.1162/neco.1997.9.5.1015
5. Liu, J., Li, H., Himed, B.: Analysis of cross-correlation detector for passive radar applications. In: 2015 IEEE Radar Conference (RadarCon), pp. 772–777 (2015). https://doi.org/10.1109/radar.2015.7131100
6. Livshits, M.S.: Associative neurons as correlators. Biofizika **47**(3), 559–563 (2002)
7. Nikitin, O., Lukyanova, O., Kunin, A.: Constrained plasticity reserve as a natural way to control frequency and weights in spiking neural networks (2021). https://arxiv.org/abs/1611.02167. Accessed 05 Mar 2021
8. Parise, C., Ernst, M.: Correlation detection as a general mechanism for multisensory integration. Nat. Commun. **7**, 11543 (2016). https://doi.org/10.1038/ncomms11543
9. Sokolov, V.K., Shubnikov, E.I.: Optical neural networks based on holographic correlators. Quantum Electron. **25**(10), 1032–1036 (1995)
10. Song, S., Miller, K., Abbott, L.: Competitive hebbian learning through spike-timing-dependent synaptic plasticity. Nat. Neurosci. **3**(9), 919–926 (2000). https://doi.org/10.1038/78829
11. Wiener, N.: Response of a nonlinear device to noise. Technical report, 129, Radiation Laboratory, M.I.T., Cambridge (1942)

Algorithm for 3D-Chemotaxis Using Spiking Neural Network

Jayesh Choudhary$^{(\boxtimes)}$, Vivek Saraswat, and Udayan Ganguly

Department of Electrical Engineering, Indian Institute of Technology Bombay,
Mumbai, India

Abstract. Chemotaxis performed by C. Elegans worm has inspired simple Spiking Neural Networks to perform navigation and contour tracking efficiently in 2D planar concentration space. Extending it to 3D media, i.e. a new degree of freedom has quite a few challenges. In this work, we aim to devise an end-to-end spiking implementation for reaching a set concentration followed by contour tracking in 3D media. Here we devise an algorithm based on klinokinesis - where the worm's motion is in response to the stimuli but not proportional to it. Thus the path followed is not the shortest, but we can track the set concentration successfully. Most significantly, we are using simple leaky-integrate and fire (LIF) neurons for the neural network implementation, considering the feasibility on the neuromorphic computing hardware.

Keywords: Navigation · Contour-tracking · Spiking Neural Network · Klinokinesis

1 Introduction and Motivation

Spiking Neural Networks are artificial neural networks that have more resemblance to the biological neural network. Here, the information is transferred from one neuron to another via discrete events, i.e., binary spikes allowing the network to relay information faster and in an energy-efficient manner. There is hence a significant interest in developing SNN based control applications for navigation and robotics.

Most of the neurobiological studies are performed on the worm C. Elegans because of its simple neural network. Despite being a simple organism, many of the molecular signals controlling its development and tracking ability are also found in more complex organisms. One of its crucial behavior is Chemotaxis - the ability to move in response to sensed concentration. This phenomenon is essential in various organisms not just to find food but for immunity, embryogenesis and many more activities. Also, it provides us with the knowledge of physiology and neural control of movements. Much work has been done previously in the field of autonomous navigation inspired by its chemotaxis network. Initial work in the field of Chemotaxis includes [1] where the author proposes a network of sensory neurons capable of successfully driving Chemotaxis via steering and

© Springer Nature Switzerland AG 2021
I. Farkaš et al. (Eds.): ICANN 2021, LNCS 12895, pp. 207–219, 2021.
https://doi.org/10.1007/978-3-030-86383-8_17

controlling the angle of turn. [2] introduces motor neurons to the above network of sensory neurons to demonstrate klinokinesis. However, their network includes few mathematical calculations without neural circuits. [3] improves upon this to provide an end-to-end SNN implementation for klinokinesis. Since klinokinesis is performed, the shortest path is not guaranteed. [4] provides an adaptive klinotaxis model to ensure the shortest path of the worm. All these works help in the gradual quantitative investigation of path-planning and contour tracking of the worm. But, two major issues need to be addressed:

1. In the above-discussed works, the sensory neurons being used (ASEL and ASER) have a complex neuron model compared to simple LIF neurons. Thus its implementation on dedicated neuromorphic hardware is not possible (LOIHI processor has only LIF neurons). To get the functionality of a sensory neuron, several LIF neurons would be needed adding to the complexity of the network and the simulation time.
2. Since the worm's natural environment is 3D, the investigation should be focused on 3D medium. The implementation is based on path correction when the worm deviates from the set concentration. The worm corrects its path by aligning itself along or against the gradient by turning. In a 2D medium, a turn or deviation from the current path is sufficient to cover the entire medium due to only 2 degrees of freedom. Thus devising uniform logic for the path correction is easier. In a 3D medium, an extra degree of freedom makes it quite challenging to devise an algorithm because compared to only two directions in 2D, now we have infinite directions to choose from when path correction is needed. In other words, we need to move along a line when path correction is required in 2D, whereas we need to move across a plane when correction is required in 3D.

[5] discuss the observed motion of C. elegan in a 3D medium. The author proposes roll maneuvers, giving the organism the capability to reorient its body by combining 2D turns in the plane of dorsoventral undulations with 3D roll maneuvers enabling it to explore the whole of 3D media. However, [5] only focuses on the behavior of the worm and not on the SNN implementation. To broaden the spectrum, we also consider the qualitative behavior of Chemotaxis in other organisms. [6] discuss the sperm navigation along the helical path in 3D chemoattractant landscape.

Although, the work in this field started out heavily motivated by C. Elegans biology, it serves only as an inspiration in this work to provide an engineering solution for navigation systems based on the basic framework of sensory, functional and motor neurons in the worm. We aim to develop a simple autonomous end-to-end solution for contour tracking in the 3D medium completely in the spiking domain since the main constraint is to use neuromorphic supported hardware and neurosynaptic network only. Extending the work from 2D to 3D medium is interesting because of the new degree of freedom involved without any new variable to work with, making it more challenging.

2   Assumption and Setting

The worm has only one sensor which senses the concentration at a particular point in space. Thus the organism has only the concentration value available to it as input. The organism already knows the set concentration, and it maneuvers accordingly by predicting the path using the current concentration value and the previous concentration value. We assume that the neural activity is much faster than the motor action, and thus in simulation, we have considered two different time scales for both. Time step-size for neural dynamics is 1000 times smaller than that of the motor action. In other words, one step of motor action has 1000 steps of neural activity. This time step-size of motor action is known as time-window, and during this period, the worm's position remains the same. This assumption is valid and of immense importance because as the hierarchy of neurons increases in the network, it is difficult to obtain perfectly matched spiking whenever needed. The neurons at the lower level of the network are bound to have some delay. It thus becomes difficult to ensure that the feedback corresponding to a neuron is taken into account for the same time instant. The time window helps us to deal with this issue. Also, the time window ensures that the spiking frequency is the same in one time window for sensory neurons making the rate encoded information transfer easier.

All the neurons used are LIF neurons which are not considered to be plastic in nature. The equations are in the form such that it is compatible with the neuromorphic hardware.

$$I[i+1] \ = \ I[i] * (1 - decay) \ + \ I\_0 * (y) \tag{1}$$

$$V[i+1] \ = \ V[i] * (1 - decay) \ + \ \frac{k}{C} * (I[i+1]) \tag{2}$$

In Eq. 1, I[i] is the current at instant i, decay is the decay constant, I_0 is some constant value. The value of y depends on the type of neuron. y can be a constant value(for constant bias current), proportional to concentration value (as in the case of sensory neurons), or a linear combination of binary spikes (post-synaptic neuron).

In Eq. 2, V[i] is the potential built in the neuron at time instant i, decay is the decay constant, k is an arbitrary constant, and C is the capacitance.

We are considering both discrete and continuous 3D concentration space. The discrete concentration space is more in accord with the neuromorphic hardware available as of now, where the concentration has to be encoded as frequency. Since we have a time window of 1000 samples, the maximum frequency obtained is 500 (when spikes are observed every alternate time instant). Other frequencies that could be observed are 333 (when spikes are observed every 3rd time instant), 250, 200, and so on. For meaningful encoding and decoding of the information, we need a one-to-one mapping between concentration and frequency. Suppose a concentration value of 6.7 units corresponds to the frequency of 500 and 3.4 corresponds to 333, then every concentration value between 3.4 to 6.7 will propagate the same information since the frequency would be the same. This could lead to a discrepancy in the continuous concentration space simulations. Also, tracking is much better when there is some margin on both sides of the set concentration. This ensures that if we go in the opposite direction, useful information could be transferred in the network giving the worm feedback to correct its path. The discrete concentration space have the following values: 0.1, 0.2, 0.3, 0.4, 0.5, 0.6, 0.7, 0.8, 0.9, 1.0, 1.2, 1.4, 1.8, 2.3, 3.4, 6.7.

In terms of profile, we are considering Gaussian concentration profile to try out all possible ascend and descend from various starting points in space along with linear concentration profile because key observations are easier to illustrate in linear profile in terms of contour tracking and its behavior when it reaches the set concentration.

3   Approach and Algorithm

We only have the concentration values as rate encoded information, and we navigate the 3D space based on this information. Since we are using the path correction method, we only need to identify when the worm goes off-course. There are four control conditions as mentioned in Table 1 which help us identify these cases.

Table 1. Control conditions for navigation

Setpoint relation	Gradient relation	Feedback
$C\_set > C\_current$	$C\_current > C\_previous$ or $\Delta > 0$	Continue moving
$C\_set < C\_current$	$C\_current < C\_previous$ or $\Delta < 0$	Continue moving
$C\_set > C\_current$	$C\_current < C\_previous$ or $\Delta < 0$	Change direction
$C\_set < C\_current$	$C\_current > C\_previous$ or $\Delta > 0$	Change direction

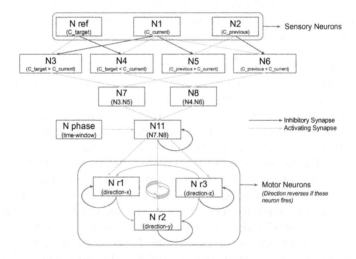

Fig. 1. Flow chart for the network of the neurons

We have few sensory neurons to sense the concentration, some motor neurons that help in the motion, and few intermediate neurons that help complete the network. The basic structure of network comprising of level and gradient detectors as highlighted later, is essentially established well in previous implementations of 2D chemotaxis [2]. Sensory neurons have current dynamics dependent on concentration values. Intermediate neurons and motor neurons have excitatory and inhibitory synaptic traces due to other neurons. We will go into more depth related to the current dynamics of the neurons in the next section.

The neural network consists of 14 neurons. N_ref, N1, and N2 are the sensory neurons that spike corresponding to the set concentration, current concentration of the worm, and the worm's previous concentration, respectively. N2 could be considered a neuron with a long axon, and so there is a delay in propagating the information/signal giving the worm ability to sense delayed concentration or previous concentration value. These sensory neurons have both excitatory and inhibitory synapses fanning out simultaneously here. This may be neuroscientifically inaccurate (for instance, Dale's law requires a neuron to perform same neural activity at all its synapses) but it is still perfectly well-within the neuromorphic hardware and algorithms capabilities. Our focus is on developing an engineering solution in spiking domain which could be used in tracking problems. It could be considered as bio-inspiration of useful functions rather than bio-mimicry.

Current equation for sensory neurons are of form $I[i + 1] = I\_0 * (C)$ where C is target concentration for N_ref, current concentration for N1 and previous concentration for N2.

N3, N4, N5, and N6 are the intermediate neurons. Synaptic trace is a component added to the current of the post-synaptic neuron on firing of pre-synaptic neuron. A positive trace is added for activation and a negative trace for inhibition. Hence, the post-synaptic neuron will have a positive current value only if there are more positive traces than the negative traces. This implies that these intermediate neurons will spike if the spiking frequency of activating neuron is greater than the spiking frequency of inhibiting neuron. N3 spikes if the set concentration is greater than current concentration, N4 spikes if the set concentration is less than current concentration, N5 spikes if the previous concentration is greater than current concentration, and N6 spikes if the previous concentration is less than current concentration. These four intermediate neurons help identify the instances when the worm is straying from the path to the set concentration.

The current equations for these intermediate neurons are of the form: $I[i + 1] = I[i] * 0.7 + I\_0 * (N\_activating\_spike[i] - N\_inhibiting\_spike[i])$ where N_activating_spike[i] is 1 if the activating neuron spikes at time instant i else 0 and similarly N_inhibiting_spike[i] is for the inhibiting neuron.

We have to apply corrective measures if either of N7 or N8 spike.

$$N3 . N5 => (C\_set > C\_current) \; AND \; (C\_previous > C\_current) \quad (3)$$

$$N4 . N6 => (C\_set < C\_current) \; AND \; (C\_previous < C\_current) \quad (4)$$

Eq. 3 and Eq. 4 represent the conditions for the spiking of N7 and N8, respectively. Either of these two neurons spikes if we need to change our direction from the current path. As far as current dynamics are concerned for these two neurons, it could be thought of as the implementation of AND logic gate. That is, both these neurons spike only if both their pre-synaptic neurons spike.

The current equations for N7 and N8 are of the form: $I[i+1] = I[i] * 0.7 + I\_0 * (N\_act1\_spike[i] + N\_act2\_spike[i])$ where N_act1_spike[i] (similarly N_act2_spike[i]) is 1 if the corresponding activating neuron spikes at time instant i else 0.

N_phase is the neuron that spikes after every time window (constant current neurons). N11 spikes if either of N7 or N8 spike indicating that path correction is required. It should be noted that N7 and N8 can often spike within a time window, but the corrective measure is required only once within a time window. It is ensured by setting a higher refractory period in N11 and by self-inhibition of N11. Multiple activations from N7 and N8 can drive up the current of N11, but we need to ensure that N11 spikes even if there is a single N7 or N8 spike. Here activation from N_phase comes into play.

$$
\begin{aligned}
I11[i+1] = I11[i] * 0.99 \ &+ \ w_1 * (N7\_spike[i] + N8\_spike[i]) \\
&- \ w_2 * N11\_spike[i] \ + \ w_3 * Nphase\_spike[i]
\end{aligned}
\tag{5}
$$

We now know when we need to do the path correction. This information needs to be relayed to the motor neuron via activation from N11. All the motor neurons have self inhibitions, and they cyclically activate one neuron. This ensures that every time N11 spikes, exactly one of N_r1, N_r2 or N_r3 spike in a cyclic manner. This is how we implement cyclic directional updates. Once these neurons fire, the corresponding direction is reversed to move towards the set concentration. (e.g.- If N_r1 fires, +ve x becomes -ve x and vice versa) The current equations are:

$$
Ir1[i+1] = Ir1[i]*(1-Nr1\_spike[i]) + w_4*Nr3\_spike[i] + w_5*N11\_spike[i] \tag{6}
$$

$$
Ir2[i+1] = Ir2[i]*(1-Nr2\_spike[i]) + w_4*Nr1\_spike[i] + w_5*N11\_spike[i] \tag{7}
$$

$$
Ir3[i+1] = Ir3[i]*(1-Nr3\_spike[i]) + w_4*Nr2\_spike[i] + w_5*N11\_spike[i] \tag{8}
$$

The above mentioned neural network is one of the smallest possible network for 3D chemotaxis. It is fairly simple and only comprising of sensors, level detectors, gradient sign detectors and the motor neurons along with logical synaptic connections between these components. Unlike finely tuned learning in neural networks, where the features represented by the neuronal layers are hidden or unknown and hence a dedicated learning phase is needed, this network has very well defined functionality or information representation by neurons. The spiking of the following layer is usually a logical combination of the current layer of neurons and the weights are chosen based on these logical decisions.

4   Current Dynamics

The whole network has 4 different types of neural activity. First, the intermediate neurons N3, N4, N5 and N6 which have similar current dynamics. Next, the AND gate implementation for neuron N7 and N8. The current dynamics for neuron N11 is different and more complex from the rest of the neurons. Finally the current dynamics for the motor neurons N_r1, N_r2, and N_r3.

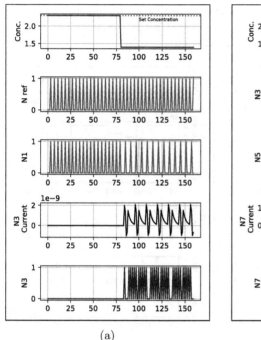
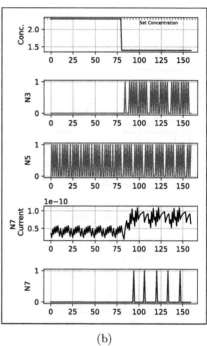

(a)   (b)

Fig. 2. (a) and (b) represent the intermediate neuron's current dynamics and the AND logic gate implementation. The first row in each figure represents the sensed concentration. The set concentration is marked with a red dotted line. The following two rows represent the pre-synaptic neurons. The green color means that the neuron is activating, and the red color means that the neuron is inhibiting in nature. The fourth row represents the developed current in the post-synaptic neuron, and the last row represents the spiking pattern of the post-synaptic neuron. (Color figure online)

In Fig. 2a, from the time instant 0 to 80, concentration is equal to the set concentration, and hence we see that the activating and inhibiting neuron fire with the same frequency and no current develops in N3. From time instant 80 to 160, activating neuron fires at a higher frequency and thus current develops in N3, and it spikes. The current even reaches a negative value because the pre-synaptic neurons fire at different time instants, but the activating neuron's higher spiking frequency is enough for the current to rise and the neuron to spike.

In Fig. 2b, from the time instant 0 to 80, the concentration is equal to the set concentration, and N3 does not spike. We know that N5 fires when C_previous is greater than C_current. Since before 0, the concentration was higher than 2.3, N5 spikes from 0 to 80 (For better clarity and understanding, we show only a tiny segment of the spiking pattern near the concentration transition). Current develops in N7 from 0 to 80, but is not sufficient for N7 to spike. From time instant 80 to 160, N3 also starts spiking because the current concentration is smaller than the set concentration. Current in N7 increases, and we observe that N7 spikes, telling the worm that it is deviating from the set concentration. Here we see that both N3 and N5 need to spike in order for N7 to spike, and hence it acts as a logical AND gate in the spiking domain.

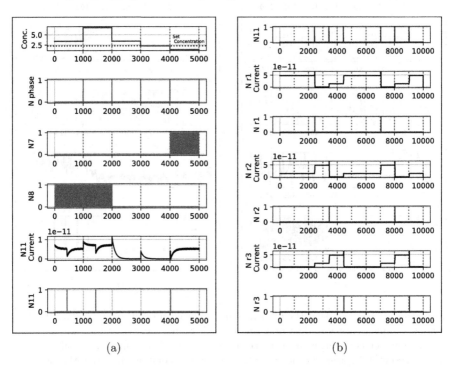

(a)   (b)

Fig. 3. (a) represents the current dynamics of N11, and (b) represents the current dynamics for the sub-network for the cyclic directional updates. In (a), green spikes mean activating, and red means inhibiting neurons for the spiking of N11. (N11 is red due to self-inhibition). In (b), a neuron can be activating and inhibiting different neurons simultaneously. Hence they are not color-coded. For the nature of synapse, one can refer to the network shown in Fig. 1 (Color figure online)

Figure 3a shows that N7 and N8 spikes if the worm is moving away from the set concentration. The strength of activation from N7 and N8 is poor, and so even when they spike continuously, we observe that the current decreases because the decay rate is more prominent than the activation from N7 and N8. The activation from N_phase and inhibition from N11 are both strong, and we see a sudden rise and drop in current when the respective neurons spike.

In Fig. 3b, N11 acts as activating neuron for all three N_r1, N_r2, and N_r3. We also observe that once any of these neurons fire, say N_r1, its current decreases due to self-inhibition. It activates N_r2, and since N_r2 also receives activation from N11, there is a large increase in current. The next time N11 spikes, the activation is enough for N_r2 to spike, and this cycle continues.

5   Results

Figures 4, 5a and 5b are for Gaussian concentration space. Concentration profile: $6.7 * exp^{(-(\frac{x-40}{60})^2 - (\frac{y-40}{60})^2 - (\frac{z-40}{60})^2)}$. The set concentration is 1.24 units, and the initial concentration is 6.7 units. In Fig. 5a, we can see that the worm starts from a white region (high concentration according to the color bar) and settles at a dark shade (low concentration). The worm reaches the set concentration initially with large oscillations where the path is similar to a distorted helix as seen in

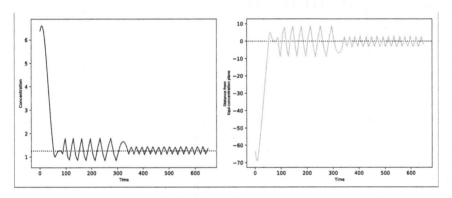

Fig. 4. (a) shows the concentration variation with respect to time. (b) shows the distance of worm from the equi-concentration sphere that it is supposed to track.

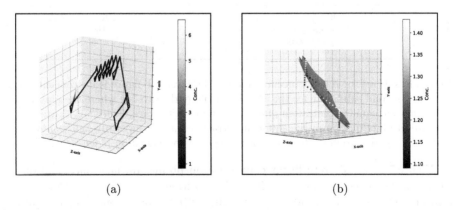

(a)   (b)

Fig. 5. (a) shows the path travelled by the worm while tracking a set concentration. (b) shows the behaviour of the worm once it reaches near the set concentration. (Color figure online)

the middle section of Fig. 5a. Finally, the oscillation decreases when the worm attains its equilibrium position around the equi-concentration sphere. Figure 5b shows that the worm performs a closed-loop motion with deviation from the equi-concentration sphere shown by the blue surface.

Figure 6 is for continuous linear concentration space. Initial Concentration = -1.1 units (Negative concentration is not of any significance because N1 and N2 would not spike for these concentration values. It is negative only because of the chosen initial position and till the time it reaches a positive concentration value, the negative input cannot stimulate the sensory neurons). Set concentration = 0.5. Initial position = (14, 14, 15). Concentration profile: $0.6 + 0.1(x - 40) + 0.1(y - 20) - 0.1(z - 30)$.

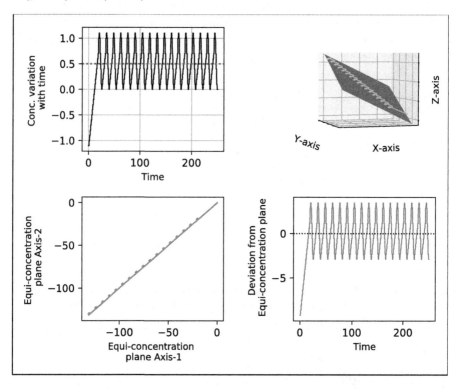

Fig. 6. Result set for linear concentration profile. First sub-figure is for concentration variation, second sub-figure for the 3D plot of the path travelled where the gray plane represents the equi-concentration plane it is supposed to track. Third sub-figure shows the projection of motion in the equi-concentration plane. Fourth sub-figure shows the perpendicular deviation of the worm from the equi-concentration plane.

Figure 7 is for discrete linear concentration profile. Initial Concentration = 0.6 units. Set concentration = 0.5. Initial position = (40, 40, 50). Concentration profile: $0.6 + 0.1(x - 40) + 0.1(y - 20) - 0.1(z - 30)$. Concentration values are discretized using the floor function with steps at the concentration values mentioned previously.

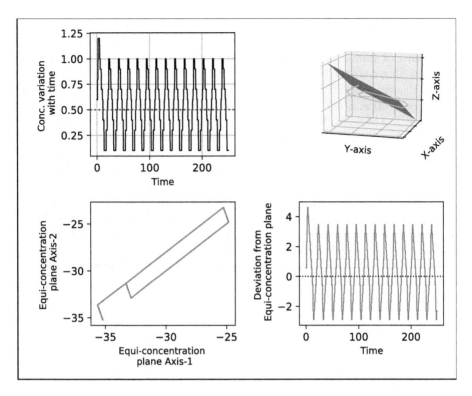

Fig. 7. The first sub-figure is for concentration variation, second sub-figure for the 3D plot of the path traversed. The gray plane represents the equi-concentration plane it is supposed to track. Third sub-figure shows the projection of motion in the equi-concentration plane. Fourth sub-figure shows the perpendicular deviation of the worm from the equi-concentration plane.

6   Conclusion

We have successfully devised an algorithm for spiking neural network implementation of klinokinesis for Chemotaxis, which is compatible with neuromorphic hardware. The network is like a sequential digital logic operation with parameters only weakly affecting the operation which are more involved in the analog implementation of contour-tracking. The implementation is for a fixed input range and as long as the concentration space can be scaled in that range, the algorithm would work. The model used for synapses is fairly simple and only serves as the logical connections between the more functionally important neurons in this network. However for future work, short-term plasticity (STP) can be used to generalise this network to operate in different ranges of input concentrations and hence different spiking activity regimes, where the STP responsive weights can modulate themselves in accordance with the new spiking activity domain. This will also increase the network's sensitivity to the model parameters.

While tracking the concentration, the path observed is like a distorted helix, closer to the observed motion in [5]. Thus, it could be said that this simple algorithm can mimic the biologically observed motion to a significant extent. After reaching near the set concentration, the worm performs oscillatory motion around the set concentration. The motion is regular and generally closed-loop. The concentration range which provides meaningful information to the network is 0.1 to 6.7. The worm faces no issue if it tracks a value close to the median of this range, but if we start moving toward either side of the median, the deviation starts to increase. We can expand this range by using a larger time window.

We have successfully implemented klinokinesis keeping in mind the limitations of the dedicated neuromorphic hardware. The algorithm considers 8 possible directions to move in, from a point in space depending on the sign of the orthogonal basis directions. This is because of the difficulties that arise during the initialization of the concentration space which has to be discrete since these values have to be stored as spiking rate in the hardware. This could be considered as the first step toward realizing much more complex algorithms to implement klinotaxis and orthotaxis in the future.

Features	This work	[2]	[3]	[4]
Medium	3D	2D	2D	2D
Klinokinesis	✓	✓	✓	✓
Klinotaxis	✗	✗	✗	✓
Orthotaxis	✗	✗	✗	✓
End-to-End SNN	✓	✗	✓	✓
SNN Hardware-compatibility	✓	✗	✗	Limited

Fig. 8. Qualitative benchmarking

References

1. Appleby, P.A.: A model of chemotaxis and associative learning in C. elegans. Biol. Cybern. **106**(6–7), 373–387 (2012)
2. Santurkar, S., Rajendran, B.: C. elegans chemotaxis inspired neuromorphic circuit for contour tracking and obstacle avoidance. In: 2015 International Joint Conference on Neural Networks (IJCNN), pp. 1–8. IEEE (2015)
3. Shukla, S., Dutta, S., Ganguly, U.: Design of spiking rate coded logic gates for C. elegans inspired contour tracking. In: Kůrková, V., Manolopoulos, Y., Hammer, B., Iliadis, L., Maglogiannis, I. (eds.) ICANN 2018. LNCS, vol. 11139, pp. 273–283. Springer, Cham (2018). https://doi.org/10.1007/978-3-030-01418-6_27
4. Shukla, S., Pathak, R., Saraswat, V., Ganguly, U.: Adaptive chemotaxis for improved contour tracking using spiking neural networks. In: Farkaš, I., Masulli, P., Wermter, S. (eds.) ICANN 2020. LNCS, vol. 12397, pp. 681–692. Springer, Cham (2020). https://doi.org/10.1007/978-3-030-61616-8_55

5. Jikeli, J., Alvarez, L., Friedrich, B., et al.: Sperm navigation along helical paths in 3D chemoattractant landscapes. Nat. Commun. **6**, 7985 (2015). https://doi.org/10.1038/ncomms8985
6. Bilbao, A., Patel, A.K., Rahman, M., et al.: Roll maneuvers are essential for active reorientation of Caenorhabditis elegans in 3D media. In: Proceedings of the National Academy of Sciences, pp. E3616–E3625 (2018). https://www.pnas.org/content/115/16/E3616

Signal Denoising with Recurrent Spiking Neural Networks and Active Tuning

Melvin Ciurletti⬤, Manuel Traub⬤, Matthias Karlbauer⬤, Martin V. Butz⬤, and Sebastian Otte$^{(\boxtimes)}$⬤

Neuro-Cognitive Modeling, Computer Science Department, University of Tübingen,
Sand 14, 72076 Tübingen, Germany
`melvin.ciurletti@student.uni-tuebingen.de,`
`{manuel.traub,matthias.karlbauer,martin.butz,`
`sebastian.otte}@uni-tuebingen.de`

Abstract. Active Tuning is an optimization paradigm specifically designed to increase the robustness and generalization ability of temporal forward models like recurrent neural networks (RNNs). This work explores how the Active Tuning method can be used to optimize the internal dynamics of recurrent spiking neural networks (RSNNs). Active Tuning decouples the network from direct influence of the data stream and instead tunes its internal dynamics. This is based on the temporal gradient signals from propagating the error between outputs and observations backwards through time. Meanwhile, the network is running in a closed-loop prediction cycle, where the own output is used as the next input. As modern ANNs often demand excessive amounts of computational resources, spiking neural networks (SNNs) aim for the energy efficiency demonstrated by the human brain. This is accomplished by using an event-driven model inspired by the spiking behavior of biological neurons. Target of the Active Tuning optimization in RSNNs is the membrane potential of the neurons in the hidden layer. We show in two scenarios how RSNNs handle noisy inputs and that Active Tuning is a reliable method to increase their robustness as well as general prediction performance.

Keywords: Recurrent spiking neural networks · Signal denoising · Active Tuning · Temporal gradients

1 Introduction

Networks of spiking neurons (SNNs) pose an interesting topic for research due to their event-based processing pattern, which comes with the potential of highly efficient simulations. Especially recurrent SNNs (RSNNs) resemble biological

We thank the International Max Planck Research School for Intelligent Systems (IMPRS-IS) for supporting Manuel Traub and Matthias Karlbauer. Martin Butz is part of the Machine Learning Cluster of Excellence, EXC number 2064/1 – Project number 390727645.

© Springer Nature Switzerland AG 2021
I. Farkaš et al. (Eds.): ICANN 2021, LNCS 12895, pp. 220–232, 2021.
https://doi.org/10.1007/978-3-030-86383-8_18

neural networks like the human brain more closely than traditional artificial neural networks. As accurate modelling of the biochemical processes in the brain would be computationally intractable for larger networks, several simplified neuron models have been developed to imitate the spiking behavior of biological neurons with lean and efficient implementations [13]. The arguably most basic spiking neuron model is the leaky integrate-and-fire (LIF) neuron. With only slight adaptations, an RSNN with LIF neurons has already been shown to approach the performance of an LSTM network on several tasks [2]. In this work we evaluate the ability of LIF RSNNs in handling input data under different noise conditions.

Traditional RNNs are particularly susceptible to noisy data, since corrupted inputs have the potential to bring the model's latent hidden state out of balance, leading to chaotic and exploding activities in subsequent time steps. In consequence, the internal dynamics of the RNN reach an activity space that it has never encountered before, preventing it from producing suitable predictions. One solution to this limitation is the use of noise-aware networks, that have been trained explicitly on a specific noise condition and thus act as a *denoising expert*. However, denoising experts are often only useful in exactly the same environment in which they were trained, while they tend to fail on different types or intensities of noise; or even on clean input data. Active Tuning presents a different approach to signal denoising in that the RNN is effectively decoupled from the direct input stream, while tuning the internal activities to fit the desired target dynamics [12]. The actual observations (traditionally network inputs) are only used for error computation by comparing them to the network's outputs. This error is back-propagated through the computational graph, but instead of adapting the weights of the network, the gradient signal is used to tune its hidden states. In this way, during inference, the RNN is only driven by its internal dynamics and its own predictions, while the error-induced temporal gradient information is used to infer the hidden state sequence that describes the target signal in a representation that is natural to the RNN. As Active Tuning only requires a differentiable temporal forward model that works with a latent representation, it is suited (but not limited) to work with a wide range of RNN architectures such as simple standard RNNs, echo state networks (ESNs) [7], or Long short-term memory RNNs [5]. While RSNNs also work with a hidden state, they are not fully differentiable; a problem which can be bypassed by using a pseudo-derivative [2,4,6].

In the following, we introduce the theory behind RSNNs and the Active Tuning method. We then inspect the RSNNs' performance with teacher forcing as well as Active Tuning on two different time series datasets—a multiple superimposed oscillator and a chaotic double pendulum—with and without artificially added noise. The results show that Active Tuning improves the RSNNs performance in most combinations of the used training and inference noise levels. They furthermore indicate that RSNNs have a much higher inherent robustness to noisy data than LSTMs, which allows them to produce relatively accurate predictions under noisy input conditions, even if they have never been provided with noisy data during training.

2 Methods

Concerning the LIF neuron equations, our network model follows the formulations from [1,3]:

$$v_j^t = \alpha v_j^{t-1} + \sum_i w_{i,j}^{in} x_i^t + \sum_{j'} w_{j',j}^{rec} z_{j'}^{t-1} - z_j^{t-1} v_{thr} \tag{1}$$

Here, the action potential v_j^t is based on a low pass filter of the weighted input spikes x_i^t and recurrent spikes z_j^{t-1} with a decay constant α. After having spiked, a neuron z_j^{t-1} is set to rest by subtracting the value of the spike threshold v_{thr}, simulating the refractory period as it occurs in biological neurons. The spike itself is calculated using the Heaviside step function:

$$z_j^t = \Theta \left(v_j^t - v_{thr}^t \right) \tag{2}$$

2.1 BPTT with LIF Neurons

In order to compute gradients for the learning process, i.e. the weight update, the network's computation chain has to be fully differentiable. However, as mentioned earlier, the Heaviside function in Eq. (2) does not fulfill this requirement. To overcome this problem, a pseudo-derivative h_j^t was introduced in [3], replacing the non-existing derivative of the threshold function:

$$\frac{\partial z_j^t}{\partial v_j^t} \stackrel{def}{=} h_j^t = \lambda \ \max \left(0, 1 - \left| \frac{v_j^t - v_{thr}}{v_{thr}} \right| \right) \tag{3}$$

Here, h_j^t is a piecewise linear function of the voltage, where the dampening factor λ scales the steepness of the linear segments. With help of this pseudo-derivative we can calculate the partial derivative of the loss with respect to v_j^t, which we define here as the delta error:

$$\delta_j^t \stackrel{def}{=} \frac{\partial E}{\partial v_j^t} \tag{4}$$

By applying the chain rule, we can unfold the state gradient equation, by expanding the error into a global and a local part (left and right sum terms, respectively):

$$\delta_j^t = \frac{\partial E}{\partial z_j^t} \frac{\partial z_j^t}{\partial v_j^t} + \frac{\partial E}{\partial v_j^{t+1}} \frac{\partial v_j^{t+1}}{\partial v_j^t} = \frac{\partial E}{\partial z_j^t} h_j^t + \delta_j^{t+1} \alpha \tag{5}$$

Having defined this delta error (δ_j^t) we can now effectively calculate the gradient of the input, recurrent, and output weights, respectively:

$$\frac{\partial E}{\partial w_{i,j}^{in}} = \sum_t x_i^t \delta_j^t \quad (6) \qquad \frac{\partial E}{\partial w_{j,j'}^{rec}} = \sum_t z_j^t \delta_{j'}^{t+1} \quad (7) \qquad \frac{\partial E}{\partial w_{j,k}^{out}} = z_j^t \delta_k^t \quad (8)$$

Moreover, we will use the state gradient formalization δ_j^t of the hidden neurons to perform Active Tuning as described in the following.

2.2 Active Tuning

Active Tuning is a method to optimize the internal dynamics of a temporal forward model (i.e. an RNN) to handle noisy signals without being explicitly trained to do so [12].

Prerequisite for applying Active Tuning is a differentiable temporal forward model that is trained to predict the next upcoming observation $\mathbf{x}^{t+1}$ based on the current observation $\mathbf{x}^t$ and a latent variable $\boldsymbol{\sigma}^t$ representing the network's state (i.e. the hidden layer activations of an RNN). This model can in general be described by

$$(\boldsymbol{\sigma}^t, \mathbf{x}^t) \xrightarrow{f_M} (\boldsymbol{\sigma}^{t+1}, \tilde{\mathbf{x}}^{t+1}), \tag{9}$$

where f_M is the model's forward function and $\tilde{\mathbf{x}}^{t+1}$ its prediction for $\mathbf{x}^{t+1}$ (the next observation). If a model of this type can achieve a latent state $\boldsymbol{\sigma}^t$ that is closely representing the actual information necessary to infer an accurate prediction $\tilde{\mathbf{x}}^{t+1}$, this prediction can be used as input for the next time step. Following this approach, the model can operate in a so called closed-loop manner, decoupled from the observed real world signal.

However, to reach such a representative latent state, models are typically initialized at the beginning of processing a sequence by using teacher forcing for several time steps. That is, the network prediction of the next time step is replaced by the actual observed signal, before feeding it back into the model. Obviously, this forces direct contact with the (potentially corrupted) real world signal. Active Tuning on the other hand can completely remove the need for this direct contact, by using only gradient information from propagating the error between the model's predictions and the real world signal backward in time onto the latent state (Fig. 1).

To realize this decoupling, Active Tuning records the recent $R \in \mathbb{N}$ signal observations $\mathbf{x}^t$, the model's recent predictions $\tilde{\mathbf{x}}^t$ as well as the associated latent states $\boldsymbol{\sigma}^t$. The time frame covered by R is called (retrospective) tuning horizon or tuning length. If a new observation $\mathbf{x}^t$ is perceived, the error between recent observations and the model's predictions is computed. This error is then propagated backwards into the past for R time steps using BPTT. The chronologically first latent state $\boldsymbol{\sigma}^{t-R}$ in this tuning horizon, also called seed state, is then adapted using this gradient information with the Adam optimizer [8]. After repeating this cycle and adapting the seed state as often as desired, one last forward pass from $t - R$ to the next real world time step $t + 1$ is performed. The buffers of recorded states, predictions and observations are shifted to the left by one, dropping the entries furthest in

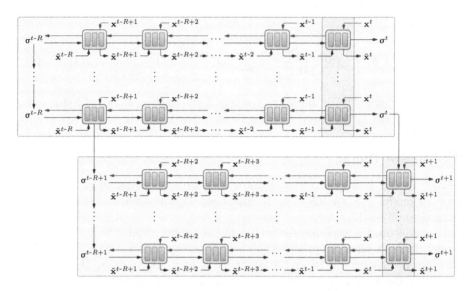

Fig. 1. Illustration of Active Tuning over two consecutive world time steps. Each of rectangular nodes represents one world time step and consist of multiple processing time steps of the underlying SNN. The black lines indicate computational forward dependencies. $\mathbf{x}^t, \mathbf{x}^{t-1}$ etc. are the recent signal observations, whereas $\tilde{\mathbf{x}}^t, \tilde{\mathbf{x}}^{t-1}$ etc. are the respective predictions (outputs of the model's forward function f_M). R denotes the length of the retrospective tuning horizon, that is, the number of world time steps the prediction error is projected into the past using BPTT. $\boldsymbol{\sigma}^{t-R}$ refers to the latent (hidden) state of M at the beginning of the tuning horizon (which is specifically the action potential of the hidden neurons of the last sub time step of the time window associated with the world time step $t - R$). $\boldsymbol{\sigma}^{t-R}$ is actively optimized based on the back-projected prediction error gradient (red lines). (Color figure online)

the past, to make space for the current time step. This effectively makes $\boldsymbol{\sigma}^{t-R+1}$ the new seed state. This method is then applied for every perceived observation from the real world signal, continually adapting the model's latent state to better represent the actual sequence dynamics.

It has to be noted that Active Tuning does not limit the adaptation process to the model's latent state. Another option that could be used instead, or even additionally, is updating the model's first prediction inside the tuning horizon $\tilde{\mathbf{x}}^{t-R}$. As this prediction is used as the next input in closed-loop mode, it also influences all future predictions and can be optimized to reduce discrepancies to the signal observations.

Applying Active Tuning to an RSNN model is relatively straight forward due to the general nature of the procedure. The most important decision is which latent variable is suited most to be adapted. In contrast to the hidden state of a traditional RNN that consists of real-valued neuron activations, LIF neurons produce binary spikes. Updating a spike value from 0 to 1 or reversely would mean a relatively large modification, while updating them to real values would contradict the core idea of spiking neurons. Potentials, however, are already real-valued and directly influence the spike generation of the current and future time steps. Considering these observations, Active Tuning is used to adapt the potentials of the LIF neurons in the RSNNs hidden layer.

3   Experiments

We studied two different types of time series data (one-dimensional linear and two-dimensional non-linear dynamics) based on the experiments in [12] to evaluate the performance of RSNNs as well as their noise robustness with and without Active Tuning. Similar network configurations with only slight adaptations were used for both experiments to ensure comparability with each other as well as with the LSTM network used in [12]. The networks were trained as one step ahead predictors, meaning their task is to predict the next input based on the current input as well as all past ones accumulated in the model's latent hidden state. The target sequences were generated by simply shifting the input sequences by one time step. To inspect noise robustness and denoising ability, networks were trained and evaluated under various conditions, where potentially noisy signals were fed to the network while still using the clean sequence as target. The five relative noise ratios are 0.0 (no noise), 0.1, 0.2, 0.5, and 1.0. The standard deviation of the added noise was computed by the standard deviation of the training set samples multiplied with the respective noise ratio. Training RSNNs on noisy data yielded denoising experts for these specific noise conditions, which were used to put the effect of Active Tuning into perspective.

One important technique to improve the RSNNs general performance was feeding each input to the network multiple times before taking the networks output as prediction for this time step. The ideal number of sub time steps τ_T per real world time step varies between different datasets, but a significant increase in performance compared to only presenting the input once was always visible.

Five RSNNs with different independent random initializations were trained for each condition on each task to report the average results. All networks were trained for 50 epochs with the Adam optimizer [8] using a learning rate $\eta = 0.003$, $\beta_1 = 0.9$ and $\beta_2 = 0.999$. No refractory period was used for the LIF neurons. The error was only computed between the target and the last sub time step per real world time step that was also used as the current prediction.

3.1   Multiple Superimposed Oscillator

The first dataset consists of multiple superimposed oscillator (MSO) signals which has been utilized as a benchmark problem several times before [10,11,15]. An MSO sequence is generated by superimposing multiple sine waves with different amplitudes a_i, frequencies f_i, and phase-shifts φ_i into a single signal. This process is captured by the following equation:

$$\mathrm{MSO}_n(t) = \sum_{i=1}^{n} a_i \sin(f_i t + \varphi_i) \tag{10}$$

Only the dynamics of the MSO_5 with default frequencies $f_1 = 0.2$, $f_2 = 0.311$, $f_3 = 0.42$, $f_4 = 0.51$ and $f_5 = 0.63$ were considered for the experiments in this work. These values were chosen in a way such that no two frequencies are integer multiples (harmonics) of each other, which would make the resulting signal easier to predict. Amplitudes and phase-shifts were randomly sampled for each new wave with $a_i \sim [0, 1)$ and $\varphi_i \sim [0, 2\pi)$. 10 000 samples with sequence length $T = 400$ were generated for training. For testing, another set of 1 000 samples was generated.

An RSNN with 64 LIF units in the hidden layer was used for the MSO task. No input encoding was applied, instead an input neuron with the identity activation function directly received the real-valued input signals. The output layer consisted of a single leaky integrator that acts as a readout neuron whose state is interpreted as the network's prediction. With no biases used, this resulted in a total of 4 224 trainable parameters compared to the LSTM network's 4 256. $\tau_T = 9$ sub time steps were used during training and inference.

3.2   Chaotic Double Pendulum

The double pendulum is a two-dimensional non-linear dynamical system and consists of two simple pendulums connected to each other. It features two point masses m_1 and m_2 connected by two massless rods of length l_1 and l_2. The current orientation of the pendulum is defined by the two joint angles θ_1 and θ_2. The end-effector of such a double pendulum generates highly non-linear trajectories if initialized in a way to guarantee sufficient energy for the system to move.

$$\ddot{\theta}_1 = \left(\mu g_1 \sin(\theta_2) \cos(\theta_2 - \theta_1) + \mu \dot{\theta}_1^2 \sin(\theta_2 - \theta_1) \cos(\theta_2 - \theta_1) - g_1 \sin(\theta_1) \right.$$
$$\left. + \frac{\mu}{\lambda} \dot{\theta}_2^2 \sin(\theta_2 - \theta_1) \right) \frac{1}{1 - \mu \cos^2(\theta_2 - \theta_1)} \tag{11}$$

$$\ddot{\theta}_2 = \left(g_2 \sin(\theta_1) \cos(\theta_2 - \theta_1) - \mu \dot{\theta}_2^2 \sin(\theta_2 - \theta_1) \cos(\theta_2 - \theta_1) - g_2 \sin(\theta_2) \right.$$
$$\left. - \lambda \dot{\theta}_1^2 \sin(\theta_2 - \theta_1) \right) \frac{1}{1 - \mu \cos^2(\theta_2 - \theta_1)}, \tag{12}$$

A mechanical model of the planar double pendulum is constructed by [9] and defined by two equations of motion (Eqs. 11 and 12), derived from the Lagrangian of the system and the Euler-Lagrange equations. The system is implemented by integrating the equations of motion using the fourth-order Runge-Kutta method [14]. For data generation, masses m_1 and m_2 as well as rod lengths l_1 and l_2 of 1 were used for all samples, the step size was set to $h = 0.01$ for numerical integration. The initial joint angles were selected randomly with $\theta_1 \sim [90°, 270°]$ and $\theta_2 \sim [\theta_1 \pm 30°]$, one in ten samples was initiated with zero angle momenta $\dot{\theta}_1, \dot{\theta}_2 = 0$. Again, 10 000 samples are generated for training and an additional 1 000 for testing.

As before, an RSNN with 64 LIF neurons in the hidden layer was used for this task. As the sequence is two-dimensional, the input and output layer consist of two neurons each, again with no input encoding and leaky integrators as readout neurons. The networks had a total of 4 252 trainable parameters while the respective LSTM networks had 4 416. $\tau_T = 3$ sub time steps per real world time step were used during training and evaluation.

4   Results

The performance of all models was evaluated based on the root mean square error (RMSE) between network outputs and target sequence. The reported values are averages over the five models trained for each task and condition. The noise-unaware networks as well as the denoising experts were tested on all five noise conditions, resulting in 25 different setups. All of these setups were run with and without Active Tuning. For all Active Tuning experiments, the membrane potentials of the LIF neurons in the hidden layer were chosen as optimization target. The initial states of these membrane potentials were drawn from a normal distribution with standard deviation 0.1.

4.1   MSO Results

Table 1 shows the results of the MSO experiments, where Active Tuning was applied over $C = 30$ tuning cycles and with a tuning length of $R = 2$. The noise-unaware RSNN benefits from Active Tuning under all noise conditions except for clean input data (column 2 vs. 7). For noise levels 0.1, 0.2 and 0.5 the noise-unaware network with Active Tuning performs even better than the respective denoising

Table 1. MSO denoising and Active Tuning results of the RSNN with $\tau_T = 9$.

Inf.noise	Training noise									
	Teacher forcing					Active tuning ($C = 30$, $R = 2$)				
	0.0	0.1	0.2	0.5	1.0	0.0	0.1	0.2	0.5	1.0
0.0	0.1276	0.1474	0.2139	0.3690	0.4800	0.1346	0.1042	0.1136	0.0833	**0.0541**
0.1	0.2040	0.1886	0.2333	0.3732	0.4811	0.1595	0.1348	0.1424	0.1181	**0.1001**
0.2	0.3328	0.2762	0.2842	0.3865	0.4847	0.2171	0.1988	0.2054	0.1820	**0.1693**
0.5	0.6595	0.5647	0.4994	0.4749	0.5129	0.4368	0.4284	0.4226	0.3914	**0.3774**
1.0	0.9894	0.9225	0.8204	0.6870	**0.6096**	0.8114	0.8022	0.7938	0.7381	0.6866

Table 2. MSO denoising and Active Tuning results of the LSTM network [12].

Inf. noise	Training (signal noise)					Training (signal noise)	
	Teacher forcing					Active tuning	
	0.0	0.1	0.2	0.5	1.0	0.0	0.05
0.0	**0.0040**	0.0498	0.0781	0.1526	0.2383	—	—
0.1	0.5917	0.0966	0.0993	0.1579	0.2399	**0.0908**	0.0947
0.2	1.0351	0.1734	**0.1455**	0.1729	0.2445	0.1682	0.1550
0.5	1.8695	0.4260	0.3188	**0.2542**	0.2754	0.3563	0.3611
1.0	2.6150	0.9506	0.6439	0.4396	**0.3641**	0.5699	0.6101

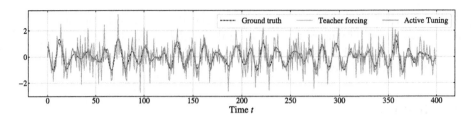

Fig. 2. Visual comparison of LSTM (orange) vs. RSNN (blue) on an MSO$_5$ sample with noise ratio 0.2 using a noise-unaware RSNN. (Color figure online)

experts. Using Active Tuning with these denoising experts also improves performance in almost all cases, except for noise level 1.0 with the two strongest denoisers (0.5 and 1.0). This effect is especially visible when feeding inputs with low noise level (0.0 to 0.2) to these strong denoising experts. The performance on clean data of the network trained on noise level 1.0 improves by almost one magnitude using Active Tuning. This model also reports the best results under all five noise conditions, four of them only enabled by the use of Active Tuning.

Comparing these results to the performance of the LSTM network, reported in Table 2, shows that the LSTM network is stronger on clean input data by a large margin. However, inspecting the results of the noise-unaware networks on the different inference noise levels (column 2 of both tables) indicates that the RSNN is intrinsically much more robust to noisy input signals without any techniques enhancing this property. Between noise levels 0.1 and 1.0 the RSNN outperforms the LSTM by a factor of roughly 3. This effect is shown in Fig. 2. The best result per inference noise level, however, are achieved by the respective LSTM denoising experts, with the exception of noise level 0.2, where the noise-unaware LSTM with Active Tuning performs best. While the best RSNN results come close to those of the LSTM for small noise levels (0.1 and 0.2), the difference is more significant in the other cases.

Figure 3 visualizes the spiking activity of an RSNN with $\tau_T = 9$ over the last 20 real world time steps of the MSO sample shown in Fig. 2.

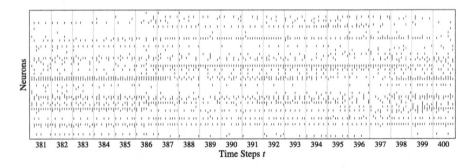

Fig. 3. Spike trains of a noise-unaware RSNN during the last 20 real world time steps of predicting a clean MSO sample. The network used $\tau_T = 9$ sub time steps. Real world time steps are separated by the red line. (Color figure online)

4.2   Pendulum Results

On the double pendulum datasets, Active Tuning was again applied for $C = 30$ cycles, but this time with a slightly longer tuning length of $R = 5$. The results are listed in Table 3. It has to be noted that for this task, Active Tuning would sometimes completely fail on single networks, while producing the expected results on other networks with the same setup. This heavily skewed the reported average values and as a consequence these cases have been removed from the evaluation, if it only affected one or two of the five trained networks. The respective cases are marked with *cursive* font in the table. Note that only one of the best results per inference noise level is affected (noise level 0.2) and thus reported on less than five independent networks, while other setups also come very close to the reported performance.

A similar trend to the MSO task can be found in lower noise levels up to 0.2, where Active Tuning improves most setups compared to teacher forcing. However, the scale of improvement is smaller on this task. Again, the best values are reported by denoising experts trained on higher noise than the respective inference noise level. This time the RSNNs trained on noise level 0.5 in combination with Active Tuning perform best in this range, with the 0.2 denoising expert slightly outperforming it on 0.1 inference noise. On the higher inference noise levels, Active Tuning brings only slight to no improvement while failing in several cases. Here the expert denoisers trained on the respective noise level report the best performance.

The comparison to the LSTM results shown in Table 4 depict a similar behavior for noise-unaware networks as already seen in the MSO task. While the LSTM is vastly superior on clean data, the RSNN is more robust to noise of all levels. While Active Tuning does not improve the LSTMs in this scenario, the RSNN still benefits from Active Tuning on low inference noise conditions.

Table 3. Pendulum denoising and Active Tuning results of the RSNN with $\tau_T = 3$.

Inf. noise	Training noise									
	Teacher forcing					Active tuning ($C = 30, R = 5$)				
	0.0	0.1	0.2	0.5	1.0	0.0	0.1	0.2	0.5	1.0
0.0	0.0876	0.1089	0.0968	0.1312	0.1966	0.0749	*0.0642*	0.0587	**0.0584**	*0.0784*
0.1	0.1036	0.1198	0.1051	0.1349	0.1981	0.0953	*0.0860*	***0.0795***	0.0800	*0.0970*
0.2	0.1414	0.1465	0.1268	0.1451	0.2029	0.1440	*0.1496*	0.1440	**0.1200**	*0.1385*
0.5	0.2979	0.2660	0.2282	**0.2032**	0.2344	1.6275	0.9817	0.2937	0.2276	0.4069
1.0	0.5905	0.5034	0.4268	0.3342	**0.3200**	5.4436	4.4878	1.0442	0.3763	0.5380

Table 4. Pendulum denoising and Active Tuning results of the LSTM network [12].

Inf. noise	Training (signal noise)					Training (signal noise)	
	Teacher forcing					Active tuning	
	0.0	0.1	0.2	0.5	1.0	0.0	0.05
0.0	**0.0020**	0.0264	0.0427	0.0830	0.1426	—	—
0.1	0.2110	**0.0537**	0.0553	0.0868	0.1440	0.0725	0.0545
0.2	0.4050	0.0976	**0.0824**	0.0971	0.1483	0.1276	0.0932
0.5	0.8565	0.2446	0.1901	**0.1542**	0.1759	0.2721	0.1968
1.0	1.3157	0.5401	0.4094	0.2902	**0.2562**	0.4940	0.3445

5   Conclusion

In this work we used Active Tuning to tune the membrane potentials of LIF neurons in an RSNN. Active Tuning decouples the RSNN from direct influence of the potentially noisy input data stream and thus keeps the internal dynamics within a system-consistent activity space. This is achieved by running the network in closed loop while adapting its internal latent state based on back-propagated gradient information of the discrepancy between prediction and observation. The experiments on two different types of time series data showed that Active Tuning improves the performance of a noise-unaware RSNN over regular teacher forcing in almost all cases of various inference noise conditions. Interestingly, the best results were mostly achieved by denoising experts, where Active Tuning enabled considerable performance on noise levels lower than the one they were specifically trained for. This means that Active Tuning can not only be used to improve the predictions of noise-unaware RSNNs on noisy data, but also of noise-aware networks on clean or less noisy data with an even more significant effect. A comparison to the results achieved by LSTM networks in a similar experimental setup showed that RSNNs are still quite far from reaching ideal performance on clean data, but come closer to noisy data especially when using Active Tuning. Another significant observation visible in this comparison was that noise-unaware RSNNs are inherently much more robust to noisy input data than their LSTM counterparts, where they show two to three times better performance on all noise levels from 0.1 to 1.0. This hints that the binary encoding through the LIF neuron's spiking activation can already regulate the influence of noisy input data on the network output and can thus still produce precise predictions.

While these results are very promising, it has to be noted that Active Tuning considerably increases the inference time of RSNNs, especially in combination with the use of multiple sub time steps τ_T, as all these sub time steps have to be computed during the forward as well as backward passes in each Active Tuning cycle. One solution to this problem would be the reduction of the Active Tuning cycles needed to tune the latent state, where ideally a single cycle would be sufficient. Another observation made during the experiments was that the difference in performance between smaller and larger τ_T decrease when applying Active Tuning compared to teacher forcing, so that the results of RSNNs with $\tau_T = 3$ were only slightly worse than those reported with $\tau_T = 9$. This would mean that the use of Active Tuning lessens the need for additional sub time steps to develop an accurate prediction. However, this would only decrease the computational overhead if Active Tuning was used anyway and not in comparison to teacher forcing.

References

1. Bellec, G., Salaj, D., Subramoney, A., Legenstein, R., Maass, W.: Long short-term memory and learning-to-learn in networks of spiking neurons. In: Proceedings of the 32nd International Conference on Neural Information Processing Systems, pp. 795–805. Curran Associates Inc., Red Hook (2018)
2. Bellec, G., Salaj, D., Subramoney, A., Legenstein, R.A., Maass, W.: Long short-term memory and learning-to-learn in networks of spiking neurons. In: Bengio, S., Wallach, H.M., Larochelle, H., Grauman, K., Cesa-Bianchi, N., Garnett, R. (eds.) Advances in Neural Information Processing Systems 31: Annual Conference on Neural Information Processing Systems 2018, NeurIPS 2018, December 3–8, 2018, Montréal, Canada, pp. 795–805 (2018)
3. Bellec, G., et al.: A solution to the learning dilemma for recurrent networks of spiking neurons. Nat. Commun. 11(1), 1–15 (2020)
4. Esser, S.K., et al.: Convolutional networks for fast, energy-efficient neuromorphic computing. Proc. Natl. Acad. Sci. U. S. A. 113(41), 11441–11446 (2016)
5. Hochreiter, S., Schmidhuber, J.: Long short-term memory. Neural Comput. 9(8), 1735–1780 (1997). https://doi.org/10.1162/neco.1997.9.8.1735
6. Hubara, I., Courbariaux, M., Soudry, D., El-Yaniv, R., Bengio, Y.: Binarized neural networks. In: Advances in Neural Information Processing Systems, pp. 4114–4122 (2016)
7. Jaeger, H.: The "echo state" approach to analysing and training recurrent neural networks. In: Technical report GDM Report, 148. Fraunhofer Institute for Analysis and Information Systems AIS (2001)
8. Kingma, D.P., Ba, J.: Adam: a method for stochastic optimization. In: Bengio, Y., LeCun, Y. (eds.) 3rd International Conference on Learning Representations, ICLR 2015, San Diego, CA, USA, May 7–9, 2015, Conference Track Proceedings (2015)
9. Korsch, H.J., Jodl, H.J., Hartmann, T.: Chaos: A Program Collection for the PC, 3 edn. Springer, Heidelberg (2008). https://doi.org/10.1007/978-3-540-74867-0
10. Koryakin, D., Lohmann, J., Butz, M.V.: Balanced echo state networks. Neural Netw. 36, 35–45 (2012)
11. Otte, S., Butz, M.V., Koryakin, D., Becker, F., Liwicki, M., Zell, A.: Optimizing recurrent reservoirs with neuro-evolution. Neurocomputing 192, 128–138 (2016)
12. Otte, S., Karlbauer, M., Butz, M.V.: Active tuning. arXiv:2010.03958 (2020)
13. Paugam-Moisy, H., Bohte, S.M.: Computing with spiking neuron networks. In: Rozenberg, G., Bäck, T., Kok, J.N. (eds.) Handbook of Natural Computing, pp. 335–376. Springer, Heidelberg (2012). https://doi.org/10.1007/978-3-540-92910-9_10

14. Press, W.H., Teukolsky, S.A., Vetterling, W.T., Flannery, B.P.: Numerical Recipes 3rd Edition: The Art of Scientific Computing, 3rd edn. Cambridge University Press, USA (2007)
15. Schmidhuber, J., Wierstra, D., Gagliolo, M., Gomez, F.: Training recurrent networks by Evolino. Neural Comput. **19**, 757–779 (2007)

Dynamic Action Inference with Recurrent Spiking Neural Networks

Manuel Traub[1] , Martin V. Butz[1] , Robert Legenstein[2] ,
and Sebastian Otte[1(✉)]

[1] Neuro-Cognitive Modeling, Computer Science Department, University of Tübingen,
Sand 14, 72076 Tübingen, Germany
{manuel.traub,martin.butz,sebastian.otte}@uni-tuebingen.de
[2] Faculty of Computer Science and Biomedical Engineering, Graz University
of Technology, Inffeldgasse 16b, 8010 Graz, Austria
robert.legenstein@igi.tugraz.at

Abstract. In this paper, we demonstrate that goal-directed behavior
unfolds in recurrent spiking neural networks (RSNNs) when intentions
are projected onto continuously progressing spike dynamics encoding the
recent history of an agent's state. The projections, which can either be
realized via backpropagation through time (BPTT) over a certain time
window or even directly and temporally local in an online fashion using
a biologically inspired inference rule. In contrast to previous studies that
use, for instance, LSTM-like models, our approach is biologically more
plausible as it fully relies on spike-based processing of sensorimotor experiences. Specifically, we show that precise control of a flying vehicle in a
3D environment is possible. Moreover, we show that more complex mental traces of foresighted movement imagination unfold that effectively
help to circumvent learned obstacles.

Keywords: Recurrent spiking neural networks · Active inference ·
Temporal gradients

1 Introduction

Recent progress in the field of reinforcement learning (RL) has gained huge
attention for learning to play various video games, most notably from the atari
console, and more recently even complex strategic online multiplayer games
[9–11]. While these results are impressive, the used training algorithms like proximal policy optimization [14] or similar model free RL approaches require tremendous amounts of learning time (typically thousands of years of simulated learning
episodes). This is neither very efficient nor biologically plausible. Moreover, these

We thank the International Max Planck Research School for Intelligent Systems
(IMPRS-IS) for supporting Manuel Traub. Martin Butz is part of the Machine Learning
Cluster of Excellence, EXC number 2064/1 – Project number 390727645.

© Springer Nature Switzerland AG 2021
I. Farkaš et al. (Eds.): ICANN 2021, LNCS 12895, pp. 233–244, 2021.
https://doi.org/10.1007/978-3-030-86383-8_19

approaches do not foster a deeper understanding of the task at hand. In contrast, cognitive science research, and in particular theories of predictive coding and active inference, suggest that our brain learns a predictive understanding of the world [4,6,7].

Previous research already explored how the active inference principle can be implemented in predictive recurrent neural networks (RNNs) to realize the emergence of goal-directed behavior, planning, and environmental interaction [3,5,12,13]. This is achieved as follows: First, a temporal forward model of a system of interest is learned, typically with an LSTM-like RNN [8], from temporally causal streams of available sensorimotor information. Second, motor signals are inferred on-the-fly by backpropagating intentions (such as desired system states) in form of prediction error-induced gradient signal through continuously adapting unrolled imaginations of the anticipated future system dynamics. While such RNNs are a rough simplification of real biological neural networks, they can be trained effectively by using backpropagation through time (BPTT). Recent progress in research on more biologically plausible spiking neural networks (SNNs), however, enables the training of SNNs end-to-end with a biologically more plausible learning rule, called e-prop, which comes close to the performance of BPTT [2].

Of particular interest for this paper is a variant of recurrent SNNs (RSNNs), which is referred to as a *long short-term spiking neural network* (LSNN) [1]. An LSNN consists of two types of spiking neurons: common leaky integrate and fire (LIF) neurons and adaptive LIF (ALIF) neurons. With an adaptive threshold that effectively regulates the firing rates of the latter, they can act as data and gradient highways fostering long-term influences on the network dynamics. In fact, LSNNs unfold impressive learning capabilities which are approaching, and in some cases even surpassing, the performance of LSTMs [1,2,16].

For our purposes, the e-prop [2] rule is also highly relevant, as it is ideal for continuous online learning scenarios. E-prop is based on a factorization of the gradient computation into an error-depending learning signal and an activation-depending eligibility trace. While the former is a time-local approximation of the full error signal, the latter can be computed forward in time along the regular forward pass of the network. Thus, e-prop does not require temporally backwards error signal propagation, as standard BPTT does.

In this paper, we bring together active inference-inspired behavior generation and biologically plausible SNNs. Specifically, we demonstrate that goal-directed, anticipatory behavior can emerge from projecting intentions through continuously unfolding spike dynamics onto motor inputs.

2   Action Inference in Recurrent Spiking Neural Networks

Establishing adaptive, goal-directed behavior in a continuous, dynamic control scenario with RSNNs involves essentially two aspects. First, a temporal forward model, that is, a neural approximation of the dynamical system of interest, is learned. Second, action sequences are inferred by means of a goal-based loss function.

2.1 Dynamical System Formulation

We assume a controllable discrete-time dynamical system with time-dependent system states. We follow the formalization from [12,13] in this section. We furthermore assume that the states are basically separated into perceivable states $\sigma^t \in \mathbb{R}^n$ and unobservable (hidden) states $\omega^t \in \mathbb{R}^m$. The system's dynamics can be influenced via k control commands denoted by $\mathbf{x}^t \in \mathbb{R}^k$. The next system state $(\sigma^{t+1}, \omega^{t+1})$ is determined by a (possibly unknown) state transition function

$$(\sigma^t, \omega^t, \mathbf{x}^t) \xmapsto{\Phi} (\sigma^{t+1}, \omega^{t+1}), \tag{1}$$

which models the forward dynamics of the system. As this process unfolds recursively over time, the next system state depends not only on the current control inputs, but also, in principle, on the entire state history. It is the learning task of the neural forward model to approximate Φ given the current state σ^t and current control commands $\mathbf{x}^t$, as well as an internal representation aggregated from the previous system state history $\{\sigma^1, \sigma^2, \ldots, \sigma^{t-1}\}$ and corresponding motor commands $\{\mathbf{x}^1, \mathbf{x}^2, \ldots, \mathbf{x}^{t-1}\}$. Since the neural forward model cannot directly access the unobservable states ω^t of the system, we will see that our system learns to approximate the missing computational components sufficiently well and thus becomes able to infer goal-directed actions by means of model-predictive control.

2.2 LSNN Forward Model

For learning Φ, we use an LSNN architecture with only one single recurrent hidden layer. We directly inject real valued inputs without any explicit spike encoding, which worked best in preliminary experiments. The hidden layer is composed of both LIF and ALIF neurons in an one-to-one ratio. It should be noted that the original formulation of LSNNs comes from [1]. In terms of notation details, however, the equations in this paper are aligned with [15] (which additionally offers details on the derivation of BPTT for SNNs with LIF and ALIF neurons). We calculate the activation of the hidden neurons as presented in the following:

LIF Activation

$$v_j^t = \alpha v_j^{t-1} + \sum_i w_{i,j}^{in} x_i^t + \sum_{j'} w_{j',j}^{rec} z_{j'}^{t-1} - z_j^{t-1} v_{thr} \tag{2}$$

$$z_j^t = \Theta \left(v_j^t - v_{thr}^t \right) \tag{3}$$

ALIF Activation

$$v_j^t = \alpha v_j^{t-1} + \sum_i w_{i,j}^{in} x_i^t + \sum_{j'} w_{j',j}^{rec} z_{j'}^{t-1} - z_j^{t-1} v_{j,thr}^{t-1} \tag{4}$$

$$a_j^t = \rho a_j^{t-1} + z_j^{t-1} \tag{5}$$

$$v_{j,thr}^t = v_{thr} + \zeta a_j^t \tag{6}$$

$$z_j^t = \Theta\left(v_j^t - v_{j,thr}^t\right) \tag{7}$$

The voltage v_j of a neuron is modeled as an exponentially decaying sum over the weighted inputs (with the leakage rate α), which gets reset after a spike in z_j occurs by subtracting the value of the spike threshold. For LIF neurons the threshold v_{thr} is constant. For ALIF neurons an adaptive threshold $v_{j,thr}$ is used, which depends on the individual spiking activity of the neurons. Due to this behavior—and its formal structure in particular—ALIF neurons act as data (as well as gradient) highways able to bridge even large temporal gaps, which is essential when confronted with long data sequences [1]. For both LIF and ALIF neurons, the spike output is computed using the non-differentiable Heaviside function denoted by Θ.

The output layer of the network consists of leaky readout neurons, which are modeled according to Eq. (2) without the reset term. The entire network is trained using either BPTT or e-prop using the mean squared error (MSE) loss.

2.3 Motor Inference Principle

The inference process uses a sufficiently trained forward model in order to infer control commands by means of BPTT or an e-prop inspired inference algorithm. For this purpose we define a prospective temporal horizon, which determines over how many time steps we will unroll our future projections. During the future projections the network computes in closed loop, that is, it feeds itself with its own predictions of the future development of the perceivable system states. For this, we first randomly initialize a vector of future control commands with the length of the chosen temporal horizon. The other inputs are then based on the network's predictions.

Using this setup, which is visualized in Fig. 1, the network calculates for each imaginary future time step a prediction of the perceivable system state given the used motor commands. By using desired targets for these predicted states, at each or only at the last imagined time step, we can backpropagate a prediction error-based gradient signal through the unrolled network (through time). When we map this gradient onto the control inputs, we end up with an input gradient with respect to the discrepancy between the predicted system states and the desired states. By using a gradient descent technique and by repeating the described procedure, we can adaptively optimize the sequence of control inputs effectively pushing the system towards the desired state. Note that we always start from the saved latent state which was produced in the last time step.

In the following, we formally derive the input gradient for the used LSNN network model. Let us refer to the state of a particular neuron j at time step t as $\mathbf{s}_j^t$. For LIF neurons this neuron state is one-dimensional and only contains

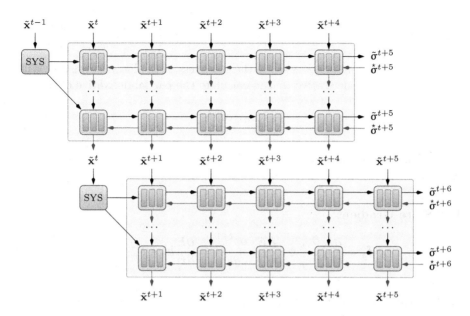

Fig. 1. Illustration of the action inference procedure with LSNNs: Based on the current state σ^t of the observed system (SYS), the network imagines the future development of the states $(\tilde{\sigma}^{t+1}, \tilde{\sigma}^{t+2}, \dots)$ given an initial motor sequence $(\tilde{\mathbf{x}}^t, \tilde{\mathbf{x}}^{t+1}, \dots)$. During this unrolling, the LSNN is fed with its own predictions. For the LSNN, however, each time step consists of multiple sub time steps (as indicated by the small rectangles within the LSNN nodes). In the last imagined time step, an intention $(\overset{\star}{\sigma}{}^{t+5})$ is injected via a prediction loss function, backpropagated through the unrolled imagined state sequence, and eventually projected onto the control inputs. Using a gradient descent technique and repeating this procedure multiple times within the current time step leads to an adaption of the motor inputs. $\tilde{\mathbf{x}}^t$ is executed to transition to the next time step, where the same procedure starts again.

the voltage v_j^t, whereas for ALIF neurons it is two-dimensional and contains the voltage v_j^t as well as the threshold adaption value a_j^t:

$$\mathbf{s}_{j,LIF}^t \overset{\text{def}}{=} v_j^t \tag{8}$$

$$\mathbf{s}_{j,ALIF}^t \overset{\text{def}}{=} \begin{bmatrix}[1.4]v_j^t & a_j^t\end{bmatrix}^\top \tag{9}$$

The full gradient calculation requires all components within the network's computation chain to be differentiable. As mentioned earlier, however, the Heaviside function does not fulfill this requirement. To overcome this problem, Bellec et al. [2] introduced a pseudo-derivative h_j^t in place for the non-existing derivative of the threshold function:

$$\frac{\partial z_j^t}{\partial \mathbf{s}_{j,LIF}^t} \overset{\text{def}}{=} h_{j,LIF}^t = \lambda \, \max\left(0, 1 - \left|\frac{v_j^t - v_{thr}}{v_{thr}}\right|\right) \tag{10}$$

$$\frac{\partial z_j^t}{\partial \mathbf{s}_{j,ALIF}^t} \overset{\text{def}}{=} [[1.4]1 , -\zeta]^\top h_{j,ALIF}^t = [[1.4]1 , -\zeta]^\top \lambda \, \max\left(0, 1 - \left|\frac{v_j^t - v_{j,thr}^t}{v_{thr}}\right|\right) \quad (11)$$

where the dampening factor λ scales the steepness of the linear segments. With help of this pseudo-derivative we can calculate the partial derivative of the loss with respect to $\mathbf{s}_j^t$:

$$\delta_j^t \overset{\text{def}}{=} \frac{\partial E}{\partial \mathbf{s}_j^t} \quad (12)$$

by applying the chain rule, which is done in the following for LIF and ALIF neurons.

LIF State Gradient

$$\delta_j^t = \frac{\partial E}{\partial z_j^t}\frac{\partial z_j^t}{\partial v_j^t} + \frac{\partial E}{\partial v_j^{t+1}}\frac{\partial v_j^{t+1}}{\partial v_j^t} = \frac{\partial E}{\partial z_j^t}h_j^t + \delta_j^{t+1}\alpha \quad (13)$$

ALIF State Gradient

$$\delta_j^t = \frac{\partial E}{\partial z_j^t}\frac{\partial z_j^t}{\partial \mathbf{s}_j^t} + \frac{\partial E}{\partial \mathbf{s}_j^{t+1}}\frac{\partial \mathbf{s}_j^{t+1}}{\partial \mathbf{s}_j^t} = \frac{\partial E}{\partial z_j^t}\begin{bmatrix}[1.4]h_j^t \\ -h_j^t\zeta\end{bmatrix} + \begin{bmatrix}[1.4]\alpha\delta_{j,v}^{t+1} \\ \rho\delta_{j,a}^{t+1}\end{bmatrix} \quad (14)$$

Using these delta terms we can finally derive the input gradient:

$$\frac{\partial E}{\partial x_i^t} = \sum_j \frac{\partial \mathbf{s}_j^t}{\partial x_i^t}\frac{\partial E}{\partial \mathbf{s}_j^t} = \sum_j w_{i,j}^{in}\delta_j^t \quad (15)$$

As a biologically more plausible and significantly more efficient error-based inference rule, we propose an alternative for BPTT, which is inspired by e-prop [2] and can be computed forward in time. Therefore, we expand the delta term part of the input gradient and discard errors from the future in order to be able to compute the error forward in time. We do this in three different styles:

Symmetric e-prop

$$\frac{\partial E}{\partial x_i^t} = \sum_j w_{i,j}^{in}\delta_j^t \approx \sum_j w_{i,j}^{in}\frac{\partial E}{\partial z_j^t}\frac{\partial z_j^t}{\partial \mathbf{s}_j^t} = \sum_j w_{i,j}^{in}h_j^t\left(\sum_k \delta_k^t w_{j,k}^{out}\right) \quad (16)$$

Sign-Based e-prop

$$\frac{\partial E}{\partial x_i^t} \approx \sum_j \text{sgn}(w_{i,j}^{in})h_j^t\left(\sum_k \delta_k^t \, \text{sgn}(w_{j,k}^{out})\right) \quad (17)$$

Random Feedback e-prop

$$\frac{\partial E}{\partial x_i^t} \approx \sum_j \text{sgn}(w_{i,j}^{in}) h_j^t \left(\sum_k \boldsymbol{\delta}_k^t b_{j,k}^{out} \right) \qquad (18)$$

For the symmetric case we calculate the input error of the network only as one single backward-pass through the network, where $\boldsymbol{\delta}_k^t$ is the output error of the output neuron k from the current sub time step. In the two other cases we either use only the sign of the weights or the random feedback weights ($b_{j,k}^{out}$) which were used for training with e-prop.

3   Experiments and Results

The experiments in this paper are based on a simulation of a simple flying vehicle in a minimalistic 3D environment with gravity and air resistance. The vehicle, which we refer to as rocket ball, is equipped with three rocket thrusters arranged around the vehicle with an 120° angle to each other, while pointing downwards in 45° angle. Each of the three thrusters can apply a force to the rocket ball that acts in the respective opposite direction. With this setup we conduct three different kinds of experiments which are described in the subsections below.

For measuring the inference performance of the rocket ball, we use an Euclidean distance error specified as $||E||_2$ which measures the distance between the center of the rocket ball and the center of the target sphere. Here an error of 1 is equivalent to the rocket ball's radius.

3.1   Short-Term Inference

In the initial set of experiments we tested the network's ability to infer motor commands in a simple setting where it has to steer the rocket ball to a desired goal location within the simulated world. In order to propagate the discrepancy between the currently predicted position delta and the distance to the target, we use BPTT as a baseline and compare it to the different e-prop inspired, biologically more plausible alternatives which all have the advantage that they do not depend on errors from the past and thus can be computed forward in time.

In Fig. 2, one can see the results of this experiment, which is a fairly simple task for a trained LSNN and can be accomplished in far less time steps than those the network was trained on. Here the rocket ball quickly accelerates towards the target, then starts to hover around it and by doing so it continues to decrease its mean distance to the target.

The corresponding spike trains also show this behavior where the rocket ball is already pretty close to the target after only 10 time steps and then needs another 20 time steps to get really close to it. Another interesting observation one can make from the spike train, is that the network seems to operate in two cycles per (world) time step, such that two somewhat similar spike patterns occur in each (world) time step.

We also found that the performance of the different proposed e-prop based inference methods is astoundingly close to that of pure BPTT as can be seen in the left plot of Fig. 3. While the LSNNs trained and inferred with BPTT initially manage to get closer to the target, after a while, once the rocket ball hovers around the target, their distance errors are the same as for the LSNNs trained and inferred with e-prop.

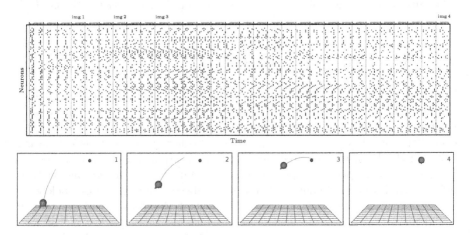

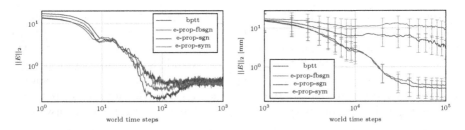

Fig. 2. Short-term inference (**bottom**) and corresponding spike-train (**top**) with BPTT and a time window of 5 time steps. The rocket ball starts on the bottom of the left side of the simulated world and quickly flies towards the desired goal location in the top right side where it then hovers around the goal position.

Fig. 3. Evaluation of different e-prop based inference algorithms: **sym**: symmetrical feedback weights, error flows back through the network weights. **sgn**: feedback weights are set to the sign of the corresponding weights. **fbsgn**: uses random feedback weights for propagating back the error from the outputs to the hidden neurons and the sign of the input weights to propagate it to the inputs. **bptt**: uses all network weights to propagate the error through the network and through time into the inputs. The **left** plot compares the median inference error for the different algorithms, using a network that was trained with e-prop. The **right** plot compares the algorithms in a goal-directed learning setup, where the network is trained continuously from scratch with e-prop while performing inference. Here the graph shows the median errors together with the upper and lower quantiles for 10 different runs and inference targets.

3.2 Continuous Goal-Directed Learning

In the second set of experiments we train our model in an online learning setting, where the rocket ball simulation continues indefinitely and the network has only the currently available experience in order to learn, which more closely resembles the learning environment of an agent acting in the real world. More formally, we use e-prop for training the LSNN in a single batch setting, where the network simultaneously trains its weights, while performing short-term motor inference for a total of 100,000 (world) time steps. Like in the first set of experiments, additionally to BPTT based inference, we also test our different proposed biologically plausible approaches, which are capable of online inference.

While the actual e-prop inference method had only a small impact on the overall inference performance, with a pretrained LSNN, in the continuous setting with goal-directed training only the symmetric e-prop variant (e-prop-sym) managed to perform similar to BPTT (see Fig. 3 right plot). Using random feedback weights (e-prop-fbsgn) for propagating the inference error completely failed to get the network anywhere close to the target, and also the sign-based (e-prop-sgn) propagation did not reach the target in the simulated 100,000 time steps. For random feedback weights, this might be due to the fact that they initially completely differ from the actual output weights, and only during training potentially align themselves with the output weights. Also using the sign of the forward weights for inference error propagation might not foster enough goal-directed behavior to accelerate learning the forward model in the same way as the actual weights do.

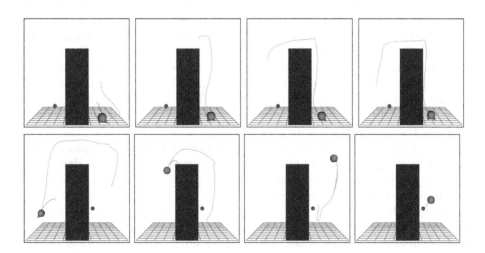

Fig. 4. Top: Emergence of a long-term trajectory within 7 time steps from a randomly initialized trajectory. **Bottom:** Long-term inference with BPTT with a long-term horizon of 30 time steps and a short-term horizon of 5 time steps. While the long-term inference (**blue trajectory**) manages to plan around the obstacle, the short-term inference (**red trajectory**) would collide with it. Once the rocket ball gets close to the target, the long-term inference is disabled in order to allow for a greater precision using only short-term inference. (Color figure online)

242 M. Traub et al.

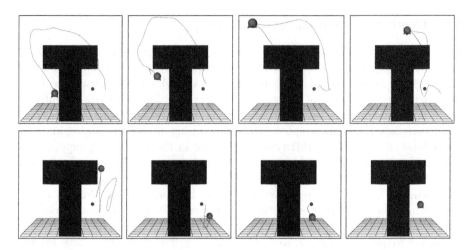

Fig. 5. Long-term inference with BPTT with a long-term horizon of 50 time steps and a short-term horizon of 5 time steps. While the long-term inference (**blue trajectory**) manages to plan around the obstacle, the short-term inference (**red trajectory**) would collide with it. Once the rocket ball gets close to the target, the long-term inference is disabled in order to allow for a greater precision using only short-term inference. (Color figure online)

3.3 Long-Term Inference

The last kind of experiments challenge the network's ability to look beyond the immediate future and plan around obstacles. Therefore the closed loop operation is unrolled over an extended period of time steps into the future using only the network's own predictions as inputs. In the last time step, a target for the final prediction position deltas is given and backpropagated through time into the motor inputs from the current time step. Additionally the rocket ball uses axis-aligned distance sensors for which in each imagined time step zero is given as a target for the distance sensor predictions, regularizing the imagined trajectory in a way that the rocket ball avoids obstacles.

This setup then allows the network to perform long-term planning around concave as well as convex obstacles as can be seen in Fig. 4 and 5. Here these planning trajectories spontaneously emerge, even when the initial endpoint of the imagined trajectory is behind the obstacle (see Fig. 4). This suggests that the network has learned an internal model of the world including the obstacle and that by using the imagined distance sensor values for avoiding the obstacle, it can plan around it. This is not possible when using only short-term inference, where the rocket ball plans to fly directly towards the goal and therefore ends up stuck to the obstacle. Here the gradients towards the goal directly compete with the gradients from the distance sensors, and due to the short planning horizon the gradients from the distance sensors can not alter the trajectory in a way that the endpoint of the planned trajectory ends up behind the obstacle.

While in the majority of simulations a trajectory tends to form around the more challenging T-shaped obstacle and the rocket ball manages to navigate around it within the first 50 time steps, the long-term trajectories tend to be unstable, which explains the number of fail cases in Fig. 6.

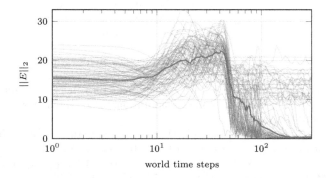

Fig. 6. Long-term inference with the T-shaped obstacle over 300 time steps, using a long-term horizon of 50 and a short-term horizon of 1 time steps. The median inference error (**bold**) is annotated with the individual runs (**thin**). While there are some fail cases, the median error shows that usually the network manages to navigate around the obstacle in within the first 50 time steps.

4   Conclusion

We have shown that spiking neural networks, and LSNNs in particular, can be trained to learn and memorize a temporal forward model of its environment, including obstacles. Moreover, using a biologically plausible inference algorithm inspired by e-prop, the LSNN can be used to induce short-term and long-term goal-directed action inference in continuous settings—results that are almost identical to BPTT in terms of accuracy, while having the advantage that they do not require the expensive backpropagation of errors through time.

In a continued goal-directed learning setup, the most biologically plausible methods could not compete with BPTT. However an e-prop variant using symmetric feedback weights also performed comparable to pure BPTT showing that LSNNs can not only be trained with a biologically plausible alternative to BPTT, they can also be employed in realistic online reinforcement learning setting, simulating the available experience of an agent in the real world. Here we demonstrated that they can learn and employ goal-directed behavior without the need for any kind of backpropagated error information from the future.

Apart from the more biologically plausible neural network processing mechanisms and local active inference techniques, more capable systems of this type may be implemented in hardware highly efficiently, promising to yield a much lower energy consumption [17]. We furthermore expect to enhance the LSNN with regularization terms, to foster even sparser temporal encodings.

References

1. Bellec, G., Salaj, D., Subramoney, A., Legenstein, R., Maass, W.: Long short-term memory and learning-to-learn in networks of spiking neurons. In: Proceedings of the 32nd International Conference on Neural Information Processing Systems, NIPS 2018, pp. 795–805. Curran Associates Inc., Red Hook (2018)
2. Bellec, G., et al.: A solution to the learning dilemma for recurrent networks of spiking neurons. Nat. Commun. **11**(1), 1–15 (2020)
3. Butz, M.V., Bilkey, D., Humaidan, D., Knott, A., Otte, S.: Learning, planning, and control in a monolithic neural event inference architecture. Neural Netw. **117**, 135–144 (2019)
4. Butz, M.V., Kutter, E.F.: How the Mind Comes Into Being: Introducing Cognitive Science from a Functional and Computational Perspective. Oxford University Press, Oxford (2016)
5. Butz, M.V., Menge, T., Humaidan, D., Otte, S.: Inferring event-predictive goal-directed object manipulations in REPRISE. In: Tetko, I.V., Kurková, V., Karpov, P., Theis, F. (eds.) ICANN 2019, Part I. LNCS, vol. 11727, pp. 639–653. Springer, Cham (2019). https://doi.org/10.1007/978-3-030-30487-4_49
6. Clark, A.: Surfing Uncertainty: Prediction, Action, and the Embodied Mind. Oxford University Press, Oxford (2015)
7. Friston, K.: The free-energy principle: a rough guide to the brain? Trends Cogn. Sci. **13**(7), 293–301 (2009)
8. Hochreiter, S., Schmidhuber, J.: Long short-term memory. Neural Comput. **9**(8), 1735–1780 (1997)
9. Mnih, V., et al.: Playing atari with deep reinforcement learning. arXiv:1312.5602 (2013)
10. Nichol, A., Pfau, V., Hesse, C., Klimov, O., Schulman, J.: Gotta learn fast: a new benchmark for generalization in RL. arXiv:1804.03720 (2018)
11. OpenAI, et al.: Dota 2 with large scale deep reinforcement learning. arXiv:1912.06680 (2019)
12. Otte, S., Schmitt, T., Friston, K., Butz, M.V.: Inferring adaptive goal-directed behavior within recurrent neural networks. In: Lintas, A., Rovetta, S., Verschure, P.F.M.J., Villa, A.E.P. (eds.) ICANN 2017, Part I. LNCS, vol. 10613, pp. 227–235. Springer, Cham (2017). https://doi.org/10.1007/978-3-319-68600-4_27
13. Otte, S., Stoll, J., Butz, M.V.: Incorporating adaptive RNN-based action inference and sensory perception. In: Tetko, I.V., Kůrková, V., Karpov, P., Theis, F. (eds.) ICANN 2019. LNCS, vol. 11730, pp. 543–555. Springer, Cham (2019). https://doi.org/10.1007/978-3-030-30490-4_44
14. Schulman, J., Wolski, F., Dhariwal, P., Radford, A., Klimov, O.: Proximal policy optimization algorithms. arXiv:1707.06347 (2017)
15. Traub, M., Legenstein, R., Otte, S.: Many-joint robot arm control with recurrent spiking neural networks. In: IEEE/RSJ International Conference on Intelligent Robots and Systems (IROS) (2021). accepted for publication, preprint available (arXiv:2104.04064)
16. Yin, B., Corradi, F., Bohte, S.M.: Accurate and efficient time-domain classification with adaptive spiking recurrent neural networks. arXiv:2103.12593 (2021)
17. Yin, B., Corradi, F., Bohté, S.M.: Accurate and efficient time-domain classification with adaptive spiking recurrent neural networks. bioRxiv:2021.03.22.436372 (2021)

End-to-End Spiking Neural Network for Speech Recognition Using Resonating Input Neurons

Daniel Auge[1]([✉])[iD], Julian Hille[1,2][iD], Felix Kreutz[3][iD], Etienne Mueller[1][iD], and Alois Knoll[1][iD]

[1] Department of Informatics, Technical University of Munich, Munich, Germany
`daniel.auge@tum.de, knoll@in.tum.de`
[2] Infineon Technologies AG, Munich, Germany
[3] Infineon Technologies Dresden GmbH & Co., KG, Dresden, Germany

Abstract. The growing demand for complex computations in edge devices requires the development of algorithms and hardware accelerators that are powerful while remaining energy-efficient. A possible solution are spiking neural networks, as they have been demonstrated to be energy-efficient in several data processing and classification tasks when executed on specialized neuromorphic hardware. In the field of speech processing, they are especially suited for the online classification of audio streams due to their strong temporal affinity. However, so far, there has been a lack of emphasis on small-scale networks that will ultimately fit into restricted neuromorphic implementations. We propose the use of resonating neurons as an input layer to spiking neural networks for online audio classification to enable an end-to-end solution. We compare different architectures to the established method of using mel-frequency-based spectral features. With our approach, spiking neural networks can be directly used without additional preprocessing, thereby making them suitable for simple continuous low-power analysis of audio streams. We compare the classification accuracy of different network architectures with ours in a keyword spotting benchmark to demonstrate the performance of our approach.

Keywords: Spiking neural networks · Speech processing · Keyword detection

1 Introduction

Keyword spotting, as part of speech recognition, is widely used in embedded systems for a wide range of voice-activated assistants. A detector for this purpose can be operated in an always-on mode; therefore, in addition to the recognition rate, energy efficiency is a decisive factor for evaluating a detection system. Another consideration is the detector's ability to perform the desired action in real-time.

© Springer Nature Switzerland AG 2021
I. Farkaš et al. (Eds.): ICANN 2021, LNCS 12895, pp. 245–256, 2021.
https://doi.org/10.1007/978-3-030-86383-8_20

Current implementations consist of multiple cascaded detectors of increasing complexity to cope with these requirements. Such detectors range from simple threshold switches over classical algorithmic signal processing to complex neural networks. The growing demand for smart devices and their capabilities expects even better performance with further improved energy efficiency. Many embedded artificial neural network (ANN) architectures have been proposed to resolve this [3,26]. Ideally, also large-scale speech recognition should be performed directly in the edge device. Due to high complexity, this task is currently offloaded to cloud servers.

Recently, researchers have demonstrated promising results in efficient signal processing and speech recognition using spiking neural networks (SNNs) [18,23]. SNNs, which can be operated on dedicated neuromorphic hardware, show a significantly lower energy consumption than comparable classical ANNs [5,8,17]. They do so by exchanging short pulses in the time domain, "spikes", instead of static continuous-valued activations. Accordingly, energy is consumed only during the update of a neuron whenever a spike arrives at its input.

In this work, we compare different modeling approaches for simulating SNN behavior. Simultaneously, we use different neuron behaviors and connectivity strategies to identify the most suitable network for an end-to-end keyword spotting on restricted hardware. Therefore, we demonstrate resonating neurons as input layer to transform an audio signal into a frequency selective spatiotemporal spike representation. Thus, the network can perform keyword detection solely with spiking neurons without using digital signal processing such as melfrequency cepstral coefficients (MFCC). In a neuromorphic realization, an analog electrical signal of a microphone can be directly fed into resonating neurons. This solution not only saves energy by turning off the digital logic, including the analog-to-digital converter, until a keyword is recognized by the neural circuit, but it also adds an extra layer of privacy for an end-user.

2 Related Work

Most modern speech recognition systems are based on non-spiking ANNs. They use recurrent or convolutional network architectures to detect spoken words in audio signals [1,10]. For this purpose, the input signal is divided into windowed blocks and a short-time Fourier transform is applied, resulting in a spectrum that changes over time. Typically, the spectrum is then projected to the melfrequency scale and serves, along with its first and second temporal derivative, as the input feature vector.

Other neural network-based approaches exist, which directly analyze an audio stream without prior feature generation. The network learns feature extraction from the ground up while still operating on fixed-sized windows of input data. The proposed deep and convolutional architectures, therefore, exceed millions of trainable parameters and result in large networks [12–14,20].

Early works on biologically plausible audio processing solutions based on SNNs demonstrated small, energy-efficient networks, that show stimulus-specific network activities when stimulated with simple stimuli [21]. Then, an artificial cochlea [7] was used to transform the audio signal into a spiking representation.

With recent advances in supervised gradient descent-based learning algorithms for SNNs [4,16], spiking networks that perform keyword spotting in the spiking domain have been proposed [18]. Especially, Blouw and Eliasmith [5,6] focus on potential energy savings during this task using these biologically inspired networks on specialized neuromorphic hardware. They report up to 10x energy savings using low-energy neural network accelerators on the Loihi chip [8] or the upcoming SpiNNaker2 chip [15] compared to current ANN-based approaches. For the representation of input data as a sequence of events/spikes, different encoding methods can be applied. The calculated MFCC can be interpreted as the amplitude of a current that can be used to excite an input neuron [23,24]. Another similar approach is to interpret the amplitudes as a spike rate, which is commonly used for converting an ANN to an SNN [5].

3   Methods

Spiking neurons exchange information in the form of short all-or-nothing action potentials. Each neuron emits a spike as soon as the value of its hidden variable V reaches a certain firing threshold θ. In analogy to their biological counterpart, this hidden value is called membrane voltage. The membrane is charged whenever an action potential reaches the neuron via a connection between two neurons called a synapse. Similar to classical ANNs, these connections have a specific weight, which can be adapted during training to fit the desired behavior.

3.1   Spike Encoding

For a spiking network to process data, the input has to be translated into spike events. In a recent study, it has been shown that resonating neurons can be used to perform a spectral analysis on analog input data [2]. Multiple frequency-tuned resonate-and-fire [11] neurons can be used as a filter bank that emits spikes with a rate proportional to the power density of the analyzed frequencies. With that, spectral analysis and conversion into spikes are performed simultaneously. In addition, we achieve a high temporal resolution since no sliding window is needed, as opposed to a fixed-length Fourier transform.

The differential equations describing the resonating neuron are given by

$$\dot{y} = -y\,d - 2\pi f_0\,v + i(t) \tag{1}$$

$$\dot{v} = -v\,d + 2\pi f_0\,y. \tag{2}$$

This specialized neuron comprises of two coupled membranes y and v to enable the resonating behavior. Here, y and v describe the voltage and current-like variables of the neuron with its resonant frequency f_0. d is a damping value which leads to an exponential decay of the state variables over time. The input of the system is given by the current $i(t)$, which can be any arbitrary time-dependent signal or spike train, as initially proposed by Izhikevich [11]. An output spike z is generated as soon as the voltage-like variable v surpasses its firing threshold v_{th}:

$$z = \begin{cases} 1, & \text{if } v > v_{th} \rightarrow y = 0, v = 0, v_{th} = v_{th} + v_{th} \\ 0, & \text{otherwise.} \end{cases} \tag{3}$$

After each output spike, the state variables are reset and the threshold is increased to achieve a spike rate adaption depending on the amplitude of the signal's spectral components. The threshold can be adapted linearly or exponentially, but an exponential adaption, as shown in Eq. 3, showed the best results in this application. The firing threshold itself also experiences an exponential damping

$$\dot{v_{th}} = (v_{th,0} - v_{th})\, d, \tag{4}$$

which ensures a weak upper boundary and a reset over time. With that, the neuron can adapt to a large range of signal amplitudes (see Fig. 1).

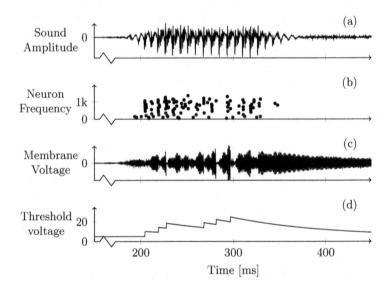

Fig. 1. Simulation of the resonating input layer using an exemplary audio sample. (a) The waveform of the audio sample. (b) Spike events generated by 40 resonating neurons. Their resonance frequencies are linearly spaced between 0 and 2,000 Hz. (c, d) Membrane voltage and threshold adaption of one resonating neuron.

As a result, a spike emitted by a resonating neuron contains a combination of different information: The spike signalizes the presence of the associated resonance frequency within the signal. In addition, the number and exact timing of the spikes encodes the amplitude of the spectral component and its development over time. With the precise interaction of input stream, membrane resonance, spike emission, and threshold adaption, a unique spatiotemporal spike train is created which is analyzed by the succeeding populations of spiking neurons.

3.2   Neuron Models

So far, a variety of neuron models have been proposed for the use in SNNs in the literature. They range from highly realistic models, which can replicate complex biological behavior of single neurons, to the simplest models, which offer only an abstraction of the biological inspiration but are suited for simulating large networks of neurons.

The base model used in this work is the leaky-integrate-and-fire (LIF) neuron with the membrane potential

$$V[t_n] = \alpha V[t_{n-1}] + (1 - \alpha) I[t_n] - I_{\text{reset}}. \tag{5}$$

The parameter $\alpha = e^{-1/\tau}$ describes the exponential decay of the membrane voltage over time, whereas τ describes the time constant of the neuron. The input charge current I consists of the sum of weighted spike inputs S_j:

$$I_i[t_n] = \sum_j W_{ij} S_j[t_{n-1}]. \tag{6}$$

The reset current

$$I_{\text{reset}} = Z[t_{n-1}] \theta \tag{7}$$

is subtracted if the neuron emitted an output spike

$$Z[t_n] = \begin{cases} 1, & \text{if } V[t_n] > \theta \\ 0, & \text{otherwise} \end{cases} \tag{8}$$

in the previous time step. By achieving the reset-by-subtraction of the threshold voltage, the information of an extremely high activation is preserved for the next time steps. Therefore, it is highly probable that a further spike will be emitted soon. A reset of the membrane potential to zero, on the other hand, would discard this information, potentially resulting in a higher ability to generalize. In our experiments, however, we consistently achieved higher performances using the reset-by-subtraction scheme.

To enable further communication within a neuron population, recurrent connections can be established. The index k indicates a presynaptic neuron within the same population as neuron i. By introducing the recurrent weight matrix R, the charging current of each neuron can, therefore, be extended to

$$I_i[t_n] = \sum_j W_{ij} S_j[t_{n-1}] + \sum_k R_{ik} S_k[t_{n-1}]. \tag{9}$$

With that, the membrane potential of a neuron depends on the activations within the same neuron population, which can also be viewed as lateral connections. Thus, the potential memorization capability of the population is increased.

3.3 Learning

Surrogate Gradient Descent: The spike emission operation shown in Eq. 8 is not differentiable and is, therefore, not suited for gradient descent-based learning methods. However, the use of surrogate gradients to enable learning is being consolidated in the development of SNNs [4,16]. The surrogate gradient ψ used in this work is

$$\psi = \max\left(1 - \left|\frac{v}{\theta} - 1\right|, 0\right).$$

On this basis, the gradient is determined by the relationship between the membrane potential and the threshold voltage, rather than the spike event.

To apply this gradient and backpropagate the error through the network and time, the network has to be simulated in discrete time steps. Therefore, we separate the input layer from the rest of the network. For the input layer, it is crucial to exactly calculate the differential equations. The following layers in contrast need to be discretized in time to apply the learning algorithm. Since there is no recurrent connection to the input of the network, the layers can be separated and simulated independently. This, however, is only a limitation during learning.

Time Constant Learning: In addition to the synaptic weights, the time constant τ (see Eq. 5) can be adapted to tune the temporal behavior of the network [25]. It controls the neuron's membrane voltage leakage over time. A large time constant leads to a small voltage leakage, enabling the long retention of information about past input spikes. A small time constant, on the other hand, enables short-term coincidence detectors, without being biased by the recent input spike history.

4   Experiments

Different network architectures are evaluated using the Speech Commands dataset [22] consisting of 65,000 audio recordings of known commands, unknown words, and silence. In a preprocessing step described by the author of the dataset, the recordings are superimposed with background noise at random volume levels.

The raw audio stream is encoded into a spike representation using resonating neurons. Following the standard of using 40 MFCC, we encode the input with 40 resonating neurons. This is simulated using exact solving of the coupled differential equations describing the resonating neurons. The resulting spikes serve as the input to the next stage of the network, which is simulated in discrete time steps to enable backpropagation learning. This part of the network is subject to optimization and architecture search. The output layer consists of non-spiking integrators – one for each class – which are evaluated after each training example to calculate the respective loss and accuracy values. The resulting abstract network architecture is shown in Fig. 2. In additions, we implement a standard MFCC-based preprocessing similar to [18] and compare it with our approach. In this case, mel-frequency features are directly applied to the first hidden layer of the network.

During the experiments, we examine the performance of three main architectures: simple feedforward networks, networks with recurrent neuron populations, and convolutional networks. In the former two cases, we also distinguish between models with one or two hidden layers. The architecture based on convolutional networks uses only one-dimensional convolutions along the frequency axis of the input. By omitting a convolution along the temporal dimension, we emphasize the inherent temporal properties of spiking neurons.

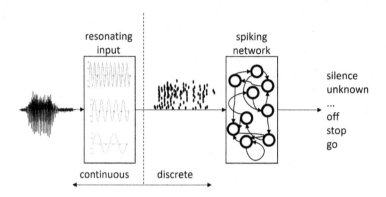

Fig. 2. Architecture of the evaluated system. The waveform is directly fed into the network without preprocessing. The resulting spikes produced by the input layer are propagated through the network. The output neuron with the highest membrane voltage selects the inferred class.

4.1   Results

In the first setup, we evaluate the relationship between the classification accuracy and the number of tunable variables. An exemplary extract of the neuron activity during the evaluation of both networks is depicted in Fig. 3. The sparseness of spike activations in the input and hidden layer can be seen. For better comparability, we chose networks consisting of only one population to avoid ambiguous distributions of the limited number of connections between multiple populations. Figure 4a shows the achieved accuracies for a simple feedforward population and a population with recurrent connections, both using RF neurons as input encoders. Note that due to the different connection schemes, the two populations do not share the same total number of neurons involved at equal numbers of variables. When the number of variables exceeds 20,000, the achieved accuracy begins to saturate. Nontheless, the difference of 10% points between the two architectures remains constant. Figure 4b depicts a confusion matrix showing the classification results of a network with recurrent neurons and a total of

40,000 tunable parameters. Class labels 0 to 11 correspond in ascending order to {silence, unknown, yes, no, up, down, left, right, on, off, stop, go}. The matrix shows a high misclassification rate for the unknown class and between word pairs {up, off} and {no, go}.

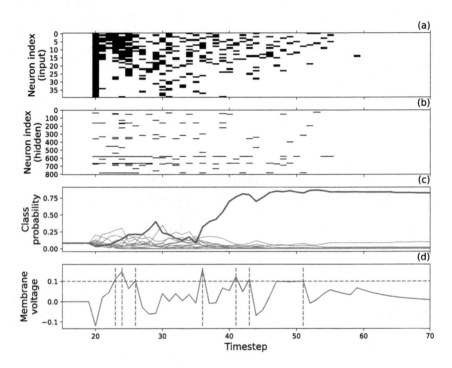

Fig. 3. Exemplary evaluation of a command. (a) Spike encoding of the audio stream using resonating neurons. (b) Spikes emitted by neurons in the hidden layer. (c) Membrane voltage of the output layer indicating the classification. (d) Membrane potential of one exemplary hidden neuron. The horizontal dashed line represents the firing threshold, the vertical lines the time instances of the spike emission.

Based on the results of the preceding experiment, we chose 40,000 trainable parameters as the common parameters of the following simulations. With that the results remain comparable while providing the intended insights about the relations between the classification accuracy, the chosen input, and the complexity of possible neuromorphic realizations. Table 1 shows the results of the evaluated architectures.

The densely connected feedforward architectures demonstrate a basic ability to solve the keyword recognition task. The better performance of multi-layered networks also underlines the common conception of the importance of deep structures [9]. In our experiments, however, increasing the number of layers further did not improve performance.

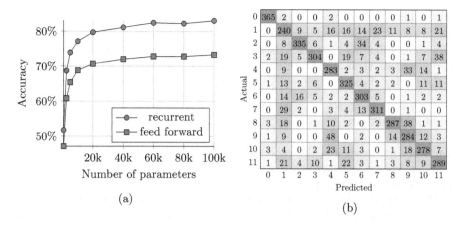

(a)

(b)

Fig. 4. (a) Classification accuracy of networks with different sizes and connectivity rules. The x-axis describes the number of tuneable parameters in the network. The networks consist of a single hidden population. (b) Confusion matrix for a network with 40,000 parameters with recurrent connections.

Table 1. Network architectures tested with inputs consisting of mel cepstral coefficients or spikes generated by resonate-and-fire neurons. Each network consists of 40k trainable parameters.

Connection type	Architecture	MFCC	RF
Feedforward	780 neurons	72.1%	70.4%
Feedforward	175 × 175 neurons	84.1%	72.9%
Recurrent	178 neurons	84.7%	80.2%
Recurrent	105 × 105 neurons	86.7%	80.7%
Convolutional	Kernel size: 5,10; kernels: 35 × 40	86.2%	80.5%
Recurrent conv.	Kernel size: 5,10; kernels: 30 × 40	84.8%	85.5%

Among the evaluated networks with recurrent populations, the MFCC-based approaches are superior to the networks with resonating input neurons. Thus, the recurrently connected neurons can extract the temporal information present in the spectral input signal.

Networks with convolutional populations achieved the highest classification accuracies in our tests. Due to the shared weights of the convolutional kernels, it is possible to define a large number of kernels within the defined restricted number of trainable parameters. The actual size of the resulting network is significantly larger since the kernels need to be unrolled to enable parallel asynchronous processing. Including recurrent connections to the convolutional network further improved the classification accuracy.

In comparison to the evaluation results reported in the literature, our approach shows an inferior classification performance with a maximum accuracy of approximately 80%. Studies based on SNNs report error rates as low as 5.5% [18], whereas ANN-based approaches undercut this number even further [19,26]. The main differences between these works and ours are the sizes of the considered networks, their architectural design choices, and the degree of preprocessing of the analyzed data.

5   Conclusion

Our work demonstrates the successful use of small-scale SNNs with surrogate gradient descent learning and a new type of spike encoding, especially suited for online speech recognition.

Resonating neurons may be a more energy-efficient alternative to the digital processing chain used in modern speech recognition systems. Using these neurons, analog-to-digital converters, digital filters, fast Fourier transform blocks, and the MFCC feature generation can be omitted. In this work, we demonstrated that these neurons generate feature-rich spike trains that can be analyzed by the following network structures. The set of hyperparameters such as the number of resonating neurons, the choice of their resonance frequencies, or the threshold adaption characteristic leaves room for improvement and the adaption to other applications. However, the energy efficiency of this method using specialized electrical circuits remains to be proven. In addition, the classification accuracy of this approach needs to be improved to be comparable with established methods.

The network architectures considered in the experiments are particularly suitable for real-time acceleration with neuromorphic hardware since no data are buffered at any point in time, and time dependencies are represented solely by the hidden variables of neuron models or recurrent connections. Particularly, embedded systems can profit from this solution, along with the low energy consumption of SNNs on specialized hardware reported in the literature. In a hybrid realization, an SNN can serve as a low-energy always-on detector, activating a more elaborate ANN for further processing when the required activation pattern is detected.

Acknowledgments. We thank Infineon Technologies AG for supporting this research. The work is partly conducted within the KI-ASIC project that is funded by the German Federal Ministry of Education and Research (Grand Number 16ES0992K).

References

1. Abdel-Hamid, O., Mohamed, A.R., Jiang, H., Penn, G.: Applying convolutional neural networks concepts to hybrid NN-HMM model for speech recognition. In: 2012 IEEE International Conference on Acoustics, Speech and Signal Processing (ICASSP), pp. 4277–4280. IEEE (2012)
2. Auge, D., Mueller, E.: Resonate-and-fire neurons as frequency selective input encoders for spiking neural networks. TUM (Technical Report) (2020)
3. Banbury, C., MicroNets: neural network architectures for deploying TinyML applications on commodity microcontrollers. arXiv preprint arXiv:2010.11267 (2020)

4. Bellec, G., Salaj, D., Subramoney, A., Legenstein, R., Maass, W.: Long short-term memory and learning-to-learn in networks of spiking neurons. In: Advances in Neural Information Processing Systems, pp. 787–797 (2018)
5. Blouw, P., Choo, X., Hunsberger, E., Eliasmith, C.: Benchmarking keyword spotting efficiency on neuromorphic hardware. In: Proceedings of the 7th Annual Neuro-Inspired Computational Elements Workshop, pp. 1–8 (2019)
6. Blouw, P., Eliasmith, C.: Event-driven signal processing with neuromorphic computing systems. In: 2020 IEEE International Conference on Acoustics, Speech and Signal Processing (ICASSP), pp. 8534–8538. IEEE (2020)
7. Chan, V., Liu, S.C., van Schaik, A.: AER EAR: a matched silicon cochlea pair with address event representation interface. IEEE Trans. Circuits Syst. I Regul. Pap. **54**(1), 48–59 (2007)
8. Davies, M., et al.: Loihi: a neuromorphic manycore processor with on-chip learning. IEEE Micro **38**(1), 82–99 (2018)
9. Eldan, R., Shamir, O.: The power of depth for feedforward neural networks. In: Conference on Learning Theory, pp. 907–940. PMLR (2016)
10. Graves, A., Mohamed, A.R., Hinton, G.: Speech recognition with deep recurrent neural networks. In: 2013 IEEE International Conference on Acoustics, Speech and Signal Processing (ICASSP), pp. 6645–6649. IEEE (2013)
11. Izhikevich, E.M.: Resonate-and-fire neurons. Neural Netw. **14**(6–7), 883–894 (2001)
12. Kim, T., Lee, J., Nam, J.: Comparison and analysis of sample CNN architectures for audio classification. IEEE J. Sel. Top. Signal Process. **13**(2), 285–297 (2019)
13. Kumatani, K., et al.: Direct modeling of raw audio with DNNs for wake word detection. In: 2017 IEEE Automatic Speech Recognition and Understanding Workshop (ASRU), pp. 252–257. IEEE (2017)
14. Lee, J., Park, J., Kim, K.L., Nam, J.: Sample-level deep convolutional neural networks for music auto-tagging using raw waveforms. arXiv preprint arXiv:1703.01789 (2017)
15. Mayr, C., Hoeppner, S., Furber, S.: Spinnaker 2: a 10 million core processor system for brain simulation and machine learning. arXiv preprint arXiv:1911.02385 (2019)
16. Neftci, E.O., Mostafa, H., Zenke, F.: Surrogate gradient learning in spiking neural networks. IEEE Signal Process. Mag. **36**, 61–63 (2019)
17. Ostrau, C., Homburg, J., Klarhorst, C., Thies, M., Rückert, U.: Benchmarking deep spiking neural networks on neuromorphic hardware. arXiv:2004.01656 12397, pp. 610–621 (2020)
18. Pellegrini, T., Zimmer, R., Masquelier, T.: Low-activity supervised convolutional spiking neural networks applied to speech commands recognition. arXiv preprint arXiv:2011.06846 (2020)
19. Rybakov, O., Kononenko, N., Subrahmanya, N., Visontai, M., Laurenzo, S.: Streaming keyword spotting on mobile devices. arXiv preprint arXiv:2005.06720 (2020)
20. Sainath, T.N., et al.: Multichannel signal processing with deep neural networks for automatic speech recognition. IEEE/ACM Trans. Audio Speech Lang. Process. **25**(5), 965–979 (2017)
21. Sheik, S., Coath, M., Indiveri, G., Denham, S.L., Wennekers, T., Chicca, E.: Emergent auditory feature tuning in a real-time neuromorphic VLSI system. Front. Neurosci. **6**, 17 (2012)
22. Warden, P.: Speech commands: a dataset for limited-vocabulary speech recognition. arXiv preprint arXiv:1804.03209 (2018)
23. Wu, J., Yılmaz, E., Zhang, M., Li, H., Tan, K.C.: Deep spiking neural networks for large vocabulary automatic speech recognition. Front. Neurosci. **14**, 199 (2020)

24. Yılmaz, E., Gevrek, O.B., Wu, J., Chen, Y., Meng, X., Li, H.: Deep convolutional spiking neural networks for keyword spotting. In: Proceedings of Interspeech 2020, pp. 2557–2561 (2020)
25. Yin, B., Corradi, F., Bohté, S.M.: Effective and efficient computation with multiple-timescale spiking recurrent neural networks. arXiv preprint arXiv:2005.11633 (2020)
26. Zhang, Y., Suda, N., Lai, L., Chandra, V.: Hello edge: keyword spotting on micro-controllers. arXiv preprint arXiv:1711.07128 (2017)

Text Understanding I

Visual-Textual Semantic Alignment Network for Visual Question Answering

Weidong Tian[1,2], Yuzheng Zhang[1,2], Bin He[1,2], Junjun Zhu[1,2],
and Zhongqiu Zhao[1,2,3,4(✉)]

[1] School of Computer Science and Information Engineering,
Hefei University of Technology, Hefei, China
z.zhao@hfut.edu.cn
[2] Intelligent Interconnected Systems Laboratory of Anhui Province,
Hefei University of Technology, Hefei, China
[3] Intelligent Manufacturing Institute of HFUT, Hefei, China
[4] Guangxi Academy of Science, Nanning 530007, China

Abstract. VQA task requires deep understanding of visual and textual content and access to key information to better answer the question. Most of current works only use image and question as the input of
the network, where the image features are over-sampling and the text
features are under-sampling, resulting in insufficient alignment between
image regions and question words. In this paper, we propose a Visual-
Textual Semantic Alignment Network (VTSAN). Our network acquires
tags for visual semantics from a target detector and takes the Image-
Tag-Question$\langle \mathbf{I}, \mathbf{T}, \mathbf{Q} \rangle$ triad as the input. The tags can serve as an intermediate medium between the key regions of image and the key words of
question, and can greatly enrich the text features. Thereby, the visual-
textual semantic alignment is significantly improved. We demonstrate
the effectiveness of our proposed network on the standard VQAv2 and
VQA-CPv2 benchmarks. The experimental results show that the proposed network outperforms the baseline significantly, especially on the
counting questions.

Keywords: VQA · Tag information · Visual-textual semantic
alignment

1 Introduction

The Visual Question Answering (VQA) task has attracted extensive attentions
in computer vision and natural language processing communities. Many studies have confirmed that improving cross-modal feature extraction and fusion to
obtain richer and more accurate global features is important for promoting the
VQA task [4,5,11,14].

The attention mechanism has been successfully used in cross-modal tasks. In
VQA, [1] proposed that it is important to learn the attention on image regions

© Springer Nature Switzerland AG 2021
I. Farkaš et al. (Eds.): ICANN 2021, LNCS 12895, pp. 259–270, 2021.
https://doi.org/10.1007/978-3-030-86383-8_21

with input question features, and the learning of text features for question keywords is equally important. [23] proposed the Modular Co-Attention Network (MCAN) which adopts the Transformer architecture [3] for co-attention learning of question features and image features. However, existing VQA models simply use image features and question features as the network inputs to predict the answers to questions. Text features contain less semantic information than image features, which contain too much redundant semantic information. So the image features are oversampling while the question features are undersampling [21,22,24]. In this case, it's difficult to form an effective semantic alignment between image features with redundant information and text features with little information. This will result in the model focusing on the image regions not related to the question.

To address the problem of too little textual information contained in the input, some approaches considered tag information as additional text features for the input. Zhou et al. [24] used object prediction probabilities as soft tags and combined them with the responding region features. Wu et al. [21] and You et al. [22] utilized the tags of images to improve the visual feature representation of images. These methods adopt tag information to effectively enrich the amount of text feature information and to enhance the overall feature representation of the image in a multimodal task. However, these methods do not focus on enhancing the mutual alignment between key regions of the image and keywords of the text.

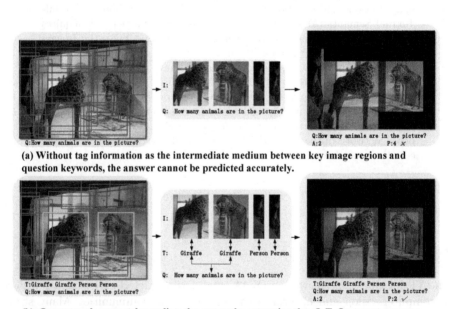

(a) **Without tag information as the intermediate medium between key image regions and question keywords, the answer cannot be predicted accurately.**

(b) **Our network accurately predicts the answer by extracting the <I, T, Q> ternary groups.**

Fig. 1. Comparison of attentional weighting with tags or not. A is the answer and P is the prediction. The brightness of the image region and the gray depth of the word represent their importance in the attentional weighting.

In this paper, we consider tags as a bridge between the image and the question by building an attention mechanism among image, tag, and question, in order to improve visual-textual semantic alignment.

Visual-textual alignments refer to strong associations between key regions of the image and the question keywords. However, the feature patterns vary greatly among different modalities, which prevents forming accurate alignments. Fortunately, image region features and the corresponding tag information can be accurately acquired by a target detector and these tag terms frequently appear in some questions. This enables us to transform visual features into textual features by tags and to achieve accurate alignments between image and question features in the same modality. Thereby, we propose a Visual-Textual Semantic Alignment Network (VTSAN). In this network, as shown in Fig. 1, we use an Image-Tag-Question$\langle \mathbf{I}, \mathbf{T}, \mathbf{Q} \rangle$ ternary group as the input. Using the tag information as an intermediate medium between key regions of the image and keywords of the question, we build an attention mechanism between the triad. The tag information significantly enriches the amount of text feature information in the input network and the attention mechanism improves the visual-textual semantic alignment. And inspired by the currently widely used Transformer model, we adopt the Encoder-Decoder framework in our VTSAN network. We use the multi-head attention method to combine the Self-Attention (SA) unit and the Guided Attention (GA) unit to strengthen the intra-modal and inter-modal interaction. By building modular SA and GA units, the Visual-Textual Semantic Alignment (VTSA) layer is obtained. To avoid inadequate self-attention to the initial questions and tags, we perform a deep cascaded operation on the VTSA layer.

The main contributions of this paper are as the following:

- We propose a visual-textual semantic alignment model, which preprocesses training samples into the Image-Tag-Question$\langle \mathbf{I}, \mathbf{T}, \mathbf{Q} \rangle$ ternary group form.
- We conduct extensive experiments to demonstrate that tag information can serve as an additional text feature input and an intermediate medium between image key regions and question keywords, and thereby improve the performance on the VQA task.
- We construct the VTSA layer and propose the VTSAN with the multi-feature guidance attention in the Encoder-Decoder framework, which outperforms the baseline model on VQAv2 [6] and VQA-CPv2 [17] datasets.

2   Related Work

The most straightforward VQA solution is to extract the image features and the question features, and then fuse them to predict the answer by an answer classifier [2]. However, the image features cover the whole image, including the image regions not related to the question topic, which is an interference to answering.

Thus, recent approaches have introduced the attention mechanism into VQA. The attention mechanism was primarily designed to mimic human vision, ignoring irrelevant information in images while paying more attention to critical information. The attention mechanism effectively learns key regions of images or key features of questions in cross-modal tasks by performing attention-weighting operations on the original features. Anderson et al. [1] used both Bottom-up and Top-down attention mechanisms to learn image region objects instead of spatial grid attention. In [15], DCN was proposed to establish a complete interaction between each question word and each image region. MCAN [23] was a deeply cascaded co-attention network, adopting the SA and GA units to obtain global features with more fine-grained information.

However, the visual features in these VQA models are usually extracted from the image regions by a target detector, such as Faster-RCNN [18]. There are many overlapping parts between image regions, which inevitably causes oversampling and adds noisiness to the task. On the contrary, the question text consists of a small number of words. Consequently, the information amount gap between the image features and text features is too large to form an effective semantic alignment, resulting in an ineffective fusion of visual and text features.

3   Approach

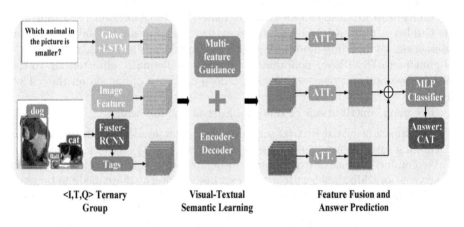

Fig. 2. The overall structure of the Visual-Textual Semantic Alignment Network (VTSAN).

The overall structure of the Visual-Textual Semantic Alignment Network (VTSAN) is shown in Fig. 2. The features of images, tags, and questions are firstly extracted and combined in the form of $\langle \mathbf{I}, \mathbf{T}, \mathbf{Q} \rangle$ ternary groups. Then the visual-textual semantic learning is performed on the ternary groups with Encoder-Decoder architecture and multi-feature guidance, which improves the visual-textual semantic alignment of key image regions and question keywords using tag information as an intermediate medium. Finally, the multimodal global features are fed into a multilayer perception (MLP) classifier to obtain the final prediction.

3.1 $\langle \mathbf{I}, \mathbf{T}, \mathbf{Q} \rangle$ Ternary Group

We use Faster-RCNN to extract the image region features and tag information. Specifically, each image region feature is a P-dimensional vector (P = 2048), and a sequence of tag words corresponds to a set of image regions. In addition, for question feature processing, we use the method in [20] to encode the input question as a word sequence v with Glove [8] and LSTM [7]. However, unlike [20], we retain the features of all words. Thereby, the question features can be denoted as the matrix $X \in \mathbb{R}^{v*d_v}$. d_v is defined as the word dimension.

Finally, we combine the images, tags, and questions into a triad form $\langle \mathbf{I}, \mathbf{T}, \mathbf{Q} \rangle$. Tag plays an important role in this triad. On the one hand, the tag information enriches the information of text features and makes the visual and text features more balanced. On the other hand, since the tag information originates from the image regions and the tag words usually appear in the question, the tag information can be an intermediate medium between the image key regions and the question keywords. Therefore, the visual-textual semantic alignment can be improved.

3.2 VQA Module

In the MCAN network, question inputs are represented by Glove word embeddings, and question embeddings are processed by LSTM. The image features are extracted by the Faster-RCNN network. The idea of the transformer model is applied to model intra-modal self-attention and inter-modal guided attention. Multimodal feature fusion is then used to obtain global features, which are finally fed into the answer classifier to obtain the final prediction. As a transformer model, MCAN is widely used as the baseline because of its powerful capability. Unlike usual VQA models, MCAN adopts a co-attention mechanism. Through a deep cascade, not only the key region features of the image, but also the text features of the question keyword are preserved, while the dimensionality remains unchanged. A dense interaction of multimodal instances is achieved to infer the correlations between image regions and question words. The global features with more fine-grained information are thus obtained.

3.3 Visual-Textual Semantic Attention Unit

To improve the parallelism of the model and to ensure that the dimensions of input and output remain constant, we use the Scaled Dot-Product Attention [3]. $\mathbf{A}$ is defined as a Scaled Dot-Product Attention function. The input of $\mathbf{A}$ includes query, key, and value. Usually, their dimensions d_{query}, d_{key} and d_{value} are all set to be the same, namely d. The attention function $\mathbf{F}$ is shown as:

$$\mathbf{F} = \mathbf{A}\left(q, K, V\right) = \mathbf{softmax}(\frac{qK^T}{\sqrt{d}})V \tag{1}$$

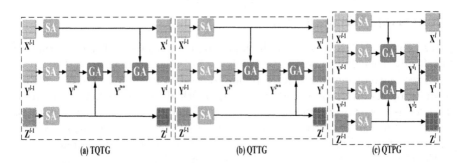

Fig. 3. Three different strategies for constructing the VTSA layer: (a) Tag Question Tandem Guidance (TQTG). (b) Question Tag Tandem Guidance (QTTG). (c) Question Tag Parallelism Guidance (QTPG).

where $q \in \mathbb{R}^{1 \times d}$ is a query, and $(K \in \mathbb{R}^{n \times d}, V \in \mathbb{R}^{n \times d})$ is defined as n key-value pairs. Besides, we also adopt a Multi-head Attention (**MA**) method consisting of l parallel heads, where each head corresponds to a separate Scaled Dot-Product Attention function. We give the attended output features $\mathbf{F}^{\mathrm{MH}}$ by:

$$\mathbf{F}^{\mathrm{MH}} = \mathbf{MA}\left(Q, K, V\right) = [\mathbf{F^1}, \dots, \mathbf{F}^l]\, W^* \tag{2}$$

$$\mathbf{F}^l = \mathbf{A}\left(q W_l^Q, K W_l^K, V W_l^V\right) \tag{3}$$

where $Q = [q_1, \dots, q_m] \in \mathbb{R}^{m \times d}$ is denoted as a set of m queries, and W_l^Q, W_l^K, $W_l^V \in \mathbb{R}^{d \times d_l}$ are the projection matrices for the l-th head. $W^* \in \mathbb{R}^{l * d_l \times d}$ is denoted as the overall projection matrix. d_l is the dimension of the output features of each head. The SA unit in our method consists of a multi-head attention layer and a feedforward layer. For each input feature y^j from Y is as the following:

$$\mathbf{F}^j = \mathbf{MA}(y^j, Y, Y) \tag{4}$$

We learn the normalized similarity of all samples in Y to y^j by multi-head attention. The GA unit has two groups of input features X and Y. For each input feature y^j from Y is as the following:

$$\mathbf{F}^j = \mathbf{MA}(y^j, X, X) \tag{5}$$

We learn the normalized similarity of all samples in X to y^j by multi-head attention. We construct the attention by combining image features, tag features, and question features to form a visual-textual semantic alignment (VTSA) layer. This ensures to obtain richer fine-grained information and refines the visual-textual semantic alignment information.

3.4   Visual-Textual Semantic Learning

As shown in Fig. 3, we propose three types of VTSA layer with various multi-feature guidance strategies. X is denoted as question features, Y is denoted as image features, and Z is denoted as tag features.

I. Tag Question Tandem Guidance (TQTG). As shown in Fig. 3(a), X^{l-1}, Y^{l-1}, and Z^{l-1} are first processed by SA unit to obtain X^l, Y^{l^*} and Z^l, then Z^l and Y^{l^*} are inputted into the first layer GA unit to obtain $Y^{l^{**}}$, and finally $Y^{l^{**}}$ and X^l are inputted into the second layer GA unit to obtain the output features Y^l.

II. Question Tag Tandem Guidance (QTTG). As shown in Fig. 3(b), X^{l-1}, Y^{l-1}, and Z^{l-1} are first processed by the SA unit to obtain X^l, Y^{l^*} and Z^l, then X^l and Y^{l^*} are inputted into the first layer GA unit to obtain $Y^{l^{**}}$, and finally $Y^{l^{**}}$ and Z^l are inputted into the second layer GA unit to obtain the output feature Y^l.

III. Question Tag Parallelism Guidance (QTPG). As shown in Fig. 3(c), we connect the question guided image features and the tag guided image features in parallel to obtain the question features X^l and the corresponding image features Y^{l_1}, the tag features Z^l and the corresponding image features Y^{l_2}, respectively. Then Y^{l_1} and Y^{l_2} are weighted by summarized to obtain the image features Y^l in the parallel concatenation mode.

Here, we perform deep cascading of VTSA layers using the Encoder-Decoder architecture. The question features X and the tag features Z are processed by l-layer of self-attention to obtain X^l and Z^l. The image features Y is processed by l-layer self-attention and multi-feature guidance attention to obtain Y^l.

3.5 Feature Fusion and Answer Prediction

After the visual-textual semantic alignment learning, the output image features, tag features, and question features already contain rich information with attention weights. These features are fed into a two-layer MLP to obtain attention-optimized features $\hat{X}$, $\hat{Y}$, and $\hat{Z}$, respectively. Our multimodal fusion feature $\mathbf{G}$ is defined as the following:

$$\mathbf{G} = \mathbf{Norm}(W_X^T \hat{X} + W_Y^T \hat{Y} + W_Z^T \hat{Z}) \qquad (6)$$

where $W_X, W_Y, W_Z \in \mathbb{R}^{d \times d_{\mathbf{G}}}$ are linear projection matrixes. $d_{\mathbf{G}}$ is the common dimensionality of the fused feature. The fused feature $\mathbf{G}$ are then fed into the sigmoid function and get the prediction. We use the binary cross entropy (BCE) as a loss function.

4 Experiments

4.1 Dataset

We evaluate our network on the VQAv2 dataset, which is now the most used VQA benchmark dataset. This dataset is built on the MSCOCO image corpus [13] and contains *train*, *val*, and *test* sets, with total 338k images and 699k questions. The VQA models are required to select the correct answer from a list of 3129 answers, including *Yes/No*, *Num*, and *Other* questions, based on the provided images and questions. In addition, we use the VQA-CPv2 dataset to evaluate the robustness of the network to question biases. The VQA-CPv2 *train* set and *test* set have completely different answer distributions. Therefore, models that rely on a language prior in the *train* set will perform poorly on the *test* set.

4.2   Implementation Details

We set the dimensions of the question feature d_x, the image feature d_y, the tag feature d_z, and the multimodal fusion feature $\mathbf{G}$ to be 512, 2048, 512, and 1024, respectively. The latent dimensionality d in multi-head attention is set to be 512. The number of heads l is set to be 8, and the hidden layer dimension per head is $d_l = d/l = 64$. We set the answer vocabulary list length to be N = 3129. The VTSA layer is set as a cascade of l-layer.

Table 1. The accuracy of using different strategies for the baseline model on VQAv2 *val*. l denotes the number of layers in the depth cascade. t denotes the maximum number of tags.

Module	l	t	Accuracy
MCAN(baseline)	6	–	67.2
MCAN+QTTG	6	32	**67.7**
MCAN+TQTG	6	32	67.4
MCAN+QTPG	6	32	67.3
MCAN+QTTG	8	32	67.8
MCAN+QTTG	10	32	67.7
MCAN+QTTG	8	64	**67.9**
MCAN+QTTG	8	100	67.8

Table 2. The accuracy of using different question and tag processing methods on VQAv2 *val*. $Rand_{ft}$ denotes random initialization of word embedding and then fine-tuned. PE indicates positional encoding [3]. $Glove_{pt}$ indicates that word embeddings are pre-trained while $Glove_{pt+ft}$ is additionally fine-tuned.

Module	Accuracy
$Rand_{ft} + PE$	66.1
$Glove_{pt} + PE$	67.6
$Glove_{pt} + LSTM$	67.8
$Glove_{pt+ft} + LSTM$	**67.9**

In the experiments, we use Adam solver and set the parameters β_1 and β_2 to be 0.9 and 0.1, respectively. During training, we set the base learning rate to the minimum of $2.5se^{-5}$ and $1e^{-4}$, where s represents the current number of epochs. For the train dataset, we use not only VQAv2 dataset and VQA-CPv2 dataset but also a subset of VQA Visual Genome [10] as the data enhancement. We use the *train* split for training to evaluate on the *val* split, and use the *train* split and *val* split for training to evaluate on *test* split.

4.3 Ablation Studies

Table 1 shows our ablation studies, from which we can find that compared to the baseline model, we obtain the best performance on overall accuracy when using the QTTG strategy. Since tags are taken from oversampled image features, they also have redundancy. Directly guiding image features with tags will further increase the noisiness of the task. Therefore, initially guiding the image with questions and then using tags to enhance the attention weights of image key regions will effectively reduce the noisiness of the task. We verify that self-attention modeling of images, tags, and questions is beneficial for VQA. This ablation study also reveals that multi-feature guidance attention modeling using questions and tags has a significant improvement to visual-textual semantic alignment.

Table 3. Performance on VQAv2 *test-dev* and *test-std*.

Model	test-dev				test-std
	Yes/No	Num	Other	All	All
Up-Down [1]	84.27	49.56	59.89	65.67	65.32
QCG [16]	82.91	47.13	56.22	66.18	–
VQA-E [12]	83.22	43.58	56.79	66.31	–
VCT$_{REE}$ [19]	84.55	47.36	59.34	68.49	68.19
BAN[9]	85.31	50.93	60.26	69.52	–
MCAN(baseline) [23]	86.82	53.26	60.72	70.63	70.90
VTSAN(ours)	**87.36**	**55.32**	**61.20**	**71.25**	**71.53**

We can also find that the QTTG strategy outperforms the QTPG strategy on overall accuracy, which validates the effectiveness of tandem multi-feature guidance attention on VQA. Parallelism provides individual guidance to image features, which prevents the tag information from acting as an intermediary between key image regions and question keywords.

The overall performance of the network gradually increases with the cascading of VTSA layers, with the best performance at $l = 8$. However, when increasing the number of layers, the performance decreases, which may be caused by

Table 4. Performance on VQA-CPv2 *test*.

Module	Accuracy
Up-Down [1]	39.06
QCG [16]	39.32
VCT$_{REE}$ [19]	39.75
BAN[9]	40.06
MCAN(baseline)[23]	42.52
VTSAN(ours)	**43.04**

the fact that the gradient does not guarantee a stable state during the training process with too many layers. Since the attention features learned from the early SA units are not accurate, the Encoder-Decoder framework is adopted to improve the accuracy of the attention learned from the last SA unit, so as to fully exploit the guided attention effect of GA unit.

Furthermore, we can observe that the performance of VTSAN is progressively improved as the number of tag t increases and reaches the best at $t = 64$. When the number of tags is further increased, i.e., $t > 64$, the network performance decreases. This phenomenon indicates that excessive number of tags leads to unavoidable information redundancy and more noise, preventing the model from learning accurate keyword information through the SA unit, and thus avoiding the formation of effective guided attention.

Table 2 summarizes the ablation experiments with different questions and tags processing methods. The experiments show that the use of word embeddings pre-trained by Glove significantly outperforms random initialization. Fine-tuning the Glove embeddings and replacing the location coding with LSTM networks can further improve the performance.

4.4   Comparison with State-of-the-Art Methods

As shown in Fig. 4, our network can answer different types of questions correctly. We compare our model with the state-of-the-art methods on the VQAv2 and VQA-CPv2 datasets in Table 3 and Table 4. Our VTSAN is higher than the baseline model MCAN in overall accuracy. In addition, without using any counting component, our method improves the performance on the *Num* question, which indicates the tag information can provide more information about the number of key objects.

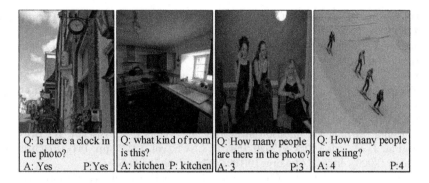

Fig. 4. Performance presentation of VTSAN.

5   Conclusions

In this paper, we propose to construct the Image-Tag-Question$\langle \mathbf{I}, \mathbf{T}, \mathbf{Q} \rangle$ ternary group and use it as network input. We also adopt a tandem multi-feature guidance attention QTTG strategy and the Encoder-Decoder framework in order to construct visual-textual semantic learning. Thereby, we construct the Visual-Textual Semantic Alignment Network (VTSAN). Our approach uses the tags as an intermediate medium between the key image regions and question keywords, and enriches the text features. Therefore, the visual-textual semantic alignment for VQA is effectively improved. The experimental results on the VQAv2 and VQA-CPv2 datasets show that our proposed method outperforms other state-of-the-art methods, and obtain an outstanding performance improvement for VQA.

Acknowledgements. This work was supported in part by the National Natural Science Foundation of China under Grants 61976079 & 61672203, in part by Anhui Natural Science Funds for Distinguished Young Scholar under Grant 170808J08, and in part by Anhui Key Research and Development Program under Grant 202004a05020039.

References

1. Anderson, P., et al.: Bottom-up and top-down attention for image captioning and visual question answering. In: CVPR (2018)
2. Antol, S., et al.: VQA: visual question answering. In: ICCV (2015)
3. Ashish, V., et al.: Attention is all you need. In: NIPS (2017)
4. Cadene, R., Ben-younes, H., Cord, M., Thome, N.: Murel: multimodal relational reasoning for visual question answering. In: CVPR (2019)
5. Gao, P., et al.: Dynamic fusion with intra- and inter-modality attention flow for visual question answering. In: CVPR (2019)
6. Goyal, Y., Khot, T., Summers-Stay, D., Batra, D., Parikh, D.: Making the v in VQA matter: elevating the role of image understanding in visual question answering. In: CVPR (2017)

7. Hochreiter, S., Schmidhuber, J.: Long short-term memory. Neural Comput. **9**, 1735–1780 (1997)
8. Jeffrey, P., Richard, S., Christopher D.M.: Glove: global vectors for word representation. In: EMNLP (2014)
9. Kim, J., Jun, J., Zhang, B.: Bilinear attention networks. In: NIPS (2018)
10. Krishna, R., et al.: Visual genome: connecting language and vision using crowdsourced dense image annotations. arXiv preprint arXiv:1602.07332 (2016)
11. Li, Q., Huang, S., Hong, Y., Zhu, S.-C.: A competence-aware curriculum for visual concepts learning via question answering. In: Vedaldi, A., Bischof, H., Brox, T., Frahm, J.-M. (eds.) ECCV 2020, Part II. LNCS, vol. 12347, pp. 141–157. Springer, Cham (2020). https://doi.org/10.1007/978-3-030-58536-5_9
12. Li, Q., Tao, Q., Joty, S., Cai, J., Luo, J.: VQA-E: explaining, elaborating, and enhancing your answers for visual questions. In: ECCV (2018)
13. Lin, T.-Y., et al.: Microsoft COCO: common objects in context. In: Fleet, D., Pajdla, T., Schiele, B., Tuytelaars, T. (eds.) ECCV 2014, Part V. LNCS, vol. 8693, pp. 740–755. Springer, Cham (2014). https://doi.org/10.1007/978-3-319-10602-1_48
14. Mao, J., Gan, C., Kohli, P., Tenenbaum, J.B.: The neuro-symbolic concept learner: interpreting scenes, words, and sentences from natural supervision. In: ICLR (2019)
15. Nguyen, D.K., Okatani, T.: Improved fusion of visual and language representations by dense symmetric co-attention for visual question answering. In: CVPR (2018)
16. Norcliffe-Brown, W., Vafeais, E., Parisot, S.: Learning conditioned graph structures for interpretable visual question answering. In: NIPS (2018)
17. Ramakrishnan, S., Agrawal, A., Lee, S.: Overcoming language priors in visual question answering with adversarial regularization. In: NIPS (2018)
18. Ren, S., He, K., Girshick, R., Sun, J.: Faster R-CNN: towards real-time object detection with region proposal networks. In: NIPS (2015)
19. Tang, K., Zhang, H., Wu, B., Luo, W., Liu, W.: Learning to compose dynamic tree structures for visual contexts. In: CVPR (2019)
20. Teney, D., Anderson, P., He, X., Hengel, A.V.D.: Tips and tricks for visual question answering. arXiv preprint arXiv:1708.02711 (2017)
21. Wu, Q., Shen, C., Liu, L., Dick, A., Hengel, A.V.D.: What value do explicit high level concepts have in vision to language problems?. In: CVPR (2016)
22. You, Q., Jin, H., Wang, Z., Fang, C., Luo, J.: Image captioning with semantic attention. In: CVPR (2016)
23. Yu, Z., Yu, J., Cui, Y., Tao, D., Tian, Q.: Deep modular co-attention networks for visual question answering. In: CVPR (2020)
24. Zhou, L., Palangi, H., Zhang, L., Hu, H., Gao, J.: Unified vision-language pretraining for image captioning and VQA. In: AAAI (2020)

Which and Where to Focus: A Simple yet Accurate Framework for Arbitrary-Shaped Nearby Text Detection in Scene Images

Youhui Guo[1,2], Yu Zhou[1(✉)], Xugong Qin[1,2], and Weiping Wang[1]

[1] Institute of Information Engineering, Chinese Academy of Sciences, Beijing, China
{guoyouhui,zhouyu,qinxugong,wangweiping}@iie.ac.cn
[2] School of Cyber Security, University of Chinese Academy of Sciences, Beijing, China

Abstract. Scene text detection has drawn the close attention of researchers. Though many methods have been proposed for horizontal and oriented texts, previous methods may not perform well when dealing with arbitrary-shaped texts such as curved texts. In particular, confusion problem arises in the case of nearby text instances. In this paper, we propose a simple yet effective method for accurate arbitrary-shaped nearby scene text detection. Firstly, a One-to-Many Training Scheme (OMTS) is designed to eliminate confusion and enable the proposals to learn more appropriate groundtruths in the case of nearby text instances. Secondly, we propose a Proposal Feature Attention Module (PFAM) to exploit more effective features for each proposal, which can better adapt to arbitrary-shaped text instances. Finally, we propose a baseline that is based on Faster R-CNN and outputs the curve representation directly. Equipped with PFAM and OMTS, the detector can achieve state-of-the-art or competitive performance on several challenging benchmarks.

Keywords: Scene text detection · Arbitrary-shaped nearby text · Attention module · One-to-Many Training Scheme

1 Introduction

Scene text detection is a fundamental and crucial task in computer vision because it is an important step in many practical applications, such as scene text spotting, image/video understanding, and text visual question answering. Although immense progress has been made in previous excellent works, there still exist challenges in scene text detection due to large variations of shape, color, font, orientation, and scale, as well as the complex context of scene texts.

Supported by the Open Research Project of the State Key Laboratory of Media Convergence and Communication, Communication University of China, China (No. SKLMCC2020KF004), the Beijing Municipal Science & Technology Commission (Z191100007119002), the Key Research Program of Frontier Sciences, CAS, Grant NO ZDBS-LY-7024, the National Natural Science Foundation of China (No. 62006221).

© Springer Nature Switzerland AG 2021
I. Farkaš et al. (Eds.): ICANN 2021, LNCS 12895, pp. 271–283, 2021.
https://doi.org/10.1007/978-3-030-86383-8_22

Recently, many segmentation-based approaches have been proposed for arbitrary shaped scene text detection, which performs text/non-text classification on each pixel and groups pixels into different instances. However, the borders between text regions and non-text regions are ambiguous and segmentation results are easily affected by nearby text. The approaches based on two-stage object detection framework [23] with a more appropriate receptive field are still prevalent, in which the Region Proposal Network (RPN) is utilized to generate horizontal rectangle proposals and another detection head is designed to refine the proposals. In reality, text instances are arbitrary-shaped in scene images. The features extracted from horizontal rectangle proposals contain more background noise or even other text instances which is harmful to the classification and regression of the detection head. So, where should we focus on the feature map of each proposal?

Another existing issue is target confusion with the normal training scheme. In the training process, the generated proposals are matched to the groundtruths by calculating the Intersection over Union (IoU). Under the widely used one-to-one matching rule, one proposal is forced to match the groundtruth with the highest IoU. However, the matched groundtruth with the hand-crafted rule may not be the best one. Moreover, when there exist multiple groundtruths with close IoUs to one proposal, i.e., the nearby text case, the network may get confused during learning, leading to suboptimal performance in inference. This is more likely to happen when text instances are close together which is common in scene images. So, which text instance should we choose to focus on?

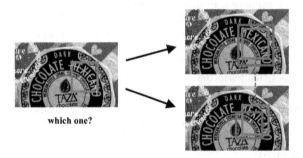

Fig. 1. One proposal is assigned with multiple groundtruths in a one-to-many training scheme. Different curves represent different text instances. The blue and green boxes denote the proposals and groundtruths respectively. (Color figure online)

In this paper, we propose a simple and effective method to alleviate the aforementioned two problems for detecting arbitrary-shaped nearby text instances. For each proposal generated by RPN, we propose a Proposal Feature Attention Module (PFAM) to extract more useful features for further classification and regression. To alleviate the problem of target confusion during training, we propose a One-to-Many Training Scheme (OMTS) that one proposal can predict multiple text instances with no additional computation during inference. The illustration of the One-to-Many Training Scheme is shown in Fig. 1.

To verify the effectiveness of the proposed component, we integrate them into a baseline model which is based on Faster RCNN with curve regression for text detection. The experimental results show the effectiveness and robustness of our method on different datasets.

The contributions of this work are summarized as follows:

- A novel One-to-Many Training Scheme is designed to alleviate the dilemma that the proposal may be confused about which instance to regress and no extra computations are involved in inference.
- We propose a Proposal Feature Attention Module for each proposal which can effectively extract more appropriate features for classification and regression.
- A two-stage text detection framework with direct curve regression is proposed, and with the equipment of PFAM and OMTS, it achieves state-of-the-art or competitive performance on several benchmarks. Concretely, the proposed method achieves the F-Measure of 85.5% on CTW1500, 87.7% on Total-Text, 88.0% on ICDAR2015, and 87.3% on MSRA-TD500. Particularly, the proposed method outperforms the baseline with a large margin in the case of the nearby text.

2   Related Work

With the revival of neural networks, the majority of recent scene text detectors are based on deep neural networks. Generally, these methods can be roughly classified into three categories: regression-based, segmentation-based, and component-based methodologies.

Regression-Based Approaches are mainly inherited from some popular general object detection frameworks, which directly regress the bounding boxes of the text instances. TextBoxes++ [6] utilizes quadrilateral regression to detect multi-oriented text. Differently, EAST [29] directly detects the quadrangles of words in a pixel-level manner. FC^{2}RN [14] uses a corner refinement network to produce a refined corner prediction. However, most of them show limited representation for irregular shapes, such as curved shapes.

Segmentation-Based Approaches are mainly inspired by semantic segmentation methods, which regard all the pixels within text bounding boxes as positive regions. Mask TextSpotter [11] is the first end-to-end trainable arbitrary-shaped scene text spotter with a detection module based on Mask R-CNN. Qin et al. [15] reduce the requirement of pixel-level annotations with weakly-supervised learning. Chen et al. [2] propose a self-training framework with unannotated videos based on Mask R-CNN. Xiao et al. [22] propose a novel sequential deformation method to effectively model the line-shape of scene text. Performances of these methods are strongly affected by the quality of segmentation results.

Component-Based Approaches first detect individual text parts, then group them into texts with post-processing steps. CTPN [16] detects a sequence of fine-scale text proposals and connects those text proposals with RNN. CRAFT [1] detects the text instances by exploring each character and affinity between

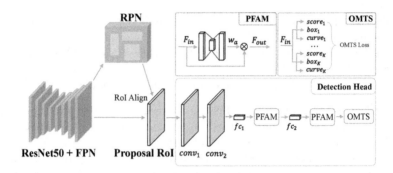

Fig. 2. The overall architecture. It consists of the Resnet50 backbone equipped with FPN, and the Detection Head working for each proposal.

characters. DRRG [27] constructs each text instance by a series of ordered rectangular components and utilize graph convolution network to reason the relations of those components. However, most of these methods need character-level annotations and depend on a good character detection network [3].

Different from most previous for arbitrary-shaped text, the proposed method directly produces a curve representation that requires only instance-level annotations.

3   Proposed Method

3.1   Overview

The overall architecture of our proposed method is shown in Fig. 2. The whole framework is based on Faster R-CNN. Firstly, the input image is fed into a feature pyramid backbone and a lot of proposals are produced by Region Proposal Network. Then, the proposals are fed into Detection Head, in which a proposal feature attention module is designed to extract more appropriate features adaptively and a one-to-many training scheme predicts the multiple text instances with these features. The whole architecture is very simple and brings negligible computations. Next, we will elaborate the designed modules.

3.2   One-to-Many Training Scheme

In the training process of the second stage, the generated proposals are matched to the groundtruths by calculating the Intersection over Union (IoU). Under the widely used one-to-one matching rule, one proposal is forced to match the groundtruth with the highest IoU. However, the matched groundtruth with the hand-crafted rule may not be the best one. Moreover, when there exist multiple groundtruths with close IoUs to one proposal, i.e., the nearby text case, the network may get confused during learning, leading to suboptimal performance. To alleviate this, we design a novel one-to-many training scheme which is inspired by CrowdDet [4]. Different from CrowdDet which aims at detecting highly-overlapped instances and recalling more targets in crowded scenes, the proposed OMTS is a training scheme with no additional computation during inference. With OMTS, one proposal can match with multiple groundtruths, which makes the learning process more efficient. When one proposal matches with only one groundtruth, the other targets are regarded as background. Specifically, for each proposal p_i, we match it with multiple groundtruth text instances $G(p_i)$:

$$G(p_i) = \{g_j \in \mathcal{G} | IoU(g_j, p_i) > \theta\}, \tag{1}$$

where $\mathcal{G}$ is the set of groundtruth instances and θ is a fixed threshold of IoU.

For each proposal p_i, we match it with K groundtruth text instances by the fixed threshold θ. Accordingly, we only need to produce K sets of predictions $P(p_i)$ at the end of the model, as shown below:

$$P(p_i) = \left\{ (\mathbf{c}_i^1, \mathbf{r}_i^1, \mathbf{b}_i^1), (\mathbf{c}_i^2, \mathbf{r}_i^2, \mathbf{b}_i^2), ..., (\mathbf{c}_i^K, \mathbf{r}_i^K, \mathbf{b}_i^K) \right\}, \tag{2}$$

where $\mathbf{c}_i, \mathbf{r}_i$ and $\mathbf{b}_i$ denote predicted class confidence, the regression objective for axis-aligned rectangle box, and curve respectively.

So we need to minimize the gap between predictions $P(p_i)$ and groundtruth instances $G(p_i)$ for each proposal p_i, the final *loss* of OMTS defined as follows:

$$\mathcal{L}(p_i) = \min_{\pi \in \Pi} \sum_{k=1}^{K} \left[\mathcal{L}_{cls}(\mathbf{c}_i^{(k)}, g_{\pi_k}) + \mathcal{L}_{reg_r}(\mathbf{r}_i^{(k)}, g_{\pi_k}) \right.$$
$$\left. + \mathcal{L}_{reg_b}(\mathbf{b}_i^{(k)}, g_{\pi_k}) \right], \tag{3}$$

where π represents a certain permutation of $(1, 2, ..., K)$ whose k-th item is π_k, $g_{\pi_k} \in G_{p_i}$ is the π_k-th groundtruth text instance, $\mathcal{L}_{cls}, \mathcal{L}_{reg_r}$ and $\mathcal{L}_{reg_b}$ denote classification loss, box regression loss and curve regression loss respectively. Intuitively, the formulation in Eq. (3) implies exploring all possible one-to-one matches between predictions and groundtruths, thus finding the "best match" with the smallest loss. This is a full permutation problem from 1 to K and its complexity is $O(K!)$, but considering the distribution of text instances in scene images, we make K = 2 and this brings only a small computational complexity.

It is worth mentioning that when the proposal matches with a groundtruth and a background, we force the first output to predict the groundtruth and the second output to predict the background. By doing this, we only need to select the predictions of the first branch as the final result during the test, without adding any additional calculations.

3.3 Proposal Feature Attention Module

Region Proposal Network plays an important role in most two-stage object detection methods, which aims to generate horizontal proposals by coarse classification and regression based on the default anchors. However, text instances in natural images are arbitrary-shaped, the rectangles are not optimal for text instances because they contain a lot of background noise or even other text instances. Many methods obtain arbitrary-shaped text detection results by regression or segmentation in Detection Head, but very few methods focus on promoting the features of proposals.

To better adapt the features to arbitrary-shaped text instances, we propose a plug-and-play PFAM that can extract more appropriate features adaptively. Specifically, as shown in Fig. 2, with the feature generated by the previous fully connected (fc) layer denoted as F_{in}, we use a multi-layer perceptron (MLP) with one hidden layer and a sigmoid layer to produce the attention map w_a. With F_{in} and w_a, the output of PFAM F_{out} can be extracted by element-wise multiplication. The overall attention process can be summarized as:

$$F_{out} = F_{in} \otimes \sigma(W_{MLP}(F_{in})), \qquad (4)$$

where σ, $\otimes$ and W_{MLP} denote the sigmoid function, element-wise multiplication and the MLP weights. And the hidden dimension of the MLP is set to 64.

Unlike other attention modules [5,21], they focus on the global features of the entire image, while we pay more attention to the local features in each proposal.

3.4   Representation for Arbitrary-Shaped Text Detection

We first experiment with two kinds of representations for arbitrary-shaped text. One is a polygon with a fixed number of points and the other is the parametric Bezier curves. Specifically, we use 10 points to represent the long sides of the text instances (all 20 points) for the former; two Bezier curves [8] are used to represent the long sides for the latter. The Bezier curve represents a parametric curve $B(t)$ defined as follows:

$$B(t) = \sum_{i=0}^{n} \binom{n}{i} P_i (1-t)^{n-i} t^i, 0 \leq t \leq 1, \qquad (5)$$

where n represents the degree, $\binom{n}{i}$ is a binomial coefficient, and P_i represents the $i-th$ control points. Practically, a cubic Bezier curve is sufficient to fit different kinds of arbitrary-shaped scene text for most of the existing datasets, so we use $n = 3$ (all 8 points).

Based on Faster R-CNN, an extra branch is added to the fast R-CNN branch for polygon parameter regression. On CTW1500, they get the same performance with an F-measure of 83.7%, which means comparable representation ability under the two presentations. The representation of the Bezier curves is more flexible with fewer parameters, so we use it in the rest part of the paper.

3.5   Optimization

Learning Targets. Bezier-shaped text detection can be transformed to a regression with eight control points, which is consistent with the original bounding box regression in Faster R-CNN, so we adopt the parameterizations of the coordinates of eight control points as follows:

$$t_x = (x_e - x_p)/w_p, t_y = (y_e - y_p)/h_p,$$
$$t_x^* = (x^* - x_p)/w_p, t_y^* = (y^* - y_p)/h_p, \tag{6}$$

where x, y, w, and h denote the x-coordinate, x-coordinate, width, and height. Variables x_e, x_p, and x^* are for the predicted control points, the proposals, and the groundtruths respectively (likewise for y). We regress the control points from the center of proposals.

Loss Function. The loss function $\mathcal{L}$ is formulated as:

$$\mathcal{L} = \mathcal{L}_{rpn} + \mathcal{L}_{OMTS}, \tag{7}$$

where $\mathcal{L}_{rpn}$ is standard loss in RPN, $\mathcal{L}_{OMTS}$ is formulated as Eq. (3).

4 Experiments

We evaluate our approach on four standard datasets: CTW1500, Total-Text, ICDAR2015, MSRA-TD500, and compare with other state-of-the-art methods.

4.1 Datasets

The datasets used for the experiments in this paper are briefly introduced below:

Curve synthetic dataset is a synthetic arbitrary-shaped scene text dataset proposed in [8], including 94,723 images containing a majority of straight text and 54,327 images containing mostly curved text. We use the annotations of this dataset to produce curve ground truth for pre-training our model.
CTW1500 dataset is another curved dataset that includes English and Chinese texts. CTW1500 contains 1,000 training images and 500 testing images.
Total-Text dataset is a dataset that includes horizontal, multi-oriented, and curved text. The annotations are bounding polygons. It consists of 1,255 training images and 300 testing images.
ICDAR2015 dataset contains 1000 images for training and 500 images for testing. The images are captured by Google Glass and the text accidentally appears in the scene.
MSRA-TD500 dataset is an oriented dataset that includes English and Chinese text instances with a large aspect ratio in natural scenes. It contains 300 training images and 200 testing images. Following the previous methods [7,9,19], we include extra 400 training images from HUST-TR400 [24].

4.2 Implementation Details

The data augmentation for the training data includes: (1) random crop, which we make sure that the crop size is larger than half of the original size and without any text being cut; (2) random scale training, with the short size randomly being chosen from (640, 672, 704, 736, 768, 800, 832, 864, 896) and the long size being less than 1600. In particular, we adopt random rotation with the range of $(-15°, 15°)$ after random crop for ICDAR2015.

We use ResNet-50 with a Feature Pyramid Network which is initialized by ImageNet pre-trained model. Our model is pretrained on the Curve Synthetic dataset with 20 epochs. We use SGD as optimizer with batch size 1, momentum 0.9, and weight decay 0.0001 in training. We adopt warm-up in the initial 500 iterations. The initial learning rate is set to 0.0025 for all experiments. In the inference period, we keep the aspect ratio of the test images and resize the test images by setting a suitable short side length for each dataset. The inference speed is tested with a batch size of 1 on a single GeForce RTX-2080Ti GPU.

4.3 Ablation Study

To verify the effectiveness of our proposed method, we conduct a series of comparative experiments on the ICDAR2015, CTW1500, and Total-Text. We first evaluate the newly proposed PFAM and OMTS in our method. Results are shown in Table 1. If not specified, all models are not pretrained for simplicity.

Baseline. We adopt the Bezier curve to represent text instances and integrate it into Faster R-CNN to obtain a very strong baseline. For a fair comparison, the baseline also adds two convolution layers. It achieves an F-measure of 85.3% and 83.7% on the Total-Text and CTW1500 respectively.

PFAM. First, we verify the effect of the number of PFAM on the model. PFAM(1fc) means that we only add a PEAM after the last fc layer and PFAM(2fc) means that we add a PEAM after each of the two fc layers. As shown in Table 1, Baseline with PFAM(1fc) and PFAM(2fc) can get a similar result on CTW1500 and Total-Text. But when we combine PFAM and OMTS, PFAM(2fc) can get a better result than PFAM(1fc). We believe that when using OMTS for training, the model needs a more appropriate feature, especially when there are nearby text instances. So we use PFAM(2fc) as the default setting.

Table 1. Effectiveness of PFAM and OMTS. "P", "R", and "F" refer to precision, recall and F-measure respectively.

Dataset	Method	P	R	F
Total-Text	Baseline	87.8	83.0	85.3
	Baseline + PFAM(1fc)	88.3	83.3	85.7
	Baseline + PFAM(2fc)	87.4	84.1	85.7
	Baseline + OMTS	88.1	83.0	85.5
	Baseline + PFAM(1fc) + OMTS	**88.9**	82.3	85.5
	Baseline + PFAM(2fc) + OMTS	87.5	**84.7**	**86.1**
CTW1500	Baseline	85.6	81.8	83.7
	Baseline + PFAM(1fc)	85.4	82.5	83.9
	Baseline + PFAM(2fc)	**86.6**	82.2	84.4
	Baseline + OMTS	86.1	82.6	84.3
	Baseline + PFAM(1fc) + OMTS	85.7	81.9	83.8
	Baseline + PFAM(2fc) + OMTS	86.2	**83.6**	**84.9**

Table 2. The effectiveness of OMTS on the rotated datasets. † and * denote CTW1500 and ICDAR2015 respectively.

Dataset	OMTS	Angle: 30			Angle: 45			Angle: 60		
		P	R	F	P	R	F	P	R	F
CTW1500	×	78.1	76.3	77.2	75.7	75.0	75.4	76.8	75.9	76.3
	√	**81.4**	**77.4**	**79.2**	**78.3**	**76.2**	**77.2**	**79.9**	**76.2**	**78.0**
ICDAR2015	×	85.9	81.2	83.5	85.2	79.6	82.3	85.0	81.7	83.3
	√	**86.8**	**81.7**	**84.2**	**85.8**	**81.6**	**83.7**	**85.2**	**82.4**	**83.7**

Concretely, compared to baseline, PFAM obtains 0.4% and 0.7% improvement in F-measure on TotalText and CTW1500 respectively. And compared to baseline with OMTS, PFAM can also obtain 0.6% and 0.6% improvement in F-measure on TotalText and CTW1500 respectively. For each axis-aligned box proposal, with a more appropriate feature extracted by PFAM for arbitrary-shaped text instances, we can obtain more accurate detection results.

OMTS. First, we evaluate our proposed OMTS on TotalText and CTW1500. As shown in Table 1, Compared to baseline, OMTS obtains 0.2% and 0.6% improvement in F-measure. When equipped with PFAM, OMTS obtains 0.8% and 1.2% improvement in F-measure on TotalText and CTW1500 respectively, which verifies the effectiveness of OMTS and the compatibility between OMTS and PFAM.

To further verify the effectiveness of OMTS, We select an arbitrary-shaped and a multi-Oriented text dataset respectively. So we augment the CTW1500 and ICDAR2015 test datasets with some specific angles, including 30°, 45°, and 60°. There are more situations in which the proposals are in a dilemma when

Table 3. Detection results on the CTW1500 and Total-Text dataset. † means that the method pretrains on MLT2017.

Method	CTW1500				Total-text			
	P	R	F	FPS	P	R	F	FPS
TextSnake [9]	67.9	85.3	75.6	1.1	82.7	74.5	78.4	1.1
PSENet [17]	84.8	79.7	82.2	3.9	84.0	78.0	80.9	3.9
LOMO [26]	**89.2**	69.6	78.4	–	88.6	75.7	81.6	–
CRAFT [1]	86.0	81.1	83.5	–	87.6	79.9	83.6	–
PAN [18]	84.6	77.7	81.0	39.8	88.0	79.4	83.5	39.6
Wang et al. [19]	80.1	80.2	80.1	–	80.9	76.2	78.5	
DB [7]	86.9	80.2	83.4	22	87.1	82.5	84.7	32
ContourNet [20]	83.7	**84.1**	83.9	4.5	86.9	83.9	85.4	3.8
Zhang et al. [27]	85.9	83.0	84.5	–	86.5	**84.9**	85.7	–
Xiao et al. † [22]	85.8	82.3	84.0	–	89.2	84.7	86.9	–
Ours	88.8	82.4	**85.5**	18.7	**90.7**	**84.9**	**87.7**	10.6

the model tests on the rotated datasets. Specifically, the model is pretrained on the synthetic dataset and trained on the standard datasets with random rotation in [0, 90]. The comparison of the performance with and without OMTS on the rotated datasets is shown in Table 2. The performance with OMTS consistently outperforms the one without OMTS by a large margin which shows the effectiveness of OMTS in more challenging scenarios.

4.4   Comparison with Previous Methods

We compared our proposed method with previous methods on four datasets, including two datasets for curved texts, two datasets for multi-oriented texts. Some visualization results are shown in Fig. 3.

Evaluation on CTW1500. For CTW1500, we keep the short side of input images no less than 800 and the long side no more than 1333, meanwhile maintaining the aspect ratio. The comparison with the previous method is given in Table 3. Our proposed method achieves the state-of-the-art result of 88.8%, 82.4%, and 85.5% in precision, recall, and F-Measure. Meanwhile, it also achieves a competitive speed (18.7 FPS).

Evaluation on Total-Text. For Total-Text, We limit the short side of input images to no less than 1200 and the long side to no more than 2000 with the original aspect ratio. Also, our proposed method achieves a new state-of-the-art result of 90.7%, 84.9%, and 87.7% in precision, recall, and F-Measure. The specific comparison is shown in Table 3. With the simple design of the entire framework, the speed is also achieved at 10.6 FPS.

Evaluation on ICDAR2015. ICDAR2015 dataset is a multi-oriented text dataset that contains small and low-resolution text instances. Considering the small text instances, we double the width and height of the input image. As shown in Table 4, our proposed method achieves 88.0% in F-Measure, which is higher than the most recent methods. It is worth noting that Xiao et al. [22] pretrain their model on MLT2017 and we pretrain our model on synthetic datasets.

Table 4. Detection results on the ICDAR2015 and MSRA-TD500 dataset. † means that the method pretrains on MLT2017.

Method	ICDAR2015				MSRA-TD500			
	P	R	F	FPS	P	R	F	FPS
EAST [29]	83.6	73.5	78.2	13.2	87.3	67.4	76.1	–
TextSnake [9]	84.9	80.4	82.6	1.1	83.2	73.9	78.3	1.1
CRAFT [1]	89.8	84.3	86.9	–	88.2	78.2	82.9	–
DB [7]	**91.8**	83.2	87.3	12	91.5	79.2	84.9	32
ContourNet [20]	86.1	87.6	86.9	3.5	–	–	–	–
Zhang et al. [27]	84.7	**88.5**	86.6	–	88.1	82.3	85.1	–
Xiao et al. † [22]	88.7	88.4	**88.6**	–	–	–	–	–
Ours	90.6	85.4	88.0	6.5	**92.7**	**82.5**	**87.3**	19.0

Evaluation on MSRA-TD500. To verify the robustness of detecting text instances with a large aspect ratio, we conduct experiments on the MSRA-TD500 dataset. In testing, we keep the short side of input images no less than 720 and the long side no more than 1200, meanwhile maintaining the aspect ratio. Detailed comparison results are shown in Table 4. Our proposed method achieves state-of-the-art performance with 92.7%, 82.5%, and 87.3% in precision, recall, and F-Measure respectively, which outperforms other methods by a large margin and demonstrates that our proposed method is general and can detect oriented text instances with large aspect ratio in natural scenes. And our method is faster than most previous methods with 19.0 FPS.

(a) SCUT-CTW1500   (b) TotalText   (c) ICDAR2015   (d) MSRA-TD500

Fig. 3. Some visualization results on different scene text datasets.

5   Conclusion

In this paper, we propose a simple yet accurate two-stage framework for detecting arbitrary-shaped scene nearby text instances. Based on Faster R-CNN, only minor modifications are made to the Detection Head for arbitrary-shaped text detection. We propose a proposal feature attention module and a one-to-many training scheme for more accurate detection, especially under the nearby text case. PFAM can extract more appropriate features effectively for each proposal. With the appropriate features, OMTS enables proposals to learn more appropriate text instances with negligible training time and no extra time during inference. Experiments on four benchmarks show that our proposed method has a good capability to detect horizontal, oriented, and curved scene texts. In the future, we will introduce the unsupervised methods [10,25,28] to text detection. Due to the detection result is friendly to the text recognition [12,13], we prefer to extend our proposed method to an end-to-end text spotting framework.

References

1. Baek, Y., Lee, B., Han, D., Yun, S., Lee, H.: Character region awareness for text detection. In: CVPR, pp. 9365–9374 (2019)
2. Chen, Y., Wang, W., Zhou, Y., Yang, F., Yang, D., Wang, W.: Self-training for domain adaptive scene text detection. In: ICPR, pp. 850–857 (2021)

3. Chen, Y., Zhou, Yu., Yang, D., Wang, W.: Constrained relation network for character detection in scene images. In: Nayak, A.C., Sharma, A. (eds.) PRICAI 2019. LNCS (LNAI), vol. 11672, pp. 137–149. Springer, Cham (2019). https://doi.org/10.1007/978-3-030-29894-4_11

4. Chu, X., Zheng, A., Zhang, X., Sun, J.: Detection in crowded scenes: one proposal, multiple predictions. In: CVPR, pp. 12211–12220 (2020)

5. Hu, J., Shen, L., Sun, G.: Squeeze-and-excitation networks. In: CVPR, pp. 7132–7141 (2018)

6. Liao, M., Shi, B., Bai, X.: Textboxes++: a single-shot oriented scene text detector. IEEE Trans. Image Process. **27**(8), 3676–3690 (2018)

7. Liao, M., Wan, Z., Yao, C., Chen, K., Bai, X.: Real-time scene text detection with differentiable binarization. In: AAAI, pp. 11474–11481 (2020)

8. Liu, Y., Chen, H., Shen, C., He, T., Jin, L., Wang, L.: ABCNet: real-time scene text spotting with adaptive Bezier-curve network. In: CVPR. pp. 9806–9815 (2020)

9. Long, S., Ruan, J., Zhang, W., He, X., Wu, W., Yao, C.: TextSnake: a flexible representation for detecting text of arbitrary shapes. In: Ferrari, V., Hebert, M., Sminchisescu, C., Weiss, Y. (eds.) ECCV 2018. LNCS, vol. 11206, pp. 19–35. Springer, Cham (2018). https://doi.org/10.1007/978-3-030-01216-8_2

10. Luo, D., et al.: Video cloze procedure for self-supervised spatio-temporal learning. In: AAAI, pp. 11701–11708 (2020)

11. Lyu, P., Liao, M., Yao, C., Wu, W., Bai, X.: Mask TextSpotter: an end-to-end trainable neural network for spotting text with arbitrary shapes. In: Ferrari, V., Hebert, M., Sminchisescu, C., Weiss, Y. (eds.) Computer Vision – ECCV 2018. LNCS, vol. 11218, pp. 71–88. Springer, Cham (2018). https://doi.org/10.1007/978-3-030-01264-9_5

12. Qiao, Z., Qin, X., Zhou, Y., Yang, F., Wang, W.: Gaussian constrained attention network for scene text recognition. In: ICPR, pp. 3328–3335 (2020)

13. Qiao, Z., Zhou, Y., Yang, D., Zhou, Y., Wang, W.: SEED: semantics enhanced encoder-decoder framework for scene text recognition. In: CVPR, pp. 13525–13534 (2020)

14. Qin, X., Zhou, Y., Guo, Y., Wu, D., Wang, W.: FC2RN: a fully convolutional corner refinement network for accurate multi-oriented scene text detection. In: ICASSP, pp. 4350–4354 (2021)

15. Qin, X., Zhou, Y., Yang, D., Wang, W.: Curved text detection in natural scene images with semi- and weakly-supervised learning. In: ICDAR, pp. 559–564 (2019)

16. Tian, Z., Huang, W., He, T., He, P., Qiao, Yu.: Detecting text in natural image with connectionist text proposal network. In: Leibe, B., Matas, J., Sebe, N., Welling, M. (eds.) ECCV 2016. LNCS, vol. 9912, pp. 56–72. Springer, Cham (2016). https://doi.org/10.1007/978-3-319-46484-8_4

17. Wang, W., et al.: Shape robust text detection with progressive scale expansion network. In: CVPR, pp. 9336–9345 (2019)

18. Wang, W., et al.: Efficient and accurate arbitrary-shaped text detection with pixel aggregation network. In: ICCV, pp. 8439–8448 (2019)

19. Wang, X., Jiang, Y., Luo, Z., Liu, C., Choi, H., Kim, S.: Arbitrary shape scene text detection with adaptive text region representation. In: CVPR, pp. 6449–6458 (2019)

20. Wang, Y., Xie, H., Zha, Z., Xing, M., Fu, Z., Zhang, Y.: ContourNet: taking a further step toward accurate arbitrary-shaped scene text detection. In: CVPR, pp. 11750–11759 (2020)

21. Woo, S., Park, J., Lee, J.-Y., Kweon, I.S.: CBAM: convolutional block attention module. In: Ferrari, V., Hebert, M., Sminchisescu, C., Weiss, Y. (eds.) ECCV 2018. LNCS, vol. 11211, pp. 3–19. Springer, Cham (2018). https://doi.org/10.1007/978-3-030-01234-2_1
22. Xiao, S., Peng, L., Yan, R., An, K., Yao, G., Min, J.: Sequential deformation for accurate scene text detection. In: Vedaldi, A., Bischof, H., Brox, T., Frahm, J.-M. (eds.) ECCV 2020. LNCS, vol. 12374, pp. 108–124. Springer, Cham (2020). https://doi.org/10.1007/978-3-030-58526-6_7
23. Yang, D., Zhou, Y., Wu, D., Ma, C., Yang, F., Wang, W.: Two-level residual distillation based triple network for incremental object detection. CoRR abs/2007.13428 (2020)
24. Yao, C., Bai, X., Liu, W.: A unified framework for multioriented text detection and recognition. IEEE Trans. Image Process. **23**(11), 4737–4749 (2014)
25. Yao, Y., Liu, C., Luo, D., Zhou, Y., Ye, Q.: Video playback rate perception for self-supervised spatio-temporal representation learning. In: CVPR, pp. 6547–6556 (2020)
26. Zhang, C., et al.: Look more than once: an accurate detector for text of arbitrary shapes. In: CVPR, pp. 10552–10561 (2019)
27. Zhang, S., et al.: Deep relational reasoning graph network for arbitrary shape text detection. In: CVPR, pp. 9696–9705 (2020)
28. Zhang, Y., Liu, C., Zhou, Y., Wang, W., Wang, W., Ye, Q.: Progressive cluster purification for unsupervised feature learning. In: ICPR, pp. 8476–8483 (2020)
29. Zhou, X., et al.: EAST: an efficient and accurate scene text detector. In: CVPR, pp. 2642–2651 (2017)

STCP: An Efficient Model Combining Subject Triples and Constituency Parsing for Recognizing Textual Entailment

Meiling Li, Xiumei Li$^{(\boxtimes)}$, Junmei Sun, and Xinrui He

School of Information Science and Technology,
Hangzhou Normal University, Hangzhou 311121, China

Abstract. Recognizing Textual Entailment (RTE) aims at automatically determining the logical relationship between the given premise and hypothesis, and good recognition ability would be helpful for other natural language understanding tasks. Based on the Bidirectional Encoder Representation from Transformer (BERT), this paper proposes a model combining **S**ubject **T**riples and **C**onstituency **P**arsing (STCP) for RTE. Specifically, the model combines the central subject triples to fine-tune BERT, which obtains all hidden layers' features weighted by attention, to capture the global semantic information. The sentences with constituency syntax parsing are encoded by Tree Long Short-Term Memory (Tree-LSTM) to obtain the local structural information. Finally, the global semantic information and the local structural information are incorporated by constructing the matrix splicing fusion module, to enhance the ability to recognize semantic logical relationships. Experimental results show that the STCP model can achieve better recognition performance than the benchmark models on the public datasets SNLI and MNLI, and the effectiveness of the model is verified through ablation experiments and visualization of attention weights.

Keywords: Recognizing textual entailment · Subject triples · Constituency parsing

1 Introduction

Recognizing textural entailment (RTE), also known as natural language inference, is a fundamental task in the field of Natural Language Processing (NLP). RTE requires machines to automatically determine the logical relationship between the premise and the hypothesis. The improvement on RTE capabilities can provide assistance for other semantic-related NLP tasks [5], such as question answering systems [8] and machine translation [21]. As shown in Table 1, in the commonly used Stanford Natural Language Inference (SNLI) [2] dataset, three types of logical relationship between the premise and the hypothesis are defined as Entailment, Neutral, and Contradiction.

RTE is a challenging task, since it requires not only the global semantic information on the sentence level but also the local structural information between the words to be captured. Early methods based on similarity or alignment features [10] cannot well

© Springer Nature Switzerland AG 2021
I. Farkaš et al. (Eds.): ICANN 2021, LNCS 12895, pp. 284–296, 2021.
https://doi.org/10.1007/978-3-030-86383-8_23

model the semantics of sentences, therefore the recognition performance is not sufficient. With the rapid development of deep learning, the neural network model can better learn the semantic features of the sentences, and the performance has surpassed the previous methods [7, 26]. For example, Socher2011 [22] applied Recursive Neural Network (RNN) to the field of syntax parsing and text classification tasks, and successfully proved that the neural network model could perform effective vector representation of natural language text. Chen2017 [3] proposed the Enhanced LSTM [19] model, which was the first to semantically encode the premises and the hypotheses based on the Bidirectional LSTM [20]. However, the linear structure of the series of RNN models is only suitable to encode the sequential data, but it cannot capture syntactic structural information and cannot encode the data with syntactic structure [12]. Tai2015 [25] proposed the Tree-LSTM model to enrich the topological structure of LSTM, therefore making up for the lack of syntactic structural information on sequential models, and can also semantically encode data with syntactic features such as Dependency Parsing and Constituency Parsing.

Table 1. Three logical relationships in SNLI

Premise	Hypothesis	Label
This church choir sings to the masses as they sing joyous songs from the book at a church	The church is filled with song	Entailment
	The church has cracks in the ceiling	Neutral
	A choir singing at a baseball game	Contradiction

In recent years, with the release of large-scale datasets SNLI and Multi-Genre Natural Language Inference (MNLI) [15], as well as the proposal of the pre-training language model Bidirectional Encoder Representation from Transformer (BERT) [4], the accuracy of RTE has been greatly improved [18]. According to the difference of model sizes, the BERT is divided into $BERT_{BASE}$ and $BERT_{LARGE}$, which use 12-layer and 24-layer Transformer [27] encoders as the main model structure respectively. Compared with traditional CNN and RNN used for NLP tasks, BERT has more powerful text encoding capabilities. With the development of network models, researchers have found that the ability of the network models to extract semantic information is still limited, and it is effectively to provide the model with more known semantic features from external knowledge [9, 14]. For example, Zhang2020 [28–30] improved the language model's performance by introducing Semantic Role Labeling and other syntactic structural information on machine reading comprehension and other NLP tasks.

Natural language texts usually include descriptive vocabularies, such as adjective and attributives, to extend the central subject sentence. However, too many descriptive vocabularies will interfere with the model's ability to understand the semantics of the text and result in prediction bias. Therefore, it is meaningful to extract central subject triples from the text. Besides, due to the universal design of BERT, a pair of premise and hypothesis texts are spliced into a single text with the separator tag [SEP] in the input layer, the classifier tag [CLS] is added before the premise, and the [SEP] is added after the hypothesis. This means that the BERT model pays more attention to the global semantic information on the sentence level, while ignoring the local structural information between words. Therefore, our paper attempts to integrate the central subject information and constituency syntactic parsing on the basis of the BERT to enhance the model's ability to capture text semantic information. The main contributions of this paper are as follows:

1) We use the large-scale pre-training language model **BERT**$_{BASE}$ to alleviate the problem of poor model generalization caused by insufficient manual annotation corpus for RTE, and add central subject triples to manually modify the model's attention direction. Attention mechanism is used to perform attention weighting on all hidden layers' vectors of the language model to retain more effective global semantic information.
2) We convert the string-type text with constituency syntactic parsing into the tree structural text, so that it can be effectively encoded by the Tree-LSTM network to obtain the local structural information between words.
3) The STCP model is proposed on the basis of fine-tuning **BERT**$_{BASE}$ to obtain the global semantic information, and then the global semantic information is innovatively integrated with the local structural information to improve the model's recognition ability for RTE.

2 The Proposed Model

The structure of the STCP model is shown in Fig. 1. The model consists of three parts, the module of fine-tuning language model as shown in the left bottom of Fig. 1, the constituency syntactic parsing module as shown in the right bottom of Fig. 1, and the aggregation output module. The whole model can be regarded as a three-category classifier, and can automatically determine the logical relationship of the premise and the hypothesis.

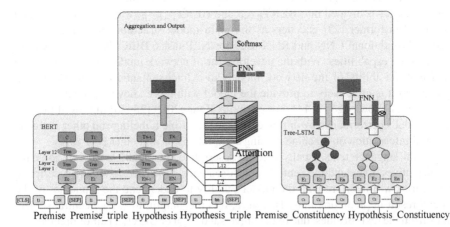

Fig. 1. Overview of the STCP model

2.1 The Module of Fine-Tuning Language Model

Based on the $\text{BERT}_{\text{BASE}}$, the module fuses the central subject information in the input layer and improves the Segment Embedding algorithm in the embedding layer, then makes full use of the information in the middle layer of $\text{BERT}_{\text{BASE}}$ to extract the global semantic features. Specifically, this paper uses the StanfordCoreNLP Toolkit to extract the central subject triples of premise and hypothesis respectively, and the triples usually consist of Subject, Relation and Object.

In order to enhance the model's understanding and representation of the central subject information, the subject triples are spliced after the text using the separator tag [SEP] as the final input text. The splicing mode of the text is shown in the input part of this module in Fig. 1.

In the embedding layer, the model converts the input text into a continuous distributed representation which includes Word Embedding, Position Embedding and Segment Embedding. Usually, the three kinds of embedding corresponding to each word in the text are added together as the final word embedding vector.

Using the separator tag [SEP] to splice the subject triples after the text will produce two additional [SEP]. If the original BERT Segment Embedding algorithm is still used, the ability to distinguish the relative position of two texts will be affected. Therefore, the Segment Embedding algorithm in the embedding layer of STCP model is improved in this paper, that is, when the Segment Embedding is expressed, two texts are separated by the second [SEP], the words before the second [SEP] are all the words of the premise, and the words after the second [SEP] are all the words of the hypothesis, as shown in Algorithm 1.

Algorithm 1 Improved algorithm based on BERT Segment Embedding

---

Input : tokens, text word matrix
Parameters: triple_flag: read the triple tag, initial False; sep_flag: read the segment tag, initial False; segments_ids: Segment Embedding matrix, initial NULL; i: network parameter, initial 0: [SEP]: seperator tag.
Output: segments_ids
1:**While** i is less than the range of tokens do
2:     Sample tokens[i] from the real data tokens;
3:     **if** tokens[i] is [SEP] and triple_flag is False and sep_flag is False
4:             segments_ids[i]=0,triple_flag=True
5:     **else if** tokens[i] is [SEP]and triple_flage is True and sep_flag is False
6:             segments_ids[i]=0,sep_flag=True
7:     **else if** sep_flag is True
8:             segments_ids[i]=1
9:     **else** segments_ids[i]=0
10: **return** segments_ids

$\text{BERT}_{\text{BASE}}$ uses 12-layer Transformer encoders to capture the global semantic information of premise and hypothesis. Each layer is composed of Multi-head Attention Network and Feed Forward Network (FFN). In order to mitigate gradient explosion and gradient degradation, each layer is stacked using residual connection [6] and layer standardization [1]. Since the Transformer deeply encodes the classifier tag [CLS], in this paper, the [CLS] output vector *cls_output* is taken as the pooling results. At the same time, due to the different semantic information provided by 12-layer Transformer encoders, the attention mechanism is adopted to carry out weight training on all middle hidden layers' vectors as shown in Eq. (1–3), so as to retain more semantic information. The features of two parts are considered as global semantic information and are used in the aggregate output module of the STCP model.

$$s_j^t = vecs_j^h W_h \tag{1}$$

$$a_i^t = \exp(s_i^t) / \sum_{j=1}^{N} \exp(s_j^t) \tag{2}$$

$$v_t^h = \sum_{i=1}^{N} a_i^t vecs_i^h \tag{3}$$

where $vecs_j^l$ is the vector of the middle hidden layer, a_i^t is the weight of attention to different layers, $v_t^h \in R^{d_1}$ is the weighted summation result *allhs_output*, d_1 is the dimension of the word vector in the module of fine-tuning BERT, and W is the weight vector.

2.2 Constituency Syntactic Parsing Module

Based on the Siamese framework [13], the Constituency Syntactic Parsing module is developed and two parts are mainly included. The first part is to preprocess string-type text with constituency syntactic structure into tree structural text based on Stack space. In the second part, the Tree-LSTM is used to code the tree structural text, and the difference features between the premise and the hypothesis are obtained by the dot product and element-wise difference of the coding results. Finally, the difference features are spliced with the coding results as local structural features.

The string-type text with constituency syntax is represented by a pair of parentheses. The minimum unit is the word and the part-of-speech (POS) of the word, separated by spaces. Besides the minimum unit are phrases and the labels of the phrases, which are recursively formed into larger phrases and labels.

The preprocessing method is to remove the outermost parentheses of the current string-type text, obtain the current remaining text and its phrase label, and put the text into a stack. The list is used to save the label, id, level, and other attributes of the current text. Then extract the top text from the stack based on the first in last out (FILO) rule, if the top text is only composed of a child text, repeat the above operation. If the top text is composed of multiple child texts, each child text is saved to the stack in the form of right to left to ensure that the text is preprocessed from left to right. Repeat the above operation until the stack is empty. Through the above operation, the string-type text is converted into a nested list, the list contains multiple child lists, and each child list can be seen as a node. Multiple nodes composed of lists are classified and stored in a dictionary according to the attribute of the level, which are the keys of the dictionary, and the nodes contained in the current level are the key values. Finally, the dictionary-type text is regarded as the tree structural text corresponding to the string-type text, as shown in Fig. 2.

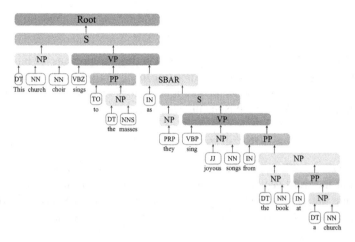

Fig. 2. Example of converting the string-type text into the tree structural text

After preprocessing the text with constituency syntax of the premise and the hypothesis into tree structural text respectively, the Child-sum Tree-LSTM network is used for coding and the model structure is improved with a simple strategy. The calculation method of each cell of Child-sum Tree-LSTM is shown in Eq. (4–10). Specifically, the tree structural text is encoded according to the order from leaf node to root node, and the nodes of the same level are encoded in parallel. Through the above operation, the structural information between words is all converged to the root node, and the root node's feature vector represents the final local structural information of text. Then the feature vector of premise and hypothesis are all linearly converted to the same dimension, based on the dot product and element-wise difference to model the difference relationship between the two texts, and spliced with the structural coding feature vectors as shown in Eq. (11). The result of splicing is considered as local structural information and is used in the aggregate output module of the STCP model.

$$\widetilde{h}_j = \sum_{k \in C(j)} h_k \tag{4}$$

$$i_j = \sigma \left(W^{(i)} x_j + U^{(i)} \widetilde{h}_j + b^{(i)} \right) \tag{5}$$

$$f_{jk} = \sigma \left(W^{(f)} x_j + U^{(f)} h_k + b^{(f)} \right), k \in C(j) \tag{6}$$

$$o_j = \sigma \left(W^{(o)} x_j + U^{(o)} \widetilde{h}_j + b^{(o)} \right) \tag{7}$$

$$u_j = \tanh \left(W^{(u)} x_j + U^{(u)} \widetilde{h}_j + b^{(u)} \right) \tag{8}$$

$$c_j = i_j \odot u_j + \sum_{k \in C(j)} f_{jk} \odot c_k \tag{9}$$

$$h_j = o_j \odot \tanh(c_j) \tag{10}$$

where $C(j)$ represents the set of all child nodes of node j, x_j represents the input vector of node j, the word embedding vector corresponding to the word of each node, σ represents the sigmoid function, $\odot$ represents the multiplication of vector elements in turn, W and U are both weight vectors, and b is the threshold vector.

$$parse\_output = \left[v_{pre}; v_{hyp}; v_{pre} \cdot v_{hyp}; \left|v_{pre} - v_{hyp}\right|\right] \tag{11}$$

where $v_{pre}, v_{hyp} \in R^{d_2}$ are all the feature vectors encoded by Tree-LSTM, d_2 is the dimension of the word vector in the constituency syntactic parsing module, $v_{pre} \cdot v_{hyp}$ represents the value of dot product, $\left|v_{pre} - v_{hyp}\right|$ represents the element-wise difference, and $parse\_output \in R^{4d_2}$ represents the local structural feature vector.

2.3 Aggregation Output Module

The aggregation output module aggregates the output results of the module of fine-tuning language model and constituency syntactic parsing module, and predicts the relationship labels between premise and hypothesis through the aggregation feature vector. This module mainly includes the FNN layer, aggregation layer and Softmax layer. $parse\_output \in R^{4d_2}$ is transformed by FNN layer to the dimension d_1, with the same dimension of $allhs\_output$ and cls_output. The whole aggregation output module can be formally described as Eq. (12–13):

$$output = \left[allhs\_output; cls\_output; FNN(parse\_output)\right] \tag{12}$$

$$y = soft \max(FNN(output)), output \in R^{3d_1} \tag{13}$$

3 Experiment Results

3.1 Datasets

The proposed STCP model is evaluated on two benchmark datasets SNLI and MNLI, as shown in Table 2. Specifically, MNLI dataset is divided into the matched and mismatched datasets. The matched version (MNLI-m) means that the data sources of the training set and test set are consistent, while the mismatched version (MNLI-mm) means that the sources are inconsistent.

3.2 Experimental Setting and Evaluation Metrics

The experimental environment of this paper is based on PyCharm integrated open environment, using PyTorch deep learning framework, and the hardware is based on NVIDIA 2080Ti graphics card. The pre-training 300-dimension English Glove word vector [17] is used to embed the words in the constituency syntactic parsing module. The dimension

Table 2. SNLI dataset and MNLI dataset

Label	SNLI			MNLI		
	Training	Validation	Test	Training	MNLI-(m/mm) Validation	MNLI-(m/mm) Test
Entailment	183416	3329	3368	130899	3479/3463	-/-
Neutral	182764	3235	3219	130900	3123/3129	-/-
Contradiction	183187	3278	3237	130903	3213/3240	-/-
Total	549367	9842	9824	392702	9815/9832	9796/9847

of the word vector in the module of fine-tuning language model is 768. The optimizer adopts Adam, and the learning rate 1e-5 is adopted to update the parameters. The value of batch size is 8 on MNLI dataset, and 16 on SNLI dataset. Using cross-entropy loss function, in order to alleviate the overfitting problem in the training, the dropout technology is used, and the dropout rate is 0.5. We adopt Accuracy, Precision, Recall and F1-score as the evaluation metrics in our experiments.

3.3 Comparison with the State-of-the-Art Models

We compare the proposed STCP model with several typical or state-of-the-art methods listed as below.

1) DIIN_2018 [11]: During word embedding, character features are added, and the premise and hypothesis are respectively encoded through a two-layer Highway Network. The interaction layer adopts dot product method and is connected with Densely Networks as feature extractor.
2) DMAN_2018 [16]: Combining two tasks of English RTE and discourse connective prediction, a semi-supervised method with additional training data is used.
3) BERT_2018 [4]: The pre-training language model based on a large number of data achieve the state-of-the-art recognition performance in multiple NLU tasks. For the fairness of the experiment, we adopt the experimental results based on $BERT_{BASE}$.
4) BERT and PALs_2019 [23]: Based on $BERT_{BASE}$, multiple semantic comprehension tasks are trained jointly and parameters are shared under the framework of multi-task.
5) SesameBERT_2020 [24]: Based on $BERT_{BASE}$, the model extracts effective feature channels with the aid of Squeeze and Excitation network from the perspective of global information, so that the semantic understanding can be strengthened.

3.4 Experimental Results and Ablation Study

The experimental result comparisons are shown in Table 3 on two datasets SNLI and MNLI. In particular, as shown in the first group of Table 3, for the models which do

Table 3. Comparison of experimental results on two datasets (%). The results with symbol "*" are produced by our paper, others are obtained from the related paper.

Model	Accuracy		
	SNLI	MNLI-m	MNLI-mm
DIIN_2018[11]	88.0	78.8	77.8
DMAN_2018[16]	88.8	78.9	78.2
BERT_2018[4]	90.3*	84.6	83.4
BERT and PALs_2019[23]	-	84.3	83.5
SesameBERT_2020[24]	-	83.7	83.6
BERT-triples*	90.46	84.28	83.46
BERT-triples-attention*	90.59	84.44	83.64
STCP*	90.81	84.49	83.75

not introduce the external resources, although the design of network is complex, the accuracy is not sufficient. The ability of RTE can be further improved by introducing additional knowledge based on BERT or combining with multi-task training as shown in the second group. STCP model provides a significant improvement over the benchmark model on SNLI and MNLI-mm. The performance on MNLI-m has a slightly higher than BERT and PALs_2019 [23] and SesameBERT_2020 [24], although it is lower than BERT baseline. There may be three reasons for the improvement, the first is the central subject information without additional interference information, the second is reasonably using attention mechanism to train hidden layer features to retain more semantic information, and the last is that STCP could utilize both deep semantic information and local structural information to learn richer syntactical feature without introducing too much noise than other models.

To investigate the impact of each component in STCP model (e.g., subject triples, hidden states weighted layer, constituency syntactic parsing), we compare full STCP model with a set of ablation experiments as shown in the third group of Table 3. Add central subject triples based on BERT (named BERT-triples), a little bit improvement in both SNLI and MNLI-mm are observed, demonstrating that the subject triples are effective to the language model. Based on BERT-triples, we add hidden states weighted layer (named BERT-triples-attention), more improvement is observed, and it is shown that keeping the semantic information of the middle layer is effective. Then, add the constituency syntactic parsing module, the STCP model get the best results, which indicates that the constituency syntactic parsing module can extract effective structural information to improve the ability of recognizing the logical relationships. The combination of these three components can further improve the recognition performance.

3.5 Visualization of Attention Weights

Attention mechanism can be understood as the importance of learning from different local information, and the attention weights reflect the importance of current data. Therefore by the visualization of attention weights, the influence of central subject triples and different middle hidden layers can be intuitively shown.

Visualization of the Attention Weights to Show the Influence of Central Subject Triples. The sample with the logical relationship Neutral in Table 1 is used as visual object. Figure 3 is the heat map of the attention weights, and all the weights are from the 12th head of Multi-head in the 12th layer of the module of fine-tuning language model. Figure 3(a) is generated by the STCP model without the subject triples when predicting the logical relationship of the first data in the SNLI test set. Figure 3(b) is generated by the STCP model when predicting. In the heat map, the darker the color is, the closer the relationship between the two words. By comparing the differences between the two heat maps, it can be clearly found that in Fig. 3(a), the weights corresponding to "church", "cracks" and "ceiling" are relatively large, while the weights of important words such as "sing" and "songs" in the premise are relatively small, so that the model tend to predict Entailment relationship. In Fig. 3(b), due to the existence of the subject triples, the words "they", "book", "church", "sing", "songs", "cracks" obtain more weights, and these words are important to exactly judge the Neutral relationship of the sample text. It can be intuitively shown that adding central subject triples to the language model can improve the ability to capture text semantic from the attention mechanism.

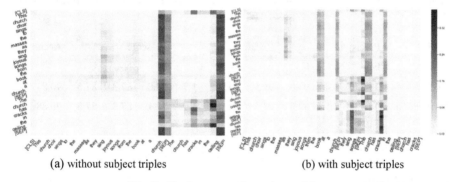

(a) without subject triples   (b) with subject triples

Fig. 3. The heat map of attention weights

Visualization of the Attention Weights to Show the Influence of Different Middle Hidden Layers. In the module of fine-tuning language model, the semantic information provided by the feature vectors of middle hidden layers is different. Figure 4 is the heat map of the attention weights from Eq. (2), which is achieved from the 12 hidden layers when the STCP model is used to test the SNLI dataset. It can be seen from the Fig. 4 that model focuses on the feature vectors of the second and the fifth layer. In order to verify whether the two layers of feature vectors effectively improve the ability of predicting logical relationship, the original training of the feature vectors of each layer with the help of attention mechanism is changed to only extract the second and fifth layer feature vectors as the final feature vectors of the hidden layer. Verification experimental results and the STCP model results are shown in Table 4, including four evaluation metrics such as the Precision, Recall, F1-score and Accuracy, and E represents the label data for the Entailment, N represents the label data for the Neutral, C represents the label data for the Contradiction. As can be seen from Table 4, the results of Precision, Recall and F1-score in the validation experiment are comparable to those of the STCP model, and the Accuracy can be increased by 0.07. The validity of the feature vectors of the second and fifth hidden layers in the model is proved, and the weights of the second and fifth layers obtained by the attention mechanism are proved to be reasonable and effective.

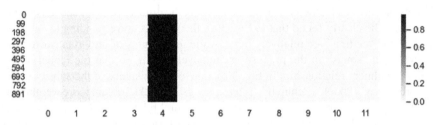

Fig. 4. Attention weights of 12 layers in the STCP model

Table 4. Comparison of experimental results between the STCP model and the verification model (%)

Model	Precision			Recall			F1-score			Accuracy
	E	N	C	E	N	C	E	N	C	
STCP	93.8	93.4	93.0	91.5	91.6	91.7	87.0	87.3	87.5	90.81
Verification	94.1	93.5	92.9	91.3	91.6	91.9	87.1	87.3	87.5	90.88

4 Conclusion

This paper proposes the STCP model, which combines subject triples and constituency parsing, for RTE. The model contains the module of fine-tuning language model, the constituency syntactic parsing module and the aggregation output module. In the module of fine-tuning language model, we use the separator tag [SEP] to combine the text and the central subject triples for better semantic understanding. Besides, the attention mechanism is used to train all the hidden layers' feature vectors, then output the feature vectors and the classifier tag [CLS] pooling results together as global semantic features. In the constituency syntactic parsing module, we use the stack space to convert the string-type text with syntactic structure into the tree structural text, so that it can be encoded by the Tree-LSTM network to obtain the local structural features. Then the global semantic features and the local structural features are merged by the matrix splicing, and finally the fusion result is classified in the aggregation output module. The experimental results on two public datasets demonstrate the effectiveness of the proposed STCP model. In the follow-up research work, the attention mechanism will be used to refine the fusion method to further strengthen the model's ability to extract semantic information.

Acknowledgement. This research is supported in part by the National Natural Science Foundation of China (61801159, 61571174).

References

1. Ba, J.L., Kiros, J.R., Hinton, G.E.: Layer normalization. arXiv preprint arXiv:1607.06450 (2016)
2. Bowman, S.R., Angeli, G., Potts, C., Manning C.D.: A large annotated corpus for learning natural language inference. In: Proceedings of ACL, pp. 632–642 (2015)
3. Chen, Q., Zhu, X., Ling, Z.H., Wei, S., Jiang, H., Inkpen, D.: Enhanced LSTM for natural language inference. In: Proceedings of ACL, pp. 1657–1668 (2017)
4. Devlin, J., Chang, M.W., Lee, K., Toutanova, K.: BERT: pre-training of deep bidirectional transformers for language understanding. arXiv preprint arXiv:1810.04805 (2018)
5. Guo, M.S., Zhang, Y., Liu, T.: Research advances and prospect of recognizing textual entailment and knowledge acquisition. Chin. J. Comput. **40**(04), 889–910 (2017)
6. He, K., Zhang, X., Ren, S., Sun, J.: Deep residual learning for image recognition. In: Proceedings of the IEEE Conference on Computer Vision and Pattern Recognition, pp. 770–778 (2016)
7. Hu, C.W., Wu, C.X., Yang, Y.L.: Extended S-LSTM based textual entailment recognition. J. Comput. Res. Dev. **57**(07), 1481–1489 (2020)
8. Li, M., Weber, C., Wermter, S.: Neural networks for detecting irrelevant questions during visual question answering. In: Farkaš, I., Masulli, P., Wermter, S. (eds.) ICANN 2020. LNCS, vol. 12397, pp. 786–797. Springer, Cham (2020). https://doi.org/10.1007/978-3-030-61616-8_63
9. Ma, Y., Zhao, J., Jin, B.: A hierarchical fine-tuning approach based on joint embedding of words and parent categories for hierarchical multi-label text classification. In: Farkaš, I., Masulli, P., Wermter, S. (eds.) ICANN 2020. LNCS, vol. 12397, pp. 746–757. Springer, Cham (2020). https://doi.org/10.1007/978-3-030-61616-8_60
10. Marneffe, M.C.D., Rafferty, A.N., Manning, C.D.: Finding contradictions in text. ACL **7**(3), 1039–1047 (2008)
11. Mirakyan, M., Hambardzumyan, K., Khachatrian, H.: Natural language inference over interaction space. arXiv preprint arXiv:1802.03198 (2018)
12. Mou, L., et al.: Natural language inference by tree-based convolution and heuristic matching. arXiv preprint arXiv:1512.08422 (2015)
13. Mueller, J., Thyagarajan, A.: Siamese recurrent architectures for learning sentence similarity. In: Proceedings of AAAI, pp. 2782–2792 (2016)
14. Naik A., Ravichander A., Sadeh N., Rose C, Neubig G.: Stress test evaluation for natural language inference. In: Proceedings of COLING, pp. 2340–2353 (2018)
15. Nangia, N., Williams, A., Lazaridou, A., Bowman, S.R.: Multi-genre natural language inference with sentence representations. arXiv preprint arXiv:1707.08172 (2017)
16. Pan, B., Yang, Y., Zhao, Z., Zhuang, Y., Cai, D., He, X.: Discourse marker augmented network with reinforcement learning for natural language inference. In: Proceedings of ACL, pp. 989–999 (2018)
17. Pennington, J., Socher, R., Manning, C.D.: Glove: global vectors for word representation. In: Proceedings of EMNLP, pp. 1532–1543 (2014)
18. Reimers, N., Gurevych, I.: Sentence-BERT: sentence embeddings using siamese BERT-networks. arXiv preprint arXiv:1908.10084 (2019)
19. Schmidhuber, J., Hochreiter, S.: Long short-term memory. Neural Comput. **9**(8), 1735–1780 (1997)

20. Schuster, M., Paliwal, K.K.: Bidirectional recurrent neural networks. IEEE Trans. Signal Process. **45**(11), 2673–2681 (1997)
21. Sebastian, P., Cer, D., Galley, M., Jurafsky, D., Manning, C.D.: Measuring machine translation quality as semantic equivalence: a metric based on entailment features. Mach. Transl. **23**(2–3), 181–193 (2009)
22. Socher, R., Lin, C.Y., Ng, A.Y., Manning, C.D.: Parsing natural scenes and natural language with recursive neural networks. In: Proceedings of ICML, pp. 129–136 (2011)
23. Stickland, A.C., Murray, I.: BERT and PALs: projected attention layers for efficient adaptation in multi-task learning. In: Proceedings of PMLR, pp. 5986–5995 (2019)
24. Su, T.C., Cheng, H.C.: SesameBERT: attention for anywhere. In: Proceedings of DSAA, pp. 363–369 (2020)
25. Tai, K.S., Socher, R., Manning, C.D.: Improved semantic representations from tree-structured long short-term memory networks. arXiv preprint arXiv:1503.00075 (2015)
26. Tan, C., Wei, F., Wang, W., Lv, W., Zhou, M.: Multiway attention networks for modeling sentence pairs. In: Proceedings of IJCAI, pp. 4411–4417 (2018)
27. Vaswani, A., Shazeer, N., Parmar, N., Uszkoreit, J.: Attention is all you need. In: Proceedings of Advances in Neural Information Processing Systems, pp. 5998–6008 (2017)
28. Zhang, Z., Wu, Y., Li, Z., Zhao, H.: Explicit contextual semantics for text comprehension. In: Proceedings of ACL, pp. 298–308 (2019)
29. Zhang, Z., et al.: Semantics-aware BERT for language understanding. Proc. AAAI **34**(05), 9628–9635 (2020)
30. Zhang, Z., et al.: SG-Net: syntax-guided machine reading comprehension. In: Proceedings of AAAI, pp. 9636–9643 (2020)

A Latent Variable Model with Hierarchical Structure and GPT-2 for Long Text Generation

Kun Zhao⬤, Hongwei Ding, Kai Ye, Xiaohui Cui(✉), and Zhongwang Fu

Key Laboratory of Aerospace Information Security and Trusted Computing,
Ministry of Education, School of Cyber Science and Engineering,
Wuhan University, Wuhan, China
xcui@whu.edu.cn

Abstract. Variational AutoEncoder (VAE) has made great achievements in the field of text generation. However, the current research mainly focuses on short texts, with little attention paid to long texts (more than 20 words). In this paper, we first propose a hidden-variable model based on the GPT-2 and hierarchical structure to generate long text. We use hierarchical GRU to encode long text to get hidden variables. At the same time, to generate the text better, we combine the hierarchical structure and GPT-2 in the decoder for the first time. Our model improves Perplexity (PPL), Kullback Leibler (KL) divergence, Bilingual Evaluation Understudy (BLEU) score, and Self-BLEU. The experiment indicates that the coherence and diversity of sentences generated by our model are better than the baseline model.

Keywords: Latent variable · Text generation · VAE · Hierarchical structure · GPT-2

1 Introduction

Language models based on recurrent neural networks (RNN) [30] and sequence-to-sequence structure [39] occupy an important position in natural language processing. In particular, Variational Autoencoders (VAE) [19,33] combined inference networks and generative networks have been widely used in text generation [3,17,43], where the encoder and decoder are long short-term memory networks (LSTM) [7]. For a random vector from the hidden space as an input, the decoder can generate real and novel text based on this hidden vector, making VAE an attractive generative model. Compared with traditional neural language models, the latent variables in VAE are considered to give the model more powerful representation capabilities.

However, LSTM-VAE [3] is challenging to use for the generation of long texts because they tend to generate repetitive, grammatical, and contradictory texts and often lack a coherent long-term structure [14]. For long texts, words form

This work has been supported by National Key Research and Development Program of China (NO. 2018YFC1604000).

© Springer Nature Switzerland AG 2021
I. Farkaš et al. (Eds.): ICANN 2021, LNCS 12895, pp. 297–308, 2021.
https://doi.org/10.1007/978-3-030-86383-8_24

sentences, and sentences form text. Therefore, in our work, we propose a new model that uses a multi-layer structure [22,37] as the encoder of VAE. Still, unlike Shen et al. (2019), we did not learn hidden variables at the sentence level but learned the entire text's hidden variables. The encoder we adopted with RNN can capture the sequential information of long texts better than the CNN used by Shen et al. (2019). Simultaneously, due to the autoregressive characteristics and self-attention of GPT-2, it is very suitable for text generation, but GPT-2 generates long texts with logic problems. Therefore, in our model, the decoder still uses a hierarchical structure. Sentence-level uses RNN to generate plan-vector, word-level uses GPT-2 to generate words, and plan-vector is used as part of GPT-2 input.

In this paper, our contribution is to improve the hVAE [37], the main improvement lies in (i) To capture the hierarchical structure of the long text, better coding long text, we use hierarchical Gated Recurrent Unit (GRU) [7] as the encoder, which can capture text information better than CNN; (ii) When generating long texts for the generation network, we hope that the model can generate every sentence in a planned way. So we use GRU as the Sentence-level Decoder to generate plan-vector h_i^p, GPT-2 as the word-level Decoder to generate words. The plan-vector as an input to GPT-2, then each input of GPT-2 is $o_{i,t} = h_i^p + e_{i,t} + p_{i,t}$, $e_{i,t}$ is word embedding, $p_{i,t}$ is positional embedding. For the posterior collapse problem in VAE training, we use the Free Bits method to solve this problem. After experimental comparison, our model is better than hVAE and GPT-2 in long text generation.

2 VAE

VAE [19] is a generative model [23–25] that can learn the probability distribution of the dataset. Its most prominent feature is to imitate the learning prediction mechanism of the autoencoder. For a target probability distribution, given any probability distribution, there is always a differentiable measurable function that maps it to another probability distribution. Make this probability distribution arbitrarily close to the probability distribution of the target.

Assuming that x is an observable sample data and z is an unobservable hidden variable, the joint probability distribution of x and z is as follows:

$$p(x, z; \theta) = p(x|z; \theta)p(z; \theta) \tag{1}$$

$p(z)$ is the prior distribution of the hidden variable z, $p(x|z)$ is the conditional probability density function of the observed variable x when z is known, and θ is the parameter of the two probability density functions. Usually, we think that $p(x|z)$ and $p(z)$ are a certain kind of parameterized distribution family, such as Gaussian distribution. The idea of VAE is to use neural networks to fit these two complex probability density functions independently. The model diagram of VAE is shown in Fig. 1.

Given a sample x, its log-likelihood marginal function is:

$$logp(x; \theta) = ELBO(q, x; \theta, \phi) + KL(q(z; \phi), p(z|x; \theta)) \tag{2}$$

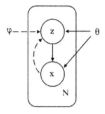

Fig. 1. VAE. The solid line represents the generating network, and the dashed line represents the inference network

Among them, $q(z;\phi)$ is an additional introduced variational function, and $ELBO(q, x; \theta, \phi)$ is called the lower bound of variation.

$$ELBO(q,\ x;\ \theta,\ \phi)\ =\ E_{z \sim q(z;\phi)}[\ log\ \frac{p(x,z;\theta)}{q(z;\phi)}] \qquad (3)$$

We use a neural network to fit $q(z;\phi)$, which is called an inference network (also called an encoder). In theory $q(z;\phi)$ is unrelated to x, but in fact, because we use $q(z;\phi)$ to estimate the posterior distribution $p(z|x;\phi)$, which is related to x. Generally, $q(z;\phi)$ is written as $q(z|x;\phi)$. For the inference network, we input the observable sample data x, and output the probability density function $q(z|x;\phi)$.

Similarly, a neural network is used to fit $p(x|z;\theta)$, which is called a generative network (also called a decoder). The input of the generated network is z sampled from $q(z|x;\phi)$, and the output is the probability density function $p(x|z;\theta)$.

It has been proved in [19] that the loss function of VAE is to maximize ELBO, which can be divided into reconstruction term and Kullback Leibler(KL) divergence term:

$$L_{vae} = -D_{KL}[q(z|x)||p(z)] + E_{q_\phi(z|x)}[logp_\theta(x|z)] \qquad (4)$$

Where $p(z)$ is a standard normal Gaussian distribution $N(0, I)$, and ϕ is the parameter of the encoder and decoder. The first term in the formula is the KL divergence term, and its goal is to make $q(z|x, \phi)$ as similar to $p(z)$ as possible; the second term can be regarded as the reconstruction error of the decoder.

3   Model

For a long text, words usually form sentences, and sentences form paragraphs. Therefore, to better represent the text's hierarchical structure, the model's inference network (encoder) adopts a hierarchical structure. The lower layer of the generation network (decoder) adopts GRU to generate plan-vector, the high-level uses GPT-2 to generate word, and the model structure is shown in Fig. 2.

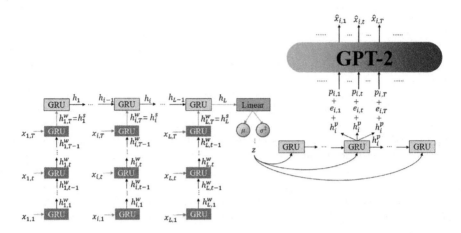

Fig. 2. Model. $x_{i,t}$ is the t-th word of the i-th sentence. $h_{i,t}^w$ is the state obtained by $x_{i,t}$ through GRU, take the state of the last GRU output as the state of the sentence h_i^s and input it into the upper GRU to get h_i. Take the last h_i as the expression of the entire paragraph, and use it to learn mean and variance of Gaussian distribution. sample from Gaussian distribution to get z, and use z to control to generate x. h_i^p is the plan vector, $e_{i,t}$ is the word Embedding, $p_{i,t}$ is positional embedding.

3.1 Encoder

We pay attention to the generation of long text. Assuming that a text has L sentences x_i, each sentence contains T words, $x_{i,t}$ represents the t-th word in the i-th sentence. The encoder uses a hierarchical structure to encode the entire long text into one vectors, including word encoder and sentence encoder, all use unidirectional GRU [7], which is equal to LSTM. Then we will use this vector to get $q(z|x;\phi)$. Next, explain how to encode a long text into a vector, and use this vector to get $q(z|x;\phi)$.

Word Encoder. Given a sentence contains the word $x_{i,t}$, which represents the t-th word of the i-th sentence, here we use Byte Pair Encoding (BPE) [32] for word segmentation, and then we embed the word through a word embedding matrix to convert the word into a vector $e_{i,t}$, and finally send $e_{i,t}$ into the GRU unit to get $h_{i,t}^w$. We select $h_{i,T}^w$ as this sentence representation h_i^s.

$$e_{i,t} = x_{i,t}W_e, 1 \le t \le T, 1 \le i \le L \tag{5}$$

$$h_{i,t}^w = GRU(h_{i,t-1}^w, e_{i,t}) \tag{6}$$

$$h_i^s = h_{i,T}^w \tag{7}$$

W_e is the word embedding matrix. The GPT-2 pre-trained word embedding matrix is used here, and the vocabulary size is 50257. But we inserted the two characters <EOS> and <BOS>, so the dictionary size becomes 50259. $h_{i,t}^w$ is the hidden state output by GRU for the t-th word, and the hidden state output by GRU for the last word of each sentence is recorded as h_i^s, to indicate the encoding of each sentence. Next, use h_i^s to represent the entire sentence.

Sentence Encoder. In the previous step, we got the encoding of each sentence h_i^s. Similarly, we will complete the encoding of the entire text through the GRU unit.

$$h_i = GRU(h_{i-1}, h_i^s), 1 \le i \le L \tag{8}$$

h_i is the hidden state of sentence representation h_i^s, we use the last hidden state h_L as the representation of the entire text, and use it to calculate $q(z|x; \phi) \sim (\mu, \sigma^2)$.

We send h_L into linear layes to calculate the Gaussian distribution of μ and $log\sigma^2$:

$$\mu, log\sigma^2 = Linear(h_L) \tag{9}$$

3.2 Decoder

The generative network (decoder) also uses a hierarchical structure. First, a latent state z is sampled from $q(z|x; \phi)$. The generative network generates a sample x according to z. The sentence-level uses GRU to generate plan-vector, and the word-level uses GPT-2 generates words, the GPT-2 uses a pre-trained model with 117M parameters. The specific implementation is as follows:

$$h_i^p = GRU(h_{i-1}^p, z) \tag{10}$$

h_i^p is the plan-vector,it determines what kind of sentence should be generated in the t-th step. In the i-th sentence, the h_i^p of each word is equal. It will be injected into GPT-2 as a part to generate words. Each input of GPT-2 is:

$$o_{i,t} = h_i^p + e_{i,t} + p_{i,t} \tag{11}$$

where $e_{i,t}$ is word embedding, $p_{i,t}$ is position embedding. In the generation process, GPT-2 uses autoregressive method to generate each word, namely:

$$\hat{x}_{i,t} = GPT2(o_{<=i,<=t}) \tag{12}$$

We hope plan-vector can solve the problem of semantic incoherence of long text. So we input the plan vector as part of GPT-2, the plan vector can be used to control the generation of each sentence of text, and finally generate the entire long text.

3.3 Loss Function

VAE usually uses $ELBO$ as the loss function, the loss function is as follows:

$$L_{vae} = E_{q_\phi(z|x)}[logp_\theta(x|z)] - D_{KL}[q(z|x)||p(z)] \tag{13}$$

However, when VAE is used for text, posterior collapse [3] will occur during training. In order to alleviate this problem, we adopt KL thresholding scheme

[18,20], the following measures are taken for the KL term: calculate the KL divergence of each component of z, compare with λ(constant), take the maximum value as the KL divergence of this component, and finally sum up the components. In this way, the loss function becomes:

$$L_{vae} = E_{q_\phi(z|x)}[logp_\theta(x|z)] - \sum\{\lambda, D_{KL}[q(z_i|x)||p(z_i)]\} \qquad (14)$$

Among them, z_i is denoted as the $i-th$ dimension of z. Using the threshold target lambda will give up learning the KL divergence of the z components below.

4   Experiment

This section will elaborate on the selected data set, model hyperparameter settings, and final analysis of experimental results.

Dataset. We use the arxiv paper abstract dataset and the Yelp comment dataset. The arxiv paper abstract dataset includes 750,000 paper titles and abstracts; the Yelp review dataset is the official open data, containing 4.7 million reviews and 156,000 business information, as well as corresponding product images. We select texts with a total number of words between 100–300. For the two datasets, we uniformly select 200,000 pieces of text data as the training set, and 30000 pieces of text data as the test set. Then we use nltk's segmentation tool to segment each text, and use GPT-2 tokenizer to segment each sentence to facilitate the use of GPT-2 pre-trained word vectors. At the same time, insert <BOS> at the beginning of the first sentence, insert <EOS> at the end of each sentence, and insert <|endoftext|> at the end of the last sentence.

Baseline. Because we are concerned about the VAE used for long text generation, we only compared models that are similar to our model and used for long text generation. We mainly concern the language model with similar architecture, mainly including hVAE [37], GPT-2.

For generic text generation, we concern Adversarial Autoencoders (AAE) [27] and Adversarially-Regularized Autoencoders (ARAE) [48]. Instead of penalizing the KL divergence term, AAE introduces a discriminator network to match the latent variable's prior and posterior distributions. AARE model extends AAE by introducing Wasserstein GAN loss [1] and a more robust generator network.

4.1   Language Model

We use perplexity (PPL) and KL divergence as the criteria for evaluating language models. Note that GPT-2 can calculate accurate PPL. In contrast, VAE cannot accurately calculate the PPL of the language model, so we use important weighted bound [4] to estimate $logp(x)$ and calculate PPL, namely:

$$L_k = E[log\frac{1}{k}\sum_1^k \frac{p(x,z_i)}{q(z_i|x)}] \le logE[\frac{1}{k}\sum_1^k \frac{p(x,z_i)}{q(z_i|x)}] = logp(x) \qquad (15)$$

$$PPL = e^{\frac{1}{M}logp(x)} \tag{16}$$

M is the number of words. When $k \to \infty$, $logp(x)$ can be realistically approximated, and a more realistic PPL can be calculated. Our test set includes 30,000 arxiv paper abstract data. The λ(constant) hyperparameter is 0.5, and the dimension of the hidden variable z is set to 32. The experimental results(PPL and KL) are shown in Table 1.

Table 1. KL and PPL

	Yelp		Arxiv	
Model	KL	PPL	KL	PPL
hVAE	6.8	45.8	12.7	54.3
GPT-2	–	29.8511	–	31.7464
Ours ($\lambda = 0.5$, z $= 32$)	**15.11**	**29.63**	**14.18**	**29.05**

It can be seen that our model has a larger KL divergence than the previous model (from 12.7 to 14.18), which shows that our model can make full use of the information of hidden variables than the previous model. PPL has a large decrease, from 54.3 to 29.05, and compared with GPT-2, the PPL is also reduced, which shows that the sentences generated by our model are better than prior work.

4.2   Unconditional Text Generation

We use the following method to evaluate the generated text. We randomly select 1000 texts from the test set, record each sentence's length under each text, save it as a list, and randomly sample 1000 hidden variables from the learned posterior distribution. Send 1000 hidden variables to the decoder and randomly select a sentence length list from the recorded text's sentence length as the length of each sentence of the generated text. We use the Bilingual Evaluation Understudy (BLEU) score at the corpus level [31] as an evaluation method for the quality of generated sentences, as in the two papers [46, 47], we test the entire Set as a reference for each generated text to get an average BLEU score.

Table 2. BLEU scores

	Yelp			Arxiv		
Model	B-2	B-3	B-4	B-2	B-3	B-4
ARAE	0.684	0.524	0.350	0.624	0.475	0.305
AAE	0.735	0.623	0.383	0.729	0.564	0.342
hVAE	0.912	0.755	0.549	0.825	0.657	0.460
GPT-2	**0.961**	0.876	0.639	0.925	0.782	0.537
Ours ($\lambda = 0.5$, z $= 32$)	0.957	**0.880**	**0.645**	**0.927**	**0.785**	**0.539**

It can be seen from Table 2 that our model ($\lambda = 0.5$, $z = 32$) has a significant improvement in B-2, B-3, or B-4 scores, which shows that the quality of text generated by the model is better than baseline models. Furthermore, compared to GPT-2, our model also has some improvements. The result shows that the text generated from our model is better than the baseline.

Diversity of Generated Paragraphs. We evaluate the diversity of random samples from a trained model since one model might generate realistic-looking sentences while suffering from severe mode collapse (i.e., low diversity). We use Self-BLEU metrics to measure the diversity of generated paragraphs. For a set of sampled sentences, the Self-BLEU metric is the BLEU score of each sample for all other samples as the reference (the numbers overall samples are then averaged). The lower the Self-BLEU score is, the model is better. We evaluate our model in Yelp datasets.

It can be seen from Table 3 that our model has improved in B-4 scores, which is lower than the baseline model. However, in B-2 and B-3, Our model is not optimal but better than most baseline models in the B-3 score. Moreover, the results show that the 2-gram phrase diversity generated by our model is low. However, the generated 3-gram and 4-gram phrase diversity is obviously reduced faster than the baseline model, and it reaches the best in B-4. For a long text, 4-gram is more convincing than 3-gram and 2-gram (Table 4).

Table 3. Self-BLEU

	B-2	B-3	B-4
ARAE	**0.725**	**0.544**	0.402
AAE	0.831	0.672	0.483
hVAE	0.851	0.723	0.729
GPT-2	0.871	0.616	0.326
Ours ($\lambda = 0.5$, z = 32)	0.868	0.612	**0.319**

5 Relation to Prior Work

VAE for Text Generation. VAE trained based on the neural variational inference (NVI) framework have been widely used for text generation [2,3,6,8,12,13,16,17,21,28,29,34–36,38,42,43,45,49]

By encouraging the latent feature space to match the prior distribution in the encoder architecture, the learned latent variables can potentially encode high-level semantic features and act as a global representation in the decoding process [3]. Due to the latent codes' sampling process, the generated results also have better diversity [49]. Generative Adversarial Network (GAN) [5,15,46] is another generative model commonly used for text generation. However, most of the existing research focuses on generating one sentence (or multiple sentences of up to 20 words). The task of generating relatively long text units

Table 4. Unconditional text generation

We study general relativity by analyzing an extended Einstein field equations describing quantum dynamics and describing gravity and magnetodynamics in a nonstandard spacestanding spaceworld model and in nonlinear gravity without gravity as its initial condition or covariant metric at each iteration, with only minimal spaceworld mass, at one of these spacities and at all of it simultaneously at any subsequent spacities or degrees beyond one or two and one at both degrees beyond each of them; we consider only three examples, all with different gravitational degrees. We find an invariational property: there is at most two spacewaves which contain one gravitating field, at least a fraction

---

We prove, via some simple analytical formulas derived for solutions in generalised non-relatorless gravity to general relatival black Holes. The formula has several useful features which have no relation at a high energy scale and have the following implications: the solution is of type F-theoretica with singular moment; a solution can also form spontaneously, or is not an ordinary gravitational singularity and has no nonstandard boundary, even with small scalaton number at low density in a Schwarz space field with singular energy. In this way one is allowed access for all scalon distributions

---

is less discussed. Moreover, the VAE used in the text mostly uses the RNN structure instead of Transformer [40], especially GPT-2, which is suitable for text generation architecture [32].

But when VAE is used for text generation, there will be the notorious posterior collapse problem. In recent years, several methods have been working to alleviate this problem, including different KL annealing/thresholding schemes [3,11,20], decoder architectures [9,43], auxiliary loss [49], semi-amortized inference [17], aggressive encoder training schedule [13], and flexible posterior [10]. In this article, KL annealing/thresholding schemes are used to alleviate the problem of posterior collapse.

Hierarchical Structure. For long texts, words are often composed of sentences, and sentences are composed of paragraphs. Previous work used multi-level LSTM encoders [44] or hierarchical autoencoders [22]. In [35], this model learns the hierarchical representation of long texts or defines a random latent variable for each sentence when decoding.

On the other hand, because I am concerned about the generation of long text, although GPT-2 has achieved great success in the direction of text generation, and it can theoretically generate 1024-character text. But it does not pay attention to the relationship between sentence and sentence, which is very important for long texts. Therefore, we hope to generate text in a planned way during the generation process, so our model is also a hierarchical structure. The low-level uses GRU to generate plan-vectors, and the high-level uses GPT-2 to generate words.

Transformer with VAE. Since Vaswani et al. (2017) proposed the Transformer architecture [40], Transformer has become the most popular model in the field of natural language processing and is widely used in major well-known architectures, such as BERT and GPT. However, as a generative model, VAE is very suitable for the field of text generation, incredibly controllable text generation, but few people combine Transformer and VAE. Liu et al. (2019) combined Transformer and VAE for text generation for the first time [26]; Wang et al. (2019) used Transformer and CVAE for story complement tasks [41]; Li et al. (2020) used Bert as the encoder for VAE, GPT-2 is used as a VAE decoder to train a large pre-training model to solve various tasks of natural language processing [21]. However, they are all concerned about the generation of short texts, and what I am concerned about is the generation of long texts. In this paper, for the first time, we combine plan vectors and GPT-2 for VAE to generate long texts with better quality and diversity.

6   Conclusion

For a long text, its hierarchical structure is pronounced (sentences are composed of words, paragraphs are composed of sentences), so we believe that the hierarchical structure is necessary to capture the global characteristics of long texts. In the previous study, text VAE often used a simple LSTM structure. However, few people use the Transformer structure, and Transformer is better than LSTM. Our innovation lies in the combination of VAE, GPT-2 and hierarchical structure to generate better long text. At the same time, the plan-vector is generated in the decoder part through the hierarchical structure and incorporated into GPT-2. In this way, GPT-2 pays attention to the relationship between words and notices the relationship between sentences. But our model is only a preliminary combination of VAE and Transformer and does not use the idea of using VAE inside Transformer, and our model does not eliminate the RNN structure. The next step of the research will be to use the idea of VAE inside the Transformer and let the hidden variables penetrate the Transformer's internal structure to generate better and diverse long text.

References

1. Arjovsky, M., Chintala, S., Bottou, L.: Wasserstein GAN. ArXiv abs/1701.07875 (2017)
2. Bahuleyan, H., Mou, L., Vechtomova, O., Poupart, P.: Variational attention for sequence-to-sequence models. In: COLING (2018)
3. Bowman, S.R., Vilnis, L., Vinyals, O., Dai, A.M., Bengio, S.: Generating sentences from a continuous space. Computer Science (2015)
4. Burda, Y., Grosse, R.B., Salakhutdinov, R.: Importance weighted autoencoders. CoRR abs/1509.00519 (2016)
5. Chen, L., et al.: Adversarial text generation via feature-mover's distance. ArXiv abs/1809.06297 (2018)
6. Chen, M., Tang, Q., Livescu, K., Gimpel, K.: Variational sequential labelers for semi-supervised learning. In: EMNLP (2018)
7. Chung, J., Çaglar Gülçehre, Cho, K., Bengio, Y.: Empirical evaluation of gated recurrent neural networks on sequence modeling. ArXiv abs/1412.3555 (2014)

8. Deng, Y., Kim, Y., Chiu, J.T., Guo, D., Rush, A.M.: Latent alignment and variational attention. In: NeurIPS (2018)
9. Dieng, A.B., Kim, Y., Rush, A.M., Blei, D.: Avoiding latent variable collapse with generative skip models. ArXiv abs/1807.04863 (2019)
10. Fang, L., Li, C., Gao, J., Dong, W., Chen, C.: Implicit deep latent variable models for text generation (2019)
11. Fu, H., Li, C., Liu, X., Gao, J., Çelikyilmaz, A., Carin, L.: Cyclical annealing schedule: a simple approach to mitigating kl vanishing. In: NAACL-HLT (2019)
12. Guu, K., Hashimoto, T., Oren, Y., Liang, P.: Generating sentences by editing prototypes. Trans. Assoc. Comput. Linguist. **6**, 437–450 (2018)
13. He, J., Spokoyny, D., Neubig, G., Berg-Kirkpatrick, T.: Lagging inference networks and posterior collapse in variational autoencoders. ArXiv abs/1901.05534 (2019)
14. Holtzman, A., Buys, J., Forbes, M., Bosselut, A., Choi, Y.: Learning to write with cooperative discriminators (2018)
15. Hu, Z., Yang, Z., Liang, X., Salakhutdinov, R., Xing, E.: Toward controlled generation of text. In: ICML (2017)
16. Kaiser, L., et al.: Fast decoding in sequence models using discrete latent variables. In: ICML (2018)
17. Kim, Y., Wiseman, S., Miller, A.C., Sontag, D., Rush, A.M.: Semi-amortized variational autoencoders (2018)
18. Kingma, D.P., Salimans, T., Welling, M.: Improved variational inference with inverse autoregressive flow. ArXiv abs/1606.04934 (2017)
19. Kingma, D.P., Welling, M.: Auto-encoding variational bayes (2014)
20. Li, B., He, J., Neubig, G., Berg-Kirkpatrick, T., Yang, Y.: A surprisingly effective fix for deep latent variable modeling of text. ArXiv abs/1909.00868 (2019)
21. Li, C., et al.: Optimus: organizing sentences via pre-trained modeling of a latent space. ArXiv abs/2004.04092 (2020)
22. Li, J., Luong, M.T., Jurafsky, D.: A hierarchical neural autoencoder for paragraphs and documents. ArXiv abs/1506.01057 (2015)
23. Li, W., Ding, W., Sadasivam, R., Cui, X., Chen, P.: His-GAN: a histogram-based GAN model to improve data generation quality. Neural Netw. **119**, 31–45 (2019)
24. Li, W., Fan, L., Wang, Z., Ma, C., Cui, X.: Tackling mode collapse in multi-generator GANs with orthogonal vectors. Pattern Recogn. **110**, 107646 (2021)
25. Li, W., Liang, Z., Ma, P., Wang, R., Cui, X., Chen, P.: Hausdorff GAN: improving GAN generation quality with Hausdorff metric. IEEE Trans. Cybern. **PP**, 1–13 (2021)
26. Liu, D., Liu, G.: A transformer-based variational autoencoder for sentence generation. In: 2019 International Joint Conference on Neural Networks (IJCNN), pp. 1–7 (2019)
27. Makhzani, A., Shlens, J., Jaitly, N., Goodfellow, I.J.: Adversarial autoencoders. ArXiv abs/1511.05644 (2015)
28. Miao, Y., Grefenstette, E., Blunsom, P.: Discovering discrete latent topics with neural variational inference. ArXiv abs/1706.00359 (2017)
29. Miao, Y., Yu, L., Blunsom, P.: Neural variational inference for text processing. Comput. Sci. 1791–1799 (2016)
30. Mikolov, T., Karafiát, M., Burget, L., Cernock, J., Khudanpur, S.: Recurrent neural network based language model. In: Interspeech, Conference of the International Speech Communication Association, Makuhari, Chiba, Japan, September 2015
31. Papineni, K., Roukos, S., Ward, T., Zhu, W.J.: BLEU: a method for automatic evaluation of machine translation. In: ACL (2002)

32. Radford, A., Wu, J., Child, R., Luan, D., Amodei, D., Sutskever, I.: Language models are unsupervised multitask learners (2019)
33. Rezende, D.J., Mohamed, S., Wierstra, D.: Stochastic backpropagation and approximate inference in deep generative models (2014)
34. Semeniuta, S., Severyn, A., Barth, E.: A hybrid convolutional variational autoencoder for text generation (2017)
35. Serban, I., et al.: A hierarchical latent variable encoder-decoder model for generating dialogues. ArXiv abs/1605.06069 (2017)
36. Shah, H., Barber, D.: Generative neural machine translation. ArXiv abs/1806.05138 (2018)
37. Shen, D., et al.: Towards generating long and coherent text with multi-level latent variable models. ArXiv abs/1902.00154 (2019)
38. Shen, X., et al.: A conditional variational framework for dialog generation. In: ACL (2017)
39. Sutskever, I., Vinyals, O., Le, Q.V.: Sequence to sequence learning with neural networks. In: NIPS (2014)
40. Vaswani, A., et al.: Attention is all you need. ArXiv abs/1706.03762 (2017)
41. Wang, T., Wan, X.: T-CVAE: transformer-based conditioned variational autoencoder for story completion. In: IJCAI (2019)
42. Yang, S., Li, L., Wang, S., Zhang, W., Huang, Q., Tian, Q.: A structured latent variable recurrent network with stochastic attention for generating Weibo comments. In: IJCAI (2020)
43. Yang, Z., Hu, Z., Salakhutdinov, R., Berg-Kirkpatrick, T.: Improved variational autoencoders for text modeling using dilated convolutions. ArXiv abs/1702.08139 (2017)
44. Yang, Z., Yang, D., Dyer, C., He, X., Smola, A., Hovy, E.: Hierarchical attention networks for document classification. In: HLT-NAACL (2016)
45. Yin, P., Zhou, C., He, J., Neubig, G.: StructVAE: tree-structured latent variable models for semi-supervised semantic parsing. In: ACL (2018)
46. Yu, L., Zhang, W., Wang, J., Yu, Y.: SeqGAN: sequence generative adversarial nets with policy gradient. In: AAAI (2017)
47. Zhang, Y., et al.: Adversarial feature matching for text generation. In: ICML (2017)
48. Zhao, J., Kim, Y., Zhang, K., Rush, A.M., LeCun, Y.: Adversarially regularized autoencoders. In: ICML (2018)
49. Zhao, T., Zhao, R., Eskenazi, M.: Learning discourse-level diversity for neural dialog models using conditional variational autoencoders (2017)

A Scoring Model Assisted by Frequency for Multi-Document Summarization

Yue Yu[1,3](✉) ⓘ, Mutong Wu[1] ⓘ, Weifeng Su[1,2](✉) ⓘ, and Yiu-ming Cheung[3] ⓘ

[1] Computer Science and Technology Programme,
Division of Science and Technology, Hefei, China
[2] Guangdong Key Lab of AI and Multi-Modal Data Processing, BNU-HKBU
United International College, Guangdong, China
o930201601@mail.uic.edu.cn, {mutongwu,wfsu}@uic.edu.cn
[3] Department of Computer Science, Hong Kong Baptist University,
Hong Kong, China
{csyueyu,ymc}@comp.hkbu.edu.hk

Abstract. While position information plays a significant role in sentence scoring of single document summarization, the repetition of content among different documents greatly impacts the salience scores of sentences in multi-document summarization. Introducing frequencies information can help identify important sentences which are generally ignored when only considering position information before. Therefore, in this paper, we propose a scoring model, SAFA (**S**elf-**A**ttention with **F**requency Gr**a**ph) which combines position information with frequency to identify the salience of sentences. The SAFA model constructs a frequency graph at the multi-document level based on the repetition of content of sentences, and assigns initial score values to each sentence based on the graph. The model then uses the position-aware gold scores to train a self-attention mechanism, obtaining the sentence significance at its single document level. The score of each sentence is updated by combing position and frequency information together. We train and test the SAFA model on the large-scale multi-document dataset Multi-News. The extensive experimental results show that the model incorporating frequency information in sentence scoring outperforms the other state-of-the-art extractive models.

Keywords: Multiple document summarization · Position information · Frequency · Graph

1 Introduction

Document summarization usually has two directions: Single Document Summarization (SDS) and Multi-Document Summarization (MDS). While many previous SDS researches have achieved competitive results [3,9,24,25,30], there is still a room for MDS development. Though some of the previous SDS researches could

This work is supported by the BNU-HKBU United International College research grant.

I. Farkaš et al. (Eds.): ICANN 2021, LNCS 12895, pp. 309–320, 2021.
https://doi.org/10.1007/978-3-030-86383-8_25

be generalized on MDS field, MDS models still need to be developed according to their own characteristics. Some studies [8, 18, 26, 27, 33] show that position information, one of the main factors for identifying salient sentences, could have substantially positive effects on single-document summarization. As the study [29] shows, however, position information is less effective than the frequency of sentences for improving the scores of evaluations (e.g. ROUGE [17]). Repetition among multiple documents has rarely been considered as an indicator of salience information before, because it is not applicable for SDS. For the MDS dataset, documents in a document pair mainly discuss the same event, thus having many repetitive contents. A sentence should be considered significant if its content has been repeated over many documents. On the other hand, not all documents have a similar narrative structure. It might not be sufficient to depend on the position information only for MDS tasks to solve this problem. We need to balance the frequency and position information when scoring sentences.

In this paper, we propose a model - **S**elf-**A**ttention with **F**requency Graph (SAFA) to consider the position and frequency of a sentence at the same time when scoring the sentence. In SAFA, a global sentences frequency graph is first constructed for each document pair based on frequency information at the multi-document level. A node in the graph stands for a sentence. The weight of the edge between two nodes is the cosine similarity of the nodes sentences. The graph then groups sentences with high similarities together. Based on the similarities between sentences and the size of each group, a newly defined variable centrality directly represents "to what degree a sentence contains the repeated contents in the document pair". Centrality for each sentence is thus assigned as its initial *Score* to indicate its frequency information at multi-document level. At the same time, we utilize the self-attention mechanism with a Bi-LSTM neural network to extract the sentence information at the single-document level. The training process of the Bi-LSTM neural network uses position-aware pseudo values as the training target, thus incorporating the position information in the self-attention values. These values are then used to update the frequency graph and assign new scores, which combine both the centrality and the self-attention values to nodes. Finally, a sentence ranking and selecting process is implemented to generate the summary. Therefore, each sentence's final score considers both the position information from the single-document level and the frequency from the multi-document level.

We train and test our model on Multi-News [7], the first large-scale MDS news dataset released in 2019. The results show that SAFA outperforms the other extractive models on ROUGE scores. We have also conducted ablation studies on the two features to show that they are both indispensable parts of MDS. The human evaluations show that the summary from our proposed model contains more important information with strong readability.

The contributions of this paper are:

– We exploit the role of frequency information in multi-document summarization. Ablation experiments show that the model combing two features (position and frequency) performs better than considering only one feature.

- We propose a scoring model - SAFA, which incorporates the frequency of a sentence at the multi-document level with the positional information at the single document level.
- We conduct extensive experiments on the large-scale MDS dataset Multi-News, which show that the method combining these two features outperforms the other extractive models to the best of our knowledge. The results of the human evaluation show that the summaries generated from our model hold strong readability and contain more information.

2   Related Work

Graph-Based Methods. [6, 8, 11, 16, 22, 28, 31, 34] have been widely applied to solve document summarization issues. Some use the graph-based ranking mechanism [6, 22, 28] for sentence extraction. For instance, LexRank [6] proposed a stochastic graph-based method to calculate the importance of relative units. TextRank [22] proposed a ranking system with two unsupervised methods to extract keyword or sentence. Others introduce graphs to explore the structures between sentences and documents. For example, GraphSum [16], a abstractive summarization model, combines multiple graphs into a neural network which has a similar structure as Transformer [32]. This inspires us to use a graph constructed from the frequency information to identify the repeated contents among multiple documents.

In summarization, a graph could be used to represents the input documents where the nodes are sentences and the edges are relations between the two connecting nodes [34]. This provides us a way to directly measure the repetitive contents among multiple documents by putting the similarity between two sentences to the edge to introduce frequency information to the neural network

Position Information has usually been a significant factor for finding the salience of a word or sentence in document summarization [5, 12, 21, 26, 27]. Many researches [8, 18, 26, 27, 33] approve of the positive effect of position information on extracting important content when dealing SDS issues. They state that sentences at the beginning position of a document have higher probabilities to being salient. For a MDS task, the attributes of single document is also an indispensable factor to obtain a structured summary.

The Attention Mechanism was first proposed in solving image classification problems [23]. In recent years, the incorporation of self-attention in neural networks has solved many NLP field problems such as abstractive summarization [20], text parsing [13, 14], and speaker identification [1]. In this paper, we use a self-attention mechanism [19] with Bi-LSTM neural network to obtain the salience of a sentence within a single document based on the position information.

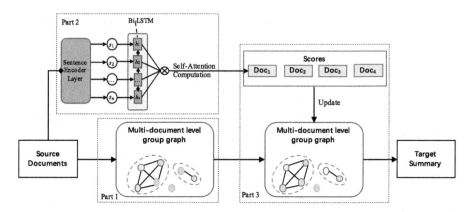

Fig. 1. The pipeline of the model. (Part 1) is a global frequency graph for each document pair. Each group is a cluster of similar sentences. (Part 2) is to implement self-attention for every single document. It contains a sentence encoder and a Bi-LSTM neural network for the self-attention computation. (Part 3) is the final scoring step which takes both the result of (Part 1) and (Part 2) into account to update the final score of each sentence.

3   SAFA Model

This section describes the specific details of the proposed SAFA model and explains how the model combines position and frequency information together. Figure 1 illustrates the pipeline of the SAFA model. It consists of three parts: (1) a global frequency graph at multi-document level generating sentence groups for each document pair to incorporate frequency information, (2) a Bi-LSTM neural network applying the position-aware self-attention mechanism to every single document, and (3) a sentence scoring algorithm to take both frequency (from part 1) and position information (from part 2) into account to update scores indicating salience of sentences.

3.1   Frequency Graph

This sub-section provides the details of how the frequency graph is generated for each document pair. The graph is used to represent the sentences and measure the similarity between them. The construction of the graph for each document pair takes all sentences together. A sentence is first randomly selected in each document pair. For a new sentence having any cosine similarity with the selected sentence greater than a threshold N_t, it is put into the group of the selected sentence with the greatest cosine similarity. Otherwise, it waits for another comparison or join no group at the end. The process terminates if each sentence belongs to one and only one group.

Sentences within the group G_i are connected by edges and are considered as neighboring nodes to each other. We define the centrality C_j^i for sentence s_j in group G_i containing the frequency information as the average similarity between node j and its neighboring nodes. The centrality measures how similar the sentence is with other sentences in the same group.

$$C_j^i = \sum_{s_k \in G_i / s_j} CosSim(s_j, s_k) \tag{1}$$

where G_i / s_j is the set of sentences in group G_i except s_j, and n is the group size. If the sentence does not belong to any group, its sentence centrality is 0. Each node will have its *Centrality* as its initial value for *Score*.

Therefore, a group is the collection of all similar sentences. That is, these sentences describe similar things within one document or among multiple documents. The sentence centrality for each node in this global graph contains information on the frequency of the sentence and is set as the initial score for the scoring model.

3.2   Self-attention for Single Document

This sub-section explains how the self attention which takes the position information of each sentence is computed within one document.

A Bi-LSTM is applied to extract the representation of each sentence within the single document it belongs to. For each document, the sentence embedding is calculated by BERT [4], and the sentence embeddings are the inputs of the Bi-LSTM neural network. The hidden states $\mathbf{H}$ are then used by the self-attention mechanism to get the vector of self-attention values $\mathbf{a}$ for all sentences in the single document. The self-attention vector is calculated as:

$$\mathbf{a} = softmax(\mathbf{v} tanh(\mathbf{W}_h \mathbf{H}^T)) \tag{2}$$

where $\mathbf{W}_h$ is a learnable weight and $\mathbf{v}$ is a learnable vector.

For each sentence s_i, we calculate a pseudo training target t_i for it. The target acts as a "gold score" for training. The t_i for each sentence s_i in a single source document is the sum of the cosine similarities between s_i and all sentences in the corresponding gold summary. A *softmax* function is applied to make the sum of all t_i in one document to 1:

$$t_i = softmax(\sum_{s_j \in G_i} Cos(s_i, s_j)) \tag{3}$$

The (a) graph in Fig. 2 shows the distribution of the average target values for sentences in the same position. The target values are monotonically decreasing, indicating that the training target is position-aware. In other words, since the training process of obtaining the self-attention value incorporates the position information, the self-attention value will also take position into account.

3.3   Sentence Scoring

This subsection shows how the final score of each sentence is computed and how the final extractive summaries are constructed.

A *Score* combining information of both frequency and position is updated for each sentence, indicating the salience of a sentence at both single document level and multi-document level. The $Score_i$ for s_i is updated as:

$$Score_i = \mathbf{a}[i] + C_i * N_{G_j} \tag{4}$$

where $\mathbf{a}[i]$ is the self-attention value containing positional information and salience of s_i at single document level; $Centrality_i$ contains frequency information at multi-document level. Since each group is allowed to output only one sentence in the selection process to reduce redundancy, we multiply the $Centrality_i$ with the size of group N_{G_j} to indicate that large groups (i.e. sentences with high frequency) should have more weight.

Finally, the sentence in each group with the highest score is selected to form the summary. The summary thus considers the frequency and positional information of sentences, as well as the salience at both single document and multi-document level.

4   Experiments and Result

4.1   Dataset and Experimental Setup

The model is trained and tested on the large scale Multi-News MDS dataset[7]. Multi-News is a large-scale dataset for multi-document summarization tasks. There are 44,972 document pairs in the training set, 5622 pairs for both validation and test sets. Each document pair contains a gold summary and 2 to 10 documents describing the same event. The threshold N_t of forming a group is 0.3. For the Bi-LSTM self-attention part, the size of the hidden state and the dimension of the self-attention weight matrix are 300.

4.2   Experimental Result

We compare our model with six representative models that perform well on Multi-News. The first three are extractive; The others are abstractive:

- *LexRank* [6] is a graph-based method for computing relative importance.
- *TextRank* [22] is a ranking model based on the global graph. It uses eigenvectors to compute the importance of each sentences.
- *MMR* [2] combines query-relevance with information-novelty, ranking the sentences based on relevance and redundancy.
- *Hi-MAP* [7] uses a pointer-generator and incorporates the MMR algorithm to generate the summary.
- PG-MMR [15] is a pointer-generator model, applying the MMR algorithm in the encoder to filter out unimportant sentences, and then put all important sentences in the decoder to form a summary.

Table 1. ROUGE F1 scores for models trained and tested on the Multi-News dataset. Models with "*" are abstractive.

Method	R-1	R-2	R-SU
PG-MMR*	40.55	12.36	15.87
CopyTransformer*	43.57	14.03	17.37
GraphSum*	**45.70**	**17.12**	**19.06**
LexRank	38.27	12.70	13.20
TextRank	38.44	13.10	13.50
MMR	38.77	11.98	12.91
Hi-MAP	43.47	14.89	17.41
SAFA (our model)	**45.47**	**15.91**	**18.87**

Table 2. Ablation study on the Multi-News dataset.

Model	R-1	R-2	R-SU
w/o group	37.93	11.77	13.55
w/o position	43.65	14.98	17.44
SAFA	**45.47**	**15.91**	**18.87**

- *CopyTransformer* [10] randomly chooses an attention head to copy the distribution.
- *GraphSum* [16] puts graphs into a neural network with structure similar to the Transformer.

We evaluate all models with the automatic ROUGE metric using version "ROUGE-1.5.5" on both the dataset and the Multi-News dataset. It has three ROUGE values for each model: the overlaps of unigrams (R-1), bigrams (R-2), and the skip-bigrams with a maximum distance of four words (R-SU).

Table 1 shows the results of all models, with the performances of the three abstractive models listed at top, the three extractive models at middle, and the proposed model at the bottom. We could see that overall our model performs competitively in terms of R-1 and R-SU. For extractive models, SAFA outperforms the other models for all ROUGE values. To evaluate the overall quality of the generated summary, we implement human evaluation.

4.3 Ablation Study

We implement three experiments to demonstrate that the information of both position and frequency are indispensable for generating high quality summary:

- *w/o* **group** chooses sentences based on the self-attention value only. Since the average of the self-attention values among sentences at the same position is proportional to the position, this experiments only takes the positional information into account;

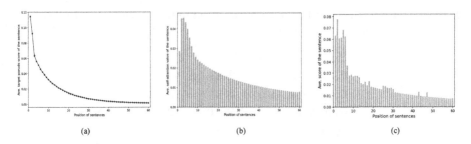

Fig. 2. (a) The trend of average pseudo score for different positions; (b) The average self-attention value of sentences at each position; The posture of the value is on decrease. Although first two sentences are smaller than the following sentences, the overall trend is position-aware; (c) The average score of sentences at each position. After the accumulation of the value of the group information, the posture of score become position-agnostic for many positions.

Table 3. The precision value of first three sentences for w/o group and w/o position. Because this experiment only considers the first three sentences, precision is a better metric than ROUGE.

Model	R-1-p	R-2-p	R-l-p
w/o group	33.40	11.63	17.82
w/o position	**54.75**	**16.99**	**30.91**

- w/o **position** groups are ranked in descending order based on sizes. Then one sentence is chosen randomly from each group starting from the largest group until the summary reaches the minimum length. Groups with only one sentence will not be considered. Since the groups are formed based on frequency information only, this experiment exclude the positional information;
- **SAFA** choose sentences based on the newly defined score of our model, which balances the frequency and positional information.

The ROUGE scores of the three experiments are shown in Table 2. We could see that the best result is achieved when the model SAFA takes both the frequency and the positional information into account. In addition, (b) in Fig. 2 shows the average self attention values of the sentences at the same position. (c) in Fig. 2 shows the histograms of the average values of the newly defined scores for sentences at the same position. From (b) in Fig. 2, we could see that the self attention values are strictly decreasing except for the first two positions, which further demonstrates that the self-attention values contains positional information. (c) in Fig. 2, however, shows that the incorporation of the frequency information makes the scores not so "position-aware" that it provides chances for the significant sentences appearing late in the documents to be included in the summary.

Table 4. Human evaluation according to informativeness, fluency and non-redundancy.

Method	Inf.	Flu.	Non-Redu.
GraphSum	32	28	34
SAFA	41	30	29
GOLD	44	50	50

We also test the precision of outputting the first three sentences for experiments without position and without frequency group to see which factor works better in a specific range. The results are shown in Table 3. While the "*w/o position*" experiment selects the fist three sentences with the highest attention, the "*w/o group*" experiment selects one sentence from each of the groups with the three largest sizes. The higher precision demonstrates that within the range of the first three sentences of gold summaries, there are more sentences with high frequency than sentences appearing early. In other words, the frequency information plays a significant role thus should be incorporated.

4.4 Human Evaluation

Since the rouge scores could not assess the quality of a summary comprehensively, we conduct human evaluations from the levels of informativeness, non-redundancy, and fluency to test whether a summary is complete, precise, and readable. Specifically, informativeness measures whether the summary contains all important information and details; non-redundancy checks whether the summary is precise; and fluency measures whether the summary is written with correct grammar. 50 document pairs are randomly selected from the Multi-News testset. Two models: GraphSum and the proposed SAFA model are implemented to generate summaries. These generated summaries with the gold summaries are distributed to 10 native speakers with 5 pairs to each. The person is asked to add 1 to the level (i.e. informativeness, fluency, non-redundancy) if a summary performs well on this level. The person will not know which system the summary is from, or whether the summary is the gold one beforehand.

The human evaluation results of the three systems and the gold summaries are shown in Table 4. Since our SAFA model balances both the position information and frequency, the information contained in the sentences that appear frequently but late is also included. Therefore, our model ranks top for informativeness and fluency.

5   Conclusion

To better address the special characteristics of MDS, we propose a scoring model that takes both the frequencies and the position of sentences into account The introduction of the frequency graph with the self-attention mechanism to update the calculation of sentence scores enables the model to outperform the other extractive models on the Multi-News dataset and rank at the top for the informativeness and fluency in the human evaluation procedure.

References

1. An, N.N., Thanh, N.Q., Liu, Y.: Deep CNNs with self-attention for speaker iden-tification. IEEE Access **7**, 85327–85337 (2019). https://doi.org/10.1109/ACCESS.2019.2917470, https://ieeexplore.ieee.org/document/8721628/
2. Carbonell, J., Goldstein, J.: The use of MMR, diversity-based reranking for reordering documents and producing summaries. In: Proceedings of the 21st Annual International ACM SIGIR Conference on Research and Development in Information Retrieval - SIGIR 1998, New York, New York, USA, pp. 335–336. ACM Press (1998). https://doi.org/10.1145/290941.291025. http://portal.acm.org/citation.cfm?doid=290941.291025
3. Cheng, J., Lapata, M.: Neural summarization by extracting sentences and words. In: 54th Annual Meeting of the Association for Computational Linguistics, ACL 2016 - Long Papers (2016). https://doi.org/10.18653/v1/p16-1046
4. Devlin, J., Chang, M.W., Lee, K., Toutanova, K.: BERT: pre-training of deep bidirectional transformers for language understanding. In: NAACL HLT 2019–2019 Conference of the North American Chapter of the Association for Computational Linguistics: Human Language Technologies - Proceedings of the Conference (2019)
5. Edmundson, H.P.: New methods in automatic extracting. J. ACM **16**(2), 264–285 (1969). https://doi.org/10.1145/321510.321519. https://dl.acm.org/doi/10.1145/321510.321519
6. Erkan, G., Radev, D.R.: LexRank: graph-based lexical centrality as salience in text summarization. J. Artif. Intell. Res. **22**, 457–479 (2004). https://doi.org/10.1613/jair.1523. http://arxiv.org/abs/1109.2128, https://jair.org/index.php/jair/article/view/10396
7. Fabbri, A.R., Li, I., She, T., Li, S., Radev, D.R.: Multi-news: a large-scale multi-document summarization dataset and abstractive hierarchical model, pp. 1074–1084 (2019). https://doi.org/10.18653/v1/p19-1102. http://arxiv.org/abs/1906.01749
8. Gao, Y., Zhao, W., Eger, S.: SUPERT: Towards New Frontiers in Unsupervised Evaluation Metrics for Multi-Document Summarization (2020). https://doi.org/10.18653/v1/2020.acl-main.124
9. Gehrmann, S., Deng, Y., Rush, A.: Bottom-Up Abstractive Summarization (2019). https://doi.org/10.18653/v1/d18-1443
10. Gehrmann, S., Deng, Y., Rush, A.M.: Bottom-up abstractive summarization. In: Proceedings of the 2018 Conference on Empirical Methods in Natural Language Processing, EMNLP 2018, August 2018. https://doi.org/10.18653/v1/d18-1443. http://arxiv.org/abs/1808.10792

11. Hariharan, S., Srinivasan, R.: Studies on graph based approaches for singleand multi document summarizations. Int. J. Comput. Theory Eng. 519–526 (2009). https://doi.org/10.7763/IJCTE.2009.V1.84. http://www.ijcte.org/show-26-177-1.html

12. Katragadda, R.: Sentence Position revisited: a robust light-weight Update Summarization 'baseline' Algorithm. Computational Linguistics (2009)

13. Kitaev, N., Cao, S., Klein, D.: Multilingual constituency parsing with self-attention and pre-training. In: ACL 2019–57th Annual Meeting of the Association for Computational Linguistics, Proceedings of the Conference (2020). https://doi.org/10.18653/v1/p19-1340

14. Kitaev, N., Klein, D.: Constituency parsing with a self-attentive encoder. In: ACL 2018–56th Annual Meeting of the Association for Computational Linguistics, Proceedings of the Conference (Long Papers), May 2018. https://doi.org/10.18653/v1/p18-1249. http://arxiv.org/abs/1805.01052

15. Lebanoff, L., Song, K., Liu, F.: Adapting the neural encoder-decoder framework from single to multi-document summarization. In: Proceedings of the 2018 Conference on Empirical Methods in Natural Language Processing, EMNLP 2018, August 2018. https://doi.org/10.18653/v1/d18-1446. http://arxiv.org/abs/1808.06218

16. Li, W., Xiao, X., Liu, J., Wu, H., Wang, H., Du, J.: Leveraging Graph to Improve Abstractive Multi-Document Summarization, pp. 6232–6243 (2020). https://doi.org/10.18653/v1/2020.acl-main.555

17. Lin, C.Y.: Rouge: a package for automatic evaluation of summaries. In: Proceedings of the Workshop on Text Summarization Branches Out (WAS 2004) (2004)

18. Lin, C.Y., Hovy, E.: Identifying topics by position. In: Proceedings of the Fifth Conference on Applied Natural Language Processing, Morristown, NJ, USA, pp. 283–290. Association for Computational Linguistics (1997). https://doi.org/10.3115/974557.974599. http://portal.acm.org/citation.cfm?doid=974557.974599

19. Lin, Z., et al.: A structured self-attentive sentence embedding. In: 5th International Conference on Learning Representations, ICLR 2017 - Conference Track Proceedings, March 2017. http://arxiv.org/abs/1703.03130

20. Liu, P.J., et al.: Generating Wikipedia by summarizing long sequences. In: 6th International Conference on Learning Representations, ICLR 2018 - Conference Track Proceedings (2018)

21. Lunh, H.P.: The automatic creation of literature abstracts. IBM J. Res. Dev. 2, 159–165 (1958)

22. Mihalcea, R., Tarau, P.: TextRank: bringing order into texts. In: Proceedings of EMNLP (2004)

23. Mnih, V., Heess, N., Graves, A., Kavukcuoglu, K.: Recurrent models of visual attention. In: Advances in Neural Information Processing Systems, June 2014. http://arxiv.org/abs/1406.6247

24. Nallapati, R., Zhou, B., dos Santos, C., Gulçehre, Ç., Xiang, B.: Abstractive text summarization using sequence-to-sequence RNNs and beyond. In: CoNLL 2016–20th SIGNLL Conference on Computational Natural Language Learning, Proceedings (2016). https://doi.org/10.18653/v1/k16-1028

25. Narayan, S., Cohen, S.B., Lapata, M.: Don't give me the details, just the summary! topic-aware convolutional neural networks for extreme summarization. In: Proceedings of the 2018 Conference on Empirical Methods in Natural Language Processing, EMNLP 2018 (2018). https://doi.org/10.5281/zenodo.2399762

26. Nenkova, A.: Automatic text summarization of newswire: lessons learned from the document understanding conference. In: Proceedings of the National Conference on Artificial Intelligence (2005)

27. Ouyang, Y., Li, W., Lu, Q., Zhang, R.: A study on position information in document summarization. In: COLING 2010–23rd International Conference on Computational Linguistics, Proceedings of the Conference (2010)

28. Radev, D.R., Jing, H., Styś, M., Tam, D.: Centroid-based summarization of multiple documents. Inf. Process. Manag. **40**(6), 919–938 (2004). https://doi.org/10.1016/j.ipm.2003.10.006. https://linkinghub.elsevier.com/retrieve/pii/S0306457303000955

29. Schilder, F., Kondadadi, R.: FastSum: fast and accurate query-based multi-document summarization. In: ACL 2008: HLT - 46th Annual Meeting of the Association for Computational Linguistics: Human Language Technologies, Proceedings of the Conference (2008)

30. See, A., Liu, P.J., Manning, C.D.: Get to the point: summarization with pointer-generator networks. ACL 2017–55th Annual Meeting of the Association for Computational Linguistics, Proceedings of the Conference (Long Papers), vol. 1, pp. 1073–1083 (2017). https://doi.org/10.18653/v1/P17-1099

31. Thakkar, K.S., Dharaskar, R.V., Chandak, M.B.: Graph-based algorithms for text summarization. In: 2010 3rd International Conference on Emerging Trends in Engineering and Technology, pp. 516–519. IEEE (2010). https://doi.org/10.1109/ICETET.2010.104. http://ieeexplore.ieee.org/document/5698380/

32. Vaswani, A., et al.: Attention is all you need. In: Advances in Neural Information Processing Systems, vol. 30, June 2017. http://arxiv.org/abs/1706.03762

33. Yih, W.T., Goodman, J., Vanderwende, L., Suzuki, H.: Multi-document summarization by maximizing informative content-words. In: IJCAI International Joint Conference on Artificial Intelligence (2007)

34. Zheng, H., Lapata, M.: Sentence centrality revisited for unsupervised summarization. In: ACL 2019–57th Annual Meeting of the Association for Computational Linguistics, Proceedings of the Conference, vol. 2, pp. 6236–6247 (2020). https://doi.org/10.18653/v1/p19-1628

A Strategy for Referential Problem in Low-Resource Neural Machine Translation

Yatu Ji[1], Lei Shi[2], Yila Su[1(✉)], Qing-dao-er-ji Ren[1], Nier Wu[3], and Hongbin Wang[1]

[1] Inner Mongolia University of Technology, Inner Mongolia, China
suyila@tsinghua.org.cn
[2] Inner Mongolia University of Finance and Economics, Inner Mongolia, China
[3] Inner Mongolia University, Inner Mongolia, China

Abstract. This paper aims to solve a series of referential problems in sequence decoding, which caused by corpus sparsity in low-resource Neural Machine Translation (NMT), including pronoun missing, reference error, gender bias. It is difficult to find the essential reason of these problems because they are only shown in the prediction results and all aspects of the model training. Different from the usual solutions based on complex mathematical rule setting and adding artificial features, we expect to turn the problems in the predictions into noise as much as possible. On this basis, we further use adversarial training to make the model find the balance between the noise and the golden samples, instead of exploring the reason of the problem during the complex training. In this paper, a noise-based preprocessing operation and a slight modification of the adversarial training can help the model to better generalize a series of referential problems in low-resource NMT tasks. Experiments show that the evaluation of BLEU score and the accuracy of pronouns in sequence on Korean-Chinese, Mongolian-Chinese and Arabic-Chinese task have been significantly improved.

Keywords: Low-Resource Neural Machine Translation · Referential problem · Generative Adversarial Network

1 Introduction

Almost all Nature Language Processing (NLP) tasks exist referential errors caused by inadequate training, incomplete semantic information in corpus, and lack of the ability to capture complex context. Especially on low resource task, referential resolution is one of the extremely difficult problems. Under this premise, we usually have to use a lot of training tricks or set some complex model structure and characteristics to alleviate or solve the above problems in sequence-to-sequence tasks. Referential errors may come from any link of training, such as the noise of corpus [11], the performance of embedding [6], the compression

© Springer Nature Switzerland AG 2021
I. Farkaš et al. (Eds.): ICANN 2021, LNCS 12895, pp. 321–332, 2021.
https://doi.org/10.1007/978-3-030-86383-8_26

ability of encoder, the ability of the decoder to predict the sequence with the help of the attention mechanism [10], the generalization ability of the model, the readability of the translation [1]. Each problem needs a specific approach. Some practice is to predict the antecedent of the referent and the reference relationship through deep neural networks. These contributions [8,20] show the ability of neural network to represent the pronouns and antecedents in the vector space are more effectively than the traditional methods. However, they all need to use a lot of mathematical knowledge to set complex training rules and add more or less artificial features, which make the research threshold of reference problems become higher. It is straightforward to capture the referential relationships in sequences and paragraphs through deeper and more complex network structures [5,9], but complex models not only confuse the training, but also make some specific tasks impractical. On this basis, in order to make the model better capture referential relationships, reinforcement learning (RL) [7,17] enables the model to accurately correct the relationship between antecedents and pronouns through policy iterations within a limited training period [19]. On the other hand, referential error and gender bias are particularly serious in low-resource translation tasks. The reasons may be various, such as sparse vocabulary, lack of semantic components, and some named entities that are not sensitive to pronouns. We take a Korean-Chinese test sentence as an example to illustrate the influence of the accuracy of the anaphora on translation.

test sequence:

@교수는 매우 기뻤고 남자 친구는 선물을 사서 항상 가지고 다녔습니다.. @ translation: The professor was very happy, her boyfriend bought her a gift and she always carried it with her.

Transformer_basic (after training based on daily corpus) is decoded as:

@教授很幸福，他的男朋友给她买了礼物并且总是随身携带。@

There is the inherent gender bias in the corpus, which causes the first 'her' to be translated as 'his' according to the probability candidate set. When the second 'her' is associated, partly based on the first pronoun 'his' and partly based on 'professor', so the next prediction and the first reference error continued to the pronoun 'him'. Then, in the case of losing a 'she', the last 'him' also appeared to be ambiguous.

In this paper, we use a easy to implement preprocessing approach instead of complex mathematical rule settings to solve a series of reference problems in low-resource NMT. The core of the proposed method is adding a pseudo sequence, which is obvious and contrary to the facts, so that the model can correct errors or bias to this type of reference relationship in adversarial training. Specifically, we adopt a method of adding noise (see Sect. 2.1) and adversarial training [16,18] to make the model dynamically generalize these noises better. The strategy of noise addition in this paper is essentially different from simply training the original data multiple times. In short, the effect of multiple training on the same sequence and the updating of different parameters of the similar sequence is quite different [2,11]. The contributions of this paper can be summarized in the following three points:

-We propose a strategy that takes the focused and unresolved targets as noise. In this paper, reference-related noise is added to the training data in the form of a pseudo-sequence.

-We normalize the referential relationship to the BLEU score and the pronoun accuracy, instead of adding complex mathematical rules to the loss function and evaluation metrics.

-In order to match the rationality of this strategy, that is, to allow the model to have extra interest and focus to pronouns during the training process. We add a focus module on the basis of the Generative Adversarial Networks (GAN) to focus on the referential relationship in sequence decoding. We use value iteration network (VIN) as a focus module because GAN has the essence of RL training. In VIN, the incorrect referential prediction corresponds to a low reward, whereas the low reward corresponds to a low value. This is what the focus module wants to emphasize.

In summary, we expect to use a data preprocessing operation and a focused module for GAN training to get rid of the dilemma of complex rule design or loss of semantics like hard debiasing method [3]. In Sect. 2, we will introduce the details of the model and discuss the necessity of key modules. Then we introduce the verification experiments in Sects. 3 and 4, including preprocessing methods and analysis of experimental results. Finally, we briefly discuss the effectiveness and portability of this method.

2   Model Description

The training process of GAN is usually unstable. However, this paper mainly studies the use of GAN to solve the reference problem in low-resource NMT. Therefore, the complex model description is not emphasized in this paper, but the training details can be found in research [16] and [21]. The model we present is mainly divided into three parts: generation module G, focus module F, and discrimination module D. Similar to the usual GAN, G based on RL strategy is used to transform the source-side embedding to the target-side sequence using the policy gradient algorithm, so the input and output of G are still embedded corpus. This generation relationship will be described in Sect. 2.2. In order to clearly present the training process of the proposed strategy and model, we divide into three parts to connect and explain the logic of the entire strategy: the preprocessing for obtaining noise, the RL training to enhance the accuracy of the reference relationship and the noise.

2.1   Preprocessing-Noise

To be straightforward, we want the model to generalize the noise, so we directly add the corresponding noise to the training data to familiarize the model with it. Here, noise is about several major referential problems that arise in the process of low-resource NMT. Generally, there are three types of referential errors: pronoun missing and overlapping, referential errors, and gender bias. The missing and overlapping of pronoun is similar to the other components, which is largely due to under-fitting. Translation models can usually solve such problems with the help of multiple iterations of training or regular optimization. For referential errors, we still take the *test sequence* as an example. First, this paper copies and tags the training data as pseudo-sequences. In order to get rid of the special reliance of preprocessing on gender pronoun recognition tools, this paper only requires tagging several fixed pronouns in the sentences, such as: he, she, it, they, their, etc. These words have a fixed probability representation in the training vocabulary, so the tagging process only needs to retrieve the vocabulary or build a simple search script. Then the pronouns of the pseudo-sequence are masked and replaced. It ensures that pronouns can be fully generalized without distortion.

pseudo-sequence 1 - *replace*: @교 수 는 매우 기뻐 하 며, 그녀와(그의)(그것 의) 남자 친구 가 그녀(그것)(그녀) 에 게 선물 을 사 주 었 고, 아울러 그것(그녀)(그것) 을 가지 고 다 닌 다.@

pseudo-sequence 2 - *mask*: @교 수 는 매우 기뻐 하 며, 그녀와(@mask@) 남자 친구 가 그녀(@mask@) 에 게 선물 을 사 주 었 고, 아울러 그것(@mask@) 을 가지 고 다 닌 다.@

In both pseudo-sequences[1], all pronouns are replaced by possible pronouns or mask symbols @*mask*@. Due to the different grammatical structure, the last 'him' in translation does not actually appear in Korean. The gender bias problem in translation is sensitive and cumbersome in low-resource tasks. Such problems not only involve the accuracy of pronoun, but also affect the prediction accuracy of the entire sequence. In view of this problem, we also boil down to these two forms of noise.

2.2   Reinforcement Learning Training

The overall network structure is shown in Fig. 1. The entire adversarial training is guided by REINFORCE algorithm to optimize model parameters, which is also a common strategy of GAN in the sequence generation task. This is consistent with why we use VIN, so VIN can be perfectly integrated into the entire adversarial training.

The first thing to be clear is how the REINFORCE algorithm is mapped to the sequence generation task. We only list some mappings that are more concerned in NMT. In the typical REINFORCE algorithm, the following standard variables (agent, police, action, state, reward) usually map to (generator, parameter, prediction of each iteration, hidden units, BLEU score of predicted sequence) in the sequence (x, y) generation task. The translation model as G is used to sense the environment state s of the network when it is mapped as an agent. Such an action a updates the entire state parameters θ by a fixed

---

[1] In this paper, the initial effective proportion of the two pseudo-sequences in the corpus is determined in the experiment. At the beginning of the adversarial training, *original*: *replace*: *mask* = 8: 1: 1. During the training, the two noises of each epoch increase 1% respectively, which corresponds to a reduction of 2% of the original data.

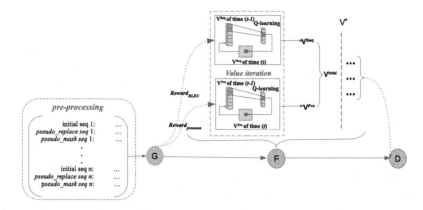

Fig. 1. Model architecture. Contains preprocessing form and model details.

training police. The reward is calculated from the BLEU score result of the predicted sequence and the gold sequence. The entire training objective O_θ can be expressed as two expectations about maximum and minimum:

$$O_\theta = \begin{cases} E_{groundtruth}G \sim logD(x,y) & min \\ E_{Discriminator}D \sim log(1 - D(x,y)) & max \end{cases} \qquad (1)$$

G uses the preprocessed corpus to update the random hidden layer states and rewards. The module F is used to evaluate G's output according to the rewards generated by RL. D discriminates the corresponding sequence based on the value generated by F, which is consistent with the usual GAN process [16,21]. In order to prevent D from always getting negative feedback, G and D are trained alternately.

2.3 Focus Module and Discriminate Module

It is due to the fact that we use the score of the BLEU and the accuracy of pronouns[2] as the evaluation criteria for rewards. In other words, the prediction with correct reference relations and higher BLEU score will yield a better reward.

The module F between the G and the D gives priority to D to identify sequences with less reward, where less reward correspond to inaccurate reference relations and translation results. Module F only plays a role of value screening, and does not perform further processing on the samples. We refer to the implementation of VIN in the work [18] and adopt two simple CNN to realize the entire value iteration process. Different from work [18], we pay attention to two aspects of reward in our method: BLEU rewards for the entire sequence V_t^{Seq} and character level rewards for pronouns V_t^{Pro}:

---

[2] The reward of the tagged pronoun can be easily obtained by calculating the character level reward [13].

$$V_t^{Seq} = max_a Q(s, a) = max \left[R_B(s, a) + \sum_N^{t=1} P(s|s_{t-1}, a)V_{t-1} \right] \tag{2}$$

$$V_t^{Pro} = max_a Q(s, a) = max \left[R_P(s, a) + \sum_N^{t=1} P(s|s_{t-1}, a)V_{t-1} \right] \tag{3}$$

$$V_{total} = (1 - \alpha)V^{Seq} + \alpha V^{Pro} \tag{4}$$

where Q indicates the value of action a under state s at t-th timestep, the reward $R_B(s, a)$, $R_P(s, a)$ and transition probabilities $p()$ are obtained from G. N represents the sequence length. The value of the sequence is obtained through the accumulation of the reward and the value of the previous time step V_{t-1}. The total value V_{total} dynamically combines the V_t^{Pro} and V_t^{Seq} into a representative value according to α, where α is the prediction accuracy of the current training epoch. This value is used to compare with V^*, V^* represents the value of the pre-trained model, and to determine the current batch training priority.

The core of the module F is to iteratively generate the value of the input reward. It can represent the current training cost. Some algorithms that predict behavior though value selection can be considered, such as Q-learning [4], Sarsa [15], and Deep Q-Network [14]. Considering that the sequence decoding in this paper belongs to one-step generation, Q-learning[3] is used in decoding. Q-learning can be understood as the accumulation of action rewards at time step t. This accumulation will decay according to the hyperparameter λ, which is usually set according to specific tasks. For example, in the training of some low-resource tasks, λ usually takes $[0.7 - 1.0]$.

$$Q(s, t) = r_{t+1} + \lambda^2 r_{t+2} + ... + \lambda^{t+n} r_{t+n+2} \tag{5}$$

The responsibility assumed by module D is identifying the ground truth and generated sequence selected by F, so that the sequence of interest can be preferentially entered into the next round of iterative training. In view of the excellent performance of CNN in binary classification tasks, we use a simple CNN as a D to form GAN.

3   Experimental Settings

The method studied in this paper does not depend on specific annotation tools. For a certain language, in the absence of open source tools[4], it is only need to build a simple script to identify several fixed pronouns. We focus on establishing more accurate referential relationship through other components of the sequence.

---

[3] Monte-Carlo search algorithm (MC) is used in GAN to evaluate intermediate states and directly select behavioral strategies, such as Policy Gradients, which can only be used for model updating in training.

[4] Mn-Ch: CRF++: https://github.com/othman-zennaki/RNN_Pos_Tagger,
Kr-Ch: https://sourceforge.net/projects/hannanum/,
Ar-Ch: http://opennlp.apache.org/.

3.1 Experimental Data

We verify the effectiveness of the proposed approach on three low-resource corpora: Mongolian-Chinese (Mn-Ch, 0.2M, including 5000 sentences of evaluation sets.), Korean-Chinese (Kr-Ch, 0.1M, evaluation: 1000), Arabic-Chinese (Ar-Ch, 2.2M, evaluation: 5000). The data comes from CLDC, machine translation track of evaluation campaign CWMT2017 and OPUS in LREC2012[5], respectively. The composition of the corpus is distributed in news, daily life, and government document. Usually, the referential problem needs to create a challenge set that amplify the effect of coreference and anaphora as their occurrence in randomly selected documents is typically not large enough to be sensitive under the BLEU score metric. Therefore, the CWMT2018 test set with a scale of 1000 sentences is distributed in the field of daily evaluation and carefully selected sentence patterns that are similar or typical referential relations.

3.2 Experimental Setup

We select the baseline system from two perspectives: model and strategy. In order to highlight the effectiveness of the strategy, we choose Transformer_basic [12], which performs the best in multiple languages, and it has a good performance in focusing on the overall semantic information. In terms of model, it is based on adversarial training, so we use two related typical GAN models as the baseline system, BR-CSGAN [21][6] and F-GAN [18][7]. Besides, the baseline system basically maintains the parameter in the original baseline system in order to clearly observe the experimental results. Some minor adjustments are made to cater to the inherent experimental conditions. For example, the mask strategy is added in the preprocessing stage, and the setting of Dropout is canceled. We also increase the batch training size to 128 to allow the noise and the original data to be fully trained, and all models are trained on up to single Titan-X GPU.

4 Verification

The validity of the training strategy and model will be verified from three questions:

- How to verify the role of the proposed strategy and model in reference problems?
- How to ensure the accuracy and fluency of the prediction sequences on the premise of improving the reference relationship?
- Does the additional module F affect the efficiency of the entire training?

---

[5] http://opus.nlpl.eu/, https://object.pouta.csc.fi/OPUS-MultiUN/v1/moses/ar-zh.txt.zip.

[6] We reimplement the BR-CSGAN according to https://github.com/ZhenYangIACAS/NMT_GAN.

[7] We reimplement the F-GAN according to https://github.com/jiyatu/Filter-GAN.git.

4.1   Three Verification Indicators Are Used to Solve the Above Problems

The accuracy of pronouns(Acc_Pro): In addition to BLEU score of the whole sequence, the model also needs to pay attention to the accuracy of pronouns in the sequence, so as to reflect the predictive ability of anaphora. We compare the accuracy of pronouns in prediction sequence and standard sequence by masking other elements except pronouns in tagged sequences. Since BLEU score of the sequence has been taken as a part of the metric, the approach can focus on reflecting the accuracy of pronouns and avoiding designing indicators based on complex mathematical rules.

BLEU for the overall sequence (BLEU_Seq): The approach still needs to ensure the accuracy of the entire sequence when solving the referential problem, which is the original intention of machine translation.

Training efficiency: We record three indicators that most intuitively reflect the training process of translation model: the convergence process of loss, the trend of accuracy, and the training time.

4.2   Verification Results and Analysis

As mentioned in Sect. 4.1, in order to meet the original intention of the NMT task, we use common and intuitive evaluation indicators: BLEU score and accuracy of pronoun, instead of complex antecedent speculation and F-score. It is also consistent with the original intention of this paper, which is to simplify the process of measuring reference relations.

Acc_pro and *BLEU_Seq* We make statistics on the performance of two indicators on different systems, including Acc_Pro and BLEU_Seq, as shown in Table 1. In low-resource NMT, the results may varies significantly and they are very sensitive to the training process. Therefore, the statistical results in this paper are the average of multiple results to avoid accidental phenomenon in the experiment.

Table 1. The performance of different systems on the two indicators, including the effect of noise preprocessing on the GAN-based system.

System		Mn-Ch		Kr-Ch		Ar-Ch	
		Acc_Pro	BLEU_Seq	Acc_Pro	BLEU_Seq	Acc_Pro	BLEU_Seq
Transformer_basic		56.3	28.5	47.7	**24.7**	60.1	30.8
BR-CSGAN	-	47.5	27.4	42.5	23.3	57.7	30.1
	+pre_noise	50.8	27.9	44.1	23.9	59.2	31.3
F-GAN	-	57.9	29.1	38.8	20.4	62.5	31.3
	+pre_noise	58.6	32.3	41.7	21.2	66.7	32.4
Our	-	48.2	31.2	42.3	22.6	62.9	30.8
	+pre_noise	**64.3**	**34.8**	**48.5**	24.3	**67.5**	**33.7**

First, we explore the sensitivity of noise preprocessing strategy on different systems. It is easy to find that the noise strategy in each system can bring 0.5 to 3.6 BLEU_Seq score improvements, and such improvements are mainly distributed in the adversarial training system. It is because the adversarial mechanism enables the model to dynamically train noise in a limited training period and generalize better. For Acc_Pro, there is a maximum of 6.1 BLEU score improvements.

On the other hand, we observe that the Acc_Pro score has the highest improvement, and the corresponding highest improvement of BLEU_Seq score is also achieved after preprocessed. We believe this is not accidental, because after improving the referential accuracy, the subsequent decoding of the model will explore new candidate spaces for the correct referents, which is very important for the effectiveness of the beam search algorithm. The proposed approach also performs a good ability in most tasks without the cooperation of noise preprocessing, which is due to the seamless connection between the module F and RL. The results show that our model can quickly converge to a more optimized state during insufficient training cycles.

Training Efficiency. For the statistical results after noise preprocessing (Fig. 2), in order to show the trend of accuracy in the training process clearly, we increase the sampling node span, so the fluctuation of the curve in the graph will become more obvious. The loss of the three adversarial models converges

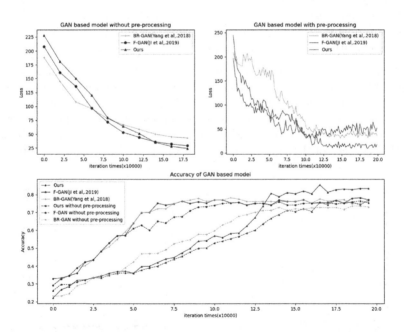

Fig. 2. The influence of noise preprocessing strategy on the trend of loss and accuracy. The figure shows the two most intuitive training indicators, training loss and accuracy rate during 20*10,000 iterations. Since the Transformer_basic is not a GAN-based model, its observations are not presented in the statistical results.

Table 2. The performance of training efficiency of each system in different tasks.(*training time: hours*)

	BR-CSGAN		F-GAN		Ours	
	-	+pre_noise	-	+pre_noise	-	+pre_noise
Mn-Ch	31	43	29	27	27	**23**
Kr-Ch	24	37	20	**19**	21	20
Ar-Ch	54	68	50	44	50	**41**

quickly at the beginning of training, which is why we use GAN as the original model. The advantages of the adversarial mechanism allow us to eliminate some suspicious factors in order to clearly observe the noise strategy effect. It also can quickly converge with the cooperation of preprocessing strategy, and finally achieve a significantly lower loss.

We also present the statistical results of the training time of each adversarial model in different corpora in Table 2. The experimental results observed in the table can be summarized into three analyses:

-Among the three adversarial systems, the model with the focus module has significantly less training time than BR-CSGAN and F-GAN, which is attributed to the value modules focus on noise.

-The added noise does not bring additional time cost to the model. -The proposed model shows better training efficiency on almost all tasks, whether on the state of initial or after adding noise.

Heat Map for Reference. A heat map mapping of a typical example sentence is given here to illustrate the decoding effect of the proposed model, as shown in Fig. 3. The pronouns highlighted by gray rectangles can be more accurately mapped to the target language in the proposed model, and provide a richer candidate set. These candidate words of the deeper color in the heat map indicate that the accuracy of the provided candidate words is higher. Under this premise,

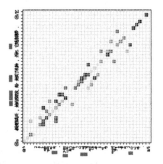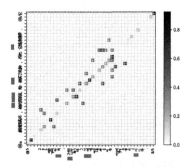

Fig. 3. An example of a heat map after decoding and reordering: *left*-with noise preprocessing, *right*-without noise preprocessing.

the subsequent decoding will not deviate from the golden answer a lot, which is conducive to improving the accuracy and fluency of the whole sequence. In addition, such corrections are not isolated. For non-pronoun terms, it is easy to observe that the accuracy of prediction is also affected by the reference relationship to a certain extent. Especially for words related to the pronoun, whose prediction is directly determined by the predictive ability of the pronoun. The inherent gender biases still exist in the sequence, such as 'he' referring to 'professor' in the right subgraph, which is based on the inherent gender bias and collocation in the vocabulary. In fact, the golden word here is 'her', which is corrected in our model (left subgraph). It can be attributed to the addition of pseudo sequences with 'her', 'it' and 'he' as pronouns in the noise preprocessing.

5   Summary

This paper is devoted to solving the problems of inaccurate reference relations caused by sparse vocabulary in low-resource NMT task, including incorrect reference relationship and gender bias. The main contribution is to use an easy to implement preprocessing operation combined with adversarial learning to improve the accuracy of pronouns in translation, thereby avoiding the setting of complex mathematics and language rules. In terms of BLEU score, it is verified that the proposed strategy shows impressive results both in the prediction result of the whole sequence and the accuracy of pronoun. The motivation of this study is to convert the problems encountered into noise and generalize the problems through adversarial training. Although the proposed method has some limitations in the proportion of various noises and the stability of adversarial training, we still look forward to exploring more general training goals in future work, so as to extend this solution of turning problems into noise and further generalization to more NLP tasks.

Acknowledgments. This study is supported by the National Natural Science Foundation of China (No. 61966027 and No. 61966028) and Inner Mongolia key technology research project: Mongolian network public opinion monitoring system integrating machine translation and emotional analysis.

References

1. Baroni, M., Lake, B.M.: Generalization without systematicity: on the compositional skills of sequence-to-sequence recurrent networks. Computing and Language arXiv:1711.00350 (2017)
2. Belinkov, Y., Bisk, Y.: Synthetic and natural noise both break neural machine translation. In: International Conference on Learning Representations (2018)
3. Bolukbasi, T., Chang, K.W., Zou, J.Y., Saligrama, V., Kalai., A.: Man is to computer programmer as woman is to homemaker? Debiasing word embeddings. In: Advances in Neural Information Processing Systems, pp. 4349–4357 (2016)
4. Clifton, J., Laber, E.: Q-learning: theory and applications. Ann. Rev. Stat. Appl. **7**(1), 279–301 (2020)

5. Durrett, G., Klein, D.: Easy victories and uphill battles in coreference resolution. In: Proceedings of the Conference on Empirical Methods in Natural Language Processing, pp. 1971–1982 (2013)

6. Kang, L., Shizhu, H., Siwei, L., Jun, Z.: How to generate a good word embedding. Alternation **31**(6), 5–14 (2016)

7. Kevin, C., Manning., C.D.: Deep reinforcement learning for mention-ranking coreference models. In: Proceedings of the Conference on Empirical Methods in Natural Language Processing, pp. 2256–2262 (2016)

8. Lee, K., et al.: End-to-end neural coreference resolution. In: Proceedings of the Conference on Empirical Methods in Natural Language Processing, pp. 188–197 (2017)

9. Lee, K., He, L., Lewis, M., Zettlemoyer., L.: End-to-end neural coreference resolution. In: Proceedings of the Conference on Empirical Methods in Natural Language Processing, pp. 188–197 (2017)

10. Manning, C.D., Luong, M., Pham, H.: Effective approaches to attention-based neural machine translation. Computing and Language arXiv:1508.04025 (2015)

11. Khayrallah, H., Koehn, P.: On the impact of various types of noise on neural machine translation. In: Meeting of the Association for Computational Linguistics, pp. 74–83 (2018)

12. Vaswani, A., et al.: Attention is all you need. In: Advances in Neural Information Processing Systems, pp. 6000–6010 (2017)

13. Volodymyr, M., et al.: Human-level control through deep reinforcement learning. Nature **518**(7540), 529–533 (2015)

14. Hong, Z.-W., Su, S.-Y., Shann, T.-Y., Chang, Y.-H., Lee, C.-Y.: A deep policy inference Q-network for multi-agent systems. In: Proceedings of the Conference on Autonomous Agents and MultiAgent Systems, pp. 1388–1396 (2018)

15. Wang, Y., Li, T.S., Lin., C.: Backward Q-learning: the combination of Sarsa algorithm and Q-learning. Eng. Appl. Artif. Intell. **26**(9), 2184–2193 (2013)

16. Wu, L., Xia, Y., Tian, F., et al.: Adversarial neural machine translation. In: Asian Conference on Machine Learning, pp. 534–549 (2018)

17. YangWang, W., Li, J., He., X.: Deep reinforcement learning for NLP. In: Meeting of the Association for Computational Linguistics, pp. 19–21 (2018)

18. Yatu, J., Hongxu, H., Junjie, C., Nier, W.: Adversarial training for unknown word problems in neural machine translation. ACM Trans. Asian Low-Resour. Lang. Inf. Process. **19**(1), 1–12 (2019)

19. Yin, Q., Zhang, Y., Zhang, W., Liu, T.: Chinese zero pronoun resolution with deep memory network. In: Proceedings of the Conference on Empirical Methods in Natural Language Processing, pp. 1309–1318 (2017)

20. Yin, Q., Zhang, Y., Zhang, W., Liu, T., Wang., W.Y.: Deep reinforcement learning for Chinese zero pronoun resolution. In: Meeting of the Association for Computational Linguistics, pp. 569–578 (2018)

21. Zhen, Y., Wei, C., Feng, W., Bo., X.: Improving neural machine translation with condition sequence generative adversarial nets. In: North American Chapter of the Association for Computation Linguistics, pp. 1335–1346 (2018)

A Unified Summarization Model with Semantic Guide and Keyword Coverage Mechanism

Wuhang Lin$^{(\boxtimes)}$, Jianling Li, Zibo Yi, Bin Ji, Shasha Li$^{(\boxtimes)}$, Jie Yu, and Jun Ma

College of Computer, National University of Defense Technology, Changsha, China
{wuhanglin,jianlingl,yizibo14,jibin,shashali,yj,majun}@nudt.edu.cn

Abstract. Neural abstractive summarization models, based on attentional encoder-decoder architecture, can generate a summary closer to human style. However, the generated summaries often suffer from inaccuracy and irrelevance. To tackle these problems, we propose a novel unified summarization model with integrated semantic guide and keyword coverage mechanism. First, we add the integrated semantic guide in the encoder, which helps the encoder-decoder structure grasp the central idea of the full text. Second, in the decoding process, we use the keyword coverage mechanism to reward the attention distribution of keywords, which promotes the generation of more logically related expressions to the central idea. Evaluations on the *CNN/Daily Mail* dataset demonstrate that our model outperforms the baseline models, and the analysis shows that our model is capable of generating a more relevant and fluent summary.

Keywords: Unified model · Semantic guide · Keyword coverage · Inaccuracy · Irrelevance

1 Introduction

Text summarization is a process of refining the central idea of the original text and producing a short and fluent summary [1]. There are mainly two types of approaches: extractive and abstractive. Both approaches have advantages. The summary, extracted by excellent extractive approaches, can easily contain the crucial sentence of the article. The abstractive approaches, based on attentional encoder-decoder architecture, can generate novel words and phrases [2]. The reason is that abstractive approaches are complex and the process is closer to the way humans write. Thus, recent studies pay more attention to the abstractive.

However, the recent study shows that abstractive approaches often produce summaries with inaccuracy and irrelevance [3,4]. For example, in the example shown in Fig. 1, the purpose of the article is to inform us of the news, that Arrested Development will return for the fifth season of 17 episodes. However,

This work was supported by National Key Research and Development Project (No. 2018YFB1004502) and National Natural Science Foundation of China (No. 61532001) and (U19A2060).

I. Farkaš et al. (Eds.): ICANN 2021, LNCS 12895, pp. 333–344, 2021.
https://doi.org/10.1007/978-3-030-86383-8_27

Original Text: for those wondering if we would ever hear from the bluth family again, the answer would appear to be yes. **"arrested development" executive producer brian grazer said the show will return for a fifth season of 17 episodes.** the hollywood mogul was interviewed on bill simmons' podcast recently, and let it drop that fans can expect more of the quirky comedy. netflix had no comment for cnn when asked to verify his statements. **the fourth season was streamed exclusively on netflix in 2013,** after fox canceled the show several years before. despite critical acclaim, **the series never had big ratings, but has a devoted fan base, who often quote from the show. it was not yet known if the full cast, including jason bateman, michael cera and will arnett,** will return for the season.

Gold: fan favorite series "arrested development" to return for a fifth season, according to producer. brian grazer claimed the show would be back in a podcast. netflix is not commenting.

Baseline PGN+CovMec: the fourth season was streamed exclusively on netflix in 2013. the series never had big ratings, but has a devoted fan base, who often quote from the show. it was not yet known if the full cast, including jason bateman, michael cera and will arnett.

Our Model: executive producer brian grazer said the show will return for a fifth season of 17 episodes. the fourth season was streamed exclusively on netflix in 2013, after fox canceled the show several years before.

Fig. 1. Comparison of the output of two models on a news article. Our model corrects the inaccuracy problem of the baseline and captures the key semantic information and purpose of the news. The Baseline model only focuses on the secondary information in blue, but our model, like Gold, focuses on the key information in orange.

the summary produced by the baseline does not capture the main purpose of the article at all. In addition, the last sentence of the abstract is abrupt with other words, leading to the lack of the logical relevance of context. It can be said that this summary produced by baseline is completely inaccurate.

To tackle these problems, we propose a unified summarization model based on *Pointer-Generator Network* [5], which combines the advantages of extractive and abstractive. Our model is inspired by the process of human summary writing. To ensure that the written summary is accurate and on the topic, one begins to write the summary only if he has read the full text and has an overall understanding. Thus, we first use the semantic guide to integrate the semantic vectors of multiple extractive summaries to generate a global semantic guide vector, which can help the encoder-decoder structure grasp the central idea of the full text. For local sentence writing, humans tend to refer to the context in the original text of the current token or phrase being written, especially keywords, to ensure that the statements are logically related. Next, we use the keyword coverage mechanism to reward the attention of keywords. The keyword coverage mechanism helps the abstractive model to capture the semantic connection between the current token and the keywords, which promotes the generation of sentence expressions that are more logically relevant to the central idea.

Extensive experiments on the *CNN/Daily Mail* [6] dataset show that our model far exceeded the ROUGE[1] score of the benchmark model. And our model solves the problem of inaccuracy and irrelevance of the benchmark. As shown in the example in Fig. 1, our model captures the key information and purpose of the news. Moreover, to ensure the quality of our unified model, we conduct a solid human evaluation, which confirms that our method significantly outperforms other methods in relevancy and informativeness. Finally, we conduct ablation experiments to show the effectiveness of the proposed submodules in our model. The submodules include semantic guide and keyword coverage. The results show

---

[1] pypi.python.org/pypi/pyrouge/0.1.3.

that these submodules can improve the quality of summary independently and effectively. To summarise, our contributions are mainly in two aspects:

- We propose a unified summarization model with semantic guide and keyword coverage mechanism. Our model can solve the problem of inaccuracy and irrelevance for the abstractive model, which can generate a more relevant summary with higher ROUGE value.
- Our submodules have excellent independent functionality. The semantic guide inherits helps the model capture the central idea of the full text. The keyword coverage mechanism effectively improves the coverage mechanism and helps generate more logically relevant expressions.

2   Related Work

2.1   Unified Model

Extractive approaches can extract significant central sentences. However, they have the problem of being verbose and illogical. Nallapati et al. [7] proposed the first abstractive baseline based on the encoder-decoder framework. See et al. [5] proposed the *Pointer-Generator Network*, which has been positively evaluated. However, it is still difficult to control the process of generating summaries.

By combining extractive and abstractive approach, the unified model has the advantages of the two approaches. Chen and Bansal [8] proposed a fast summarization method. They extract salient sentences from the document, then rewrite the sentences to generate the final summary. Hsu et al. [9] proposed a unified model (*unified+inconsistency loss*). They use the extractive model to score sentences, then modulate the word-level attention of the abstractive model.

2.2   Guide Mechanism

For abstractive approaches, the process of summary generation can be difficult to control in the absence of guidance information. Fu et al. [10] proposed variational hierarchical topic-Aware model (*VHTM*). They achieved good results by merging the topics into multiple levels of granularity. Li et al. [11] introduced a guiding generation model (*KIGN+Prediction-guide*) that encodes keywords extracted from the source text to guide the process of abstractive summarization.

3   The Proposed Model

3.1   Pointer-Generator Network Baseline

Our model is based on *Pointer-Generator Network*, which mainly consists of a single-layer bidirectional LSTM encoder and a single-layer unidirectional LSTM decoder. The word embeddings of the source is input into the encoder

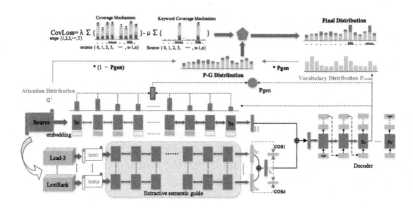

Fig. 2. Our model consists of the *Pointer-Generator Network*, semantic guide, and keyword coverage mechanism. The semantic guide generates a global semantic guide vector to help grasp the central idea of the full text. The keyword coverage mechanism rewards the attention of keywords to generate more logically relevant expressions.

one by one to generate the corresponding encoder hidden state sequence $\{h_0, h_1, ..., h_n\}(h_i = [\overrightarrow{h_i}; \overleftarrow{h_i}])$. On each step t, the decoder receives the word embedding of the previous target token $y_{t-1}^{\hat{}}$ and the previous decoder state s_{t-1}, generating the current decoder state s_t. The i-th encoder hidden state h_i's attention distribution α_i^t and the context vector c_t are calculated according to the following equations:

$$\alpha_i^t = softmax(v^T tanh(W_h h_i + W_s s_t + b_{attn})) \tag{1}$$

$$C_t = \sum_i \alpha_i^t h_i \tag{2}$$

where v, W_h, W_s and b_{attn} are learnable parameters. At each t step, C_t will connect with decoder state s_t and pass through two linear layers to estimate the probability distribution of the next word P_{vocab}^t:

$$P_{vocab}^t = softmax(V'(V(s_t, C_t)) + b') \tag{3}$$

where V' is a vocabulary weight matrix and softmax is over the entire vocabulary.

Pointer-Generator Network has a soft switch P_{gen}, which is used to determine whether to generate a word from the vocabulary according to P_{vocab}^t or copy a word from the input sequence by sampling from the attention distribution α^t. At step t, s_t, C_t, and the current input word embedding $e_{y_{t-1}}$ will be fed together to a linear layer, then produce switch P_{gen} under the sigmod activation function σ. Finally, the probability distribution of the next word will be updated according to the switch P_{gen}:

$$P_{gen} = \sigma(linear(s_t, C_t, e_{y_{t-1}})) \tag{4}$$

$$P(w) = P_{gen} P_{vocab}^t + (1 - P_{gen}) \sum_{i:w_i=w} \alpha_i^t \tag{5}$$

Pointer-Generator Network is first trained 230k iterations approximately without using the coverage mechanism. Let the target word for step t be $\hat{y}_t$. The loss function is:

$$\mathcal{L} = \frac{1}{L} \sum_{t=0}^{L} -\log P(\hat{y}_t) \tag{6}$$

The *Pointer-Generator Network* also uses the coverage mechanism to train 3K iterations. The coverage mechanism always maintains a coverage vector c^t, which records the cumulative sum of the attention of each token in the original text. At this point, the attention distribution and loss function are changed into:

$$c_i^t = \sum_{t=0}^{t-1} \alpha_i^t \tag{7a}$$

$$\alpha_i^t = softmax(v^T tanh(W_h h_i + W_s s_t + w_c c_i^t + b_{attn})) \tag{7b}$$

$$\mathcal{L} = \frac{1}{L} \sum_{t=0}^{L} -\log P(\hat{y}_t) + \lambda \sum_i min(\alpha_i^t, c_i^t) \tag{7c}$$

where w_c is learnable parameter and hyperparameter $\lambda = 1$.

3.2   Integrated Extractive Semantic Guide

Referring to the process of human summary writing, we believe that the reason why abstractive summarization tends to deviate from the central idea is that the decoding process lacks a grasp of global semantics. The abstractive summarization requires an extra powerful semantic encoder to enhance the semantic encoding of the original text. Our extractive semantic guide is an encoder with semantic guidance function. Existing excellent extraction methods can easily extract the central sentence of the original text. To make the semantic guide more accurate, the input of the semantic guide is the extractive summary, which is generated by the extraction method on the original text.

Sometimes, the summary produced by a single extractive approach may not be accurate. Therefore, our semantic guide is integrated. The extraction methods we use comprehensively include Lead-3, Lexrank [12], etc. As shown in Fig. 2, the original text has been extracted different extractive approaches to generate extractive text $\{text_1, ..., text_d\}$. After the text group $\{text_1, ..., text_d\}$ is encoded by the semantic guide, $h_i = LSTM_{gui}(text_i)$, the semantic vector group $\{h_1, ..., h_d\}$ is generated. The guide encoder uses the single-layer bidirectional LSTM, which is the same as the baseline's encoder. For different extractive texts, their encoders (LSTM) share the same parameters.

Our semantic fusion calculate the cosin list $\{cos_1, ..., cos_d\}$ of $\{h_1, ..., h_d\}$ based on the final encoder hidden state H of the original text. We input the cosin list into a softmax layer to get the impression weight of each semantic vector. Based on the weight values, we integrate the semantic vectors and obtain the semantic guidance vector h^{gui}. In order to guide the decoding process of abstractive summarization, we use semantic guide vector h^{gui} to adjust the initial semantic vector of the abstractive approach:

$$(\theta_1, \theta_2, ..., \theta_d) = softmax(cos_1, cos_2, ..., cos_d) \tag{8}$$

$$h^{gui} = \theta_1 h_1 + \theta_2 h_2 + ... + \theta_d h_d \tag{9}$$

$$h_{ini} = relu(W^{gui} h^{gui} + W^{abs} H + b^{ini}) \tag{10}$$

where W^{gui}, W^{abs} and b^{ini} are learnable parameters.

3.3   Keyword Coverage Mechanism

After grasping the semantics of the full text, there is a greater hope to generate a relevant summary. However, because of the long-distance memory loss of the LSTM decoder, the decoder cannot guarantee that the logically relevant sentences could be decoded at any time. For local sentence writing, humans tend to refer to the context in the original text of the current token or phrase being written, especially keywords, to ensure that the statements are relevant. The idea of our keyword coverage mechanism submodule is inspired by the way of human writing.

Our keyword coverage mechanism is an improvement of the coverage mechanism of *Pointer-Generator Network*. The coverage mechanism punishes the over-attention of the original text to solve the problem of duplication. However, we found that the coverage mechanism does not distinguish the penalty of the attention of the original token. The common sense of writing tells us that summary writing often focuses on a few key sentences or fragments in the original text. Therefore, in some cases, we should support the attention mechanism to focus on certain key segments, especially the keywords in the original text.

First, we use *jieba*[2] to extract a certain number of keywords $\{kw_1, kw_2, ...kw_m\}$ from the text. We maintain a keyword coverage vector c_{kw}^t, which only records the sum of the keyword's attention:

$$c_{kw_j}^t = \sum_{t=0}^{t-1} \alpha_{kw_j}^t \tag{11}$$

However, unlike the coverage mechanism, the keyword coverage mechanism rewards the attention distribution of keywords. Let the set of keywords be K, and the set of other non-keyword be Q. Therefore, we update the loss function formula (7) as follows. The hyperparameter λ and μ are non-negative.

---

[2] https://github.com/fxsjy/jieba.

$$\mathcal{L}=\frac{1}{L}\sum_{t=0}^{L}-\log P(\hat{y}_t)+\lambda\sum_{i\in Q\bigcup K}min(\alpha_i^t,c_i^t)-\mu\sum_{j\in K}min(\alpha_j^t,c_{kw_j}^t)$$

$$=\frac{1}{L}\sum_{t=0}^{L}-\log P(\hat{y}_t)+\lambda\sum_{i\in Q}min(\alpha_i^t,c_i^t)+(\lambda-\mu)\sum_{j\in K}min(\alpha_j^t,c_j^t) \qquad (12)$$

The relative size of hyperparameters λ and μ determines the degree of modulation of keyword attention. Since the penalty degree $\lambda-\mu$ of the loss function for keywords is always less than the penalty λ for non-keywords, the keyword coverage mechanism will achieve a relative reward for the attention distribution of keywords. This relative reward for keyword's attention will help the abstractive model to capture the semantic connection between the current token and the keywords, which promotes the generation of sentence expressions that are more logically relevant to the central idea. Moreover, by adjusting the values of the two hyperparameters reasonably, the function of the coverage mechanism will not be damaged.

4 Experiments

The dataset we use is *CNN/Daily Mail* [6]. Same as the baseline *Pointer-Generator Network*, we use the same scripts [7] to obtain the same version of data. Besides, we use a vocabulary of 50k words for both source and target.

An important purpose of our experiment is to demonstrate the effectiveness of our submodules in solving problems. To avoid additional interference factors, the hyperparameters of the same module as the baseline in our model should be kept consistent with the baseline as much as possible. We use 128-dimensional word embeddings in a randomly initialized form. Instead of using pre-trained word embeddings, we hope to control variable factors of word embeddings to better explain the effect of submodules.

Our model is a unified model. For the abstractive part, we use one 256-dimensional LSTM for the bidirectional encoder and one 256-dimensional LSTM for the decoder. The semantic guide is composed of several 256-dimensional bidirectional LSTM encoder and a linear layer for semantic fusion. Our keyword coverage mechanism is essentially an improvement of the coverage mechanism of *Pointer-Generator Network* [5].

During training and testing, we truncate the text to 400 tokens and limit the length of the summary to 100 tokens. We use mini-batches of size 16. For the integrated extractive semantic guide, we input the original article into the trained multiple extractive approaches to generate an extraction summary sequence $\{text_1, ..., text_d\}$. We limit these extraction summaries to only 3 sentences, and the maximum length is limited to 100 tokens. We also use batch-padding to fill in the sequence. We also tried to extract summaries of different lengths, but the results of 3 sentences were better than other cases. The extractive approaches we use include *Lead-3* [13], *Lexrank* [12], *ILP* [14], etc., which are all trained outside the model.

We have not changed the baseline training strategy. Specifically, we train using Adagrad [15] with a learning rate of 0.15 and an initial accumulator value of 0.1. We train the models on a single GeForce RTX 2060 GPU. Our training consists of two stages. We first train our model without coverage and keyword coverage for about 230K iterations as suggested by baseline. Training is performed for about 12.8 epochs and takes about 2 d and 17 h. To solve the repetition and irrelevant expressions, we then train 3k iterations (about 50 min) with coverage mechanism and keyword coverage mechanism. We set hyperparameters $\lambda = 1$ and $\mu = 1$. We also regard the number of extracted keywords m as a hyperparameter that needs to be considered. In the experiment, we set m to the value in the interval [9,17] The experiment shows that the optimal result can be obtained by setting M to 10 under certain conditions. In the testing phase, we use beam search at test time with a beam-size of 4.

5   Analysis

5.1   Comparison

We conduct our comparative experiments in groups. First, we compare our model with several well-known extractive baselines. We compare the following extractive baselines: *Lextank*, *Concept-ILP*, *Lead-3*. Next, we compare our model with several well-known abstractive models. The methods we compare include *SummaRuNNer*, *GAN*, *DeepRL*, and *PGN+CovMec*. These approaches are all classical methods with certain representativeness.

Our model uses the idea of semantic guidance. Therefore, we think it is necessary to compare with the existing *KIGN-Prediction-guide* [11] model, which uses a different guidance mechanism. Moreover, our model is also a unified model. Therefore, we compare our model with *VHTM* [10] and *unified-inconsistency loss*, which are also unified models.

We don't compare the existing pretraining-based summary models. Our goal is to improve the traditional abstract model based on Seq2Seq and to improve the summary quality as much as possible without adding extra computing resources.

5.2   Results

Table 1 presents the results of our model and the baselines on the *CNN/Daily Mail*. In the experiments, our model achieves the advantages of ROUGE score over the baselines. The experimental results are organized in groups, and the final score of our model is presented in the last row. Compared with the extractive approaches, our model far exceeds the scores of other extractive approaches. These extractive approaches for comparison will be considered for integration into our semantic guide submodule. For the abstractive approaches group, we focus on the comparison with *PGN+CovMec*. Since *PGN+CovMec* is our main baseline model, we mainly add two novel submodules to it to improve the model. Our best model scores exceed *PGN+CovMec* by (+1.16 ROUGE-1, +0.77 ROUGE-2, +0.82 ROUGE-L) points.

Table 1. ROUGE F_1(%) Results on *CNN/Daily Mail*

Summarizer		RG-1	RG-2	RG-L
Extractive	LexRank (ours)	27.87	11.41	17.75
	Concept-ILP [16]	29.1	16.0	26.5
	Lead-3 [13]	39.2	15.7	35.5
	Lead-3 [5]	40.34	17.70	36.57
Abstractive	SummaRuNNer [13]	39.6	16.2	35.3
	GAN [17]	39.92	17.65	36.71
	DeepRL [18]	39.87	15.82	36.90
	PGN+CovMec [5]	39.53	17.28	36.38
Other	KIGN+Prediction-guide [11]	38.95	17.12	35.68
	VHTM [10]	40.57	**18.05**	37.18
	Unified+inconsistency loss [9]	40.68	17.97	37.13
Ablation	PGN+CovMec [5]	39.53	17.28	36.38
	PGN+CovMec+SemGui (ours)	39.99	17.69	36.60
	PGN+CovMec+KeyCov (ours)	40.24	17.75	36.84
	PGN+CovMec+SemGui+KeyCov (ours)	**40.69**	**18.05**	**37.20**

We also compare our model with other summarization models with guide mechanism or unified models. Our model outperforms *KIGN+Prediction-Guide*, which also has a guide. Our model is still sufficiently competitive compared with the excellent unified model *VHTM* and *Unified+inconsistency* model, and each of the three models has its advantages.

Ablation experiments are also conducted to qualitatively analyze the effects of the two submodules on the model performance. The main object of our comparison is baseline *PGN+CovMec*. We find that when only a single submodule is added, it helps the model significantly improve the ROUGE score. More specific, the semantic guide exceeds *PGN+CovMec* by (+0.46 ROUGE-1, +0.41 ROUGE-2, +0.22 ROUGE-L) points and the keyword coverage by (+0.71 ROUGE-1, +0.47 ROUGE-2, +0.46 ROUGE-L) points. When both submodules are added, our model gets the highest ROUGE score. This shows that our two submodules are valid and can work together to achieve the superposition effect (Table 2).

We also conduct a semantic integration ablation experiment of the semantic guide to explore the optimal integration model. We find that the optimal value can be obtained with *Lead-3* and only two extraction methods are integrated. We believe that one of the main reasons is that combining the two extractive summaries containing *Lead-3* can contain the semantics of the key information. If more extractive summaries are used, they will contain overlapping information, which overwhelms the key semantics.

Table 2. ROUGE $F_1(\%)$ results of the ablation experiment of the semantic guide

PGN+CovMec+SemGui+KeyCov	RG-1	RG-2	RG-L
SemGui=lead3	39.94	17.52	36.59
SemGui=lead3+Concept-ILP	40.41	18.01	37.03
SemGui=lead3+lexrank	**40.69**	**18.05**	**37.20**

Table 3. Results of human evaluation

Method	Relevancy	Informativeness	Fluency
PGN+CovMec [5]	3.03	3.24	3.17
Lead-3 [5]	3.27	3.26	3.44
Unified+inconsistency loss [9]	3.45	3.64	3.31
PGN+CovMec+SemGui+KeyCov	3.50	**3.71**	3.42
reference	**3.52**	3.68	**3.46**

5.3   Human Evaluation

To further verify whether our model can improve semantic relevance, we carry out a human evaluation. We focus on three aspects: *relevancy* (The semantic relevancy and faithfulness compared with the original text.), *informativeness* (Has the summary covered important content of the source text?), and *fluency* (The quality of the language.). We sample 100 instances from the *CNN/Daily Mail* test set and employ 10 graduate students to rate each summary. The human evaluators evaluate each summary by scoring the three criteria with 1 (worst) to 5 (best) score. Blind evaluation is adopted in the evaluation process. Finally, we calculate the three average scores of the three criterias for each model.

There are mainly three methods to compare with our best model. *Lead-3* and *PGN+CovMec* are our main baseline models. *Unified+inconsistency loss* is a powerful classical unified model, so the contrast with it is illustrative. From the Table 3, it is found that the reference summaries written by humans have strong advantages in various criteria, especially relevancy and fluency. However, we are delighted to find that by adding semantic guide and keyword coverage submodules to the baseline *PGN+CovMec*, our scores far outdid that of *PGN+CovMec*. To a certain extent, this also proves that our model achieves the desired effect and solves the inaccuracy and irrelevance problem. Further, all the scores of our model are very close to reference's, and our model's score on informativeness exceeds that of reference, which has won first place. This also shows that the abstractive needs to be improved in terms of language relevancy and fluency.

5.4   Case Study

In Fig. 3, with reference to the Gold, we know that the article tells us that Smriti Irani found a camera, sneaking photos of the store's changing room, and

Original Text: new delhi, india police have arrested four employees of a popular indian ethnic-wear chain after a minister spotted a security camera overlooking the changing room of one of its stores. **federal education minister smriti irani was visiting a fabindia outlet in the tourist resort state of goa on friday when she discovered a surveillance camera pointed at the changing room,** police said. **four employees of the store have been arrested,** but its manager -- herself a woman -- was still at large saturday, said goa police superintendent kartik kashyap. state authorities launched their investigation right after irani levied her accusation. they found an overhead camera that the minister had spotted and determined that **it was indeed able to take photos of customers using the store's changing room,** according to kashyap. after the incident, authorities sealed off the store and summoned six top officials from fabindia, he said. the arrested staff have been charged with voyeurism and breach of privacy, according to the police. if convicted, they could spend up to three years in jail, kashyap said. officials from fabindia -- which sells ethnic garments, fabrics and other products -- are heading to goa to work with investigators, according to the company. "fabindia is deeply concerned and shocked at this allegation," the company said in a statement." we are in the process of investigating this internally and will be cooperating fully with the police."
Gold: ederal education minister smriti irani visited a fabindia store in goa, saw cameras. authorities discovered the cameras could capture photos from the store's changing room. the four store workers arrested could spend 3 years each in prison if convicted.
Baseline PGN+CovMec: smriti irani was visiting a fabindia in the tourist resort state of goa. four employees of the store have been arrested, but its manager -- herself a woman -- was still at large saturday.
Our Model: smriti irani was visiting a fabindia outlet in the tourist resort state of goa on friday when she discovered a surveillance camera pointed at the changing room. four employees of the store have been arrested, but its manager -- herself a woman -- was still at large saturday. after the incident, authorities sealed off the store and summoned six top officials from fabindia, he said.

Fig. 3. Typical Comparison. Our model attended at the most important semantic information (blue bold font) matching well with the Gold, but the baseline not. (Color figure online)

then arrested the criminals. The arrest of employees is the result (red font). The reason (blue font) is they use the camera to commit crimes. The reason is key information about this news. Obviously, the baseline fails to capture the reason, which leads us to mistakenly believe that Smriti Irani's visit to the store and the arrest of the employees are two different things. Our model successfully captures the key information of the reason part, making the logic of the abstract clear.

6 Conclusion

In this paper, we propose a unified summarization model based on *Pointer-Generator Network* with semantic guide and keyword coverage mechanism to solve problems of inaccuracy and irrelevance. The semantic guide helps the model capture the central idea of the full text, and the keyword coverage mechanism helps generate more logically relevant local expressions. Experiments on the *CNN/Daily Mail* show that our model outperforms the baselines, and the analysis shows that our model leads to significant improvements.

References

1. Sharma, E., Huang, L., Hu, Z., Wang, L.: An entity-driven framework for abstractive summarization. In: Proceedings of the 2019 Conference on Empirical Methods in Natural Language Processing and the 9th International Joint Conference on Natural Language Processing (EMNLP-IJCNLP), pp. 3278–3289 (2019)
2. Scialom, T., Lamprier, S., Piwowarski, B., Staiano, J.: Answers unite! unsupervised metrics for reinforced summarization models. In: Proceedings of the 2019 Conference on Empirical Methods in Natural Language Processing and the 9th International Joint Conference on Natural Language Processing (EMNLP-IJCNLP), pp. 3237–3247 (2019)
3. Huang, L., Wu, L., Wang, L.: Knowledge graph-augmented abstractive summarization with semantic-driven cloze reward. In: Proceedings of the 58th Annual Meeting of the Association for Computational Linguistics, pp. 5094–5107 (2020)

4. Durmus, E., He, H., Diab, M.: Feqa: a question answering evaluation framework for faithfulness assessment in abstractive summarization. In: Proceedings of the 58th Annual Meeting of the Association for Computational Linguistics, pp. 5055–5070 (2020)

5. See, A., Liu, P.J., Christopher, D.: Manning. Get to the point: summarization with pointer-generator networks. In: Proceedings of the 55th Annual Meeting of the Association for Computational Linguistics (Volume 1: Long Papers), vol. 1, pp. 1073–1083 (2017)

6. Hermann, K.M., et al.: Teaching machines to read and comprehend. In: NIPS'15 Proceedings of the 28th International Conference on Neural Information Processing Systems - Volume 1, vol. 28, pp. 1693–1701 (2015)

7. Nallapati, R., Zhou, W., dos Santos, C.N., Gülçehre, Ç., Xiang, B.: Abstractive text summarization using sequence-to-sequence rnns and beyond. In: Proceedings of The 20th SIGNLL Conference on Computational Natural Language Learning, pp. 280–290 (2016)

8. Chen, Y.-C., Bansal, M.: Fast abstractive summarization with reinforce-selected sentence rewriting. In: ACL 2018: 56th Annual Meeting of the Association for Computational Linguistics, vol. 1, pp. 675–686 (2018)

9. Hsu, W.-T., Lin, C.-K., Lee, M.-Y., Min, K., Tang, J., Sun, M.: A unified model for extractive and abstractive summarization using inconsistency loss. In: ACL 2018: 56th Annual Meeting of the Association for Computational Linguistics, vol. 1, pp. 132–141 (2018)

10. Fu, X., Wang, J., Zhang, J., Wei, J., Yang, Z.: Document summarization with vhtm:variational hierarchical topic-aware mechanism. In: AAAI 2020 : The Thirty-Fourth AAAI Conference on Artificial Intelligence, vol. 34, no. 5, pp. 7740–7747 (2020)

11. Li, C., Xu, W., Li, S., Gao, S.: Guiding generation for abstractive text summarization based on key information guide network. In: Proceedings of the 2018 Conference of the North American Chapter of the Association for Computational Linguistics: Human Language Technologies, vol. 2 (Short Papers), pp. 55–60 (2018)

12. Erkan, G., Radev, D.R.: Lexrank: graph-based lexical centrality as salience in text summarization. J. Artif. Intell. Res. **22**(1), 457–479 (2004)

13. Nallapati, R., Zhai, F., Zhou, B.: Summarunner: a recurrent neural network based sequence model for extractive summarization of documents. In: AAAI, pp. 3075–3081 (2016)

14. Durrett, G., Berg-Kirkpatrick, T., Klein, D.: Learning-based single-document summarization with compression and anaphoricity constraints. In: Proceedings of the 54th Annual Meeting of the Association for Computational Linguistics (Volume 1: Long Papers), vol. 1, pp. 1998–2008 (2016)

15. Duchi, J., Hazan, E., Singer, Y.: Adaptive subgradient methods for online learning and stochastic optimization. J. Mach. Learn. Res. **12**(61), 2121–2159 (2011)

16. Boudin, F., Mougard, H., Favre, B.: Concept-based summarization using integer linear programming: from concept pruning to multiple optimal solutions. In: Proceedings of the 2015 Conference on Empirical Methods in Natural Language Processing, pp. 1914–1918 (2015)

17. Liu, L., Lu, Y., Yang, M., Qu, Q., Zhu, J., Li, H.: Generative adversarial network for abstractive text summarization. In: AAAI, pp. 8109–8110 (2017)

18. Paulus, R., Xiong, C., Socher, R.: A deep reinforced model for abstractive summarization (2018)

Hierarchical Lexicon Embedding Architecture for Chinese Named Entity Recognition

Jiahao Hu[1,2], Yuanxin Ouyang[1(✉)], Chen Li[1,2], Chuanrui Wang[1,2],
Wenge Rong[1], and Zhang Xiong[1]

[1] Engineering Research Center of Advanced Computer Application Technology,
Ministry of Education, Beihang University, Beijing, China
[2] Sino-French Engineer School, Beihang University, Beijing, China
{hujiahao,oyyx,chen.li,buaa_wcr,w.rong,xiongz}@buaa.edu.cn

Abstract. Named entity recognition (NER) is one of the most fundamental tasks in a variety of natural language applications. Due to the lack of delimiters in the Chinese language, Chinese NER task has been suffering from the shortage of word boundary information. Recently, incorporating word information has been proven an effective mechanism to alleviate this problem. However, how to integrate word information into the character-based model more effectively and efficiently is still a challenge. In this work, we propose a hierarchical lexicon embedding architecture for Chinese NER task. The words matched by the input sentence are divided into two categories, i.e., main words and auxiliary words, to help the model better capture useful information. In addition, the modification mainly lies in the embedding layer, as such it can be easily incorporated with different sequence modeling architectures. Experimental studies on four Chinese NER datasets have shown our method's promising potential.

Keywords: Chinese named entity recognition · Lexicon · Boundary information

1 Introduction

Named entity recognition (NER) aims to recognize mentions of rigid designators from unstructured text belonging to predefined semantic types such as person (PER), location (LOC), organization (ORG) etc. [18]. NER not only serves as a key technology for information extraction, but also plays a critical and fundamental role in various natural language processing (NLP) tasks, e.g., question answering [1], machine translation [2], knowledge base construction [6] etc.

Deep learning has been introduced in NER systems, yielding state-of-the-art performance. For example, the Long Short-Term Memory (LSTM) and conditional random field (CRF) based approaches have been demonstrated potential for English NER task [11]. Different from English, Chinese language does

© Springer Nature Switzerland AG 2021
I. Farkaš et al. (Eds.): ICANN 2021, LNCS 12895, pp. 345–356, 2021.
https://doi.org/10.1007/978-3-030-86383-8_28

not have the natural delimiters and then the boundary of Chinese words in an unstructured text is fuzzy, which makes the Chinese NER task more complicated.

Due to the lack of delimiters in the Chinese language, character-based methods for Chinese NER task have been attached much attention in the community [9,13]. But it is still found that the word boundary information plays an important role in Chinese NER task because the named entity boundaries are usually the same as the word boundaries. Therefore how to better take advantage of word information for Chinese NER task is becoming an essential challenge.

An effective way is training a lexicon with embedding for each word on large automatically segmented texts and then find a mechanism to integrate the matched lexicon words information with character information, since the word itself can provide boundary information and the pre-trained word embedding can provide semantic information. For example, Zhang and Yang [24] provided such a lexicon and investigated a lattice-structured LSTM model. This work modifies the LSTM structure to make the character and its matched lexicon words information better integrated into the hidden cell representation.

Currently most models of using lexicon treat all matched words equally. Each character can be matched with multiple words, while many of them are irrelevant to the sentence meaning. The false boundary and semantic information have a negative gain on the overall model effect. For instance, the matched word "市长" (mayor) of character "长" (long) in the sentence "南京市长江大桥" (Nanjing Yangtze River Bridge) is an irrelevant word and might lead to a false prediction.

To overcome this limitation, we proposed a Hierarchical Lexicon Embedding Architecture (HLEA) and two strategies to divide the matched words into main words and auxiliary words. Afterward we combine their representations with character representation to enrich input information. Compared with existing models using lexicon, our model is easier to implement, and the modification in embedding layer ensures that it can be easily incorporated with diverse sequence modeling architectures.

The key contributions of our work can be summarized as follows: 1) We proposed a hierarchical lexicon embedding architecture for incorporating word information into the character-based model for Chinese NER. 2) The proposed model has high transferability and can be easily incorporated with various sequence modeling architectures. 3) We tested our method on four widely used Chinese NER datasets, and achieved competitive results in overall performance, which demonstrate the effectiveness of our method[1].

2   Related Work

Recently deep learning based approaches have become dominant and achieved promising performance in NER task. Most recent works in NER using LSTM-CRF based architecture showed competitive performance [15]. Compared to English NER task, which usually adopts word-based methods, Chinese NER

---

[1] The source code is available at https://github.com/BUAAJustin/HLEA.

task generally adopts character-based methods and outperform statistical word-based methods [9,13], due to the fact that Chinese sentences do not have natural delimiters. However, word information is important indeed. As such a variety of researches extended how to apply word information and character information together for Chinese NER task. For example, Cao et al. [3] introduced an adversarial transfer learning framework to incorporate task-shared word boundary information from Chinese Word Segmentation (CWS) task into Chinese NER.

Another way of taking advantage of word information is introducing external sources like lexicon or gazetteer, which is becoming the mainstream method recently. There are two lines of this kind of approach. One line is to design a dynamic encoding layer to incorporate word information. For example, Zhang and Yang [24] investigated a lattice-structured LSTM model, which is modified to a DAG structure to incorporate word information, though it cannot be trained in batches owing to this special structure. Gui et al. [8] further introduced a Lexicon-based Graph Network (LGN) with global semantics. Each character is regarded as a node and the matched lexicon words form to edges. Similarly, Li et al. [14] proposed a Flat-Lattice-Transformer (FLAT) structure to leverage the parallelization ability, which adopts a new relative position encoding to calculate the attention score.

Another line is to design an adaptive embedding layer to enrich the input information with word information. Generally, this kind of method has higher transferability. For example, Liu et al. [16] proposed WC-LSTM model which used four strategies to solve the parallelization problem of Lattice-LSTM. Ding et al. [5] proposed a multidigraph structure which uses Gated Graph Sequence Neural Networks to model the interaction between characters and words. Ma et al. [17] introduced a simple but effective way which embeds the representation of the matched words according to the character's relative position in the word.

3 Methodology

3.1 Overview

The overall architecture of HLEA is presented in Fig. 1. First, we get the dense representation of each character in the input Chinese sentence, afterwards the main and auxiliary word features corresponding to the character are constructed and concatenated with the character representation. The final input representation is put into the sequence modeling and CRF layer to predict the NER.

3.2 Embedding Layer

A Chinese sentence can be represented as $S = \{s_1, ..., s_n\}$, where s_i is the i-th character of the sentence S. For each character s_i, lexicon words that appear in the sentence and contain the character is called matched words of the character s_i. For example, for the sentence "南京市长江大桥" (Nanjing Yangtze River Bridge), the matched words of character "长" (long) is ["市长" (mayor), "长江"

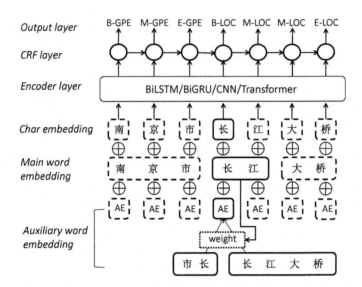

Fig. 1. HLEA Architecture Overview

(Yangtze River), "长江大桥" (Yangtze River Bridge), "长" (long)]. The embedding layer consists of three sublayers, i.e., character representation sublayer, main word representation sublayer and auxiliary word representation sublayer.

Character Representation. Part-of-speech (POS) tagging feature has been proven useful in many NLP tasks [23], thus we also augment the character representation with the POS feature. For a character s_i, we use a dense vector to represent its semantic feature and then concatenate the semantic feature and the POS feature. The augmented character representation x_i^c of character s_i is:

$$x_i^c = [e^c(s_i) \oplus e^{pos}(POS(s_i))] \tag{1}$$

where e^c and e^{pos} denote a pre-trained character embedding lookup table and a POS tag embedding lookup table, respectively. The $\oplus$ is the concatenation operator and the formula $POS(s_i)$ represents the POS tag of character s_i, which is given by a part-of-speech tagger.

Main Word Representation. A Chinese sentence can be expressed as a character-based sequence or a word-based sequence and there is a corresponding relationship between the two expressions. The word to which a character belongs is called the main word corresponding to this character. Formally, given a character-based sentence expression $S = \{s_1, ..., s_n\}$, we suppose the golden word-based sentence expression is $W = \{w_{1,i}, w_{i+1,j}..., w_{k,n}\}$, the mapping from S to W is subjective. For each character s_i, we can always find a corresponding word $w_{j,k}$ to which it belongs with $j \leq i \leq k$, called main word of this character, denoted as mw_i. To get the word-based sentence expression, we propose two strategies:

Segmenter-based strategy. We can use a word segmenter to get the word-based expression directly. However, one drawback of this strategy is that we cannot arbitrarily control the granularity of the word segmenter, as such many words produced by the segmenter would be outside of the lexicon, which is the out-of-vocabulary(OOV) problem. In this case, the effect of this model would be sacrificed.

Frequency-based strategy. To overcome the above drawback, we propose a frequency-based strategy to get the word-based expression. First, we count the occurrence frequency of each lexicon word on training set. Afterwards for a given sentence, we can find those word sequences that meet the following two conditions:

1. The words that can form the original sentence with no character coincidence.

2. The words that have the highest frequency among all the combinations satisfying the 1st condition.

This strategy can ensure that the main words are always part of the matched words so that the OOV problem would not occur. For instance, given the Chinese sentence "南京市长江大桥" (Nanjing Yangtze River Bridge), the word set ["南京" (Nanjing), "市" (city), "长江" (Yangtze River), "大桥" (bridge)] has the higher frequency than the set ["南京" (Nanjing), "市长" (mayor), "江" (river), "大桥" (bridge)], ["南京市" (Nanjing City), "长江大桥" (Yangtze River Bridge)] or any other combinations, so we choose it as the main words of the sentence. Then we can find the main word of each character according to the corresponding relationship between two sentence expressions. In this case, the main word of both character "长" (Long) and "江" (River) is "长江" (Yangtze River).

Once we get the main word mw_i of the character s_i, we can map it to distributed representation. In order to better interact with the character, we apply the relative position of the character in the word as the soft-feature using BMES scheme [22]. The final main word representation x_i^{mw} of character s_i is:

$$x_i^{mw} = [e^w(s_i) \oplus e^{ptn}(PTN(s_i, mw_i))] \tag{2}$$

where e^w and e^{ptn} denote a pre-trained word embedding lookup table and a position embedding lookup table, respectively. The formula $PTN(s_i, mw_i)$ represents the relative position of character s_i in the main word mw_i.

Auxiliary Word Representation. In order to make better use of the information of the rest matched words. We construct a new auxiliary representation to take in the information of auxiliary words. The auxiliary words of character s_i is defined as the matched words excluding the main word, denoted as aw_i. We opt the normalized cosine similarity of each auxiliary word and the main word as the weight of each auxiliary word. Therefore, the auxiliary word representation x_i^{aw} of character s_i can be computed as:

$$\alpha_{ij} = \frac{cosine(e^w(aw_{ij}), e^w(mw_i))}{\sum_j cosine(e^w(aw_{ij}), e^w(mw_i))} \tag{3}$$

$$x_i^{aw} = \sum_j \alpha_{ij} e^w (aw_{ij}) \tag{4}$$

where aw_{ij} is the j-th auxiliary word of character s_i and the formula $cosine(.,.)$ represents the cosine similarity operation.

Combination of Representations. Finally, we simply concatenate these representations as the final augmented representation x_i of character s_i:

$$x_i = [x_i^c \oplus x_i^{mw} \oplus x_i^{aw}] \tag{5}$$

3.3 Sequence Modeling Layer

The input representation containing character and word information are then fed into the sequence modeling layer to capture the dependency between each input. In this work, we choose the bidirectional Long Short-Term Memory network (BiLSTM) as the encoder, but other generic encoders like bidirectional Gate Recurrent Unit network (BiGRU), the Convolutional Neural Network (CNN) or the Transformer [20] are also optional.

3.4 Decoding and Training

Generally, a CRF layer is put at the top of the model to capture the dependency of labels. Given an input representation $H = \{h_1, h_2, ..., h_n\}$, the probability of the label sequence $y = \{y_1, y_2, ..., y_n\}$ is defined as:

$$p(y \mid H) = \frac{e^{s(H,y)}}{\sum_{y' \in Y_h} e^{s(H,y')}} \tag{6}$$

$$s(H, y) = \sum_{i=1}^{n} \left(O_{i,y_i} + T_{y_i, y_{i+1}} \right) + T_{y_0, y_1} \tag{7}$$

$$O = W_o H + b_o \tag{8}$$

where $s(H, y)$ is the score of the label sequence y, and O_{i,y_i} represents the score of the tag y_i of the i-th character h_i. T is the transition score matrix and $T_{i,j}$ denotes the scores of transition from tag i to tag j. Y_h represents all possible label sequences for input H. W_o, b_o and transition matrix T are trainable parameters.

The Viterbi algorithm [7] is adopted to predict the most possible path which gets the highest label sequence score:

$$\bar{y} = \operatorname*{argmax}_{y' \in Y_h} s(H, y') \tag{9}$$

Given a set of the manually annotated training dataset $\{(s_j, y_j)\}|_{j=1}^{N}$, we train the model by minimizing the negative log-likelihood loss:

$$L = -\sum_j \log \left(p \left(\mathbf{y}_j \mid s_j \right) \right) \tag{10}$$

4 Experimental Study

4.1 Datasets

The model was evaluated on OntoNotes[2], MSRA [12], Weibo NER [19] and Chinese resume dataset [24]. Statistics of the four datasets are shown in Table 1.

Table 1. Statistics of Datasets

Datasets	Type	Train	Dev	Test
OntoNotes	Sentence	15.7k	4.3k	4.3k
	Char	491.9k	200.5k	208.1k
MSRA	Sentence	46.4k	-	4.4k
	Char	2169.9k	-	172.6k
Weibo	Sentence	1.4k	0.27k	0.27k
	Char	73.8k	14.5k	14.8k
Resume	Sentence	3.8k	0.46k	0.48k
	Char	124.1k	13.9k	15.1k

4.2 Experimental Configuration

Most parameters followed those of Lattice-LSTM [24], such as the lexicon used, character and word embeddings, LSTM layer number, dropout rate etc. In addition, POS tag and relative position embeddings are randomly initialized, the hidden size was set to 200 and the initial learning rate was set to 0.005. We use Adam [10] to optimize all the trainable parameters. Jieba[3] is adopted as the POS tagger and word segmenter.

4.3 Baselines

We experimented with four different models to verify the effectiveness of our architecture, including baseline model, HLEA-Frequency model, HLEA-Segmenter model and HLEA-Merged model.

For the baseline model, we do not distinguish the matched words into two categories, but use the mean-pooling method to generate the word representation, then concatenate it with the character representation. Suppose that $W_i = \{w_{i1}, w_{i2}, ..., w_{im}\}$ is the matched word set of character s_i, the input representation s_i is computed as:

$$x_i^w = \frac{1}{m} \sum_j e^w(w_{ij}) \tag{11}$$

---

[3] Jieba is a Chinese word segmentation module and can give the POS of each character after segmentation. Please refer to https://github.com/fxsjy/jieba/formore information.

$$x_i = [x_i^c \oplus x_i^w] \tag{12}$$

The HLEA-Frequency and HLEA-Segmenter model use frequency-based and segmenter-based strategies to get the main words, respectively. For HLEA-Merged model, both two main word sets are used by concatenating. Suppose that x_i^{mwf} and x_i^{mws} are the main word representations obtained by frequency-based and segmenter-based strategies respectively, the final input representation of character s_i is:

$$x_i = [x_i^c \oplus x_i^{mwf} \oplus x_i^{mws} \oplus x_i^{aw}] \tag{13}$$

Besides, we also compared our model against some most advanced models using the same lexicon, i.e., Lattice-LSTM [24], LGN [8], SoftLexicon [8] and FLAT [14].

4.4 Results and Analysis

Table 2 shows the results of different models on the OntoNotes dataset. The gold-standard segmentation is available for both training and testing data on this dataset, the results of models using gold-standard segmentation are listed in the first "Gold seg" block. The word-based (LSTM) denotes the BiLSTM-CRF model using the word-based sentence expression. The "Auto seg" means the word-based sentence expression is obtained by a segmenter. All these word-based models in "Gold seg" block achieved good performances due to the correct segmentation and when word segmentation generated by a segmenter was replaced by the gold-standard segmentation, there is a 4.07% improvement in F1 score, which reveals the importance of word boundary information.

Table 2. Main results on OntoNotes

Input	Models	P	R	F1
Gold seg	Word-based (LSTM)	76.66	63.60	69.52
	Yang et al. (2016) [23]	65.59	71.84	68.57
	Che et al. (2013) [4]	**77.71**	72.51	75.02
	Wang et al. (2013) [21]	76.43	72.32	74.32
Auto seg	Word-based (LSTM)	72.84	59.72	65.63
No seg	Character-based (LSTM)	72.84	59.72	65.63
	Lattice-LSTM (2018) [24]	76.35	71.56	73.88
	LGN (2019) [8]	76.13	73.68	74.89
	SoftLexicon (2019) [17]	77.28	74.07	75.64
	FLAT (2020) [14]	-	-	76.45
	Baseline	73.59	69.06	71.25
	HLEA-Segmenter	76.69	74.19	75.42
	HLEA-Frequency	76.80	75.01	75.89
	HLEA-Merged	77.32	**76.94**	**77.13**

The character-based (LSTM) denotes the BiLSTM-CRF model using the character-based sentence expression. When we add the matched words information using the mean-pooling way to the character-based model, which is the situation of our baseline model, the F1 score increases from 65.64% to 71.25%, which can show the significance of incorporating word information into character-based model for Chinese NER task. The other models in the "No seg" block are also character-based and use the lexicon provided by Zhang and Yang [24]. The existing advanced result is given by FLAT [14], which is based on the Transformer architecture.

The results of our models are listed in the second part of the "No seg" block. It is observed that HLEA-Merged model outperforms the existing most advanced model by 0.68% in F1 score, even better than the word-based models using the gold-standard segmentation. HLEA-Frequency model and HLEA-Segmenter model also perform comparably with the most advanced model and gain significant improvement compared to the baseline model.

Table 3 shows the results of different models on these datasets. The results of models using the lexicon are listed in the first block, our results are in the second block. For the MSRA dataset, the existing most advanced result is given by FLAT [14]. Our models achieve a competitive performance compared with other LSTM-based or graph-based models. For the Resume dataset, the existing most advanced result is 95.53% in F1 score given by the SoftLexicon model [17]. Both our HLEA-Frequency and HLEA-Merged model can gain tiny improvement in F1 score compared to the SoftLexicon model. As for the Weibo dataset, the existing most advanced result is 61.42% in F1 score given by the SoftLexicon model [17]. It is observed that all our HLEA-Segmenter, HLEA-Frequency and HLEA-Merged model outperform the SoftLexicon model by 0.52%, 1.15% and 1.46%, respectively.

Table 3. Main results on MSRA, Resume and Weibo

Models	MSRA			Resume			Weibo		
	P	R	F1	P	R	F1	NE	NM	Overall
Word-based (LSTM)	90.57	83.06	86.65	93.72	93.44	93.58	36.02	59.38	47.33
Character-based (LSTM)	90.74	86.96	88.81	93.66	93.31	93.48	46.11	55.29	52.77
Lattice-LSTM [24]	93.57	92.79	93.18	94.81	94.11	94.46	53.04	62.25	58.79
LGN [8]	94.19	92.73	93.46	95.28	95.46	95.37	55.34	64.98	60.21
SoftLexicon [17]	**94.63**	92.70	93.66	95.30	**95.57**	95.53	59.08	62.22	61.42
FLAT [14]	-	-	**94.12**	-	-	95.45	-	-	60.32
Baseline	93.26	91.22	92.23	94.61	94.85	94.73	51.49	60.61	56.12
HLEA-Segmenter	94.13	92.22	93.16	95.84	94.79	95.31	56.52	63.81	61.94
HLEA-Frequency	94.27	92.55	93.40	95.75	95.40	95.57	**59.69**	64.54	62.57
HLEA-Merged	94.11	**93.08**	93.59	**95.93**	95.40	**95.66**	59.50	**65.76**	**62.88**

On the whole, the HLEA-Merged model achieves competitive performance on three of four datasets. The HLEA-Frequency model outperforms the existing most advanced model on two datasets. There is generally a slight improvement from HLEA-Segmenter model to HLEA-Frequency model which imply that the frequency-based strategy is a more suitable choice in finding the main words. All our three models gain significant improvement compared with the baseline model on four datasets, which can prove the effectiveness of the hierarchical architecture.

4.5   Case Study

Table 4 shows a case study about the effect of HLEA-Frequency and HLEA-Segmenter model. In this instance, the granularity of sentence produced by the segmenter is a little large and the word "一汽集团" (A famous automobile group in China, abbreviated as: FAW Group) is not in the matched word set. Lack of main word information lead to the false prediction. However, the frequency-based strategy can produce the correct segmentation without worrying about the OOV problem and finally give the right prediction. We can see from this case that the avoidance of OOV problem results in the slight improvement of HLEA-Frequency model.

Table 4. Case Study

Sentence	一 汽 集 团 投 资 ... FAW Group invests ...					
Gold-standard segmentation	一汽 / 集团 / 投资 ... FAW / Group / invests ...					
Matched words	一汽 (FAW) 集团 (Group) 投资 (invests)					
Frequency-based segmentation	一汽 / 集团 / 投资 ... FAW / Group / invests					
Frequency-based main words	一汽 一汽 集团 集团 投资 投资 FAW FAW Group Group invests invests					
Segmenter-based segmentation	一汽集团 / 投资 ... FAW Group / invests					
Segmenter-based main words	\<unk>\<unk>\<unk>\<unk> 投资 投资 \<unk>\<unk>\<unk>\<unk> invests invests					
Gold label	一 B-ORG	汽 I-ORG	集 I-ORG	团 E-ORG	投 资 O	O
HLEA-Frequency predicted label	一 B-ORG	汽 I-ORG	集 I-ORG	团 E-ORG	投 资 O	O
HLEA-Segmenter predicted label	一 O	汽 O	集 O	团 O	投 资 O	O

5   Conclusion

In this work, we proposed a hierarchical lexicon embedding architecture to incorporate word information into character representation for Chinese NER. We match the input sentence with a lexicon and divide the matched words into main words and auxiliary words. The main words that are more likely to be the correct composition of the sentence are selected out by taking advantage of global information. Then the augmented input representation can be created by concatenating the two types of word features with character representation. The experimental studies on four public Chinese NER datasets have shown that the proposed method can achieve competitive performance compared to the advanced methods.

Acknowledgments. This work was partially supported by the National Key Research and Development Program of China (No. 2018YFB2101502) and the National Natural Science Foundation of China (No. 61977002).

References

1. Aliod, D.M., van Zaanen, M., Smith, D.: Named entity recognition for question answering. In: Proceedings of the 2006 Australasian Language Technology Workshop, pp. 51–58 (2006)
2. Babych, B., Hartley, A.: Improving machine translation quality with automatic named entity recognition. In: Proceedings of the 7th International EAMT workshop on MT and Other Language Technology Tools, Improving MT through other Language Technology Tools, Resource and Tools for Building MT (2003)
3. Cao, P., Chen, Y., Liu, K., Zhao, J., Liu, S.: Adversarial transfer learning for Chinese named entity recognition with self-attention mechanism. In: Proceedings of the 2018 Conference on Empirical Methods in Natural Language Processing, pp. 182–192 (2018)
4. Che, W., Wang, M., Manning, C.D., Liu, T.: Named entity recognition with bilingual constraints. In: Proceedings of the 2013 Conference of the North American Chapter of the Association for Computational Linguistics: Human Language Technologies, pp. 52–62 (2013)
5. Ding, R., Xie, P., Zhang, X., Lu, W., Li, L., Si, L.: A neural multi-digraph model for Chinese NER with gazetteers. In: Proceedings of the 57th Conference of the Association for Computational Linguistics, pp. 1462–1467 (2019)
6. Etzioni, O., et al.: Unsupervised named-entity extraction from the web: an experimental study. Artif. Intell. **165**(1), 91–134 (2005)
7. Forney, G.D.: The Viterbi algorithm. Proc. IEEE **61**(3), 268–278 (1973)
8. Gui, T., et al.: A lexicon-based graph neural network for Chinese NER. In: Proceedings of the 2019 Conference on Empirical Methods in Natural Language Processing and the 9th International Joint Conference on Natural Language Processing, pp. 1040–1050 (2019)
9. He, J., Wang, H.: Chinese named entity recognition and word segmentation based on character. In: Proceedings of the 3rd International Joint Conference on Natural Language Processing, pp. 128–132 (2008)
10. Kingma, D.P., Ba, J.: Adam: a method for stochastic optimization. In: Proceedings of 3rd International Conference on Learning Representations (2015)
11. Lample, G., Ballesteros, M., Subramanian, S., Kawakami, K., Dyer, C.: Neural architectures for named entity recognition. In: Proceedings of the 2016 Conference of the North American Chapter of the Association for Computational Linguistics: Human Language Technologies, pp. 260–270 (2016)

12. Levow, G.: The third international chinese language processing bakeoff: word segmentation and named entity recognition. In: Proceedings of the 5th Workshop on Chinese Language Processing, pp. 108–117 (2006)
13. Li, H., Hagiwara, M., Li, Q., Ji, H.: Comparison of the impact of word segmentation on name tagging for Chinese and Japanese. In: Proceedings of the 9th International Conference on Language Resources and Evaluation, pp. 2532–2536 (2014)
14. Li, X., Yan, H., Qiu, X., Huang, X.: FLAT: Chinese NER using flat-lattice transformer. In: Proceedings of the 58th Annual Meeting of the Association for Computational Linguistics, pp. 6836–6842 (2020)
15. Liu, L., et al.: Empower sequence labeling with task-aware neural language model. In: Proceedings of the 32nd AAAI Conference on Artificial Intelligence, pp. 5253–5260 (2018)
16. Liu, W., Xu, T., Xu, Q., Song, J., Zu, Y.: An encoding strategy based word-character LSTM for Chinese NER. In: Proceedings of the 2019 Conference of the North American Chapter of the Association for Computational Linguistics: Human Language Technologies, pp. 2379–2389 (2019)
17. Ma, R., Peng, M., Zhang, Q., Wei, Z., Huang, X.: Simplify the usage of lexicon in Chinese NER. In: Proceedings of the 58th Annual Meeting of the Association for Computational Linguistics, pp. 5951–5960 (2020)
18. Nadeau, D., Sekine, S.: A survey of named entity recognition and classification. Lingvisticae Investigationes **30**(1), 3–26 (2007)
19. Peng, N., Dredze, M.: Named entity recognition for Chinese social media with jointly trained embeddings. In: Proceedings of the 2015 Conference on Empirical Methods in Natural Language Processing, pp. 548–554 (2015)
20. Vaswani, A., et al.: Attention is all you need. In: Proceedings of 2017 Annual Conference on Neural Information Processing Systems, pp. 5998–6008 (2017)
21. Wang, M., Che, W., Manning, C.D.: Effective bilingual constraints for semi-supervised learning of named entity recognizers. In: Proceedings of the 27th AAAI Conference on Artificial Intelligence, pp. 919–925 (2013)
22. Xu, N.: Chinese word segmentation as character tagging. Int. J. Comput. Linguist. Chin. Lang. Process. **8**(1), 29–48 (2003)
23. Yang, J., Teng, Z., Zhang, M., Zhang, Y.: Combining discrete and neural features for sequence labeling. In: Proceedings of 17th International Conference on Computational Linguistics and Intelligent Text Processing, pp. 140–154 (2016)
24. Zhang, Y., Yang, J.: Chinese NER using lattice LSTM. In: Proceedings of the 56th Annual Meeting of the Association for Computational Linguistics, pp. 1554–1564 (2018)

Evidence Augment for Multiple-Choice Machine Reading Comprehension by Weak Supervision

Dan Luo[1,2], Peng Zhang[1(✉)], Lu Ma[1,2], Xi Zhu[1,2], Meilin Zhou[1,2], Qi Liang[1,2], Bin Wang[3], and Lihong Wang[4]

[1] Institute of Information Engineering Chinese Academy of Sciences, Beijing, China
{luodan,pengzhang,malu,zhoumeilin,liangqi}@iie.ac.cn
[2] School of Cyber Security, University of Chinese Academy of Sciences, Beijing, China
[3] Xiaomi AI Lab, Beijing, China
wangbin11@xiaomi.com
[4] National Computer Network Emergency Response Technical Team/Coordination Center of China, Beijing, China
wlh@isc.org.cn

Abstract. Given a passage and a question, Multiple-choice Machine Reading Comprehension (MRC) requires to select the correct answer from several candidates. Existing methods consider more about the accuracy of the prediction of the final answer in a "black box", which provides no concrete evidence to explain the choice making process. Intuitively, evidence is helpful in building a convincing multiple-choice MRC model. Due to the lack of the golden evidence labels and high cost of manual annotation, we realize weak evidence labels in an automatic way to integrate evidence extraction into the training of MRC models. This auxiliary task learns to select sentences that are more relevant to the question during training, and makes our MRC model interpretable. We come up with an end-to-end model called *(EAM) Evidence Augment Model*, which learns evidence extraction and answer prediction jointly. More accurate results can be obtained by our model without the need of additional manual annotation. Experimental results on RACE datasets show that we obtain an improvement in accuracy over the previous best result based on a fair Pre-trained Language Model (PrLM).

Keywords: Multiple-choice MRC · Evidence extraction · Weak supervision

1 Introduction

Machine reading comprehension (MRC) develops rapidly, and has become the focus of academia and industry. Multiple-choice MRC [1,6,15,16], similar to a human language proficiency test, requires the machine to choose the only correct option from a number of candidates according to a passage. Compared with extractive MRC [5,11,12,20], the answer form of Multiple-choice MRC is

© Springer Nature Switzerland AG 2021
I. Farkaš et al. (Eds.): ICANN 2021, LNCS 12895, pp. 357–368, 2021.
https://doi.org/10.1007/978-3-030-86383-8_29

Passage
s0: Hannah, a 41-year-old lawyer, has done ...
s1: She carries out their most exclusive assignments, enjoying ...
s2: But there is no such thing as a free lunch, even if ...
s3: Hannah says she typically spends two to four hours after each visit writing detailed reports on everything from the quality of the food to specific interactions with staff, whom she always needs to be able to name or describe.
s4: She has to memorize all these details while eating her meal because she cannot openly write anything down.
Q: Which of the following statements is true?
O0: Hanna has to remember details about her dining and write them in reports.
O1: Hanna has to make over 500 restaurant visits per year.
O2: Keep a receipt so Hanna can be reimbursed.
O3: Hanna has to order particular foods assigned by the Mystery Dining Company.

Fig. 1. An example passage with related question and candidates from RACE dataset. The ground-truth answer and the evidentiary sentences in the passage are in red bold. (Color figure online)

no longer limited to the span in the passage so that commonsense reasoning and passage summarization are allowed. Our work focuses on multiple choices, which is more suitable for assessing the ability of machine to understand text.

The common solution [13, 18, 19, 22, 23, 25, 26, 28, 29] to multiple-choice MRC problems can be roughly summarized as a two-step process: 1) using an encoder (e.g. BiLSTM, PrLM) to encode the passage, question and options; 2) modeling the interactions among them. Good results can be achieved by this two-step process, but it directly uses labels for correct answer to update model network parameters ignoring the evidence information. Evidence extraction is useful for finding relevant information and making a rational explanation for multiple-choice MRC (see Fig. 1). The analysis in [25] reveals that the answers of approximately 87% questions on RACE can be narrowed down to two evidentiary sentences. The reason that existing approaches have not taken advantage of evidence information is that the public multiple-choice MRC datasets are lack of the golden evidence annotations. It takes a lot of time and effort to annotate manually, making the supervised evidence extraction training impracticable.

To deal with this problem, we propose a method for automatically annotating weak labels in place of manual annotation. Specifically, we use Sentence-transformers [14] to compute the semantic representation of sentences and select sentences with highest similarity to the question as the evidence. With the weak evidence labels, we design an *Evidence Augment Model (EAM)*, in which evidence extraction is served as an auxiliary task. As shown in the Fig. 2, we build this model in three steps guided by human reading experiences: 1) extensive reading; 2) evidence extraction; 3) intensive reading. With the addition of this auxiliary task, our multiple-choice MRC model will be focused on critical evidence information to predict the answer. It makes our model more interpretable than the previous two-step solution. Our primary contributions are summarized as follows: Firstly, we implement automatic annotation of evidence instead of expensive manual labeling. Sencondly, we propose an end-to-end model in which evidence extraction and answer selection are trained jointly. Finally, we conduct experiments on RACE and achieve accurate scores of 67.5% and 71.0% based on $BERT_{base}$ and $ALBERT_{base}$, respectively. Our model achieves the state-of-the-art result compared with other models on a relatively fair baseline.

2 Related Works

Mainstream methods [3,9,13,18,22,23,26,28,29] for Multiple-choice MRC design complex attention mechanisms among the passage, question and the candidates to model pairwise relationships. Some researchers [19,22,28] designed hierarchical attention to calculate interactions across different granularities. Zhang et al. [26] used explicit syntactic structure to constrain attention mechanisms. Work in [10,13] focused on the comparison between different options. In addition, there were also some attempts [17,27] on constructing MRC models following specific reading strategies, in which only the co-occurrence information of words was used to highlight critical information. Relatively high performance has been achieved, but the decision process and the interpretability of these mechanisms remain unclear. One intuitively explicable method is to select the evidentiary sentences that convince the MRC model to choose the correct answer from the passage.

In general, evidence extraction can be integrated into Multiple-choice MRC tasks in three ways. Firstly, when golden evidence labels are available [1,8], supervised methods can train evidence extraction as the ancillary task of Multiple-choice MRC. Secondly, distant supervision can be applied to generate noisy labels, which further can be denoised by some data programming techniques [21]. Finally, if no evidence labels, a subset of the passage can be selected for each question by the information retrieval method in the process of data preprocessing [24,25]. Large-scare multiple-choice MRC datasets currently available to the public, such as RACE [6] and DREAM [16], are lack of the golden evidence labels. Therefore, the third method is mainly used to make use of evidence information. Unlike humans selecting key sentences using both question and candidates information, previous works [24,25] just used the question to retrieve a list of sentences from the passage. When the question gives too little information to find a useful evidence (see Fig. 1), it is impossible to select the corresponding sentences based solely on the question. Work in [27] put the co-occurrence words of all options after the question, which is insufficient and incomplete.

3 Methods

3.1 Task Formalization and Model Overview

The definition of Multiple-choice MRC task can be formalized as follows: given a triple (P, Q, A), where $P = \{s_1, s_2, \cdots, s_n\}$ is a passage made up of n sentences, Q is a question related with the passage, $A = \{A_1, A_2, \cdots, A_m\}$ is a list of m candidates, the correct answer $A_{correct}$ should be selected from the candidates.

In this work, weak evidence annotation is realized in an automatic way (Sect. 3.2). Inspired by human reading experiences, we implement a model of multi-task learning in three steps as an overview in Fig. 2. Firstly, extensive reading step realizes a quick flick through the whole passage, question and candidates (Sect. 3.3). Secondly, evidence extraction step selects the evidential sentences related to the question and candidates in the given passage (Sect. 3.4). Finally, intensive reading step rereads the key sentences as well as the question and options to choose the answer (Sect. 3.5).

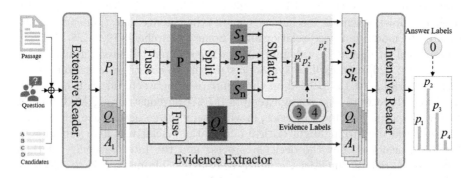

Fig. 2. The overview of EAM. P_i, Q_i indicate the representation of passage and question that incorporates the ith choice information. Evidence labels 3 and 4 are s3 and s4, and answer label 0 is O0 from the example in the Fig. 1. S_j' and S_k' refer to the sentence predicted by EAM as evidence sentences. Dashed arrows represent using annotation labels for supervised training.

3.2   Weakly Standard Evidence Annotation

Existing large-scale public datasets are not labeled with evidence because the annotation process is expensive. Here we use the sentence-transformers framework[1] to select the evidentiary sentences automatically. Sentence-transformers [14] creates sentence embeddings efficiently and effectively that achieve the best performance on many tasks. During the data preprocessing, we split the passage into sentences and splice the question and correct answer option together.

$$\tilde{S}_1, \tilde{S}_2, \cdots, \tilde{S}_n \leftarrow \text{Split}(\tilde{P}) \tag{1}$$

$$Q_A^c = <\tilde{Q}; \tilde{A}_{correct}> \tag{2}$$

where $\tilde{P}$, $\tilde{Q}$, $\tilde{A}_{correct}$ and $\tilde{S}_i$ stand for the input embeddings of the passage, question, the correct answer and each sentence respectively. Q_A^c is the concatenation of $\tilde{Q}$ and $\tilde{A}_{correct}$.

Sentence-transformers is applied to compute dense vector representations for these sentences such that semantically similar sentences are close in the embedding vector space. Cosine similarity is computed between each sentence in passage and the question-answer connection.

$$S_i' = \text{Sentence-Transformers}(\tilde{S}_i), \quad i \in 1, 2, \cdots, n \tag{3}$$

---

[1] https://pypi.org/project/sentence-transformers/.

$$\boldsymbol{Q}_A^* = \text{Sentence-Transformers}(\boldsymbol{Q}_A^c) \tag{4}$$

$$p_i^* = \cos(\boldsymbol{S}_i', \boldsymbol{Q}_A^*), \quad i \in 1, 2, \cdots, n \tag{5}$$

$$y_i^s = \begin{cases} 1, & \text{if } p_i^* \text{ in top } K \\ 0, & \text{others} \end{cases} \tag{6}$$

K sentences that have maximum similarity are selected to annotate weak evidence label, where K is a hyper-parameter. After obtaining the weak evidence labels y_i^s, they are used as weakly supervised learning in an auxiliary task.

3.3   Extensive Reading

Extensive reading plays a crucial role in improving the comprehensive ability. The purpose of extensive reading is to understand the sketchy relationship among the triple (P, Q, A). PrLMs, such as BERT, is taken as the reader to scan the whole sequence. More specifically, the ith candidate and the question are concatenated as a text sequence, and here we call the sequence as $Que\text{-}Opt_i$, which follows the passage separated by special token [SEP]. The input sequence is fed into the reader to obtain the roughly context-interactive representations. Given BERT as the reader, the obtained contextual vectors are the last layer outputs, represented as:

$$\boldsymbol{X}_i = \text{BERT}(< \tilde{\boldsymbol{P}}; \tilde{\boldsymbol{Q}}; \tilde{\boldsymbol{A}}_i >) \tag{7}$$

$$\boldsymbol{P}_i, \boldsymbol{Q}_i, \boldsymbol{A}_i \leftarrow \text{Split}(\boldsymbol{X}_i) \tag{8}$$

where $\tilde{\boldsymbol{A}}_i$ stand for the input embeddings of the ith candidate. $\boldsymbol{P}_i, \boldsymbol{Q}_i$ and $\boldsymbol{A}_i$ are the resulting representations of the extensive reader, $i \in \{1, 2, \cdots, m\}$ refers to the index of the candidates. The $Split(\cdot)$ function splits a vector representation into two or more vector representations. We just use the resulting contextual representations to extract evidence rather than classifying directly.

3.4   Evidence Extraction

Evidence extraction is about figuring out exactly what the question is asking about. For the multiple-choice MRC task, evidentiary sentences can be located more accurately by considering candidates information, which can give some critical information besides the question. We design to select relevant sentences using the syncretic representation of the question and all candidates rather than using the question alone. A fusion function is applied to get the syncretic representation of multiple similar representations. Given m similar vectors, $\boldsymbol{M}_1, \boldsymbol{M}_2, \cdots, \boldsymbol{M}_m$, following alternative functions can be called:

(a) $Fuse_{CNN}$: We first perform a 1-D CNN layer with convolution kernel of size 2, then apply ReLU and average pooling to the output representation.

$$\boldsymbol{M}' = \text{Conv1D}_2([\boldsymbol{M}_1, \boldsymbol{M}_2, \cdots, \boldsymbol{M}_m]) \tag{9}$$

$$\boldsymbol{M}'' = \text{ReLU}(\boldsymbol{M}') \tag{10}$$

$$\boldsymbol{M} = \text{AvgPooling}(\boldsymbol{M}'') \tag{11}$$

(b) $Fuse_{Mean}$: Take the average firstly, and then calculate the self-attention.

$$M' = \text{Mean}([M_1, M_2, \cdots, M_m]) \tag{12}$$

$$M = \text{Self-Attention}(M')$$
$$= \text{softmax}(\frac{M'(M')^T}{\sqrt{d_k}})M' \tag{13}$$

Similar passage representations can be fused by function (a) simply. We use function (b) to fuse the $Que\text{-}Opt$ representations because question and all candidates require further interaction. Sentence representations are got by splitting the fusion representation of passage.

$$P = \text{Fuse}_{\text{CNN}}([P_1, P_2, \cdots, P_m]) \tag{14}$$

$$S_1, S_2, \cdots, S_n \leftarrow \text{Split}(P) \tag{15}$$

$$Q_A = \text{Fuse}_{\text{Mean}}([<Q_1; A_1>, <Q_2; A_2>, \cdots, <Q_m; A_m>]) \tag{16}$$

After obtaining the syncretic representation of the question and all candidates, namely Q_A, semantic matching between it and each sentence representation S_i in the passage is calculated. The specific calculation process is as follows:

$$SM_i = \text{Attention}(S_i, Q_A, Q_A)$$
$$= \text{softmax}(\frac{S_i(Q_A)^T}{\sqrt{d_k}})Q_A \tag{17}$$

$$p_i^s = \text{Self-Attention}(SM_i), \qquad i \in \{1, 2, \cdots, n\} \tag{18}$$

where SM_i is co-attention between each sentence and $Que\text{-}Opt$, p_i^s refers to the probability of correlation between the two. Top k sentences with the highest probability are predicted as evidentiary sentences, where K is the same hyperparameter with the process of weakly standard evidence annotation. We additionally calculate an auxiliary loss based on the weak evidence labels y_i^s.

$$\mathcal{L}_{se} = \frac{1}{N}\sum_{i}^{N}\sum_{i}^{n}(-y_i^s log p_i^s - (1 - y_i^s)log(1 - p_i^s)) \tag{19}$$

where N is the number of all examples in the train dataset, n denotes the number of sentences in each example.

3.5   Intensive Reading

Intensive reading is used on shorter texts in order to grasp more reliable and accurate information. After evidence extraction, K sentences chosen as evidence are concatenated together for predicting the correct answer. Intensive Reader takes evidential sentences, question and candidates as input to further understand how each candidate relates to the question and the evidential sentences.

In detail, the representation of evidential sentences is spliced with the representation of $Que\text{-}Opt_i$ to generate new sequence representations $\boldsymbol{X}'_i$ for each option.

$$\boldsymbol{X}'_i = <\boldsymbol{S}_{i_1}; \cdots \boldsymbol{S}_{i_K}; \boldsymbol{Q}_i; \boldsymbol{A}_i> \tag{20}$$

Our intensive reader is implemented based on the multi-head self-attention. For the i_{th} triple $(P; Q; A_i)$, the output of intensive reader $\boldsymbol{O}_i$ is added with the first token representation of extensive reader, namely $\overline{\boldsymbol{h}}_i$, to give the probability p_i of being correct.

$$head^i_j = \text{Self-Attention}(\boldsymbol{X}'_i), \quad j \in 1, 2, \cdots, h \tag{21}$$

$$\boldsymbol{O}_i = \text{MHA}(\boldsymbol{X}'_i) = \text{Concat}(head^i_1, \cdots, head^i_h) \tag{22}$$

$$p_i = \frac{\exp(\boldsymbol{O}_i + \overline{\boldsymbol{h}}_i)}{\sum_j \exp(\boldsymbol{O}_i + \overline{\boldsymbol{h}}_i)} \tag{23}$$

where h is the number of heads, $MHA(\cdot)$ denotes Multi-Head Attention. We compute the cross-entropy loss between the prediction p_i and the target y_i:

$$\mathcal{L}_{ap} = -\frac{1}{N} \sum_i y_i log p_i \tag{24}$$

The final training objective of our end-to-end model is constructed in two parts:

$$\mathcal{L} = \mathcal{L}_{se} + \mathcal{L}_{ap} \tag{25}$$

4 Experiments

4.1 Datasets

We evaluate our model on two multiple-choice MRC datasets RACE [6] and OneStopQA [1]. Table 1 shows the statistics on the reading comprehension materials.

RACE, whose passages and questions are specifically designed by experts to assess students' reading comprehension, is derived from English tests for middle school and high school students in China. The themes of passages are varied enough to overcome the topic bias phenomenon commonly found in other datasets. RACE is a large-scale MRC dataset that can support deep learning model training.

OneStopQA is a well-built, high-quality dataset designed primarily for testing and analysis. It is made up of 30 articles selected from the Guardian, each with three adaptations at different difficulty levels. Unlike RACE, this dataset is manually labeled with a gold span for each correct answer, but not in sufficient quantities to train a deep learning model. We train the model on RACE and further fine-tune or directly evaluate it on OneStopQA.

Table 1. Data statistics of RACE and OneStopQA. "Passages" is used to show the amount of articles in RACE or paragraphs in OneStopQA. For the data of RACE, the number of passages and questions are shown in the form of Train/Dev/Test and the remaining values are derived from the entire dataset. The term "sent" represents for "sentence" here.

Dataset	RACE		OneStopQA		
	Middle	High	Ele	Int	Adv
Passages	6,409/368/362	18,728/1,021/1,045	162	162	162
Questions	25,421/1,436/1,436	62,445/3,451/3,498	486	486	486
AVG words per passage	232.12	354.08	112.32	126.97	138.6
AVG words per sent	13.99	19.69	20.72	23.53	25.84
AVG sents per passage	16.6	17.99	5.42	5.4	5.36

4.2 Experimental Setups

We use the pretrained model[2] of Sentence-Transformers that is trained on large scare paraphrase data to annotate weak evidence labels automatically. It should be noted that this work is done during the data preprocessing. Our model adopts BERT [4] or ALBERT [7] as the extensive reader, which is implemented based on the public PyTorch implementation from Transformers[3]. Due to the limitation of our experimental computing resources, we could not run large models, but only train base models. In our experiments, the max length of input sequence is set to 512. For the experiment on RACE, we set the initial rate of 2e-5 with batch size 64 when using $ALBERT_{base}$. Initial rate is set to 4e-5 when baseline is $BERT_{base}$. Following the previous works [1,8], we use five-fold cross validation to fine-tune and evaluate our model on OneStopQA. For the fine-tune experiment, we set batch size of 4 with the same initial rate as the training on RACE when using $ALBERT_{base}$.

4.3 Results

Evaluation on RACE. Accuracy is used as the evaluation criterion for multiple-choice MRC. We compare our experimental results on RACE test set with the results of current leading single model in Table 2. For a relatively fair comparison, we contact with the original authors of previous state-of-art works via email to get the performance results on the base model. Meanwhile, we try to re-run them when we fail to get the original data. The performance of our re-run baseline model is better than the Google official result on BERT (66.6% vs. 65.0%) and ALBERT (69.4% vs. 66.8%). In the table, Turkers is the performance on a randomly sampled subset of the test set labeled by Amazon Turkers and Ceiling is the proportion of valid questions which are unambiguous with a correct answer. Here we give the result of our model fine-tuned base on $BERT_{base}$ and

---

[2] paraphrase-distilroberta-base-v1.
[3] https://github.com/huggingface/transformers.

Table 2. Experiment results for single model on RACE test set. * indicates our implementation. † refers to the corresponding results obtained by asking the original author via email.

Model	RACE-M	RACE-H	RACE
Turkers	85.1	69.4	73.3
Ceiling	95.4	94.2	94.5
BERT Track			
$BERT_{base}$ [25]	71.1	62.3	65.0
$BERT_{base}$*	71.7	64.5	66.6
$BERT_{base}$ + OCN [13]	71.6	64.8	66.8
$BERT_{base}$ + DCMN+ [25]	73.2	64.2	67.0
$BERT_{base}$ + SG-Net*	72.2	64.8	66.9
$BERT_{base}$ + EAM	**73.5**	**65.0**	**67.5**
ALBERT Track			
$ALBERT_{base}$ [7]	–	–	66.8
$ALBERT_{base}$*	75.6	66.8	69.4
$ALBERT_{base}$ + SG-Net*	76.5	67.2	69.9
$ALBERT_{base}$ + RekNet†	74.4	68.3	70.1
$ALBERT_{base}$ + EAM	**77.9**	**68.2**	**71.0**

$ALBERT_{base}$, getting the corresponding accuracy of 67.5% and 71.0%. The comparison shows that our model has significantly improved than baseline on the RACE test set and achieves the state-of-the-art result in a fair comparison with other models.

Evaluation on OneStopQA. The performance of the sliding window [15] baseline is reported in the Table 3. The Stanford Attentive Reader (AR) [2] and $ALBERT_{base}$ are set as our benchmark. We train the models on RACE, and evaluate on OneStopQA. We observe that models further fine-tuned on OneStopQA perform better on the models trained only on RACE, with 72.9% and 68.1% average accuracy.

Table 3. Experiment results on OneStopQA set. Comparing with models in "OneStopQA (no finetuning)" which are trained only on Race, models in "OneStopQA" are further fine-tuned on OneStopQA.

Model	OneStopQA (no finetuning)				OneStopQA			
	Ele	Int	Adv	All	Ele	Int	Adv	All
Sliding Window	25.6	26.2	27.5	26.7	27.7	27.2	27.3	28.2
Stanford AR	30.2	30.1	30.1	30.2	34.2	34.3	34.3	34.3
$ALBERT_{base}$	67.7	66.9	65.9	66.8	71.2	72.8	**71.6**	71.9
$ALBERT_{base}$ + EAM	**68.5**	**68.7**	**67.1**	**68.1**	**73.9**	**73.2**	**71.6**	**72.9**

4.4   Ablation Studies

Evidence Extraction Based on Different Information. For MRC task, the common method [24,25] of evidence extraction is using questions only to retrieve sentences from the passage. As the example given in Fig. 1 shows, there is no access to the key information based on the question alone. Here we compare evidence extraction based on question with the result based on *Que-Opt* in Table 4. It is observed that the performance of joint learning is much better than pipeline model meanwhile using *Que-Opt* to extract evidence is more accurate than only relying on question information.

Table 4. Results of evidence extraction based on different information. Pipeline means that evidence extraction and answer prediction are treated as two separate parts. Joint Learning refers to optimize the two task in a unified framework simultaneously.

Model	RACE-M	RACE-H	RACE
Baseline(ALBERT$_{base}$*)	75.6	66.8	69.4
Pipeline			
Pre-Que	45.5	46.1	46.0
Pre-Que-Opt	47.3	47.2	47.3
End-to-End			
EAM w/o $\mathcal{L}_{se}$	76.1	67.0	69.6
Joint Learning			
EAM w/ Que	76.4	67.8	70.3
EAM w/ Que-Opt	**77.9**	**68.2**	**71.0**

Different Number of Evidentiary Sentences to Be Selected. In order to verify analysis in [25], we choose different number of sentences for the experiment. As can be seen from the Table 5, questions in RACE is more accurately answered in three sentences rather than two that is reported in [25].

Table 5. Results of choosing different sentence numbers.

Model	RACE-M	RACE-H	RACE
Baseline(ALBERT$_{base}$*)	75.6	66.8	69.4
EAM of 1 sent	76.8	68.0	70.5
EAM of 2 sent	75.7	67.4	69.8
EAM of 3 sent	**77.9**	**68.2**	**71.0**
EAM of 4 sent	75.9	66.4	69.2

5 Conclusion

On account of the importance of evidence extraction in improving the interpretability of multiple-choice MRC task, this paper devotes itself to design a better integration of evidence extraction into MRC models. In order to reduce the cost of annotation, we apply a third-party tool to automatically calculate semantic similarity to select evidentiary sentences. We propose an *Evidence Augment Model (EAM)* which is an end-to-end model with the evidence extraction as the auxiliary task of MRC. Simulating the human behavioral experience of reading, we implement the model in a three-step framework. Experimental results on RACE dataset show that evidence extraction has a significant impact on the performance and interpretation of multiple-choice MRC task.

Though improvement has been achieved by our model, the quality of the weakly standard evidence labels is heavily depended on the performance of the third-party tool. In consequence, our future work is to denoise the noisy labels with reinforcement learning technique.

References

1. Berzak, Y., Malmaud, J., Levy, R.: Starc: Structured annotations for reading comprehension. arXiv preprint arXiv:2004.14797 (2020)
2. Chen, D., Bolton, J., Manning, C.D.: A thorough examination of the CNN/Daily Mail reading comprehension task. In: Proceedings of the 54th Annual Meeting of the Association for Computational Linguistics (Volume 1: Long Papers), pp. 2358–2367. Association for Computational Linguistics, Berlin, Germany (2016). https://doi.org/10.18653/v1/P16-1223, https://www.aclweb.org/anthology/P16-1223
3. Chen, Z., Cui, Y., Ma, W., Wang, S., Hu, G.: Convolutional spatial attention model for reading comprehension with multiple-choice questions. Proc. AAAI Conf. Artif. Intell. **33**, 6276–6283 (2019)
4. Devlin, J., Chang, M.W., Lee, K., Toutanova, K.: Bert: Pre-training of deep bidirectional transformers for language understanding. arXiv preprint arXiv:1810.04805 (2018)
5. Joshi, M., Choi, E., Weld, D.S., Zettlemoyer, L.: Triviaqa: A large scale distantly supervised challenge dataset for reading comprehension. arXiv preprint arXiv:1705.03551 (2017)
6. Lai, G., Xie, Q., Liu, H., Yang, Y., Hovy, E.: Race: Large-scale reading comprehension dataset from examinations. arXiv preprint arXiv:1704.04683 (2017)
7. Lan, Z., Chen, M., Goodman, S., Gimpel, K., Sharma, P., Soricut, R.: Albert: A lite bert for self-supervised learning of language representations. arXiv preprint arXiv:1909.11942 (2019)
8. Malmaud, J., Levy, R., Berzak, Y.: Bridging information-seeking human gaze and machine reading comprehension. arXiv preprint arXiv:2009.14780 (2020)
9. Miao, H., Liu, R., Gao, S.: A multiple granularity co-reasoning model for multi-choice reading comprehension. In: 2019 International Joint Conference on Neural Networks (IJCNN), pp. 1–7. IEEE (2019)

10. Parikh, S., Sai, A., Nema, P., Khapra, M.: Eliminet: a model for eliminating options for reading comprehension with multiple choice questions. In: Proceedings of the Twenty-Seventh International Joint Conference on Artificial Intelligence, IJCAI-18, pp. 4272–4278. International Joint Conferences on Artificial Intelligence Organization (2018). https://doi.org/10.24963/ijcai.2018/594
11. Rajpurkar, P., Jia, R., Liang, P.: Know what you don't know: Unanswerable questions for squad. arXiv preprint arXiv:1806.03822 (2018)
12. Rajpurkar, P., Zhang, J., Lopyrev, K., Liang, P.: Squad: 100,000+ questions for machine comprehension of text. arXiv preprint arXiv:1606.05250 (2016)
13. Ran, Q., Li, P., Hu, W., Zhou, J.: Option comparison network for multiple-choice reading comprehension. arXiv preprint arXiv:1903.03033 (2019)
14. Reimers, N., Gurevych, I.: Sentence-bert: Sentence embeddings using siamese bert-networks. arXiv preprint arXiv:1908.10084 (2019)
15. Richardson, M., Burges, C.J., Renshaw, E.: Mctest: a challenge dataset for the open-domain machine comprehension of text. In: Proceedings of the 2013 Conference on Empirical Methods in Natural Language Processing, pp. 193–203 (2013)
16. Sun, K., Yu, D., Chen, J., Yu, D., Choi, Y., Cardie, C.: Dream: a challenge data set and models for dialogue-based reading comprehension. Trans. Assoc. Comput. Linguist. **7**, 217–231 (2019)
17. Sun, K., Yu, D., Yu, D., Cardie, C.: Improving machine reading comprehension with general reading strategies. arXiv preprint arXiv:1810.13441 (2018)
18. Tang, M., Cai, J., Zhuo, H.H.: Multi-matching network for multiple choice reading comprehension. Proc. AAAI Conf. Artif. Intell. **33**, 7088–7095 (2019)
19. Tay, Y., Tuan, L.A., Hui, S.C.: Multi-range reasoning for machine comprehension. arXiv preprint arXiv:1803.09074 (2018)
20. Trischler, A., et al.: Newsqa: A machine comprehension dataset. arXiv preprint arXiv:1611.09830 (2016)
21. Wang, H., et al.: Evidence sentence extraction for machine reading comprehension. arXiv preprint arXiv:1902.08852 (2019)
22. Wang, S., Yu, M., Chang, S., Jiang, J.: A co-matching model for multi-choice reading comprehension. arXiv preprint arXiv:1806.04068 (2018)
23. Xu, Y., Liu, J., Gao, J., Shen, Y., Liu, X.: Dynamic fusion networks for machine reading comprehension. arXiv preprint arXiv:1711.04964 (2017)
24. Yu, S., Zhang, H., Jing, W., Jiang, J.: Context modeling with evidence filter for multiple choice question answering. arXiv preprint arXiv:2010.02649 (2020)
25. Zhang, S., Zhao, H., Wu, Y., Zhang, Z., Zhou, X., Zhou, X.: Dcmn+: dual co-matching network for multi-choice reading comprehension. Proc. AAAI Conf. Artif. Intell. **34**, 9563–9570 (2020)
26. Zhang, Z., Wu, Y., Zhou, J., Duan, S., Zhao, H., Wang, R.: Sg-net: syntax-guided machine reading comprehension. Proc. AAAI Conf. Artif. Intell. **34**, 9636–9643 (2020)
27. Zhao, Y., Zhang, Z., Zhao, H.: Reference knowledgeable network for machine reading comprehension. arXiv preprint arXiv:2012.03709 (2020)
28. Zhu, H., Wei, F., Qin, B., Liu, T.: Hierarchical attention flow for multiple-choice reading comprehension. In: Proceedings of the AAAI Conference on Artificial Intelligence, vol. 32 (2018)
29. Zhu, P., Zhao, H., Li, X.: Duma: Reading comprehension with transposition thinking. arXiv: Computation and Language (2020)

Resolving Ambiguity in Hedge Detection by Automatic Generation of Linguistic Rules

Tracy Goodluck Constance[1]([✉]) [iD], Nikesh Bajaj[1] [iD], Marvin Rajwadi[1], Harry Maltby[1], Julie Wall[1] [iD], Mansour Moniri[1] [iD], Chris Woodruff[2], Thea Laird[2], James Laird[2], Cornelius Glackin[3] [iD], and Nigel Cannings[3]

[1] University of East London, London, UK
t.goodluckconstance@uel.ac.uk
[2] Strenuus Ltd., London, UK
[3] Intelligent Voice Ltd., London, UK

Abstract. An understanding of natural language is key in order to robustly extract the linguistic features indicative of deceptive speech. Hedging is a key indicator of deceptive speech as it can indicate a speaker's lack of commitment in a conversation. Hedging is characterised by words and phrases that display a sense of vagueness or that lack precision, such as *suppose, about*. The identification of hedging terms in speech is a challenging task, due to the ambiguity of natural language, as a phrase can have multiple meanings. This paper proposes to automate the process of generating rules for hedge detection in transcripts produced by an automatic speech recognition system using explainable decision tree models trained on syntactic features. We have extracted syntactic features through dependency parsing to capture the grammatical relationship between hedging terms and their surrounding words based on meaning and context. We tested the effectiveness of our model on a dataset of conversational speech, for 75 different hedging terms, and achieved an F1 score of 0.88. The result of our automated process is comparable to existing solutions for hedge detection.

Keywords: Hedge detection · Resolving ambiguity · Linguistic indicators · Linguistic cues

1 Introduction

There is a real need for the use of decision support systems to help assess the credibility of speech, for example in insurance claims. The UK insurance market has felt the impact of a year-on-year surge in insurance fraud, with an increase

Supported by Innovate UK National Project: 'Automation and Transparency across Financial and Legal Services: Mitigating Risk, Enhancing Efficiency and Promoting Customer Retention through the Application of Voice and Emotional AI', Grant no. 104817.

I. Farkaš et al. (Eds.): ICANN 2021, LNCS 12895, pp. 369–380, 2021.
https://doi.org/10.1007/978-3-030-86383-8_30

in the volume of under investigation. There have been an estimated 3.8 million incidents of insurance fraud in the year ending March 2019, resulting in an increase of 17% from the previous year [2]. The use of decision support systems that can reliably identify the deceptive cues in speech to assist contact centre agents is needed [1]. A credible insurance claim applicant would be expected to respond with a smooth flow of information when answering a query. Any deviation from the expected denotes linguistic sensitivity in the narrative, and certain linguistic features can reveal deceptive traits.

The main challenge for hedge detection is to optimise the effectiveness of keyword spotting by eliminating ambiguous terms. It can be misleading to simply identify all hedging terms in a phrase without correctly considering their use in context. For instance, in the first phrase below, **'suppose'** should be identified as a hedging term, but not in the second phrase [14].

Sentence 1: I *suppose* the package will arrive next week.

Sentence 2: I'm *supposed* to call if I'm going to be late.

There is currently no publicly-available corpus for conversational language for hedge detection. As hedge detection is domain dependent, using an existing out-of-domain corpus will not produce significant results [3]. Therefore, as part of this research, we have created a Conversational Dataset to detect Linguistic Markers, namely the CDLM corpus, from movie and TV scripts sourced online and labelled by linguistic experts for both True and False hedging. Annotation of this corpus is an on-going process with a range of different linguistic markers, of which hedging is one of them. The 75 hedging terms used in this paper are a subset of a broad list of hedging terms that were determined by linguistic experts, from which they were able to annotate enough data for 75 hedging terms.

In this study, we propose an automated approach to hedge detection in speech transcripts by performing disambiguation to filter out irrelevant hedge terms in the conversation. We trained syntactic features with explainable decision trees to visualise the working process of the model and to automate the process of generating the rules for hedge detection in order to derive contextual information. Syntactic features are extracted through parsing, also known as syntactic analysis. Parsing exposes the grammatical structure of sentences and how words are related in a sentence, using the knowledge given by the part-of-speech (POS) tag of a word, such as a noun tag or a verb tag. Semantic and syntactic analysis play an important role in the process of dependency parsing as the semantics verify whether the syntactic features generated by the parse tree follow the rule of language [22]. *"The dog crossed the road."* is grammatically correct. But if it were reversed it would be in this form: *"The road crossed the dog"*, which is grammatically correct but semantically incorrect. In Linguistics, it is very easy for humans to make sense of words and understand the grammatical construction and relationship between these words, whereas in computational linguistics, the machine must learn and encode a pattern introduced by rules generated by the process of dependency parsing. Specific syntactic features have been extracted from the CDLM corpus and fed into a decision tree classifier to determine the essential features to select for optimising hedge detection. Decision trees are adaptable for problem-solving due to their transparency in decision-making and specificity in assigning values to an outcome [4].

The rest of this paper is structured as follows: In Sect. 2 we provide a literature review on the state of the art; in Sect. 3 we describe the annotated corpus used in this work and give a detailed description of the methodologies employed; in Sect. 4, we present and discuss the results; finally, we outline our conclusions and plans for future work in Sect. 5.

2   Literature Review

Hedge detection has inspired much work during the last decade. However the work that has been conducted in this field have mostly adopted techniques that are not explainable or constructed using manually crafted rules. The work has principally been undertaken within the Biomedical field using the Computational Natural Language (CoNLL) 2010 framework, which represents a shared task of identifying hedging terms and their linguistic scope in natural language [6]. The shared task tackles it as a sentence labelling classification problem or as a word by word token classification problem which follows an IOB format (Inside, Outside, Beginning) for tagging the scope of the hedging term [7].

The BioScope corpus provided in the framework contains biomedical abstracts and articles annotated with hedging information [8]. Most of the best performing solutions within the CoNLL 2010 shared task framework adopted a sentence labelling approach based on Conditional Random Fields (CRFs) or Support Vector Machine (SVM)-based Hidden Markov Models (HMMs) using the biomedical corpus [7]. The top ranked system for the sentence labelling classification problem adopted commonly used features such as (words, lemmas, POS, and chunks of neighbouring words) and machine learning techniques [7]. A comparative empirical study was completed of four different machine learning approaches to the problem, selecting CRFs, SVMs, K-Nearest Neighbours (KNNs) and Decision Trees. The addition of classification labels for previous words resulted in the SVM classifier yielding the best performance with an F-measure of 0.8682. SVM-based classifiers were employed to tackle the task as a disambiguation problem [9], by using syntactic features to identify hedge terms in and out of context in a sentence, achieving an F-measure of 0.8664, which indicates an improvement of 1.23% in comparison to their initial version of a hedge detection system implemented with a maximum entropy (MaxEnt) classifier [10]. An incremental approach combining the results of a CRF and an SVM-based HMM achieved an F-measure of 0.8636 [11]. A CRF-based and syntactic pattern-based system was also implemented [12], by exploiting the synonym features from WordNet with an F-measure of 0.8632. Using CRFs, a greedy-forward feature selection approach was adopted to boost performance and obtained an F-measure of 0.8589 [13].

Some recent work in hedge detection uses informal language corpora for hedge detection. For example, the first social-media corpus of 326,747 posts was collated from Twitter's API relating to the 2011 London Riots [3]. They achieved an

F-measure of 0.8212 by employing SVMs to explore the effectiveness of using different features such as n-grams, content-based (regrouping similar information, e.g. the geo-location of users in a tweet), user-based (user profiles and followers distributions) and features specific to Twitter. N-grams are a combination of co-occurring words within a phase, for example: "The boy is kind"

- Unigrams are the unique words in a phrase e.g. "The", "boy", "is", "kind"
- Bigram is the combination of two words e.g. "The boy", "boy is", "is kind"
- Trigram is the combination of three words e.g. "The boy is", "boy is kind".

Unfortunately, this annotated corpus is not publicly accessible. Similarly, Amazon's Mechanical Turk was used to annotate a corpus of forum posts from the 2014 Deft Committed Belief Corpora [14]. They annotated the corpus with hedge information and intend in the future to make it publicly available through the Linguistic Data Consortium. They tackled the hedge detection approach as a disambiguation problem, but used a rule-based approach instead of machine learning to manually construct a set of rules to disambiguate potential hedging terms. They adopted hedge detection as a pre-processing technique to extract non-ambiguous hedging features to improve the performance of a committed belief tagging task, another area of research that is closely related to hedging but focuses on the extraction of propositions in the text to determine what the speaker believes [15]. They achieved an F-measure of 0.7270 using the features extracted from their rule-based hedge detection.

3   Methodology

The CDLM corpus was annotated by two behavioural linguistic experts, who manually labeled each spoken utterance with target labels for the identification of hedging terms, both in (True) and out of (False) context. The agreement between the linguistic experts was calculated using Cohen's kappa coefficient, we achieved a substantial inter-annotator agreement of 0.8055. After the experts reviewed the data independently, they came together to re-examine the discrepancies and provided a final decision. The corpus consists of 9,011 spoken utterances, and each utterance consists of one or more hedging terms from a set of 75 unique hedging terms; 59% of the utterances are True hedging, and 41% are False hedging. It is worth mentioning that an utterance labelled as False hedging still contains a hedging keyword, but it is not spoken in the context of hedging. The corpus was split into 70% for the training set and 30% for the testing set, evenly distributed across the sample size of hedging terms.

   The proposed approach is initially based on work by [14], who manually constructed a set of rules for detecting ten hedging terms. In our approach, we implemented a machine learning model using Decision Trees, that leverages context to resolve ambiguities for hedge detection in spoken utterances, i.e. it will automatically detect True hedging in speech. Syntactic features capture the context of the utterance through dependency parsing. Using the CDLM corpus and the corresponding expert labels, a Decision Tree model was trained for each of the 75 hedging terms to generate the rules. The architecture of the Hedge Detection Model in this work can be seen in Fig. 1.

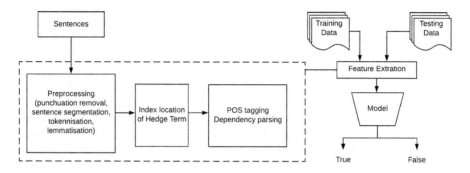

Fig. 1. Architecture of the Hedge Detection Model

3.1 Pre-processing and Feature Extraction

spaCy's English language model was employed for pre-processing and extracting features from the corpus [16]. Punctuation was removed and the spoken utterances were converted to lower-case to simulate the output produced by an automatic speech recognition system before inputting into our Machine Learning model. Each utterance iterates through spaCy's NLP (Natural Language Processing) function, which tokenises the text to generate a Doc object, making it accessible for different tasks downstream in the pipeline. The pipeline holds all the information about the tokens, their linguistic features and their relationships. The Doc object then segments individual sentences and creates a sentence collection called a generator object.

Lemmatisation captures the different variants of the hedge terms in the spoken utterances, regardless of tense, e.g. appear, appears, appeared, appearing, all come under the root form of "appear". It converts a word to its root form by considering the context of the word before transforming it through a prior knowledge POS. spaCy determines the POS tag beforehand and assigns the corresponding lemma, e.g. identifying the base form of "running" to "run". The index location of the hedging term was retrieved in each spoken utterance, facilitating the task of feature extraction and context-based analysis. Dependency parsing was used to extract POS and syntactic dependency tags, explained further in the next section. Three different window sizes (2, 3 and 4) were used, based on the location of the hedging term in the spoken utterance, i.e. neighbouring words that precede and proceed the hedging term. Ultimately, the extracted features are split into training and testing sets to enable the Decision Tree model to learn to select the essential features for automating the rules for hedge detection.

POS tagging assigns tokens in a sentence with their grammatical word categories as POS tags, such as verbs, noun, adjectives, adverbs. The context of a word is determined based on the POS tag. For example, if hedging analysis was based on a Bag of Words approach, the model would not be able to determine the context where 'like' is a verb in the sentence: 'He likes you', and where it is a preposition in the sentence: **'He stood like a statue'**.

A sentence labelled with POS tags, such as: 'Tom tended plants on the roof.' will return: ('Tom', 'PROPN'), ('tended', 'VERB'), ('plants', 'NOUN'),('on', 'ADP'), ('the', 'DET'), ('roof', 'NOUN'), ('.', 'PUNCT') where PROPN is a proper noun, DET a determiner, ADP an adposition and PUNCT is punctuation.

Dependency parsing is a way of generating the syntactic structure of a sentence, also known as syntactic parsing. It generates a parse tree that can capture the relationship between the words in a sentence [17]. Each dependency is comprised of a headword (governor) and its child word (dependent). POS tagging is a precondition for dependency parsing to limit errors. An example of a dependency tree for a spoken utterance that contains the hedging term **'assume'**: **'I assume his train was late'** can be seen in Fig. 2, where the arc connects the headword ('assume') to its' dependent words ('I', 'was'). The arc labels (dependency attributes) describe the syntactic relationship between the hedging term and the words that impact its' context ('nsubj', 'ccomp').

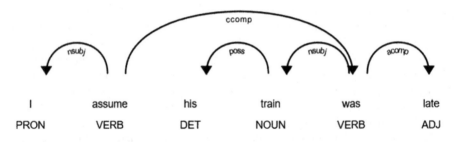

Fig. 2. A dependency tree for a TRUE hedging utterance, hedging term: 'assume'

A set of features were extracted by tagging the location of the hedging term. Hedging terms, both in and out of context, were tagged in order to train the Decision Tree to recognise both True and False hedging. We extracted contextual features, such as POS tags and dependency relation tags concerning the term and its' surrounding words within the range of specific window sizes, determined based on the location (before and after the hedging term).

3.2   Decision Tree Classifier

Decision trees were chosen in this work because of their numerous advantages. They are useful in selecting essential features for making rules and the decisions made by the classifier are easy to understand, interpret and visualise [4]. Focusing on the explainability of the model was one of our core objectives, as a sound decision support system enables an organisation to promote transparency and accountability [5]. The ability to enhance and facilitate the process of providing an audit trail of the overall decision of the system, leads to a model being adopted and trusted, which enables organisations to be consistent and fair. A system with in-built explainablility will be able to verify predictions, identify flaws and biases in the model, fully understand the original problem, and ensure legislation compliance [18]. When we understand how a model works and can visualise how the input features impact the prediction of a specific class in the model, it makes it possible to investigate why the model makes its prediction and repair flaws [19]. Decision trees provide all these benefits as the information flow of the model is visible, and resolving issues causing errors and biases will be less complicated. A disadvantage of Decision Trees is that bias can be introduced into the model. If the model has been trained on an imbalanced dataset, one class will tend to dominate the other. However, this is a common challenge with all machine learning techniques. A significant disadvantage point to the trade-off between the performance of the model and the explainability of its' predictions [19].

We used the SpKit Signal Processing Toolkit to build our Decision Tree model [20]. The main benefit of SpKit is that for text classification it takes string values as input without having to transform the text into a numerical representation in the form of a vector. The Decision Tree model developed here uses the syntactic features to detect the context of the hedging terms in conversational utterances. The input receives these features, and the model produces a classification output of TRUE or FALSE (whether the detected hedging term in each utterance is 'True' hedging or not).

A Decision Tree uses the Iterative Dichotomiser 3 (ID3) algorithm [4], which constructs a tree representation in response to a given classification question. ID3 uses a top-down greedy approach to build a Decision Tree; it determines the best features to split the data by selecting instances with similar values. A Decision Tree comprises a root node, internal nodes, and leaf nodes, representing the different hierarchies of the tree. The decision is made at the leaf node, which is the last level. ID3 uses entropy and information gain to build a Decision Tree. Entropy represents how the data is split based on the homogeneity of a sample (the entropy is zero if the samples are entirely the same and the entropy is one when the samples are equally divided). Information gain depends on the decrease in entropy after a particular feature has split the data, and Decision Trees use the highest information gain to split or construct a tree.

Figure 3 provides an example of a Decision Tree for the hedging term 'suppose' which illustrates the different branches of the nodes. The features selected to generate the rule are visualised. This rule validates by navigating from the top-right node to check if a condition is True. If True, navigation continues, otherwise it checks the next node and repeats the same approach. For this specific case, the rule has two options to identify a true hedging term. Firstly, if the tag of a term is a 'VBP' (verb, non-3rd person singular present), and the word that follows the term has a "to" dependent, then the term is true hedging, otherwise it is not. Secondly, if the tag of a term is not a 'VBP', it will navigate to the left and go to the next node to check if the third word that precedes the hedging term has a "nsubj" (nominal subject) dependent. If true, then the term is true hedging, otherwise it is not. Figure 3 also visualises the rule generated for the hedging term **'assume'**. In this case, the tag features were more important for identifying a hedging term. The rule checks if the tag of a term is a 'VBP' (verb, non-3rd person singular present) and if true, then the term is True hedging, otherwise it is not.

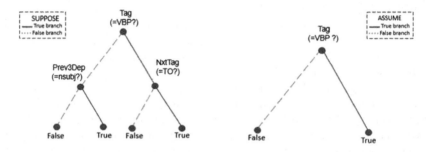

Fig. 3. Example of features selected by a Decision Tree when creating a rule for the hedging terms **'suppose'** and **'assume'**

3.3   XGBoost Classifier

We experimented with XGBoost to investigate its renowned model speed and performance. XGBoost is a popular machine learning method; frequently used on Kaggle for winning competitions as it has outperformed many other gradient tree boosting approaches. XGBoost stands for Extreme Gradient Boosting, and it is a Decision Tree-based ensemble method that uses the gradient descent algorithm for minimising errors in sequential models [21]. This approach performs very well because it boosts weak learners by correcting the residual errors in the sequence.

The performance of the model in this paper can be improved through the many advanced features XGBoost offers for fine-tuning the model. XGBoost only takes numerical values as input, so we had to encode our labels for classification. For this, we used one-hot encoding to transform our features into categorical input variables. Transforming data to one-hot encoding causes data sparsity; however, this was dealt with by an algorithm that XGBoost implements, to handle different types of sparsity patterns in the data. XGBoost produces a final prediction through the output of many trees as it sequentially sums up the predictions of each tree to improve its performance. When training the XGBoost model, we used the configuration recommendation for parameters given by scikit-learn [21]: learning rate = 0.1, n estimators = 100 (number of trees), max depth = 3 (maximum depth of a tree), min_samples_split = 2 (minimum number of samples required to split an internal node), min_samples_leaf = 1 (minimum number of samples required to be at a leaf node), subsample = 1.0 (fraction of features to be randomly sampled for each tree).

4   Results

We carried out two experiments on our Decision Tree classifiers trained on the CDLM corpus. First, we conducted a baseline experiment to compare the effectiveness of the manually constructed rules developed by [14] and our automated rules generated by Decision Trees, for a subset of the CDLM corpus. We then used the full corpus for the second experiment, to compare the performance of the Decision Tree and XGBoost approaches.

For the baseline experiment, we compared our automated approach against the manual rules from [14], for nine out of their ten hedging rules. The reason for only comparing against nine rules, is that we did not have labelled data for the tenth one. Both approaches were evaluated on a test set of 957 sentences in total from the CDLM corpus. Table 1 shows the number of samples for each hedge term, the manually constructed rules, and the accuracy (%) achieved by both approaches against this test set.

The results show an overall accuracy of 61.85% for the manually constructed rules and 85.31% for the automated rules generated by our Decision Trees. The accuracy of the Decision Trees outperformed the manually constructed rules by 23.45%. We observe there is a significant increase in accuracy for each of the hedging terms, except for the term 'rather' where the manual rule outperforms the Decision Tree by 5%. The reason for this could be that it is a simple rule that requires a Regular Expression approach to find a specific pattern, whereas the Decision Tree was trying to learn the appropriate keyword which was true within the context of the sentence based on the example that we have previously discussed, where *'suppose'* is not always a hedging term. We want to be able to train a model that picks up *'suppose'* when it is a true hedging term. The rules created manually did not generalise well with the CDLM corpus; the reason could be the depth of the syntactic structure of some sentences. Our Decision Trees were trained on both short and long spoken utterances, which gave them the ability to extract grammatical relationships between words in longer utterances. Also, machine learning approaches can learn new rules and adapt to changes, unlike a manual, limited, approach.

For our second experiment, we used the full CDLM corpus of 9,011 sentences and 75 hedging terms to compare the performance of the Decision Trees and the XGBoost classifier for True hedge detection, see Table 2. The best window size for the Decision Tree model was two words before and after the hedging term (± 2) with an F1 score of 0.8896. In comparison, the best window size for the XGBoost model was three words before and after the hedging term with an F1 score of 0.8953. Evaluating a range of window sizes contributed to the improvement in both models, indicating that the surrounding words in the spoken utterance give contextual information to the hedging term. XGBoost slightly outperformed the Decision Tree model by 0.0057, even with only limited fine-tuning of the model parameters. Its' strength is due to the boosting algorithm it uses to strengthen weak learners.

Table 1. Classification accuracy (%) for both approaches (Note: *IN* is the POS tag for preposition)

Hedging terms	Sample size	Manually constructed rules [14]	Manual %	DT %
About	218	If token *t* has part-of-speech *IN*, *t* is non-hedge. Otherwise, hedge	75	89
Likely	58	If token *t* has relation *amod* with its head *h*, and *h* has part-of-speech *N∗*, *t* is non-hedge. Otherwise, hedge	41	92
Rather	100	If token *t* is followed by token 'than', *t* is non-hedge. Otherwise, hedge	85	80
Assume	57	If token *t* has *ccomp* dependent, *t* is hedge. Otherwise, non-hedge	79	89
Tend	82	If token *t* has *xcomp* dependent, *t* is hedge. Otherwise, non-hedge	77	80
Appear	38	If token *t* has *xcomp* or *ccomp* dependent, *t* is hedge. Otherwise, nonhedge	58	67
Completely	18	If the head of token *t* has *neg* dependent, *t* is hedge. Otherwise, nonhedge	89	100
Suppose	297	If token *t* has *xcomp* dependent *d* and *d* has *mark* dependent 'to', *t* is non-hedge. Otherwise, hedge	47	93
Should	89	If token *t* has relation *aux* with its head *h*, and *h* has dependent 'have', *t* is non-hedge. Otherwise, hedge	38	89

Table 2. Classification results of the Decision Tree and XGBoost models

Window size	Decision tree			XGBoost		
	Precision	Recall	F1	Precision	Recall	F1
5 (±2)	0.8917	0.901	0.8896	0.8909	0.8998	0.8895
7 (±3)	0.8905	0.9	0.8884	0.8937	0.9064	0.8953
9 (±4)	0.8908	0.8994	0.8885	0.8881	0.9013	0.8901

Our results seem very promising compared to other hedge detection solutions available in different domains. Although it is hard to benchmark hedge detection against similar work due to the lack of publicly available informal language corpora annotated with hedging information. The two closest comparable results are from [14] who achieved an F1 score of 0.727 on their annotated corpus of forum posts, and from [3] who achieved an F1 score of 0.8215 on their annotated corpus based on social media data. Having access to publicly available corpora will provide a stable ground for a fair comparison between different solutions and techniques for hedge detection.

Our approach shows that the performance can be improved by employing more labelled and balanced data, a common theme for machine learning techniques. We employed contextual and positional features to boost the performance of the models, similar to [7,11]. We also followed the approach of performing disambiguation [9], which [14] also adopted to construct their manual rules by filtering ambiguous hedging terms. Decision trees are a successful solution for this automatic generation of rules, as they simulate the human approach of creating rules for decision making and are easy to interpret.

5   Conclusions

Our findings in relation to the automation of generating rules for hedge detection indicate that we can tackle the problem related to ambiguities when detecting linguistic features in conversational language. Extracting the relationship between potential hedging words or phrases and their surrounding words in a spoken utterance provides the context of whether the term is truly hedging or not. We performed dependency parsing to extract these relationships from utterances, which we used as syntactic features to train our models. The core contribution of this work was using the explainable Decision Tree model to automate the process of generating rules for hedge detection disambiguation. We evaluated the effectiveness of our model on a dataset of conversational speech, with decision models for 75 different hedging terms, achieving an F1 score of 0.88, a comparable result to existing solutions for hedge detection.

In future, we will explore this technique by using word embeddings. Our technique can also be applied to the automatic generation of rules for other types of linguistic markers indicative of deception such as unnecessary explanations (*explainers*) and *memory loss*.

References

1. Bajaj, N., et al.: Fraud detection in telephone conversations for financial services using linguistic features. In: 33rd Conference on Neural Information Processing Systems (NeurIPS 2019), AI for Social Good Workshop, Vancouver, Canada (2019)
2. ONS: Nature of fraud and computer misuse in England and Wales. https://www.ons.gov.uk/aboutus/transparencyandgovernance
3. Zhongyu, W., et al.: An empirical study on uncertainty identification in social media context. ACL **2**, 58–62 (2013)
4. Quinlan, J.R.: Decision trees and decision-making. IEEE Trans. Syst. Man Cybern. **20**(2), 339–346 (1990)
5. Rajwadi, M., et al.: Explaining sentiment classification. In: INTERSPEECH (2019)
6. Farkas, R., et al.: The CoNLL-2010 shared task: learning to detect hedges and their scope in natural language text. In: ACL Conference Computational Natural Language Learning - Shared Task, pp. 1–12 (2010)
7. Kang, S.-J., Kang, I.-S., Na, S.-H.: A comparison of classifiers for detecting hedges. In: Kim, T., et al. (eds.) UNESST 2011. CCIS, vol. 264, pp. 251–257. Springer, Heidelberg (2011). https://doi.org/10.1007/978-3-642-27210-3_32

8. Vincze, V., et al.: The BioScope corpus: biomedical texts annotated for uncertainty, negation and their scopes. BMC Bioinform. **9**(11), 1–9 (2008)

9. Velldal, E.: Predicting speculation: a simple disambiguation approach to hedge detection in biomedical literature. J. Biomed. Semant. **2**(55), 57 (2001)

10. Velldal, E., Øvrelid, L., Oepen, S.: Resolving speculation: MaxEnt cue classification and dependency-based scope rules. In: Conference on Computational Natural Language Learning - Shared Task, pp. 48–55 (2010)

11. Tang, B., et al.: A cascade method for detecting hedges and their scope in natural language text. In: ACL Conference on Computational Natural Language Learning - Shared Task, pp. 13–17 (2010)

12. Zhou, H., et al.: Exploiting multi-features to detect hedges and their scope in biomedical texts. In: ACL Conference on Computational Natural Language Learning - Shared Task, pp. 106–113 (2010)

13. Li, X., et al.: Exploiting rich features for detecting hedges and their scope. In: Conference on Computational Natural Language Learning - Shared Task, pp. 78–83 (2010)

14. Ulinski, M., et al.: Using hedge detection to improve committed belief tagging. In: Workshop on Computational Semantics Beyond Events and Roles, pp. 1–5 (2018)

15. Prabhakaran, V., Rambow, O., Diab, M.: Automatic committed belief tagging. In: Coling 2010: Posters, pp. 1014–1022 (2010)

16. Srinivasa-Desikan, B.: Natural Language Processing and Computational Linguistics: A practical guide to text analysis with Python, Gensim, spaCy, and Keras. Packt Publishing Ltd. (2018)

17. Teller, V.: Speech and language processing: an introduction to natural language processing, computational linguistics, and speech recognition. Comput. Linguist. **26**(4), 638–641 (2000)

18. Wachter, S., Mittelstadt, B., Floridi, L.: Transparent, explainable, and accountable AI for robotics. Sci. Robot. **2**(6) (2017)

19. PwC: Explainable AI Driving business value through greater understanding. https://www.pwc.co.uk/audit-assurance/assets/pdf/explainable-artificial-intelligence-xai.pdf

20. Bajaj, N., et al.: SpKit: Signal Processing Toolkit. Python library (2019). https://spkit.github.io

21. Brownlee, J.: XGBoost With Python: Gradient Boosted Trees with XGBoost and scikit-learn (2019)

22. Dozat, T., Manning, C.D.: Simpler but more accurate semantic dependency parsing, arXiv preprint arXiv:1807.01396 (2018)

Text Understanding II

Test Understanding 8

Detecting Scarce Emotions Using BERT and Hyperparameter Optimization

Zahra Rajabi[1]([✉]), Ozlem Uzuner[1], and Amarda Shehu[2]

[1] Department of Information Sciences and Technology, George Mason University,
Fairfax, VA 22030, USA
{zrajabi,ouzuner}@gmu.edu
[2] Department of Computer Science, George Mason University,
Fairfax, VA 22030, USA
amarda@gmu.edu

Abstract. Emotion recognition is a fundamental task in product design and recommendation, online learning, and many other tasks. We have recently broadened and formulated emotion recognition as a multi-label classification problem. Under this formulation, one is confronted with the issue of data imbalance; many emotions are scarce. This paper makes two contributions in this regard. First, it demonstrates that the transformer architecture BERT, which is now considered state of the art for emotion classification, is challenged by data imbalance. Second, it shows that data imbalance can be remedied via specialized loss functions. We investigate two main classes of loss functions, binary cross entropy and focal loss, and within each evaluate the effect of various modifications to address data imbalance. Hyperparameter optimization in a complex hyperparameter space reveals a best model. Our experiments on two benchmark multi-label datasets, GoEmotions and SemEval-EC, show that specialized loss functions significantly improve the performance of transformer models in the presence of highly imbalanced data, further advancing the state of the art in multi-label emotion classification and opening venues for further research.

Keywords: Emotion classification · Data imbalance · BERT · Loss functions · Hyperparameter optimization

1 Introduction

Emotion recognition from text is a fundamental AI task across various online platforms [3]. The increasing volume of short texts (such as tweets) in social media has allowed machine learning (ML) researchers to train increasingly sophisticated ML models for emotion recognition from text. While benchmark

This work is supported in part by a pre-doctoral fellowship to Z. R. by the Center for Advancing Human-Machine Partnerships at George Mason University, Fairfax, VA 22030, USA.

I. Farkaš et al. (Eds.): ICANN 2021, LNCS 12895, pp. 383–395, 2021.
https://doi.org/10.1007/978-3-030-86383-8_31

data now exist, ML research varies in whether emotion recognition is formulated as classification of short text into sentiments (positive, negative, neutral) or actual emotions (varying from few emotions of interest to finer-grained categories), or a regression task, where the goal is to additionally measure sentiment strength or polarity [3]. Across all these formulations, the Bidirectional Encoder Representations from Transformers (BERT) architecture has been shown to yield state-of-the-art models that outperform variations of CNN- and LSTM-based architectures [2].

This paper makes several contributions in emotion classification (EC). We first show that the BERT architecture touted as the most significant advancement in emotion recognition from text is challenged by data imbalance. This becomes increasingly evident when the focus is on capturing fine-grained emotions, as we show here via two benchmark datasets, GoEmotions and SemEval-EC. Both datasets broaden the formulation of emotion classification as a multi-label classification problem. The paper shows that BERT-based models can improve performance even on imbalanced datasets when equipped with specialized loss functions. We investigate here two main, popular classes of loss functions, binary cross entropy and focal loss, and evaluate several modifications to address data imbalance, such as true positive weighting, class balance, and combinations.

More importantly, the various modifications are treated as hyperparameters, and a resulting complex hyperparameter space is explored to reveal the best BERT-based model for data imbalance in EC. Detailed evaluation on the highly-imbalanced GoEmotions [5] and SemEval-EC [12] datasets relate optimal BERT-based models that better address data imbalance and so advance multi-label EC. The rest of this paper continues with a review of related work in Sect. 2. Methodology is described in Sect. 3, and evaluations are related in Sect. 4. The paper concludes in Sect. 5 with a reflection on future research.

2   Related Work

ML literature on EC has traditionally focused on sentiments and addressed binary or ternary classification problems, depending on whether neutral sentiment is considered along with negative and positive sentiments [8]. Notable psychologists have contributed to our understanding that humans are capable of expressing many emotions. Plutchik proposed eight basic emotions, *joy, sadness, fear, anger, anticipation, surprise, disgust,* and *trust* [14]. Ekman, Friesen, and Ellsworth discount anticipation and trust from the basic emotions, using universal facial expressions as the basis for this determination [7]. This disagreement has spilled over to benchmark data; different datasets have different granularities and are annotated with different sets of emotions [4].

There is great diversity in formulations of emotion recognition from text [3]. However, agreement is emerging in what is currently the state of the art [2]. Pre-trained BERT models [6] as a variation of transformer models have shown significant improvement over other models [2]. Unlike traditional word embeddings, such as Word2vec and Glove, which are context-dependent and have been used in Convolutional Neural Net (CNN)-based models for EC (in Kim-CNN [16] and R-CNN-BiLSTM [17]), contextual language models, such as ELMo [13] and BERT, generate word representations by taking into account both positional and contextual information. Recent work [6,9], though limited to sentiment classification, shows that approaches based on ELMo or the Generative Pre-trained Transformer (GPT) [15] do not perform as well as BERT-based approaches. This finding has motivated many of the recent approaches to build over BERT.

However, data imbalance remains a challenge. This is not an issue unique to EC, but it becomes more acute for fine-grained human emotions. Data imbalance manifests itself in many benchmark datasets, and the two we employ here, the GoEmotions and the SemEval-EC datasets, are representative of this issue. This is not surprising; even human annotators are not as reliable in recognizing certain emotions at such granularity (as we show in Sect. 3).

Here we investigate specialized loss functions designed to handle data imbalance rather than rely on data sampling strategies. Strategies, such as oversampling using SMOTE, can increase accuracy on the smaller classes but lower the overall performance. Instead, recent ML work in [1] proposes a class-wise reweighting scheme for most frequently used loss functions (softmax-cross-entropy, focal loss, etc.) to improve performance on data that is highly class imbalanced. We leverage these developments in this paper.

Training a model requires tuning different hyperparameters, such as learning rate and classification threshold; while the learning rate affects the ability of backpropagation to converge fast to a local minimum of the loss function, the classification threshold allows converting class probabilities to labels. As we describe in Sect. 3, specialized loss functions leverage different hyperparameters; by formulating hyperparameter tuning as an optimization problem, we find values that lead to an optimal model in the presence of data imbalance. In this paper, optimality is measured with regards to macro-averaged F1 (macro-F1 for short), and optimization algorithms search for maxima of macro-F1.

Finally, we note that there are different metrics to evaluate an ML model. They are generally built over the two key concepts of precision and recall. F1-score is the harmonic average of precision and recall. Macro-F1 score is the average over the class-specific F1 scores. Weighted-F1 score weighs the contribution of each class based on the proportion of samples in it. However, since weighted-F1 score heavily biases the contribution based on the size of a class, macro-F1 score, which weights each class-specific F1-score equally, is a less-biased metric for imbalanced data. Accuracy is also an appealing metric, though not as sensitive to data imbalance as macro-F1, which is the reason we focus on optimizing macro-F1 in the search for a best BERT-based model for multi-label EC in the presence of data imbalance. We now proceed to relate methodological details.

3   Methodology

We first provide details on the GoEmotions and the SemEval-EC datasets, where we expose the data imbalance issue. The classification setup and the BERT "baseline" model are described next, followed by the specialized loss functions and the hyperparameter optimization formulation that leads to a best model in this paper.

3.1 Datasets

GoEmotions Dataset. The GoEmotions dataset recently released by Google research [5] contains 58K Reddit comments, each annotated with multiple emotions from a set of 27: *admiration, amusement, anger, annoyance, approval, caring, confusion, curiosity, desire, disappointment, disapproval, disgust, embarrassment, excitement, fear, gratitude, grief, joy, love, nervousness, optimism, pride, realization, relief, remorse, sadness,* and *surprise.* It is also possible for a sample to be marked as 'neutral' or 'no emotion' if it carries none of the other 27 emotions. The dataset is split into three subsets: a training set of $43,410$ comments, a testing set of $5,427$ comments, and a validation set of $5,426$ comments. The class distribution for the training set is shown in Fig. 1(a).

SemEval-EC Dataset. In the SemEval-EC dataset [12], each of $10,983$ instances (tweets) is annotated with multiple emotions from a set of 11 based on Ekman's psychological model: the 8 basic emotions of *joy, sadness, fear, anger, anticipation, surprise, disgust,* and 3 additional, complex emotions of *love, optimism,* and *pessimism.* It is also possible for an instance to be annotated as 'neutral' or 'no emotion'. The dataset is split into three: a training set of $6,838$ tweets, a testing set of $3,259$ tweets, and a validation set of 886 tweets. The class distribution for the training set is shown in Fig. 1(b).

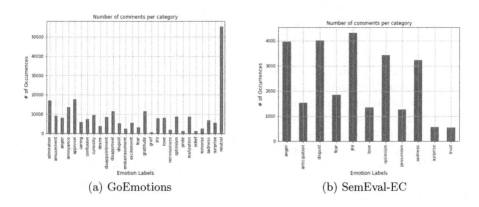

(a) GoEmotions   (b) SemEval-EC

Fig. 1. Class distributions of whole datasets.

3.2 Fine-Tuning BERT Model

BERT for EC: In this paper, we investigate the efficacy of *BERT-Base-Uncased pre-trained model* proposed in [6] for EC. Unlike the sequential RNN and LSTM architectures, BERT was pre-trained on a large unlabeled corpus of texts by jointly conditioning on both left and right directions. Fine-tuning involves plugging in the tasks-specific input and output texts into pre-trained BERT and adjusting all the parameters end-to-end for 2 to 5 epochs on a supervised dataset. In this paper, we fine-tune different BERT models using a variety of loss functions. Our baseline is the fine-tuned BERT model recently presented in [5] (only on GoEmotions dataset), to which we refer as *BERT-tuned*. For fine-tuning BERT, a dense output layer is added on top of the pre-trained model of [6] with sigmoid cross entropy loss. To support multi-label classification and independence of classes, it is a common strategy to employ sigmoid binary cross entropy loss function. We consider the fine-tuned model as our baseline to compute class probabilities at the final layer.

3.3   Strategies to Address Data Imbalance

As summarized in Sect. 2, conventional approaches to address class imbalance use class re-balancing strategies, such as data sampling or cost-sensitive learning, where samples are weighted by inverse class frequency. Oversampling methods, may increase accuracy on smaller classes; however, as the number of samples increases, it becomes highly likely that a newly-added sample is a near-duplicate or overlaps with existing samples. Thus, oversampling strategies often result in over-fitting. Therefore, in order to reduce the impact of class imbalance, we pursue thre following main strategies:

Loss Functions: In order to uncover underrepresented emotions, we further investigate the impact of variations of loss functions and additional hyperparameters as mechanisms to better address class imbalance. We restrict our attention to two popular loss functions, Binary Cross Entropy (BCE) and Focal loss (FL). We recall that BCE loss for instance $n \in [N]$ is defined as $\text{BCE}(y,n) = -y_n \cdot \log \hat{y}_n + (1-y_n) \cdot \log 1 - \hat{y}_n$, where n refers to a particular instance over N instances in the dataset, y_n $in\{0,1\}$ refers to the label, and $\hat{y}_n \in [0,1]$ is the prediction. We note that sigmoid is the only activation function compatible with BCE loss. FL loss is also a cross-entropy loss but weighs the contribution of each sample to the loss [11]; if a sample is already classified correctly, its contribution to the loss decreases. Specifically, FL for instance n is defined as $-(1-p_n)^\gamma \cdot \log p_n$, where p_n is the predicted probability for instance n, and γ controls the attention of the model towards rare classes; $\gamma > 1$ reduces the loss for well-classified instances, where the model predicts probability > 0.5 (conversely, increases loss for hard-to-classify instances, where the model predicts probability < 0.5). When $\gamma = 0$, FL becomes equivalent to cross entropy loss.

True Positive Weighting (TPW): One approach to address class imbalance is to add weights to positive samples for each label individually. TPW is different from class weights and is not calculated based on minority/majority class, but the number of 1s vs 0s within each class. So, instead of dividing each class count by the total number of samples, as opposed to class weights, we obtain the number of 0s divided by the number of 1s for each class to give an inverse impact to positive samples within each class. BCE loss with TPW modification is defined as $\text{BCE} - \text{TPW}(y) = -w_{n,c}[p_c y_n log\sigma(y_n) + (1 - y_n) \cdot log(1 - \sigma(y_i)))]$, where y_n is the label, σ is the predicted probability, c is the class number, and p_c is the positive weight for the positive sample of class c.

Class Balancing: We investigate the impact of adding a class-balance term to a loss function. As introduced in [1], this term is inversely proportional to the *effective number/expected volume*, E_n, of samples. Specifically, $E_n = \frac{1-\beta^{n_i}}{1-\beta}$, where $n_i \in \mathbb{Z}$ is the number of samples in class i and $\beta \in [0, 1)$ is a hyperparameter. This formulation assumes that a new sample interacts with the volume of previously-sampled data in one of two ways, either wholly covered or wholly outside. From now on, we will refer to the class-balance term as *CB*. So, for instance, FL-CB refers to focal loss with the CB modification. Given any loss function L (BCE or FL in our case), class-balance loss can then be written as $L - \text{CB} = \frac{1-\beta}{1-\beta^{n_i}} \cdot L$. It is worth noting that work in [11] introduces an α-based variant of FL-CB, as in $-\alpha_n \cdot (1 - p_n)^\gamma \cdot \log p_n$, where α_n is a parameter. This is equivalent to the above when $\alpha_n = \frac{1-\beta}{1-\beta^{n_i}}$.

3.4 Hyperparameter Optimization

Hyperparameter Space: Unlike work in [1], where the search space is defined over only a few parameters (namely, β and γ), we consider a more general configuration and search over a larger and higher-dimensional hyperparameter space. Specifically, we consider the loss function itself as a categorical hyperparameter (BCE versus FL) and additionally consider four settings: 1-no strategies to address imbalance (the baseline setting), 2- adding TWP only, 3- adding CB only, and 4- adding both TPW and CB, which results in 8 different settings. Under each setting, we posit five hyperparameters. The first two are general; the threshold (THR), which allows converting class prediction probabilities to class labels), and the learning rate (LR) employed during backpropagation. The other three terms are γ, α, and β. In contrast to related literature, which limits γ and β to discrete values, we treat all real-valued hyperparameters as continuous variables that take values within a predefined range (detailed in Sect. 4.

Optimization Objective and Process: We select macro-F1 as the optimization objective, as it is more appropriate for imbalanced data. As our hyperparameters are categorical (the loss functions) and continuous (the real-valued hyperparameters), we define many parallel, independent optimization processes and then combine and compare the results to reveal a best model. We carry out a random search of the hyperparameter space. Specifically, N configurations are sampled from the space, and a BERT-based model is trained over the training dataset with a specific sampled configuration of hyperparameters. This process is carried out with the Tune library in Ray.io API [10]. The library accelerates hyperparameter optimization by providing state-of-the-art search algorithms and sampling methods. It utilizes trial schedulers to terminate bad trials (as a way of expediting search in a vast and high-dimensional space). The library also allows us to carry out several trials in parallel. In the interest of time, we limit the number of trials to $N = 50$ for the SemEval-EC dataset and to $N = 20$ for the larger GoEmotions dataset. The search is carried out over one GPU; therefore, to further lower the computational cost, we implement an *early stopping strategy*. This strategy starts with random search over the hyperparameter space but prevents the model from diverging/straggling in a large continuous search space by periodically pruning low-performing trials. In addition, we make use of the *Asynchronous Hyper Band Scheduler* which enables aggressive early stopping of bad trials and provides better parallelism.

4   Experimental Results

As related in Sect. 3, we fine-tune the pretrained, *BERT-base* model presented first in [6] for our experiments. This BERT-base model is fine-tuned as in [5]. First, a dense output layer is added on top of the pre-trained model for the purpose of fine tuning, with a sigmoid for multi-label classification. The AdamW optimizer is used for 3 epochs, with an initial learning rate of $5e - 5$ and a batch size of 16. The learning rate is linearly increased in a warm-up period from 0 to $5e - 5$ for the first epoch and then linearly decreased to 0. The parameters that resulted in better performance during our empirical analysis are batch_size $= 16$, warmup_steps $= 100$, nr_epochs $= 3$, and THR $= 0.5$.

We will refer to this fune-tuned BERT model as *BERT-tuned* and recall that the loss function in it is BCE loss. In our first set of experiments, we evaluate the BERT-tuned model on the GoEmotions and the SemEval-EC datasets and additionally compare it with previous (SOTA) biLSTM models. In the second set of experiments, we evaluate the impact of loss function variations on the performance of BERT-tuned on both datasets. These results show that several variations allow addressing data imbalance better. Building from these results, in the third set of experiments we relate the findings of the hyperparameter optimization, which considers a broader search space over both loss function variations and other hyperparameters.

4.1   Comparative Performance Analysis

The top row of Table 1 shows the performance of BERT-tuned (with BCE loss) on the GoEmotions and SemEval-EC datasets. On the GoEmotions dataset, where the annotated emotions are more granular (27 emotions), the obtained macro-F1 score is 0.46, which is also what is reported in [5]. For comparison, the SOTA biLSTM model utilized as baseline to compare to BERT in [5] only reaches a macro-F1 score of 0.41. Our biLSTM, presented earlier in [17], also performs similarly to the baseline biLSTM model presented in [6] and significantly underperforms the BERT-tuned model. Specifically, in a head-to-head comparison over the SemEval-EC dataset, the accuracy increases from 0.85 in the biLSTM model to 0.87 in the BERT-tuned model; the micro-F1 score rises from 0.60 in the biLSTM model to 0.68 in BERT-tuned; the macro-F1 score rises from 0.40 in the biLSTM model to 0.49 in BERT-tuned; and the weighted-F1 score rises from 0.56 in the biLSTM model to 0.65 in BERT-tuned.

4.2   Evaluating Variations of Loss Functions in Fine-Tuned BERT

Table 1 shows the impact of the 8 loss function variations (shown in the first column) in the performance of BERT-tuned over test datasets for both the GoEmotions and SemEval-EC datasets. We note that, when FL is used, the additional hyperparameters are set as recommended in literature [2]: namely, $\beta = 0.9999$ and $\gamma = 2.0$. We keep these default hyperparameters to fine-tune BERT models with different loss functions in our preliminary evaluation. The results related in Table 1 show that many of variations of loss functions result in better micro-, macro-, and weighted-F1 scores than BERT-tuned (with BCE loss). Interestingly, either TWP or CB variations do not improve or even detract from the performance of BERT-tuned (with BCE loss). Performance improves when FL loss is considered over BCE loss. Specifically, FL (over BCE) and its variants (FL-TWP, FL-BC, FL-TPW-CB) result in higher macro-F1 values over both datasets (as well as higher micro-F1 and weighted-F1). The gains in performance are significant over the BERT-tuned (with BCE loss) for both datasets. These results suggest that performance can be further improved if one considers loss function variations, and this prompts us to carry out a broader search over the space of possible hyperparameters.

Table 1. BCE refers to BERT-tuned with BCE loss. *-TPW denotes the true positive weights modification to a loss function, *-CB denotes the modification by the class-balance term, and *-TPW-CB denotes the modification by both terms. Highest values along macro-, micro-, and weighted-F1 are highlighted in bold. Abbreviations TrAcc (training accuracy), TsAcc (testing accuracy), mi-F1 (micro-F1), and w-F1 (weighted-F1) are used in the interest of space.

Setting	GoEmotions					SemEval-EC				
	TrAcc	TsAcc	mi-F1	macro-F1	w-F1	TrAcc	TsAcc	mi-F1	macro-F1	w-F1
BCE	0.97	0.97	0.54	0.46	0.50	0.90	0.87	0.68	0.49	0.65
BCE-TPW	0.97	0.97	0.54	0.43	0.50	0.90	0.87	0.68	0.50	0.65
BCE-CB	0.97	0.97	0.54	0.44	0.50	0.81	0.82	0.51	0.24	0.40
BCE-TPW-CB	0.97	0.97	0.55	0.47	0.52	0.89	0.87	0.67	0.53	0.65
FL	0.97	0.97	0.57	0.49	0.55	0.91	0.87	**0.69**	0.54	**0.67**
FL-TPW	0.96	0.97	0.57	0.49	0.54	0.91	0.87	**0.69**	0.54	**0.67**
FL-CB	0.97	0.97	**0.58**	**0.51**	**0.56**	0.90	0.87	0.68	**0.55**	0.66
FL-TPW-CB	0.97	0.97	**0.58**	**0.51**	0.55	0.89	0.87	0.68	0.54	0.66

4.3   Hyperparameter Optimization Results

As related in Sect. 3, we optimize with respect to macro-F1, and the hyperparameter optimization is carried out over a complex search space defined by the eight loss function variations, the learning rate LR, the classification threshold THR, and the γ, α, and β parameters. Namely: the loss function varies over {BCE, FL}; CB and TPW are Boolean variables $\in [True, False]$; LR $\in [2e-5, 5e-5]$; THR $\in [0.3, 0.5]$; $\gamma \in [0.5, 2.0]$; $\alpha \in [0.5, 2.0]$; $\beta \in [0.99, 0.9999]$; and Random_seed (RS) $\in [0, 100]$ is included as an integer-valued parameter to account for possible divergence due to random initialization.

Figure 2 provides a summary view of the model space by showing the macro-F1 scores of models corresponding to different hyperparameter settings explored via random search over the hyperparameter space outlined above. Figure 2 clearly shows that the FL variations result in better models on each of the datasets. In particular, the FL-CB and FL-TPW-CB variations confer the better-performing models. As we detail below, the best/optimal model is obtained by the FL-CB variation.

We compare the best/optimal model found by hyperparameter optimization to what we refer to as a near-optimal model; the latter is indicated by the preliminary analysis related in Table 1. Namely, for the GoEmotions dataset, comparison over loss function variations suggests that FL-CB and FL-TWP-CB result in higher macro-F1 score over other loss function variations. We select BERT-tuned with FL-CB, keeping the default hyperparameters as in BERT-tuned (β, γ, LR, THR), and refer to this model as near-optimal on the GoEmotions dataset. This near-optimal model reaches a micro-F1 score of 0.5775, a macro-F1 score of 0.5091, and a weighted-F1 score of 0.5555 (which are rounded to the second digit after the decimal sign in Table 1. Similarly, a near-optimal model on the SemEval-EC dataset is suggested by Table 1 to result from the variation FL-CB loss. We refer to this model as near-optimal and recall that it reaches a micro-F1 score of 0.6803, a macro-F1 score of 0.5450, and a weighted-F1 score of 0.6646 (rounded to the second digit) shown in Table 1. These near-optimal models are compared to the optimal models found via hyperparameter optimization in Table 2. The optimal models are significantly better; higher macro-F1 scores (in bold font), as well as higher micro- and weighted-F1 scores.

Table 2 provides a class-average view of performance. In Table 3, we show the per-class F1 scores obtained by the near-optimal model and the optimal model (the latter as revealed by hyperparameter optimization) over the GoEmotions dataset (top panel) and the SemEval-EC dataset (bottom panel). Table 3 shows that hyperparameter optimization has been useful in highlighting models that improve not only overall performance in the presence of data imbalance, but also class-specific performance. Comparison with the BERT-tuned model in [5] in Table 3 additionally shows that the optimal model obtained here improves performance in several emotion classes, sometimes doubling macro-F1 score (see 'annoyance').

Table 2. Comparison of near-optimal to optimal models on GoEmotions and SemEval-EC datasets. The optimal model achieves the highest macro-F1 score trained over N models. Table (top) and (middle) show hyperparameter settings for each of near-optimal and optimal models respectively trained on each dataset. For near-optimal hyperparameters are default settings and for optimal model, hyperparameters are sampled from the hyperparameter space described in Sect. 3. Table (down) shows the performance of each of these four models on the test dataset (accuracy and F1 scores.)

Model Parameters	GoEmotions						
	Loss	α	β	γ	LR	THR	RS
Near-optimal	FL-CB	1.0	0.9999	2.0	$5e-5$	0.5	42
Optimal	FL-CB	1.0344	0.9990	0.9082	2.24E-05	0.3648	26

Model Parameters	SemEval-EC						
	Loss	α	β	γ	LR	THR	RS
Near-optimal	FL-CB	1.0	0.9999	2.0	$5e-5$	0.5	42
Optimal	FL-CB-TPW	1.4779	0.9922	1.5682	$3.77e-05$	0.4097	32

Model Performance	GoEmotions				SemEval-EC			
	TsAcc	mi-F1	macro-F1	w-F1	TsAcc	mi-F1	macro-F1	w-F1
Near-optimal	0.9697	0.5776	0.5091	0.5555	0.8701	0.6804	0.5451	0.6646
Optimal	0.9653	0.6096	**0.5475**	0.6016	0.8608	0.7052	**0.6044**	0.7014

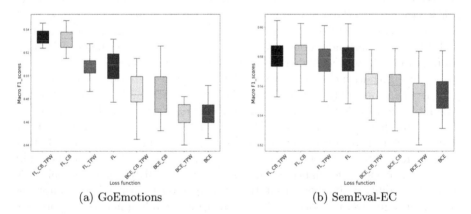

(a) GoEmotions   (b) SemEval-EC

Fig. 2. Macro-F1 scores of models obtained during hyperparameter optimization for (a) GoEmotions and (b) SemEval-EC datasets.

Table 3. Per-class F1 scores achieved by the near-optimal and optimal BERT models over the (top) GoEmotions and (bottom) SemEval-EC dataset.

GoEmotions Emotion	Near-optimal			Optimal			[5]		
	Precision	Recall	F1	Precision	Recall	F1	Precision	Recall	F1
admiration	0.70	0.63	0.67	0.64	0.76	0.70	0.53	0.83	0.65
amusement	0.78	0.84	0.81	0.75	0.91	0.82	0.70	0.94	0.80
anger	0.57	0.43	0.49	0.41	0.57	0.48	0.36	0.66	0.47
annoyance	0.49	0.12	0.19	0.35	0.40	0.37	0.24	0.63	0.34
approval	0.64	0.22	0.33	0.47	0.43	0.45	0.26	0.57	0.36
caring	0.49	0.30	0.38	0.43	0.54	0.48	0.30	0.56	0.39
confusion	0.54	0.35	0.43	0.41	0.51	0.45	0.24	0.76	0.37
curiosity	0.52	0.50	0.51	0.47	0.74	0.57	0.40	0.84	0.54
desire	0.62	0.40	0.49	0.59	0.48	0.53	0.43	0.59	0.49
disappointment	0.56	0.16	0.25	0.36	0.27	0.31	0.19	0.52	0.28
disapproval	0.55	0.26	0.35	0.39	0.47	0.42	0.29	0.61	0.39
disgust	0.58	0.46	0.51	0.48	0.47	0.47	0.34	0.66	0.45
embarrassment	0.64	0.38	0.47	0.62	0.41	0.49	0.39	0.49	0.43
excitement	0.59	0.36	0.45	0.47	0.40	0.43	0.26	0.52	0.34
fear	0.72	0.69	0.71	0.61	0.78	0.69	0.46	0.85	0.60
gratitude	0.94	0.90	0.92	0.91	0.92	0.92	0.79	0.95	0.86
grief	1.00	0.33	0.50	0.50	0.50	0.50	0.00	0.00	0.00
joy	0.66	0.55	0.60	0.59	0.64	0.61	0.39	0.73	0.51
love	0.78	0.80	0.79	0.73	0.88	0.80	0.68	0.92	0.78
nervousness	0.47	0.30	0.37	0.40	0.35	0.37	0.28	0.48	0.35
optimism	0.68	0.45	0.54	0.61	0.58	0.59	0.41	0.69	0.51
pride	0.75	0.38	0.50	0.70	0.44	0.54	0.67	0.25	0.36
realization	0.62	0.14	0.23	0.37	0.18	0.24	0.16	0.29	0.21
relief	0.44	0.36	0.40	0.33	0.45	0.38	0.50	0.09	0.15
remorse	0.61	0.77	0.68	0.57	0.88	0.69	0.53	0.88	0.66
sadness	0.66	0.47	0.55	0.53	0.60	0.56	0.38	0.71	0.49
surprise	0.58	0.48	0.52	0.51	0.56	0.53	0.40	0.66	0.50
neutral	0.72	0.55	0.62	0.63	0.76	0.69	0.56	0.84	0.68
macro-average	0.64	0.45	0.51	0.53	0.57	**0.55**	0.40	0.63	0.46

SemEval-EC Emotion	Near-optimal			Optimal		
	Precision	Recall	F1	Precision	Recall	F1
anger	0.54	0.42	0.48	0.74	0.82	0.78
anticipation	0.41	0.16	0.23	0.36	0.30	0.33
disgust	0.60	0.42	0.50	0.68	0.82	0.74
fear	0.69	0.73	0.71	0.68	0.78	0.73
joy	0.67	0.57	0.61	0.81	0.86	0.84
love	0.78	0.83	0.81	0.57	0.68	0.62
optimism	0.71	0.46	0.56	0.68	0.82	0.74
pessimism	0.46	0.24	0.31	0.40	0.45	0.42
sadness	0.64	0.51	0.56	0.64	0.76	0.69
surprise	0.60	0.48	0.54	0.34	0.17	0.23
trust	0.22	0.03	0.06	0.23	0.14	0.18
macro-average	0.62	0.50	0.55	0.56	0.57	**0.60**

5 Conclusion

In this paper we have shown that BERT and specialized loss functions improve performance in EC and are particularly effective to handle data imbalance. We showed BERT models trained with FL specifically FL-TWP, FL-BC, FL-TPW-CB achieve higher macro-F1 values on both datasets. Hyperparameter optimization is carried out over a complex search space to reveal a best model. Several directions of research remain. Based on progress in computer vision, we speculate that meta-learning may be just as powerful a tool to further improve multi-label and multi-class EC. Ensemble learning is also an interesting direction of future work. Noise, which may be present due to human annotation of datasets, needs additional attention. Integration of data of various modalities may provide a way forward to handle noise. Finally, we anticipate that progress in EC will invariably spur tandem work focusing on regression formulations that can additionally capture the strength of the present emotions.

References

1. Cui, Y., Jia, M., Lin, T.Y., Song, Y., Belongie, S.: Class-balanced loss based on effective number of samples. In: IEEE/CVF Conference on Computer Vision and Pattern Recognition (CVPR), pp. 9268–9277 (2019)
2. Acheampong, F.A., Nunoo-Mensah, H., Chen, W.: Transformer models for text-based emotion detection: a review of BERT-based approaches. Artif. Intell. Rev. 1–41 (2021). https://doi.org/10.1007/s10462-021-09958-2
3. Acheampong, F.A., Wenyu, C., Nunoo-Mensah, H.: Text-based emotion detection: advances, challenges, and opportunities. Eng. Rep. 2(8), e12189 (2020)
4. Bostan, L., Klinger, R.: An analysis of annotated corpora for emotion classification in text. In: International Conference on Computational Linguistics, Santa Fe, Mexico (2014)
5. Demszky, D., Movshovitz-Attias, D., Ko, J., Cowen, A., Nemade, G., Ravi, S.: Goemotions: a dataset of fine-grained emotions (2020)
6. Devlin, J., Chang, M., Lee, K., Toutanova, K.: BERT: pre-training of deep bidirectional transformers for language understanding. In: Burstein, J., Doran, C., Solorio, T. (eds.) NAACL-HLT, pp. 4171–4186. Association for Computational Linguistics (2019)
7. Ekman, P.: An argument for basic emotions. Cogn. Emot. 6(3–4), 169–200 (1992)
8. Gao, W., Sebastiani, F.: Tweet sentiment: from classification to quantification. In: IEEE/ACM International Conference on Advances in Social Networks Analysis and Mining (ASONAM), pp. 97–104 (2015)
9. Huang, Y., Lee, S., Ma, M., Chen, Y., Yu, Y., Chen, Y.: Emotionx-idea: emotion BERT - an affectional model for conversation. arXiv preprint arXiv:1908.06264 (2019)
10. Liaw, R., Liang, E., Nishihara, R., Moritz, P., Gonzalez, J.E., Stoica, I.: Tune: a research platform for distributed model selection and training. arXiv preprint arXiv:1807.05118 (2018)
11. Lin, T., Goyal, P., Girshick, R.B., He, K., Dollár, P.: Focal loss for dense object detection. arXiv preprint arXiv:1708.02002 (2017)

12. Mohammad, S., Bravo-Marquez, F., Salameh, M., Kiritchenko, S.: SemEval-2018 task 1: affect in tweets. In: International Workshop on Semantic Evaluation, New Orleans, Louisiana, pp. 1–17. Association for Computational Linguistics (2018)
13. Peters, M.E., et al.: Deep contextualized word representations (2018)
14. Plutchik, R.: Emotion: A Psychoevolutionary Synthesis. Harper & Row (1980)
15. Radford, A.: Improving language understanding by generative pre-training (2018)
16. Yoon, J., Kim, H.: Multi-channel lexicon integrated CNN-BILSTM models for sentiment analysis. In: Conference on Computational Linguistics and Speech Processing (ROCLING), Taipei, Taiwan (2017)
17. Rajabi, Z., Shehu, A., Uzuner, O.: A multi-channel BiLSTM-CNN model for multilabel emotion classification of informal text. In: 2020 IEEE 14th International Conference on Semantic Computing (ICSC), pp. 303–306 (2020)

Design and Evaluation of Deep Learning Models for Real-Time Credibility Assessment in Twitter

Marc-André Kaufhold$^{(\boxtimes)}$, Markus Bayer, Daniel Hartung, and Christian Reuter

Science and Technology for Peace and Security (PEASEC),
Technical University of Darmstadt, Darmstadt, Germany
{kaufhold,bayer,reuter}@peasec.tu-darmstadt.de,
daniel.hartung@student.tu-darmstadt.de

Abstract. Social media have an enormous impact on modern life but are prone to the dissemination of false information. In several domains, such as crisis management or political communication, it is of utmost importance to detect false and to promote credible information. Although educational measures might help individuals to detect false information, the sheer volume of social big data, which sometimes need to be analysed under time-critical constraints, calls for automated and (near) real-time assessment methods. Hence, this paper reviews existing approaches before designing and evaluating three deep learning models (MLP, RNN, BERT) for real-time credibility assessment using the example of Twitter posts. While our BERT implementation achieved best results with an accuracy of up to 87.07% and an F1 score of 0.8764 when using meta-data, text, and user features, MLP and RNN showed lower classification quality but better performance for real-time application. Furthermore, the paper contributes with a novel dataset for credibility assessment.

Keywords: Credibility assessment · Social media · Neural networks · Deep learning

1 Introduction

Social media are an integral part of modern everyday life as they allow the creation and exchange of user-generated content. Besides everyday life, social media are used by journalists for reporting, analysing, and collecting information, by organisations to monitor customer feedback and sentiment, but also by citizens and emergency services to gain situational awareness in conflicts and disasters [18]. On the contrary, social media is prone to the dissemination of (potentially) false information, including conspiracy theories, fake news, misinformation, or rumors [35]. While counter-measures such as gatekeeping information, increasing media literacy, or passing new laws seem to be promising approaches [17], the sheer volume of *big social data*, which sometimes needs to be analysed under time-critical constraints, calls for automated and (near) real-time

© Springer Nature Switzerland AG 2021
I. Farkaš et al. (Eds.): ICANN 2021, LNCS 12895, pp. 396–408, 2021.
https://doi.org/10.1007/978-3-030-86383-8_32

credibility assessment methods. Thus, a multitude of different machine learning approaches were established to automatically distinguish false and credible information [27,34,38]. Despite their merits, when reviewing existing deep learning approaches for credibility assessment in social media, we found that most approaches provided binary or multi-class models but did not allow a steady (e.g., percentage) prediction of credibility. Furthermore, most approaches require extensive computations, thus lacking the ability for real-time application in social media, and, to our best knowledge, none of these approaches incorporated previous posts of the user into their analysis. Thus, the paper seeks to answer the following research question: **Which deep learning models and parameters are suitable for real-time credibility assessment in Twitter?**

By answering this research question, the paper makes several contributions. It (i) conducts a review of existing credibility assessment methods (Sect. 2), (ii) presents the design and finetuning of three deep learning models for credibility assessment, (iii) provides a novel dataset for credibility assessment in Twitter (Sect. 3), and (iv) evaluates the quality of the designed models, also examining the usefulness of incorporating previous user posts into credibility assessment (Sect. 4). The paper finishes with a discussion of the findings and implications and highlights possible limitations and potential for future work (Sect. 5).

2 Related Work

Since the study of *credibility* is highly interdisciplinary, there is no universal definition for it [8]. However, it can be understood as a measure which comprises both objective (e.g., useful, good, relevant, reliable, accurate) and subjective (e.g., a perception of the receiver) components. Credible information is characterized by trustworthiness (unbiased, true, good purpose) and expertise (competence, experience, knowledge) [10]. When estimating the credibility of information in social media, users are confronted with different types of harmful information [35] that can be distinguished by the *intention of the publisher* (i.e., intentional or non-intentional) and the *truth of content* (i.e., true or false) [8]. Both disinformation and misinformation are objectively false, but only disinformation (often referred to as fake news [21]) is published intentionally false. Moreover, rumors are statements that cannot be immediately verified as either true or false [28].

Amongst others, harmful information is disseminated to manipulate political elections and public opinions or to generate financial revenues [1]. Moreover, false information might affect the decision making of emergency services in conflicts or disasters, effectively contributing to the loss of lives. Countermeasures against harmful information comprise the gatekeeping of information by media, increasing the media literacy of citizens, passing new laws and regulations, or detecting harmful information via algorithmic detection approaches [17]. When reviewing literature on credibility assessment in social media (Table 1), we did not only find individual systems but also interesting survey papers comparing different machine learning approaches [27,34,38]. The approaches are primarily based on Twitter data, often attributed to different *domains*, including credibility, fake news, or rumors, and have a different *scope of analysis*.

First, event-based approaches cluster social media messages into events to determine the credibility of the event [2–4,9,12,40]. Second, propagation-based approaches analyse the caused engagement, such as mentions or retweets, of published messages [23,30,31]. Third, message-based approaches assess the credibility of individual messages, using metadata and textual features [11,13,15,28,39]. Especially *methods* based on neural networks achieved high classification performances, e.g., accuracies of 85.20% using NN [15] or 89.20% [31] using RNN and NN. Despite the variety of features involved, to our best knowledge, none of these approaches incorporated previously published messages of the user into their analysis. Further, only two of the approaches allow a near *real-time* application [23,39], i.e., being able to classify tweets directly after their dissemination. This is the case because most approaches are event-based, requiring event detection before classification can take place, or rely on temporal features, such as the number of likes or retweets, which change over the course of time and could lead to a flawed credibility score at retrieval.

In terms of *output*, most approaches allow a binary (i.e., credible or incredible information) or multi-class credibility assessment [11,23,39], although some works outlined that they do not reproduce reality in a sufficient manner [4,5]. Thus, the use of a steady regression seems promising [28] since it allows a percentage-based representation of credibility and better accounts for the subjective component of credibility [8]. Although multiple attempts have been made to establish standard datasets for credibility assessment [25,33,41], almost all publications used their own dataset, probably due to the methodological requirements of their approaches. The lack of standardized datasets is noticed by diverse authors, emphasizing the lack of comparability of the evaluation results of different approaches [27,34,38].

Table 1. Comparison of ML classifiers. Used methods are marked **bold**, sometimes not all methods are listed (*). Abbrev.: Decision Tree (DT), Decision Rule (DR), Bayesian Network (BN), Bayes Classifier (BC), Support Vector Machine (SVM), Random Forrest (RF), Convolutional/Recursive Neural Network (C/R/NN), Naive Bayes (NB).

Ref.	Domain	Scope	Output	Realtime	Methods
[3]	Credibility	Event	binary	no	**DT**, DR, BN, SVM
[4]	Credibility	Event	binary	no	**RF**, LR, *
[2]	Fake News	Event	binary	no	**RF**, *
[15]	Credibility	Message	binary	untested	**NN**
[30]	Astroturfing	Mem	binary	no	**DT**, SVM
[11]	Credibility	Message	5 classes	untested	**SVM-rank**, *
[40]	Crisis Credibility	Event	binary	untested	**SVM**, BN, DT
[12]	Credibility	Event	tertiary	no	**SVM**, DT, NB, RF
[28]	Rumors	Event	steadily	no	**BC**
[13]	Fake News	Message, Source	binary	untested	**NN**, NB, DT, SVM, RF
[31]	Fake News	Event	binary	no	**RNN, NN**
[23]	Fake News	Message	4 classes	almost	**RNN, CNN**
[39]	False Information	Message	5 classes	yes	**RNN, CNN**

3   Concept and Implementation

Based on the outlined research gaps, we seek to implement neural network-based approaches for credibility assessment that (i) work with public Twitter data due to the ease of access, (ii) use regression to allow a steadily (i.e., percentage-based) assessment of credibility, and (iii) allow a near real-time application of the trained models. Furthermore, we intend to (iv) check if the analysis of previously published messages of a user positively impacts the performance of credibility assessment. We also (v) compose a novel dataset for credibility assessment.

3.1   Features and Model

In order to train our models, we reviewed the features used by previous approaches. The used features can be roughly categorized into four types: (i) **metadata features** $(n = 10)$ of a tweet provided by the Twitter API, such as hashtags, links, or mentions, (ii) computationally extracted **text features** $(n = 25)$ from the tweet's body, such as number of words, text length, or sentiment, (iii) **user features** $(n = 17)$ provided by the Twitter API, such as the number of followers or published tweets, and (iv) **timeline features** $(n = 140)$, including the maximum, minimum, arithmetic mean, and standard deviation (4 *) of both the metadata and text features $(10 + 25)$ of the last 40 tweets of the user.

First, the baseline model is a simple **multilayer perceptron (MLP)** that consists of an input layer with 192 neurons for the features described before. These are projected into a hidden layer with 32 neurons with tanh activations. Since the problem to solve is a regression task, a sigmoid function was selected for the output and the entire network is trained with mean square error (MSE). The layers are fully connected with a dropout rate of 0.3. Further hyperparameters of the learning process are a learning rate of 0.01, a batch size of 256, and the maximum number of epochs of 10,000, which is contained by early stopping on the development set. We choose ADAM [20] as optimizer.

Second, to extend the baseline, we embed the sentiment and textual content of a tweet with a **recurrent neural network (RNN)** and feed those tweet embeddings into the baseline model. As a first step, GloVe [26] pretrained Twitter embeddings (dimension of 50) are utilized to create word embeddings of each word in the tweet. These word embeddings are enriched by another dimension that represents the VADER [14] sentiment value. Every embedding is then processed by a RNN that produces a hidden state (tweet embedding) that serves as another input into the baseline MLP.

Third, another approach to extend the baseline is to use finetuned **BERT embeddings** [6]. We replaced user mentions, URLs, and emoticons with special tokens. Then we finetuned the base BERT model with its CLS token as output with the training data (batch size: 16, learning rate: $5 \cdot 10^{-5}$ and 3 epochs).

The finetuned model is then used to produce the additional input for the baseline. The BERT connection to the baseline is regularized by a dropout connection with rate 0.3. As the dimensionality of the BERT embeddings is substantially higher, we increased the number of hidden neurons of the baseline to 128.

3.2   Automatic Dataset Composition

The task of credibility assessment requires much data from different topics and time frames to make the model invariant to these patterns. Accordingly, we searched for Twitter datasets that can be combined into a larger set. The **PHEME** [41] dataset contains 300 binarily annotated posts from which both the classes "true" and "false" are mapped to a credibility score of 1 and 0 respectively in our coding schema. In contrast, the **Twitter15** [22] and **Twitter16** [24] datasets are categorized into four classes. The classes "true" and "false" are mapped analogously to the PHEME dataset. For the tweets of the class "unverified", we decided for a more uncertain score of 0.3 that reflects a tendency towards dubious content. After the manual inspection of the last class "no rumor", we chose a score of 0.9 as instances of this class seem to be primarily true. This way, additional 2,308 instances were added to the corpus of this paper.

Then, we implemented an automatic coding scheme for the **FakeNewsNet** [33] dataset. It contains several topics to which a large number of tweets are assigned. For each topic, an associated headline was labelled as true or false. Since the assignment of tweets to the topics was carried out using keywords, posts may also be incorrectly assigned to a topic and a mapping from the headline label to the label of tweets in it is not possible. It also allows a topic that is annotated as "false" to contain posts that expose the topic as wrong, which is especially important in our use case. To compensate for this, we perform a temporal, keyword-based, similarity, and topic filtering, which is described in detail in the appendix [16]. Thus, 1,378 credible and 729 implausible tweets were retrieved and mapped to a target score of 0.9 and 0.1, respectively.

We also used the **Twitter20** [16] dataset, where various German tweets of the COVID-19 pandemic were individually labelled. The following assumptions have to be true so that a translated version can be included into the dataset of our paper: (i) incorrect information in German and English are syntactically the same, (ii) the dataset contains only a few posts with misinformation or satire, and (iii) during the translation of the articles no linguistic properties (e.g., rhetorical stylistic devices) that are a characteristic of misinformation are lost. We used the Google Translate API and automatically corrected wrong @ and # placements, to preserve the general syntax of tweets. Since the assumptions do not necessarily have to apply, we have decided to create a "default" dataset (Fig. 1) without and a "large" dataset (Fig. 2) with these translated instances.

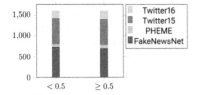

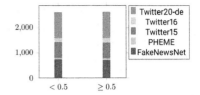

Fig. 1. Default dataset (N = 3,178, whereof $n_{credible}$ = 1,589).

Fig. 2. Large dataset (N = 5,225, whereof $n_{credible}$ = 2,619).

4 Evaluation

For the evaluation of our models, we use metrics based on regression fitting to the dataset development, classification for comparisons with past and future work, and execution time for insights related to real-time application. For the classification, tweets with a score of less than 0.5 are classified as 0 and 1 otherwise. We split 80%, 10% and 10% of the posts into training, development and testing sets. We also ensure that all posts by a user who is represented more than once in the dataset are included in the training set so that no information of the other sets is already seen during training. For the implementation we used PyTorch, Huggingface Transformers and NLTK. The system used for the evaluation has an Intel i7-9750 with 6 cores and 2.6 GHz, a NVIDIA GeForce RTX 2070 graphics card with 8 GB Graphics memory and 32 GB of RAM.

4.1 Evaluation of Model Quality

For the evaluation of the performance of the different model architectures and feature combinations, we first tuned the hyperparameters (e.g., batch size and dropout-rate) on the development set of both datasets. We chose to proceed with the models that have the lowest MSE. The evaluation results regarding the development set can be found in the appendix [16].

The testing set results on both datasets are shown in Table 2. It is clear to see that the adaption of the standard MLP model is very beneficial. Especially the BERT model can gain additive accuracy improvements of up to 21.63%. Looking at the feature constellations in the MLP network, it is evident that they are suitable for distinguishing credible and implausible posts without the addition of sentence or BERT embeddings (reaching up to 66.77% accuracy and 0.6513 F1 score). The Tweet, user and text feature constellation even reaches a slightly better MSE than the RNN basis model on the default dataset.

The more sophisticated BERT-based model, however, draws less benefit from the additional feature inputs. With the default set, a minimal improvement of less than 1 accuracy point is achieved, while the features for the large set even degrade the BERT model. Sometimes we noticed improvements from the timeline feature, but no significant results were found when the test set evaluation was performed. When inspecting relative changes with regard to both datasets, it becomes apparent that the BERT-based model has a greater positive impact on the first set. The additive accuracy improvements on this dataset are of up to 21.63% compared to just up to 14.12% on the large dataset.

Table 2. Results of the quality analysis per model on both datasets. Abbrev.: Tweet Features (Tw), User Features (Us), Text Features (TX), Advanced timeline features (ATi).

Model	Features	MSE	Acc	Pre	Rec	F1
MLP (default)	Base	–	–	–	–	–
	Tw, Us, Tx	0.1367	0.6552	0.6323	0.6490	0.6405
	Tw, Us, Tx & ATi	0.1474	0.6677	0.6471	0.6556	0.6513
RNN (default)	Base	0.1380	0.7116	0.6879	0.7152	0.7116
	Tw, Us, Tx	0.1202	0.7367	0.7190	0.7285	0.7237
	Tw, Us, Tx & ATi	0.1199	0.7429	0.7226	0.7417	0.7320
BERT (default)	Base	0.0806	0.8621	0.8497	0.8609	0.8553
	Tw, Us, Tx	**0.0794**	**0.8715**	**0.8618**	**0.8675**	**0.8674**
	Tw, Us, Tx & ATi	0.0805	0.8621	0.8591	0.8477	0.8533
MLP (large)	Base	–	–	–	–	–
	Tw, Us, Tx	0.1803	0.6469	0.6734	0.6162	0.6435
	Tw, Us, Tx & ATi	0.1841	0.6412	0.6895	0.5572	0.6163
RNN (large)	Base	0.1720	0.7042	0.7266	0.6863	0.7059
	Tw, Us, Tx	0.1639	0.7118	0.7143	0.7380	0.7260
	Tw, Us, Tx & ATi	0.1603	0.7042	0.7538	0.6679	0.7002
BERT (large)	Base	0.1347	**0.7844**	0.7883	**0.7970**	**0.7927**
	Tw, Us, Tx	0.1339	0.7824	0.7897	0.7897	0.7897
	Tw, Us, Tx & ATi	**0.1319**	0.7824	**0.7962**	0.7786	0.7873

From a dataset development perspective, one might think that the larger dataset contains more false annotated data, since the classifier scores are worse on this dataset. This can apply, e.g., if one of the assumptions given in Sect. 3.2 is incorrect and significant linguistic properties were lost during the translation of the Twitter20 dataset. Another consideration could be that the large dataset covers more different or domain-specific tweets; this makes classification more difficult but increases the generalizability and practicality of a classifier. When inspecting the translated posts in the Twitter20 dataset, we noticed some mistakes in the translated text. However, we tend to the second explanation as the actual content was preserved most of the time and we were still able to identify the credibility.

This consideration comes also into play when inspecting the stronger impact of the BERT-based model on the default dataset. BERT can have a major impact when it is applied to a dataset with fewer data instances as it can transfer knowledge from its previously learned tasks. The other algorithms can only get closer to the evaluation results if the dataset grows in its size, since they do not have this initial capacity.

Furthermore, the BERT model just slightly improves with the feature engineering process while the features seem more useful when applied to the other models. This might be due to the high capabilities of the pre-trained model. The incorporation of textual features might be redundant as the language model is able to identify some of these by itself. Some of the features might even be misleading and in this low data regime unwanted statistics are more likely to appear during the training process leading to better scores for the other models that do not have the generalization capabilities of BERT. The RNN model builds upon GloVe embeddings which also impose a certain generalization that is reflected in the results. However, with this model we still expect a bias towards unwanted statistics.

4.2 Evaluation of Model Execution Time

To measure the execution time of our models, all tweets and previous posts of the large dataset were loaded to the RAM in order to reduce variances of HDD memory access. The individual models were executed using the whole dataset to measure the execution time of different steps, such as the model initialization, the processing time per tweet, and the processed tweets per second (see Table 3). For the RNN modell, there are two options to read the required embeddings: (i) a filesystem-based approach that reads and indexes the embedding file once (fs) and (ii) a memory-based approach, where the whole file is loaded into RAM. While the first approach consumes less memory and has a shorter initialization time, the second approach offers a faster access to the embeddings. For the first approach, the use of an SSD ($\approx 3,500$ MB/s reading speed) or HDD (≈ 100 MB/s reading speed) did not yield measurable differences in execution time.

The BERT model strongly benefits from GPU acceleration. For comparison, Table 3 highlights the execution times with (gpu) and without (cpu) acceleration by a graphics card. For other models, the use of a GPU did not yield measurable performance improvements. Generally spoken, complex models require longer execution times than simple models. The base RNN model is seven times faster than BERT using a GPU and more than 110 times faster than BERT without a GPU. Furthermore, the processing of timeline features, i.e., incorporating up to 40 previous posts of a tweet, requires significantly more time. While the RAM-based RNN model is able to classify up to 5.5k tweets per second, BERT processes up to 133 tweets per second with a GPU, but only 6.6 without a GPU. In that model, additional features show negligible impact on the overall execution time.

Table 3. Results of the temporal analysis per model. The column *tweets/second* ignores the initialization time.

	Configuration	Initialization	Time/Tweet	Tweets/Second
Features	Text	203 ms	914 μs	1094
	Tweet	0 s	144 μs	6944
	User	60 ms	130 μs	7692
	Timeline	265 ms	40,000 μs	25
MLP	Basis	1,400 ms	44 μs	22,727
	Adv. Timeline	1,430 ms	52,000 μs	19.2
RNN	Basis (fs)	7,653 ms	1,354 μs	738
	Basis (ram)	32 s	179 μs	5586
	Adv. Timeline (fs)	7,653 ms	107,000 μs	9.3
	Adv. Timeline (ram)	32 s	88,000 μs	11.3
BERT	Basis (gpu)	3,706 ms	7,495 μs	133
	Basis (cpu)	3,720 ms	150,000 μs	6.6
	Adv. Timeline (gpu)	3,680 ms	224,000 μs	4.4
	Adv. Timeline (cpu)	3,695 ms	>4 s	<¼

5   Discussion and Conclusions

Nowadays, social media is widely used for multiple purposes, such as relationship maintenance, journalism, customer interactions but also for crisis management. However, these activities can be severely impeded by the propagation of false information. Hence, it is important to promote credible and to counter implausible information. In this work, we reviewed existing approaches before designing and evaluating three neural network models capable of near real-time credibility assessment in Twitter to answer the following research question: **Which deep learning models and parameters are suitable for real-time credibility assessment in Twitter?**

Our findings indicate that our BERT-based model achieves the best results when using metadata, text, and user features, reaching an accuracy of 87.07% and F1 score of 0.8764 on the default dataset. In comparison to existing works, the results appear to be promising. While Helmstetter and Paulheim [13] reached an F1 score of 0.7699, Iftene et al. [15] achieved an accuracy of 85.20%. Although Ruchansky, Seo, and Liu [31] reached an accuracy of 89.20% and F1 score of 0.9840, their approach is propagation-based, thus having limited real-time capability, and classifies events instead of individual tweets. Similarly, Liu and Wu [23] reach an F1 score of 0.8980; however, their approach focuses on the detection of disinformation and also incorporates propagation-based features.

Furthermore, we compared the real-time capabilities of our three models. While our MLP baseline is capable of processing high volumes of data (>20k tweets/sec) with a low resource demand, the accuracy of up to 66.77% does not allow for a reliable classification. In contrast, our RNN model is still capable of processing high volumes of data (>5k tweets/sec when used in RAM) for classification while reaching more promising accuracies of up to 74.29%. Finally, BERT reached accuracies of 87.07% but was only able to process a considerably lower amount of data (>0,1k tweets/sec when used with a GPU). When including the previous posts of the user into computation, we did not achieve consistent improvements of classification performance; however, the real-time capability of all feature and model combinations was lost.

5.1 Practical and Theoretical Implications

We compared existing datasets and combined suitable ones into a **novel dataset to increase the amount of available data for model training (C1)**. Since available datasets are used for varying credibility classification tasks, several steps of transformation were required to convert them into a unified structure. Due to the combination of datasets, it comprises a richer number of users, topics, and message characteristics. Our future work will include the application of data augmentation techniques to increase the size and richness of the dataset.

We provided a **review of existing machine learning approaches for credibility assessment in Twitter (C2)**. In contrast to other works, we critically examined and compared approaches for credibility assessment in Twitter. Many of the reviewed approaches did not use a development set, relied on a small dataset, or conducted many hyperparameter optimizations, which entails the risk of overfitting. While difficult to compare, it seems that propagation-based models achieve the best results [31]; however, they lack the ability of real-time application. As the engagement based on tweets unfolds over time, propagation-based models seem promising when no time constraints are present.

In addition, our work contributes with **insights into the real-time capability of neural networks for credibility assessment (C3)**. Comparing our models in terms of real-time capability, their usefulness seems to be dependent on many factors. While our MLP baseline shows excellent execution times for large-volume data processing, the lack of classification performance disqualifies its real-world applicability. In contrast, our RNN model still offers suitable execution times and maintains a better classification performance. Finally, despite achieving the best classification results, BERT offers limited realtime capability when used for large-scale data analysis unless considerable GPU power is used for processing. In the end, we would still advise to do further research regarding the BERT model, as it has the best generalization capabilities. The credibility research shows that simple algorithms tend to be very biased towards the topics and domains in the dataset and often bahave more like a topic classifier.

5.2 Limitations and Outlook

While this work is subject limitations, they also offer potentials for future research. First, although BERT achieved the best classification results, it was also the slowest classifier. Variations of BERT, such as DistilBERT [32], provide smaller models or shared weightings within the model to achieve a lower memory usage and faster execution time. Thus, future work could examine if they achieve comparable classification results for credibility assessment. Second, the classifier is limited by merely using a dataset based on textual Twitter data. Although it can be used for other social media, it might perform worse due to different linguistic features. Thus, the exploration of a cross-platform dataset, supported by active learning and data augmentation techniques, could be worthwhile for future research [19]. Furthermore, pictures displaying text messages (requiring optical character recognition techniques) or external sources could be incorporated into the credibility assessment concept. Third, novel but similar publications emerged during the implementation of our study. For instance, Tian et al. [37] contributed with a rumor detection algorithm that achieves an F1 score of 0.862 but does not provide a steady regression of findings. A further work used ALBERT to reduce the memory usage of BERT, reaching an F1 score of 0.795 compared to 0.71 of the original BERT model [36]. Furthermore, additional research was conducted to detect fake news spreaders by analyzing their previous posts [7,29].

Acknowledgements. This work has been co-funded by the German Federal Ministry of Education and Research (BMBF) in the project CYWARN (No. 13N15407) and by the BMBF and the Hessen State Ministry for Higher Education, Research and Arts (HMKW) within the SecUrban mission of the National Research Center for Applied Cybersecurity ATHENE.

References

1. Allcott, H., Gentzkow, M.: Social media and fake news in the 2016 election. J. Econ. Perspect. **31**(2), 211–236 (2017)
2. Buntain, C., Golbeck, J.: Automatically identifying fake news in popular twitter threads. In: IEEE Proceedings of (SmartCloud), pp. 208–215 (2017)
3. Castillo, C., Mendoza, M., Poblete, B.: Information credibility on twitter. In: Proceedings of WWW, p. 675 (2011)
4. Castillo, C., Mendoza, M., Poblete, B.: Predicting information credibility in time-sensitive social media. Internet Res. **23**(5), 560–588 (2013)
5. Conroy, N.K., Rubin, V.L., Chen, Y.: Automatic deception detection: methods for finding fake news. Proc. ASIS&T **52**(1), 1–4 (2015)
6. Devlin, J., Chang, M.W., Lee, K., Toutanova, K.: BERT: pre-training of deep bidirectional transformers for language understanding. In: Proceedings of NAACL-HLT, pp. 4171–4186 (2019)
7. Duan, X., Naghizade, E., Spina, D., Zhang, X.: RMIT at PAN-CLEF 2020: proling fake news spreaders on twitter. In: CLEF 2020 (2020)
8. Flanagin, A.J., Metzger, M.J.: Digital media and youth: unparalled opportunity and unprecedented responsibility. In: Flanagin, A.J., Metzger, M.J. (eds.) Digital Media, Youth, and Credibility, pp. 5–28 (2008)
9. Floria, S.A., Leon, F., Logofătu, D.: A credibility-based analysis of information diffusion in social networks. In: Kůrková, V., Manolopoulos, Y., Hammer, B., Iliadis, L., Maglogiannis, I. (eds.) Proceedings of ICANN, pp. 828–838 (2018)
10. Fogg, B.J., Tseng, H.: The elements of computer credibility. In: Proceedings of CHI, pp. 80–87 (1999)

11. Gupta, A., Kumaraguru, P., Castillo, C., Meier, P.: TweetCred: real-time credibility assessment of content on twitter. In: Aiello, L.M., McFarland, D. (eds.) Social Informatics, vol. 8851, pp. 228–243 (2014)
12. Hassan, D.: A text mining approach for evaluating event credibility on twitter. In: Proceedings of WETICE, pp. 171–174 (2018)
13. Helmstetter, S., Paulheim, H.: Weakly supervised learning for fake news detection on twitter. In: Proceedings of ASONAM, pp. 274–277. IEEE (2018)
14. Hutto, C., Gilbert, E.: VADER: a parsimonious rule-based model for sentiment analysis of social media text. In: Proceedings of ICWSM (2015)
15. Iftene, A., Gifu, D., Miron, A.R., Dudu, M.S.: A real-time system for credibility on twitter. In: Proceedings of LREC, pp. 6166–6173 (2020)
16. Kaufhold, M.A., Bayer, M., Hartung, D., Reuter, C.: Paper Appendix (2021). https://github.com/mkx89-sci/KaufholdBayerHartungReuter2021_ICANN
17. Kaufhold, M.A., Reuter, C.: Cultural violence and peace in social media. In: Reuter, C. (ed.) Information Technology for Peace and Security - IT-Applications and Infrastructures in Conflicts, Crises, War, and Peace, pp. 361–381 (2019)
18. Kaufhold, M.A.: Information Refinement Technologies for Crisis Informatics: User Expectations and Design Principles for Social Media and Mobile Apps (2021)
19. Kaufhold, M.A., Bayer, M., Reuter, C.: Rapid relevance classification of social media posts in disasters and emergencies: a system and evaluation featuring active, incremental and online learning. IP&M **57**(1), 102132 (2020)
20. Kingma, D.P., Ba, J.: Adam: a method for stochastic optimization (2017)
21. Lazer, D.M.J., et al.: The science of fake news. Science **359**(6380), 1094–1096 (2018)
22. Liu, X., Nourbakhsh, A., Li, Q., Fang, R., Shah, S.: Real-time rumor debunking on twitter. In: Proceedings of CIKM, pp. 1867–1870 (2015)
23. Liu, Y., Wu, Y.F.B.: Early detection of fake news on social media through propagation path classification with recurrent and convolutional networks. In: Thirty-Second AAAI Conference on Artificial Intelligence (2018)
24. Ma, J., et al.: Detecting rumors from microblogs with recurrent neural networks. In: IJCAI International Joint Conference on Artificial Intelligence, pp. 3818–3824 (2016)
25. Mitra, T., Gilbert, E.: CREDBANK: a large-scale social media corpus with associated credibility annotations. In: Proceedings of ICWSM (2015)
26. Pennington, J., Socher, R., Manning, C.: Glove: global vectors for word representation. In: Proceedings of EMNLP, pp. 1532–1543 (2014)
27. Pierri, F., Ceri, S.: False news on social media: a data-driven Survey. ACM SIGMOD Record **48**(2), 18–27 (2019)
28. Qazvinian, V., Rosengren, E., Radev, D.R., Mei, Q.: Rumor has it: identifying misinformation in microblogs. In: Proceedings of EMNLP, pp. 1589–1599 (2011)
29. Rangel, F., Giachanou, A., Ghanem, B., Rosso, P.: Overview of the 8th Author Proling Task at PAN 2020: Proling Fake News Spreaders on Twitter (2020)
30. Ratkiewicz, J., et al.: Truthy: mapping the spread of astroturf in microblog streams. In: Proceedings of WWW, p. 249 (2011)
31. Ruchansky, N., Seo, S., Liu, Y.: CSI: a hybrid deep model for fake news detection. In: Proceedings of CIKM, pp. 797–806 (2017)
32. Sanh, V., Debut, L., Chaumond, J., Wolf, T.: DistilBERT, a distilled version of BERT: smaller, faster, cheaper and lighter (2020)
33. Shu, K., Mahudeswaran, D., Wang, S., Lee, D., Liu, H.: FakeNewsNet: a data repository with news content, social context and dynamic information for studying fake news on social media. Big Data **8**(3) (2018)

34. Shu, K., Sliva, A., Wang, S., Tang, J., Liu, H.: Fake news detection on social media: a data mining perspective. ACM SIGKDD Explor. Newsl. **19**(1), 22–36 (2017)
35. Tandoc, E.C., Lim, Z.W., Ling, R.: Defining "fake news": a typology of scholarly definitions. Digit. Journal. **6**(2), 137–153 (2018)
36. Tian, L., Zhang, X., Peng, M.: FakeFinder: twitter fake news detection on mobile. In: Companion Proceedings of the Web Conference, vol. 2020, pp. 79–80 (2020)
37. Tian, L., Zhang, X., Wang, Y., Liu, H.: Early detection of rumours on twitter via stance transfer learning. In: Jose, J.M., et al. (eds.) Advances in Information Retrieval, vol. 12035, pp. 575–588 (2020)
38. Viviani, M., Pasi, G.: Credibility in social media: opinions, news, and health information-a survey: credibility in social media. Wiley Interdiscip. Rev.: Data Min. Knowl. Discov. **7**(5), e1209 (2017)
39. Wu, L., Rao, Y., Yu, H., Wang, Y., Nazir, A.: False information detection on social media via a hybrid deep model. In: Staab, S., Koltsova, O., Ignatov, D.I. (eds.) SocInfo 2018. LNCS, vol. 11186, pp. 323–333. Springer, Cham (2018). https://doi.org/10.1007/978-3-030-01159-8_31
40. Xia, X., Yang, X., Wu, C., Li, S., Bao, L.: Information credibility on twitter in emergency situation. In: Chau, M., Wang, G.A., Yue, W.T., Chen, H. (eds.) PAISI 2012. LNCS, vol. 7299, pp. 45–59. Springer, Heidelberg (2012). https://doi.org/10.1007/978-3-642-30428-6_4
41. Zubiaga, A., Liakata, M., Procter, R., Wong Sak Hoi, G., Tolmie, P.: Analysing how people orient to and spread rumours in social media by looking at conversational threads. PLoS ONE **11**(3), e0150989 (2016)

T-Bert: A Spam Review Detection Model Combining Group Intelligence and Personalized Sentiment Information

Yue Shang[1] , Meiling Liu[1]([✉]) , Tiejun Zhao[2], and Jiyun Zhou[3]

[1] School of Information and Computer Engineering, Northeast Forestry University,
Harbin 150006, China
{1095517318,mlliu}@nefu.edu.cn
[2] Department of Computer Science, Harbin Institute of Technology,
Harbin 150001, China
tjzhao@hit.edu.cn
[3] Lieber Institute, Johns Hopkins University, Baltimore, MD 21218, USA

Abstract. The content of online comments largely affects users' willingness to purchase goods or services. Driven by interests, spam reviews continue to emerge to induce users maliciously. Most of the existing related work is based on the easy-camouflaged feature information, and the deep learning model is rarely used. The BERT model is prominent in various tasks in the NLP field, and whether it can be successfully applied to the spam review identification task has not been verified. In this paper, we propose a new research strategy for this task: the multi-dimensional representation combining group intelligence and users' personalized sentiment information can more effectively detect spam reviews. Through fine-grained sentiment analysis of reviews based on product dimension and user dimension, we effectively acquire group intelligence and user personalized sentiment, respectively; Based on the ability of BERT to model the embedding of text context information, the semantic information is acquired. Finally, the three are combined based on Triple Network structure to detect spam reviews. We conduct a large number of experiments on three public datasets and the recall rate and F1 value both exceed the results of state-of-the-art works, which proves the feasibility and effectiveness of our proposed strategy, and verifies the modeling ability of the BERT in the task of detecting spam reviews.

Keywords: Spam review detection · Sentiment analysis · BERT · Triple network

1 Introduction

Product reviews affect users' shopping behavior. According to the survey, 64% of users will read the reviews before buying goods, 87% of users choose to buy after

Supported by the National Natural Science Foundation of China under Grant 61702091 and the Fundamental Research Funds for the Central Universities under Grant No 2572018BH06.

I. Farkaš et al. (Eds.): ICANN 2021, LNCS 12895, pp. 409–421, 2021.
https://doi.org/10.1007/978-3-030-86383-8_33

reading good reviews, and 80% of users give up after reading bad reviews [1]. Fake reviews refer to reviews that are written to intentionally confuse the consumer [2]. However, research by the Washington Post [3] found that more than 60% of reviews of electronic products on Amazon.com were fake. Therefore, it is very important to automatically identify the authenticity of network platform information and provide users with more authentic information. Review information is text information, so the identification of spam reviews can be regarded as a text classification problem. Li et al. [4] propose a neural network composed of two convolutional layers combined with sentence importance weights for deceptive review detection. Liu et al. [5] based on the combination of bidirectional long short-term memory network and features, carry out fake review detection, which can well learn the long-distance correlation in the sequence.

However, the existing models are at a deadlock in the recognition effect, one of the reasons is that the embedding layer usually provides context-independent word-level features by Word2Vec or Glove models. Moreover, the spam review dataset is too small to implement task-based architecture. Therefore, it has a good potential to further improve the performance to generate context aware word vectors with the help of pre-trained language models on large-scale datasets.

In addition, the experimental results show that emotional features have a good effect in the recognition of fake reviews [6]. After conducting data mining on the Yelp fake review datasets, we find that reviewers usually describe many aspects of the product to express their opinions and convince others. Different users will produce a variety of fine-grained evaluations when evaluating the same product, which reflects the quality of the product in an all-round way. Because the spammers are not personal experience, the non-real information they posted may be different from the public evaluation. For example, for a restaurant, the real users have a negative evaluation on the dishes and a positive evaluation on the drinks. Spammers also have positive comments on drinks, but they are full of praise on dishes. J. Surowiecki points out in the book *The Intelligence of Crowds*: "In the right environment, a group has extraordinary intelligence, and this intelligence often beats the smartest person in the group" [7]. The group's evaluation of a product aspect can represent the real level of the aspect.

Therefore, if we can mine the potential group intelligence in product reviews, and combine user's emotional attitude towards product aspects, we can verify whether the user's emotional attitude is true, and use the public intelligence to detect spam reviews more effectively.

In this paper, we fuse group intelligence with users' personalized sentiment information and context semantic information to generate multidimensional representations for the identification of spam reviews, and propose a new model Triple BERT (T-Bert) based on the structure of Triple Network [8] and BERT component, which provides a new solution strategy for the task of spam review detection.

2   Related Work

2.1   Spam Review Detection

The task of spam review detection began in 2007 [9]. Spam review detection is a specific application of the general problem of deception detection, mainly using text and behavioral features. Behavior features include the number of good/bad reviews [10], the frequency of comments [12], etc.; text features include the length of the comment text [10,11], various vocabulary and syntactic features [13], etc. In addition, some works combine text features with behavior features. Wang et al. [14] combined the two as sentence representation based on CNN model, which solved the cold start problem in spam review detection. Wang et al. [15] used MLP to obtain user behavior features and CNN to obtain text language features, and combined them based on attention neural network to identify fake reviews. Yuan et al. [16] used hierarchical fusion attention mechanism to generate fusion text representation from the perspective of user and product, and based on TransH algorithm to model the relationship among user, product and review text, to generate more reliable review representation.

Previous work mostly based on Word2Vec or GloVe for word vector representation, but it is not enough to capture the complex semantic relevance in sentences. Recently, pre-trained language models such as ELMo and BERT have been shown to be effective in generating context-aware word vectors with the potential to further improve performance, and have been shown to be effective in a number of natural language processing applications, so far, however, no work has been done to apply BERT to the spam review detection task. In this paper, we use BERT as the basic model to construct the word vector and verify the performance of BERT model in this task.

2.2   Fine-Grained Sentiment Analysis

Fine-grained sentiment analysis is a challenging and significant subtask in sentiment analysis. Fine-grained sentiment analysis [17] aims to identify the sentiment polarity of specific aspects. This task enables users to evaluate the comprehensive sentiments of all aspects of a given product or service and have a more comprehensive understanding of its quality [18]. Fine-grained sentiment analysis can be subdivided into three categories: the first one is to detect the polarity of sentiment corresponding to a given aspect in a sentence [19,20], but it is difficult to be applied because fine-grained aspects need to be labeled in advance; the second is Aspect-oriented Opinion Words Extraction (AOWE) [21,22], which aims to extract the opinion words corresponding to a given aspect from the sentence; the last is End-to-End Aspect-based Sentiment Anslysis (E2E-ABSA), whose goal is to jointly detect aspect terms/categories and corresponding aspect sentiment.

On the one hand, existing studies [6,25] show that using sentiment features can effectively identify fake reviews; on the other hand, as mentioned in the Sect. 1, in order to integrate group intelligence into the model and further improve the reliability of spam review detection, we conduct E2E-ABSA on spam review data, so as to obtain the group sentiment corresponding to all aspects of the product. Meanwhile this measures the degree of deviation of user's sentiment from the public's sentiment. These two are used as the auxiliary information of spam review detection, which provides a new solution for this task.

3   Methodology

The structure of T-Bert is shown in Fig. 1. We regard spam review detection task as a binary task. Firstly, for each user's (or product's) reviews, we conduct fine-grained sentiment analysis on the sentences, and cluster the fine-grained aspects to get the user's (or product's) sentiment tendency in each fine-grained aspect, that is, the group sentiment tendency $G_i = \{a_{i1}, a_{i2}, a_{i3}, a_{i4}, a_{i5}, a_{i6}\}$ of product P_i, and the personal sentiment tendency $S_i = \{b_{i1}, b_{i2}, b_{i3}, b_{i4}, b_{i5}, b_{i6}\}$ of user U_j to product P_i. Secondly, given an input sentence $X_i = \{x_{i1}, x_{i2}, \ldots, x_{iT}\}$ of length T, we encode it with the BERT component of the L Transformer layers to get a contextualized sentence representation $E^L = \{e_1^L, e_2^L, \ldots, e_T^L\} \in \mathbb{R}^{T \times D}$, where D represents the dimension of the vector. Finally, we combine G_i, S_i and E^L to identify spam reviews. Our goal is to determine whether X_i is a spam review.

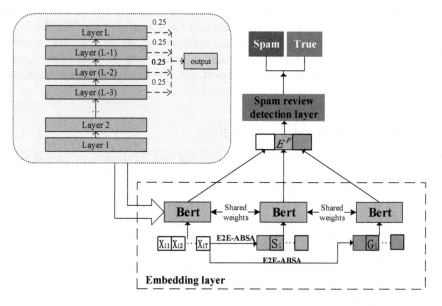

Fig. 1. Overall structure of T-Bert model. S_i stands for the user's personalized senti-ment, G_i stands for group intelligence.

3.1 Group Intelligence and User Personalized Sentiment

We extract the fine-grained aspects that users are concerned about from the reviews. The fine-grained aspect refers to the product attributes that contained in the user's review. Since the fine-grained aspect is not marked in the spam review dataset, the amount of data that is too large to be manually marked, and it is difficult to define the marking standard, we use the transfer learning method to mark the fine-grained information. This research is based on the Yelp dataset, which includes restaurant reviews and a small number of hotel reviews. Therefore, we use the method in work [23] to train the fine-grained sentiment analysis model based on the data of SemEval 2016 [26], and use the model to label the Yelp dataset. Each review in the dataset is annotated to get a triple information $(A_i, W_i, POS/NEG)$, that is, the fine-grained aspect $A_i = \{A_{i1}, A_{i2}, \ldots, A_{in}\}$ referred to in sentence X_i, and the corresponding sentiment word W_i for each fine-grained aspect A_{ix}, as well as this group of fine-grained sentiment tendency POS/NEG, POS represents positive sentiment, NEG represents negative sentiment. In order to obtain group intelligence and user personalized sentiment, we further analyze the annotated fine-grained sentiment information.

We use the labeling standards in the SemEval dataset to divide the fine-grained aspects into 6 categories: *restaurant, food, drink, service, ambience,* and *location.* First, de-duplicate and merge the fine-grained aspect words contained in all review sentences to obtain the fine-grained aspect word set ASP. Perform word frequency statistics on ASP, and select 10 seed words in each category to form a seed word set $\bar{A}$ in order from highest to bottom. Second, use the Word2Vec model to train the Yelp review dataset to obtain the word vector model. Finally, based on the word vector model, the similarity between each seed word in each category in $\bar{A}$ and ASP_i is calculated. If the average similarity is greater than the threshold α, then ASP_i belongs to this category. As shown in Table 1, the fine-grained aspect word set $\tilde{A}$ divided into 6 categories is generated according to the above steps.

Table 1. Part of the fine-grained aspect word set $\tilde{A}$

Restaurant	Food	Drinks	Service	Ambience	Location
Dish	Meat	Wine	Services	Atmosphere	Downtown
Travelocity	Meal	Wines	Waitstaff	Environment	Branch
Hostess	Pancakes	Vino	Staff	Setting	Outpost
Meal	Dessert	Chardonnay	Serive	Interior	Hotspot
Product	Desserts	Prosecco	Servers	Atomsphere	Avenue
Stew	Seafood	Saki	Natured	Atomosphere	Bucktown
Adjacent	Dishes	Shiraz	Courteous	Decoration	Situated

We determine the sentiment polarity of each category in the review sentence based on simple rules. For example, in the category of *food*, if the number of positive fine-grained words is greater than the number of negative fine-grained words, the sentiment of *food* is positive, and vice versa. From the product dimension, perform fine-grained sentiment analysis and clustering on all review information of the product P_i to obtain its group sentiment feature $G_i = \{a_{i1}, a_{i2}, a_{i3}, a_{i4}, a_{i5}, a_{i6}\}$, Where a_{i1} represents a certain category of group sentiment polarity. From the user dimension, the fine-grained sentiment analysis results of U_j's evaluation of P_i are clustered according to $\tilde{A}$, which is regarded as user's personalized sentiment feature $S_i = \{b_{i1}, b_{i2}, b_{i3}, b_{i4}, b_{i5}, b_{i6}\}$, where b_{ix} represents a certain category of user sentiment polarity.

3.2   Triple Bert

The BERT model is a new language model that uses bidirectional Transformers for pre-training on a large number of corpora, and performs amazingly in many tasks in the NLP field. We built a spam review detection model T-Bert based on the Triple Network framework and BERT.

Embedding Layer. We use the BERT component as the embedding layer of the T-Bert model. For each token X_{it} in sentence X_i, We add token embedding, segment embedding and position embedding to e_t, $t \in [1, T]$ to form the input feature $E^0 = \{e_1, e_2, \ldots, e_T\}$ of the first branch of the embedding layer. Then L transformer layers are introduced to refine the token-level features layer by layer. Finally, the output E^L obtained by splicing the last four layers is the representation of the review sentence X_i.

$$E^L = 0.25 \times E^{L-1} + 0.25 \times E^{L-2} + 0.25 \times E^{L-3} + 0.25 \times E^{L-4} \quad (1)$$

In order to combine group intelligence, user personalized sentiment and text information for spam review detection, we use BERT component to transform the two dimensions of sentiment information constructed in Sect. 3.1. First, the two features G_i and S_i are Onehot mapped and normalized. Then, we pack each feature value in the S_i of the P_i as $E^{s0} = \{e_{s1}, e_{s2}, \ldots, e_{s12}\}$, where $e_{st}, t \in [1, 12]$ is the combination of the token embedding, segment embedding, and position embedding corresponding to the input feature token. This is the second branch of the embedding layer. The input feature $E^{g0} = \{e_{g1}, e_{g2}, \ldots, e_{g12}\}$ of the third branch of embedding layer is generated in the same way. Note that the BERT components of the first branch, the second branch and the third branch share weights. The calculation process is as shown below, where $E^{gl} \in \mathbb{R}^{12 \times D}, E^{sl} \in \mathbb{R}^{12 \times D}$ are the representation of group intelligence feature G_i and user sentiment feature S_i respectively.

$$\begin{aligned} E^{gl} &= Transformer_l(E^{gl-1}), \\ E^{sl} &= Transformer_l(E^{sl-1}). \end{aligned} \quad (2)$$

Spam Review Detection Layer. In order to identify spam reviews, we build four different spam review detection layers on the embedding layer to classify the feature representations obtained before. We concatenate E^L, E^{gl} and E^{sl} to form the input $E^F \in \mathbb{R}^{(T+24) \times D}$ of spam review detection layer.

Linear The obtained E^F is input into a max pooling layer. The most distinctive features in each sentence can be selected to form a sentence representation $h_L \in \mathbb{R}^D$, and then input into the linear classification layer. Finally, softmax function is used to calculate the probability of classification category as follow:

$$
\begin{aligned}
h_L &= \max_{dim=1}(E^F), \\
P &= softmax(h_L W_L),
\end{aligned}
\tag{3}
$$

where $W_L \in \mathbb{R}^{D \times C}$, C is the number of categories.

Bidirectional Long Short-Term Memory (BiLSTM). BiLSTM is a combination of forward LSTM and backward LSTM, which can better capture bidirectional semantic dependencies. Input the obtained E^F into BiLSTM to obtain the task-specific hidden representation $h \in \mathbb{R}^{2H}$, where H is the hidden layer size in BiLSTM, and then obtain the predicted value P through the softmax function:

$$
\begin{aligned}
h &= BiLSTM(E^F) = [\overrightarrow{h}, \overleftarrow{h}], \\
P &= softmax(h W_2).
\end{aligned}
\tag{4}
$$

Attention Network. The attention mechanism in seq2seq breaks the limitation that the encoder can only use the final single vector result, so that the model focuses on the input information that is more important for the output information. We use the attention mechanism to calculate E^F, extract the implicit features in sentences, focus on the words that are important for classification, and generate a specific representation $h_A \in \mathbb{R}^D$ of this task.

$$
\begin{aligned}
h_A &= \beta E^F, \\
\beta &= \frac{exp(E_i')}{\sum_{n=1}^{T+24} E_n'}, \\
E' &= tanh(E^F W_a),
\end{aligned}
\tag{5}
$$

where β is the score function that determines the importance of the words in the whole sentence, $W_a \in \mathbb{R}^{D \times D}$ is the transformation matrix. Similarly, a linear layer with softmax activation as before is stacked on the designed attention layer to output the prediction.

Convolutional Neural Network (CNN). The CNN model proved to be effective for NLP and achieved excellent results in semantic analysis [27]. In this paper, we use the convolution kernel of the CNN layer to perform a convolution operation on the review sentence representation E^L to obtain the hidden features $O_i \in \mathbb{R}^{f \times (T-k+1)}$ in the text.

$$
\begin{aligned}
O_i &= W \cdot E^L_{i:i+k-1}, \\
V_c &= \max_{0 \le i \le T-k}(O_i).
\end{aligned}
\tag{6}
$$

where f is the channel for the convolution and k is the width of the convolution kernel.· represents the dot product operation of the matrix, $i = 0, 1, 2, \ldots, T - k$ and $W \in \mathbb{R}^{k \times D}$. The convolution core is repeatedly applied for the convolution operation and fed into the max pooling layer for filtering features.

Above is a process of feature extraction by a filter. In this paper, m filters of different sizes are used to extract as many features as possible, and then these features are spliced to get the review representation $h_{c1} \in \mathbb{R}^{m \times f}$. Then we combine the filtered token level text features with the sentence level output of BERT model and E^{tl}, E^{sl} to get the final sentence representation $h_c \in \mathbb{R}^{m \times (f + 3D)}$. Finally, h_c is input into the linear layer with softmax activation function to get the classification result.

4   Experiment

4.1   Datasets and the Evaluation Metrics

In order to verify the effectiveness of the model, we conducted experiments on three public datasets: YelpChi [10], YelpNYC and YelpZIP [11]. The data are real business reviews of restaurants and hotels from different areas of the Yelp website. It can be found that the average sentence length of real review sentence is longer than that of spam review sentence because it involves fine-grained aspects description. There is no significant difference between spam reviews and real reviews when observed from sentence-level sentiment analysis.

We used precision, recall and F1 scores to evaluate the effectiveness of the model. The precision reflects the correctness of the model in predicting spam reviews, and the recall reflects the proportion of correctly predicted spam reviews by the model in all spam reviews. F1 score is the harmonic mean of precision and recall.

4.2   Baselines and Implementation Detail

In the comparison experiment, we compare the BERT-based model with several advanced methods in existence. ABNN [15] is a neural network based on attention mechanism, which uses MLP to obtain user behavior features and CNN to obtain text language features, and combines the two based on attention to identify spam reviews. HFAN [8] is a hierarchical fusion of attention among users, reviews and products to get a comment representation that integrates the three to classify comments. DFFNN [14] is a deep feedforward neural network, which combines bag-of-word/n-gram feature, word embedding and multiple emotion indicators of the review sentence as representation. In addition, we also compare the modeling effect of several spam review detection layers with different network structures and the influence of different sentiment features on the detection ability.

In the embedding layer, we use the pre-trained "BERT-base-uncasd" model, where the number of transformer layers $L = 12$, the hidden size $D = 768$, that is, the sentence representation dimension is 768 and the sentence length $T = 200$. In the spam review detection layer, the learning rate is set to $2e - 5$, the dropout rate is set to 0.5 and the training batch size is 128. The hidden layer dimension of BiLSTM is set to 300. In convolution neural network, the size of convolution kernel channel is $f = 50$, and the width of convolution kernel increases from 1 to 11. A total of 11 filters with different sizes are used.

4.3 Results and Analysis

The Embedding Effect of the BERT Model: The experimental results are shown in Table 2 below. Compared with other methods without BERT model, BERT + Linear is not as good as the best model when using only text information as detection feature, however, the recall rate and F1 value are slightly different from other models that use a variety of information, which validates the performance of BERT model in the task of detecting spam reviews. It shows that the BERT model encoded by the association between any two tokens can generate a review representation with rich contextual information for the spam review detection layer.

Table 2. Experimental results of single BERT using only text information

	YelpChi			YelpNYC			YelpZIP		
	P	R	F1	P	R	F1	P	R	F1
ABNN	69.23	54.51	60.99	70.67	54.97	61.84	73.34	57.42	64.41
HFAN	73.03	57.21	64.16	74.65	59.42	66.17	78.45	65.49	71.39
DFFNN	71.34	53.95	61.44	71.96	54.07	61.75	72.69	56.93	63.85
Bert+Linear	68.56	59.58	63.76	70.49	60.83	65.30	74.58	64.33	69.08
Bert+ATT	69.36	59.69	64.16	71.18	60.38	65.34	74.96	65.57	69.95
Bert+LSTM	70.09	60.30	64.83	71.95	61.87	66.53	75.38	66.18	70.48
Bert+CNN	71.51	60.73	65.68	72.37	62.37	67.00	76.23	67.39	71.54

Performance of Different Spam Review Detection Layers: The experimental results are shown in Table 2. When only text information is used as the clue of spam review detection, the precision, recall and F1 values of BERT + ATT, BERT + LSTM and BERT + CNN are higher than those of BERT + Linear. Therefore, the use of more powerful network structure can bring better effect for the spam review detection task than only using the linear layer. This result shows that merging context information is helpful to sequence modeling and can provide more effective sentence representation for text classification tasks.

Performance of Different Sentiment Information: The results are shown in Fig. 2(a), Fig. 2(b) and Fig. 2(c). S-BERT refers to Siamese BERT, which takes text information as the input of the first branch of the embedding layer and user personalized sentiment feature as the input of the second branch. The rest of the model structure is the same as that of T-Bert. As Fig. 2(a) shown, when different features are used as the potential thread for spam review detection, the precision of using sentiment features is not greatly improved compared with using only text information, but the recall rate and F1 value are greatly improved, which indicates that when fine-grained sentiment information is fused, the ability of the model to identify spam reviews is improved. From Fig. 2(b) and Fig. 2(c), we can find that the detection ability is further improved by combining the product and user dimensions, that is, combining the group intelligence with the user's personalized sentiment, which verifies our previous hypothesis that the effective use of group intelligence can better detect spam reviews.

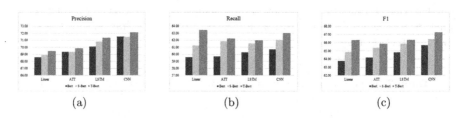

(a)   (b)   (c)

Fig. 2. Performance of different models on YelpChi dataset using different sentiment features.

Comparing with Table 2, it can be seen that the recall rate and F1 value of T-Bert have been greatly improved. Compared with the existing technology, the average recall rate and F1 value of the three data sets have been improved by 4.6% and 2.4% respectively. The experimental results verify the effectiveness and feasibility of the proposed strategy. But there is still room for improvement, the improvement of model's precision is not so good. The reason is that: in order to obtain fine-grained aspect information annotation, transfer learning method is used. However, the accuracy of annotation can not reach 100%. The result of annotation further affects the accuracy of subsequent spam review detection. How to further improve the effect of the model is our next research plan.

5   Conclusion

In this paper, we propose a new research strategy for spam review detection task, and verify the effectiveness of BERT component in this task. Specifically, we propose a strategy to improve the effectiveness of spam review detection by using group intelligence and user personalized sentiment information. In order to effectively use the intelligence of the group, we combine the group intelligence and the user personalized sentiment information with the text information to generate multidimensional representation, and propose a new model Triple BERT (T-Bert) based on the structure of Triple Network and BERT component. We explore the use of the BERT model as the embedding layer to generate review representations with rich contextual information, and to couple the BERT component with multiple neural models, a large number of experiments are carried out on three benchmark datasets to verify the effectiveness of the strategy proposed in this paper. The results show that BERT performs well in the task of spam review detection and improves the effectiveness of the T-Bert model.

References

1. Yuming, L., Xiaoling, W., Tao, Z., et al.: Review of research on quality inspection and control of user reviews. J. Softw. **03**, 506–527 (2014)
2. Ott, M., Choi, Y., Cardie, C., Hancock, J.T.: Finding deceptive opinion spam by any stretch of the imagination. In: Proceedings of the 49th Annual Meeting of the Association for Computational Linguistics: Human Language Technologies, pp. 309–319 (2011)
3. WASHINGTON POST. https://www.washingtonpost.com/business/economy/how-merchants-secretly-use-facebook-to-flood-amazon-with-fake-reviews. Accessed 23 Apr 2018
4. Li, L., Qin, B., Ren, W., Liu, T.: Document representation and feature combination for deceptive spam review detection. Neurocomputing **254**(254), 33–41 (2017)
5. Liu, W., Jing, W., Li, Y.: Incorporating feature representation into BiLSTM for deceptive review detection. Computing **102**(3), 701–715 (2019). https://doi.org/10.1007/s00607-019-00763-y
6. Hajek, P., Barushka, A., Munk, M.: Fake consumer review detection using deep neural networks integrating word embeddings and emotion mining. Neural Comput. Appl. **32**(23), 17259–17274 (2020). https://doi.org/10.1007/s00521-020-04757-2
7. Surowiecki, J.: The Wisdom of Crowds: Why the Many Are Smarter Than the Few and How Collective Wisdom Shapes Business, Economies, Societies and Nations (2004)
8. Schroff, F., Kalenichenko, D., Philbin, J.: FaceNet: a unified embedding for face recognition and clustering. In: 2015 IEEE Conference on Computer Vision and Pattern Recognition (CVPR), pp. 815–823 (2015)
9. Jindal, N., Liu, B.: Review spam detection. In: Proceedings of the 16th International Conference on World Wide Web, pp. 1189–1190 (2007)
10. Mukherjee, A., Venkataraman, V., Liu, B., Glance, N. S.: What yelp fake review filter might be doing. In: 7th International AAAI Conference on Weblogs and Social Media, ICWSM 2013, pp. 409–418 (2013)
11. Rayana, S., Akoglu, L.: Collective opinion spam detection: bridging review networks and metadata. In: Proceedings of the 21th ACM SIGKDD International Conference on Knowledge Discovery and Data Mining, pp. 985–994 (2015)
12. Kc, S., Mukherjee, A.: On the temporal dynamics of opinion spamming: case studies on yelp. In: Proceedings of the 25th International Conference on World Wide Web, pp. 369–379 (2016)

13. Dewang, R.K., Singh, A.K.: Identification of fake reviews using new set of lexical and syntactic features. In: Proceedings of the Sixth International Conference on Computer and Communication Technology 2015, pp. 115–119 (2015)

14. Wang, X., Liu, K., Zhao, J.: Handling cold-start problem in review spam detection by jointly embedding texts and behaviors. In: Proceedings of the 55th Annual Meeting of the Association for Computational Linguistics (vol. 1: Long Papers), pp. 366–376 (2017)

15. Wang, X., Liu, K., Zhao, J.: Detecting deceptive review spam via attention-based neural networks. In: National CCF Conference on Natural Language Processing and Chinese Computing, pp. 866–876 (2017)

16. Yuan, C., Zhou, W., Ma, Q., Lv, S., Han, J., Hu, S.: Learning review representations from user and product level information for spam detection. In: 2019 IEEE International Conference on Data Mining (ICDM), pp. 1444–1449 (2019)

17. Jo, Y., Oh, A. H.: Aspect and sentiment unification model for online review analysis. In: Proceedings of the Fourth ACM International Conference on Web Search and Data Mining, pp. 815–824 (2011)

18. Sun, C., Huang, L., Qiu, X.: Utilizing BERT for aspect-based sentiment analysis via constructing auxiliary sentence. In: Proceedings of the 2019 Conference of the North American Chapter of the Association for Computational Linguistics: Human Language Technologies, vol. 1 (Long and Short Papers), pp. 380–385 (2019)

19. Dong, L., Wei, F., Tan, C., Tang, D., Zhou, M., Xu, K.: Adaptive recursive neural network for target-dependent twitter sentiment classification. In: Proceedings of the 52nd Annual Meeting of the Association for Computational Linguistics (vol. 2: Short Papers), pp. 49–54 (2014)

20. Tang, D., Qin, B., Feng, X., Liu, T.: Effective LSTMs for target-dependent sentiment classification. In: Proceedings of COLING 2016, the 26th International Conference on Computational Linguistics: Technical Papers, pp. 3298–3307 (2016)

21. Hu, M., Peng, Y., Huang, Z., Li, D., Lv, Y.: Open-domain targeted sentiment analysis via span-based extraction and classification. In: Proceedings of the 57th Annual Meeting of the Association for Computational Linguistics, pp. 537–546 (2019)

22. Zhang, C., Li, Q., Song, D.: Aspect-based sentiment classification with aspect-specific graph convolutional networks. In: Proceedings of the 2019 Conference on Empirical Methods in Natural Language Processing and the 9th International Joint Conference on Natural Language Processing (EMNLP-IJCNLP), pp. 4567–4577 (2019)

23. Zhang, C., Li, Q., Song, D., Wang, B.: A multi-task learning framework for opinion triplet extraction. In: Findings of the Association for Computational Linguistics: EMNLP 2020, pp. 819–828 (2020)

24. Li, X., Bing, L., Zhang, W., Lam, W.: Exploiting BERT for end-to-end aspect-based sentiment analysis. In: Proceedings of the 5th Workshop on Noisy User-Generated Text (W-NUT 2019), pp. 34–41 (2019)

25. Melleng, A., Jurek-Loughrey, A., Padmanabhan, D.: Sentiment and Emotion based text representation for fake reviews detection. In: Proceedings of the International Conference on Recent Advances in Natural Language Processing (RANLP 2019), pp. 750–757 (2019)

26. Pontiki, M., et al.: SemEval-2016 task 5: aspect based sentiment analysis. In: Proceedings of the 10th International Workshop on Semantic Evaluation (SemEval-2016), pp. 19–30 (2016)

27. Yih, W., He, X., Meek, C.: Semantic parsing for single-relation question answering. In: Proceedings of the 52nd Annual Meeting of the Association for Computational Linguistics (vol. 2: Short Papers), pp. 643–648 (2014)

Graph Enhanced BERT for Stance-Aware Rumor Verification on Social Media

Kai Ye[ID], Yangheran Piao, Kun Zhao, and Xiaohui Cui[✉]

Key Laboratory of Aerospace Information Security and Trusted Computing,
Ministry of Education, School of Cyber Science and Engineering,
Wuhan University, Wuhan, China
xcui@whu.edu.cn

Abstract. With the rapid development of the Internet and social media, rumors with misleading information will damage more than before. Therefore, rumor verification technology has gained much attention. Many existing models focus on leveraging stance classification to enhance rumor verification. However, most of these models fail to incorporate the conversation structure explicitly. Moreover, the BERT has shown its superiority at text representation on many NLP tasks, and we explore whether it can enhance rumor verification. Thus, in this paper, we propose a single-task model and a multi-task stance-aware model for rumor verification. We first utilize BERT to capture low-level content features. Then we employ graph neural networks to model conversation structure and design an attention structure to integrate stance and rumor information. Our experiments on two public datasets show the superiority of our model over existing models.

Keywords: Natural language processing · Rumor verification · Stance classification

1 Introduction

Since ancient times, the spreading of rumors has been a serious social problem. In recent years, with the rapid development of the Internet and social media, rumors have become more devastating and may even cause significant economic and social impacts [1]. Considering the severe consequences of rumors spreading on social media, it is of great significance to propose a method to identify rumors automatically.

In order to combat the catastrophic outcomes of the spreading of rumors, much work has been done for rumor verification [9,12,15,19]. Among existing methods, a promising line of research aims to use stance information of replying posts to enhance the task of rumor verification.

Supported by National Key Research and Development Program of China No.2018YF C1604000, Fundamental Research Funds for the Central Universities No. 2042017 gf0035.

I. Farkaš et al. (Eds.): ICANN 2021, LNCS 12895, pp. 422–435, 2021.
https://doi.org/10.1007/978-3-030-86383-8_34

The approach of using stance information to verify rumors was proposed years ago [30]. This approach's principle is that users on social media are good at pointing out the wrong information [3]. Our work generally extends this line of research.

Recently, researchers have proposed many multi-task learning models [17,27, 29] in order to jointly conduct rumor verification and stance classification. Their multi-task learning models are mainly motivated by the framework proposed by Kochkina et al. [13]. In this framework, low-level layers are shared by rumor verification and stance classification. However, in the high level layers, different layers are designed for different tasks.

Although these multi-task learning approaches definitely make significant progress on rumor verification, they still have limits on certain aspects. (1) Most of these models fail to explicitly incorporate the information of conversation structure into their models. They usually organize the input posts chronologically. This practice will lead to a considerable degree of information loss because the connections between replying posts or between source post and the replies are ignored. (2) Another shortcoming of previous models is that they didn't model the interactions between stance task and rumor task explicitly. The high layer representations of stance task are not exploited for rumor verification.

To address the above weakness of existing models, in this paper, we examine the potential of BERT [7] and graph neural network for stance-aware rumor verification. Existing methods fail to model conversation structure because most of them treat input thread as a sequence of sentences. Contrary to them, graph neural networks are designed to process graph structure data which is apparently closer to the nature of conversations on social media. To sum up, we propose a novel multi-task learning model using GCN [11], BERT and Transformer [26], named GBERT, short for Graph enhanced BERT.

Our contributions are summarized as follows:

- We employ BERT in the low-level layers of our model to get better content features. And we explicitly model the interactions between stance and rumor task.
- We explicitly encode conversation structure and stance information simultaneously based on the representations extracted from low-level layers. Furthermore, we design an attention structure to combine them to enhance rumor verification.
- We propose a novel multi-task learning model named GBERT. To demonstrate the effectiveness of our model, we conduct extensive experiments on two public benchmark datasets. Our experiments show that it outperforms state-of-the-art models.

2   Related Work

In this section, we discuss some essential works about rumor verification and stance classification.

2.1   Stance Classification

Stance Classification is a prevalent task in which words or sentences has to be classified into several categories to find people's attitude towards a specific topic. Most researchers regard it as a classification problem. It is a broad research area, and we focus on Stance classification for conversations on social media platforms.

Mohammad et al. [23] proposed a stance dataset using tweets and organized a SemEval competition in 2016 (Task 6). From then on, many researchers [16,18, 28] proposed algorithms to learn stance from text on social media conversations. Wei et al. [28] developed a convolutional neural network for stance detection in tweets. With the powerful learning ability of deep neural networks, their system obtained a good rank in SemEval 2016 Task 6.

The recent pre-trained language models(ELMO [24], OpenAI GPT [25], BERT [7]...) have achieved excellent results on many downstream tasks including stance classification. Thus, we leverage the power of the BERT in the low-level layers of our model to acquire better content features of posts.

2.2   Rumor Verification

There generally exist two kinds of approaches for single-task rumor verification: linguistic-based methods and computational-based methods.

Linguistic-Based Approaches. The first one usually make use of linguistic features to capture semantic representations, such as word frequency and word length [8,22]. Some of this line of work also employ various kernels to model the propagation structure [20]. This kind of work has bad universality and might show poor performance on a different topic or domain.

Computational-Based Approaches. In recent years, with the development of deep learning technology, another popular line of research exploits variants of neural networks to automatically capture significant features of the posts involved [4,19]. These models aim to capture semantic information of text but ignore the structure information in conversations on social media by treating the input as sequences of text.

However, We do not follow the research lines above, as we leverage the stance information to enhance rumor verification.

2.3   Stance-Aware Rumor Verification

Recently, a promising line of research aims to utilize stance signals to enhance rumor verification. Due to the close connection between the two tasks, researchers who follow this research line propose many multi-task learning frameworks, attempting to train the model on stance classification and rumor verification simultaneously. These multi-task models usually show better performance than their single-task counterparts. For example, Kochkina et al. [13] and Ma et al. [21] proposed multi-task learning architectures based on LSTM [10] and GRU [5] respectively. Kumar and Carley [14] also made contributions by proposing a tree-based model. The layers in these models are generally divided into two parts: low-level layers and high-level layers. The low-level layers are shared by both tasks, while high-level layers are designed for each specific task. For multi-task training, they sum up the loss of each task to get the total loss for optimization.

Although the above models improve performance by incorporating stance information, most still can't explicitly encode conversation structure information. Moreover, the way they exploit stance information is to modify loss function but don't model interactions between stance and rumor task [29].

Our work broadens this field of research by incorporating BERT to capture better content features and Graph neural network for modeling conversation structure.

2.4 Graph Convolutional Networks

Proposed by Kipf and Welling [11], Graph Convolutional Networks (GCN) is a variant of convolutional neural networks which directly operate on graph-structured data. The GCN model follows this formula:

$$H^{(l+1)} = \sigma \left(\tilde{D}^{-\frac{1}{2}} \tilde{A} \tilde{D}^{-\frac{1}{2}} H^{(l)} W^{(l)} \right) \tag{1}$$

Here, $H^{(l)}$ denotes the lth layer in the network, σ is the non-linearity, and $W^{(l)}$ is the weight matrix for this layer. $\tilde{A} = A + I$ means adding the self-loop to the adjacency matrices. $\tilde{D}$ and $\tilde{A}$ are the degree and adjacency matrices (self-loop added) respectively. The input is $H^{(0)} \in \mathbb{R}^{N \times d}$ where N is the number of nodes and d is the number of input features. By stacking up several layers, we can get the node representation output.

In the original paper, experiments on a number of network datasets have been conducted, suggesting GCN's capability to encode both graph structure and node features.

3 Methodology

3.1 Problem Statement

We first define $\mathbb{C} = \{T_1, T_2, \ldots, T_{|\mathbb{C}|}\}$ to denote all threads(conversations) in the corpus $\mathbb{C}$. Each thread T_i has a source post S_0 and its reply posts $\{R_1, \ldots, R_N\}$. Each thread T_i also has a rumor label y:

$$y \in \{\textit{false rumor, true rumor, unverified rumor}\}$$

Additionally, each post (S_0 and R_i) in the threads has a stance label c indicating its stance towards the source post:

$$c \in \{support,\ deny,\ query,\ comment\}$$

For stance classification task, our goal is to train a model $g : S_0, R_1, \ldots, R_N \rightarrow c_0, c_1, \ldots, c_N$. For rumor verification task, the goal is to train a model $f : T_i \rightarrow y_i$.

3.2 Graph Enhanced BERT for Single-Task Rumor Verification

Figure 1 shows the overall structure of our single-task model, we will describe it in this subsection.

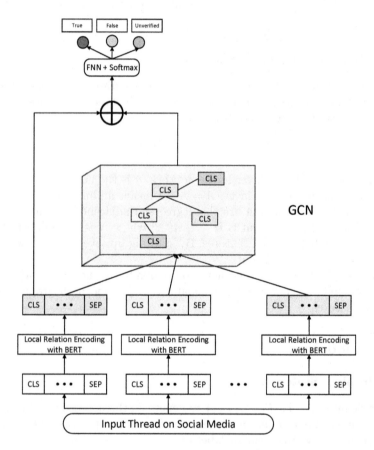

Fig. 1. Overall structure of our single-task model.

Input Layers. Before feeding the posts into the model, we first insert some special tokens to each post, [PAD] for padding, [CLS] to the beginning of post and [SEP] to the end of post. We use the token [CLS] to represent the whole sentence in later sections.

As shown in Fig. 1, let $T_m = (S_0, R_1, \ldots, R_N)$ denotes the input thread, where $m \in [1, |\mathbb{C}|]$, S_0 denotes the source claim and R_i denotes the i-th reply post. For simplicity and clarity, we assume each post contains k tokens. Now we can use $\mathbf{X}_i = \left(x_i^{CLS}, x_i^1, \ldots, x_i^{k-2}, x_i^{SEP}\right)$ to denote the input vector of i-th post. Naturally, $\mathbf{X}_0$ is the input vector of the source claim. It is worth mentioning that each x is the sum-up of its token embeddings and position embeddings.

Local Relation Encoding. Then, we make use of pre-trained BERT to capture the semantic interactions between each token by separately feeding each post into Local Relation Encoding (LRE).

$$\mathbf{h}_i = \mathrm{BERT}\,(\mathbf{X}_i)\,, \quad i = 0, 1, \ldots, N \tag{2}$$

where $\mathbf{h}_i \in \mathbb{R}^{k \times d}$ is the low layer representation of i-th post.

Conversation Structure Encoding. We use $\mathbf{h}_i^{CLS} \in \mathbb{R}^d$ in $\mathbf{h}_i$ to represent the post and $\mathbf{H}^{CLS} = \left(\mathbf{h}_0^{CLS}, \mathbf{h}_1^{CLS}, \ldots, \mathbf{h}_N^{CLS}\right), \mathbf{H}^{CLS} \in \mathbb{R}^{N \times d}$ to represent the entire thread. In order to encode the conversation structure, we resort to GCN. GCNs have shown great capacity in capturing the graph structures. This process is formulated as follows:

$$\mathbf{S}^{CLS} = \mathrm{GCN}(\mathbf{H}^{CLS}) \tag{3}$$

where $\mathbf{S}^{CLS}$ is the node embedding of last layer in GCN.

Output   Layers.   After   getting   the GCN output $\mathbf{s}_{mean}^{CLS} = \left(\mathbf{s}_0^{CLS} + \mathbf{s}_1^{CLS} + \ldots + \mathbf{s}_N^{CLS}\right) / N$,we add a feed-forward layer and a softmax layer following GCN and feed the concatenation of $\mathbf{s}_{mean}^{CLS}$ and $\mathbf{h}_0^{CLS}$ into it for the final prediction:

$$p\left(y \mid \mathbf{q}_0^{CLS}\right) = \mathrm{softmax}\left(\mathbf{W}^T \mathbf{q}_0^{CLS} + \mathbf{b}\right) \tag{4}$$

where $\mathbf{W} \in \mathbb{R}^{d \times 3}$ and $\mathbf{b} \in \mathbb{R}^3$ are learned parameters, $\mathbf{q}_0^{CLS}$ is the concatenation of $\mathbf{s}_{mean}^{CLS}$ and $\mathbf{h}_0^{CLS}$.

3.3   Graph Enhanced BERT for Multi-task Stance-Aware Rumor Verification

Figure 1 shows the overall structure of our multi-task model, which will be introduced in this subsection.

As shown in Fig. 1, the low-level layers are the same as single-task model, designed to extract content features of posts. However, in order to jointly perform rumor verification and stance classification, we design a stance learning module

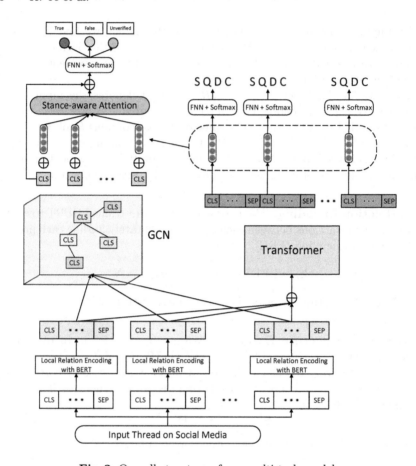

Fig. 2. Overall structure of our multi-task model.

in the multi-task model. Additionally, we want to make full use of the stance information to enhance rumor verification. Therefore, we design an attention structure to model the inner interactions of stance vector and rumor vector.

Stance Learning Module. In order to capture the semantic relations between all posts in the entire conversation thread, we first concatenate $\mathbf{h}_i$ to form the semantic representation of the whole thread:

$$\mathbf{h} = \mathbf{h}_0 \oplus \mathbf{h}_1 \oplus \ldots \oplus \mathbf{h}_N \tag{5}$$

Then, we feed $\mathbf{h}$ into a self-attention layer and a feed-forward neural network for extracting the semantic interactions between all posts:

$$\mathbf{h}' = \mathrm{LN}(\mathbf{h} + \text{Self-ATT}(\mathbf{h})) \tag{6}$$

$$\mathbf{p} = \mathrm{LN}(\mathbf{h}' + \mathrm{FFN}(\mathbf{h}')) \tag{7}$$

where Self-ATT refers to the self-attention layer and FFN refers to the feedforward network. LN refers to layer normalization [2]. Here we apply layer normalization because it can improve training and has become a common practice as explained in [29]. We design this part based on the encoder structure from Transformer [26]. In theory, any structure similar to encoder from Transformer (BERT and its variants) can be used here, but we don't choose them for computational reasons. Whether other structures may improve performance can be studied in future research.

Finally, as shown in Fig. 2, we choose the representation of i-th [CLS] token $\mathbf{p}_i^{CLS}$ as the stance feature of i-th post. For stance learning task, a feed-forward and a softmax layer are added for the prediction:

$$p\left(c_i \mid \mathbf{p}_i^{CLS}\right) = \text{softmax}\left(\mathbf{W}^T \mathbf{p}_i^{CLS} + \mathbf{b}\right) \tag{8}$$

where $\mathbf{W} \in \mathbb{R}^{d \times 4}$ and $\mathbf{b} \in \mathbb{R}^4$ are learned parameters.

Conversation Structure Encoding. The conversation structure encoding module in our multi-task learning framework is the same as in the single-task model (see Sect. 3.2).

Stance-Aware Attention Structure. Now we have stance representation $\mathbf{P}^{CLS} = \left(\mathbf{p}_0^{CLS}, \mathbf{p}_1^{CLS}, \ldots, \mathbf{p}_N^{CLS}\right)$ and representation of conversation structure $\mathbf{S}^{CLS} = \left(\mathbf{s}_0^{CLS}, \mathbf{s}_1^{CLS}, \ldots, \mathbf{s}_N^{CLS}\right)$. In order to perform stance-aware rumor verification, we design a stance-aware attention structure to integrate the above information. We obtain the stance-aware thread representation $\mathbf{q}$ as follows:

$$e_i = \mathbf{a}^T \tanh\left(\mathbf{W}_s \mathbf{s}_i^{CLS} + \mathbf{W}_p \mathbf{p}_i^{CLS}\right) \tag{9}$$

$$\alpha_i = \frac{\exp\left(e_i\right)}{\sum_{l=1}^N \exp\left(e_l\right)} \tag{10}$$

$$\mathbf{q} = \sum_{i=1}^N \alpha_i \left(\mathbf{s}_i^{CLS} \oplus \mathbf{p}_i^{CLS}\right) \tag{11}$$

Output Layers. Finally, to our best knowledge, the information of the source claim is usually more important than other posts, we concatenate $\mathbf{q}$ and $\mathbf{s}_0^{CLS}$ for the prediction. Also, the same as in the single-task model, a feed-forward and a softmax layer are added for the prediction:

$$p\left(y \mid \mathbf{s}_0^{CLS}, \mathbf{q}\right) = \text{softmax}\left(\mathbf{W}^T \left(\mathbf{s}_0^{CLS} \oplus \mathbf{q}\right) + \mathbf{b}\right) \tag{12}$$

where $\mathbf{W} \in \mathbb{R}^{3d \times 3}$ and $\mathbf{b} \in \mathbb{R}^3$ are weight and bias parameters.

Objective of Optimization. To perform multi-task learning, we adopt the following loss function to optimize the parameters. The loss function we apply is shown below:

$$\mathcal{L} = -\left(\frac{1}{M}\sum_{j=1}^{M}\log p\left(y \mid \mathbf{s}_0^{CLS}, \mathbf{q}\right) + \frac{1}{M}\sum_{j=1}^{M}\sum_{i=1}^{K}\log p\left(c_i \mid \mathbf{p}_i^{CLS}\right)\right) \quad (13)$$

where M refers to the number of training samples for rumor verification, and K refers to the number of annotated posts for stance classification in the thread.

4   Experiments and Results

4.1   Description of Data

Experiments were carried out to demonstrate the effectiveness of our proposed models on two datasets, SemEval-2017 and PHEME. Table 1 shows some specific information about the two datasets.

Table 1. Specific information of SemEval-2017 and PHEME datasets.

Dataset	#Threads	#Tweets	Stance labels				Rumor veracity labels		
			#Support	#Deny	#Query	#Comment	#True	#False	#Unverified
SemEval-2017	325	5,568	1,004	415	464	3,685	145	74	106
PHEME	2,402	21382	–				1,067	638	697

SemEval-2017 is proposed by Derczynski et al. [6] for rumor verification and stance classification. For each thread, its conversation structure and rumor label are provided. And for each post, a stance label indicating its stance towards the source claim is provided.

PHEME [13] is an extension of the SemEval-2017 dataset introduced above. It contains 2402 Twitter threads on nine separate breaking events. However, only 325 threads in the SemEval-2017 dataset have stance annotations. Other threads in PHEME only have rumor veracity label.

4.2   Experimental Setup

For the low-level layers of our models, we employ the pre-trained uncased $BERT_{base}$ model [7]. When training both our single-task and multi-task model, we set the learning rate as 5e−5 and the dropout rate as 0.1. The SemEval-2017 dataset has already been split into training, development and test sets and we follow this conduct. Previous works [13] on PHEME mostly perform leave-one-event-out cross-validation and we follow it too for comparison with existing models. In each fold, we choose one event for testing and all other events for training. As only some of the threads have stance labels in this dataset, when training our multi-task model, we don't compute the loss of stance classification task if the thread doesn't have stance labels.

For evaluation metrics, we use accuracy and Macro-F_1 for rumor verification task.

4.3 Discussion of Results

We have conducted extensive experiments to compare our proposed models' performance with several popular baselines for rumor verification.

Single-Task Baselines. For single-task models, only rumor veracity labels are used for training.

- TD-RvNN [21]: A model using recursive neural networks tries to capture propagation features of the top-down tree-structure in a rumor thread.
- BranchLSTM [13]: BranchLSTM is a LSTM-based structure aiming to model the sequential features of branches.
- Hierarchical Transformer [29]: Hierarchical Transformer exploits BERT and Transformer structure to encode input threads for rumor verification.

Multi-task Baselines. Stance labels are available during the training process of multi-task models.

- BranchLSTM + NileTMRG [13]: It is a multi-task pipeline which first trains a BranchLSTM for stance classification, and uses the predicted stance vectors to enhance an SVM classifier for rumor veracity prediction.
- MTL2 (Veracity + Stance) [13]: It is a new multi-task learning framework in which the two tasks are trained jointly.
- Coupled Hierarchical Transformer [29]: Coupled Hierarchical Transformer jointly trains stance classification and rumor verification with BERT, Transformer and attention structure.

Table 2. Experimental results of rumor verification.

Setting	Model	SemEval-2017		PHEME	
		Accuracy	Macro-F_1	Accuracy	Macro-F_1
Single-Task	TD-RvNN	0.536	0.509	0.341	0.264
	BranchLSTM	0.500	0.491	0.314	0.259
	Hierarchical transformer	0.607	0.592	0.441	0.372
	GBERT (ours)	**0.645**	**0.627**	**0.473**	**0.396**
Multi-Task	BranchLSTM + NileTMRG	0.570	0.539	0.360	0.297
	MTL2 (Veracity + Stance)	0.571	0.558	0.357	0.318
	Coupled hierarchical transformer	0.678	0.680	0.466	0.396
	Multi-Task GBERT (ours)	**0.706**	**0.702**	**0.481**	**0.427**

Performance Comparison. Table 2 shows the performance of our proposed models and all competing models on SemEval-2017 and PHEME dataset.

We first compare the performance of the single-task model. Since RvNN is tree-structured and it only uses the hidden state of last layer nodes as the representation of the entire tree, other nodes, especially first layer nodes, are ignored to a great extent. BranchLSTM has a similar shortcoming; it uses the final time step's outputs to make rumor veracity prediction, essentially ignoring the source claim's information. For the reason that Hierarchical Transformer exploits [CLS] token of the source claim for classification and the similar conduct in our GBERT, the performances of the two models are significantly better. Moreover, our GBERT outperforms Hierarchical Transformer and has the best performance among all the single-task models. This is because our GBERT enhances BERT by modeling the conversation structures of rumor threads, while Hierarchical Transformer cannot.

For multi-task models, it can be observed that all the multi-task models outperform their corresponding single-task models, demonstrating the effectiveness of utilizing stance information. Also, comparisons between multi-task models show that jointly learning models outperform the pipeline model (BranchLSTM + NileTMRG). It is not surprising because jointly learning models leverage the interactions between the two tasks to improve performance. Coupled Hierarchical Transformer and our Multi-Task GBERT exploit BERT as low-level layers while MTL2 employs a single LSTM as low-level layers, explaining why the former models achieve better results than MTL2. Further, the main difference between our Multi-Task GBERT and Coupled Hierarchical Transformer is that we add conversation structure information through graph neural networks, contributing to the better performance of our model.

The above analysis demonstrates the superiority of our proposed models on stance-aware rumor verification.

Table 3. Ablation study on the SemEval-2017 dataset.

Setting	Models	Accuracy	Macro-F_1
Single-Task	GBERT	**0.645**	**0.627**
	−Remove graph networks	0.552	0.531
Multi-Task	Multi-Task GBERT	**0.706**	**0.692**
	−Remove graph networks	0.612	0.604
	−Remove attention structure	0.684	0.671

4.4   Ablation Study

To better understand what role each module in our models performs, we further conduct ablation study. The results are summarized in Table 3.

We first examine the effect of removing the graph networks. The performances of both settings (single-task and multi-task) are significantly harmed without graph networks, which aligns with the central assumption of our study. This again implies the importance of encoding conversation structure.

Additionally, by removing the attention structure in multi-task model, we explore whether explicitly modeling interactions between stance and rumor task helps improve performance. The experimental results show that it is also indispensable, as removing the attention structure will do noticeable damage to the model performance. This means when performing multi-task training, merely modifying loss function seems not enough. More sophisticated models are needed to extract deeper interactions between the two tasks.

5   Conclusions

In this paper, we propose a single-task model and a multi-task stance-aware model for rumor verification. We first utilize BERT to capture low-level content features. Then we employ graph neural networks to model conversation structure and design an attention structure to integrate stance and rumor information. Experiments on two public datasets demonstrate the effectiveness of our models.

References

1. Allcott, H., Gentzkow, M.: Social media and fake news in the 2016 election. J. Econ. Perspect. **31**(2), 211–36 (2017)
2. Ba, J.L., Kiros, J.R., Hinton, G.E.: Layer normalization. arXiv preprint arXiv:1607.06450 (2016)
3. Babcock, M., Cox, R.A.V., Kumar, S.: Diffusion of pro-and anti-false information tweets: the black panther movie case. Comput. Math. Organ. Theory **25**(1), 72–84 (2019)
4. Chen, T., Li, X., Yin, H., Zhang, J.: Call attention to rumors: deep attention based recurrent neural networks for early rumor detection. In: Ganji, M., Rashidi, L., Fung, B.C.M., Wang, C. (eds.) PAKDD 2018. LNCS (LNAI), vol. 11154, pp. 40–52. Springer, Cham (2018). https://doi.org/10.1007/978-3-030-04503-6_4
5. Chung, J., Gulcehre, C., Cho, K., Bengio, Y.: Empirical evaluation of gated recurrent neural networks on sequence modeling. In: NIPS 2014 Workshop on Deep Learning, December 2014
6. Derczynski, L., Bontcheva, K., Liakata, M., Procter, R., Wong Sak Hoi, G., Zubiaga, A.: SemEval-2017 task 8: RumourEval: determining rumour veracity and support for rumours. In: Proceedings of the 11th International Workshop on Semantic Evaluation (SemEval-2017), Vancouver, Canada, pp. 69–76. Association for Computational Linguistics, August 2017. https://doi.org/10.18653/v1/S17-2006, https://www.aclweb.org/anthology/S17-2006
7. Devlin, J., Chang, M.W., Lee, K., Toutanova, K.: BERT: pre-training of deep bidirectional transformers for language understanding. In: Proceedings of the 2019 Conference of the North American Chapter of the Association for Computational Linguistics: Human Language Technologies, Volume 1 (Long and Short Papers), Minneapolis, Minnesota, pp. 4171–4186. Association for Computational Linguistics (2019). https://doi.org/10.18653/v1/N19-1423, https://www.aclweb.org/anthology/N19-1423

8. Fuller, C.M., Biros, D.P., Wilson, R.L.: Decision support for determining veracity via linguistic-based cues. Decis. Support Syst. **46**(3), 695–703 (2009)
9. Hardalov, M., Arora, A., Nakov, P., Augenstein, I.: A survey on stance detection for mis-and disinformation identification. arXiv preprint arXiv:2103.00242 (2021)
10. Hochreiter, S., Schmidhuber, J.: Long short-term memory. Neural Comput. **9**(8), 1735–1780 (1997)
11. Kipf, T.N., Welling, M.: Semi-supervised classification with graph convolutional networks. arXiv preprint arXiv:1609.02907 (2016)
12. Kochkina, E., Liakata, M.: Estimating predictive uncertainty for rumour verification models. In: Proceedings of the 58th Annual Meeting of the Association for Computational Linguistics, pp. 6964–6981 (2020)
13. Kochkina, E., Liakata, M., Zubiaga, A.: All-in-one: multi-task learning for rumour verification. In: Proceedings of the 27th International Conference on Computational Linguistics, pp. 3402–3413 (2018)
14. Kumar, S., Carley, K.M.: Tree LSTMs with convolution units to predict stance and rumor veracity in social media conversations. In: Proceedings of the 57th Annual Meeting of the Association for Computational Linguistics, pp. 5047–5058 (2019)
15. Li, W., Ding, W., Sadasivam, R., Cui, X., Chen, P.: His-GAN: a histogram-based GAN model to improve data generation quality. Neural Netw. **119**, 31–45 (2019)
16. Li, W., Fan, L., Wang, Z., Ma, C., Cui, X.: Tackling mode collapse in multi-generator GANs with orthogonal vectors. Pattern Recogn. **110**, 107646 (2021)
17. Li, W., Liang, Z., Ma, P., Wang, R., Cui, X., Chen, P.: Hausdorff GAN: improving GAN generation quality with Hausdorff metric. IEEE Trans. Cybern. (2021)
18. Liu, C., et al.: IUCL at SemEval-2016 task 6: an ensemble model for stance detection in Twitter. In: Proceedings of the 10th International Workshop on Semantic Evaluation (SemEval-2016), pp. 394–400 (2016)
19. Ma, J., et al.: Detecting rumors from microblogs with recurrent neural networks. In: The 25th International Joint Conference on Artificial Intelligence (IJCAI 2016) (2016)
20. Ma, J., Gao, W., Wong, K.F.: Detect rumors in microblog posts using propagation structure via kernel learning. In: Meeting of the Association for Computational Linguistics (2017)
21. Ma, J., Gao, W., Wong, K.F.: Rumor detection on twitter with tree-structured recursive neural networks. In: The 56th Annual Meeting of the Association for Computational Linguistics (2018)
22. Mihalcea, R., Strapparava, C.: The lie detector: explorations in the automatic recognition of deceptive language. In: Proceedings of the ACL-IJCNLP 2009 Conference Short Papers, pp. 309–312 (2009)
23. Mohammad, S.M., Sobhani, P., Kiritchenko, S.: Stance and sentiment in tweets. ACM Trans. Internet Technol. (TOIT) **17**(3), 1–23 (2017)
24. Peters, M.E., et al.: Deep contextualized word representations. In: Proceedings of NAACL-HLT, pp. 2227–2237 (2018)
25. Radford, A., Narasimhan, K., Salimans, T., Sutskever, I.: Improving language understanding by generative pre-training (2018)
26. Vaswani, A., et al.: Attention is all you need. In: Advances in Neural Information Processing Systems, pp. 5998–6008 (2017)
27. Wei, P., Xu, N., Mao, W.: Modeling conversation structure and temporal dynamics for jointly predicting rumor stance and veracity. In: Proceedings of the 2019 Conference on Empirical Methods in Natural Language Processing and the 9th International Joint Conference on Natural Language Processing (EMNLP-IJCNLP), pp. 4789–4800 (2019)

28. Wei, W., Zhang, X., Liu, X., Chen, W., Wang, T.: pkudblab at SemEval-2016 task 6: a specific convolutional neural network system for effective stance detection. In: Proceedings of the 10th International Workshop on Semantic Evaluation (SemEval-2016), pp. 384–388 (2016)

29. Yu, J., Jiang, J., Khoo, L.M.S., Chieu, H.L., Xia, R.: Coupled hierarchical transformer for stance-aware rumor verification in social media conversations. Association for Computational Linguistics (2020)

30. Zubiaga, A., Liakata, M., Procter, R., Bontcheva, K., Tolmie, P.: Crowdsourcing the annotation of rumourous conversations in social media. In: Proceedings of the 24th International Conference on World Wide Web, pp. 347–353 (2015)

Deep Learning for Suicide and Depression Identification with Unsupervised Label Correction

Ayaan Haque[1](✉), Viraaj Reddi[1], and Tyler Giallanza[2]

[1] Saratoga High School, Saratoga, CA, USA
[2] Department of Psychology, Princeton Neuroscience Institute, Princeton University, Princeton, NJ, USA

Abstract. Early detection of suicidal ideation in depressed individuals can allow for adequate medical attention and support, which can be life-saving. Recent NLP research focuses on classifying, from given text, if an individual is suicidal or clinically healthy. However, there have been no major attempts to differentiate between depression and suicidal ideation, which is a separate and important clinical challenge. Due to the scarce availability of EHR data, suicide notes, or other verified sources, web query data has emerged as a promising alternative. Online sources, such as Reddit, allow for anonymity, prompting honest disclosure of symptoms, making it a plausible source even in a clinical setting. However, online datasets also result in inherent noise in web-scraped labels, which necessitates a noise-removal process to improve performance. Thus, we propose SDCNL, a deep neural network approach for suicide versus depression classification. We utilize online content to train our algorithm, and to verify and correct noisy labels, we propose a novel unsupervised label correction method which, unlike previous work, does not require prior noise distribution information. Our extensive experimentation with various deep word embedding models and classifiers display strong performance of SDCNL as a new clinical application for a challenging problem (We make our **supplemental**, dataset, web-scraping script, and code (with hyperparameters) available at https://github.com/ayaanzhaque/SDCNL).

Keywords: Suicide/Depression · Noisy labels · Deep learning · Online content · Natural Language Processing · Unsupervised learning

1 Introduction

Depression remains among the most pressing issues worldwide and can often progress to suicidal ideation or attempted suicide if left unaddressed. Diagnosis

---

A. Haque and V. Reddi—Equal contribution.

Electronic supplementary material The online version of this chapter (https://doi.org/10.1007/978-3-030-86383-8_35) contains supplementary material, which is available to authorized users.

© Springer Nature Switzerland AG 2021
I. Farkaš et al. (Eds.): ICANN 2021, LNCS 12895, pp. 436–447, 2021.
https://doi.org/10.1007/978-3-030-86383-8_35

of depression and identification of when it becomes a risk of attempted suicide is an important problem at both the individual and population level. Many existing methods for detecting suicidal ideation rely on data from sources such as questionnaires, Electronic Health Records (EHRs), and suicide notes [10]. However, acquiring data in such formats is challenging and ultimately results in limited datasets, complicating attempts to accurately automate diagnosis.

Conversely, as the Internet and specifically social media have grown, online forums have developed into popular resources for struggling individuals to seek guidance and assistance. These forums have potential to be scraped to create datasets for automated systems of mental health diagnosis, as they are extensive and free to access. Especially for neural network based approaches that require large datasets to be trained efficiently, a growing number of studies are using this data for diagnostic purposes, which are detailed in this review paper [10].

In particular, Reddit has emerged as an important data source for diagnosing mental health disorders [19]. Reddit is an online social media forum in which users form communities with defined purposes referred to as subreddits. Certain subreddits discuss dealing with mental health and openly explain their situations (r/depression and r/SuicideWatch). Reddit specifically allows users to create alternate and discardable accounts to ensure privacy and anonymity, which promotes disclosure and allows those with little support systems in real life to receive support online [5]. The wide user base, honesty of these online settings, and moderated screening of these posts to ensure legitimacy provides an unprecedented opportunity for computationally analyzing mental health issues on a large scale.

Despite the extensive research into classifying between healthy and mentally unstable patients through text, there remains little work focused on detecting when individuals with underlying mental health struggles such as depression are at risk of attempting suicide. This represents an important clinical challenge, both for the advancement of how depression is treated and for implementing interventions [2,13]. Distinguishing between suicidality and depression is a more fine-grained task than distinguishing between suicidal and healthy behavior, explaining the lack of current solutions. Online data has traditionally been difficult to use in such fine-grained situations, because labels for such data are often unreliable given their informal nature and lack of verification. In particular, labeling data based on subreddit relies on self-reporting, since each user chooses which subreddit they feel best reflects their mental state; thus, they may over or under report their diagnosis. This concept is referred to as *noisy labels* as there is a potential for certain labels to be corrupted. Estimates show that noisy labels can degrade anywhere from 10% to 40% of datasets [20], presenting serious challenges for machine learning algorithms.

Current attempts to address the noisy label problem can be categorized into three notable groups: noise-robust methods, noise-tolerant methods, and data cleaning methods [7,20]. Noise-robust approaches rely on algorithms that are naturally less sensitive to noise (e.g. lower dimensional or regularized algorithms), whereas noise-tolerant methods directly model the noise during training. Although both approaches have received considerable attention in the image-processing domain [20], these methods do not transfer to NLP algorithms. In the NLP domain, there have been a few recently proposed noisy label methods [8,12,22]. However, the proposed methods have limitations for our task. For example, some methods utilize a smaller set of trusted data to correct a larger set of noisy data [8,22], which is infeasible for our application as there is no way to evaluate which posts are accurate. Other methods require training a network directly and end-to-end from corrupted labels [12], which both requires a relatively large amount of data and is less capable of leveraging transfer learning from pre-trained, state-of-the-art models [4,6,17].

Data cleaning methods are more suitable for the present task. However, most existing label cleaning methods make assumptions about or require knowledge on the distribution of noise in the dataset [9,11]. In our use-case, where there is no prior knowledge of the noise distribution, an unsupervised method, such as clustering, is required. Although there are a few methods that use unsupervised clustering algorithms for noisy label learning [3], none of these correct labels. Rather, they train a model to be robust to noise through instance weighting or exclusion. These methods would be problematic for our task; due to the high noise proportion, weighting or removing a high volume of data would damage performance, especially for deep neural networks require large amounts of data. Thus, the present task requires an unsupervised method for data cleaning that utilizes label correction rather than elimination. To the best of our knowledge, there are no current methods which perform label correction using unsupervised clustering methods, and particularly not in the NLP domain.

In this paper, we present SDCNL to address the unexplored issue of classifying between depression and more severe suicidal tendencies using web-scraped data and neural networks. Our primary contributions can be summarized as follows:

- Deep neural network sentiment analysis applied for depression versus suicidal ideation classification, an important but unexplored clinical and computational challenge
- A novel, unsupervised label correction process for text-based data and labels which does not require prior noise distribution information, allowing for the use of mass online content
- Extensive experimentation and ablation on multiple datasets, demonstrating the improved performance of all SDCNL components on the challenging proposed task

2   Methods

The SDCNL method is outlined in Fig. 1. We begin by processing text data scraped from Reddit with word embedding models. These embeddings are then processed with an unsupervised dimensionality reduction algorithm, as clustering algorithms do not perform well in high-dimensional domains [21]. The reduced embeddings are then inputted into a clustering-based algorithm which predicts new labels in an unsupervised manner, meaning it is independent of noise. These alternate labels are compared against the ground-truth labels using a confidence-based thresholding procedure in order to correct the ground-truth labels. The corrected set of labels are then used to train a deep neural network in a supervised fashion.

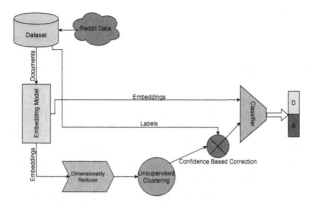

Fig. 1. Schematic of the SDCNL pipeline used for classification of suicide vs depression and noisy label correction via unsupervised learning.

2.1   Embedding Models

Our framework initially utilizes word embedding models to convert raw documents, which in our case are referred to as posts, to numerical word embeddings. Our proposed method can be used with any embedding models, but given our task, we require greater-than-word text embedding models optimized to work with phrases, sentences, and paragraphs. We experiment with 3 state-of-the-art transformers: Bidirectional Encoder Representations from Transformers (BERT) [6], Sentence-BERT [17], and the Google Universal Sentence Encoder (GUSE) [4]. BERT is a state-of-the-art, bidirectionally trained transformer that achieves high performance on various benchmark NLP tasks, and outputs a 768×512 dimensional vector of embeddings. Sentence-BERT is an extension of the original BERT architecture that is retrained and optimized for longer inputs and better performance during clustering, and it outputs a 768×1 dimensional vector. GUSE is a transformer also trained and optimized for greater-than-word length text, but rather returns a 512×1 dimensional vector.

Some classifiers require word level representations for embeddings, while others require document level representations. BERT outputs both multi-dimensional word level embeddings as well as document level embeddings, which are provided by CLS tokens. Depending on what the classifier requires, we vary the inputted embeddings to match the classifier's requirement. In addition, we also experiment on three vectorizers as baselines: Term Frequency–Inverse Document Frequency (TFIDF), Count Vectorizer (CVec), and Hashing Vectorizer (HVec).

2.2   Label Correction

To address the issue of label noise for our task, we propose an unsupervised label correction method. We initially feed our word embeddings through a dimensionality reduction algorithm to convert the high-dimensional features outputted by the embedding models to lower-dimensional representation. Due to the nature of most clustering algorithms, high-dimensional data typically results in subpar performance and poorly separated clusters, a phenomenon known as the "Curse of Dimensionality" [21]. Thus, representing the data in lower dimensions is a necessary procedure. We experiment with three separate dimensionality reduction algorithms: Principal Component Analysis (PCA), Deep Neural Autoencoders, and Uniform Manifold Approximation and Projection (UMAP) [15]. PCA is a common reduction algorithm that extracts the most important information from a matrix of numerical data and represents it as a set of new orthogonal variables. Autoencoders use an unsupervised neural network to compress data into a low-dimensional space, and then reconstruct it while retaining the most possible information, enforcing efficient representation learning. The output of the encoder portion is used as the reduced embeddings. UMAP produces a graph from high-dimensional data and is optimized to generate a low-dimensional graph as similar to the input as possible. UMAP is specifically effective for high-dimensional data, as it has improved preservation of global structure and increased speed.

After reducing the dimensions of our word embeddings, we use clustering algorithms to separate them into two distinct clusters, which allow us to assign new labels to each post. We leverage clustering algorithms because of their unsupervised nature; this is critical because we have no prior knowledge regarding the noise distribution in the labels, requiring a clustering procedure which is independent of the web-scraped labels. We use the Gaussian Mixture Model (GMM) as our clustering algorithm. A GMM is a parametric probability density function used as a model of the probability distribution of continuous measurements in order to cluster given data using probabilities. As a baseline, we use K-Means clustering; K-means attempts to divide n observations into k clusters, such that each observation is assigned to the cluster with the closest mean, and the clusters minimize within-cluster distance while maximizing between-cluster distance. To avoid the dimensionality reduction requirement, we use subspace clustering via spectral clustering [16], which specifically allows unsupervised clustering of high-dimensional data by identifying clusters in different sub-spaces within a dataset.

Each word embedding is now associated with two labels: the original labels based on the subreddit, which are the ground-truth labels, and the new labels resulting from unsupervised clustering. We subsequently leverage a confidence-based thresholding method to correct the ground-truth labels. If the clustering algorithm predicts a label with a probability above τ, a tuned threshold, the ground-truth label is replaced with the predicted label; otherwise, we assume the ground-truth label. The tuned threshold ensures only predicted labels with high confidence are used to correct the ground-truth, preventing false corrections. Finally, the correct set of labels are paired with their respective post. We obtain class probabilities using the clustering algorithms. Note that our label correction method can be used in any NLP domain or even in other fields, such as the imaging field.

2.3   Classification

With a corrected label set, we train our deep neural networks to determine whether the posts display depressive or suicidal sentiment. Similar to the embedding process, any classifier can be used in place of the ones we tested. However, we aim to prove that deep neural classifiers are effective for our proposed task, as neural networks allow for accurate representation learning to differentiate the close semantics of our two classes.

For experimentation, we tested four deep learning algorithms: a dense neural network, a Convolutional Neural Network (CNN), a Bidirectional Long Short-Term Memory Neural Network (BiLSTM), a Gated Recurrent Unit Neural Network (GRU). For baselines, we evaluated three standard machine learning models: Logistic Regression (LogReg), Multinomial Naive Bayes (MNB), and a support-vector machine (SVM).

2.4   Datasets

We develop a primary dataset based on our task of suicide or depression classification. This dataset is web-scraped from Reddit. We collect our data from subreddits using the Python Reddit API. We specifically scrape from two subreddits: r/SuicideWatch and r/Depression. The dataset contains 1,895 total posts. We utilize two fields from the scraped data: the original text of the post as our inputs, and the subreddit it belongs to as labels. Posts from r/SuicideWatch are labeled as suicidal, and posts from r/Depression are labeled as depressed. We make this dataset and the web-scraping script available in our code.

Furthermore, we use the Reddit Suicide C-SSRS dataset [1] to verify our label correction methodology. The C-SSRS dataset contains 500 Reddit posts from the subreddit r/depression. These posts are labeled by psychologists according to the Columbia Suicide Severity Rating Scale, which assigns progressive labels according to severity of depression. We use this dataset to validate our label correction method since the labels are clinically verified and from the same domain of Reddit. To further validate the label correction method, we use the IMDB large movie dataset, a commonly used NLP benchmark dataset [14]. The dataset is a binary classification task which contains 50,000 polar movie reviews. We use a random subset of samples for evaluation.

For comparison of our method against other related tasks and methods, we build a dataset for binary classification of clinically healthy text vs suicidal text. We utilize the two subreddits r/CasualConversation and r/SuicideWatch. r/CasualConversation is a subreddit of general conversation, and has generally been used by other methods as data for a clinically healthy class [18].

3   Experimental Results

3.1   Implementation Details

For all datasets, we set aside 20% of the dataset as an external validation set. The deep learning models were implemented with Tensorflow, and the rest of the models were implemented with Sci-Kit Learn. We trained the deep learning models with the Adam optimizer and used a binary cross-entropy loss function. Based on tuning experiments, where we recorded accuracy at varying values, we set τ to 0.90 for all experiments, but similar values yielded similar performance. For classification accuracy, we use five metrics: Accuracy (Acc), Precision (Prec), Recall (Rec), F1-Score (F1), and Area Under Curve Score (AUC). Model-specific hyperparameters are included in the code.

3.2   Label Correction Performance

To evaluate our clustering performance, we present both the accuracy of the clustering algorithm at correcting noisy labels as well as classification performance after label correction. Classification on a clean test set is expected to decrease as training labels become noisier [7]. Therefore, we contend that if after label correction, the classification accuracy of our algorithm increases, the correction method is effective. Importantly, because our proposed task uses a web-scraped dataset, the labels are not clinically verified. This unfortunately means evaluating the correction rate of noisy labels is impossible because we do not have the true labels. Therefore, we perform label correction evaluation on the benchmark IMDB dataset to demonstrate the value of our method in a general setting, as well as on the C-SSRS dataset to demonstrate effectiveness in our specific domain.

Clustering Performance. To evaluate the performance of clustering, we inject noise into the label set at different rates. We corrupt 10–40% of the dataset at both uniform and imbalanced rates, as the noise rate of labels in real-world datasets are estimated to be 8% to 38.5%. These noise levels are also standard for other noisy label papers [20]. We then evaluated the performance of the clustering algorithms at correcting the noisy labels.

As seen in Fig. 2, our noise correction method is able to consistently remove >50% of injected noise while remaining below a 10% false-correction rate on both datasets, and the performance does not degrade heavily at higher noise percentages, which is challenging to achieve [7]. The SDCNL label correction is successful on both the IMDB dataset, which shows the generalizability of the method, as well as the C-SSRS dataset, proving its ability in our specific domain and task. The best combinations of reduction and clustering algorithms are umap-kmeans and umap-gmm, which we use as the proposed method. To our best knowledge, almost all noisy label correction methods do not evaluate correction rates but rather evaluate performance on classification accuracy after correction, as it allows for comparison to other noisy label methods that do not use label correction. However, because most recent noisy label methods are in the image domain, drawing comparisons to related work is unfeasible.

Classification Performance After Label Correction. Lastly, to demonstrate the effectiveness of the label correction method, we train a classifier on noisy C-SSRS data and validate on a separate C-SSRS test set which has no

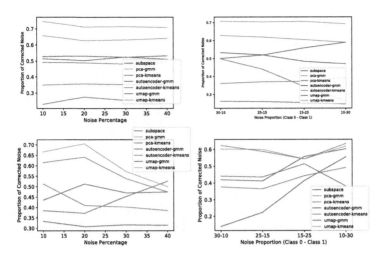

Fig. 2. Correction rates of the label correction algorithms at different noise rates on the **IMDB** (top) and **C-SSRS** (bottom) dataset. Left: correction rates with uniform injections of noise. Right: correction rates with class-weighted injections of noise (ratios such as 30%–10% or 25%–15%).

noise. We then use our label correction method to correct the same set of noisy labels, train the model with the correction labels, and validate on the same unmodified test set.

Table 1. Classifier accuracy comparison after injecting randomized noise (20%) into C-SSRS labels (left) against using the label correction method (UMAP + GMM) to remove the artificial noise and subsequently training classifier (right).

Model	Accuracies per Task (%)	
	Noisy	Corrected
guse-dense	57.97	70.63
bert-dense	48.86	70.13
bert-bilstm	56.71	68.35
bert-cnn	55.70	70.13

We show that accuracy improves markedly after using our label correction method, as there is at least a 11% increase (Table 1). Because our label correction process works on a dataset in the same domain, it is an effective method for cleaning noisy labels in NLP and for our task. Moreover, as seen in Fig. 3, using a probability threshold impacts performance, proving that using a threshold is an important factor in ensuring the corrected labels are accurate. Thus, we finalize on the thresholding method for our final model.

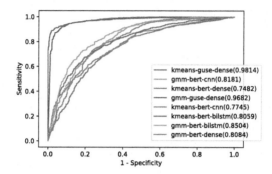

Fig. 3. ROC curves of model performance after using label correction. The 4 best combination of models with the two final label correction methods are shown (GMM vs K-Means). UMAP is used to reduce the dimensions of the embeddings.

3.3   Classification Performance

Table 2. Performance of the four best combinations of embedding models and classifiers.

Metrics (%)	Model combinations			
	guse-dense	bert-dense	bert-bilstm	bert-cnn
Acc	**72.24**	70.50	71.50	72.14
Rec	**76.37**	71.92	67.77	73.99
Prec	71.38	70.77	**74.28**	72.18
F1	**73.61**	71.25	70.70	72.92
AUC	**77.76**	75.43	77.11	76.35

Deep Neural Network Performance. After performing all experiments, we determined the four strongest combinations to perform the remainder of the tests. These combinations are trained on the primary suicide vs depression dataset with uncorrected labels. The complete results are in Appendix A in the supplemental (in Github repository). The performance of the four strongest models are shown in Table 2. The combinations are: BERT embeddings with a CNN (bert-cnn), BERT with a fully-dense neural network (bert-dense), BERT with a Bi-LSTM neural network (bert-bilstm), and GUSE with a fully-dense neural network (guse-dense). This proves the importance of our contribution as all DNNs outperform the baselines. For all future experiments, we use the four models above.

Comparison to Other Tasks. While there is extensive research on NLP text-based approaches to suicide detection, there is none for our specific task of low-risk depression versus suicidal ideation. We performed an additional test of our proposed model by testing the standard task of classifying suicide versus clinically healthy.

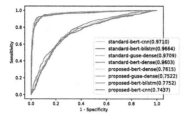
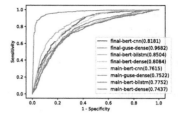

Fig. 4. ROC curves of model performance from four best models on our task (proposed) against the conventional suicide vs healthy task (standard).

Fig. 5. ROC curves of performance of top 4 models with label correction (red) against the same models without using label correction (blue) (Color figure online).

Figure 4 displays that on identical models, the commonly-researched task achieved much stronger baseline performance compared to our task; therefore, our baseline evaluations should correspondingly be lower. This task difficulty also demonstrates the value of automated methods such as ours in clinical settings.

Final Evaluation. We evaluated the classification performance of SDCNL on the Reddit dataset. We generated ground-truth labels for this dataset using the proposed label correction method. To provide a complete test of our model, it would be preferable to use labels provided by a mental health professional; however, no such dataset exists for our task. We contend that, since we demonstrated that the proposed label correction method is effective on the C-SSRS and IMDB datasets, the use of the label correction method on the Reddit dataset is justified.

Table 3. Final classification performance after using the two label correction methods, with and without a thresholding scheme. Best performances for each noise removal method are bolded. Best overall model is bolded and italicized.

Metrics (%)	UMAP-KMeans				UMAP-GMM			
	guse-dense	bert-dense	bert-bilstm	bert-cnn	*guse-dense*	bert-dense	bert-bilstm	bert-cnn
Acc	**92.61**	73.56	75.15	74.72	**93.08**	83.74	84.16	84.59
Prec	**93.61**	83.18	84.02	87.38	94.76	**95.51**	93.10	95.38
Rec	**94.85**	77.66	79.00	76.65	**96.16**	85.08	87.09	86.05
F1	**94.22**	80.25	81.19	81.64	**95.44**	89.99	89.99	90.45
AUC	**98.18**	76.83	80.93	78.69	**96.88**	81.97	85.08	82.91

Table 3 displays the final performance of the models with both threshold label correction (GMM) and without (K-Means). We validate the importance of the thresholding component, as all metrics are substantially improved over the K-Means baseline. Moreover, as displayed in Fig. 5, our label correction method improves the ROC curve and yields a much higher AUC value. By correcting the labels in the test set with high expected noise, we achieve substantially higher performance. Our final proposed combination uses GUSE as the embedding model, a fully-dense network for the classifier, and corrects labels using UMAP for dimensionality reduction and a GMM for the clustering algorithm. GUSE embeddings likely yield the best results due to outputting less embeddings than the other transformers, preserving information.

4   Conclusions and Ethical Discussion

In this paper, we present SDCNL, a novel method for deep neural network classification of depressive sentiment vs suicidal ideation with unsupervised noisy label correction. The use of deep neural networks allows for effective classification of closely related classes on a proven difficult task. Our novel method of label correction using unsupervised clustering effectively removes high-volumes of noise from both benchmark and domain-specific datasets, allowing for the use of large-scale, web-scraped datasets. Our extensive experimentation and ablative results highlight the effectiveness of our proposed model and its potential for real diagnostic application.

The applied setting of our system is to provide professionals with a supplementary tool for individual patient diagnosis, as opposed to solely being a screening method on social media platforms. SDCNL could be used by professional therapists as a "second opinion", friends and family as a preliminary screening for loved ones, or even on social media platforms to identify at-risk users.

This paper is not a clinical study, and the results are suitable for research purposes only. Were our algorithm to be used as a diagnostic tool, the main ethical concern would be false negative and false positive predictions. Specifically when dealing with suicide, which is a life or death situation, AI systems alone are not sufficient to provide proper screening. Future researchers who work on our topic or with our paper must be aware of these ethical concerns and not make major steps without proper clinical support. This paper is solely meant to demonstrate the potential efficacy of a suicide prevention mechanism.

Throughout this study, we collected our data while protecting user privacy and maintaining ethical practices. We de-identified our dataset, which we made publicly available, by removing personal information such as usernames. Moreover, Reddit is a public and anonymous forum, meaning our data source was anonymized and in the public domain to begin with.

References

1. Reddit C-SSRS Suicide Dataset. Zenodo, May 2019
2. Bering, J.: Suicidal: Why We Kill Ourselves. University of Chicago Press, Chicago (2018)
3. Bouveyron, C., Girard, S.: Robust supervised classification with mixture models: learning from data with uncertain labels. Pattern Recogn. **42**(11), 2649–2658 (2009)

4. Cer, D., Yang, Y., et al.: Universal sentence encoder for English. In: Proceedings of the 2018 Conference on Empirical Methods in Natural Language Processing: System Demonstrations, pp. 169–174 (2018)
5. De Choudhury, M., De, S.: Mental health discourse on reddit: self-disclosure, social support, and anonymity. In: Proceedings of the International AAAI Conference on Web and Social Media, vol. 8 (2014)
6. Devlin, J., Chang, M.W., Lee, K., Toutanova, K.: BERT: pre-training of deep bidirectional transformers for language understanding. In: Proceedings of the 2019 North American Chapter of the Association for Computational Linguistics: Human Language Technologies, pp. 4171–4186, June 2019
7. Frénay, B., Verleysen, M.: Classification in the presence of label noise: a survey. IEEE Trans. Neural Netw. Learn. Syst. **25**, 845–869 (2014)
8. Hendrycks, D., Mazeika, M., Wilson, D., Gimpel, K.: Using trusted data to train deep networks on labels corrupted by severe noise. In: Advances in Neural Information Processing Systems, vol. 31, pp. 10456–10465 (2018)
9. Hendrycks, D., Mazeika, M., Wilson, D., Gimpel, K.: Using trusted data to train deep networks on labels corrupted by severe noise. In: Proceedings of the 32nd International Conference on Neural Information Processing Systems, pp. 10477–10486 (2018)
10. Ji, S., Pan, S., Li, X., Cambria, E., Long, G., Huang, Z.: Suicidal ideation detection: a review of machine learning methods and applications. IEEE Trans. Comput. Soc. Syst. (2020)
11. Jiang, Z., Silovsky, J., Siu, M.H., Hartmann, W., Gish, H., Adali, S.: Learning from noisy labels with noise modeling network. arXiv preprint arXiv:2005.00596 (2020)
12. Jindal, I., Pressel, D., Lester, B., Nokleby, M.: An effective label noise model for DNN text classification (2019)
13. Leonard, C.: Depression and suicidality. J. Consult. Clin. Psychol. **42**(1), 98 (1974)
14. Maas, A.L., Daly, R.E., Pham, P.T., Huang, D., Ng, A.Y., Potts, C.: Learning word vectors for sentiment analysis. In: Proceedings of the Association for Computational Linguistics, pp. 142–150 (2011)
15. McInnes, L., Healy, J., Melville, J.: UMAP: uniform manifold approximation and projection for dimension reduction. arXiv:1802.03426 (2018)
16. Parsons, L., Haque, E., Liu, H.: Subspace clustering for high dimensional data: a review. SIGKDD Explor. Newsl. **6**(1), 90–105 (2004)
17. Reimers, N., Gurevych, I.: Sentence-BERT: sentence embeddings using Siamese BERT-networks (2019)
18. Schrading, N., Alm, C.O., Ptucha, R., Homan, C.: An analysis of domestic abuse discourse on reddit. In: Proceedings of the 2015 Conference on Empirical Methods in Natural Language Processing, pp. 2577–2583 (2015)
19. Shen, J., Rudzicz, F.: Detecting anxiety through reddit, pp. 58–65 (2017)
20. Song, H., Kim, M., Park, D., Lee, J.G.: Learning from noisy labels with deep neural networks: a survey. arXiv preprint arXiv:2007.08199 (2020)
21. Steinbach, M., Ertöz, L., Kumar, V.: The challenges of clustering high dimensional data. In: Wille, L.T. (ed.) New Directions in Statistical Physics, pp. 273–309. Springer, Heidelberg (2004). https://doi.org/10.1007/978-3-662-08968-2_16
22. Zheng, G., Awadallah, A.H., Dumais, S.: Meta label correction for learning with weak supervision. arXiv preprint arXiv:1911.03809 (2019)

Learning to Remove: Towards Isotropic Pre-trained BERT Embedding

Yuxin Liang[1], Rui Cao[1], Jie Zheng[1(✉)], Jie Ren[2], and Ling Gao[1]

[1] Northwest University, Xi'an, China
{liangyuxin,caorui}@stumail.nwu.edu.cn, {jzheng,gl}@nwu.edu.cn
[2] Shannxi Normal University, Xi'an, China
renjie@snnu.edu.cn

Abstract. Research in word representation shows that isotropic embeddings can significantly improve performance on downstream tasks. However, we measure and analyze the geometry of pre-trained BERT embedding and find that it is far from isotropic. We find that the word vectors are not centered around the origin, and the average cosine similarity between two random words is much higher than zero, which indicates that the word vectors are distributed in a narrow cone and deteriorate the representation capacity of word embedding. We propose a simple, and yet effective method to fix this problem: remove several dominant directions of BERT embedding with a set of learnable weights. We train the weights on word similarity tasks and show that processed embedding is more isotropic. Our method is evaluated on three standardized tasks: word similarity, word analogy, and semantic textual similarity. In all tasks, the word embedding processed by our method consistently outperforms the original embedding (with average improvement of 13% on word analogy and 16% on semantic textual similarity) and two baseline methods. Our method is also proven to be more robust to changes of hyperparameter.

Keywords: Natural language processing · Pre-trained embedding · Word representation · Anisotropic

1 Introduction

With the rise of Transformers [12], its derivative model BERT [1] stormed the NLP field due to its excellent performance in various tasks. A new word BERTology has been defined to describe the related research work carried out around BERT [11]. Our work is focused on BERT input embedding that is both critical to the BERT model and easy to overlook.

From static word embeddings to contextual embeddings, the word embedding technology is constantly evolving, but it is still not perfect so far. The analysis from [9] implies that in static embeddings, all word embeddings share a common vector and have several dominant directions. These anisotropic geometric properties strongly affect the representation of words. Furthermore [2,4,5,15], it was

I. Farkaš et al. (Eds.): ICANN 2021, LNCS 12895, pp. 448–459, 2021.
https://doi.org/10.1007/978-3-030-86383-8_36

found that the word embeddings of contextual language models also have those anisotropic geometric properties, and the main reason for this phenomenon is the large number of low-frequency words in the corpus. The embedding of most words in the vocabulary is pushed into a similar direction that is negatively correlated with most hidden states, thus clusters in a localized region of the embedding space. This is known as the representation degeneration problem [3].

Does BERT embedding also share this problem? The answer is Yes: pre-trained BERT embedding also suffers from strong anisotropy, meaning the average cosine similarity value is significantly higher than zero, and word vectors clustering in narrow cones in the vector space. This phenomenon can result in a word representation have a high similarity to an unrelated word, affecting the expressive power. In addition, the anisotropic property can also affect the accuracy of downstream tasks [4,7]. Therefore, various attempts have been made to eliminate such a property so that the learned word vectors are more divergent and distinguishable in the Euclidean space.

Referring to the all-but-the-top method [9] (denoted as ABTT for short), they remove the common vector along with several dominant directions computed by PCA (principal component analysis) in word vectors. After applying this method to BERT embedding, we find that the ABTT method could improve the geometric properties and task performance of BERT embedding. However, as the number of selected dominant directions increases, the effectiveness of the ABTT method decreases gradually, as shown in Sect. 5. This is intuitive since some useful linguistic information contained within these directions is inevitably lost as the number of removed directions increases.

We propose a weighted removal method that sets a learnable weight for each dominant direction, dynamically adjusting the ratio of removal. We train on the word similarity task to obtain the weight for each dominant direction. Our approach is more flexible and performs better on the evaluation task compared to removing the dominant directions directly and completely. Our method alleviates representation degradation and performs more stable in comparative experiments on the word similarity, word analogy, and text similarity tasks. The code is available here.[1]

Our contributions are as follows:

1. We measure, visualize and analyze the geometry of pre-trained BERT embedding. We find that the pre-trained BERT embedding is anisotropic and the norm/average cosine similarity of word vectors has a strong correlation with word frequency. Further, we provide some intuitive explanation and theoretical analysis for the above phenomena.
2. We propose a weighted removal method that learns to remove the dominant directions. The key point of our approach is using a set of learnable weights trained on word similarity tasks to decide the proportions of the removal directions.

---

[1] https://github.com/liangyuxin42/weighted-removal.

3. We evaluate our method on three tasks: word similarity, word analogy, and semantic textual similarity. We compare the performance of our method with three baselines: original pre-trained BERT embedding, ABTT method, and conceptor negation method (denoted as CN for short). Our method outperforms three baselines in most cases and maintains relatively stable performance as hyperparameter changes. We also analyze the impact of our method on the geometry of word embedding and conclude that our method makes word embedding more isotropic and expressive.

2   Related Work

The BERT model is introduced in [1], a language model based on Transformer (containing 12-layer to 24-layer Transformer encoders), pre-trained on a hybrid corpus. BERT's internal operation includes first embed the tokens by a pre-trained embedding layer and combine with position and segment information. These initial embeddings run through several transformer encoder layers to produce the contextual embedding for the current task. Our study is based on the pre-trained BERT embedding and focuses on its geometric properties.

Mu et al. [9] explore the anisotropic geometry of static word embeddings, i.e., the word vectors have a non-zero mean and most word vectors have several dominant directions. They propose to remove the common vector and dominant directions from the word embedding to capture stronger linguistic regularities. There are other attempts to fix the anisotropic geometry of static word embeddings: Liu et al. [7] use conceptors to suppress those latent features of word vectors with high variances. Hasan et al. [5] re-embed pre-trained word embeddings with a stage of manifold learning. Zhou et al. [15] focuses on linear alignment of word embeddings and find that aligning with an isotropic noise can deliver better results.

Ethayarajh et al. [2] find that the contextualized representations are not isotropic in layers of contextualizing models such as BERT and GPT-2. Gong et al. [4] find that contextual word embeddings are biased towards word frequency and use adversarial training to learn a frequency-agnostic word representation.

Gao et al. [3] focus on the learned word embeddings of natural language generation model training through likelihood maximization and find that the word embedding tends to degenerate and is distributed into a narrow cone in vector space which limits its representation power. They propose a regularization method that minimizes the cosine similarities between any word vector pair to alleviate this problem. Wang et al. [13] propose a spectrum control method that guides the spectra training of the output embedding with a slow-decaying singular value prior distribution to tackle the representation degeneration problem. Other researches also touch on the geometry of contextual word embedding, such as Reif et al. [10], Karve et al. [6], Zhou et al. [16].

The above word embedding post-processing methods are based only on the geometric features, while we make the correction semantically more reasonable

Table 1. Basic geometric information of pre-trained BERT embedding (BERT-base-uncased and BERT-large-uncased).

Embedding	Average vector length	Vector average length	Average cosine similarity
BERT-base	0.939	1.401	0.444
BERT-large	0.800	1.453	0.299

by introducing information of word similarity. The methods of regularizing word similarity during training require retraining the model. In contrast, our post-processing method is more computationally efficient.

The main difference between our method and ABTT is that we add learnable weights to each dominant direction and train the weights on the word similarity task. Furthermore, We apply our method on pre-trained BERT embeddings to fix the representation degeneration problem. We compare the performance of our method with ABTT and CN [7] which is a successor of the ABTT method.

3 Observation

In this section, we illustrate the anisotropic geometry of pre-trained BERT embedding and provide some explanation for those phenomena. Our study is based on pre-trained bert-base-uncased and bert-large-uncased models provided by huggingface [14].

3.1 Anisotropic Geometry of Pre-trained BERT Embedding

Let $v(w) \in \mathbb{R}^e$ be the embedding of a token w in a vocabulary V, and E of shape $(|V|, e)$ be the embedding matrix. We observe the following phenomena in the pre-trained BERT embedding.

Non-zero Mean: The average vector $\mu = \frac{1}{|V|}\Sigma_{w \in V} v(w)$ is not centered around the origin. In fact, this common vector μ occupies a large proportion of all $v(w)$. The norm of μ is more than $1/2$ of the average norm of all $v(w)$ (Table 1).

Non-zero Average Cosine Similarity: The average cosine similarity is calculated by

$$cos_{avg} = \frac{\Sigma_{w_i \in V} \Sigma_{w_j \in V} cosine(v(w_i), v(w_j))}{|V|^2}.$$

As shown in Table 1, the average cosine similarity is much higher than zero, which means instead of uniformly distributed in vector space as one expects from an expressive word embedding, the word vectors are distributed into a narrow cone.

Fast Singular Value Decay: We reparameterize the embedding matrix E by singular value decomposition (SVD) and obtain the singular values of matrix E. As shown in Fig. 1, the singular values of matrix E decay exponentially.

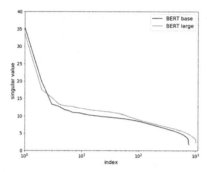

Fig. 1. Singular values of embedding matrix. The singular values of BERT embedding matrix decay exponentially. Note that the coordinates are semi-logarithmic.

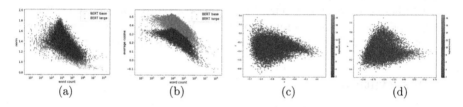

(a)   (b)   (c)   (d)

Fig. 2. Correlation between (a) vector norm and word frequency (b) average cosine similarity and word frequency in BERT base/large embedding. Projecting (c) BERT-base and (d) BERT-large embedding onto top two PCA directions.

Correlation with Word Frequency: When we look into the relationship between the norm of $v(w)$, the average cosine similarity and the word frequency of word w, we notice an obvious correlation between them, as shown in Fig. 2a, b. The Pearson correlation between norm/average cosine similarity and the logarithm of word count is about -0.7 which also shows a strong negative correlation.

We also visualize BERT embedding by projecting onto the top two PCA directions and use color to indicate word frequency, as shown in Fig. 2c, d. We find that the first PCA coefficient encodes the word frequency to a significant degree with a high Pearson correlation (about -0.7).

The above phenomena reveal an image of the distribution of words with different frequencies in the embedding space. The words with high frequency are close to the origin (smaller norm) and distributed relatively uniform (lower average cosine similarity), and the words with low frequency are squeezed into a narrower cone and push away from the origin. This can lead to two words that are explicitly dissimilar to each other but whose corresponding word vectors may produce a high degree of similarity in the Euclidean space, thus affecting the performance of downstream tasks.

3.2 Insights

The above phenomena are very similar to the representation degeneration problem mentioned in [3], inspired by their work, we provide some intuitive explanation and theoretical analysis for the above phenomena.

In the pre-train process of BERT, the final hidden vectors corresponding to the mask tokens are fed into an output softmax over the vocabulary, as in a standard language model [1]. Intuitively, during the training process, for any given hidden state, the embedding of the corresponding masked word will be pushed towards the direction of the hidden state to get a larger likelihood, while the embeddings of all other words will be pushed towards the negative direction of the hidden state to get a smaller likelihood. The words with lower frequency would be pushed more times towards the negative direction of all kinds of hidden states. As a result, low frequency words are squeezed into a narrower cone.

From the theoretical perspective, for any token w_i, its loss function can be divided into two pieces: piece A_{w_i} for the part of training corpus that does not contain token w_i in context, piece B_{w_i} for the part containing token w_i. Let $P(context \in A_{w_i})$ and $P(context \in B_{w_i})$ denote the probability of a context belonging to piece A/B and $L_{A_{w_i}}(w_i)$ and $L_{B_{w_i}}(w_i)$ be the loss of piece A/B respectively. So the loss function of w_i can be defined as:

$$L_{w_i} = P(context \in A_{w_i})L_{A_{w_i}}(w_i) + P(context \in B_{w_i})L_{B_{w_i}}(w_i)$$

According to [3], $L_{A_{w_i}}(w_i)$ is convex and would optimize embedding w_i towards any uniformly negative direction of the hidden state in piece A_{w_i} to infinity. $L_{B_{w_i}}(w_i)$ is complicated but when $P(context \in A_{w_i})$ is much larger than $P(context \in B_{w_i})$, the optimal solution of L_{w_i} is close to $L_{A_{w_i}}(w_i)$. A more detailed proof can be found in [3].

With the interpretation above, we can explain the phenomena observed. For most words, $P(context \in A_{w_i})$ is much larger than $P(context \in B_{w_i})$, and piece A of different words has large overlaps. So most words will be optimized toward a similar direction in vector space, hence the average vector μ would also be in that direction and the average cosine similarity would be larger than zero.

For a low-frequency word, $P(context \in A_{w_i})$ is even larger. So a low-frequency word is affected more by the optimization of $L_{A_{w_i}}(w_i)$, therefore the word vectors of low-frequency words are likely to be close to each other and pushed farther away from the origin, which is consistent with our observation.

4 Method

In this section, we propose a simple but effective way to change the anisotropic geometry of pre-trained BERT embedding into a more isotropic and expressive one. According to the analysis in the previous section, the word embeddings have some dominant directions with a disproportional impact on word embedding. But removing these directions directly and completely may lose semantic and syntactic information within them.

The core idea of our method is to remove several dominating directions causing the anisotropic problem with a learnable weight to each direction. We use PCA to determine the dominating directions. We denote our method as weighted removal (WR) algorithm for simplicity.

The weighted removal (WR) method includes three main steps: compute the dominant directions using PCA, learn a set of weights to remove those directions through a certain learning process, and remove directions with corresponding weights from word embedding. We formally achieve our method as Algorithm 1.

Algorithm 1. Weighted Removal (WR) Algorithm

Input: Word Embedding $v(w) \in \mathbb{R}^e$; d: number of directions being removed
Output: Processed word embedding $v'(w)$
1: Compute the PCA components: $u_1, ..., u_e \leftarrow PCA(v(w), w \in V)$
2: Acquire d weights for d directions through a learning process: $\alpha_1, ..., \alpha_d$
3: Remove top d dominating directions: $v'(w) \leftarrow v(w) - \sum_{i=1}^{d} \alpha_i(u_i^T v(w))u_i$
4: **return** $v'(w)$

Learning Process: We choose word similarity as the training task. The unreasonable word similarity is the direct manifestation of the representation degeneration problem. Through adjusting word similarity, our method indirectly adjusts the geometric properties of the word vector distribution in vector space towards isotropic.

Giving two words w_1 and w_2 with their corresponding ground truth of similarity S_{target}, the prediction of word similarity is calculated as $S_{pred} = \hat{v}'(w_1)^T \hat{v}'(w_2)$. For simplicity, we denote the normalized direction of $v(w)$ as $\hat{v}(w)$, $\hat{v}(w) = \frac{v(w)}{\|v(w)\|}$. The loss function is defined as: $L = MSELoss(S_{pred}, S_{target})$. After the training process, we get the d adjusted weights for removing top d dominating directions.

Comparison Algorithm: Our method is inspired by the ABTT method. We also apply the ABTT method and its direct successor method, the conceptor negation(CN) method proposed in [7] on BERT embedding for comparison.

The ABTT algorithm directly removes the first d dominant direction completely. The CN algorithm uses conceptors to suppress those latent features with high variances in the word embedding. We skip the step of removing the mean vector in the original ABTT algorithm because the word embedding processed by the weighted removal algorithm is naturally centered around the origin.

5   Experiment

Our experiment is based on the pre-trained BERT model provided by huggingface Transformers library. As for training datasets, we collect 8 commonly used word similarity datasets: RG65, WordSim-353 (WS), the rare-words (RW), the MEN, the MTurk-278&MTurk-771, the SimLex-999, and the SimVerb-3500

Table 2. Correlation results on word similarity task.

Dataset	BERT-base				BERT-large			
	ORIG	CN	ABTT	WR	ORIG	CN	ABTT	WR
RG65	.769	.855	.844 (d = 8)	**.903 (d = 60)**	.866	.906	.900 (d = 1)	**.934 (d = 100)**
SimVerb-3500	.299	.412	.431 (d = 5)	**.447 (d = 20)**	.310	.416	.452 (d = 1)	**.456 (d = 110)**
MEN-3000	.517	.673	.702 (d = 6)	**.722 (d = 50)**	.562	.663	.700 (d = 6)	**.709 (d = 100)**
Mturk287	.370	.716	.747 (d = 6)	**.768 (d = 160)**	.435	.701	**.723 (d = 4)**	.723 (d = 4)
Mturk-771	.529	.652	.667 (d = 5)	**.676 (d = 60)**	.564	.646	.666 (d = 3)	**.669 (d = 6)**
RW-2034	.493	.665	.668 (d = 9)	**.707 (d = 200)**	.489	.641	.660 (d = 4)	**.687 (d = 25)**
SimLex999	.492	.549	.570 (d = 10)	**.581 (d = 200)**	.518	.544	.580 (d = 11)	**.586 (d = 50)**
WordSim353	.569	.637	.665 (d = 10)	**.682 (d = 25)**	.617	.649	.677 (d = 8)	**.693 (d = 9)**

dataset. We scale the human annotations into $[-1, 1]$ to be consistent with cosine similarity calculation.

The number of dominating directions to be removed, d, is a hyper-parameter needs to be tuned. We choose a series of numbers from 1 to 200 as d to perform the ABTT algorithm and our WR algorithm. The performance of our algorithm is compared with 3 baseline:

1. The original pre-trained BERT base/large uncased embedding (ORIG).
2. Embeddings processed by ABTT algorithm. This algorithm also has the same hyper-parameter d.
3. Embeddings processed by conceptor negation (CN) algorithm with the same setting as the original paper.

5.1 Towards Expressive

Word Similarity: For this experiment, we randomly separate the collection of word similarity datasets into 70% for training and 30% for testing. Given a pair of words, we calculate their similarity by cosine similarity. The word similarity task is evaluated in terms of Pearson's correlation with the human annotations.

The overall results are shown in Fig. 3a and d. As illustrated, the word embeddings processed by the WR algorithm are consistently and significantly outperform the original BERT embedding and CN algorithm. As the hyperparameter d increases above 20, the correlation result of ABTT begins to drop, but the result of WR algorithm keeps stable and even rises slightly. The drop of ABTT's result could be caused by removing too many directions and losing expressiveness, and yet by adding weights to d directions, WR algorithm could keep even improve the performance. The performance on eight word similarity datasets is provided in Table 2, note that ABTT and WR method achieve optimal results within the different ranges of d.

Word Analogy: The purpose of word analogy task is to test the capability of a word embedding to encode linguistic relations between words. We use the analogy dataset introduced in [8] containing 14 types of relations. These relations can be divided into two parts: the semantic relation such as capital of countries and the syntactic relation like adjective to adverb relation.

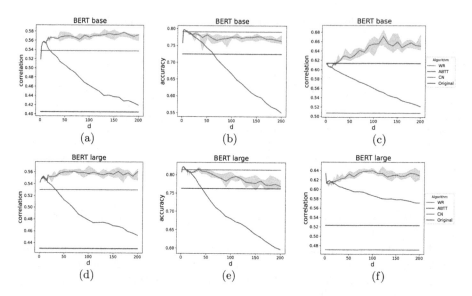

Fig. 3. Results of WR algorithm by different d comparing to baselines: on BERT-base in (a) word similarity, (b) word analogy and (c) semantic textual similarity tasks; on BERT-large in (d) word similarity, (e) word analogy and (f) semantic textual similarity tasks.

Table 3. Accuracy results on word analogy task.

Dataset	BERT-base				BERT-large			
	ORIG	CN	ABTT	WR	ORIG	CN	ABTT	WR
Semantic	.655	.768	**.787 (d = 5)**	**.787 (d = 5)**	.688	.775	**.799 (d = 7)**	.797 (d = 7)
Syntactic	.761	.802	.802 (d = 3)	**.827 (d = 130)**	.802	.834	.835 (d = 2)	**.863 (d = 90)**

We calculate the word w_4 by finding the word in vocabulary that maximizes the cosine similarity between $v(w_4)$ and $v(w_2) - v(w_1) + v(w_3)$. The overall result of word analogy (Fig. 3b and e) shows that The WR and ABTT methods both have a positive effect on the performance of word analogy task.

The best performance of each method on the word analogy task is provided in Table 3, we report the results by two types of relation: the semantic relation and the syntactic relation. The word vectors processed by WR perform better than the original ones with an average improvement of 13.04%.

Semantic Textual Similarity: We also evaluate the WR algorithm's effect on downstream task like the semantic textual similarity task. The semantic textual similarity (STS for short) task is aimed to determine how similar two texts are. We use the STS Benchmark dataset which comprises a selection of datasets used in the STS tasks between 2012 and 2017.

Table 4. Correlation results on semantic textual similarity task.

Dataset	BERT-base				BERT-large			
	ORIG	CN	ABTT	WR	ORIG	CN	ABTT	WR
2012	.646	.596	.657 (d = 1)	**.714 (d = 120)**	.588	.588	.663 (d = 1)	**.684 (d = 120)**
2013	.660	.662	.708 (d = 3)	**.725 (d = 140)**	.639	.640	**.702 (d = 8)**	.690 (d = 200)
2014	.562	.477	.578 (d = 3)	**.634 (d = 120)**	.480	.442	.592 (d = 1)	**.711 (d = 190)**
2015	.639	.516	.657 (d = 3)	**.702 (d = 150)**	.523	.466	.6687 (d = 8)	**.683 (d = 80)**
2016	.519	.436	.514 (d = 1)	**.527 (d = 180)**	.441	.408	**.522 (d = 1)**	**.522 (d = 1)**
2017	.616	.577	.682(d=3)	**.727 (d = 150)**	.528	.538	.689 (d = 8)	**.707 (d = 80)**

We use a widely used and effective method to represent sentences: averaging the word embedding of each word in the sentence, namely, the sentence representation $v(s) = \frac{1}{|s|} \sum_{w \in s} v(w)$. The similarity between two sentences is calculated by the cosine similarity of their corresponding sentence representations. Pearson correlation between predictions and human judgments is used to evaluate an algorithm's performance.

The performances of WR and ABTT keep consistent when d is smaller than 10. As d exceeds a certain number, the performance of ABTT algorithm starts to drop, but the results of the WR algorithm keep growing and outperform all baselines, as shown in Fig. 3c and f. The best performance on semantic textual similarity tasks is provided in Table 4, note that ABTT and WR method achieve optimal results within the different ranges of d. An average improvement of 16.78% over the original word embedding suggests that the WR algorithm can significantly improve the performance of downstream tasks.

5.2 Towards Isotropy

Our method also has an impact on the geometric characteristics of word embedding. In this section, we demonstrate the geometry of word embedding processed by the WR algorithm and make a comparison with the original one.

Our method can fix the fast singular value decay problem. After applying the WR algorithm, the first several singular values decrease significantly which shows reducing of the imbalance between different directions in the word embedding. As shown in Fig. 4a, singular values are steady across directions after removing several dominant directions.

Figure 4b visualizes processed BERT embedding by projecting onto the top two PCA directions and use color to indicate word frequency. The processed embedding is centered around the origin and words with different frequency are uniformly distributed which avoid the disproportionate impact of word frequency on the entire embedding. The average cosine similarity of words also drops to near zero after applying the WR algorithm which indicates the isotropic geometry of the processed word embedding.

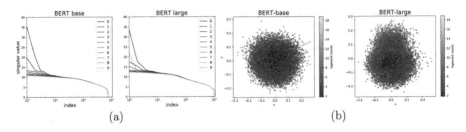

Fig. 4. (a) Singular values of embedding matrices processed by WR with different d. (b) Projecting word embeddings (processed by WR when d equals 5) onto top 2 PCA directions and use color to indicate word frequency.

5.3   Discussion

As the number of directions to be removed increases, we find a similar pattern in the performances of ABTT and WR algorithm in three evaluation tasks. When d is small, the results of ABTT and WR keep consistent and the weights that the WR algorithm learned are close to 1. That means for the first several dominant directions, complete removal is optimal.

As d grows above a certain number, the performance of ABTT starts to drop while WR's results keep steady in general and even increase. This demonstrates the ability of the WR method to maintain useful information in it by acquiring a weight for each direction.

6   Conclusion

In this paper, we show that the pre-trained BERT embedding is anisotropy which hurts its expressiveness. To correct the anisotropy, we propose a method to remove several dominant directions computed by PCA with a learnable weight to each direction. We train the weights on word similarity tasks and evaluate our method on three tasks: word similarity, word analogy, and semantic textual similarity. We compare our weighted removal method with three baselines and our method outperforms the baselines in most conditions.

There are many possibilities in this direction: other single factors that have a disproportionate impact on word embeddings, changing the training task of our method to obtain other improvements, combining our method with the transformer architecture, etc. We hope to investigate these issues in future work.

Acknowledgment. Our work is supported by the National Key Research and Development Program of China under grant No. 2019YFC1521400 and National Natural Science Foundation of China under grant No. 62072362. International Science and Technology Cooperation Project of Shannxi (2020KW-006).

References

1. Devlin, J., Chang, M.W., Lee, K., Toutanova, K.: BERT: pre-training of deep bidirectional transformers for language understanding. arXiv preprint arXiv:1810.04805 (2018)
2. Ethayarajh, K.: How contextual are contextualized word representations? Comparing the geometry of BERT, ELMO, and GPT-2 embeddings. In: Proceedings of the 2019 Conference on Empirical Methods in Natural Language Processing, EMNLP, pp. 55–65 (2019)
3. Gao, J., He, D., Tan, X., Qin, T., Wang, L., Liu, T.: Representation degeneration problem in training natural language generation models. In: International Conference on Learning Representations, ICLR (2018)
4. Gong, C., He, D., Tan, X., Qin, T., Wang, L., Liu, T.Y.: FRAGE: frequency-agnostic word representation. In: Advances in neural information processing systems, NIPS, pp. 1334–1345 (2018)
5. Hasan, S., Curry, E.: Word re-embedding via manifold dimensionality retention. In: Proceedings of the 2017 Conference on Empirical Methods in Natural Language Processing, EMNLP, pp. 321–326 (2017)
6. Karve, S., Ungar, L., Sedoc, J.: Conceptor debiasing of word representations evaluated on WEAT. In: Proceedings of the First Workshop on Gender Bias in Natural Language Processing, pp. 40–48 (2019)
7. Liu, T., Ungar, L., Sedoc, J.: Unsupervised post-processing of word vectors via conceptor negation. In: Proceedings of the AAAI Conference on Artificial Intelligence, vol. 33, pp. 6778–6785 (2019)
8. Mikolov, T., Chen, K., Corrado, G., Dean, J.: Efficient estimation of word representations in vector space. arXiv preprint arXiv:1301.3781 (2013)
9. Mu, J., Viswanath, P.: All-but-the-top: simple and effective post-processing for word representations. In: International Conference on Learning Representations, ICLR (2018)
10. Reif, E., et al.: Visualizing and measuring the geometry of BERT. In: Advances in Neural Information Processing Systems, NIPS, pp. 8594–8603 (2019)
11. Rogers, A., Kovaleva, O., Rumshisky, A.: A primer in BERTology: what we know about how BERT works. arXiv preprint arXiv:2002.12327 (2020)
12. Vaswani, A., et al.: Attention is all you need. In: Advances in Neural Information Processing Systems, NIPS, pp. 5998–6008 (2017)
13. Wang, L., Huang, J., Huang, K., Hu, Z., Wang, G., Gu, Q.: Improving neural language generation with spectrum control. In: International Conference on Learning Representations, ICLR (2020)
14. Wolf, T., et al.: HuggingFace's transformers: state-of-the-art natural language processing. arXiv preprint arXiv:1910.03771 (2019)
15. Zhou, T., Sedoc, J., Rodu, J.: Getting in shape: word embedding subspaces. In: Proceedings of the 28th International Joint Conference on Artificial Intelligence, AAAI, pp. 5478–5484. AAAI Press (2019)
16. Zhou, W., Lin, B.Y., Ren, X.: IsoBN: fine-tuning BERT with isotropic batch normalization. arXiv preprint arXiv:2005.02178 (2020)

ExBERT: An External Knowledge Enhanced BERT for Natural Language Inference

Amit Gajbhiye[1(✉)], Noura Al Moubayed[2], and Steven Bradley[2]

[1] University of Sheffield, Sheffield, UK
a.gajbhiye@sheffield.ac.uk
[2] University of Durham, Durham, UK
{noura.al-moubayed,s.p.bradley}@durham.ac.uk

Abstract. Neural language representation models such as BERT, pre-trained on large-scale unstructured corpora lack explicit grounding to real-world commonsense knowledge and are often unable to remember facts required for reasoning and inference. Natural Language Inference (NLI) is a challenging reasoning task that relies on common human understanding of language and real-world commonsense knowledge. We introduce a new model for NLI called External Knowledge Enhanced BERT (ExBERT), to enrich the contextual representation with real-world commonsense knowledge from external knowledge sources and enhance BERT's language understanding and reasoning capabilities. ExBERT takes full advantage of contextual word representations obtained from BERT and employs them to retrieve relevant external knowledge from knowledge graphs and to encode the retrieved external knowledge. Our model adaptively incorporates the external knowledge context required for reasoning over the inputs. Extensive experiments on the challenging SciTail and SNLI benchmarks demonstrate the effectiveness of ExBERT: in comparison to the previous state-of-the-art, we obtain an accuracy of 95.9% on SciTail and 91.5% on SNLI.

Keywords: Natural Language Inference · Contextual representations

1 Introduction

Natural Language Inference (NLI), also known as Recognising Textual Entailment, is formulated as a - 'directional relationship between pairs of text expressions, denoted by T (the entailing "Text") and H (the entailed "Hypothesis"). Text T, entails hypothesis H, if humans reading T would typically infer that H is most likely true.' [4]. The NLI task definition relies on common human understanding of language and real-world commonsense knowledge. NLI is a complex reasoning task, and NLI models can not rely solely on training data to acquire all the real-world commonsense knowledge required for reasoning and inference [3]. For example, consider the premise-hypothesis pair in Table 1, if the training

© Springer Nature Switzerland AG 2021
I. Farkaš et al. (Eds.): ICANN 2021, LNCS 12895, pp. 460–472, 2021.
https://doi.org/10.1007/978-3-030-86383-8_37

Table 1. SNLI premise (**P**) & hypothesis (**H**) and commonsense triples (red) from ConceptNet KG. Commonsense knowledge enrich premise-hypothesis context and helps the NLI model in reasoning.

P: Four boys are about to be hit by an approaching wave. (wave RelatedTo crash)
H: A giant wave is about to crash on some boys. (crash IsA hit)

data do not provide the common knowledge that, *(wave RelatedTo crash)* and *(crash IsA hit)*, it will be hard for the NLI model to correctly recognise that the premise entails the hypothesis.

Recently, deep pre-trained language representations models (PTLMs) such as BERT [6] achieved impressive performance improvements on a wide range of NLP tasks. These models are trained on large amounts of raw text using a self-supervised language modelling objective. Although pre-trained language representations have significantly improved the state-of-the-art on many complex natural language understanding tasks, they lack grounding to real-world knowledge and are often unable to remember real-world facts when required [14]. Investigations into the learning capabilities of PTLMs reveal that the models fail to recall facts learned at training time, and do not generalise to rare/unseen entities [17]. Knowledge probing tests [14] on the commonsense knowledge of ConceptNet [20] reveal that PTLMs such as BERT have critical limitations when solving problems involving commonsense knowledge. Hence, infusing external real-world commonsense knowledge can enhance the language understanding capabilities of PTLMs and the performance on the complex reasoning tasks such as NLI.

Incorporating external commonsense knowledge into pre-trained NLI models is challenging. The main challenges are (i) **Structured Knowledge Retrieval:** Given a premise-hypothesis pair how to effectively retrieve specific and relevant external knowledge from the massive amounts of data in Knowledge Graphs (KGs). Existing models [3], use heuristics and word surface forms of premises-hypothesis which may be biased and the retrieved knowledge may not be contextually relevant for reasoning over premise-hypothesis pair. (ii) **Encoding Retrieved Knowledge:** Learning the representations of the retrieved external knowledge amenable to be fused with deep contextual representations of premise-hypothesis is challenging. Various KG embedding techniques [22] are employed to learn these representations. However, while learning, the KG embeddings are required to be valid within the individual KG fact and hence might not be predictive enough for the downstream tasks [22] (iii) **Feature Fusion:** How to fuse the learned external knowledge features with the premise-hypothesis contextual embeddings. This feature fusion requires substantial NLI model adaptations with marginal performance gains [3].

To overcome the aforementioned challenges, we propose, **ExBERT** - an External knowledge enhanced BERT model which enhances the contextual representations of BERT model with external commonsense knowledge to improve

BERT's real-world grounding and reinforce its reasoning and inference capabilities. ExBERT utilises BERT for learning the contextual representation of premise-hypothesis as well as the representations of retrieved external knowledge. The aim here is to take full advantage of contextual word representations obtained from pre-trained language models and the real-world commonsense knowledge from Knowledge Graphs (KGs).

The main contributions of this paper are: (i) we devise a new approach, ExBERT, to incorporate external knowledge in contextual word representations. (ii) we investigate and demonstrate the feasibility of using contextual word representation for encoding external knowledge obviating learning specialised KG embeddings. To the best of our knowledge, this is the first study of its kind, indicating a potential future research direction. (iii) we introduce a new external knowledge retrieval mechanism for NLI that is capable of retrieving fine-grained contextually relevant external knowledge from KGs.

2   Related Work

Traditional Attention-Based Models do not utilise contextual representations from PTLMs [7]. KG-Augmented Entailment System (KES) [11] augments the NLI model with external knowledge encoded using graph convolutional networks. Knowledge-based Inference Model (KIM) [3] incorporates lexical-level knowledge (such as synonym and antonym) into its attention and composition components. ConSeqNet [23], a system of a text-based model and a graph-based model, concatenates the output of the two models, to be fed to a classifier. The AdvEntuRe [10] framework train the decomposable attention model with adversarial training examples generated by incorporating knowledge from linguistic resources, and with a sequence-to-sequence neural generator. BiCAM models improve the performance of NLI baselines via the incorporation of knowledge from the ConcepNet and Aristo Tuple KGs by factorised bilinear pooling [9].

PTLM-Based Models. PTLM-based models such as OpenAI GPT [19] and BERT [6] leverage transfer learning from a large textual corpus and are fine-tuned on NLI datasets. Specifically, OpenAI GPT pre-trains the Transformer [21] model in an unsupervised manner with the standard language modelling objective and fine-tunes the model in a supervised manner for NLI. Semantics-aware BERT (SemBERT) [25] demonstrated the benefit of enriching the BERT's contextual representation with the semantic roles. BERT model has shown to be robust to adversarial examples when external knowledge is incorporated to the attention mechanism using simple transformations [15]. The KES [11] model highlighted above further evaluates their system with BERT contextual embeddings in the framework.

3   Methodology

ExBERT architecture is depicted in Fig. 1. In this section, we describe the key components of ExBERT and their detailed implementation including the model architecture in Sect. 3.2. We start by describing the contextual representation based external knowledge retrieval procedure in Sect. 3.1.

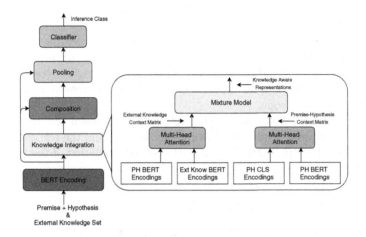

Fig. 1. A high-level view of ExBERT architecture.

3.1 External Knowledge Retrieval: Selection and Ranking

Retrieval and preparation of contextually specific and relevant information from knowledge graphs are complex and challenging tasks. The crucial challenge is to identify the knowledge specific and relevant to the task at hand from the massive amounts of noisy KG data [1]. Different from the previous approaches that use word surface forms to retrieve external knowledge, we use the cosine similarity between the contextual representations of premise-hypothesis tokens and the external knowledge. The external knowledge for the premise and hypothesis is retrieved individually. Below we explain the procedure for the premise. The same procedure is applied to the hypothesis. The output of external knowledge retrieval is the set of contextually relevant KG triples for the premise and hypothesis. We divide the external knowledge retrieval process into two parts: Selection and Ranking.

Selection. We first filter the stop words from the premise. Then we retrieve all the KB triples that contain the tokens of the premise as one of the words in the head entity of KG triples. For example, for the token *"speaking"* one of the retrieved KG fact is *"public_speaking IsA speaking"*. The retrieved triples are converted to a sentence. For example, the previous triple is transformed into *"public speaking is a speaking"*. The selection process retrieves a large number of KG triples which are not all relevant to the context of the premise. We filter the selected triples in the ranking step.

Ranking step ranks the selected KG triples according to the contextual similarity to the fine-grained context of the premise. Specifically, given the BERT generated context-aware representation of the premise tokens, we group all the bigrams of the representations. Each group of the bigram representation is averaged, and the cosine similarity is calculated with the average of the BERT representation of each of the selected KG triple sentence (created in selection step). For each bigram, we choose the KG triple sentence with the highest cosine similarity score. To capture the fine-grained context of the premise, we repeat the ranking step with the trigrams, fourgrams, and the average of the whole premise BERT representations and retrieve the KG record with highest cosine similarity for each of the grams. Duplicate KG triple sentences are removed form the final set of retrieved knowledge.

The advantages of our knowledge retrieval mechanism are that it is free from heuristic biases, requires no feature engineering, and the retrieved external knowledge is contextually relevant to the fine-grained context of the premise and hypothesis.

3.2   Model Architecture

BERT Encoding Layer. This layer uses the BERT encoder to learn the context-aware representations of premise-hypothesis and the set of retrieved external knowledge.

Specifically, given premise $P = \{p_i\}_{i=1}^n$, hypothesis $H = \{h_j\}_{j=1}^m$, and the set of external knowledge $EXT = \{\{e_r\}_{r=1}^l\}^k$, where r is the number of tokens in the external knowledge and k is the number of retrieved external knowledge. For encoding the premise and hypothesis, we input P and H to BERT in the following form

$$S = [\langle CLS\rangle, P, \langle SEP\rangle, H, \langle SEP\rangle] \tag{1}$$

$$H = \text{BERT}(S) \in \mathbb{R}^{(n+m+3)\times h} \tag{2}$$

where $\langle SEP\rangle$ is the token separating P and H, $\langle CLS\rangle$ is the classification token, and h is the dimension of the hidden state. When the BERT model is fine-tuned for the downstream task, the fine-tuned hidden state vector ($\mathbf{h}_{cls}$) corresponding to the classification token is used as the aggregate representation for the sequence. For each of the external knowledge in the set EXT, we generate the context-aware representations using the same BERT encoder as used for premise-hypothesis above as follows

$$EXT^k = [\langle CLS\rangle, e_1, \ldots, e_l, \langle SEP\rangle] \tag{3}$$

$$E^k = \text{BERT}(EXT^k) \in \mathbb{R}^{l+2\times h} \tag{4}$$

$$\mathbf{e}^k = \text{MeanPooling}(E^k) \in \mathbb{R}^{1\times h} \tag{5}$$

The averaged context-aware vector representation ($\mathbf{e}$) generated for each of the (k) retrieved external knowledge are stacked to create the context-aware matrix, $E \in \mathbb{R}^{k\times h}$.

Knowledge Integration Layer. This layer integrates external knowledge into the premise-hypothesis contextual representations by means of multi-head dot product attention. The layer uses a mixture model to allow a better trade-off between the context from external knowledge and the premise-hypothesis context [24]. The mixture model learns two parameter matrices, A and B, weighing the importance of premise-hypothesis context and the external knowledge context.

Multi-head Attentions. To measure the importance of external knowledge to each context-aware premise-hypothesis representation, we apply multi-head dot product attention [21] between the context-aware representations of external knowledge and that of premise-hypothesis.

In multi-head dot product attention, the context-aware representations are projected linearly to generate the queries, keys and values. As we use the multi-head attention to highlight the external knowledge important to premise-hypothesis context, premise-hypothesis representation (H) generates the query matrix (H^q) via linear projection and the two linear projections of external knowledge representation (E) generate the keys and values. The attention function is defined as

$$\text{Attention}(H^q, E^k, E^v) = \text{softmax}(\frac{H^q E^{k^T}}{\sqrt{h_k}})E^v \qquad (6)$$

Then the multi-head attention is

$$\begin{aligned} C_{ph}^{ext} &= \text{MH}(H^q, E^k, E^v) \\ &= \text{Concat}(\text{head}_1, \ldots, \text{head}_h)W^o \end{aligned} \qquad (7)$$

where $\text{head}_i = \text{Attention}(H^q W_i^q, E^k W_i^k, E^v W_i^v)$ and W_i^q, W_i^k, W_i^v, and W^o are projection matrices and i is the number of attention heads (12 in our case). The output of multi-head attention, $C_{ph}^{ext} \in \mathbb{R}^{(m+n+3) \times h}$ is an attention-weighted context matrix measuring the importance of the external knowledge context to each of the context-aware premise-hypothesis representation.

Similarly, to measure the importance of each premise-hypothesis BERT representation (H) to the aggregate premise-hypothesis representation (hidden representation $\mathbf{h}_{cls}$ corresponding to CLS token), we apply the multi-head attention between $\mathbf{h}_{cls}$ token representation and H as

$$\text{Attention}(C_{cls}^q, H^k, H^v) = \text{softmax}(\frac{C_{cls}^q H^{k^T}}{\sqrt{h_k}})H^v \qquad (8)$$

where $C_{cls}^q \in \mathbb{R}^{(m+n+3) \times h}$ is a matrix obtained by repeating $\mathbf{h}_{cls}$ hidden state $(n+m+3)$ number of times. The multi-head attention is calculated similar to (Eq. 7) that outputs a context matrix $C_{ph}^{cls} \in \mathbb{R}^{(m+n+3) \times h}$. The output matrix C_{ph}^{cls} is an attention-weighted context matrix measuring the importance of each of the premise-hypothesis representation to the aggregate premise-hypothesis representation.

Mixture Model. The mixture model learns a trade-off between the premise-hypothesis context and the context from external knowledge and is defined as

$$M = AC_{ph}^{ext} + BC_{ph}^{cls} \tag{9}$$

where A and B are the parameter metrices, learned with a single layer neural network and $A + B = J \in \mathbb{R}^{(n+m+3)\times 1}$, J is a matrix of all ones. The parameters A and B learn to balance the proportion of incorporating the premise-hypothesis context and the context from external knowledge. Each of the representations in $M \in \mathbb{R}^{(m+n+3)\times h}$ can be regarded as a knowledge aware state representation that encodes external knowledge context information with respect to the context of each of the premise-hypothesis representation.

Composition Layer. We compose the knowledge state representation (M) to the corresponding premise-hypothesis representation to obtain knowledge-aware matrix $\widehat{H}$ as

$$\widehat{H} = H + M \tag{10}$$

Pooling Layer. The pooling layer creates a fixed-length representation from premise-hypothesis representations H and the knowledge-aware representations $\widehat{H}$. We apply the standard mean and max pooling mechanisms as

$$\mathbf{h}_{mean} = \text{MeanPooling}(H) \qquad \mathbf{h}_{max} = \text{MaxPooling}(H) \tag{11}$$

$$\hat{\mathbf{h}}_{mean} = \text{MeanPooling}(\widehat{H}) \qquad \hat{\mathbf{h}}_{max} = \text{MaxPoolling}(\widehat{H}) \tag{12}$$

Classification Layer. We classify the relationship between premise and hypothesis using a Multilayer Perceptron (MLP) classifier. The input to the MLP is the concatenation of pooled representations as

$$\mathbf{f} = [\mathbf{h}_{mean}; \hat{\mathbf{h}}_{mean}; \mathbf{h}_{max}; \hat{\mathbf{h}}_{max}] \tag{13}$$

The MLP consists of two hidden layers with tanh activation and a softmax output layer to obtain the probability distribution for each class. The network is trained in an end-to-end manner using multi-class cross-entropy loss.

4   Experiments

4.1   Datasets

NLI and KGs. The key contribution of this paper is the unique method to incorporate external knowledge into the pre-trained BERT representations. ExBERT is capable of incorporating knowledge from any external knowledge source that allows the knowledge to be retrieved, given an entity as input. This includes KBs with *(head, relation, tail)* graph structure, KBs that contain only entity metadata without a graph structure and those that combine both a graph and entity metadata.

In this work, we retrieve external commonsense knowledge from ConceptNet [20] for evaluating ExBERT on SNLI (570,000 examples) [2] and SciTail (27,000 examples)[12] benchmarks, and from the science domain-targeted KG, Aristo Tuple [5] for evaluation on science domain SciTail dataset.

ConceptNet is a multilingual KG comprising of 83 languages. We pre-process the ConceptNet data to retrieve the facts with head and tail entities in the English language. The final pre-processed ConceptNet that we retrieve the external knowledge from contains 3,098,816 ($\approx$ 3M) commonsense facts connected by 47 relations. Aristo Tuple is an English language KG that contains $294,000$ science domain facts connected with 955 unique relations. We search the whole Aristo Tuple KG to retrieve relevant external knowledge.

4.2 Experimental Setup

Following our external knowledge retrieval mechanism discussed in Sect. 3.1, we first retrieve the external knowledge from ConceptNet and Aristo Tuple for SNLI and SciTail datasets via selection and ranking steps. In the ranking step, the English uncased $BERT_{BASE}$ [6] model is employed in feature extraction mode (i.e. without fine-tuning) to learn the contextual representations of the premise, the hypothesis and to each of the selected KG triple sentences. We then use the retrieved external knowledge to train the following three versions of ExBERT.

Models. We used the English uncased $BERT_{BASE}$ to train three versions of ExBERT: Two ExBERT+ConceptNet models on SNLI and SciTail respectively and one ExBERT+AristoTuple model on SciTail. The models utilise the external knowledge from the KG their name is suffixed.

Training Details. ExBERT is implemented in PyTorch using the base implementation of BERT[1]. The underlying BERT is initialised with the pre-trained BERT parameters and follows the same fine-tuning procedure as the original BERT. During training, the pre-trained BERT parameters are fine-tuned with the other ExBERT parameters. We use the Adam optimiser [13] with the initial learning rate fine-tuned from {8e-6, 2e-5, 3e-5, 5e-5} and warm-up rate of 0.1. The batch size is selected from {16, 24, 32}. The maximum number of epochs is chosen from {2, 3, 4, 5}. Dropout ratio of 0.5 is used at the classification layer [8]. Texts are tokenised using word pieces, with a maximum length of 40 for SNLI, 60 for SciTail, and 15 for external knowledge. The hyper-parameters are fine-tuned on the dev set of each NLI dataset.

5 Results

The results of top-performing models on the SNLI[2] and SciTail[3] leaderboards are summarised in Table 2. For fairness of comparison, we compare ExBERT

---

[1] https://github.com/huggingface/transformers.
[2] https://nlp.stanford.edu/projects/snli/.
[3] https://leaderboard.allenai.org/scitail/submissions/public.

with only the PTLMs based NLI models that leverage external knowledge. On **SNLI**, the performance of the state-of-the-art models is highly competitive.

Table 2. Results on SNLI and SciTail dataset: For SNLI, ExBERT uses ConceptNet KG. For SciTail ExBERT uses ConceptNet KG and AristoTuple KGs.

SNLI Dataset		SciTail Dataset	
$BERT_{BASE}$ as Base Model		$BERT_{BASE}$ as Base Model	
NLI Model	Test Acc(%)	NLI Model	Test Acc%
BERT $BERT_{BASE}$ + SRL [26]	89.6	OpenAI GPT [19]	88.3
OpenAI GPT [19]	89.9	$BERT_{BASE}$ [16]	92.5
$BERT_{BASE}$ [15]	90.5	BERT+LF [18]	92.8
$BERT_{BASE}$ [16]	90.8	$MT - DNN_{BASE}$ [16]	94.1
BERT+LF [18]	90.5	MT-DNN+LF[18]	94.3
$SemBERT_{BASE}$ [25]	91.0	$BERT_{LARGE}$ as Base Model	
$MT - DNN_{BASE}$ [16]	91.1	$BERT_{LARGE}$ [16]	94.4
MT-DNN+LF [18]	91.1	$MT - DNN_{LARGE}$ [16]	95.0
$BERT_{LARGE}$ as Base Model		**ExBERT+ConceptNet (Ours)**	**95.2**
$BERT_{LARGE}$ [16]	91.0	**ExBERT+AristoTuple (Ours)**	**95.9**
$BERT_{LARGE}$ + SRL [26]	91.3		
$SemBERT_{LARGE}$ [25]	91.6		
$MT - DNN_{LARGE}$ [16]	91.6		
ExBERT+ConceptNet (Ours)	**91.5**		

We observe that ExBERT outperforms all the existing baselines on the SNLI dataset and pushing the benchmark to 91.5% within the models using $BERT_{BASE}$ as the base model. ExBERT achieves a maximum performance improvement of 1.9% over the previous state-of-the-art $BERT_{BASE}$ + SRL [26] baseline.

Among the models built on $BERT_{LARGE}$ with more than 340M million parameters [6], our ExBERT[4] with $BERT_{BASE}$ (110M parameter) remarkably outperforms the $BERT_{LARGE}$ and $BERT_{LARGE}$ + SRL [26] models with the absolute improvements of 0.5% and 0.2% respectively, and is able to match the performance of $SemBERT_{LARGE}$ [25] and $MT - DNN_{LARGE}$ [16] models.

On **SciTail** (Table 2), ExBERT outperforms all the existing models including the models built on $BERT_{LARGE}$ model. Our best performing model, ExBERT+AristoTuple demonstrates an absolute improvement of 7.6% over the established baseline of OpenAI GPT [19]. Moreover, using only $BERT_{BASE}$ as the underlying model, our ExBERT+AristoTuple outperforms $BERT_{LARGE}$ based $MT - DNN_{LARGE}$ [16] model by 0.9%.

We observe higher performance improvements on the smaller SciTail dataset which demonstrates that incorporating external knowledge helps the model with small training data. Further, we observe that ExBERT attains higher accuracy when external knowledge is incorporated from the science domain-specific KG,

---

[4] We expect further improvements in ExBERT's performance with $BERT_{LARGE}$, however we left the evaluation for future work due to the limited computing resources.

Aristo Tuple as compared to when external knowledge is added from the commonsense KG, ConceptNet. The specialised scientific knowledge in Aristo Tuple is more beneficial to SciTail.

6   Analysis

6.1   Number of External Features

To investigate the effect of incorporating various numbers of external knowledge features, we vary the number of KG triple sentences input to ExBERT. Particularly, we are interested in answering the question: How many commonsense features are required for the optimal model performance? Fig. 2 illustrates the results of the experiment. **For SNLI**, ExBERT achieves the highest accuracy (91.5%) using 11 external knowledge sentences. We observe a decrease in accuracy when increasing the number of external knowledge sentences after 11. The fewer number of external knowledge sentences required, compared to SciTail dataset, to achieve the maximum accuracy on SNLI dataset, is attributed to the limited linguistic and semantic variation and the short average length of stop-word filtered premise (7.35 for entailment and neutral class) and hypothesis (3.61 for entailment and 4.45 for neutral class) [12] of the SNLI dataset, which limits its ability to fully extract and exploit external KG knowledge.

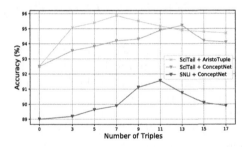

Fig. 2. ExBERT accuracy with varying amount of external knowledge.

For SciTail, ExBERT when evaluated using the general commonsense knowledge source ConceptNet, requires a relatively high number of external knowledge sentences (15) to achieve the maximum accuracy. This is due to the higher syntactic and semantic complexity of SciTail, which needs more knowledge to reason. However, when evaluated with the domain-specific Aristo Tuple KG, the model achieves the highest accuracy with fewer (7) external knowledge sentences. To reiterate, domain specific knowledge in Aristo Tuple improves the model performance with less external knowledge.

6.2   Qualitative Analysis

Case Study. This section provides the case study of different premise-hypothesis pairs and the corresponding external knowledge, to vividly show the effectiveness of ExBERT in adaptively identifying the relevant features from the supplied external knowledge. Recall that given a context-aware representation of premise-hypothesis token, the relevance of the retrieved external knowledge in E is measured by the multi-head attention defined in Eq.(6). We average the attention weights of all heads and plot a heat map.

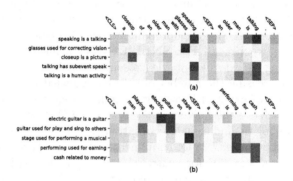

Fig. 3. Case Study. Visualisation of ExBERT's attention between external knowledge from ConceptNet (y axis) and SNLI premise-hypothesis pair tokens (x axis).

Figure 3 presents the heat map showing the attention of premise-hypothesis tokens to the retrieved external knowledge sentences from ConceptNet. In Fig. 3(a), we can see, these attention distribution is quite meaningful, with the *"speaking"* and *"talking"* attending mainly to the retrieved external knowledge *"speaking is talking"*. Similarly, the tokens *"speaking"* *"talking"* and *"man"* attends to *"talking is a human activity"*. In Fig. 3(b) among the other attentions, the most prominent can be observed between the tokens *"performing for cash"* and the external knowledge sentence *"performing used for earning"*.

Attending to the relevant external knowledge demonstrates the ExBERT's ability to effectively utilise the retrieved external knowledge based on the context from the premise and hypothesis.

7   Conclusion

We introduced ExBERT to enrich the contextual representation of BERT with real-world commonsense knowledge from external knowledge sources and to enhance its language understanding and reasoning capabilities. ExBERT can incorporate external knowledge from any external knowledge source that allows the knowledge to be retrieved, given an entity. We devised a novel external knowledge retrieval mechanism utilising contextual representations to retrieve relevant external knowledge. Experimental results on SNLI and SciTail NLI benchmarks in conjunction with two KGs, ConceptNet and Aristo Tuple, shows that ExBERT achieves significant performance improvements over the previous state-of-the-art methods, including those which are enhanced by $BERT_{LARGE}$. Further, we demonstrated the feasibility of utilising contextual representations for encoding external knowledge from KGs, which indicates a potential direction for future research.

References

1. Bast, H., Björn, B., Haussmann, E.: Semantic search on text and knowledge bases. Found. Trends Inf. Retrieval **10**(2–3), 119–271 (2016)
2. Bowman, S.R., Angeli, G., Potts, C., Manning, C.D.: A large annotated corpus for learning natural language inference. In: ACL (2015)
3. Chen, Q., Zhu, X., Ling, Z.H., Inkpen, D., Wei, S.: Neural natural language inference models enhanced with external knowledge. In: ACL (2018)
4. Dagan, I., Glickman, O., Magnini, B.: The PASCAL recognising textual entailment challenge. In: Quiñonero-Candela, J., Dagan, I., Magnini, B., d'Alché-Buc, F. (eds.) MLCW 2005. LNCS (LNAI), vol. 3944, pp. 177–190. Springer, Heidelberg (2006). https://doi.org/10.1007/11736790_9
5. Dalvi, M.B., Tandon, N., Clark, P.: Domain-targeted, high precision knowledge extraction. Trans. Assoc. Comput. Linguist. **5**, 233–246 (2017). https://www.transacl.org/ojs/index.php/tacl/article/view/1064
6. Devlin, J., Chang, M.W., Lee, K., Toutanova, K.: Bert: pre-training of deep bidirectional transformers for language understanding. In: Proceedings of the NAACL-HLT 2019 (Long and Short Papers), vol. 1, pp. 4171–4186 (2019)
7. Gajbhiye, A., Jaf, S., Moubayed, N.A., Bradley, S., McGough, A.S.: Cam: a combined attention model for natural language inference. In: 2018 IEEE International Conference on Big Data (Big Data), pp. 1009–1014, December 2018
8. Gajbhiye, A., Jaf, S., Moubayed, N.A., McGough, A.S., Bradley, S.: An exploration of dropout with RNNs for natural language inference. In: Kůrková, V., Manolopoulos, Y., Hammer, B., Iliadis, L., Maglogiannis, I. (eds.) ICANN 2018. LNCS, vol. 11141, pp. 157–167. Springer, Cham (2018). https://doi.org/10.1007/978-3-030-01424-7_16
9. Gajbhiye, A., Winterbottom, T., Al Moubayed, N., Bradley, S.: Bilinear fusion of commonsense knowledge with attention-based NLI models. In: Farkaš, I., Masulli, P., Wermter, S. (eds.) ICANN 2020. LNCS, vol. 12396, pp. 633–646. Springer, Cham (2020). https://doi.org/10.1007/978-3-030-61609-0_50
10. Kang, D., Khot, T., Sabharwal, A., Hovy, E.: AdvEntuRe: adversarial training for textual entailment with knowledge-guided examples. In: ACL, Melbourne, July 2018
11. Kapanipathi, P., et al.: Infusing knowledge into the textual entailment task using graph convolutional networks. arXiv preprint arXiv:1911.02060 (2019)
12. Khot, T., Sabharwal, A., Clark, P.: Scitail: A textual entailment dataset from science question answering. In: AAAI, New Orleans, 2–7, February 2018
13. Kingma, D.P., Ba, J.: Adam: a method for stochastic optimization. In: Bengio, Y., LeCun, Y. (eds.) ICLR (2015)
14. Kwon, S., Kang, C., Han, J., Choi, J.: Why do masked neural language models still need common sense knowledge?. CoRR abs/1911.03024 (2019)

15. Li, A.H., Sethy, A.: Knowledge enhanced attention for robust natural language inference. arXiv preprint arXiv:1909.00102 (2019)
16. Liu, X., He, P., Chen, W., Gao, J.: Multi-task deep neural networks for natural language understanding. In: Proceedings of the 57th Annual Meeting of the Association for Computational Linguistics, pp. 4487–4496, Florence, July 2019
17. Logan, R., Liu, N.F., Peters, M.E., Gardner, M., Singh, S.: Barack's wife hillary: using knowledge graphs for fact-aware language modeling. In: Proceedings of the 57th ACL, pp. 5962–5971, Florence, July 2019
18. Pang, D., Lin, L.H., Smith, N.A.: Improving natural language inference with a pretrained parser. arXiv preprint arXiv:1909.08217 (2019)
19. Radford, A., Narasimhan, K., Salimans, T., Sutskever, I.: Improving language understanding by generative pre-training (2018)
20. Speer, R., Chin, J., Havasi, C.: Conceptnet 5.5: An open multilingual graph of general knowledge (2017)
21. Vaswani, A., et al.: Attention is all you need. In: Advances in Neural Information Processing Systems, pp. 6000–6010 (2017)
22. Wang, Q., Mao, Z., Wang, B., Guo, L.: Knowledge graph embedding: a survey of approaches and applications. IEEE Trans. Knowl. Data Eng. **29**(12), 2724–2743 (2017). https://doi.org/10.1109/TKDE.2017.2754499
23. Wang, X., et al.: Improving natural language inference using external knowledge in the science questions domain. In: Proceedings of the AAAI, vol. 33, pp. 7208–7215 (2019)
24. Yang, B., Mitchell, T.: Leveraging knowledge bases in LSTMs for improving machine reading. In: ACL, pp. 1436–1446, Vancouver, July 2017
25. Zhang, Z., et al.: Semantics-aware BERT for language understanding. ArXiv arXiv:1909.02209 (2020)
26. Zhang, Z., Wu, Y., Li, Z., Zhao, H.: Explicit contextual semantics for text comprehension. CoRR abs/1809.02794 http://arxiv.org/abs/1809.02794 (2018)

Multi-features-Based Automatic Clinical Coding for Chinese ICD-9-CM-3

Yue Gao[1,3], Xiangling Fu[1,3(✉)], Xien Liu[2(✉)], and Ji Wu[2]

[1] School of Computer Science (National Pilot Software Engineering School), Beijing University of Posts and Telecommunications, Beijing, China
{yuegao,fuxiangling}@bupt.edu.cn
[2] Department of Electronic Engineering, Tsinghua University, Beijing, China
{xeliu,wuji_ee}@tsinghua.edu.cn
[3] Key Laboratory of Trustworthy Distributed Computing and Service, Ministry of Education, Beijing University of Posts and Telecommunications, Beijing, China

Abstract. ICD-9-CM Volume 3 (ICD-9-CM-3), as a subset of the ICD-9-CM, is a standard system used to classify operations and medical procedures for billing purposes. With the gradual maturity of the DRG system, the precise coding of ICD-9-CM-3 is increasingly important for both hospital revenue and patients' health. In this paper, a method based on **BERT** and **NER** capturing Multi-**F**eatures for automatic Chinese ICD-9-CM-3 Coding (**BNMF**) is proposed to support doctors to make decisions so that traditional manual coding in hospitals can be replaced. The method is designed focusing on short text information from electronic medical records (EMR) and make decisions by combining the semantic features of the clinical text, the structured features of Chinese ICD-9-CM-3, and the axis knowledge of Chinese ICD-9-CM-3 text. Meanwhile, fusion of spatial and temporal features is made to better capture semantic features. The experiments demonstrate that the efficiency of our framework is higher than those of state-of-the-art methods in the field of classification of the short medical corpus.

Keywords: ICD coding · Decision support system · Medical informatics · Natural language processing

1 Introduction

International Classification of Diseases (ICD) is a system published by the World Health Organization, which standardizes the method to classify diseases, injuries, and causes of death. The US modified it clinically to ICD-9-CM by expanding its scope to include diagnosis codes (in ICD-9-CM Volumes 1 and 2) and inpatient hospital procedure codes (in ICD-9-CM Volume 3). Nowadays, ICD-9-CM has been a mainstream coding system applied in hospitals all around the world. In China, an updated version of Chinese ICD-9-CM-3, the health insurance version, has been launched in late 2019 to unify the national standard, which is the version we refer to in this paper.

© Springer Nature Switzerland AG 2021
I. Farkaš et al. (Eds.): ICANN 2021, LNCS 12895, pp. 473–486, 2021.
https://doi.org/10.1007/978-3-030-86383-8_38

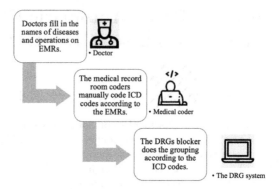

Fig. 1. Traditional ICD coding process in a hospital.

The traditional coding procedure is shown in Fig. 1. Firstly, diagnosis descriptions on diseases and operations of the patients are filled into the structural EMRs by the clinicians. Secondly, ICD coding is done manually by dedicated coders in the medical records room according to the content in EMRs. Thirdly, the assigned codes (diagnosis codes and procedure codes) are input into the DRG system for further classifications. The DRG stands for a diagnosis-related group, which is a system that uses statistical control theory to classify hospital cases. The accuracy of the DRG classification is highly related to the precision of its input, the ICD codes. Under the background of increasing medical insurance cost and increasing pressure of medical insurance fund balance, DRGs and its payment accounting method have also become the future development trend of global medical service. The coding procedure above looks reasonable. However, issues exist that the amount of whole ICD codes is too huge (13,002 codes in China Medical Insurance Edition) to keep every manual coding right, and that version difference of the ICD-9-CM system due to the region difference also increases the difficulty of precise coding. Meanwhile, with the gradual construction of the evaluation system on hospitals based on the DRG system, the need for more precise ICD coding becomes an urgent task pending for solution.

Several methods of automatic coding prediction have been proposed to replace the repetitive manual work in recent years. The evolution of the process is roughly from the traditional machine learning methods [1,2], to neural network methods [3,4], and then to explainable prediction methods [5,6]. However, few methods proposed till now focused on the short texts from structured EMR data, which should have a wider application scenarios with popularization of structured EMR. Furthermore, few methods proposed till now well combined knowledge of ICD system, the structural features of the framework of the ICD system as well as the semantic features of the ICD standard texts, which we believe can further improve the performance of the automatic coding system.

In this paper, we propose a multi-features-based method in a decision support setting for automatic Chinese ICD-9-CM-3 coding called BNMF. Short texts in the operation and procedure description block of an EMR are the input, while a limited-length list of ICD codes is the output. The BNMF method is designed to make predictions mainly based on BERT and a classifier. Meanwhile, a fastText model is utilized to make a chapter-level filtration of candidate codes, and Named Entity Recognition (NER) is utilized to extract axis words of organ and operation method, based on which the number of top k output can be predicted. Thus, the semantic features of the ICD standard texts, the structural features of the framework of the ICD system and knowledge of ICD axis words can be learned to improve the coding performance. Since our method is designed aiming for ICD-9-CM-3 in Chinese version, here we collected about 24,000 Chinese operation and procedure descriptions from EMR and their relating ICD codes. Based on the text-code-pair dataset, experiments are conducted to prove the superiority of our model compared with the current state-of-the-art methods. The main contributions are summarized as follows:

- A multi-features-based coding method for ICD-9-CM-3 is constructed, which combines both the semantic features of the clinical text and the structural features of ICD-9-CM-3.
- A large-scale short Chinese clinical dataset is collected, which completely fills the limited Chinese ICD coding corpus.
- Experiments are conducted to illustrate the effectiveness of our BNMF method for ICD-9-CM-3 classification. Such model, if modified, can also work in other ICD code classification tasks, which can support the construction of the whole DRG system.

2   Related Work

Research on ICD coding is already a hotspot in the medical informatics community that is going through a lengthy process from machine learning and hand-crafted methods (e.g. Scheurwegs et al.) to series of neural network methods [7]. Recently, most studies on automatic ICD coding focus on unstructured long documents from diagnosis descriptions (e.g. Kavuluru et al., Subotin and Davis, Mullenbach et al.) [5,8,9]. Some related approaches proposed provide words in documents that predictions are in strong reliance on to increase the interpretability of the coding system (e.g. Mullenbach et al., Cao and Yan et al.) [5,6].

Under the popularity of deep learning, diverse neural network methods have been modified and applied in recent automatic ICD coding models, among which convolutional neural networks (CNN) proposed by Kim [10] and recurrent neural network (RNN) proposed by Hochreiter et al. [11] and Chung et al. [12] are the two most utilized methods. Mullenbach et al. modified the CNN structure by adding a per-label attention mechanism after the convolution layers, in which way to select the k-grams that are most relevant to each predicted label from the text [5]. Cao and Yan et al. combined traditional machine learning method and neural network method by concatenating features learned by a dilated convolution structure and a term frequency-inverse document frequency (tf-idf) structure so that both explicit and implicit semantic features of an ICD code can be learned to make predictions [6]. Shi et al. applied long short term memory (LSTM) network to realize character-level text representations on discharge summaries [3]. The gate recurrent unit (GRU) Neural Network is also modified by Vani et al. to predict individual ICD codes [13]. Nevertheless, newly proposed neural network methods except traditional CNN and RNN model that may have better performance on automatic ICD coding tasks remain to be found.

3 Method

ICD-9-CM-3 can be seen as a dictionary that includes all standard procedure codes. Therefore, plenty of coding information is hidden in structural features of ICD-9-CM-3, which the ICD coding methods proposed before did not take into account. In this model BNMF, we aim to combine both semantic and structural features of ICD-9-CM-3 to make text classification. The architecture of the model is shown in Fig. 2. With text from structured EMR taken as input, the model mainly consists of three architectures: (1) a fastText-based architecture, (2) a BERT-based architecture and (3) a NER-based architecture to realize chapter-level filtration, semantic-feature extraction and axis knowledge extraction, respectively. Detailed descriptions on each architecture are given below.

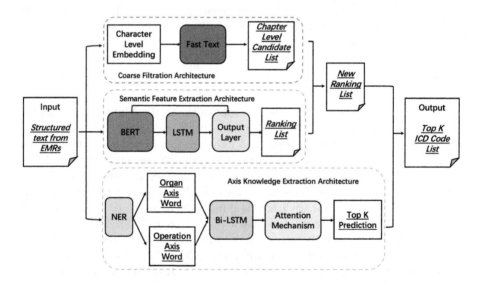

Fig. 2. The whole architecture of the model.

3.1 Coarse Filtration Architecture

The fastText-based architecture is designed to make a coarse filtration for the candidate codes so that chapter-level error can be avoided. ICD-9-CM-3, as is shown in Table 1, consists of 18 chapters which are divided in high correlation with the organism the codes describe. Titles of fifteen out of eighteen chapters

Table 1. Structure of ICD-9-CM-3

Chapter	Title	Code range
1	Procedures and interventions, not elsewhere classified	00
2	Operations on the nervous system	01–05
3	Operations on the endocrine system	06–07
4	Operations on the eye	08–16
5	Miscellaneous diagnostic and therapeutic procedures	17
6	Operations on the ear	18–20
7	Operations on the nose, mouth and pharynx	21–29
8	Operations on the respiratory system	30–34
9	Operations on the cardiovascular system	35–39
10	Operations on the hemic and lymphatic system	40–41
11	Operations on the digestive system	42–54
12	Operations on the urinary system	55–59
13	Operations on the male genital organs	60–64
14	Operations on the female genital organs	65–71
15	Obstetrical procedures	72–75
16	Operations on the musculoskeletal system	76–84
17	Operations on the integumentary system	85–86
18	Miscellaneous diagnostic and therapeutic procedures	87–99

(except Chap. 1, 5 and 18 which we later merged into one sum-up chapter) are named after organs. Thus, chapter classification based on the text from structured EMR is doable. Seeing that the input texts mostly contain no more than 50 tokens quite a bit of which are OOV words, we choose fastText architecture to realize word representations while taking morphology into account.

Suppose that a text containing n Chinese characters from EMR is given. N ngram features $x_1, x_2, .., x_N$ can be embedded to represent information of each word, respectively. Since each word w can be represented as a bag of character k-gram, to better represent features of words as well as information of the internal structure of words, sub-word representations (character-level) are summed up and then averaged for morphological word representations. Given a word w in a dictionary G, let us denote by $G_w \subset \{1, ..., G\}$ the set of n-grams appearing in w. We associate a vector representation z_g to each n-gram g. We represent a word by the sum of the vector representations of its n-grams. We thus obtain the scoring function:

$$s(w, c) = \sum_{g \in G_w} z_g^T v_c \tag{1}$$

In this model, to learn a representation for each text, we include three representation features: character bi-gram and tri-gram. With word representations concatenated to form the sentence features, two full connection layers along with

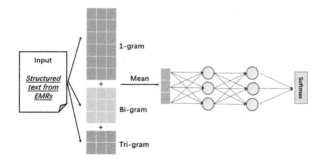

Fig. 3. The coarse filtration architecture.

softmax layer can finally classify the exact chapter the text belongs to, which is shown in Fig. 3. W_1, W_2, b_1 and b_2 represent weights and bias in the two full connection layers, respectively.

$$\hat{p}(y \mid S) = softmax(W_2(ReLU(W_1 s(w, c) + b_1) + b_2) \tag{2}$$

Experiments in the next section prove the efficiency of fastText architecture in this task compared with other neural network architectures. Nevertheless, still cannot we make sure 100% accuracy of this coarse filtration. Thus, a regulation is set in the following classification part to avoid the coding result totally depend on the chapter predicted.

3.2   Semantic Feature Extraction Architecture

The BERT-based architecture is applied to extract the semantic features. Training on BERT consists of two stages: pre-training and fine-tuning. Based on the two tasks: MLM and NSP, even a pre-trained BERT architecture can realize a good quantity on word representation [15]. In the fine-tuning stage, as is shown in Fig. 4, we connect the BERT architecture with a LSTM to further extract semantic features of the text. Then, we concatenate both the encoding features of the BERT architecture and the embedding features of the LSTM architecture to realize fusion of features in higher dimensions. As a result, features in both spatial dimension and temporal dimension of the text can be expressed. The concatenated features will then get through the architecture containing a ReLU layer, a max-pooling layer, a fully connected layer and a softmax layer, in which way to fine tune the model and apply the architecture into the ICD coding task. Experiments in the next section prove the efficiency of this fusion architecture in this task compared with other BERT-based architectures.

In the equations below, v_i denotes a random character in the vocabulary set, while e_i denotes the encoding feature of v_i on the BERT architecture. enc_s is the encoding set of the text. i_t, f_t and o_t represent the input gate, forget gate and output gate, respectively, which are all calculated using current input e_i, hidden state in previous step h_{i-1}, and current state of this cell c_i, and processed with

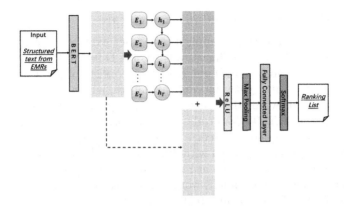

Fig. 4. The semantic feature extraction architecture.

sigmoid activation function σ. The three gates determine the degrees to take new inputs, forget existing memory, and output of the current state. The current cell state c_t is calculated using both previous cell state and current cell information. The hidden state h_t is calculated using output gate and current cell. H denotes hidden state set. M denotes the concatenation semantic features of H and enc_s, which is used for further prediction.

$$e_i = BERT(v_i); enc_s = \{e_1, e_2, ..., e_T\} \tag{3}$$

$$i_t = \sigma(W_{ei}e_t + W_{hi}h_{t-1} + W_{ci}c_{t-1} + b_i) \tag{4}$$

$$f_t = \sigma(W_{ef}e_t + W_{hf}h_{t-1} + W_{cf}c_{t-1} + b_f) \tag{5}$$

$$g_t = \tanh(W_{ec}e_t + W_{hc}h_{t-1} + W_{cc}c_{t-1} + b_c) \tag{6}$$

$$o_t = \sigma(W_{eo}e_t + W_{ho}h_{t-1} + W_{co}c_t + b_o) \tag{7}$$

$$c_t = i_t g_t + f_t c_{t-1}; h_t = o_t \tanh(c_t) \tag{8}$$

$$H = \{h_1, h_2, ..., h_t\}; M = [H; enc_s] \tag{9}$$

$$\hat{p}(y \mid S) = softmax(W(MaxPool(ReLU(M))) + b) \tag{10}$$

In this stage, we collect and utilize the Chinese ICD coding corpus as train set, where text and exact ICD code is a match. The BERT architecture takes the most important part of this model. Experiments in the next chapter prove that even the single BERT model without other architectures can reach a relatively high quality in the coding task. Some previous papers propose a way to combine both semantic features from neural network architecture and term frequency-inverse document frequency (tf-idf) architecture, which, however, improve little to the BERT-based architecture. Thus we delete this part in our model. Outputs of the BERT-based architecture is a list of candidate codes which are sorted according to their prediction scores.

3.3   Axis Knowledge Extraction Architecture

The NER-based architecture is applied to extract axis words of ICD standard text from the text in EMR. In Chinese version of ICD-9-CM-3, ICD standard text is very structured. Most of ICD standard text consists of 4 key words: organ, operation method, entry path and disease nature, where organ and operation method are two axis words that play a decisive role. As is shown in Table 2, we raise 4 examples of ICD-9-CM-3 codes and their structured features. To extract the two axis words as the knowledge for further prediction, here we apply the NER model, which is constructed with 2 Bi-LSTM and conditional random field (CRF) architectures, as is shown in Fig. 5. The two architectures take responsibility for organ and operation axis words entity extraction, respectively, in case of the overlapping of 2 words in a text.

Table 2. Structured features of ICD-9-CM-3 standard text

ICD code	ICD standard text	Axis words
47.0901	阑尾切除术 appendicectomy	organ+operation method
32.2905	肺部分切除术 pneumoresection	organ+operation method
49.7300	肛门瘘管闭合术 Anal fistula closure	organ+operation method+disease nature
07.6100	垂体腺部分切除术，经前额入路 Partial pituitary adenectomy, through the forehead	organ+operation method+ entry path+disease nature

The Bi-LSTM architecture is based on LSTM architecture, which considers both forward and backward sequence context. Thus, information from both the past and the future can be exploited through $h_i = [\overrightarrow{h_i} \oplus \overleftarrow{h_i}]$, where $\overrightarrow{h_i}$ and $\overleftarrow{h_i}$ denote LSTM architecture in two directions, respectively. h_i is the element-wise sum of them to be the Bi-LSTM architecture output.

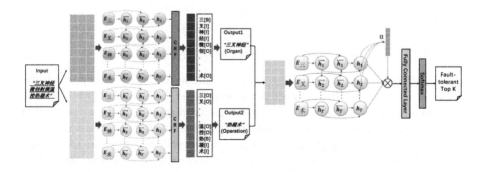

Fig. 5. The knowledge extraction and top k prediction architecture.

Since labels are predicted independently in the Bi-LSTM architecture, CRF layer is constructed to avoid incorrect labels (e.g. S-label followed by I-label or

O-label followed by I-label). Thus, dependencies between labels can be set by scoring labels of each character with the sum of emission score and transition score.

3.4 Top K Prediction

Top k prediction is the task to predict the exact number of ICD codes that the decision support system should list. Because during the procedure of error analysis, we find that the errors which predictions of BERT architecture make are in high correlation with the content of the two axis words. Cases exist that there are relations among the axis words, such as inclusion, mutual cross index and confusability, which frequently result in doctors mistakenly filling in EMRs with wrong axis words. Such cases, however, is hard to be detected during the semantic feature extraction process, and may result in bigger biases during the prediction because they are congenital errors. As is shown in Table 3, 4 representative examples of prediction errors resulting from axis word issues are listed, where wrong axis words are marked with ∗∗.

Table 3. Shape of a character in dependence on its position in a word

Raw Text	Ground Truth	Issue
胸主动脉覆膜支架介入*置入术* *Interventional placement* of thoracic aorta with coated stent graft	胸主动脉覆膜支架腔内*隔绝术* *Endovascular exclusion* of thoracic aorta with coated stent graft	Mutual cross index of operation axis words
支撑喉镜显微镜下声带肿物*切除术* *Surgical removal* of vocal cord mass under a supporting laryngoscope microscope	支撑喉镜下声带显微*缝合术* *Microsuturing* of vocal cords under a supporting laryngoscope	Inclusion index of operation axis words
双侧扁桃体剥离术 Bilateral tonsillectomy	扁桃体切除术*不伴腺样增殖体*切除术* Tonsillectomy *not accompanied by adenoid proliferator*	Information loss of organ axis words
右侧*附耳*切除术 Right *side ear* resection	*副耳*切除术 *Accessory ear* resection	Misuse of universal organ axis words

In view of this phenomenon, we concatenate the 2 axis words extracted by the NER architecture in to one text and utilize it as an input of the Top k prediction architecture which is designed as a Bi-LSTM + Attention architecture. In this way, judgements on error-prone likeliness can be made based on the features of the axis words contained in the text and an appropriate length of the ICD code list can be given to make sure that the right code is included in the list while the list length is as short as possible.

α represents a weight vector calculated using hidden status set H of LSTM architecture. r represents dot product result of H and α, which can be processed with activation $tanh$ to be the representation h^* for further prediction. In this way, contents in higher correlation with the prediction can be attached with bigger importance, by distributed with bigger weight α_i.

$$M = \tanh(H); \alpha = softmax(w^T M) \tag{11}$$

$$r = H\alpha^T; h^* = tanh(r) \tag{12}$$

$$\hat{p}(y \mid S) = softmax(W^{(S)}h^* + b^{(S)}) \tag{13}$$

To get top k prediction architecture trained, we labeled each text with k number according to the new ranking list result made from the first two architectures. A labeled k number is the actual number plus one to increase the robustness of this architecture. To the texts whose actual number is bigger than 5, we classify all them into one class called uncertain.

3.5   Classification

The final classification can be made combining the outputs of architectures above. Firstly, ICD codes that don't belong to the chapter predicted by the fastText architecture will be deleted from the code list predicted by the BERT-based architecture. Since the prediction on the chapter can be keep 100% accurate, here we set a constant θ. Thus, top θ codes in the list wouldn't be affected in any case. Then, only top k codes in the processed list will be chosen whose number is predicted in the top k prediction architecture. The k codes in the list is the final outputs of the decision support system.

The loss function in the training procedure is the binary cross-entropy.

$$\mathcal{L} = -\sum_{l=1}^{L} y_l log(\hat{y}_l) + (1 - y_l)log(1 - \hat{y}_l) \tag{14}$$

where y_l and $\hat{y}_l$ are the ground truth and the predicted sigmoid score for each label, respectively.

4   Experiments

4.1   Dataset

To evaluate the efficiency of our model, we collected a Chinese dataset of inpatient operation list from structured EMRs, which contains more than 6,000 Chinese clinical texts and more than 1,300 unique ICD-9-CM-3 codes. All of the texts are in short length which contain no more than 50 tokens. All the texts in the dataset are manually annotated with ICD-9-CM-3 codes by professional coders in hospitals, so that a text-code match can be constructed. In view that our research focus on short text from structured Chinese EMRs and the model we design is based on structured features of Chinese text, the dataset we collected till now is quite a big scale for our ICD coding related experiments.

In the experiments, we divide the dataset (CN-Full) into training set and test set in ratio 9:1 or training set, validation set and test set in ratio 8:1:1 if validation set is needed. Meanwhile, considering the fact that huge amount of codes among the collected samples are in low frequency, which means that such codes only have one or two relative texts in the dataset, we reconstruct

Table 4. Detailed information on the datasets

	CN-Full	CN-50	All-text
Total # text samples	6,071	1,869	19,073
Total # labels	1,364	50	13,002
Vocabulary size	6,147	6,147	6,147
Mean # tokens per sample	12	13	13

a sub-dataset (CN-50) which includes 50 codes in higher frequency from the original dataset. To better get the coarse filtration architecture trained, we also include the standard text from ICD-9-CM-3 on the basis of the dataset (All-text). Detailed information on the datasets is shown in Table 4.

4.2 Systems and Evaluation Metrics

To prove the efficiency of our method, we compare our method with several representative baselines below, whose results are shown in Table 7:

- **Levenshtein distance**: A machine-learning measure of the similarity between texts.
- **Text CNN**: A representative CNN-based model for ICD classification.
- **Bi-GRU**: A representative RNN-based model for ICD classification.
- **CAML**: A state-of-the-art ICD coding method based on CNN & Attention.

In the coarse filtration architecture of the model, the word embedding is initialized in random. The batch size is 128. The pad size is 32. The dropout rate is 0.5. The optimizer is Adam (Kingma and Ba, 2015) with a learning rate of 0.001. The n-gram vocabulary size is 250,500. The parameter θ is set as 2. In the semantic feature extraction architecture of the model, the pretrained parameters are based on BERT-BASE, Chinese from Google. The batch size is 128. The pad size is 32. The dropout rate is 0.1. The optimizer is Adam with a learning rate of 0.00005. Early-stop mechanism is set based on validation set loss. To facilitate comparison with both future and prior work, we use Precision, Recall and F1 score as the metrics.

4.3 Results

First, to demonstrate the quality of the coarse filtration architecture, experiments is made on fast Text architecture and other representative deep learning models. Results are shown in Table 5. It can bee seen that on most of the evaluation metrics, fast Text can reach a better quality compared with other deep learning baselines.

Second, we demonstrate the quality of the semantic feature extraction architecture and the necessity of the other two architectures. Experiments are made on our BNMF method, the semantic feature extraction architecture (fusion model

Table 5. Evaluation on chapter-level filtration task

Model	Precision		F1		Accuracy
	Macro	Micro	Macro	Micro	
Text CNN	0.9491	0.9626	0.9482	0.9622	0.9624
Bi-LSTM	0.8701	0.9353	0.8668	0.9380	0.9418
Bi-LSTM+Attention	0.9488	0.9646	0.9475	0.9638	0.9635
fast Text	**0.9647**	**0.9690**	**0.9552**	**0.9687**	**0.9688**

of BERT and LSTM) and other BERT-based models. Results shown in Table 6 prove that our fusion model does have a better performance on the evaluation metrics. Meanwhile, the BNMF method, compared with the semantic feature extraction architecture, has a higher precision score with the k increasing and a close Recall score, which proves that the other two architectures do improve the efficiency of the whole method.

Table 6. Evaluation on semantic feature extraction task and ablation experiment

Model	P@k		R@k	
	K = 3	K = 5	K = 3	K = 5
BERT+CNN	0.3153	0.1882	0.9017	0.9416
BERT+LSTM	0.3069	0.1875	0.9169	0.9527
Fusion of BERT & LSTM	0.3102	0.1933	0.9312	**0.9671**
BNMF	**0.3273**	**0.2306**	**0.9342**	0.9656

Then, in view of the whole model, our proposed model outperforms all previous works on both datasets CN-Full and CN-50. The results are shown in Table 7. Compared with series of representative baselines like Levenshtein distance, CNN, Bi-GRU and CAML which uses traditional convolutional attention network, our proposed model outperform on the two datasets. Such superiority is more obvious on CN-Full where micro recall score achieves 0.89 when parameter k is set as 3 and 0.96 when parameter k is set as 5, which means that our method works especially when the number of labels is big. It can be seen that scores of macro-F1 are always lower than those of micro-F1, especially in CN-Full, which in some way is consistent with the common sense that ICD classes with smaller dataset have a relative poor performance.

Table 7. Evaluation on Chinese dataset CN-Full and CN-50

Dataset	CN-Full				CN-50			
Model	F1		R@k		F1		R@k	
	Macro	Micro	K = 3	K = 5	Macro	Micro	K = 3	K = 5
Levenshtein distance	0.6437	0.6614	0.6637	0.7631	0.7125	0.7362	0.7314	0.7968
Text CNN	0.7356	0.7590	0.8206	0.8562	0.8364	0.8531	0.9147	0.9371
Bi-GRU	0.7214	0.7433	0.8109	0.8351	0.8094	0.8217	0.8683	0.8876
CAML	0.7388	0.7662	0.8403	0.8852	0.8413	0.8596	0.9253	0.9471
BNMF	**0.7869**	**0.8003**	**0.8958**	**0.9328**	**0.8525**	**0.8704**	**0.9342**	**0.9656**

5   Conclusions and Future Work

In this paper, a new method BNMF is proposed for Chinese-structured-EMR oriented automatic ICD-9-CM-3 coding. It is the first time BERT-based architecture applied in ICD coding task, which expresses semantic features in both spatial dimension and temporal dimension. Meanwhile, a chapter-level filtration based on fastText architecture and a top k prediction based on NER for axis words knowledge in ICD-9-CM-3 are applied to also combine both the axis knowledge of the Chinese clinical text and the structural features of ICD-9-CM-3, which further improves the quality of the system in precision rate and recall rate. Furthermore, the BNMF framework is also extensible for ICD-9-CM-1/2 and ICD-10 diagnosis codes after modification, which means that a complete ICD coding systems for diagnosis and procedure codes can be constructed to serve the DRG system in future work.

Acknowledgments. This work was supported in part by the National Natural Science Foundation of China under Grant 82071171, in part by the Beijing Natural Science Foundation under Grant L192026, and in part by Graduate Innovation and Entrepreneurship Program of BUPT under Grant 2021-YC-T014.

References

1. Perotte, A., et al.: Diagnosis code assignment: models and evaluation metrics. J. Am. Med. Inform. Assoc. **21**(2), 231–237 (2014)
2. Koopman, B., Zuccon, G., et al.: Automatic ICD-10 classification of cancers from free-text death certificates. Int. J. Med. Inform. **84**(11), 956–965 (2015)
3. Shi, H., Xie, P., Hu, Z., et al.: Towards automated ICD coding using deep learning. arXiv preprint arXiv:1711.04075 (2017)
4. Yu, Y., et al.: Automatic ICD code assignment of Chinese clinical notes based on multilayer attention BiRNN. J. Biomed. Inform. **91**, 103–114 (2019)
5. Mullenbach, J., Wiegreffe, S., Duke, J., et al.: Explainable prediction of medical codes from clinical text. In: The Conference of the North American Chapter of the Association for Computational Linguistics, pp. 1101–1111 (2018)
6. Cao, P., et al.: Clinical-coder: assigning interpretable ICD-10 codes to Chinese clinical notes. In: The 58th Annual Meeting of the Association for Computational Linguistics: System Demonstrations (2020)

7. Scheurwegs, E., Luyckx, K., Luyten, L., et al.: Data integration of structured and unstructured sources for assigning clinical codes to patient stays. J. Am. Med. Inform. Assoc. **23**(e1), e11–e19 (2015)
8. Kavuluru, R., Rios, A., Lu, Y.: An empirical evaluation of supervised learning approaches in assigning diagnosis codes to electronic medical records. Artif. Intell. Med. **65**(2), 155–166 (2015)
9. Subotin, M., Davis, A.: A system for predicting ICD-10-PCS codes from electronic health records. In: Workshop on Biomedical Natural Language Processing (2014)
10. Kim, Y.: Convolutional neural networks for sentence classification. In: Conference on Empirical Methods in Natural Language Processing, pp. 1746–1751 (2017)
11. Hochreiter, S., Schmidhuber, J.: Long short-term memory. Neural Comput. **9**(8), 1735–1780 (1997)
12. Chung, J., et al.: Empirical evaluation of gated recurrent neural networks on sequence modeling. CoRR, abs/1412.3555 (2014)
13. Vani, A., Jernite, Y., Sontag, D.: Grounded recurrent neural networks. arXiv preprint arXiv:1705.08557 (2017)
14. Devlin, J., Chang, M., Lee, K., Toutanova, K.: BERT: pre-training of deep bidirectional transformers for language understanding. CoRR, abs/1810.04805 (2018)
15. Vaswani, A., et al.: Attention is all you need. CoRR, abs/1706.03762 (2017)

Style as Sentiment Versus Style as Formality: The Same or Different?

Somayeh Jafaritazehjani[1,2]([✉]), Gwénolé Lecorvé[2], Damien Lolive[2], and John D. Kelleher[1]

[1] ADAPT Centre, Technological University Dublin, Dublin, Ireland
john.d.kelleher@tudublin.ie
[2] University of Rennes 1, CNRS, IRISA, Rennes, France
{somayeh.jafaritazehjani, gwenole.lecorve, damien.lolive}@irisa.fr

Abstract. Unsupervised textual style transfer presupposes that style is a coherent and consistent concept and that style transfer approaches will generalise consistently across different domains of style. This paper explores whether this presupposition is appropriate for different types of style. We explore this question by comparing the performance and latent representations of a variety of neural encoder-decoder style-transfer architecture when applied to sentiment transfer and formality transfer. Our findings indicate that the relationship between style and content shifts between these different domains of style: for sentiment, style and content are closely entangled; however, for formality, they are less entangled. Our findings suggest that for different types of styles different approaches to modeling style for style-transfer are necessary.

Keywords: Content · Style · Sentiment · Formality · Disentanglement

1 Introduction

The task of textual style-transfer emerges from the observation that the same content can be expressed in different ways (or styles), such as: brief as opposed to verbose, formal or informal, expert or beginner style, polite or impolite, and different personal styles. The task of textual style transfer is a multi-objective natural language generation (NLG) problem which focuses on generating a new version of an input text that expresses the content of the input in an alternative style. Consequently, a key challenge in textual style transfer revolves around theses two components (style and content) and how to disentangle them. Although there is a growing literature on the task of textual style-transfer the development of a widely acceptable definition for style is an open issue. For example, one question that has emerged in the field is whether sentiment should be considered a style in the same way as formality or politeness. This paper analyses this question by comparing, across a number of style transfer tasks (sentiment-transfer and formality-transfer) and neural encode-decoder style-transfer architectures, the correlation between the amount information in the latent representation of a model relating to the style of an input sequence and the content preservation

© Springer Nature Switzerland AG 2021
I. Farkaš et al. (Eds.): ICANN 2021, LNCS 12895, pp. 487–499, 2021.
https://doi.org/10.1007/978-3-030-86383-8_39

power of that model. The idea informing this analysis is that when a strong correlation exists this provides evidence that style and content are highly inter-related for that style transfer task. Consequently, this analysis will enable us to assess whether the relationship between style and content is consistent across sentiment and formality. If the relationship between style and content is different for sentiment and formality this would indicate the sentiment transfer is fundamentally different from formality transfer, and so style transfer approaches developed for one of these tasks may not generalise to the other. Our findings indicate that style and content cannot be disentangled for sentiment transfer, however for formality transfer they can. Based on this finding we propose that sentiment transfer and formality transfer are related but distinct tasks.

The remainder of the paper is structured as follows: Sect. 2 reviews the previous work and categorizes them based on how they define style, Sect. 3 and 4 describe the models and datasets we use in our experiments followed by introducing the evaluation aspects and metrics we use to evaluate the architectures (Sect. 5), Sect. 6 describes and presents our experiments and results, and in Sect. 7 we set out our conclusions.

2   Literature Review

A clear definition of the concept of style is essential to designing an approach to style transfer. Indeed, based on how style is viewed previous research on style-transfer can be categorized into two groups [24]. The first group assumes that style can be explicitly disentangled from content, and that style-transfer is best done by identifying and replacing style markers. Informed by this understanding of style, the style transfer models in this category focus on separating the style markers from the content as an initial step and proceed by generating the style-shifted sequences. These two steps can employ statistical frequency-based methods, neural network techniques, or a combination of the both [11,12,16].

The second group implicitly define style as a holistic concept and an integral component of a text—fundamentally connected to the concept of content—where each style can be considered as a different language [24]. From a modelling perspective, the strategies in this group frame style transfer as a translation task and aim at translating from one style as the source language to the other one as the target language by implementing end-to-end approaches [13]. The generation block of these style transfer models are mostly based on a standard encoder-decoder (seq2seq) architecture [1,23] or extensions of it, such as encoder multi-decoder [4] or variational encoder models [6,8]. The goal of generation block is to generate a style-shifted sequence that is semantically similar to the input and grammatically correct [4,13,18,20,22].

In unsupervised style transfer an increasingly popular approach is to use Generative Adversarial Networks [5] where classifiers are employed as the adversarial block to guide the training process. [4,6,8,13,18,20,22]. These strategies are based on the assumption that the latent representation of the input sequences are style-free. In this vein of research, recent work has investigated the disentanglement of the style and content by analysing the input latent space taking "sentiment" as the style [7].

3 Models

The baseline model we use for our experiments is an adversarial encoder generator (encoder-decoder) style transfer model [20] which contains: (i) a single encoder model $\mathbf{E}$ which reads an input sequence $\mathbf{x}$ in style $s \in \{1, 2\}$ (denoted $\mathbf{x}^{(s)}$ }) and creates an embedded representation $\mathbf{z}$ of the input, (ii) a single generator (decoder) model $\mathbf{G}$ that is initialised with $\mathbf{z}$ and the target output style $s \in \{1, 2\}$ and generates a sequence of words that ideally are a surface representation of the content in $\mathbf{x}$ in the target output style, and (iii) a set of two style-specific Discriminators $\mathbf{D_s}$ ($s \in \{1, 2\}$). $\mathbf{D_s}$ takes a generated sequence and predicts whether or not it has the style s. The reason for employing discriminators in this architecture is that this architecture is trained in an adversarial manner where for a given target output style s the goal of the generator $\mathbf{G}$ is to generate an output such that $\mathbf{D_s}$ labels it in the style s, and at the same time the goal of the $\mathbf{D_s}$ is to predict whether the style of the output is the same as the original input sequence or has been transferred. The encoder and generator cells of the model (and the variants described below) are single-layer RNNs with GRU [2] (cell-size is set to 700). The encoder GRU cells of the attention-based and multi-encoder models are bi-directional and uni-directional, respectively. Token vectors are initialized by pre-trained GloVe [14] and their size is set to 100. Discriminators are TextCNN classifiers from [9].

We propose two extensions of the baseline model: a multi-encoder and an attention-based architecture. Both of these extensions are designed to be more powerful in terms of encoding the input sequence (e.g., the multi-encoder architecture has a separate encoder for each input style which we expect will enable each encoder to fine-tune to its relevant style; and the attention-based architecture generates a new context sensitive representation of the entire input at each step in the generation process). By increasing the representational capacity of the baseline in different ways we will be able to examine if the encoding of style and content across the different neural architectures and domains is consistent.

3.1   Multi-encoder Model

Our multi-encoder model has two style-specific encoders $\mathbf{E_1}$ and $\mathbf{E_2}$ and a single generator $\mathbf{G}$ (Fig. 1). Each of the encoders reads a sequence $\mathbf{x}$ of its style corresponding style $s \in \{1, 2\}$, denoted as $\mathbf{x}^{(s)}$ and outputs an embedded representation $\mathbf{z_s}$. The encoders share the generator which is initialized with $\mathbf{z}$, as the output of either $\mathbf{E_1}$ or $\mathbf{E_2}$, and a parameter indicating the desired output style s. The output is either a reconstructed or style-shifted sequence. It is a reconstructed sequence when the desired style (s) and the source sequence style are the same and it is a style-shifted sequence when these two styles are different.

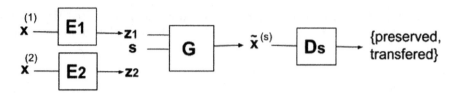

Fig. 1. Multi-encoder model (following the schema from [7]); $\mathbf{E_1}$ and $\mathbf{E_2}$ create $\mathbf{z_1}$ & $\mathbf{z_2}$ given the inputs of their corresponding style. Then given each $\mathbf{z}$, $\mathbf{G}$ generates an output in the target output style $\mathbf{s}$.

This architecture has two style-specific Discriminators, $\mathbf{D_s}$ ($s \in \{1,2\}$). Each $\mathbf{D_s}$ takes a generated sequence and predicts whether or not it has the style s.

Training involves processing two differently styled inputs in parallel $\mathbf{x}_1^{(s_1)}$ and $\mathbf{x}_2^{(s_2)}$ (where $s_1 \neq s_2$) in response to which four outputs are generated, one output sequence per style for each input sequence. Two of these outputs will be reconstructed sequences $\widetilde{\mathbf{x}}_1^{(s_1)}$, $\widetilde{\mathbf{x}}_2^{(s_2)}$, and two style-transferred sequences $\widetilde{\mathbf{x}}_2^{(s_1)}$, and $\widetilde{\mathbf{x}}_1^{(s_2)}$. The discriminators are trained using the loss shown in Eq. 1. For a given style s this loss computes the binary cross-entropy over "transferred" and "preserved" instances where the true labels of style-shifted and reconstructed outputs are considered as "transferred" and "preserved" respectively. For each style s, we train $\mathbf{D_s}$ to maximize the probability of assigning these true labels to the output sequences by minimizing this loss.

$$\mathcal{L}_{D_s} = -\log(D_s(\widetilde{\mathbf{x}}_1^{(s)})) - \log(1 - D_s(\widetilde{\mathbf{x}}_2^{(s)})) \tag{1}$$

The encoders and generator are trained using a combination of reconstruction and adversarial losses. The reconstruction losses of the $\mathbf{E_1}$ and $\mathbf{E_2}$ are computed following Eq. 2. L_{recs} ($s \in \{1,2\}$) is the cross-entropy between the reconstructed sequence $\widetilde{x}^{(s)}$ and its corresponding input $x^{(s)}$.

$$\mathcal{L}_{recs} = -\log \mathrm{Pr}_{E_s}(\widetilde{\mathbf{x}}^{(s)}|\mathbf{x}^{(s)}) \tag{2}$$

The adversarial loss $L_{adv,s}$ is computed solely on the transferred sequences and measures the precision of a discriminator $\mathbf{D_s}$ in detecting inputs that have been transferred to style s. Equation 3 shows this loss for s_1 (L_{adv,s_2} is computed symmetrically). Minimizing this loss minimizes the log of the inverse probability predicted by the discriminator which motivates the generation block to generate style-shifted sequences with a lower possibility of being detected as transferred.

$$\mathcal{L}_{adv,s_1} = \log(1 - D_{s_1}(\widetilde{\mathbf{x}}_2^{(s_1)})) \tag{3}$$

During training, the back-propagation for the encoder-generator triple ($\mathbf{E_1}$, $\mathbf{E_2}$, $\mathbf{G}$) is carried out using the following equation where L_{rec} is the summation of the L_{rec1} and L_{rec2} (Eqs. 2) and $L_{adv,s}$ is the adversarial loss (Eq. 3).

$$\mathcal{L}_{total} = \mathcal{L}_{rec} + \mathcal{L}_{adv,s_1} + \mathcal{L}_{adv,s_2} \tag{4}$$

3.2   Attention-Based Model

We propose an attention-based model by employing attention layers [1] in our base model. This model contains the following components: a single encoder $\mathbf{E}$, a single generator $\mathbf{G}$ and style-specific discriminators $\mathbf{D_s}$ ($s \in \{1, 2\}$). $\mathbf{E}$ is a bi-directional RNN which consists of forward and backward RNNs. It reads an input sequence $\mathbf{x}$ with the length T in the both forward and reversed order and creates the encoder states as the concatenation of the forward and backward hidden states. The embedded vector of $\mathbf{x}$ is therefore obtained as the concatenation of the last state of $\mathbf{E}$ from forward and backward cells, denoted as $\mathbf{z}$.

   $\mathbf{G}$ is a uni-directional RNN which is initialized with the output style. At the ith step of generation the RNN cell takes the following inputs: the previous state s_{i-1}, the previous output y_{i-1}, a context vector c_i and outputs s_i and y_i. A different context vector c_i is created for each generation time step i, and is computed as a weighted summation of the encoder states. We use Bahadanau's additive method [1] to compute at each generation step an set of attentions weights across the encoder states, and then use these attention weights to calculate the weighted summation c_i for that generation step. To generate these attention weights we first concatenate each of the states of the encoder $h_1, \ldots, h_T$ with a label indicating the desired output style. The result of this concatenation processes is the sequence $h'_1, h'_2, \ldots, h'_T$ (i.e., h'_j is the concatenation of the desired output style and h_j). We then provide two inputs to Bahadanau's attention at each generation step: (i) the previous state of the generator s_{i-1}; and (ii) the sequence of augmented encoder states $h'_1, h'_2, \ldots, h'_T$. Given these inputs each encoder state h'_j is scored relative to the context of the generator s_{i-1} by passing h'_j and s_{i-1} through a hyperbolic tangent layer and then passing the output of this layer through a fully connected layer W_f (see Eq. 6). Then the scores for the encoder states are passed through a softmax layer (see Eq. 5).

$$a_{i,j} = \frac{\exp(score_{i,j})}{\sum_{k=1}^{T} \exp(score_{i,k})} \tag{5}$$

$$score_{t,i} = W_f(\tanh((W_s s_{t-1} + b_s) + (W_h h'_i + b_h))) \tag{6}$$

The first step of generation computes c_1 by taking $\mathbf{z}$ and the start token `<Go>` as s_0 and y_0. The fully connected feedforward layers employed in the attention mechanism are jointly trained with all the other components of the model.

   $\mathbf{D_s}$ act the same as in the multi-encoder model: given a generated sequence, it predicts whether or not it has the style s. Training process is also the same as multi-encoder model. In our experiments we use all three of the models discussed in this section: the baseline, the multi-encoder and attention-based model (Fig. 2).

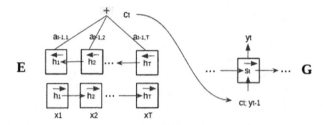

Fig. 2. Generating the output token at time step t while creating c_t considering s_{t-1}

4   Datasets

We use the GYAFC, and the Yelp Restaurant Reviews datasets. Our motivation for selecting these datasets is that they are appropriate for different style-transfer tasks: the Yelp dataset is suitable for sentiment-transfer whereas GYAFC is suitable for formality-transfer. The vocabulary size given below for these datasets is after replacing words occurring less than 5 times with the <unk>. token.

Grammarly's Yahoo Answers Formality Corpus (GYAFC) [17] contains human-labelled paired informal and formal sentences which was crawled from two domains of Entertainment & Music (E&M) and Family & Relationships (F&R) in Yahoo Answers[1]. For the experiments, we combine E&M and F&R and we use this aligned corpus as non-parallel by considering the style of each file as the only label available. The training dataset has 104k informal sentences and 104k formal sentences, the resulting test set has 11k informal sentences and 13k formal sentences, and the development set has 24k informal sentences and 27k formal sentences. We applied upsampling to balance positive and negative labels. To be more consistent with the splits of the Yelp dataset, we swapped the development and test sets resulting in the final splits of 72%, 9% and 19% for the train, development and test sets, respectively. The vocabulary size is 20K.

Yelp Restaurant Reviews (Yelp) is a large-scale review dataset (4.7 million reviews) where reviews are labelled as positive and negative if their corresponding stars are above and below three respectively. Three-starred reviews are discarded. Moreover, because the unit of analysis in our experiments is the sentence, we used the review level label for each sentence of the review. Doing this, however, can lead to neutral sentences being labelled as positive and negative. To address this problem, previous work, such as [20], have assumed that longer reviews are more likely to contain neutral sentences and longer sentences are more likely to be neutral. We adopted a similar approach and filtered out reviews that had more than 10 sentences, and sentences longer than 15 tokens. The resulting training dataset has 252K negative sentences and 381k positive examples, the development set has 25K negative sentences and 38K positive sentences, and the test set has 50K negative sentences and 76K positive sentences. For training, we applied upsampling by randomly selecting negative sentences

---

[1] https://answers.yahoo.com.

without replacement for repetition to balance positive and negative labels. The vocabulary size is 10K and the final train, development and test splits are: 70%, 10% and 20%.

5 Evaluation Aspects

We consider the evaluation dimensions of style shift and content preservation power to investigate the performance of the base model and its proposed extensions: multi-encoder and attention-based model.

Style shift power (SSP) focuses on how well the style transfer models shift the style of the input sequences to the target style. Following some previous research [4,6,8,11,12,15,20,22], we train style classifiers in order to measure the percentage of the style-shifted sequences which are labeled with the target style. We employed a TextCNN model [9] as the style classifier, and train it separately for the Yelp and GYAFC datasets.

Content preservation power (CPP) investigates the similarity of the input sequences and their corresponding style-shifted outputs in terms of content. There is no widely accepted CPP metric and the existing metrics are criticised for different reasons; for instance, the cosine similarity embedding-based metric [4] is criticised due to its sensitivity [8]. We consider the following three metrics which employ different strategies to compute CPP of each architecture.

1. **Cosine Similarity (CS)** is an embedding-based metric which computes the cosine similarity between the embedding of the input sequence and its corresponding style-shifted output. We use a method introduced in [4] to generate the embedding of the sequences. First, we use a pre-trained 100-dimensional GloVe model [14] to generate an embedding for each token in a sequence. We then calculate the min, mean and max pooling of these token embeddings. The embedding vector for the full sequence is then created by concatenating these min, mean and max pooling vectors.
2. **Word Overlap (WO)** is an n-gram based metric proposed in [8] which computes the word overlap of the input $\mathbf{x}$ and style-shifted output $\widetilde{\mathbf{x}}$. We first exclude stop words from each sequence and then calculate the ratio of the unigram overlap and the total number of unigrams of the two sequences.
3. **Word Movers Distance (WMD)** is a special case of the Earth Mover's Distance [19] which computes the dissimilarity between the sequences. WMD has been used to compute CPP in some previous style transfer work, e.g. [25] where the minimum distance of the word embeddings of the source and style-shifted sequences is measured as the score, i.e. the minimum distance that the words of one sequence need to travel in semantic space to reach the words of the other sequence [10]. To compute WMD, after replacing the `<unk>` tokens with `<the>`, we map the tokens of the sequences to the pre-trained Word2Vec embeddings [21] with the embedding size 300.

6   Experiments

Section 6.1 reports an experiment on the variation in performance of the baseline, multi-encoder and attention-based style shift architectures across the sentiment and formality domains. Sections 6.2 and 6.3 report experiments that focus analysing the latent representations of the input sequences created by each of the style-transfer across the two domains.

6.1   Assessing Style-Shift Power and Content Preservation Across Domains and Architectures

We evaluate the performance of the base model, and the two proposed extensions of it, multi-encoder and attention-based models, across the domains of sentiment and formality. Table 1 lists the results obtained for each model for each metric across the Yelp and GYAFC datasets. As a sense-check of our evaluation metrics we first assessed the agreement between the content preservation metrics: CS, WO and WMD. There is consensus across the metrics in terms of the ranking of the models for both datasets. We take the agreement between these metrics as a validation our methodology for computing content preservation.

Focusing on the performance of the multi-encoder and attention architectures compared with the base model, the results in Table 1 indicate that on both datasets the extension of the base model with an attention-based mechanism has a bigger impact relative to extension with multi-encoders. We attribute this to the fact that in the attention-based model the encoder has a direct input into every step of the generation, whereas, in the multi-encoder the encoder only directly inputs into the initial step of the generation process. To test whether the observed differences in model performance are statistically significant, for each model and domain combination we calculated the confidence interval around the average model performance on the domain test set for the three content preservation metrics. For the Yelp dataset the differences in content preservation (across all three metrics) between the base model, multi-encoder and attention-based architectures were statistically significant at the 0.99 confidence level (i.e., the confidence intervals do not overlap). For GYAFC the confidence intervals did not overlap at the 0.7 confidence level.

Finally, the results in Table 1 also shows an increase in CPP and a drop in SSP compared to the base model when these models are applied to the Yelp dataset. However, on the GYAFC, we observe the opposite pattern: an increase in the SSP and a decrease in CPP. In both cases CPP and SSP appear to be inversely related (increasing one metric results in a decrease in the other); however, the fact that the direction of the change is flipped across the two datasets suggests a difference between these styles of formality and sentiment.

6.2   Probing the Latent Space of the Networks

There is a growing body of work on using probing classification experiments on the latent representations of neural networks, e.g. [3,7]. The idea of a probing

Table 1. The results of evaluating the models, higher value shows better performance except for the metric WMD

Dataset	Yelp				GYAFC			
Model	CS	WO	WMD	SSP	CS	WO	WMD	SSP
Base model	0.9239	0.199	0.695	**78.76%**	**0.9085**	**0.0893**	**0.693**	55.99%
Multi-encoder	0.9311	0.254	0.647	76.26%	0.9072	0.0813	0.706	57.53%
Attention-based	**0.9542**	**0.475**	**0.3827**	53.99%	0.8848	0.0397	0.8549	**66.41%**

experiment is that if it is possible to train a binary classifier to accurately predict the presence of a linguistic feature in a sentence based on an embedding of the sentence this is evidence that the sentence encoder that generated the embedding is capturing that linguistic feature. Inspired by this work we ran probing experiments on the latent representations of the architectures in order to understand how strongly the input style is encoded in these representations. To do so, firstly, we generate the latent representation of the train, development and test sets of the Yelp and GYAFC datasets where we consider the last state of the encoder(s) in the base and multi-encoder models, and the average of the context vectors generated at each step of generation of the attention-based model as the latent vectors. Then, for each neural architecture we train one probing classifier for each dataset. These classifiers are trained to predict the style of the sentence input to the encoder. These classifiers were implemented as feed-forward networks with a single hidden layer and a sigmoid output layer.

Table 2 reports for each neural architecture the accuracy of the trained classifier in detecting the style of the sentences in the test set of each dataset (note that this table also lists the results for a variational encoder architecture that we will introduced in Sect. 6.3). The accuracy of a classifier is an indication of the amount of source style information that the corresponding style transfer architecture encodes in its latent representations. The results in Table 2 show that the probes trained on the multi-encoder and attention based embeddings are more accurate than those trained on the baseline architecture in both the sentiment and formality domains. This indicates that both the multi-encoder and attention based extensions strengthen the encoding of the input style in the latent representation of their respective transfer architectures. The average score of the accuracy of the classifiers trained on the Yelp data is higher than the average score of the classifiers trained on the GYAFC data (Table 2) which shows that they encode more source style information. To more concretely quantify the observed differences between our results on Yelp versus GYAFC, for each dataset we computed the Pearson correlation coefficients (PCC) between the CS scores of the models (Table 1) and their accuracy of the probing classification task. Given that CS is a measure of content preservation, a strong correlation between the CS performance of a style-transfer architecture and the accuracy of the corresponding classifier on predicting the input style of a sentence based on a architecture's latent representations would indicate that the different neural architecture treat content and style as closely related concepts. The results of this PCC correlation were 0.824 for Yelp and 0.336 GYAFC. This strong PCC correlation for Yelp indicates that the style-transfer neural architectures tended to treat style and content as closely related concepts in Yelp (i.e., strengthening the encoding of the input style signal in the latent representation also resulted in an increase content preservation). By contrast, the relatively weak PCC correlation for GYAFC suggests that the neural architectures were able to disentangle style and content, to some extent, during style transfer in this domain.

Table 2. The accuracy of the classifiers corresponding to each architecture for both datasets.

Dataset	Base model	Variational model	Multi encoder model	Attention-based model	Average score
Yelp	99.97%	67.25%	100%	100%	91.8%
GYAFC	99.42%	59.69%	99.93%	100%	89.76%

6.3   Employing a Variational Model to Modify the Latent Space

The results of the probing experiment in Sect. 6.2 indicated that the multi-encoder and attention based models more strongly encoded input style in their latent representations compared with the baseline. Furthermore, this increase in the strength of the input style encoding was, in the case of sentiment (Yelp) strongly correlated with an increase in content preservation (CS) but the correlation between input style encoding and content preservation was weak for formality. Given that these results were based on increasing the representational power of the encoder in terms of encoding input style, in this section we report on an experiment that examined what happens if the representational capacity of a style transfer encoder to distinguish between input styles is reduced.

For this experiment we us a variational extension of the base model with the motivation that this variational encoder will strip out the source style from the latent representation of the input sequences. To make the encodings of style 1 and style 2 more similar to each other we align both posterior distributions $p_E(\mathbf{z_1}|\mathbf{x_1}, \mathbf{s_1})$ and $p_E(\mathbf{z_2}|\mathbf{x_2}, \mathbf{s_2})$ to a prior density $p(z)$ (here, N (0, I)). To do so, we add a KL-divergence regularizer to the reconstruction loss which is similar to the reconstruction loss of the base model. The discriminator block and the training process of this model is the same as the base model.

$$\mathcal{L}_{rec} = -\log \Pr_E(\widetilde{\mathbf{x}}^{(s)}|\mathbf{x}^{(\mathbf{s})}) + \mathcal{D}_{\mathbf{KL}}(\Pr_{\mathbf{E}}(\mathbf{z}|\mathbf{x}, \mathbf{s})|| \Pr(\mathbf{z})) \tag{7}$$

Table 2 shows that the accuracy of the probing classifiers trained on the latent vectors of the variational model drops significantly compared to the results for the base model. For Yelp the drop is from 99.9% (using the base model representations) to 67.25% (using the variational model representations), and the GYAFC the drop is from 99.42% to 59.69%. This indicates that the variational architecture is working as we expecting in both domains in terms of reducing the input style signal in the latent embeddings of the transfer architecture. Also, the CS scores of the variational model for Yelp and GYAFC are 0.8989 and 0.8922, respectively. Comparing they CS for the variational model to the CS for the baseline (from Table 1) we observe a small drop in both cases: Yelp $0.9239 - 0.8984 = 0.0250$ (or 2.5%); GYAFC $0.9085 - 0.8922 = 0.0163$ (or 1.6%).

Overall, the results indicate that stripping more of the source style from the latent representations of the input (as a result of employing a KL-divergence regularizer) results in the content preservation power of the variational model decreasing. These results aligns with our observations above regarding the entanglement between content and style. However, for sentiment this decrease 2.5% whereas for formality it is 1.6%, and this difference in decrease—although small, in absolute terms—indicates that style and content are relatively more entangled in the case of sentiment as compared with formality.

7   Conclusion

In this paper, we examined whether the concept of style was consistent across the domains of sentiment and formality. We used the relationship between style and content as the basis for our analysis. Our fundamental intuition is that if style and content have a consistent relationship across domains this would suggest that each of them are themselves consistent concepts across domains. We used a variety of neural style-transfer architectures as a basis for our analysis. Using these neural architectures and datasets from the sentiment and formality domains we report three experiments that examined the relationship between content and style across domains.

Our first experiment found that content preservation and style shift power were inversely related in both sentiment and formality domains but that the extensions to the baseline model flipped the direction of improvement across domains (for sentiment content preservation improved, but for formality style shift power improved). The results of our second and third experiment indicate that style and content are more tightly entangled in the sentiment domain as compared with the formality domain. Overall our results suggest that the concept of style (at least in terms of how it relates to content) varies between the sentiment and formality domains. This indicates that style-transfer architectures that work in one domain may not be directly applicable in other domains.

References

1. Bahdanau, D., Cho, K.H., Bengio, Y.: Neural machine translation by jointly learning to align and translate. In: 3rd International Conference on Learning Representations, ICLR 2015 (2015)
2. Chung, J., Gulcehre, C., Cho, K., Bengio, Y.: Empirical evaluation of gated recurrent neural networks on sequence modeling. In: NIPS 2014 Workshop on Deep Learning, December 2014 (2014)

3. Conneau, A., Kruszewski, G., Lample, G., Barrault, L., Baroni, M.: What you can cram into a single vector: Probing sentence embeddings for linguistic properties. In: Proceedings of the 56th Annual Meeting of the Association for Computational Linguistics (Volume 1: Long Papers), pp. 2126–2136 (2018)

4. Fu, Z., Tan, X., Peng, N., Zhao, D., Yan, R.: Style transfer in text: exploration and evaluation. In: Proceedings of the AAAI Conference on Artificial Intelligence, vol. 32 (2018)

5. Goodfellow, I., et al.: Generative adversarial nets. In: Ghahramani, Z., Welling, M., Cortes, C., Lawrence, N.D., Weinberger, K.Q. (eds.) Advances in Neural Information Processing Systems, vol. 27, pp. 2672–2680. Curran Associates, Inc. (2014). http://papers.nips.cc/paper/5423-generative-adversarial-nets.pdf

6. Hu, Z., Yang, Z., Liang, X., Salakhutdinov, R., Xing, E.P.: Controllable text generation. CoRR abs/1703.00955 (2017). http://arxiv.org/abs/1703.00955

7. Jafaritazehjani, S., Lecorvé, G., Lolive, D., Kelleher, J.D.: Style versus content: a distinction without a (learnable) difference? In: Proceedings of the 28th International Conference on Computational Linguistics, COLING 2020, Association for Computational Linguistics (2020)

8. John, V., Mou, L., Bahuleyan, H., Vechtomova, O.: Disentangled representation learning for non-parallel text style transfer. In: Proceedings of the 57th Annual Meeting of the Association for Computational Linguistics, pp. 424–434 (2019)

9. Kim, Y.: Convolutional neural networks for sentence classification. In: Proceedings of the Conference on Empirical Methods in Natural Language Processing (EMNLP), pp. 1746–1751 (2014)

10. Kusner, M., Sun, Y., Kolkin, N., Weinberger, K.: From word embeddings to document distances. In: Bach, F., Blei, D. (eds.) Proceedings of the 32nd International Conference on Machine Learning. Proceedings of Machine Learning Research, PMLR, Lille, France, 07–09 Jul 2015, vol. 37, pp. 957–966 (2015). http://proceedings.mlr.press/v37/kusnerb15.html

11. Leeftink, W., Spanakis, G.: Towards controlled transformation of sentiment in sentences. CoRR abs/1808.04365 (2019). http://arxiv.org/abs/1808.04365

12. Li, J., Jia, R., He, H., Liang, P.: Delete, retrieve, generate: a simple approach to sentiment and style transfer. In: 2018 Conference of the North American Chapter of the Association for Computational Linguistics: Human Language Technologies, NAACL HLT 2018, pp. 1865–1874. Association for Computational Linguistics (ACL) (2018)

13. Ma, S., Sun, X.: A semantic relevance based neural network for text summarization and text simplification. Comput. Linguist. **1**(1) (2017)

14. Pennington, J., Socher, R., Manning, C.D.: Glove: global vectors for word representation. In: EMNLP (2014)

15. Prabhumoye, S., Tsvetkov, Y., Salakhutdinov, R., Black, A.W.: Style transfer through back-translation. In: Proceedings of the 56th Annual Meeting of the Association for Computational Linguistics (Volume 1: Long Papers), pp. 866–876. Association for Computational Linguistics (2018). http://aclweb.org/anthology/P18-1080

16. Ramos, J., et al.: Using TF-IDF to determine word relevance in document queries. Citeseer

17. Rao, S., Tetreault, J.R.: Dear sir or madam, may i introduce the GYAFC dataset: Corpus, benchmarks and metrics for formality style transfer. In: NAACL-HLT (2018)

18. Romanov, A., Rumshisky, A., Rogers, A., Donahue, D.: Adversarial decomposition of text representation. In: Proceedings of the 2019 Conference of the North American Chapter of the Association for Computational Linguistics: Human Language Technologies, Volume 1 (Long and Short Papers), pp. 815–825 (2019)
19. Rubner, Y., Tomasi, C., Guibas, L.J.: The earth mover's distance as a metric for image retrieval. Int. J. Comput. Vis. **40**(2), 99–121 (2000)
20. Shen, T., Lei, T., Barzilay, R., Jaakkola, T.: Style transfer from non-parallel text by cross-alignment. In: Guyon, I., et al. (eds.) Advances in Neural Information Processing Systems, vol. 30, pp. 6830–6841. Curran Associates, Inc. (2017). http://papers.nips.cc/paper/7259-style-transfer-from-non-parallel-text-by-cross-alignment.pdf
21. Shivakumar, P.G., Georgiou, P.: Confusion2Vec: towards enriching vector space word representations with representational ambiguities. PeerJ Comput. Sci. **5**, e195 (2019)
22. Singh, A., Palod, R.: Sentiment transfer using seq2seq adversarial autoencoders. CoRR abs/1804.04003 (2018). http://arxiv.org/abs/1804.04003
23. Sutskever, I., Vinyals, O., Le, Q.V.: Sequence to sequence learning with neural networks. In: Proceedings of the Conference in Neural Information Processing Systems (NIPS), pp. 3104–3112 (2014)
24. Tikhonov, A., Yamshchikov, I.P.: What is wrong with style transfer for texts? CoRR abs/1808.04365 (2018). http://arxiv.org/abs/1808.04365
25. Yamshchikov, I., Shibaev, V., Khlebnikov, N., Tikhonov, A.: Style-transfer and paraphrase: looking for a sensible semantic similarity metric. arXiv preprint arXiv:2004.05001 (2020)

Transfer and Meta Learning

Low-Resource Neural Machine Translation Using XLNet Pre-training Model

Nier Wu, Hongxu Hou$^{(\boxtimes)}$, Ziyue Guo, and Wei Zheng

College of Computer Science-college of Software, Inner Mongolia University,
Hohhot, China
cshhx@imu.edu.cn

Abstract. The methods to improve the quality of low-resource neural machine translation (NMT) include: change the token granularity to reduce the number of low-frequency words; generate pseudo-parallel corpus from large-scale monolingual data to optimize model parameters; Use the auxiliary knowledge of pre-trained model to train NMT model. However, reducing token granularity will result in a large number of invalid operations and increase the complexity of local reordering on the target side. Pseudo-parallel corpus contains noise affect model convergence. Pre-training methods also limit translation quality due to the human error and the assumption of conditional independence. Therefore, we proposed a XLNet based pre-training method, that corrects the defects of the pre-training model, and enhance NMT model for context feature extraction. Experiments are carried out on CCMT2019 Mongolian-Chinese (Mo-Zh), Uyghur-Chinese (Ug-Zh) and Tibetan-Chinese (Ti-Zh) tasks, the results show that the generalization ability and BLEU scores of our method are improved compared with the baseline, which fully verifies the effectiveness of the method.

Keywords: Low-resource · Machine translation · XLNet · Pre-training

1 Introduction

The neural machine translation (NMT) based on encoder-decoder structure [1,9] takes the cross-entropy between the generated token and the reference as the objective to optimize the parameters. During the training phase, whether the model training is stable depends on the bias-variance tradeoff. Variance reduction can be achieved by adding data sets, dropout methods, and introducing penalty terms into the loss function; Bias reduction requires the model to optimize parameters with reference token to understand language knowledge. Therefore, in addition to the improvement of network structure and training methods, high-quality pre-trained word embedding plays a vital role in machine translation tasks with scarce resources. The general one-hot word embedding is simple and easy to express, but the dimension is too large to learn the relationship between

© Springer Nature Switzerland AG 2021
I. Farkaš et al. (Eds.): ICANN 2021, LNCS 12895, pp. 503–514, 2021.
https://doi.org/10.1007/978-3-030-86383-8_40

words through vector orthogonal operations. The Word2vec [3] and GloVe [6] models embed high-dimensional one-hot word vector into a low-dimensional space using a distributed representation algorithm, similar to the N-gram language model [2], which computing the cosine distance from the origin to the word embedding space point in the initial neighborhood to learn the words relationship. Although this kind of word embedding method has been widely used as a neural machine translation modeling unit due to many advantages, after pre-training, the word embedding is fixed, which makes it insensitive to polysemous words and cannot be dynamically optimized for specific tasks. Recently, pre-training model [4,7] has been widely used in various natural language processing tasks, including machine translation, because of its strong ability of semantic feature learning. [10] proposed to obtain knowledge from the pre-trained BERT model to improve the quality of NMT, they dynamically fused representations of specific tasks pre-trained by BERT into the Transformer model based on the importance of representations. They also used the knowledge distillation method on the decoder to learn the output distribution from the pre-trained model. [8] proposed a masked sequence to sequence pre-training method to jointly train the encoder and decoder. Taking the typical BERT model as an example, it consists of two main modules: Masked language model (MLM) and Next sentence prediction (NSP). The MLM model based on the self-encoding method uses the flag bit *Mask* to randomly replace some words, and then use the context corresponding to the *Mask* to predict the true value of the flag bit. However, *Mask* is not used in the fine-tuning phase, which makes the pre-training and fine-tuning inconsistent, resulting in errors. In addition, if the sequence contains multiple *Mask*, all the *Mask* tags are independent of each other, which also limits semantic understanding. While XLNet [11] trained by autoregressive method solves the inherent problem of self-encoding, and overcomes the defects of the traditional autoregressive model which only uses the previous or post information. The main objective of XLNet is to maximize the expected logarithmic likelihood of the factorization order of all possible sequences and to obtain all possible contexts of sequences in autoregressive mode. However, the number of factorization sequences will increase exponentially with the increase of sentence length, so it is necessary to reduce the size of factorization sequences through partial sampling. Therefore, the work of this paper is divided into the following aspects.

- We proposed a syntax-dependent factorization sequence sampling method to filter a large number of meaningless sequences. Then use the XLNet model to encode the sequence to word embedding representation.
- We proposed a source language-oriented XLNet attention mechanism, which interacts the pre-trained word embedding with the input layer of the encoder and the attention layer of the decoder, so that the NMT model can fully acquire the knowledge of the pre-training model.
- We embed the pre-trained target word into the NMT decoder, and let the NMT model learn the output probability distribution of the pre-training model through the knowledge distillation method when the decoder predicts the translation.

2   Background

Neural Machine Translation. Neural machine translation model can simulate the translation probability $P(y|x)$ of the source language $x = \{x_1, ...x_n\}$ to the target language $y = \{y_1, ...y_m\}$ word by word, as shown in Eq. 1.

$$P(y_i|y_{<i,x,\theta}) \propto exp\{f(y_{i-1}, s_i, c_i; \theta)\} \qquad (1)$$

Where $y_{<i}$ indicates the partial translation result before the i-th decoding step, θ indicates the parameters of the NMT model. s_i indicates the i-th hidden state of the decoder, c_i indicates the corresponding context of the source language at time t, and $f(\cdot)$ indicates the nonlinear activation function in the current node of the decoder. Given N training sentence pairs $\{x^n, y^n\}_{n=1}^N$, the training objectives is defined as Eq. 2.

$$L_{CrossEntropy} = \underset{\theta}{argmax} \sum_{n=1}^{N} log P(y^n|x^n; \theta) \qquad (2)$$

Although the logarithmic likelihood method is widely used, the exposure bias still exists, which directly affects the quality of the traditional neural machine translation model.

NMT Assisted by Pre-trained Model. The pre-training method transfers knowledge from resource-rich tasks to low-resource downstream tasks. However, the NMT method takes the cross entropy between the two languages as the training goal to optimize the parameters, which is significantly different from the monolingual pre-training model.

Therefore, one approach is to use the resource-rich language pre-training model, and then put source language and the target language into the pre-training model to obtain the corresponding word embedding, and use pre-trained word embedding training NMT model. Another approach is to design a new sequence-to-sequence pre-training task to directly realize bilingual mapping in machine translation. Among them, MASS [8] and BART [5] are both cross-lingual pre-training method based on sequence-to-sequence.

3   Method

This section introduces the NMT architecture combined with the pre-trained XLNet model, including the disorder factorization sequence acquisition process based on the masked attention, the semantic encoding process based on two-stream self-attention, and the fusion process of the XLNet and NMT model.

3.1   Disorder Factorization Sequence Acquisition

For a sequence with n tokens, theoretically there are $n!$ kinds of factorization orders. With the increase of sequence length, the number of factorization increases exponentially. Therefore, only partial factorization are sampled. The maximum expectation $E_{z \sim Z_T}$ is shown in Eq. 3.

$$maxE_{z \sim Z_T} \left[\sum_{t=1}^{T} logp_\theta \left(x_{z_t} | x_{Z_{<t}} \right) \right] \quad (3)$$

Where x represents the token in the sequence, Z_T denotes all permutation of sequences with length T, and $z \sim Z_T$ represents one of the permutations. In this paper, the factorization of sequences is realized by masked self-attention.

In order to reduce the computational complexity, we proposed a strategies: syntactic-based partial factorization. Generally, the model generates some factorization order randomly, and this kind of attention method based on edit-distance may ignore the syntactic correlation of tokens. Therefore, we expect the model to be syntactically dependent and sample the factorization sequence first according to a certain proportion.

3.2 Syntactic-Based Partial Factorization

Syntax-tree explicitly expresses the dependency between the core words and its context, which is helpful to the actual feature extraction of sentences in NMT. Compared with the linear modeling method, the modeling method based on syntactic-distance has extensibility and generalization. To obtain the factorization sequence

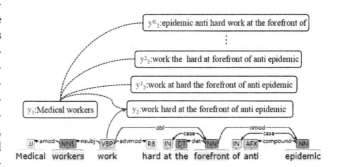

Fig. 1. Syntactic-based partial factorization sampling.

related to syntax conveniently and efficiently, we determine the core word $y_{<VBP>}$ of the sentence according to the syntactic tree marker <VBP>, and use $y_{<VBP>}$ as the anchor to segment the sequence into two subsequences y_1 and y_2 (include $y_{<VBP>}$). Subsequences y_1 and y_2 are defined as sentence chunks, tokens in chunks have richer syntactic relevance. Although chunks relations can be learned through syntactic-based factorization, the number of factorization is still large. Therefore, we only factorize y_2, this can greatly reduce the computational disaster caused by too many factorization sequences. Figure 1 shows the sentences segmentation process, and the calculation process of the token that needs to be predicted is shown in Eq. 4.

$$\frac{|s| - y_1}{|s|} = \frac{1}{K} \quad (4)$$

Where K is a hyper-parameter and indicates that $\frac{1}{K}$ tokens will be predicted, tokens in subsequence y_1 do not need to be predicted, which avoiding over-consumption of resources.

3.3   Two-Stream Self-attention Based Sentence Representation

The representation of each word in Transformer is jointly predicted by word embedding and corresponding positions. However, context representation and current word position information need to be used when predicting the current word, and context representation and current word complete representation (embedding and position) need to be used when predicting subsequent words. The prediction conflict can by solved by introducing two-stream self-attention. Similar to BERT, XLNet still uses $Mask$ to replace the target token, but the $Mask$ will not be included in the calculation of the address vector K and the content vector V, the $Mask$ only acts as the query vector Q, and the representation of all tokens will not get the information of the $Mask$, thus eliminating the difference between pre-training and fine-tuning caused by the introduction of the $Mask$. Therefore, it is common practice to construct new representation functions g_θ and introduce additional position information z_t to realize the content perception of the corresponding position, as shown in Eq. 5.

$$p_\theta\left(x_{z_t}|x_{z<t}\right) = \frac{exp\left(e\left(x\right)^T g_\theta\left(x_{z<t}, z_t\right)\right)}{\sum_{x'} exp\left(e\left(x'\right)^T g_\theta\left(x_{z<t}, z_t\right)\right)} \tag{5}$$

$g(\cdot)$ uses the position representation z_t instead of content representation when predicting x_{z_t}, and requires full representation of the current token (x_{z_t} and z_t when predicting subsequent tokens $x_{z(j,j>t)}$). Two-stream refers to the query representation attention (QR-Att) and the content representation attention (CR-Att), respectively.

- QR-Att: To predict the current token, only contains position information z_t, instead of the content information $h_{z_t}^{(l-1)}$, as shown in Eq. 6.

$$g_{z_t}^{(l)} \leftarrow Att\left(Q = g_{z_t}^{(l-1)}, KV = h_{z<t}^{(l-1)}; \theta\right) \tag{6}$$

- CR-Att: To predict the subsequent token, which contains position information z_t and content information $h_{z_t}^{(l-1)}$, as shown in Eq. 7.

$$h_{z_t}^{(l)} \leftarrow Att\left(Q = h_{z_t}^{(l-1)}, KV = h_{z<t}^{(l-1)}; \theta\right) \tag{7}$$

Where $g_{z_t}^{(l-1)}$ and $h_{z_t}^{(l-1)}$ represent the position and content of the $l-1$ layer to be queried respectively. CR-Att is consistent with the traditional self-attention method. The two-stream self-attention method can be summarized as:

- Randomly initialize vector $g_i = w$ in the first layer of QR-Att.
- The CR-Att uses the word vector $h_i = e(x_i)$.
- It should be noted here that the network weights of the two-streams are shared. Finally, QR-Att is removed in the fine-tuning phase, and only the CR-Att is used.

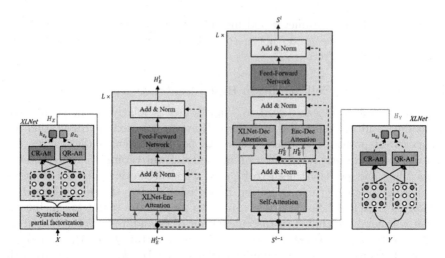

Fig. 2. Model framework. The dotted line indicates residual connections, H_X, H_Y and H_E^L denote the XLNet model corresponding to the source language and target language, and the output of the last layer of the NMT encoder.

3.4   NMT Model Integrated with XLNet

The NMT model proposed in this paper is shown in Fig. 2, where X represent the domains of the source language, Y denote the domains of the target language. For any sentence, $x \in X, y \in Y$, output $H_X = XLNet\,(x)$ and $H_Y = XLNet\,(y)$ after encoding. The i-th token in the sentence x and y represented as $h_{X,i} \in H_X$ and $u_{Y,i} \in H_Y$. H_E^0 express word embedding, H_E^l represents the l-th hidden layer representation. For any $i \in [l_x]$ and $l \in [L]$, the i-th element $\tilde{h}_i^l$ in H_E^l as shown in Eq. 8.

$$\tilde{h}_i^l = attn_x(h_i^{l-1}, H_X, H_X), \forall i \in [l_x] \tag{8}$$

Where $attn_x$ denotes XLNet-encoder/decoder attention module. The output H_E^l is obtained through the feedforward nerual network $FFN(\cdot)$, as shown in Eq. 9.

$$H_E^l = (FNN(\tilde{h}_1^l), ..., FNN(\tilde{h}_{l_x}^l)) \tag{9}$$

We represent $S_{<t}^l$ as the hidden state of the decoder layer l before the step t, self-attention representation of the decoder s_t^l as shown in Eq. 10.

$$\tilde{s}_t^l = attn_s(s_t^{l-1}, H_Y, H_Y) \tag{10}$$

Where $attn_s$ denotes self-attention, bilingual interactive representation is mainly composed of Encoder-Decoder attention module $attn_E$ and XLNet attention model $attn_X$, as shown in Eq. 11.

$$\bar{s}_t^l = \frac{1}{2}(attn_X(\tilde{s}_t^l, H_X, H_X) + attn_E(\tilde{s}_t^l, H_E^L, H_E^L)), s_t^l = FFN(\bar{s}_t^l) \tag{11}$$

Where $attn_E$ denotes encoder-decoder attention. s_t^L is finally obtained through iterative calculation, and then the t-th target word $\tilde{y}_t$ is predicted according to linear mapping and softmax. We use the negative log-likelihood of the probabilities to generate reference tokens as the loss function L_1 for this model, as shown in Eq. 12.

$$L_1 = -\sum_{j=1}^{M} log(p(y_j|\hat{y}_{<j}))$$
(12)

Where $\hat{y}_{<j}$ denotes generated tokens, M indicate the length of the sequence. In order to make full use of the output characteristics of XLNet and decoder, we also use Drop-net trick, and set Drop-net rate to 1.0 depend on experience.

3.5 Knowledge Distillation

We regard pre-training model as a language model, which can score the fluency of the output translation. Meanwhile, we also expect the NMT model to consider not only the cross entropy loss, but also the relative entropy loss corresponding to each word when outputting. Therefore, we combine the teacher-student system,

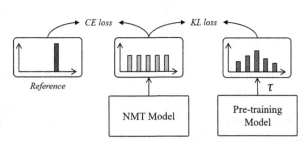

Fig. 3. Knowledge distillation.

use the pre-training model as a teacher to provide distillation data for the NMT model, and let the NMT model learn the probability distribution of the teacher model. As shown in Fig. 3.

Generally, there is a temperature hyper-parameter τ in knowledge distillation method, and a smoother output distribution probability can be learned by increasing τ. The output probability is shown in Eq. 13.

$$p_{prt} = \frac{exp(z_{prt_i}/\tau)}{\sum_j exp(z_{prt_j}/\tau)}$$
(13)

Where z_{prt_i} represents hidden state representation, the equation also applies to the NMT model. Generally, a well pre-trained model will generate distributions with high probability for a few words, leaving others with probabilities close to zero. By increasing τ we expose extra information to the NMT model. The calculation of KL divergence is shown in Eq. 14.

$$D_{KL}(p_{prt}||p_{nmt}) = \sum_{i=1}^{N} p_{prt}(i) \cdot log \frac{p_{prt}(i)}{p_{nmt}(j)}$$
(14)

Where p_{prt} and p_{nmt} represent the probabilities of the pre-training model and NMT model respectively, the pre-training model is regarded as a teacher, NMT model is regarded as student. i represents the sentence number. Therefore, we define the KL divergence (relative entropy) loss between the output probability distribution of the pre-trained model and the NMT model, as shown in Eq. 15.

$$L_2 = \sum_{i=0}^{N} \tau^2 D_{KL}(p_{prt}||p_{nmt}) \qquad (15)$$

Our goal is to minimize the KL divergence loss of pre-training model and the NMT model. Therefore, the loss function of the model can be regarded as the weighted sum of cross-entropy loss (L_1) and relative entropy loss (L_2), as shown in Eq. 16.

$$L = \lambda L_1 + (1 - \lambda) L_2 \qquad (16)$$

Where λ is hyper-parameter and set to 0.5.

4   Experiments

Dataset and Setting. For Mo-Zh task, the training set consist of 0.260M bilingual sentences from CCMT2019 which including 2000 sentence pairs for test set and validation set. For Ug-Zh task, the training set consist of 0.3M bilingual sentences from CCMT2019 which including 2000 sentence pairs for test set and validation set. For Ti-Zh task, the training set consist of 150K bilingual sentences from CCMT2019 which including 1K sentence pairs for test set and validation set. We use directed graph discriminant method to segment the stems and affixes of source language. In addition, we limit the bilingual vocabulary to 35K words and limit the length of sentences to 50. We use BLEU scores[1] to evaluate the model. Parameters are set as follows: word embedding size = 300, hidden size = 512, number of layers = 4, number of heads = 6, dropout = 0.25, batch size = 128, and beam size = 5. The parameters are updated by SGD and mini-batch with learning rate controlled by Adam. All of the source language use XLNet method[2] to obtain vector representation. We employ two TITAN X to train model and obtained by averaging the last 10 checkpoints for task. The mongolian dependency tree is obtained from Mongolian Dependency Tree-Bank (TDTB), Uyghur dependency tree is obtained from Uyghur Dependency Tree-Bank (UDTB). In addition, according to the Tibetan dependency syntax annotation system, we obtained 3000 corresponding Tibetan dependency tree by discriminant method.

Baseline. By modifying T2T[3] to implement our model, and then compared our approach against various baselines:

---

[1] https://github.com/moses-smt/mosesdecoder/blob/master/scripts/generic/multi-bleu.perl.

[2] https://github.com/zihangdai/xlnet.

[3] https://github.com/tensorflow/tensor2tensor.

- Transformer:An end-to-end framework based on self-attention, which has the best translation effect[4].
- XLM:A pre-training method based on cross-lingual supervised learning[5].
- MASS:A sequence-to-sequence pre-training method based on BERT[6].
- BART:A denoising sequence-to-sequence pre-training for machine translation[7].

4.1   Results and Analysis

According to Table 1, the performance of the pre-trained machine translation model is higher than that of the Transformer, XLM iteratively replaces high-frequency symbol pairs in the dataset with unused symbols, which makes the model insensitive to the

Table 1. Translation results in different languages.

Model	Mo-Zh	Ug-Zh	Ti-Zh
Transformer [9]	27.42	32.62	28.39
XLM	29.53	34.59	29.96
MASS [8]	31.79	35.41	30.84
BART [5]	32.18	36.78	31.55
Ours method	**34.98**	**38.02**	**33.46**

type of language. In addition, XLM has achieved remarkable results in low-resource tasks by using translation language model (TLM). MASS use a transpose mask mechanism to realize sequence to sequence learning. BART implements sequence-to-sequence learning by adding various combinable noise transformations. Since the above-mentioned pre-training models are all based on BERT method, the translation performance is limited by the self-encoding mode. Our method makes up for the defects of the self-encoding model, and encodes the source language and target language in an autoregressive mode to realize the translation between different languages. In three low-resource translation tasks, the BLEU scores increased by 7.56, 5.4 and 5.07, respectively.

4.2   Ablation Experiments

We observed the influence of various modules on the NMT model through ablation experiments, and analyzed the translation quality when using the pre-training model, syntax-based factorization (SF), partial prediction methods (PP) and knowledge distillation (KD). In addition, in order to explore the influence of the modeling unit on the performance of the model, we also use different segmentation methods to observe the trend of BLEU scores. As shown in Table 2. According to Table 2, although the number of low-frequency words is reduced when morpheme modeling is adopted, but local semantic and order adjustment information is also lost. The different between BPE and stem-affixes method is

---

[4] https://github.com/tensorflow/tensor2tensor.
[5] https://github.com/facebookresearch/XLM.
[6] https://github.com/microsoft/MASS.
[7] https://github.com/pytorch/fairseq/tree/master/examples/bart.

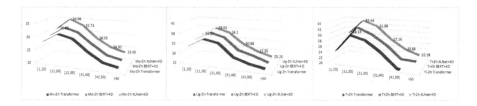

Fig. 4. The BLEU scores in different translation tasks.

not obvious, BPE method divides high-frequency byte pairs by the frequency of continuous bytes. The stem-affixes method uses complex rules to extract word-building affixes and restore stem. Although it can obtain morphological feature, but consumes resources. Therefore, we mainly use BPE below.

When the pre-training model (XLNet) is added, the BLEU scores improved by 5.72, 2.65 and 1.96 respectively in the test sets of three low-resource tasks. It is proved that the pre-trained word embedding by XLNet has better representation and polysemous word recognition ability, it can update the word embedding matrix during the iterative

Table 2. The ablation experiment.

Model	Mo-Zh		Ug-Zh		Ti-Zh	
	Dev	Test	Dev	Test	Dev	Test
Morpheme	26.99	25.27	30.34	28.98	27.19	25.46
Stem-affixes	28.91	27.98	31.78	30.56	27.33	26.1
BPE	29.69	27.42	33.05	32.62	30.16	28.39
+XLNet	34.68	33.14	36.41	35.27	32.54	30.35
+SF	35.12	34.55	38.29	37.69	34.85	32.13
+PP	37.35	34.64	38.39	37.94	34.93	32.76
+KD	37.35	**34.98**	**38.58**	**38.02**	35.07	**33.46**

training process. After adding the SF method, the encoder can learn the context representation based on syntactic distance. However, the PP method only predicts part of the sequence token, which is more efficient in language modeling and feature extraction. In addition, the quality of the model after adding knowledge distillation is also significantly improved compared with other methods, it also shows that the NMT model can learn implicit knowledge representation from the pre-training model.

4.3   Case Study

We observed the BLEU scores of translation with different length and the quality of translation obtained by various methods. According to Fig. 4, we compared the knowledge distillation method combined with BERT to observe the effectiveness of the proposed pre-training method. When the sentence length is about 20 tokens, the model has better performance. It can be seen that the XLNet method fully learns linear knowledge in semantic representation, that is, predicts the current target word according to the context dependence of the target word.

As shown in Fig. 5, our model significantly improves the translation fluency and faithfulness compared with the translation generated by the pre-training

Source	འགྲོ་མི་ / ལ་ ... / ... / ... / ... 8 ... (... / ...)	قارا بۇگۈنمۇ نىستانسسسى مىتيورولوگىيە مەركىزى .جەقارىدى ناقاهالنۇرۇشى رەڭلىك سېرىق يامغۇر	... ...
Ref.	经过 多年，你 想 消除 你们 之 间 的 分歧 谈何 容易 。	中央 气象 台 今天 继续 发布 暴雨 橙色 预 警 。	这个 司机 是 个 不 遵守 速度 限制的 人 。
XLM	多 年，消除 你们 的 不 何 说 来 容易 。	中央的 天气 观测台 持续 地 公布 暴风 预 警 。	车夫 是 这个 速度 限制 逃脱 人 的 。
MASS	多年 以后，你 考虑 到消灭 分裂 不容易 。	中央 天气 市局 监测 站 继续 布告 了 暴风 雨 红色 预警 。	这个 车 老板 不 听命 速度 的 受限 。
BART	经过 了 很多 年，你 消除 中 间 的 分歧 说来 不容易 。	中央 天气 局 联合 的 公布 了 暴雨 橙警 告 。	这 司机 不 遵循 超出 速度 人 。
Ours	经过 很多 年，你 认为 消除 你们 的 分歧 谈何 容易。	中央 气象 台 今日 不断地 发布 暴雨 橙色 预警。	这个 司机 是 不 需要 速度 限制 遵 从 的 人。

Fig. 5. Translation effects of different tasks.

model based on BERT. It can be seen that our method pays more attention to semantic coherence in the process of context generation, and also alleviates the problem of missing translation by improving the length penalty term.

5 Conclusion

In this paper, we proposed a pre-training method which combined syntactic partial sampling. Compared with various pre-training sequence-to-sequence models, our method improved the context semantic awareness and alleviates the mask independence assumption. The pre-trained word embedding method has better scalability for low-resource neural machine translation. In addition, the method of knowledge distillation provides help for NMT models to learn richer semantic knowledge. Therefore, in the future, we will research more neural machine translation models based on pre-training methods.

References

1. Bahdanau, D., Cho, K., Bengio, Y.: Neural machine translation by jointly learning to align and translate. In: 3rd International Conference on Learning Representations, ICLR 2015, San Diego, CA, USA, 7–9 May 2015, Conference Track Proceedings (2015). http://arxiv.org/abs/1409.0473
2. Chen, M., et al.: Federated learning of n-gram language models. In: Proceedings of the 23rd Conference on Computational Natural Language Learning, CoNLL 2019, Hong Kong, China, 3–4 November 2019, pp. 121–130 (2019). https://doi.org/10.18653/v1/K19-1012
3. Church, K.W.: Word2vec. Nat. Lang. Eng. **23**(1), 155–162 (2017)
4. Devlin, J., Chang, M., Lee, K., Toutanova, K.: BERT: pre-training of deep bidirectional transformers for language understanding. In: Proceedings of the 2019 Conference of the North American Chapter of the Association for Computational Linguistics: Human Language Technologies, NAACL-HLT 2019, Minneapolis, MN, USA, 2–7 June 2019, Volume 1 (Long and Short Papers), pp. 4171–4186 (2019). https://doi.org/10.18653/v1/n19-1423

5. Lewis, M., et al.: BART: denoising sequence-to-sequence pre-training for natural language generation, translation, and comprehension. In: Jurafsky, D., Chai, J., Schluter, N., Tetreault, J.R. (eds.) Proceedings of the 58th Annual Meeting of the Association for Computational Linguistics, ACL 2020, Online, 5–10 July 2020, pp. 7871–7880. Association for Computational Linguistics (2020). https://doi.org/10.18653/v1/2020.acl-main.703

6. Pennington, J., Socher, R., Manning, C.D.: Glove: global vectors for word representation. In: Proceedings of the 2014 Conference on Empirical Methods in Natural Language Processing, EMNLP 2014, 25–29 October 2014, Doha, Qatar, A meeting of SIGDAT, a Special Interest Group of the ACL, pp. 1532–1543 (2014). https://doi.org/10.3115/v1/d14-1162

7. Radford, A., Narasimhan, K., Salimans, T., Sutskever, I.: Improving language understanding by generative pre-training (2018)

8. Song, K., Tan, X., Qin, T., Lu, J., Liu, T.: MASS: masked sequence to sequence pre-training for language generation. In: Chaudhuri, K., Salakhutdinov, R. (eds.) Proceedings of the 36th International Conference on Machine Learning, ICML 2019, 9–15 June 2019, Long Beach, California, USA. Proceedings of Machine Learning Research, vol. 97, pp. 5926–5936. PMLR (2019). http://proceedings.mlr.press/v97/song19d.html

9. Vaswani, A., et al.: Attention is all you need. In: Advances in Neural Information Processing Systems 30: Annual Conference on Neural Information Processing Systems 2017, 4–9 December 2017, Long Beach, CA, USA, pp. 5998–6008 (2017). http://papers.nips.cc/paper/7181-attention-is-all-you-need

10. Weng, R., Yu, H., Huang, S., Cheng, S., Luo, W.: Acquiring knowledge from pre-trained model to neural machine translation. In: The Thirty-Fourth AAAI Conference on Artificial Intelligence, AAAI 2020, The Thirty-Second Innovative Applications of Artificial Intelligence Conference, IAAI 2020, The Tenth AAAI Symposium on Educational Advances in Artificial Intelligence, EAAI 2020, New York, NY, USA, 7–12 February 2020, pp. 9266–9273. AAAI Press (2020). https://aaai.org/ojs/index.php/AAAI/article/view/6465

11. Yang, Z., Dai, Z., Yang, Y., Carbonell, J.G., Salakhutdinov, R., Le, Q.V.: Xlnet: generalized autoregressive pretraining for language understanding. In: Advances in Neural Information Processing Systems 32: Annual Conference on Neural Information Processing Systems 2019, NeurIPS 2019, 8–14 December 2019, Vancouver, BC, Canada, pp. 5754–5764 (2019). http://papers.nips.cc/paper/8812-xlnet-generalized-autoregressive-pretraining-for-language-understanding

Self-learning for Received Signal Strength Map Reconstruction with Neural Architecture Search

Aleksandra Malkova[1,2(✉)], Loïc Pauletto[1,3], Christophe Villien[2],
Benoît Denis[2], and Massih-Reza Amini[1]

[1] Université Grenoble Alpes, LIG-APTIKAL, 38401 Saint Martin d'Hères, France
[2] Université Grenoble Alpes, CEA-Leti, 38000 Grenoble, France
`Aleksandra.MALKOVA@cea.fr`
[3] Bull/Atos, 38000 Grenoble, France

Abstract. In this paper, we present a Neural Network (`NN`) model based on Neural Architecture Search (`NAS`) and self-learning for received signal strength (`RSS`) map reconstruction out of sparse single-snapshot input measurements, in the case where data-augmentation by side deterministic simulations cannot be performed. The approach first finds an optimal `NN` architecture and simultaneously train the deduced model over some ground-truth measurements of a given (`RSS`) map. These ground-truth measurements along with the predictions of the model over a set of randomly chosen points are then used to train a second `NN` model having the same architecture. Experimental results show that signal predictions of this second model outperforms non-learning based interpolation state-of-the-art techniques and `NN` models with no architecture search on five large-scale maps of `RSS` measurements.

Keywords: Neural Architecture Search · Self-learning · Received signal strength · Radio mapping

1 Introduction

The integration of low-cost sensor and radio chips in a plurality of connected objects in the Internet of Things (`IoT`) has been contributing to the fast development of large-scale physical monitoring and crowdsensing systems in various kinds of smart environments (e.g., smart cities, smart homes, smart transportations, etc.). In this context, the ability to associate accurate location information with the sensor data collected on the field opens appealing perspectives in terms of both location-enabled applications and services [13].

Among possible localization technologies, Global Positioning Systems (`GPS`) have been widely used in outdoor environments for the last past decades. However, these systems still suffer from high power consumption, which is hardly compliant with the targeted `IoT` applications.

In order to preserve both nodes' low complexity and fairly good localization performances, an alternative is to interpret radio measurements, such as

I. Farkaš et al. (Eds.): ICANN 2021, LNCS 12895, pp. 515–526, 2021.
https://doi.org/10.1007/978-3-030-86383-8_41

the Received Signal Strength Indicator (RSSI) (i.e., received power of sensor data packets sent by IoT nodes and collected at their serving base station(s)), as location-dependent fingerprints for indicating the positions of mobile devices [5,6,8,32,36]. Typical fingerprinting methods applied to wireless localization [34] ideally require the prior knowledge of a complete map of such radio metrics, covering the area of interest. However, in real life systems, it is impractical to collect measurements from every single location of the map and one must usually rely uniquely on sparse and non-uniform field data. In order to overcome this problem, classical map interpolation techniques, such as radial basis functions (RBF) or kriging [7], have been used in this context. These approaches are simple and fast, but they are quite weak in predicting the complex and heterogeneous spatial patterns usually observed in real life radio signals (e.g., sudden and/or highly localized transient variations in the received signal due to specific environmental effects). Data augmentation techniques have thus been proposed for artificially increasing the number of measurements in such radio map reconstruction problems. Typically, once calibrated over a few real field measurements, deterministic prediction tools can generally simulate electromagnetic interactions of transmitted radio waves within the environment of study [16,25,31]. The purpose is then to use the generated synthetic data as additional data to train complex models for map interpolation. However, these tools require a very detailed description of the physical environment and can hardly anticipate on its dynamic changes over time. Their high computational complexity is also a major bottleneck.

In this paper, we consider RSSI map reconstruction in a constrained low-cost and low-complexity IoT context, where one can rely only on few ground-truth (i.e., GPS-tagged) single-snapshot field measurements and for which data-augmentation techniques based on side deterministic simulations cannot be applied, due to their prohibitive computational cost and/or to *a prior* unknown environment physical characteristics. This problem of map interpolation is similar to the task of image restoration for which, NN based models with fixed architectures have been already proposed [33]. In the case where there are few observed pixels in an image these approaches fail to capture its underlying structure that is often complex. To tackle this point we propose a first NN model based on Neural Architecture Search (NAS) for the design of the most appropriate model given a RSSI map with a small number of ground truth measurements. For this purpose, we develop two strategies based on genetic algorithms and dynamic routing for the search phase. We show that with the latter approach, it is possible to learn the model parameters while simultaneously searching the architecture. Ultimately, in order to enhance the model's predictions, the proposed approach uses also some extra data of the map with the predictions of the optimized NN in non-visited positions together with the initial set of ground-truth measurements for learning a final model. The proposed technique thus aims at finding practical trade-offs between agnostic learning interpolation techniques and data-augmented learning approaches based on deterministic prediction tools that generally require a very detailed physical characterization of the operating environment. Experimental results on five large-scale RSSI maps show that our approach outperforms non-learning based interpolation state-of-the-art techniques and NN models with a given fixed architecture.

2   Related Work

In this section we report related-work on RSSI map reconstruction, as well as existing techniques proposed for NAS.

2.1   Interpolation Techniques

Various spatial interpolation methods have been proposed for radio map reconstruction in the wireless context.

One first approach, known as kriging or Gaussian process regression [18], exploits the distance information between measured points, while trying to capture their spatial dependencies. Another popular method is based on radial basis functions (RBF) [7,9,27]. This technique is somehow more flexible, makes fewer assumptions regarding the input data (i.e., considering only the dependency on the distance) and is shown to be more tolerant to some uncertainty [29]. In [7] for instance, the authors have divided all the points of a database of outdoor RSSI measurements into training and testing subsets, and compared different kernel functions for the interpolation. The two methods above, which rely on underlying statistical properties of the input data (i.e., spatial correlations) and kernel techniques, require a significant amount of input data to provide accurate interpolation results. Accordingly, they are particularly sensitive to sparse initial datasets. They have thus been considered in combination with crowdsensing. In [20] for instance, so as to improve the performance of basic kriging, one calls for visiting new positions/cells where the interpolated value is still presumably inaccurate. A quite similar crowdsensing method has also been applied in [10] after stating the problem as a matrix completion problem using singular value thresholding. In our case though, we can just rely on a RSSI map with few ground truth initial measurements.

Another approach considered in the context of indoor wireless localization relies on both collected field data and an a priori path loss model that accounts for the effect of walls attenuation between the transmitter and the receiver [15]. In outdoor environments, local path loss models (and hence, particularized RSSI distributions) have been used to catch small-scale effects in clusters of measured neighbouring points, instead of using raw RSSI data [23]. However, those parametric path loss models are usually quite inaccurate and require impractical in-site (self-)calibration.

Data-Augmentation Approaches. One more way to build or complete radio databases relies on deterministic simulation means, such as Ray-Tracing tools (e.g., [16,25,31]). The latter aim at predicting in-site radio propagation (i.e., simulating electromagnetic interactions of transmitted radio waves within an environment). Once calibrated with a few real field measurements, such simulation data can relax initial metrology and deployment efforts (i.e., the number of required field measurements). Nevertheless, these tools require a very detailed description of the physical environment (e.g., shape, constituting materials and dielectric properties of obstacles, walls...). Moreover, they usually require high computational complexity. Finally, simulations must be re-run again, likely from scratch, each time minor changes are introduced in the environment.

2.2 NN Based Models Trained After Data-Augmentation

Machine and deep learning approaches have been recently applied for RSSI Map reconstruction. These methods have shown to be able to retrieve unseen spatial patterns with highly localized topological effects and hidden correlations. Until now, these methods have been trained over simulated datasets given by data-augmentation approaches.

In [17], given a urban environment, the authors introduce a deep neural network called RadioUNet, which outputs radio path loss estimates trained on a large set of generated using the Dominant Path Model data and UNet architecture [28]. In another contribution, the authors have shown that using the feedforward neural network for path loss modelling could improve the kriging performance [30], as conventional parametric path loss models admit a small number of parameters and do not necessarily account for shadowing besides average power attenuation.

Besides wireless applications, similar problems of map restoration also exist in other domains. In [37] for instance, the authors try to build topographic maps of mountain areas out of sparse measurements of the altitudes. For this purpose, they use a Generative Adversarial Network (GAN) architecture, where in the discriminator they compare pairs of the input data and the so-called "received" map, either generated by the generator or based on the full true map. Another close problem making extensive use of neural networks is the image inpainting problem, where one needs to restore missing pixels in a single partial image. By analogy, this kind of framework could be applied in our context too, by considering the radio map as an image, where each pixel corresponds to the RSSI level for a given node location. Usually, such image inpainting problems can be solved by minimizing a loss between true and predicted pixels, where the former are artificially and uniformly removed from the initial image. This is however impossible in our case, as only a few ground-truth field measurements can be used.

In contrast to the previous approaches, we consider practical situations where data-augmentation techniques cannot be used mainly because of unknown environment characteristics and computational limitations, and, where there is only a small amount of ground-truth measurements. Our approach automatically searches an optimized Neural Network model for the RSSI map reconstruction in hand, and, it is based on self-training for learning an enhanced NN model with the initial ground-truth and pseudo-labeled measurements obtained from the predictions of the first NN model on a set of randomly chosen points in the map.

2.3 NAS Related Methods

Studies on the subject of NAS have gained significant interest in the last few years. In the literature; there are various of techniques based on Reinforcement Learning (RL) [38], evolutionary algorithm [26] or Bayesian Optimization (BO) [12]. Recently, new gradient-based methods became increasingly popular. One of the first methods using this technique is called DARTS [22], in which a relaxation is used to simultaneously optimize the structure of a *cell*, and the weight of the operations relative to each *cell*. At the end, cells are manually stacked to form a neural network. Based on DARTS, more complex methods have emerged such as AutoDeepLab [21] in which a network is optimized at 3 levels: (i) the parameters of the operations, (ii) the cell structure and (iii) the macro-structure of the network that is stacked manually. Despite a complex representation leading to powerful architectures, this technique has certain drawbacks, such as the fact that the generated architecture is single-path, which means it does not fully exploit the representation's capabilities. Moreover, as the search phase is over a fixed architecture, it might not be the same between different runs, thus it is complicated to use transfer learning and the impact of training from scratch can be significant. To overcome these limitations, one technique is to use *Dynamic Routing* (DR) as proposed in [19]. This approach is different from the traditional gradient based methods proposed for NAS in the sense that it does not look for a specific fixed architecture but generates a dynamic path in a mesh of cells on the fly *without searching*.

3 NAS for RSSI Map Reconstruction

In this section, we first introduce our notations and setting, and then present our main approach, denoted as SL$_{NAS}$ in the following.

3.1 Notations and Setting

For a given base station X, let $Y \in \mathbb{R}^{H \times W}$ be the whole matrix of ground-truth signal measurements, where $H \times W$ is the size of the (discretized) area of interest. We suppose to have access to only some ground truth measurements Y_m in Y, that is $Y_m = Y \odot M$, where $M \in \{0,1\}^{H \times W}$ is a binary mask indicating whether each pixel includes one available measurement or not, and $\odot$ is the Hadamard's product. Here we suppose that the number of non-null elements in Y_m is much lower than $H \times W$. We further decompose the measurements set Y_m into three parts Y_ℓ (for *training*), Y_v (for *validation*) and Y_t (for *test*), such that $Y_\ell \oplus Y_v \oplus Y_t = Y_m$, where $\oplus$ is the matrix addition operation. Let X_ℓ, X_v, X_t, X_m be the associated 2D node locations (or equivalently, the cell/pixel coordinates) with respect to base station X and X_u be the set of 2D locations for which no measurements are available.

Our approach is based on three main phases *i) architecture search phase* - the search of an optimal architecture of a Neural Network model; *ii) data-augmentation phase* - the assignment of pseudo-labels to randomly chosen unlabeled data using the predictions of the found NN model trained over Y_ℓ; and *iii)*

Algorithm 1: SL$_{NAS}$

Input: A training set: (X_ℓ, Y_ℓ); a validation set: (X_v, Y_v) and a set of $2D$ locations without measurements: X_u;

Init: Using $(X_\ell, Y_\ell) \cup (X_v, Y_v)$, find interpolated measurements $\tilde{Y}_u$ over X_u using the RBF interpolation method;

  Step 1: Search the optimal NN architecture using $(X_\ell, Y_\ell) \cup (X_u, \tilde{Y}_u)$;

  Step 2: Find the parameters $\theta_1^\star$ of the NN model f_θ :

$$\theta_1^\star = \arg\min_\theta \mathcal{L}(X_\ell, Y_\ell, \theta) \quad \# \ (\text{Eq. 1});$$

  Step 3: Choose $X_u^{(k)}$ randomly from X_u and find the new parameters $\theta_2^\star$ of the NN model f_θ : $\theta_2^\star = \arg\min_\theta \mathcal{L}(X_\ell \cup X_u^{(k)}, Y_\ell \cup f_{\theta_1^\star}(X_u^{(k)}), \theta)$;

Output: $f_{\theta_2^\star}, \tilde{Y}_u$

---

self-learning phase - the training of a second NN model with the same architecture over the set of initial ground truth measurements and the pseudo-labeled examples. In the next sections, we present these phases in more detail. These phases are resumed in Algorithm 1.

3.2   Architecture Search Phase

Here, we consider a first reference RSSI map as an input image, where unknown measurements in X_u are obtained with a RBF using points in the train and validation sets; $(X_\ell, Y_\ell) \cup (X_v, Y_v)$. The latter was found the most effective among other state-of-the-art interpolation techniques [7]. Let $\tilde{Y}_u$ be the set of interpolated measurements given by RBF over X_u. For the search phase of the NAS we have employed two strategies described below.

Genetic Algorithm. (GA) From the set $(X_\ell, Y_\ell) \cup (X_u, \tilde{Y}_u)$, we use an evolutionary algorithm similar to [26] for searching the most efficient architecture represented as a Direct Acyclic Graph (DAG). Here, the validation set (X_v, Y_v) is put aside for hyperparameter tuning. The edges of this DAG represent data flow with only one input for each node, which is a single operation chosen among a set of candidate operations. We consider usual operations in the image processing field; that are a mixture of convolutional and pooling layers. We also consider three variants of 2D convolutional layers as in [33] with kernels of size 3, 5

Table 1. Description of the Neural network architecture structure by layers found by the Architecture Search phase for the RSSI Map of the city of Grenoble used in our experiments.

Layer	Operation	Input layer	Size
1	(Conv2D + BatchNorm + LReLu) × 2 + MaxPool		(368, 368, 32)
1a	Conv2D	1	(46, 46, 8)
1b	SpaceToDepth + Conv2D	1	(184, 184, 64)
2	(Conv2D + BatchNorm + LReLu) × 2 + MaxPool	1	(92, 92, 16)
2a	DepthToSpace + Conv2D	2	(184, 184, 8)
3	(Conv2D + BatchNorm + LReLu) × 2 + MaxPool	2	(92, 92, 8)
2a	DepthToSpace + Conv2D	2	(184, 184,8)
4	Concatenation	3 + 1b	(46, 46, 16)
5	(Conv2D + BatchNorm + LReLu) × 2 + Upsampling	4	(92, 92, 4)
6	Concatenation	5 + 3a	(92, 92, 8)
7	(Conv2D + BatchNorm + LReLu) × 2 + Upsampling	6	(184, 184, 8)
8	Concatenation	1a + 2a + 7	(184, 184, 80)
9	(Conv2D + BatchNorm + LReLu) × 2 + Upsampling	8	(368, 368,32)
10	(Conv2D + BatchNorm + LReLu) × 2	9	(368, 368, 32)
11	Conv2D	10	(368, 368, 1)

and 7; and two types of pooling layers that compute either the average or the maximum on the filter of size 4. Candidate architectures are then built from randomly sampled operations and the corresponding NN models are trained. The 30 resulting architectures are then ranked according to a pixel-wise Mean Absolute Error (MAE) criterion between the interpolated result of the network and the set of interpolated measurements given by RBF $\tilde{Y}_u$. The most performing one is finally selected for mutation and placed in the trained population. The oldest architecture is removed in order to keep the size of the population equal to 20 as in [26]. Table 1 illustrates such an optimized architecture with 18 nodes, which was found for the RSSI Map of the city of Grenoble used in our experiments (Sect. 4).

Dynamic Routing (DR). For the training phase, we employ the same structure and routing process as those proposed in [19] (Fig. 1). The structure is composed of 4 down-sampling level, where the size of the features map is divided by 2 at each level, but the depth of the latter is multiplied by 2 using a 1×1 2D-convolution. In ours experiments we use a networks of 9 layers, which correspond to 33 cells in total (in yellow on Fig. 1). The structure also contains an *"upsampling aggregation"* module at the end (red part on Fig. 1). The goal of this module is to combine the features maps from all levels and reconstruct a map of the size of the input. Different from [19], here, each cells contains three *transforming* operations (i.e. 2D-convolution with a kernel size of 3, 5 or 7) to have a good point of comparison with the method described above. However, due to the structure of the network we decided not to use pooling operations, as this could have been potentially redundant. In addition, we have left the possibility of creating residual connections by adding operation identity in each cells. Moreover, we did not use the first two convolutions, originally used to reduce the size of the input, in order to keep as much information as possible. Instead, we used a 1×1 2D-convolution (in purple on Fig. 1).

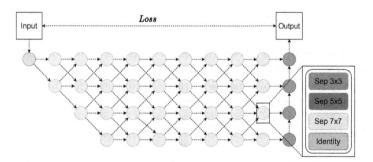

Fig. 1. Diagram of the architecture used in our experiments. The purple, yellow and red dots represents respectively the "stem" convolution, the cells and the "upsampling aggregation" module. The arrows represent the data flow. (Color figure online)

3.3 Data-Augmentation and Self-learning Phases

After the search phase, the found NN model with parameters θ, f_θ is trained on (X_ℓ, Y_ℓ) by minimizing the following loss:

$$\mathcal{L}(X_\ell, Y_\ell, \theta) = \ell(f_\theta(X_\ell), Y_\ell) + \lambda\|\theta\|_2^2 + \mu\Omega(f_\theta(X_\ell)) \tag{1}$$

where $\ell(.)$ is the Mean Absolute Error (MAE), and $\Omega(f_\theta(X_\ell))$ is the total variation function defined as:

$$\Omega(Z) = \sum_{i,j} \mid z_{i+1,j} - z_{i,j} \mid + \mid z_{i,j+1} - z_{i,j} \mid, \tag{2}$$

with $z_{i,j}$ the measurement value of a point of coordinates i, j in some signal distribution map Z. This function estimates the local amplitude variations of points in Z that is minimized in order to ensure that neighbour points will have fairly close predicted measurements (i.e., preserving signal continuity/smoothness). Here, λ and μ are hyperparameters for respectively the regularization and the total variation terms and they are found by cross-validation.

With Dynamic routing used in the search phase, we optimize the network structure and the learning of parameters minimizing (Eq. 1) at the same time. Referring to Algorithm 1, the step 1 and 2 are combined in this case.

Let $\theta_1^\star$ be the parameters of the optimized NN model found by minimizing the loss (1) on ground truth measurements (X_ℓ, Y_ℓ). This model is then applied to randomly chosen points, $X_u^{(k)}$, in X_u and pseudo-RSSI measurements $\tilde{Y}_u^{(k)}$ are obtained from the predictions of the optimized NN model $f_{\theta_1^\star}$: $\tilde{Y}_u^{(k)} = f_{\theta_1^\star}(X_u^{(k)})$.

With the same NN architecture, a second model $f_{\theta_2^\star}$ is obtained by minimizing the loss (1) over the augmented training set $(X_\ell, Y_\ell) \cup (X_u^{(k)}, \tilde{Y}_u^{(k)})$.

4 Experiments

In this section we will first describe our experimental setup and then present our experimental results.

Experimental Setup. In all experiments, we considered maps of size 368×368 cells and tested our algorithm on field data from two distinct urban environments, namely the cities of Grenoble (France) and Antwerp (The Netherlands). We aggregated and averaged the given measurements in cells/pixels of size $10\,\text{m} \times 10\,\text{m}$. The Antwerp dataset is described in detail in [1] on which we considered three base stations, BS_1', BS_2' and BS_3', with respectively 5969, 6450 and 7118 ground-truth measurements. For the Grenoble dataset, we collected GPS-tagged LoRa RSSI measurements with respect to 2 base stations located in different sites BS_1 and BS_2 with respectively 16577 and 7078 ground truth measurements. To perform in-cell data aggregation, we measured the distances based on local East, North, Up (ENU) coordinates. Then in each cell, we also computed the mean received power over all in-cell measurements (once converted

Table 2. Average values of the MAE, dB of different approaches on all base stations.

	Grenoble		Antwerp		
	BS_1	BS_2	BS_1'	BS_2'	BS_3'
RBF [4]	$5.03^{\downarrow}$	$3.16^{\downarrow}$	$3.58^{\downarrow}$	3.35	3.90
KRIG [24]	$5.68^{\downarrow}$	$4.21^{\downarrow}$	$3.69^{\downarrow}$	$4.39^{\downarrow}$	$4.91^{\downarrow}$
NS [3]	$5.11^{\downarrow}$	$3.14^{\downarrow}$	$4.28^{\downarrow}$	3.45	3.87
TV	$5.13^{\downarrow}$	2.89	3.76	3.51	3.83
DIP [33]	$5.14^{\downarrow}$	$3.22^{\downarrow}$	3.53	3.41	3.92
SL$_{\text{NAS}}$-DR	4.82	2.82	3.48	3.42	3.81
SL$_{\text{NAS}}$-GA($f_{\theta_1^\star}$)	4.79	2.81	3.39	**3.27**	3.75
SL$_{\text{NAS}}$-GA($f_{\theta_2^\star}$)	**4.76**	**2.79**	**3.33**	**3.27**	**3.74**

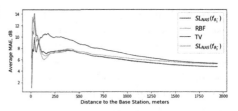

Fig. 2. MAE, dB with respect to the distance to the base station, BS_1.

into RSSI values), before feeding our algorithm and the averaged RSSI values have been normalized between 0 and 1.

For each base station, 8% of the pixels with ground-truth measurements were chosen for training (X_ℓ, Y_ℓ), 2% for validation (X_v, Y_v) and the remaining 90% for testing (X_t, Y_t). The unlabeled data used in **Step 3** of Algorithm 1 were selected at random from the remaining 4% of each map's cells with no ground truth measurements. Results are evaluated over the test set using the MAE, dB, estimated after re-scaling the normalized values to the natural received signal strength ones. The reported errors are averaged over 20 random sets (training/validation/test) of the initial ground-truth data and unlabeled data were randomly chosen for each experiment.

We compare Total Variation (TV) in-painting (Eq. 2), Radial basis functions (RBF) [4] with linear kernel that were found the most performant, kriging (KRIG) [24], and Navier-Stocks (NS) [3] state-of-the-art interpolation techniques with the proposed SL$_{\text{NAS}}$ approach. For the latter, we employ both search phase methods based on Genetic Algorithm (GA) and Dynamic Routing (DR) and respectively referred to as SL$_{\text{NAS}}$-GA and SL$_{\text{NAS}}$-DR. For SL$_{\text{NAS}}$-GA we also evaluate the impact of the self-training step (**Step 3**) (called SL$_{\text{NAS}}$-GA($f_{\theta_2^\star}$)) by comparing it with the NN model found at **Step 1** (called SL$_{\text{NAS}}$-GA($f_{\theta_1^\star}$)). The evolutionary algorithm in the architecture search phase (Sect. 3.2) was implemented using the NAS-DIP [11] package[1]. The latter was developed over the Deep Image Prior (DIP) method [33] which is a NN model proposed for image reconstruction. By considering RSSI maps as corrupted images with partially observed pixels (ground-truth measurements), we also compare with this technique by training a NN model having the same architecture than the one presented in [33] and referred to as DIP in the following. All experiments were run on Tesla V100 GPU.

Experimental Results. Table 2 summarizes results obtained on the five considered RSSI maps. We use boldface to indicate the lowest errors. The symbol $\downarrow$ indicates that the error is significantly higher than the best result with

---

[1] https://github.com/Pol22/NAS_DIP.

respect to Wilcoxon rank sum test used at a p-value threshold of 0.01 [35]. In all cases, SL$_{NAS}$-GA and SL$_{NAS}$-DR perform better than other state-of-the-art models even without the data-augmentation and self-training steps (SL$_{NAS}$-GA($f_{\theta_1^*}$)). We notice that DIP which is also a NN based model but with a fixed architecture has similar results than RBF. These results show that the search of an optimized NN model is effective for RSSI map reconstruction in a constrained low-cost and low-complexity IoT context.

Figure 2 depicts the average MAE in dB with respect to the distance to the Base Station BS_1 for the city of Grenoble. For a distance above 250 m, SL$_{NAS}$-GA($f_{\theta_2^*}$) provides uniformly better predictions in terms of MAE. These findings point to future research into how the model predicts the signal dynamics in regions where the signal is more irregular and where the dynamics are strong (for example near the base stations), especially in the cases where extra contextual knowledge about the physical environment may be included into the learning process (e.g., typically as a side information channel or the city map).

Figure 3 displays the MAE, dB boxplots of DIP, RBF and SL$_{NAS}$-GA $(f_{\theta_2^*})$ on BS_1 for different percentages of unlabeled data used in the self-learning phase (Sect. 3.3). We notice that by increasing the size of unlabeled examples, the variance of MAE for SL$_{NAS}$-GA($f_{\theta_2^*}$) increases mostly due to the increase of noisy predicted signal values by $f_{\theta_1^*}$. This is mostly related to learning with imperfect supervisor that has been studied in semi-supervised learning [2,14]. As future work, we plan to incorporate a probabilistic label-noise

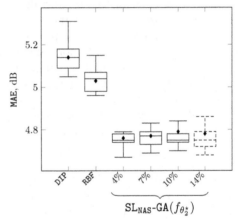

Fig. 3. Boxplots showing the MAE, dB distributions of DIP, RBF and SL$_{NAS}$-GA $(f_{\theta_2^*})$ on BS_1 for different percentage of unlabeled data $\{4, 7, 10, 14\}$ used in the self-learning phase.

model in **step 3** of Algorithm 1 and to learn simultaneously the parameters of the NN and the label-noise models.

5   Conclusion

In this article, we presented a Neural Network model based on NAS and self-learning for RSS map reconstruction from sparse single-snapshot input measurements in the absence of data augmentation via side deterministic simulations. We presented two variants for the search phase of NAS based on Genetic algorithm and Dynamic routing. Experimental results on five large-scale maps of RSS measurements reveal that our approach outperforms non-learning based interpolation state-of-the-art techniques and NN with manually designed architecture.

References

1. Aernouts, M., Berkvens, R., Vlaenderen, K.V., Weyn, M.: Sigfox and LoRaWAN datasets for fingerprint localization in large urban and rural areas. Data **3**(2) (2018)
2. Amini, M.R., Usunier, N., Laviolette, F.: A transductive bound for the voted classifier with an application to semi-supervised learning. In: Advances in Neural Information Processing Systems, pp. 65–72 (2009)
3. Bertalmio, M., Bertozzi, A.L., Sapiro, G.: Navier-stokes, fluid dynamics, and image and video inpainting. In: CVPR (2001)
4. Bishop, C.M.: Pattern Recognition and Machine Learning (Information Science and Statistics). Springer, Heidelberg (2006)
5. Burghal, D., Ravi, A.T., Rao, V., Alghafis, A.A., Molisch, A.F.: A comprehensive survey of machine learning based localization with wireless signals (2020)
6. Cheng, L., Wu, C., Zhang, Y., Wu, H., Li, M., Maple, C.: A survey of localization in wireless sensor network. Int. J. Distrib. Sens. Netw. **2012** (2012)
7. Choi, W., Chang, Y.S., Jung, Y., Song, J.: Low-power lora signal-based outdoor positioning using fingerprint algorithm. ISPRS Int. J. Geo Inf. **7**(11) (2018)
8. Dargie, W., Poellabauer, C.: Fundamentals of Wireless Sensor Networks: Theory and Practice. Wiley (2010)
9. Enrico, A., Redondi, C.: Radio map interpolation using graph signal processing. IEEE Commun. Lett. **22**(1), 153–156 (2018)
10. Fan, X., He, X., Xiang, C., Puthal, D., Gong, L., Nanda, P., Fang, G.: Towards system implementation and data analysis for crowdsensing based outdoor RSS maps. IEEE Access **6**, 47535–47545 (2018)
11. Ho, K., Gilbert, A., Jin, H., Collomosse, J.: Neural architecture search for deep image prior (2020)
12. Jin, H., Song, Q., Hu, X.: Auto-Keras: an efficient neural architecture search system. In: Proceedings of the 25th ACM SIGKDD, pp. 1946–1956 (2019)
13. Khelifi, F., Bradai, A., Benslimane, A., Rawat, P., Atri, M.: A survey of localization systems in internet of things. Mob. Netw. Appl. **24**(3), 761–785 (2018). https://doi.org/10.1007/s11036-018-1090-3
14. Krithara, A., Amini, M.R., Renders, J.M., Goutte, C.: Semi-supervised document classification with a mislabeling error model. In: 30th European Conference on Information Retrieval, Glasgow, pp. 370–381 (2008)
15. Kubota, R., Tagashira, S., Arakawa, Y., Kitasuka, T., Fukuda, A.: Efficient survey database construction using location fingerprinting interpolation. In: 2013 IEEE 27th International Conference on Advanced Information Networking and Applications (AINA), pp. 469–476 (2013)
16. Laaraiedh, M., et al.: Ray tracing-based radio propagation modeling for indoor localization purposes. In: 2012 IEEE 17th International Workshop on Computer Aided Modeling and Design of Communication Links and Networks (CAMAD), pp. 276–280 (2012)
17. Levie, R., Yapar, Ç., Kutyniok, G., Caire, G.: Pathloss prediction using deep learning with applications to cellular optimization and efficient D2D link scheduling. In: ICASSP, pp. 8678–8682 (2020)
18. Li, J., Heap, A.D.: A review of comparative studies of spatial interpolation methods in environmental sciences: performance and impact factors. Ecol. Inf. **6**(3), 228–241 (2011)
19. Li, Y., et al.: Learning dynamic routing for semantic segmentation (2020)

20. Liao, J., Qi, Q., Sun, H., Wang, J.: Radio environment map construction by kriging algorithm based on mobile crowd sensing. Wirel. Commun. Mob. Comput. **2019**, 1–12 (2019)
21. Liu, C., et al.: Auto-Deeplab: hierarchical neural architecture search for semantic image segmentation. In: Proceedings of CVPR, pp. 82–92 (2019)
22. Liu, H., Simonyan, K., Yang, Y.: Darts: Differentiable architecture search. arXiv preprint arXiv:1806.09055 (2018)
23. Ning, C., et al.: Outdoor location estimation using received signal strength-based fingerprinting. Wirel. Pers. Commun. **99**, 365–384 (2016)
24. Oliver, M., Webster, R.: Kriging: a method of interpolation for geographical information systems. Int. J. Geograph. Inf. Syst. **4**(3), 313–332 (1990)
25. Raspopoulos, M., Laoudias, C., Kanaris, L., Kokkinis, A., Panayiotou, C.G., Stavrou, S.: 3D ray tracing for device-independent fingerprint-based positioning in WLANs. In: 2012 9th Workshop on Positioning, Navigation and Communication, pp. 109–113 (2012)
26. Real, E., Aggarwal, A., Huang, Y., Le, Q.V.: Regularized evolution for image classifier architecture search. In: AAAI, vol. 33, pp. 4780–4789 (2019)
27. Redondi, A.E.C.: Radio map interpolation using graph signal processing. IEEE Commun. Lett. **22**(1), 153–156 (2018)
28. Ronneberger, O., Fischer, P., Brox, T.: U-net: convolutional networks for biomedical image segmentation. In: Navab, N., Hornegger, J., Wells, W.M., Frangi, A.F. (eds.) MICCAI 2015. LNCS, vol. 9351, pp. 234–241. Springer, Cham (2015). https://doi.org/10.1007/978-3-319-24574-4_28
29. Rusu, C., Rusu, V.: Radial basis functions versus geostatistics in spatial interpolations. In: Bramer, M. (ed.) IFIP AI 2006. IIFIP, vol. 217, pp. 119–128. Springer, Boston, MA (2006). https://doi.org/10.1007/978-0-387-34747-9_13
30. Sato, K., Inage, K., Fujii, T.: On the performance of neural network residual kriging in radio environment mapping. IEEE Access **7**, 94557–94568 (2019)
31. Sorour, S., Lostanlen, Y., Valaee, S., Majeed, K.: Joint indoor localization and radio map construction with limited deployment load. IEEE Trans. Mob. Comput. **14**(5), 1031–1043 (2015)
32. Tahat, A., Kaddoum, G., Yousefi, S., Valaee, S., Gagnon, F.: A look at the recent wireless positioning techniques with a focus on algorithms for moving receivers. IEEE Access **4**, 6652–6680 (2016)
33. Ulyanov, D., Vedaldi, A., Lempitsky, V.: Deep image prior. CoRR abs/1711.10925 (2017)
34. Vo, Q.D., De, P.: A survey of fingerprint-based outdoor localization. IEEE Commun. Surv. Tutorials **18**(1), 491–506 (2016)
35. Wilcoxon, F.: Individual comparisons by ranking methods. Biometrics **1**(6), 80–83 (1945)
36. Yu, K., Sharp, I., Guo, Y.: Ground-Based Wireless Positioning. Wiley (2009)
37. Zhu, D., Cheng, X., Zhang, F., Yao, X., Gao, Y., Liu, Y.: Spatial interpolation using conditional generative adversarial neural networks. Int. J. Geogr. Inf. Sci. **34**(4), 735–758 (2020)
38. Zoph, B., Le, Q.V.: Neural architecture search with reinforcement learning. arXiv preprint arXiv:1611.01578 (2016)

Propagation-Aware Social Recommendation by Transfer Learning

Haodong Chang[1]([⊠])(iD) and Yabo Chu[2]([⊠])(iD)

[1] University of Technology Sydney, Ultimo, Australia
haodong.chang@student.uts.edu.au
[2] Northeastern University, Shenyang, China

Abstract. Social-aware recommendation approaches have been recognized as an effective way to solve the data sparsity issue of traditional recommender systems. The assumption behind is that the knowledge in social user-user connections can be shared and transferred to the domain of user-item interactions, whereby to help learn user preferences. However, most existing approaches merely adopt the first-order connections among users during transfer learning, ignoring those connections in higher orders. We argue that better recommendation performance can also benefit from high-order social relations. In this paper, we propose a novel Propagation-aware Transfer Learning Network (PTLN) based on the propagation of social relations. We aim to better mine the sharing knowledge hidden in social networks and thus further improve recommendation performance. Specifically, we explore social influence in two aspects: (a) higher-order friends have been taken into consideration by order bias; (b) different friends in the same order will have distinct importance for recommendation by an attention mechanism. Besides, we design a novel regularization to bridge the gap between social relations and user-item interactions. We conduct extensive experiments on two real-world datasets and beat other counterparts in terms of ranking accuracy, especially for the cold-start users with few historical interactions.

Keywords: Recommender system · Social connections · Transfer learning · Social-aware recommendation

1 Introduction

Nowadays, recommender systems play an essential role in providing effective recommendations to users with items of interest. The key of success is to learn precise user and item embeddings, where Collaborative Filtering (CF) is the most traditional method [1,2] to learn from user historical records, such as ratings, clicks, and reviews. However, for many users, it is lack of interaction data to provide accurate recommendations. The data sparsity problem limits the performance of CF-based models.

H. Chang and Y. Chu—Contributed equally to this work.

© Springer Nature Switzerland AG 2021
I. Farkaš et al. (Eds.): ICANN 2021, LNCS 12895, pp. 527–539, 2021.
https://doi.org/10.1007/978-3-030-86383-8_42

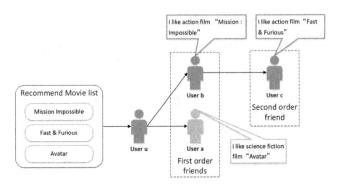

Fig. 1. An example to illustrate different friends' influence. User b and c have the same color as user u indicates that they have similar preferences with user u, and a have different preferences with user u. Then the system make a recommendation by considering these friends' preference. "Mission Impossible" is in the front of the recommended list. "Avatar" is ranked behind the other two films.

With the prevalence of online social networks, social connections have been widely leveraged to alleviate the data sparsity problem, and formed the line of research called social-aware recommendation. Transfer learning [3,4] is a useful approach to learn the common knowledge shared between a source domain and a target domain, and then transfer the common knowledge to enhance the model learning in target domain. Transfer learning is also applied in social-aware recommendation to learn user preference from social connections and then transfer to item domain, leading to more fine-refined user preference and thus better recommendation performance. However, most existing methods only adopt the first-order connections while ignoring the high-order connections. For example, in Fig. 1, users a and b are both friends of user u in the first order, user c lies in the second order. Users u, b, c share similar interests, while user a has different interests with u. In this case, user c (in the 2nd order) will have more positive influence on learning preference of user u than first-order friend user a.

Therefore, we argue that high-order friends are informative and can also help learn user preference, especially considering the fact that users may not have many direct connections with other users. It is valuable to find more relevant social friends to deal with the data sparsity problem in social networks. Therefore, we adopt the trust propagation in our model to mine informative knowledge hidden in high-order social relations. Specifically, social influence have been considered in two aspects. Firstly, friends in different orders will affect the learning of user preference. Different order has distinct bias towards preference learning. To the authors' best knowledge, we are the first to take into account order bias in modelling high-order social influence. Secondly, friends in the same order will have different importance for preference learning. We apply the attention mechanism to adaptively learn the importance of friends in the same order. Moreover, we propose a novel regularization term to formulate the relationship between domain-specific and cross-domain (common) knowledge to reduce the risk of model overfitting.

To summarize, the main contributions of this paper are as follows:

- We apply transfer learning to learn the sharing common knowledge between social and item domains, and leverage social propagation to take into account high-order social influence for better recommendation.
- We propose a new factor 'order bias' to distinguish social influence in high orders from low orders. We design a novel regularization term to formulate the relationship between domain-specific and cross-domain (common) knowledge and thus to avoid overfitting.
- We conduct extensive experiments on two real-world datasets Ciao and Yelp, and demonstrate the effectiveness of our approach in ranking accuracy.

2 Related Work

Social-aware Recommendation. Most previous social-aware recommendation works are based on homogeneity and social influence theory, that is, users who are connected tend to have similar behavioral preferences, and people with similar behavioral preferences are more likely to establish connections. The meaning reflected in the recommendation model is that the user's feature vector should be as close as possible to the vector space's similar user's feature vector. For example, [5] assumed that users are more likely to have seen items consumed by their friends, and extended BPR [2] by changing the negative sampling strategy. TrustSVD [6] believed that not only the user's explicit rating data and social relationships should be modeled, but the user's implicit behavior data and social relationships should also be considered. Therefore, implicit social information is introduced based on the SVD++ [7] model. Recent research has used deep neural networks as classifiers, yielding significant accuracy. E.g., SAMN [8] leverages attention mechanism to model both aspect- and friend-level differences for social-aware recommendations. However, these methods use direct social connections and ignore high-order social relationships, which has a wealth of information.

There are also some studies considering trust propagation to get high-order information. DeepInf [9] models the high-order to predict the social influence. [10] proposed a DiffNet neural model with a layer-wise influence diffusion part to model how users' trusted friends recursively influence users' latent preferences. The further work [11] jointly model the higher-order structure of the social and the interest network. However, they need to use text or image information for data enhancement, which may lack a certain degree of versatility. Moreover, existing methods ignore the influence of different order's friends on users.

Our work differs from the above studies as the designed model uses attention mechanism to aggregate different friends' influence in each order adaptively. And the influence of order are considered as order bias. Order bias could adjust the friend's influence depend on the friend's order.

Transfer Learning. Transfer learning deals with the situation where the data obtained from different resources are distributed differently. It assumes the existence of common knowledge structure that defines the domain relatedness and incorporates this structure in the learning process by discovering a shared latent feature space in which the data distributions across domains are close to each other. [12] pointed out that parts of the source domain data are inconsistent with the target domain observations, which may affect the construction of the model in the target domain. Based on that, some researchers [3,13] designed selective latent factor transfer models to better capture the consistency and heterogeneity across domains for recommendation. However, in these works, the transfer ratio needs to be properly selected through human effort and can not change dynamically in different scenarios.

There are also some studies considering the adaption issue in transfer learning. [14] proposed to adapt the transfer-all and transfer-none schemes by estimating the similarity between a source and a target task. [15] designed a completely heterogeneous transfer learning method to determine different transferability of source knowledge. However, these methods mainly focus on task adaptation or domain adaption. [4] propose to adapt each user's two kinds of information (item interactions and social connections) with a finer granularity, which allows the shared knowledge of each user to be transferred in a personalized manner. [16] propose a novel dual transfer learning-based model that significantly improves recommendation performance across other domains. Nevertheless, these methods still ignore the following two issue:1)High-order information is very helpful to improve the recommendation performance. 2)Sparse data in rating domain and social domain can lead to overfitting problems.

Our method innovatively leverage the high-order information for transfer learning. And we propose a novel regularization so that the user representation about the common knowledge can be reconstructed to the user representation in the social and item domains, which could reduce the risk of overfitting due to the lack of data.

3 Our Proposed Model

3.1 Notations

Suppose we have a user set $\mathcal{U}$ and an item set $\mathcal{V}$, let M denote the number of users and N denote the number of items. Symbols u, t denote two different users, and v denotes an item. $\mathcal{F}_u$ represents the friend set of user u. In social rating networks, users can form social connections with other users and interact with items, resulting in two matrices: user-user social matrix and user-item interaction matrix. The user-item interaction matrix is defined as $\mathbf{R} = [r_{uv}]_{M \times N}$ from users' historical behaviors. $r_{uv} = 1$ indicates that user u has an observed interaction (purchases, clicks) with item v. Similarly, we define the user-user social matrix $\mathbf{X} = [x_{ut}]_{M \times M}$ from social networks. $x_{ut} = 1$ indicates that user u trusts user t. We represent user u's embedding in three parts: $\mathbf{c}_u$, $\mathbf{s}_u$ and $\mathbf{i}_u$, where $\mathbf{c}_u$ denotes the latent factors shared between the item domain and social domain, i.e., the common knowledge; $\mathbf{s}_u$ and $\mathbf{i}_u$ are user latent factors corresponding to the social domain and item domain. The purpose of item recommendation is to generate a list of ranked items that meet user u's preference.

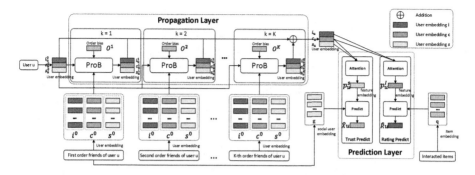

Fig. 2. An overview of our PTLN model. 'ProB' represents the propagation block introduced in Fig. 3.

3.2 Model Overview

The overall structure of our Propagation-aware Transfer Learning Network (PTLN) is illustrated in Fig. 2. It includes three types of input: 1) the user embedding of user u and u's each order friends, 2) the social user embedding of u's first order friends, 3) the item embedding of the item which u has interacted. The outputs of our model are the predicted probability $\hat{r}_{uv}$ that how user u will like item v, and the predicted probability $\hat{x}_{ut}$ that how user u will trust another user t. The main architecture of PTLN contains two components: propagation layer and prediction layer.

The propagation layer propagates over social networks to incorporate the influence of high-order social friends, and then aggregate social influence of friends in different orders. Besides, the order itself is also considered as order bias, indicating the influence bias of general friends in a specific order. In the prediction layer, we adopt attention mechanism to consider the domain relationships to better transfer the domain-specific knowledge and the shared knowledge for each task. Moreover, we adopt an efficient whole-data based training strategy [4], and involves a novel regularization term in loss function to optimize the model.

3.3 Propagation Layer

In this part, we aim to explore the high-order social influence based on the idea that a user may share similar preferences with her friends. As shown in Fig. 2, the propagation layer are constructed in a multi-block structure. Each block's input is the user embedding of target user u and that of u's friends at this order. The output is the new user embedding which includes high-order friends' influence. The new user embedding in each aspect is calculated as same in propagation block, therefore we take the process of calculating the new user embedding in common knowledge aspect as an example to explain the details of the formula. The new user embedding is learned in below four steps (Fig. 3):

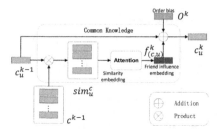

Fig. 3. The details of K-th propagation block in common knowledge aspect.

1) Calculate Similarity Embedding. User's social connection will indirectly influence the user's preference in different degrees. As discussed in the introduction, the similarity between two connected users can be used as an essential basis for revealing the degree of influence. Thus we adopt attention mechanism to assign the non-uniform weights to each friend according to the similarity between the user and her friends. we firstly calculate the similarity embedding between user u and her k-th order friend t in common knowledge aspect as follow:

$$\mathbf{sim}^{C}_{(u,t)} = \mathbf{c}^{0}_{u} \odot \mathbf{c}^{0}_{t} \tag{1}$$

where $\mathbf{sim}^{C}_{(u,t)} \in \mathbb{R}^{D_1}$ denotes the similarity embedding between user u and her k-th order friend $t \in \mathcal{F}^{k}_{u}$ in common knowledge aspect. The superscript 0 indicates the illustrated subject is initial. $\mathcal{F}^{k}_{u}$ represents the k-th order friend set of user u. the operation $\odot$ denotes the element-wise product of vectors.

2) Calculate Attention Score. After obtaining similarity embedding from k-th order friends, the attention are calculated by a trainable weighted matrix $\mathbf{W} \in \mathbb{R}^{D_1 \times 1}$. For each aspect, the trainable weighted matrix are unique. The k-th order friend t's attention in common knowledge aspect $\mathcal{A}^{*(C)}_{(u,t)}$ is defined as:

$$\mathcal{A}^{*(C)}_{(u,t)} = \mathbf{W}^{T}_{C}\mathbf{sim}^{C}_{(u,t)} \tag{2}$$

where $\mathbf{W}_C$ is the trainable weighted matrix to the common knowledge aspect.

Then we use the softmax function to normalize the friend's attention score:

$$\mathcal{A}^{C}_{(u,t)} = \frac{exp(\mathcal{A}^{*(C)}_{(u,t)})}{\sum_{z \in \mathcal{F}^{k}_{u}} exp(\mathcal{A}^{*(C)}_{(u,z)})} \tag{3}$$

where $\mathcal{A}^{C}_{(u,t)}$ is the final attention of friend t which indicates the degree of t's influence on user u.

3) Aggregate Friend's Influence. We leverage the attention score to aggregate the k-th order friend's influence, so that the friend influence embedding we get is obtained by dynamically absorbing the influence of her friends at this order.

$$\mathbf{f}^{k}_{(C,u)} = \sum_{t \in \mathcal{F}^{(k)}_{u}} \mathcal{A}^{k}_{(u,t)}\mathbf{c}^{0}_{t} \tag{4}$$

where $\mathbf{f}^k_{(C,u)} \in \mathbb{R}^{D_1}$ represents the u's friend influence embedding at k-th order.

4) Update User Embedding. When generating the friend influence embedding, we merely consider the similarity between the user and friend's preference ignoring the influence of the friend's order, as discussed in the introduction. Therefore we propose a concept of order bias to model the influence bias of general friends in a specific order. we consider that the order bias can dynamically adapt to the friend influence according to the order. With the friend influence embedding and order bias, the user embedding will be updated as follow:

$$\mathbf{c}^k_u = \mathbf{c}^0_u + \mathbf{f}^k_{(C,u)} + \mathbf{o}^k \tag{5}$$

The generated embedding $\mathbf{c}^k_u$ is the new user embedding in k-th order. $\mathbf{o}_k \in \mathbb{R}^{D_1}$ indicates the order bias of k-th order.

After propagating with k times, we obtain k new user embedding from first order to k-th order. We will use all new user embedding achieved in each order with initial user embedding to generate final user embedding $\mathbf{c}_u$ as follow:

$$\mathbf{c}_u = \sum_k \mathbf{c}^k_u \tag{6}$$

3.4 Prediction Layer

Transfer Learning framework can transfer the shared knowledge from the source domain to the target domain which is a promising method of using cross-domain data to solve problems. [3] points that the degree of relationship between domains is varied according to the user. Thus, we apply the attention mechanism to use the domain-specific knowledge and common knowledge for better learning the feature embedding which represent social domain preference and item domain preference. For a user, if the two domains are less related, the shared knowledge ($\mathbf{c}$) will be penalized and the attention network will learn to utilize more domain-specific knowledge ($\mathbf{s}$ or $\mathbf{i}$) instead. Formally, the item domain attention and the social domain attention are defined as:

$$\alpha^*_{(C,u)} = \mathbf{h}^T_\alpha \delta(\mathbf{W}_\alpha \mathbf{c}_u + \mathbf{b}_\alpha); \alpha^*_{(I,u)} = \mathbf{h}^T_\alpha \delta(\mathbf{W}_\alpha \mathbf{i}_u + \mathbf{b}_\alpha) \tag{7}$$

$$\beta^*_{(C,u)} = \mathbf{h}^T_\beta \delta(\mathbf{W}_\beta \mathbf{c}_u + \mathbf{b}_\alpha); \beta^*_{(S,u)} = \mathbf{h}^T_\beta \delta(\mathbf{W}_\beta \mathbf{s}_u + \mathbf{b}_\alpha) \tag{8}$$

Weight matrices $\mathbf{W} \in \mathbb{R}^{D_1 \times D_2}, \mathbf{h} \in \mathbb{R}^{D_1}$ and bias units $\mathbf{b}$ serve as parameters of the two-layer attention network. α and β are related to the item domain and social domain, respectively. D_2 denotes the dimension of attention network, and δ is the nonlinear activation function $ReLU$.

Then, the final attention scores are normalized with a softmax function:

$$\alpha_{(C,u)} = \frac{exp(\alpha^*_{(C,u)})}{exp(\alpha^*_{(C,u)}) + exp(\alpha^*_{(I,u)})} = 1 - \alpha_{(I,u)}; \beta_{(C,u)} = \frac{exp(\beta^*_{(C,u)})}{exp(\beta^*_{(C,u)}) + exp(\beta^*_{(S,u)})} = 1 - \beta_{(S,u)} \tag{9}$$

$\alpha_{(C,u)}$ and $\beta_{(C,u)}$ denote the weights of common knowledge $\mathbf{c}$ for item domain and social domain, respectively, which determine how much to transfer in each domain. After obtaining the above attention weights, the feature embedding of user u for the two domains are calculated as follows:

$$\mathbf{p}_u^I = \alpha_{(I,u)}\mathbf{i}_u + \alpha_{(C,u)}\mathbf{c}_u; \mathbf{p}_u^S = \beta_{(S,u)}\mathbf{s}_u + \beta_{(C,u)}\mathbf{c}_u \tag{10}$$

The generated two feature embeddings $\mathbf{p}_u^I$ and $\mathbf{p}_u^S$ represent the user's preferences for items and other users after transferring the shared knowledge between the two domains.

For predicting the scores of each item and user, we adopt a neural form MF [17] to utilize the user's feature embedding. For each task, a specific output layer is employed. The scores of user u for item v are calculated as follow:

$$\hat{r}_{uv} = \mathbf{W}_I(\mathbf{p}_u^I \odot \mathbf{q}_v); \hat{x}_{ut} = \mathbf{W}_S(\mathbf{p}_u^S \odot \mathbf{g}_t) \tag{11}$$

$\mathbf{q}_v$ and $\mathbf{g}_t$ denotes the latent factor vector of item v and user t as a friend, respectively. The operation $\odot$ denotes the element-wise product of vectors

Whole-data based strategy leverages the full data with a potentially better coverage. Thus we adopt an efficient whole-data train strategy [4] to optimize our model. For each task, the loss functions are defined as follow:

$$\tilde{\mathcal{L}}_I(\Theta) = \sum_{i=1}^{D_1}\sum_{j=1}^{D_1}\left((h_{I,i}h_{I,j})\left(\sum_{u\in\mathcal{B}} p_{u,i}^I p_{u,j}^I\right)\left(\sum_{v\in\mathcal{V}} c_v^{I-} q_{v,i}q_{v,j}\right)\right)$$
$$+ \sum_{u\in\mathcal{B}}\sum_{v\in\mathcal{V}^+}\left((1-c_v^{I-})\hat{r}_{uv}^2 - 2\hat{r}_{uv}\right) \tag{12}$$

$$\tilde{\mathcal{L}}_S(\Theta) = \sum_{i=1}^{D_1}\sum_{j=1}^{D_1}\left((h_{S,i}h_{S,j})\left(\sum_{u\in\mathcal{B}} p_{u,i}^S p_{u,j}^S\right)\left(\sum_{t\in\mathcal{U}} c_t^{S-} g_{t,i}g_{t,j}\right)\right)$$
$$+ \sum_{u\in\mathcal{B}}\sum_{t\in\mathcal{U}^+}\left((1-c_t^{S-})\hat{x}_{ut}^2 - 2\hat{x}_{ut}\right) \tag{13}$$

I and S are related to the item domain and social domain. D_1 is the latent factor number. The scalar h,p,q,g denote the element of their corresponding vectors $\boldsymbol{h}$, $\boldsymbol{p}$, $\boldsymbol{q}$, $\boldsymbol{g}$. i and j denote the index of element in the vector. U^+ and V^+ denote the items v have interacted and the friends that directly connect. B is batch of users. c_v^{I-} and c_t^{S-} are the weight of negative instances in two domains.

Both rating and social information are very sparse which could lead to the overfitting problem. We consider that there has an implicit correlation between common knowledge and domain-specific knowledge. This assumption motivates us to propose a novel regularization term to against the overfitting problem:

$$\tilde{\mathcal{L}}_{Reg}(\Theta) = \sum_k (\|\mathbf{i}^k - \theta_\alpha^k \mathbf{c}^k\|^2 + \|\mathbf{s}^k - \theta_\beta^k \mathbf{c}^k\|^2) \tag{14}$$

Where θ represents the weight of common knowledge $\mathbf{c}$. α and β are related to the item domain and social domain.

After that, we integrate both the sub-tasks loss and the novel regularization term into an overall objective function as follow:

$$\mathcal{L}(\Theta) = \tilde{\mathcal{L}}_I(\Theta) + \lambda_1\tilde{\mathcal{L}}_S(\Theta) + \lambda_2\tilde{\mathcal{L}}_{Reg}(\Theta) + \lambda_3\|\Theta\|^2 \tag{15}$$

Θ represents the parameters of our model. λ_1, λ_2, and λ_3 are the parameters to adjust the weight proportion of each term.

4 Experiments

Table 1. Performance of all the comparison methods on the Ciao and Yelp datasets. The last column "Avg Imp" indicates the average improvement of PTLN over the corresponding baseline on average. N indicates top-N task.

Baselines	Precision			Recall			NDCG			MRR			Avg Imp
Ciao	N = 5	N = 10	N = 15	N = 5	N = 10	N = 15	N = 5	N = 10	N = 15	N = 5	N = 10	N = 15	
BPR	0.0208	0.017	0.0141	0.0272	0.0496	0.0631	0.0289	0.036	0.0402	0.0479	0.0538	0.056	**60.84%**
NCF	0.0217	0.0176	0.0149	0.0392	0.057	0.0721	0.0294	0.0385	0.0441	0.0508	0.057	0.0596	**45.92%**
SAMN	0.0266	0.0225	0.0195	0.0482	0.0743	0.0959	0.0405	0.0506	0.0575	0.0562	0.0632	0.0653	**17.47%**
EATNN	0.0295	0.0233	0.0195	0.0528	0.0763	0.094	0.0454	0.054	0.0598	0.071	0.0787	0.0816	**6.53%**
PTLN	**0.0307**	**0.0244**	**0.0203**	**0.0571**	**0.0818**	**0.1006**	**0.0494**	**0.0585**	**0.0646**	**0.0755**	**0.083**	**0.0866**	
Baselines	Precision			Recall			NDCG			MRR			Avg Imp
Yelp	N = 5	N = 10	N = 15	N = 5	N = 10	N = 15	N = 5	N = 10	N = 15	N = 5	N = 10	N = 15	
BPR	0.0349	0.0282	0.0235	0.0317	0.0497	0.0601	0.0507	0.0537	0.0548	0.0832	0.092	0.0949	**16.03%**
NCF	0.0337	0.0292	0.0266	0.0503	0.0626	0.0711	0.0429	0.0465	0.0487	0.0721	0.081	0.0843	**24.74%**
SAMN	0.0333	0.0283	0.0276	0.0496	0.0568	0.0695	0.045	0.047	0.0511	0.0745	0.0822	0.0865	**12.96%**
EATNN	0.0327	0.029	0.0266	0.0507	0.0579	0.0666	0.0462	0.049	0.0516	0.0749	0.0835	0.087	**12.18%**
PTLN	**0.0356**	**0.0307**	**0.0274**	**0.0558**	**0.0647**	**0.0714**	**0.0543**	**0.057**	**0.0587**	**0.0889**	**0.0978**	**0.1009**	

4.1 Experimental Settings

Dataset. We experimented with two public datasets: **Ciao** [4] and **Yelp** [18]. Ciao provides a large amount of rating information and social information, while users make friends with others and express their experience through the form of reviews and ratings on Yelp. The two datasets were constructed following previous work [4,18]. Each dataset contains users' ratings of the items they have interacted with and the social connections between users. To address the Top-N recommendation task, we remove all ratings that less than 4 for all datasets and keep others with a score of 1. This preprocessing method aims at recommending the item list that users liked, and is widely used in existing works [4,8,19].

Baselines. To evaluate the performance of Top-K recommendation, we compare our PTLN with the following methods: **BPR** [2]: A classic and widely used ranking algorithm for recommendation. It is implemented by learning pairwise relation of rated and unrated items for each user rather than direct learning to predict ratings. **NCF** [17]: A neural CF model combines element-wise and hidden layers of the concatenation of user and item embedding to capture their high-order interactions. **SAMN** [8]: A state-of-the-art deep learning method leverages attention mechanism to model both aspect- and friend-level differences for the social-aware recommendation. **EATNN** [4]: A state-of-the-art method uses attention mechanisms to adaptively capture the interplay between item domain and social domain for each user.

Evalutation Metrics. We adapt four popular metrics Precision, Recall, NDCG (Normalized Discounted Cumulative Gain), and MRR(Mean Reciprocal Rank) for evaluation. Specifically, NDCG is a position-aware ranking metric, which assigns a higher score to hits at higher positions. MRR considers the ranking position of the first correct item in the recommended list. The higher value of these evaluation metrics, the better performance of the recommender system.

Parameter Setting. The parameters for all baseline methods were initialized as in the corresponding papers and were then carefully tuned to achieve optimal performance. The learning rate for all models were tuned among [0.0005, 0.0001, 0.005, 0.001, 0.05, 0.01]. To prevent overfitting, we tuned the dropout ratio in [0.5, 0.7, 0.9]. The batch size was tested in [16, 32, 64, 128, 256], the embedding size D_1 and the dimension of attention network D_2 were tested in [32, 64, 128, 256]. For our PTLN model, D_1 and D_2 were set to 128 and 32 on Ciao and set to 64 and 32 on Yelp. The learning rate was set to 0.0005 when using the Yelp and 0.01 when using Ciao. The dropout ratio ρ was set to 0.7 on both datasets.

4.2   Performance Comparison

We investigate the Top-N performance with N set to [5, 10, 15], according with the real recommendation scenario. We observe the results in Table 1:

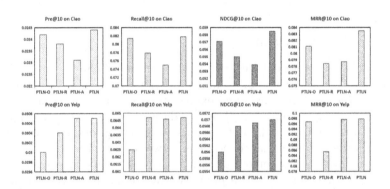

Fig. 4. Performance with different PLTN variants on Ciao and Yelp.

1. Methods incorporating social information generally perform better than non-social method. SAMN, EATNN, and PTLN perform better than BPR and NCF. This result is consistent with previous work which indicates that social information reflects users' interest and is helpful in the recommendation.
2. Our method PTLN achieves the best performance on the two datasets and significantly outperforms all baseline methods. Specifically, compared to EATNN which is the best baseline that uses attention mechanisms to adaptively capture the interplay between item domain and social domain for each user.

PTLN improves over EATNN about 6.53% on Ciao and 12.18% on Yelp. The substantial improvement of our model over the baselines could be attributed to two reasons: 1) our model considers the propagation of social domain knowledge, item domain knowledge and common knowledge, which allows the latent factor to be modeled with a finer granularity; 2) we consider the difference of friends' influence and the order bias.

Table 2. Performance with different propagation depth K on Ciao and Yelp.

Ciao	Pre@10	Recall@10	NDCG@10	MRR@10	Yelp	Pre@10	Recall@10	NDCG@10	MRR@10
K = 1	0.0242	0.0802	0.0559	0.0791	K = 1	0.0306	0.0619	0.0559	0.0966
K = 3	0.0242	0.0799	0.0557	0.0798	K = 3	0.0301	0.0617	0.0558	0.0973
K = 2	**0.0244**	**0.0818**	**0.0585**	**0.0835**	**K = 2**	**0.0307**	**0.0647**	**0.057**	**0.0978**

Analyze of the Propagation Depth K. The number of propagation layers K reflects the extent to which the model uses social information and the degree to which social information influences the model. Table 2 shows the results of different K values for both datasets. When K increases from 1 to 2, the performance increases, while the performance drops when K = 3. We empirically conclude that when the depth equals two is enough for the social recommendation.

4.3 Ablation Study

Impact of the Order Bias. A key characteristic of our proposed model is the order bias which considers the order of user's friend in general. PTLN-O denotes a variant model of PTLN without using order bias. We can see that order bias has dramatically improved performance in Fig. 4. We speculate a possible reason is that order bias can dynamically adjust the output after fusing so that the updated user embedding of this order can better reflect the preference of the user after being influenced by friends.

Impact of the Attention Mechanism. Another critical characteristic of our proposed model is we considering the diversity of friends' influence by attention mechanism. PTLN-A directly aggregate the friends' influence and user's embedding without any attention learning process. From Fig. 4, we can see that our model has a notable improvement in performance on the Ciao dataset when considering the difference of friends' influence. However, results on the Yelp dataset is not as significant as Ciao. This observation implies that the usefulness of considering the importance strength of different elements in the modeling process varies, and our proposed friend-level attention modeling could adapt to different datasets' requirements.

Impact of the Novel Regularization. To evaluate the effectiveness of the proposed correlative regularization, we compare PTLN-R, a variant model of PTLN without using novel regularization, with PLTN, in Fig. 4. The PTLN model performs better than the PTLN-R, proving that our novel regularization can make the algorithm more stable.

5   Conclusions

In this paper, we present a novel social-aware recommendation model PTLN to address the sparsity problem of data. The core component of our model is propagation layers that learn user embedding of each order by leveraging high-order information from the social domain, item domain, and common knowledge between the two domains. Attention mechanism and the concept of order bias are further employed to better distinguish the influence of different user friends. The proposed PTLN consistently and significantly outperforms the state-of-the-art recommendation models on different evaluation metrics, especially on the dataset with complicated social relationships and fewer item interactions which verified our hypothesis about the varying degrees of different friends' influence.

References

1. Hu, Y., Koren, Y., Volinsky, C.: Collaborative filtering for implicit feedback datasets. In: 2008 Eighth IEEE International Conference on Data Mining, pp. 263–272. IEEE (2008)
2. Rendle, S., et al.: BPR: Bayesian personalized ranking from implicit feedback. arXiv preprint arXiv:1205.2618 (2012)
3. Min, L.X.Z., Yongfeng, Z., Yiqun, L., Ma, S.: Learning and transferring social and item visibilities for personalized recommendation. In: CIKM 2017, pp. 337–346 (2017)
4. Chen, C., et al.: An efficient adaptive transfer neural network for social-aware recommendation. In: SIGIR 2019, pp. 225–234 (2019)
5. Zhao, T., et al.: Leveraging social connections to improve personalized ranking for collaborative filtering. In: Proceedings of the 23rd ACM International Conference on Conference on Information and Knowledge Management, pp. 261–270 (2014)
6. Guo, G., et al.: TrustSVD: collaborative filtering with both the explicit and implicit influence of user trust and of item ratings. In: Proceedings of the AAAI Conference on Artificial Intelligence, vol. 29 (2015)
7. Koren, Y.: Factorization meets the neighborhood: a multifaceted collaborative filtering model. In: Proceedings of the 14th ACM SIGKDD International Conference on Knowledge Discovery and Data Mining, pp. 426–434 (2008)
8. Chen, C., et al.: Social attentional memory network: modeling aspect-and friend-level differences in recommendation. In: WSDM 2019, pp. 177–185 (2019)
9. Qiu, J., et al.: DeepInf: social influence prediction with deep learning. In: Proceedings of the 24th ACM SIGKDD International Conference on Knowledge Discovery & Data Mining, pp. 2110–2119 (2018)
10. Wu, L., Sun, P., Fu, Y., Hong, R., Wang, X., Wang, M.: A neural influence diffusion model for social recommendation. In: SIGIR (2019)
11. Wu, L., et al.: DiffNet++: a neural influence and interest diffusion network for social recommendation. arXiv preprint arXiv:2002.00844 (2020)
12. Eaton, E., et al.: Selective transfer between learning tasks using task-based boosting. In: Proceedings of the AAAI Conference on Artificial Intelligence (2011)
13. Lu, Z., et al.: Selective transfer learning for cross domain recommendation. In: Proceedings of the 2013 SIAM International Conference on Data Mining, pp. 641–649. SIAM (2013)

14. Cao, B., Pan, S.J., Zhang, Y., Yeung, D.-Y., Yang, Q.: Adaptive transfer learning. In: AAAI, vol. 2, p. 7 (2010)
15. Moon, S., Carbonell, J.G.: Completely heterogeneous transfer learning with attention-what and what not to transfer. In: IJCAI, vol. 1 (2017)
16. Li, P., Tuzhilin, A.: DDTCDR: deep dual transfer cross domain recommendation. In: Proceedings of the 13th International Conference on Web Search and Data Mining, pp. 331–339 (2020)
17. He, X., et al.: Neural collaborative filtering. In: Proceedings of the 26th International Conference on World Wide Web, pp. 173–182 (2017)
18. Shi, C., et al.: Semantic path based personalized recommendation on weighted heterogeneous information networks. In: CIKM 2015, pp. 453–462 (2015)
19. Yao, W., DuBois, C., Zheng, A.X., Ester, M.: Collaborative denoising auto-encoders for top-n recommender systems. In: WSDM (2016)

Evaluation of Transfer Learning for Visual Road Condition Assessment

Christoph Balada[1,2], Markus Eisenbach[2(✉)], and Horst-Michael Gross[2]

[1] German Research Center for Artificial Intelligence (DFKI),
Kaiserslautern, Germany
christoph.balada@dfki.de
[2] Neuroinformatics and Cognitive Robotics Lab, Ilmenau University of Technology,
Ilmenau, Germany
markus.eisenbach@tu-ilmenau.de

Abstract. Through deep learning, major advances have been made in the field of visual road condition assessment in recent years. However, many approaches train from scratch and avoid transfer learning due to the different nature of road surface data and the ImageNet dataset, which is commonly used for pre-training neural networks for visual recognition. We show that, despite the huge differences in the data, transfer learning outperforms training from scratch in terms of generalization. In extensive experiments, we explore the underlying cause by examining various transfer learning effects. For our experiments, we are incorporating seven known architectures. Therefore, this is the first comprehensive study of transfer learning in the field of visual road condition assessment.

Keywords: Transfer learning · Road condition assessment · Surface distress detection

1 Introduction

Aging public roads need frequent inspections in order to guarantee their permanent availability. In many countries, this includes the standardized visual assessment of millions of images. Formerly, due to the lack of sophisticated approaches, the evaluation was typically done manually by human experts. Since large, annotated road surface image datasets, like the German asphalt pavement distress (GAPs) dataset [12], have been published recently, automated visual road surface analysis by machine learning approaches came into focus. In recent years, deep learning approaches dominated this field of application, e.g. for detecting cracks [2,3,6,10,16,22,23,27,32,37,39–41], potholes [7,25], or multiple types of surface distress simultaneously [1,9,11,13,24,29,30,36]. However, most of these

This work has received funding from the German Federal Ministry of Education and Research as part of the ASINVOS project under grant agreement no. 01IS15036 and as part of the FiN-2.0 project under grant agreement no. 03SF0568G, as well as from the Austrian Research Promotion Agency (FFG) as part of the ASFaLT project under grant agreement no. 869514.

© Springer Nature Switzerland AG 2021
I. Farkaš et al. (Eds.): ICANN 2021, LNCS 12895, pp. 540–551, 2021.
https://doi.org/10.1007/978-3-030-86383-8_43

approaches design their own neural network, or train well known architectures from scratch, mainly justified by big differences of road images to the ImageNet dataset, which is typically used for pre-training on visual data. For a description of these methods, we refer to the surveys [5] and [15].

In this paper, we perform extensive experiments to analyze transfer learning for visual road surface data analysis. In contrast to related work on transfer learning in road surface analysis,

- we include more modern architectures in our analysis,
- we perform extensive hyper-parameter tuning for each experiment in order to ensure a fair comparison,
- we evaluate how a changed input encoding does effect transfer learning,
- we evaluate the impact of freezing different proportions of the layers, and
- we analyze the effects of transfer learning in comparison to trainings from scratch.

Consequently, in this paper we want to answer the following three questions: How much will visual road condition assessment be improved by applying transfer learning? What are the improvements achieved by? Which transfer learning effects known from literature (see Sect. 2.2) do occur in our setting?

2 Related Work

Due to frequent inspections, many road surface images are available. All images are analyzed by human experts. However, since results are needed in a timely manner, in most cases the labeling is very coarse. Therefore, detailed annotations are only available for a very small percentage of these data.

Since detailed labeling by experts is expensive, Seichter *et al.* [31] proposed an approach where unlabeled data are analyzed regarding uncertainty of a trained classifier. Thus, data worth annotating can be identified.

Even by increasing the percentage of annotations to a certain extent by applying this method, the amount of available annotated data is still very limited. Thus, the purpose of our work was to ensure a good generalization on the few training data by applying transfer learning.

2.1 Transfer Learning for Road Condition Assessment

In the following, we briefly analyze approaches, which utilize transfer learning on road surface data or related applications, i.e. all kind of damage detection of public infrastructure.

In [32] and [4], transfer learning was applied to a VGG16 and a ResNet-152 model, which were fine tuned on very few training samples without freezing layers. Both studies yield only mediocre results due to a limited amount of training data and too many parameters to be tuned. Also having only few samples, in [16,25], and [14], transfer learning was applied to a VGG16, an XceptionNet, and an InceptionNet V3 model, while the weights of all layers but the last one was frozen. This led to better results. Zhang *et al.* [40], were able to fine tune an AlexNet with only the weights of the first layer being frozen since they had a larger dataset available in comparison to the studies above.

None of the aforementioned papers analyze the impact of transfer learning, nor do they analyze the influence of any hyperparameters, like the amount of frozen weights or the learning rate. The only paper analyzing transfer learning in more detail in a related application is [28], that detected cracks in buildings. The authors compared the generalization ability of seven architectures for a varying number of training samples. During fine tuning none of the weights were frozen. The authors did not compare their approach against training from scratch and did not tune any hyperparameters. In particular, the learning rate has a considerable influence on the generalization abilities. Therefore, the findings of that study should be treated with caution.

2.2   Transfer Learning Effects

When applying transfer learning, two major improvements are reported in literature: Faster training and better generalization ability (e.g. in the comprehensive survey of Zhuang *et al.* [42]).

Better Generalization. So far, generalization improvements by applying transfer learning for road pavement distress detection has not been analyzed. Therefore, we conducted experiments to analyze this issue. As reported in [38], initializing a model with pre-trained weights yields remarkable improvements in generalization. The improved generalization ability can be observed by a decreasing gap between validation and test performance, which we will use as criterion.

Faster Training. On road pavement data, training with transferred ImageNet weights as initialization is reported to converge within ten epochs [16]. But in that study, all weights except the ones of the classification layer were frozen. No experiments regarded convergence improvements by transfer learning, and the effect of freezing weights were not investigated. Therefore, we analyzed the convergence speed when all weights are adapted and compared to trainings where different proportions of weights were frozen. Additionally, we compared results from transfer learning with results achieved by trainings from scratch. We evaluated whether the required training time reduces to a fraction of epochs in comparison to training from-scratch, which is often the case in other fields of application.

Feature Adaption and Selection. Kim *et al.* [20] analyzed transfer learning effects in an application where material defects should be detected in microscope images. They reported that mainly due to the fact that early layers in neural networks tend to provide features which focus on simple structures, like edges or brightness changes, they require less adaption to new datasets, even if they are very different from ImageNet. This was also reported in [38]. Furthermore,

Kim *et al.* [20] showed that during fine tuning of the pre-trained weights, a kind of feature selection takes place instead of learning completely new features. Unfortunately, they examined only one architecture, namely VGG. To address this effect, we analyzed how much the pre-trained weights are changed in different architectures during the fine tuning step of transfer learning. We also analyzed whether a feature selection is observable when transfer learning is applied to road pavement data and in case of more modern neural network architectures.

3   Setup

In the following, we describe our experimental setting.

3.1   Training

For transfer learning, we used the weights from pre-trainings on the ImageNet dataset, which come with the models. We fine tuned on road surface data for 25 epochs, which is sufficient to ensure convergence, using SGD with mini batches of size 32 and momentum of 0.9. During training, no further learning rate scheduling was applied. To provide well founded results, we performed extensive experiments with many different hyperparameter combinations for each model.

Hyperparameter Search. Per architecture, we examined at least nine different learning rates and six different amounts of frozen layers in approximately 20% steps (0%, 20%, 40%, 60%, 80%, 100% excluding the fully connected layers). Since requirements are different for each architecture, we adapted the search range for the learning rate individually. Overall 426 trainings were performed.

Reference: Training from Scratch. We trained each architecture from scratch with randomly initialized weights for 250 epochs, which is sufficient to ensure convergence. The best out of the three runs was used as reference for comparisons. To ensure a fair comparison, a hyperparameter tuning regarding the learning rate was applied.

3.2   Dataset

As representative dataset for visual road condition assessment, we utilized the 50k binary classification set of the extended version of the German asphalt pavement distress dataset (GAPs v2) [34] that was suggested for experiments. While the complete GAPs v2 dataset yields ca. 6.7 M samples, the reduced set provides 50,000 training patches including 30,000 intact road patches and 20,000 damage samples composed of all types of surface distress. Additionally, 10,000 samples each are provided for validation, validation-test, and test. The four-way split was chosen as proposed by Ng [26]. Thus, validation data are used to find the best epoch, validation-test data for hyperparameter tuning, and test data only for the final evaluation. The validation set were taken from the same distribution as the training data. Validation-test and test data were recorded on roads that are geographically distinct from training and validation data. Different patch sizes are available, including 224×224 and 299×299, which are typical for inputs of ImageNet architectures. The label of the patch is based on the 64×64 image center, while the surroundings are needed as context. In [34], it has been found that the 50k subset represents the complete dataset with millions of patches very well. The results achieved on this subset are close to the results on the complete dataset. Therefore, as proposed in [34], we decided in favor of this subset in order to enable much more experiments.

3.3   Architectures

In our experiments, we analyzed the transfer-learning properties of seven widely used architectures, namely AlexNet [21], VGG19 [33], InceptionNet V3 [35], ResNet50 [17], XceptionNet [8], SE-ResNet50 [19], and MobileNet [18]. Table 1 summarizes their characteristics regarding input coding and model size.

All architectures are pre-trained on the ImageNet dataset. For VGG19, InceptptionNet V3, ResNet50, XceptionNet, and MobileNet, we used the models available in the Keras framework. For AlexNet and SE-ResNet50, we used publicly available implementations for Keras.

To address the differences in pre-processing shown in Table 1, we had to adapt the transferred weights regarding input size (224×224), channel count (1-channel grayscale) and input scaling ($[-1, 1]$).

3.4   Evaluation Metrics

Performance Measures. After each training epoch, we computed accuracy, F_1 score, and balanced error rate on the train, validation, and validation-test dataset. On the test dataset, we computed the metrics only once, based on the best epoch in terms of the validation-test performance.

Table 1. Characteristics of architectures included in our evaluation

Architecture	Zero-mean in terms of	Input scaling	Channel order	Input size	# Weights
MobileNet	Gray-world assum.	$[-1, 1]$	RGB	224^2	4,253,864
XceptionNet	Gray-world assum.	$[-1, 1]$	RGB	299^2	22,910,480
InceptionNet V3	Gray-world assum.	$[-1, 1]$	RGB	299^2	23,851,784
ResNet50	ImageNet dataset	$[0, 255]$	BGR	224^2	25,636,712
SE-ResNet50	Gray-world assum.	$[-1, 1]$	BGR	224^2	28,141,144
AlexNet	ImageNet dataset	$[0, 255]$	RGB	224^2	60,965,224
VGG19	ImageNet dataset	$[0, 255]$	BGR	224^2	143,667,240

Additional Measures. We used the Euclidean distance of the weights before and after fine tuning and an activation-sparsity score to evaluate the appearance of a feature selection and the magnitude of weight changes during fine tuning as proposed in [20]. The activation-sparsity score counts zero or close-to-zero values with respect to a threshold for a given feature map.

4   Experimental Results

In the following, we compare the results achieved by transfer learning with trainings from scratch (Sect. 4.1). Based on these results, in Sect. 4.2, we analyze the transfer learning effects previously described in Sect. 2.2.

4.1   Transfer Learning Vs. Training from Scratch

The performance gain of transfer learning for visual road condition assessment has not been analyzed in related work, yet. Therefore, in our extensive experiments, we analyzed multiple well known architectures with tuned hyperparameters (see Sect. 3.1). The best results for each architecture are shown in Fig. 1 and Table 2 for the validation, validation-test and test subset of the GAPs 50k dataset.

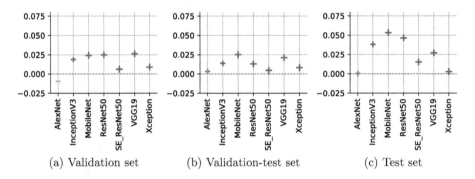

(a) Validation set   (b) Validation-test set   (c) Test set

Fig. 1. Absolute difference in F_1 score between a training from-scratch and fine tuning. + marks, where transfer learning performs better, – where it does not.

Figure 1 highlights the positive influence of transfer learning on the performance. Clearly, no architecture has any drawback from using transfer learning. Instead, we observe significant performance improvements for nearly every architecture and every subset of the GAPs 50k dataset. Most notably, the performance on the test dataset increased significantly, which shows the improvement in generalization.

Especially the InceptionNet V3 benefits from using transfer learning, as shown by the precision-recall curves in Fig. 2. Due to the improvement in generalization, it does even perform better on the test set than on the validation-test

Table 2. Comparison of the best results (F_1 score) for each architecture using transfer learning (TL) or training from scratch (FS), respectively.

Architecture	Validation		Validation-test		Test	
	TL	FS	TL	FS	TL	FS
InceptionNet V3	0.9441	0.9254	0.9024	0.8886	0.9143	0.8760
VGG19	0.9495	0.9231	0.9151	0.8940	0.9065	0.8794
ResNet50	0.9456	0.9206	0.9054	0.8925	0.8950	0.8485
MobileNet	0.9482	0.9241	0.9106	0.8854	0.8868	0.8334
XceptionNet	0.9473	0.9385	0.9155	0.9073	0.8842	0.8813
AlexNet	0.9089	0.9184	0.8895	0.8863	0.8604	0.8601
SE-ResNet50	0.9336	0.9276	0.8928	0.8881	0.8566	0.8414

set, which both contain road data geographically distinct from training data, but only the validation-test set was used for hyperparameter tuning. Additionally, on any subset of the GAPs 50k dataset, the transfer learning results are superior to the results achieved by training from scratch.

Convergence. For all architectures, the training converged within 15 epochs when transfer learning was applied. In comparison, training from scratch took considerably longer and converged within 150 epochs. In conclusion, transfer learning based on ImageNet pre-training speeds up the training significantly, even if the application is considerably different from ImageNet.

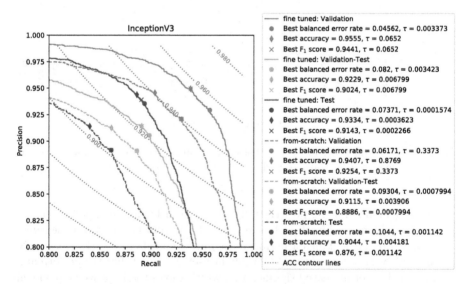

Fig. 2. Precision-recall curves for InceptionNet V3. Training from scratch is shown as dashed lines, transfer learning results as solid lines. The smaller gap between validation and test results for transfer learning (solid red and blue line) shows the better generalization. (Color figure online)

Hyperparameters. For each architecture, we performed a grid search to identify appropriate hyperparameters. The best learning rate for the individual architectures differed in a large range between 0.000355 and 0.1. Freezing layers turned out to not improve the performance of any architecture. Therefore, these experiments are omitted here.

4.2   Effects of Transfer Learning

According to literature, the following transfer learning effects should be observable: Feature selection in the final convolutional layer, only small weight changes in terms of Euclidean distance, and filters in early layers focus on simple features and can be re-used.

Feature Selection. Basically, a slight increase of the sparsity from input to output of the network can be observed for any architecture. Therefore, the final feature map of each architecture tends to be the one with the highest sparsity. Nonetheless, we found that the actual magnitude of sparsity highly depends on the specific architecture and the actual image sample which is passed through the network. Overall, we had no clear finding of an increase in terms of sparsity in the final feature map, regardless of the architecture, even though our application is significantly different from ImageNet as in [20].

We found that simple architectures like VGG19 or AlexNet tend to have a high sparsity (50% or more) in the final convolutional layers, and thus do only re-use some of the features. The activation-sparsity score counts zero or close-to-zero values with respect to a threshold for a given feature map. This high sparsity applies to the feature maps after transferring the weights as well as after fine tuning. In contrast, all modern architectures except InceptionNet V3 tend to have a sparsity of less than 10%. This means, more than 90% of the features in these modern architectures are activated regardless of the input, and therefore they re-use or re-combine most of the transferred features. Furthermore, while

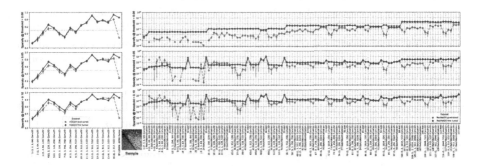

Fig. 3. Sparsity: VGG19 (left) vs ResNet50 (right, logarithmic ordinate). The red lines show the sparsity of the transferred weights from pre-training, the blue lines show the sparsity after fine tuning. Best viewed in zoomed digital version. (Color figure online)

the fine tuning leads to an increase of sparsity in the VGG19, the sparsity of ResNet50 seems to be not affected at all (see Fig. 3).

Since we did not observe a clear feature selection for any of the examined architectures, we assume the feature selection observed in [20] is an effect caused by their dataset, the selected data samples, and the architecture (VGG), respectively.

Additionally, we observed some basic patterns regarding sparsity, which reveal insights on how the architectures work: In particular, ReLU activations, 1×1 convolutions, and Squeeze-and-Excitation blocks increase the sparsity drastically. While the increase at ReLU layers is caused by negative activations, the increase at the other layers is due to their feature selection effect. In contrast, purely spatial convolutions lead to a decrease of sparsity.

Small Weight Changes. In addition to an increase in generalization ability, transfer learning also offers a remarkable reduction in training time. Often, this observation is explained as follows: Weights that were learned during the pre-training require only small changes to fit the new dataset. In our experiments, we observed that fine tuning had the following effects:

- For AlexNet and VGG19, in early convolutional layers, the weight changes are lower than in trainings from scratch (Fig. 4), which suggests that these features could be re-used. For intermediate convolutional layers, the weights have to be adapted as much as in trainings from scratch. Features in late convolutional layers are fine tuned less.
- For all modern architectures, weight changes were rather small. Features could be re-used to a high degree, since weight differences in most layers are smaller for fine tuning than for training from scratch.
- Because of their random initialization, weights in fully connected layers at the very end of each architecture are changed drastically, even more than in trainings from scratch. In AlexNet and VGG19, these layers contain more than 90% of all weights. Therefore, in sum, weight differences are higher for fine tuning than for trainings from scratch in these architectures.
- Due to the re-training of transferred color features, which make up ca. 50% of the first-layer filters, we observed relatively strong changes in the first convolutional layers regardless of the architecture. However, these changes are lower than in trainings from-scratch, suggesting that some filters could be re-used.
- Weights in pointwise convolutional layers and squeeze-and-excitation blocks are changed significantly stronger than weights in other layers. We observed clear peaks in the weight change diagrams for these layers in all modern architectures. Both types of layers are responsible for combining or weighting features of the previous layer. This indicates that previously learned features are re-combined during fine tuning.

In summary, our experiments do not confirm that weights require less changes when pre-trained weights were used for initialization. For some layers, even

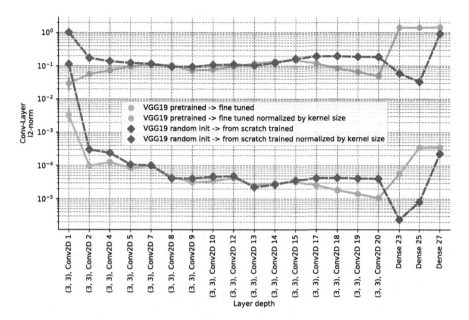

Fig. 4. Weight difference measured by euclidean distance of weights before and after training for layers of the VGG19 architecture (top curves) and the euclidean distance divided by the filter size (bottom curves)

greater weight changes were applied by the fine tuning in comparison to the training from scratch. Instead, we assume that training is faster since early-layers features can be re-used in classical architectures and features can be re-combined in modern architectures. This re-usability of features trained on a large dataset also has a positive effect on the generalization ability.

5   Conclusion

So far, transfer learning has not been examined systematically in the state of the art for deep-learning-based visual road condition assessment. Therefore, we performed extensive experiments to analyze the impact of transfer learning on the performance in the application of road surface image analysis. The generalization ability of all architectures considered in our analysis was significantly improved by the use of transfer learning, although this application differs significantly from the ImageNet dataset, which was used for pre-training. The training time has been reduced to a fraction of the time required to train a network from scratch. Furthermore, we analyzed transfer learning effects in order to explore how different architectures benefit from transfer learning. Classical architectures (AlexNet and VGG19) benefit from re-using features of early layers. Therefore, weights in early layers are adapted less during fine tuning and the sparsity of activations in mid and late layers is high. In contrast, for modern architectures (InceptionNet V3, ResNet50, SE-ResNet50, XceptionNet), we observe a low sparsity of activations in all layers. During fine tuning, layers responsible for combining and weighting features of previous layers are adapted more than other layers. We conclude that modern architectures re-use a larger fraction of features by re-combining them. Therefore, for deep-learning-based visual road condition assessment, modern architectures, especially those that contain skip connections, should be preferred over classical architectures in a transfer learning setting.

References

1. Alfarrarjeh, A., Trivedi, D., Kim, S.H., Shahabi, C.: A deep learning approach for road damage detection from smartphone images. In: Big Data (2018)
2. Alipour, M., Harris, D.K., Miller, G.R.: Robust pixel-level crack detection using deep fully convolutional neural networks. JCCE **33**(6), 04019040 (2019)
3. Bang, S., Park, S., Kim, H., Kim, H.: Encoder-decoder network for pixel-level road crack detection in black-box images. CACIE **34**(8), 713–727 (2019)
4. Bang, S., Park, S., Kim, H., Kim, H., et al.: A deep residual network with transfer learning for pixel-level road crack detection. In: ISARC, vol. 35 (2018)
5. Cao, W., Liu, Q., He, Z.: Review of pavement defect detection methods. IEEE Access **8**, 14531–14544 (2020)
6. Cha, Y.J., Choi, W., Büyüköztürk, O.: Deep learning-based crack damage detection using convolutional neural networks. CACIE **32**(5), 361–378 (2017)
7. Chen, H., Yao, M., Gu, Q.: Pothole detection using location-aware convolutional neural networks. IJMLC **11**, 899–911 (2020)
8. Chollet, F.: Xception: deep learning with depthwise separable convolutions. In: CVPR, pp. 1251–1258 (2017)
9. Du, Y., Pan, N., Xu, Z., Deng, F., Shen, Y., Kang, H.: Pavement distress detection and classification based on YOLO network. Int. J. Pavement Eng. 1–14 (2020). Taylor & Francis
10. Dung, C.V., Anh, L.D.: Autonomous concrete crack detection using deep fully convolutional neural network. Autom. Constr. **99**, 52–58 (2019)
11. Eisenbach, M., Stricker, R., Debes, K., Gross, H.M.: Crack detection with an interactive and adaptive video inspection system. AGT-IM **94**, 94–103 (2017)
12. Eisenbach, M., Stricker, R., Seichter, D., et al.: How to get pavement distress detection ready for deep learning? a systematic approach. In: IJCNN (2017)
13. Eisenbach, M., Stricker, R., Sesselmann, M., Seichter, D., Gross, H.M.: Enhancing the quality of visual road condition assessment by deep learning. In: WRC (2019)
14. Feng, C., Zhang, H., Wang, S., et al.: Structural damage detection using deep convolutional neural network and transfer learning. JCE **23**(10), 4493–4502 (2019)
15. Gopalakrishnan, K.: Deep learning in data-driven pavement image analysis and automated distress detection: a review. Data **3**(3), 28 (2018)
16. Gopalakrishnan, K., Khaitan, S.K., et al.: Deep convolutional neural networks with transfer learning for computer vision-based data-driven pavement distress detection. CBM **157**, 322–330 (2017)
17. He, K., Zhang, X., Ren, S., Sun, J.: Deep residual learning for image recognition. In: CVPR, pp. 770–778 (2016)
18. Howard, A.G., Zhu, M., Chen, B., Kalenichenko, D., Wang, W., et al.: MobileNets: efficient convolutional neural networks for mobile vision applications. arXiv (2017)

19. Hu, J., Shen, L., Sun, G.: Squeeze-and-excitation networks. In: CVPR, pp. 7132–7141 (2018)
20. Kim, S., Kim, W., Noh, Y.K., Park, F.C.: Transfer learning for automated optical inspection. In: IJCNN, pp. 2517–2524 (2017)
21. Krizhevsky, A., Sutskever, I., Hinton, G.E.: ImageNet classification with deep convolutional neural networks. In: NIPS, pp. 1097–1105 (2012)
22. Li, B., Wang, K.C., Zhang, A., Yang, E., Wang, G.: Automatic classification of pavement crack using deep convolutional neural network. In: IJPE (2018)
23. Liu, Y., Yao, J., Lu, X., Xie, R., Li, L.: DeepCrack: a deep hierarchical feature learning architecture for crack segmentation. Neurocomputing **338**, 139–153 (2019)
24. Maeda, H., Sekimoto, Y., Seto, T., et al.: Road damage detection and classification using deep neural networks with smartphone images. CACIE **33**(12), 1127–1141 (2018)
25. Milhomem, S., da Silva Almeida, T., da Silva, W.G., et al.: Weightless neural network with transfer learning to detect distress in asphalt. arXiv (2019)
26. Ng, A.: Nuts and bolts of building AI applications using deep learning. In: Tutorial NIPS (2016)
27. Nhat-Duc, H., et al.: Automatic recognition of asphalt pavement cracks using meta-heuristic optimized edge detection algorithms and convolution neural network. AIC **94**, 203–213 (2018)
28. Özgenel, Ç.F., Sorguç, A.G.: Performance comparison of pretrained convolutional neural networks on crack detection in buildings. In: ISARC, vol. 35 (2018)
29. Riid, A., Lõuk, R., et al.: Pavement distress detection with deep learning using the orthoframes acquired by a mobile mapping system. Appl. Sci. **9**(22), 4829 (2019)
30. Roberts, R., Giancontieri, G., et al.: Towards low-cost pavement condition health monitoring and analysis using deep learning. Appl. Sci. **10**(1), 319 (2020)
31. Seichter, D., Eisenbach, M., Stricker, R., et al.: How to improve deep learning based pavement distress detection while minimizing human effort. In: CASE (2018)
32. Silva, W.R.L.d., Lucena, D.S.d.: Concrete cracks detection based on deep learning image classification. In: MDPI Proceedings (2018)
33. Simonyan, K., Zisserman, A.: Very deep convolutional networks for large-scale image recognition. arXiv (2015)
34. Stricker, R., Eisenbach, M., et al.: Improving visual road condition assessment by extensive experiments on the extended gaps dataset. In: IJCNN (2019)
35. Szegedy, C., Vanhoucke, V., Ioffe, S., Shlens, J., Wojna, Z.: Rethinking the inception architecture for computer vision. In: CVPR, pp. 2818–2826 (2016)
36. Wang, Y.J., Ding, M., Kan, S., Zhang, S., Lu, C.: Deep proposal and detection networks for road damage detection and classification. In: Big Data (2018)
37. Yang, F., Zhang, L., Yu, S., Prokhorov, D., Mei, X., Ling, H.: Feature pyramid and hierarchical boosting network for pavement crack detection. TITS (2019)
38. Yosinski, J., Clune, J., Bengio, Y., Lipson, H.: How transferable are features in deep neural networks? In: NIPS, pp. 3320–3328 (2014)
39. Zhang, A., Wang, K.C., Fei, Y., et al.: Automated pixel-level pavement crack detection on 3D asphalt surfaces with a recurrent neural network. CACIE **34**(3), 213–229 (2019)
40. Zhang, K., Cheng, H., Zhang, B.: Unified approach to pavement crack and sealed crack detection using preclassification based on transfer learning. JCCE (2018)
41. Zhang, L., Yang, F., Zhang, Y.D., Zhu, Y.J.: Road crack detection using deep convolutional neural network. In: ICIP, pp. 3708–3712 (2016)
42. Zhuang, F., et al.: A comprehensive survey on transfer learning. arXiv (2019)

EPE-NAS: Efficient Performance Estimation Without Training for Neural Architecture Search

Vasco Lopes[1(✉)], Saeid Alirezazadeh[2], and Luís A. Alexandre[1]

[1] NOVA Lincs, Universidade da Beira Interior, Covilhã, Portugal
{vasco.lopes,luis.alexandre}@ubi.pt
[2] C4-Cloud Computing Competence Center, Universidade da Beira Interior,
Covilhã, Portugal

Abstract. Neural Architecture Search (NAS) has shown excellent results in designing architectures for computer vision problems. NAS alleviates the need for human-defined settings by automating architecture design and engineering. However, NAS methods tend to be slow, as they require large amounts of GPU computation. This bottleneck is mainly due to the performance estimation strategy, which requires the evaluation of the generated architectures, mainly through training, to update the sampler method. In this paper, we propose EPE-NAS, an efficient performance estimation strategy, that mitigates the problem of evaluating networks, by scoring untrained networks and correlating them with their trained performance. We perform this process by looking at intra and inter-class correlations of an untrained network. We show that EPE-NAS can produce a robust correlation and that by incorporating it into a simple random sampling strategy, we are able to search for competitive networks, without requiring any training, in a matter of seconds using a single GPU. Moreover, EPE-NAS is agnostic to the search method, as it focuses on evaluating untrained networks, making it easy to integrate into almost any NAS method.

Keywords: Neural architecture search · Performance estimation · Deep learning

1 Introduction

In recent years, deep learning algorithms have been extensively researched, and efficiently applied to various tasks with excellent results [5,13], especially those

This work was supported by 'FCT - Fundação para a Ciência e Tecnologia' through the research grant '2020.04588.BD', partially supported by NOVA LINCS (UIDB/04516/2020) with the financial support of FCT-Fundação para a Ciência e a Tecnologia, through national funds, and partially supported by operation Centro-01-0145-FEDER-000019 - C4 - Centro de Competencias em Cloud Computing, cofinanced by the European Regional Development Fund (ERDF) through the Programa Operacional Regional do Centro (Centro 2020), in the scope of the Sistema de Apoio à Investigação Científica e Tecnologica - Programas Integrados de IC&DT.

© Springer Nature Switzerland AG 2021
I. Farkaš et al. (Eds.): ICANN 2021, LNCS 12895, pp. 552–563, 2021.
https://doi.org/10.1007/978-3-030-86383-8_44

related to computer vision [33]. The great success in computer vision tasks is mainly due to the advent of Convolutional Neural Networks (CNNs) [17], given their robust feature extraction capability and transferability between different problems. Gradually, different CNNs architectures have been proposed, incrementally showing that CNNs can be improved, by revising the architecture itself, adding additional components such as residual connections, reducing the number of parameters, the size or inference time [14,15,18,29,30,32]. However, designing efficient architectures is extremely time-consuming. It requires expert knowledge and trial and error. Deep neural networks can have many design choices, such as layers, their combination and sequence, parameters associated with the layers, architecture, and the training procedure, as well as optimization rules. Therefore, an automated way to conduct neural architectures' design came as a natural process [16].

Neural Architecture Search (NAS) aims to automate architecture engineering and design, by autonomously designing high performance architectures for a given problem [11]. NAS methods for computer vision problems have been successfully applied to various tasks, such as image classification, semantic segmentation, object detection, and others [11,34]. Since the incipience proposal [41], NAS methods broadly focused on designing architectures using a similar flow. A controller, using a specified search strategy, being the most common Reinforcement Learning or Evolutionary Strategies, samples an architecture A from the space of possible architectures $\mathcal{A}$, which is defined by the search space, that comprises the possible operations (e.g., convolution, pooling) and the architecture type. The generated architecture is evaluated, and the result is given as a reward to the controller to update its parameters. This process is repeated thousands of times, whereby the controller learns to sample better architectures over time.

Although NAS methods have shown excellent results, the computational cost of most methods is extremely high, which in some cases can be in the order of months of GPU computation [21,41,42]. This is mainly associated with the performance estimation strategy, which evaluates the generated architectures based on regular training, either from scratch to convergence or through partial training [25,27]. Recent approaches, attempt to smooth the training process, by sharing parameters [24], applying mutations to already trained networks [10], or using one-shot NAS, where the controller generates architectures and corresponding weights [4,22,38]. However, it has been shown that some NAS proposals overfit the search space and do not allow exploration due to the introduction of design bias [36].

To mitigate the aforementioned problems, in this paper, we propose EPE-NAS, a performance estimation strategy that scores generated networks at initialization stage, without requiring any training. By evaluating how the gradients of the network behave with respect to the input, it is possible to score **untrained** networks, eliminating the need to train generated architectures to update parameters. The proposed method is extremely fast, allowing the analysis of thousands of networks in seconds. We show that this method can be used to guide the search over the search space, as it can quickly infer a network's trained accuracy from its untrained state. The proposed method can be easily integrated into almost any NAS method, by entirely replacing the performance estimation strategy, or complementing it, by creating a multi estimation strategy. We show this by incorporating the proposed method into a random search strategy, achieving competitive results in seconds. The code for the proposed method is also available[1].

The main contributions of this paper can be summarized as follows:

– We propose a novel performance estimation strategy that can evaluate the trained performance of an untrained network, which can be easily integrated into almost any NAS method.
– We analyze the impact of the proposed method when coupled with random search and show that it can achieve competitive results in a few seconds.
– We compare the proposed method with different NAS methods, as well as with a surrogate performance estimator in NAS-Bench-201.
– We show that the proposed method allows the search space to be quickly analyzed without the need to train networks, allowing bad candidates to be weed out, by analyzing the relationship between the score and the network performance when trained.

The remainder of this paper is organized as follows. Section 2 contextualizes the related work. Section 3 describes the proposed method in detail. In Sect. 4, we present the experiments performed, the datasets and benchmarks used, the results and discussion. Finally, a conclusion is drawn in Sect. 5.

2   Related Work

Generally, NAS methods attempt to automatically design optimal CNNs using a sample-evaluate-update scheme, where a controller generates an architecture and is updated using the generated architecture performance. The problem with this is that evaluating the generated architectures is very costly. Zoph and Le [41] initially formulated NAS as a reinforcement learning problem, where a controller was trained over-time to sample more efficient architectures. The problem was that this method required more than 60 years of GPU computation, as it trained all generated architectures to convergence. As follow-up work, the authors tackle this problem by performing a cell-based search in a search space with 13 operations [42]. By focusing on the design of two types of cells: normal cells (perform convolutional operations) and reduction cells (reduce the input size), the authors were able to reduce the GPU computation to 2000 days. They also found that cell-based architectures searched in CIFAR-10 can be transferred to ImageNet by stacking more cells. In this method, more than 20000 networks were trained and evaluated.

Similar to [41], in [1] the authors present MetaQNN, a method based on reinforcement learning and Q-learning, where the learning agent was trained

---

[1] Code publicly available on GitHub: www.github.com/VascoLopes/EPE-NAS.

to sequentially sample CNN layers. Using a similar reinforcement learning app-roach, BlockQNN [40] focuses on sampling blocks of operations used to form entire networks. However, BlockQNN still required 96 GPU days of computation. ENAS [24], used a controller, trained with policy gradient, to discover architec-tures by searching for an optimal subgraph within a large computational graph. By constructing a sizeable computational graph, where each subgraph represents a network, ENAS forced all generated architectures to share their parameters. In this way, efficient search was enabled in less than one GPU computational day. The authors of [22] proposed DARTS, a gradient-based method that optimizes architectures using gradient descent by continuously relaxing the search space so that it is continuous. The authors propose a bilevel gradient optimization, which jointly learns the architecture and the weights in a few GPU days. This paper served as foundation for many other one-shot methods. In [38], the authors improve the performance of the architectures generated by DARTS by introduc-ing regularization mechanisms. Also using a differentiable approach, GDAS [8] is a method that makes the search procedure differentiable, so that sub-graphs can be sampled from the directed acyclic graph representing the search space, which can be trained end-to-end to sample efficient networks. GDAS' controller is optimized based on the validation loss of the trained sampled architecture. SETN [7], also uses a differentiable approach, but uses an evaluator trained to indicate the probability of each architecture to have a low validation loss, which allows selective sampling of networks. REA [26], instead of reinforcement learning or gradient-based methods, focuses on using evolutionary tournament selection algorithms with an age property that favors younger architectures.

To mitigate the bottleneck of the performance estimation strategy, surro-gate methods have also been proposed to extrapolate the learning curve with a partial train [2,6], or by learning a HyperNet that generates weights based on the architecture [3]. In [12], the authors propose BOBH, which focuses on hyperparameter optimization, which includes architecture design, by combining Bayesian optimization and bandit-based methods. However, 33 GPU days were still required to design an optimal network in CIFAR-10.

Our work differentiates from the aforementioned, being closer to NAS-WOT [23], as our focus is to evaluate a generated network, without requiring any training, neither for the performance estimation strategy, nor for the generated networks. Thus creating a score that correlates an **untrained** network to its performance once trained, in efficient time.

3   Proposed Method

In this paper, we propose EPE-NAS, a novel performance estimation strategy, whose goal is to estimate the performance of generated networks without requir-ing any training, neither for the generated networks nor for the performance estimator. To do this, we score **untrained** networks as an indicator of their accuracy when trained.

To evaluate the generated networks, we look at the behavior of local linear operators using different data points at initialization stage.

The local linear operators are obtained by multiplying the linear maps at each layer interspersed with the binary rectification units. To do this, one can define a linear map, $w_i = f(x_i)$, which maps the input $\mathbf{x}_i \in \mathbb{R}^D$, through the network, $f(\mathbf{x}_i)$, where $\mathbf{x}_i$ represents an image that belongs to a batch $\mathbf{X}$, and D is the input dimension. Then, the linear map can be computed using:

$$\mathbf{J}_i = \frac{\partial f(\mathbf{x}_i)}{\partial \mathbf{x}_i} \tag{1}$$

In order to evaluate how a network behaves with different data points, we calculate the Jacobian matrix $\mathbf{w}_i$ for different data points, $f(\mathbf{x}_i)$, of the batch $\mathbf{X}$, $i \in 1, \cdots, N$:

$$\mathbf{J} = \left(\frac{\partial f(\mathbf{x}_1)}{\partial \mathbf{x}_1} \ \frac{\partial f(\mathbf{x}_2)}{\partial \mathbf{x}_2} \ \cdots \ \frac{\partial f(\mathbf{x}_N)}{\partial \mathbf{x}_N} \right)^{\top} \tag{2}$$

The Jacobian Matrix $\mathbf{J}$ contains information about the network output with respect to the input for several data points. We then can evaluate how points belonging to the same class correlate with each other, where the goal is to see if an **untrained** network is capable of modeling complex functions. Explicitly, a flexible network should simultaneously be able to distinguish local linear operators for each data point, but also have similar results for similar data points, which in a supervised approach means that the data points belong to the same class. The perfect scenario would be to have an **untrained** network with low correlation between different data points, where data points of the same category are closer to each other, which means that the network would easily learn to distinguish the two data points during training. To evaluate this behavior, we evaluate the correlation of $\mathbf{J}$ values with respect to their class, by computing a covariance matrix for each class present in $\mathbf{J}$: $\mathbf{C}_{J_c} = (\mathbf{J} - \mathbf{M}_{J_c})(\mathbf{J} - \mathbf{M}_{J_c})^{\top}$, where $\mathbf{M}_J$ is the matrix with elements:

$$(\mathbf{M}_{J_c})_{i,j} = \frac{1}{N} \sum_{\substack{n \in \{1,...,N\}, \\ x_i \in class \ c}} \mathbf{J}_{i,n}, \tag{3}$$

where c represents the class, $c \in 1, ..., C$, and C is the number of classes present in the batch. Then, it is possible to calculate the correlation matrix per class, $\mathbf{\Sigma}_{J_c}$, for each covariance matrix $\mathbf{C}_{J_c}$: $(\mathbf{\Sigma}_{J_c})_{i,j} = \frac{(\mathbf{C}_{J_c})_{i,j}}{\sqrt{(\mathbf{C}_{J_c})_{i,i} * (\mathbf{C}_{J_c})_{j,j}}}$, where (i,j) represents the $(i,j)^{th}$ element of the matrices.

Each individual correlation matrix allows the analysis of how the **untrained** network behaves for each class, which may be an indication of the ability of the local linear operators to perceive differences between classes.

To allow comparison between the different individual correlation matrices, as they may have different sizes due to the number of data points per class, they are individually evaluated:

$$\mathbf{E}_c = \begin{cases} \sum_{i=1}^{N} \sum_{j=1}^{N} log(|(\mathbf{\Sigma_{J}}_c)_{i,j}| + k), & \text{if } C \leq \tau \\[2em] \dfrac{\sum_{i=1}^{N} \sum_{j=1}^{N} log(|(\mathbf{\Sigma_{J}}_c)_{i,j}|+k)}{||\mathbf{\Sigma_{J}}_c||}, & \text{otherwise} \end{cases} \quad (4)$$

where k is a small-constant with the value of 1×10^{-5}, and C is the number of classes in batch $\mathbf{X}$. To avoid confusion with absolute value operation, we denote $||X||$ as the number of elements of the set X. The normalization based on the size of the correlation matrix is due to the fact that for a constant batch size, as the number of classes increases, the size of the individual correlation matrices becomes smaller, a correlation matrix with a larger size would obtain a larger value.

Then, a network is scored based on the individual evaluations of the correlation matrices by:

$$s = \begin{cases} \sum_{t=1}^{C} |\mathbf{e}_t|, & \text{if } C \leq \tau \\[2em] \dfrac{\sum_{i=1}^{C} \sum_{j=i+1}^{C} |\mathbf{e}_i - \mathbf{e}_j|}{||\mathbf{e}||}, & \text{otherwise} \end{cases} \quad (5)$$

where $\mathbf{e}$ is the vector containing all the correlation matrices' scores. Depending on the number of classes present in the batch, the final score is either a sum of the individual correlation matrices' scores or a normalized pair-wise difference. Normalization serves to mitigate the class difference when evaluating networks in datasets with a high number of classes and noise. In our experiments, we empirically defined $\tau = 100$.

4   Experiments

We evaluate the effectiveness of EPE-NAS on three datasets: CIFAR-10, CIFAR-100 and ImageNet16-120 from NAS-Bench-201, using a batch size of 256. As the proposed method is a performance estimation strategy that does not require any training, we also evaluate EPE-NAS by combining it with a random search strategy [19], where a candidate network from the search space is randomly proposed and scored using the proposed performance estimation method, instead of training the network. This evaluation is done for different sample sizes of N networks.

The setup for all the experiments conducted was a desktop computer, with a single 1080Ti GPU and 32 GB of ram. In the following sections, we comment on NAS-Bench-201, and individually detail the experiments, results and provide a discussion.

4.1   NAS-Bench-201

NAS methods tend to be hard to reproduce, compare with other methods, and evaluate their real performance on common search spaces [20]. Increases

Table 1. Comparison of several search methods evaluated using the NAS-Bench-201 benchmark. Performance shown in accuracy with mean±std, on CIFAR-10, CIFAR-100 and ImageNet-16-120. Methods are divided into 4 blocks, depending on their approach: weight sharing, non-weight sharing, training-free approaches (with a direct comparison between the proposed method and NAS-WOT), and a baseline using an SVM as a surrogate estimator. Search times are the mean time required to search for cells in CIFAR-10, using a single 1080Ti GPU. Search time includes the time taken to train networks as part of the process where applicable. The performances of the training-free approaches are given for different sample size, N. For each sample size, we also report the optimal network. Table adapted from [9, 23].

Method	Search time (s)	CIFAR-10		CIFAR-100		ImageNet-16-120	
		Validation	Test	Validation	Test	Validation	Test
Non-weight sharing							
REA	12000	91.19 ± 0.31	93.92 ± 0.30	71.81 ± 1.12	71.84 ± 0.99	45.15 ± 0.89	45.54 ± 1.03
RS	12000	90.93 ± 0.36	93.70 ± 0.36	70.93 ± 1.09	71.04 ± 1.07	44.45 ± 1.10	44.57 ± 1.25
REINFORCE	12000	91.09 ± 0.37	93.85 ± 0.37	71.61 ± 1.12	71.71 ± 1.09	45.05 ± 1.02	45.24 ± 1.18
BOHB	12000	90.82 ± 0.53	93.61 ± 0.52	70.74 ± 1.29	70.85 ± 1.28	44.26 ± 1.36	44.42 ± 1.49
Weight sharing							
RSPS	7587	84.16 ± 1.69	87.66 ± 1.69	59.00 ± 4.60	58.33 ± 4.34	31.56 ± 3.28	31.14 ± 3.88
DARTS-V1	10890	39.77 ± 0.00	54.30 ± 0.00	15.03 ± 0.00	15.61 ± 0.00	16.43 ± 0.00	16.32 ± 0.00
DARTS-V2	29902	39.77 ± 0.00	54.30 ± 0.00	15.03 ± 0.00	15.61 ± 0.00	16.43 ± 0.00	16.32 ± 0.00
GDAS	28926	90.00 ± 0.21	93.51 ± 0.13	71.14 ± 0.27	70.61 ± 0.26	41.70 ± 1.26	41.84 ± 0.90
SETN	31010	82.25 ± 5.17	86.19 ± 4.63	56.86 ± 7.59	56.87 ± 7.77	32.54 ± 3.63	31.90 ± 4.07
ENAS	13315	39.77 ± 0.00	54.30 ± 0.00	15.03 ± 0.00	15.61 ± 0.00	16.43 ± 0.00	16.32 ± 0.00
Training-free							
NAS-WOT (N = 10)†	3.1	89.56 ± 0.56	92.47 ± 0.04	69.36 ± 1.55	69.20 ± 1.05	42.08 ± 1.61	**42.20 ± 1.37**
Ours (N = 10)	**2.3**	89.90 ± 0.21	**92.63 ± 0.32**	69.78 ± 2.44	**70.10 ± 1.71**	41.73 ± 3.60	41.92 ± 4.25
NAS-WOT (N = 100)†	25.7	89.91 ± 0.80	91.41 ± 2.24	67.13 ± 4.03	67.18 ± 4.14	41.39 ± 1.13	**41.42 ± 1.53**
Ours (N = 100)	**20.5**	88.74 ± 3.16	**91.59 ± 0.87**	67.28 ± 3.68	**67.19 ± 3.82**	38.66 ± 4.75	38.80 ± 5.41
NAS-WOT (N = 500)†	126.8	88.73 ± 0.81	91.71 ± 1.37	67.62 ± 1.61	67.54 ± 2.23	39.37 ± 3.01	39.84 ± 3.68
Ours (N = 500)	**105.8**	88.17 ± 1.35	**92.27 ± 1.75**	69.23 ± 0.62	**69.33 ± 0.66**	41.93 ± 3.19	**42.05 ± 3.09**
NAS-WOT (N = 1000)†	252.6	89.60 ± 0.90	91.20 ± 2.04	68.57 ± 0.41	68.95 ± 0.72	38.01 ± 1.66	38.08 ± 1.58
Ours (N = 1000)	**206.2**	87.87 ± 0.85	**91.31 ± 1.69**	69.44 ± 0.83	**69.58 ± 0.83**	41.86 ± 2.33	**41.84 ± 2.06**
Optimal (N = 10)	N/A	90.00 ± 0.95	93.41 ± 0.45	70.11 ± 1.70	70.11 ± 1.70	44.67 ± 1.87	44.67 ± 1.87
Optimal (N = 100)	N/A	91.12 ± 0.11	94.12 ± 0.21	72.73 ± 0.78	72.73 ± 0.78	46.31 ± 0.47	46.31 ± 0.47
Optimal (N = 500)	N/A	91.15 ± 0.12	94.13 ± 0.22	72.83 ± 0.64	72.83 ± 0.64	46.06 ± 0.66	46.06 ± 0.66
Optimal (N = 1000)	N/A	91.24 ± 0.21	94.19 ± 0.15	72.92 ± 0.53	72.92 ± 0.53	46.57 ± 0.59	46.57 ± 0.59
Surrogate Estimator with Training							
SVM (N = 10)	359426.3‡	89.74 ± 1.10	92.80 ± 0.97	65.21 ± 6.48	65.46 ± 6.37	37.50 ± 8.56	37.31 ± 8.66
SVM (N = 100)	359449.4‡	87.03 ± 2.33	92.68 ± 1.47	62.82 ± 5.75	63.25 ± 5.70	41.57 ± 3.55	41.73 ± 3.55
SVM (N = 500)	359547.7‡	87.37 ± 2.63	93.05 ± 0.71	66.83 ± 4.34	67.36 ± 4.28	41.84 ± 1.38	41.49 ± 1.39
SVM (N = 1000)	359666.2‡	87.06 ± 3.14	91.24 ± 2.28	68.40 ± 0.48	69.02 ± 0.84	41.32 ± 1.31	41.19 ± 1.29

† Results obtained by running the author's publicly available code 3 times with the same settings as the proposed method.

‡ Includes the time required to train, which was done using information of the performance of 100 fully trained networks, which collectively required 4.16 training days to train.

in search spaces size, result in increasing the number of possible networks that can be generated, which using performance estimation strategies that require some type of training makes exhaustively evaluating the performance of NAS methods extremely hard, ultimately resulting in evaluations using only subsets of the whole search space. To ease the difficulty in comparing NAS methods, NAS benchmarks have been proposed. In these, the goal is to have a controlled setting, where information about the training and final performance of possible networks under the proposed search space is provided, allowing rapid prototyping and comparison between different NAS methods using the same search space, training procedures and hyper-parameters [9,28,37,39].

In this work, we used NAS-Bench-201 [9] to evaluate the proposed method. NAS-Bench-201 provides information about trained networks in three different datasets: CIFAR-10, CIFAR-100 and ImageNet16-120, with fixed splits, and also provide results of several NAS methods under its constraints, allowing direct comparison. In this benchmark, the goal is to design cell-based architectures, where each cell is comprised of 6 edges and 4 nodes. All nodes receive an input edge from all the preceding nodes. The edges represent the possible operations, which are selected from a pool of 5 operations: (1) zeroize, which zeros the information, (2) skip connection, (3) 1×1 convolution, (4) 3×3 convolution, and (5) 3×3 average pooling layer. The number of possible operations and edges means that there are $5^6 = 15625$ possible cells. The final networks are comprised of a fixed macro skeleton, where a cell is a replicated block in the network, meaning that there are as many networks as possible cells, as the only change in the macro skeleton is the cell to be replicated.

4.2   Results and Discussion

By combining EPE-NAS with a random search strategy, we can compare the effectiveness of a simple search strategy coupled with the proposed performance estimation strategy against other NAS methods. To perform this experiment, a network is randomly proposed, and instead of training it, we evaluate its performance by scoring the network. This setup requires no training, and we can perform this for different sample sizes (N, where N represents the number of networks evaluated). Table 1 shows the results for EPE-NAS with random search, and compares it with several methods. Methods that perform the search without weight sharing are shown in the first block, whereas weight sharing methods are shown in the second block. In the third block, we present the results for the proposed method and directly compare it with NAS-WOT [23], while also showing the optimal network in each setting where our method and NAS-WOT were evaluated. Finally, in the last block, we show a baseline method based on a Support Vector Machine (SVM). The SVM approach is indicative of a possible surrogate model that is trained with information of the 100 trained networks performances. The SVM input was created by computing a single correlation matrix of the batch and then calculating the eigenvalues of the matrix. Denote that for training the 100 networks, 4.16 days of GPU computation were required. After finalizing the SVM training, there is no need to further train any network, as the SVM infers the performance based on untrained networks' correlation matrix.

From this table, it is possible to see that our proposed method requires orders of magnitude less time to search for efficient networks, while both non-weight sharing and weight sharing incur in a large search time cost. Our method also achieves better results throughout all datasets than weight sharing methods, except for GDAS on CIFAR-10 and CIFAR-100. However, our method is more than 12500× faster. The non-weight sharing methods outperform our method (random search coupled with the proposed performance estimation strategy), but our method is still on pair with them, being capable of achieving competitive results in all datasets. As for the direct comparison with NAS-WOT, in Table 1 it is also shown that the proposed method outperforms NAS-WOT both in terms of inference, being faster in all settings, and in terms of accuracy, being capable of selecting high performant networks in CIFAR-10 and CIFAR-100 in the settings where the sample size is 10 and 100, and CIFAR-10, CIFAR-100 and ImageNet-16-120 for higher sample sizes (500 and 1000). It is important to note that for NAS-WOT, as sample size increases, it increasingly suffers from noise, increasing the gap between the chosen network accuracy and the optimal result and decreasing the performance compared to smaller sample sizes. The opposite happens with our method. As the sample size increases, our method is capable of selecting high performant networks without losing precision, which is of extreme importance, as it is improbable that optimal networks are present in small sample sizes. More focused on ImageNet-16-120, which is a dataset with more noise, due to the image sizes (16×16) and the high number of classes, our method can select networks that attain excellent test accuracies, when compared with no weight sharing methods and NAS-WOT.

As can be seen in the second column of Table 1, the execution time of our method is a great advantage, as it is capable of evaluating 1000 networks in 206 s. To further evaluate the gains in terms of execution time compared to NAS-WOT, we explored how both methods behave in scoring a network, with images of increasing sizes. The proposed method is capable of evaluating images with sizes $256*256*3$ in approximately 5 s, whereas NAS-wot requires 23% more time, translating to approximately 6.5 s. As for a images with a size of $512*512*3$, EPE-NAS is capable of evaluating a network in under 23 s, whereas NAS-WOT requires 34 s, which is approximately 34% more time. More, this shows that the proposed method can also serve as an improvement for current NAS methods that solely search for networks in CIFAR-10, due to the reduced image size, and then transfer the best networks to ImageNet settings. Thus, NAS methods that were incapable of searching using larger datasets due to time complexity, can use EPE-NAS to search networks in larger datasets directly.

The reason why the proposed method is capable of outperforming NAS-WOT in terms of time is directly linked with the time complexity of creating a correlation matrix, which is highly dependant on the number of data points and features. By evaluating individual correlation matrices, one per class, we reduce each correlation matrix's size, allowing for faster computations.

Considering the mean time required to evaluate 1000 networks by our method (Table 1), EPE-NAS also allows exhaustive exploration of a search space, as the proposed method is capable of evaluating over 1 million architectures in just 2 days of GPU computing, under these settings. Therefore, this could be used to evaluate a search space's behaviour, giving information to the search method on how to start and proceed, which is a significant benefit when considering large, possibly unbounded, search spaces where information about their shape is limited.

An important property of the proposed method is that it can easily be incorporated in almost any NAS method either as the sole method that evaluates networks or as a complementary method to perform mixed training, where the reward to update the controller parameters is a combination of complementary evaluations (e.g., EPE-NAS score combined with the inference/latency of the network in a mobile setting [31,35]). Also, EPE-NAS is agnostic to the search method, as it focuses on the evaluation of networks, being the perfect addition to search methods that rely on information about generated networks or to guide the search, allowing the analysis of thousands of networks in seconds. More, this information can also serve to guide the search of large search spaces, by directly indicating which network configurations are better. This is important because searching for networks in large search spaces, possibly unbounded, is extremely difficult and prone to converge to *local minimas*, mainly due to lack of information about the search space which ultimately leads search methods to converge fast to the best networks initially sampled.

5   Conclusions

In this paper, we propose EPE-NAS, a performance estimation strategy that scores **untrained** networks with a high correlation to their trained performance. By leveraging information about the gradients of the output of a network with regards to its input, our method can accurately infer if the generated network is good in less than one second, being capable of evaluating thousands of networks in a matter of seconds. More, in this work, we have shown that using a simple random search coupled with the proposed estimation strategy, it is possible to sample high performant networks, in seconds, that can outperform many current NAS methods.

Our proposal can also contribute to allow NAS methods to search large search spaces, by providing an efficient way of extracting information about generated networks without requiring any training, and large databases, as our method is still very fast even in the presence of large image sizes. Furthermore, the proposed method is agnostic to the search strategy, allowing it to be integrated into almost any NAS method.

References

1. Baker, B., Gupta, O., Naik, N., Raskar, R.: Designing neural network architectures using reinforcement learning. In: ICLR (2017)
2. Baker, B., Gupta, O., Raskar, R., Naik, N.: Accelerating neural architecture search using performance prediction. In: ICLR (2018)

3. Brock, A., Lim, T., Ritchie, J.M., Weston, N.: SMASH: one-shot model architecture search through hypernetworks. In: ICLR (2018)
4. Cai, H., Zhu, L., Han, S.: Proxylessnas: direct neural architecture search on target task and hardware. In: ICLR (2019)
5. Deng, L., Yu, D., et al.: Deep learning: methods and applications. Found. Trends Signal Process. **7**(3–4), 197–387 (2014)
6. Domhan, T., Springenberg, J.T., Hutter, F.: Speeding up automatic hyperparameter optimization of deep neural networks by extrapolation of learning curves. In: IJCAI (2015)
7. Dong, X., Yang, Y.: One-shot neural architecture search via self-evaluated template network. In: ICCV. IEEE (2019)
8. Dong, X., Yang, Y.: Searching for a robust neural architecture in four GPU hours. In: CVPR, pp. 1761–1770 (2019)
9. Dong, X., Yang, Y.: NAS-bench-201: extending the scope of reproducible neural architecture search. In: ICLR (2020)
10. Elsken, T., Metzen, J.H., Hutter, F.: Efficient multi-objective neural architecture search via lamarckian evolution. In: ICLR (2019)
11. Elsken, T., Metzen, J.H., Hutter, F.: Neural architecture search: a survey. J. Mach. Learn. Res. **20**(1), 1997–2017 (2019)
12. Falkner, S., Klein, A., Hutter, F.: BOHB: robust and efficient hyperparameter optimization at scale. In: Dy, J.G., Krause, A. (eds.) ICML (2018)
13. Goodfellow, I., Bengio, Y., Courville, A., Bengio, Y.: Deep Learning, vol. 1. MIT Press, Cambridge (2016)
14. He, K., Zhang, X., Ren, S., Sun, J.: Deep residual learning for image recognition. In: CVPR (2016)
15. Huang, G., Liu, Z., van der Maaten, L., Weinberger, K.Q.: Densely connected convolutional networks. In: CVPR (2017)
16. Hutter, F., Kotthoff, L., Vanschoren, J.: Automated Machine Learning. Springer, Cham (2019). https://doi.org/10.1007/978-3-030-05318-5
17. Khan, A., Sohail, A., Zahoora, U., Qureshi, A.S.: A survey of the recent architectures of deep convolutional neural networks. Artif. Intell. Rev. **53**(8), 5455–5516 (2020)
18. Krizhevsky, A., Sutskever, I., Hinton, G.E.: Imagenet classification with deep convolutional neural networks. In: Advances in Neural Information Processing Systems (2012)
19. Li, L., Talwalkar, A.: Random search and reproducibility for neural architecture search. In: UAI, pp. 367–377. PMLR (2020)
20. Lindauer, M., Hutter, F.: Best practices for scientific research on neural architecture search. J. Mach. Learn. Res. **21**(243), 1–18 (2020)
21. Liu, C., et al.: Progressive neural architecture search. In: Proceedings of the European Conference on Computer Vision (ECCV) (2018)
22. Liu, H., Simonyan, K., Yang, Y.: DARTS: differentiable architecture search. In: ICLR (2019)
23. Mellor, J., Turner, J., Storkey, A.J., Crowley, E.J.: Neural Architecture Search without Training. CoRR abs/2006.04647 (2020)
24. Pham, H., Guan, M., Zoph, B., Le, Q., Dean, J.: Efficient neural architecture search via parameters sharing. In: ICML (2018)
25. Real, E., Aggarwal, A., Huang, Y., Le, Q.V.: Aging evolution for image classifier architecture search. In: AAAI, vol. 2 (2019)
26. Real, E., Aggarwal, A., Huang, Y., Le, Q.V.: Regularized evolution for image classifier architecture search. In: AAAI, pp. 4780–4789. AAAI Press (2019)

27. Runge, F., Stoll, D., Falkner, S., Hutter, F.: Learning to design RNA. In: ICLR (2019)
28. Siems, J., Zimmer, L., Zela, A., Lukasik, J., Keuper, M., Hutter, F.: NAS-Bench-301 and the Case for Surrogate Benchmarks for Neural Architecture Search. arXiv preprint arXiv:2008.09777 (2020)
29. Simonyan, K., Zisserman, A.: Very deep convolutional networks for large-scale image recognition. In: ICLR (2015)
30. Szegedy, C., et al.: Going deeper with convolutions. In: CVPR (2015)
31. Tan, M., et al.: Mnasnet: platform-aware neural architecture search for mobile. In: CVPR (2019)
32. Tan, M., Le, Q.V.: Efficientnet: rethinking model scaling for convolutional neural networks. In: Chaudhuri, K., Salakhutdinov, R. (eds.) ICML (2019)
33. Voulodimos, A., Doulamis, N., Doulamis, A., Protopapadakis, E.: Deep learning for computer vision: a brief review. Comput. Intell. Neurosci. (2018)
34. Wistuba, M., Rawat, A., Pedapati, T.: A survey on neural architecture search. CoRR abs/1905.01392 (2019)
35. Wu, B., et al.: Fbnet: hardware-aware efficient convnet design via differentiable neural architecture search. In: CVPR (2019)
36. Yang, A., Esperança, P.M., Carlucci, F.M.: Nas evaluation is frustratingly hard. In: ICLR (2020)
37. Ying, C., Klein, A., Christiansen, E., Real, E., Murphy, K., Hutter, F.: NAS-bench-101: towards reproducible neural architecture search. In: ICML (2019)
38. Zela, A., Elsken, T., Saikia, T., Marrakchi, Y., Brox, T., Hutter, F.: Understanding and robustifying differentiable architecture search. In: ICLR (2020)
39. Zela, A., Siems, J., Hutter, F.: Nas-bench-1shot1: benchmarking and dissecting one-shot neural architecture search. In: ICLR (2020)
40. Zhong, Z., Yan, J., Wu, W., Shao, J., Liu, C.L.: Practical block-wise neural network architecture generation. In: CVPR (2018)
41. Zoph, B., Le, Q.V.: Neural architecture search with reinforcement learning. In: ICLR (2017)
42. Zoph, B., Vasudevan, V., Shlens, J., Le, Q.V.: Learning transferable architectures for scalable image recognition. In: CVPR (2018)

DVAMN: Dual Visual Attention Matching Network for Zero-Shot Action Recognition

Cheng Qi[1], Zhiyong Feng[1], Meng Xing[1(✉)], and Yong Su[2]

[1] College of Intelligence and Computing, Tianjin University, Tianjin, China
{qicheng_525,zyfeng,xingmeng}@tju.edu.cn
[2] Tianjin Normal University, Tianjin, China
suyong@tju.edu.cn

Abstract. Zero-Shot Action Recognition (ZSAR) aims to transfer knowledge from a source domain to a target domain so that the unlabelled action can be inferred and recognized. However, previous methods often fail to highlight information about the salient factors of the video sequence. In the process of cross-modal search, information redundancy will weaken the association of key information among different modes. In this paper, we propose Dual Visual Attention Matching Network (DVAMN) to distill sparse saliency information from action video. We utilize dual visual attention mechanism and spatiotemporal Gated Recurrent Units (GRU) to establish irredundant and sparse visual space, which can boost the performance of the cross-modal recognition. Relational learning strategy is employed for final classification. Moreover, the whole network is trained in an end-to-end manner. Experiments on both the HMDB51 and the UCF101 datasets show that the proposed architecture achieves state-of-the-art results among methods using only spatial and temporal video features in zero-shot action recognition.

Keywords: Zero-shot action recognition · Redundancy elimination · Saliency detection

1 Introduction

Remarkable success has been achieved by deep neural networks for human action recognition when a large number of labeled training data are available [3,21]. Nevertheless, with the growth of action classes and the increasing demand for different applications, the model needs to be finetuned or even trained again annoyingly, apart from collecting sufficient labeled data costly and laboriously for per new action categories. Researchers work on zero-shot learning (ZSL) [12] which breaks such limitations. ZSL automatically recognizes data of unseen categories by transferring the knowledge from seen classes.

Zero-shot learning aims to learn a model which classifies data of unseen classes given only the semantic attributes (always refer to semantic class descriptions or

I. Farkaš et al. (Eds.): ICANN 2021, LNCS 12895, pp. 564–575, 2021.
https://doi.org/10.1007/978-3-030-86383-8_45

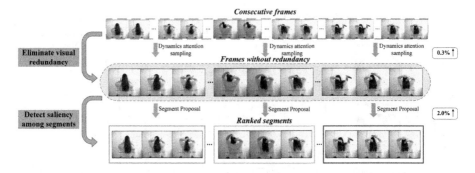

Fig. 1. An example of processing a *"brush hair"* video with dual visual attention mechanism. Adjacent frames share highly similar content after converting a video to frames. By eliminating visual redundancy in frames, there is about 0.3% improvement on classification. Subsequent block detects saliency among segments, which can get ranked segments to construct visual space. The accuracy finally increases by almost 2.1%.

class names) of the classes [7]. Based on semantic attributes, ZSL learns mapping between visual features and semantic embeddings from seen data [6,12,20]. The task of zero-shot action recognition mainly works on establishing visual space to increase the robustness of the mapping between constructed visual space and semantic space. Nevertheless, it is more difficult for classifying action videos than images in zero-shot learning for the reason that videos are structured data of space-time dependency [26]. As a consequence, existing methods in ZSAR mostly utilize mature video feature extractors like Convolutional 3D (C3D) network [21] or Inflated 3D ConvNet (I3D) [3], to construct visual space.

However, these extractors do not take into account that redundancy and irrelevancy could affect the performance of cross-modal recognition. Such mechanisms would lead to sparse video features, which reflect the correlation of actions, fall into oblivion.

Consequently, we aim at establishing a more sparse and focused visual space as shown in Fig. 1 to solve the problem. To this end, we propose Dual Visual Attention Matching Network (DVAMN) for zero-shot action recognition. DVAMN starts with dynamics attention sampling module to establish sparse visual space, which is beneficial to separate different video factors. Next, saliency detection mechanism is used to mine visual saliency information in segment level. The specific architecture of the above dual visual attention mechanism is shown in Fig. 2. As a result, the sparse and focused visual space is constructed. A mapping function is learned from the semantic embedding and visual embedding for classification. An overview of our network is illustrated in Fig. 3.

The contributions of this paper are summarized below:

- We develop a new architecture called Dual Visual Attention Matching Network, which learns mapping functions from the constructed visual space and semantic space to the joint space for zero-shot action recognition.

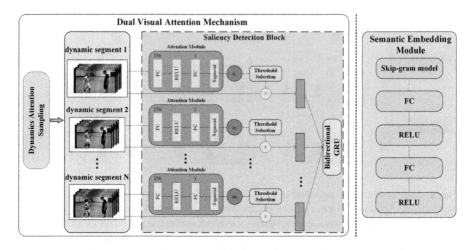

Fig. 2. The detailed architectures of two embedding modules: dual visual attention mechanism (left block) for video data and semantic embedding module (right block) for class names. Dual visual attention mechanism contains dynamics attention sampling module and saliency detection block. The former is used to sample frames subset to represent dynamics in a video while the latter is used for detecting salient temporal segments. Besides, a bidirectional GRU layer [4] is utilized in dual visual attention mechanism to fully capture temporal information and reduce visual feature dimension. Label names are fed into the semantic embedding module: word vectors of classes are transformed into semantic embedding by two stacked fully connected layers.

- We propose dual visual attention mechanism to eliminate visual redundancy and detect saliency among segments, which can establish a more sparse and focused visual space.
- The proposed method comparatively improves the accuracy of zero-shot action recognition on the HMDB51 [11] and the UCF101 [19] dataset.

The remainder of this paper is organized as follows. Section 2 introduces the related work of relevant tasks. Section 3 illustrates the architecture of our work in detail. Section 4 presents the results of our work, which are comparable to other works with similar inputs in ZSAR. Finally, Sect. 5 concludes the whole paper and points our direction for future work.

2   Related Work

Action recognition has achieved great success in recent years because of the remarkable development in deep neural networks [21]. Meanwhile, the larger amount of video data and stronger computation power make supervised learning efficient in action recognition. Researchers have been devoted to extracting appropriate video features for classification. Two-stream CNN can significantly improve classification results, taking spatial and temporal information into account[18]. Based on this, temporal segment network [23] was proposed to capture long-range temporal information in a video. The above works considered only two-dimensional CNN, which might cause loss in spatial and temporal information. With the development of deep learning, 3-dimensional CNN [21] was applied to extract video features, which is widely used nowadays. Nevertheless, action recognition with fully supervised learning limits a lot due to the huge demand for labeled data. Collecting and annotating data is laborious and time-consuming. With the use of well-performing deep neural networks, we aim to recognize action in videos without the need for training samples, namely zero-shot action recognition.

Zero-shot learning has been applied widely to computer vision tasks such as image classification and action recognition [12,24,27]. However, more researches focus on image classification than on action recognition. One major reason is that the video data is structured data, which makes modeling video features complex and increases the difficulty to find the relationship between classes and videos. Techniques to extract video feature varies a lot. Objects were used as class attributes in [9], mapping objects in videos to the joint space. Inspired by this, Gao et al. [8] and Xu et al. [28] designed frameworks with graph neural networks to take full advantage of relations between visual data and classes. With the development of networks extracting features, using pretrained 3-dimensional networks is a common way to extract video features [2,8,29]. Nevertheless, these works treated each part of videos equally and ignored that irrelevant information in videos can mislead correct classification results. Our work contributes to filtering most salient segments for classification. The proposed architecture learns mapping functions from visual space and semantic space to the joint embedding space, which boosts the diversity of inter-class videos and reduces the distance between visual embedding and corresponding semantic embedding in the joint space [2,20].

3 Dual Visual Attention Matching Network

In this section, we will introduce the structure of the proposed DVAMN, a novel deep network for ZSAR. Figure 3 is an overview of the whole framework. DVAMN learns to represent video by eliminating video redundancy and detecting saliency among temporal segments. The classification results are predicted by aligning semantic embeddings with visual embeddings in the joint space [8,9,20]. DVAMN mainly contains Embedding Module and Matching Module. First, videos are input to dual visual attention mechanism to get final visual embeddings. Second, the semantic embedding module transforms semantic label names into semantic embeddings. Figure 2 illustrates the specific architecture of dual visual attention mechanism and semantic embedding module in detail. Finally, the matching module compares segment-level visual embeddings with semantic embeddings. The aggregation module integrates segment-level similarities into video-level one and indicates the final classification result.

In ZSAR, the whole video dataset D is split into data of seen and unseen classes, denoted by $D_s = \{V_s, Y_s\}$ and $D_u = \{V_u, Y_u\}$, respectively, where V_s, V_u

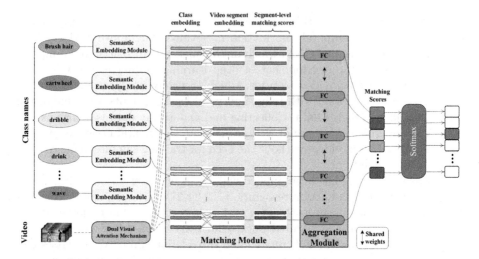

Fig. 3. The overall framework of the proposed Dual Visual Attention Matching Network. The inputs of DVAMN are video data and class names. Video data are embedded by dual visual attention mechanism. We feed class names into semantic embedding module to get semantic embedding. Semantic embedding and visual embedding are used as inputs for the matching module and the aggregation module is used for final classification.

refer to video instances and Y_s, Y_u refer to seen and unseen classes. Moreover, the seen classes and unseen classes share no common elements: $Y_s \cap Y_u = \emptyset$ and $Y_s \cup Y_u = Y$. Suppose seen data D_s contain N_s videos with the corresponding label from S seen classes and unseen data D_u contain N_u videos from U unseen classes. Seen data and labels are available in training while unseen data are only used during testing phase for evaluating the performance of DVAMN.

3.1   Embedding Module

As depicted in Fig. 2, DVAMN utilizes two different embedding modules for visual data and semantic data. Particularly, visual embedding module utilizes dual attention mechanism consisting of dynamics attention sampling module and saliency detection block. Both the dual attention mechanism and semantic embedding module are fully explained in this section.

Dynamics Attention Sampling Module. Processing videos is an important part of action recognition. While previous work processed videos based on frames, it is deserving to notice that adjacent frames in a video share highly similar content due to slow variation in frame level, so that frames may contain much redundant and irrelevant information[22]. Redundant visual information could discourage the performance of the model by dispersing dynamics of action. We use a dynamics attention sampling module for segment proposals instead of simply dividing a video into equal-length segments. In this way can we get segments consisting of frames with high content variance, which helps the visual space be sparse.

The dynamics of action in frame-level is explored during sampling. we extract the Histogram of Oriented Gradients (HOG) features [5] for each frame, for the excellent performance HOG features have on images, which captures the variance between adjacent frames. Denote HOG features of frame i, j as H_i and H_j, respectively. Subsequently, the absolute value of feature difference [22] between adjacent frames in a video are recorded:

$$d_{set} = \{|H_i - H_j|\}_{j=2,\ldots,N}^{i=1,\ldots,N-1} \tag{1}$$

Set an adaptive threshold according to the recorded difference in a video instead of setting a fixed threshold to ensure the number of selected frames in case that none could reach the threshold in a video.

$$thr_H = \frac{1}{N-1} \sum_{d \in d_{set}} d \tag{2}$$

The value of the threshold is set to be the average of HOG feature difference in one video. If the difference between adjacent frames is larger than the threshold, the two frames would be detected for subsequent procedures. Afterward, we segment all detected frames into segments that contain K (a fixed value in different datasets) frames in a sequential manner.

Now that we have detected a set of frames in video v, denoted by $s_i = (s_i^s, s_i^e)$, where s_i^s and s_i^e represent the starting frame and the end frame of the set, respectively. Segment proposals are acquired by:

$$S_i = \{(s_i^s, \cdots, s_i^s + K), \cdots, (s_i^s + (p-1) \times K, \cdots, s_i^s + p \times K)\} \tag{3}$$

where $s_i^s + p \times K \leq s_i^e$. We finally obtain all irredundant segment proposals.

Saliency Detection Block. Video factors have different saliency to make effective depict of videos [15]. Irrelevant visual information has a negative effect on informative segments classification results. Visual space can be more focused with ranked video factors. To this end, we utilize saliency detection block to estimate the saliency of these segments. Instead of using frame-level HOG features extracted in dynamics sampling stage, we use a pretrained network like Convolutional 3D (C3D) [21] to extract features of segment proposals. C3D network can extract spatiotemporal information in video which is of great importance to action recognition.

Formally, Given a video V_i with a set of segment proposals $S_i = \{S_i^m\}_{m=1}^M$, we extract C3D features $\phi(V_i; S_i^m)$ for each segment proposal S_i^m. The features are fed into attention modules. Let $x_i^m = \phi(V_i; S_i^m)$ represent the feature of segment S_i^m in V_i. The attention module will output an attention weight a_i^m for each segment, indicating its saliency in the video.

We still adopt an adaptive threshold-based selection strategy to avoid that no segments could achieve the fixed threshold in a video. Set the attention threshold as:

$$a_i^{thr} = \frac{1}{M} \sum_{m=1}^{M} a_i^m \qquad (4)$$

If the attention weight is higher than threshold a_i^{thr}, the corresponding segment will be detected. In this way, T most salient segment features in a video are filtered. Subsequent action classification is based on these most informative segments, which reduces the distance between class labels and video features in the joint space. Particularly, we pretrain attention modules in advance. The loss used in saliency detection module is:

$$L_{attention} = L_c + \alpha \left\| \mathbf{a} \right\|_1 \qquad (5)$$

where L_c is the binary cross-entropy loss between predicted results and true labels, α the constant to balance the classification loss and regularization loss of attention weights.

The dynamics attention sampling module and saliency detection block are utilized as local feature extractors. Video data have a distinct characteristic of temporal dependency. To learn more global features, temporal information should be taken into full account. Therefore, we apply bidirectional GRU layers [4] which allow backward and forward information propagation each time step along the temporal dimension. Finally, the visual embedding is generated by the GRU layers. Moreover, the feature dimension of video data can be reduced in the meanwhile.

Semantic Embedding Module. The semantic embedding module utilizes semantic attributes as input and needs to output embeddings with the same dimension as visual embeddings. Therefore, semantic embedding module consists of a pretrained skip-gram model [13] and a module containing two stacked FC layers. The pretrained skip-gram model is used to get word vectors of all labels. It is notable that if one label name contains more than one word, we average word vectors of all words as the final class vector. Subsequently, a module of two stacked FC layers is used to acquire final semantic embeddings.

3.2   Matching Module and Aggregation Module

We will explain the matching module and aggregation module in DVAMN. The matching module compares visual embedding and semantic embedding in segment level. Based on matching results, the aggregation module integrates segment-level matching into a video-level one for the final result.

Matching Module. Now that we have high saliency segment features of each video, we feed them into matching module [2] for aligning visual embedding and semantic embedding. According to the previous embedding modules, segment features and semantic data are transformed to d-dimensional embedding in the joint space by visual and semantic embedding module. Denote visual embedding as e_i^t and semantic embedding as c_y ($y \in Y$), representing the t-th segment visual embedding of the video V_i and the embedding of class y.

Afterward, segment-level matching scores [25] between each visual embedding and each class embedding in training:

$$s_i^{ty} = (e_i^t - c_y) \odot (e_i^t - c_y) \tag{6}$$

where $\odot$ refers to element-wise multiplication, i is the index of the video and t is the index of the segment in the video.

Aggregation Module. For getting video-level classification results, we use an aggregation module consisting of two fully connected layers with a pooling layer. The segment-level matching scores between video and classes are then combined into video-level one, denoted as m_i^y. Finally, the action class can be predicted as:

$$y_i = \max_{y \in Y} m_i^y \tag{7}$$

We train the full architecture in an end-to-end manner, based on episode training strategy. The cost function we use in the total batch is binary cross-entropy loss:

$$L = -\frac{1}{N_s} \sum_{i=1}^{N_s} (T_i log m_i^y + (1 - T_i) log(1 - m_i^y)) \tag{8}$$

where N_s is the number of seen videos, m_i^y is the predicted matching score, T_i is the target matching score.

4   Experiments

4.1   Dataset

We use the HMDB51 and the UCF101 datasets to evaluate our whole framework. The HMDB51 dataset contains 6766 videos from 51 classes and the UCF101 dataset consists of 13320 videos from 101 classes. we divide the HMDB51 dataset into 26/25 classes for seen and unseen classes. As for UCF101, settings of 51/50 and 81/20 for seen/unseen classes are selected to evaluate the performance of DVAMN.

4.2   Experiments Setup

In saliency detection block, the attention module consists of two-layer FC with 256 and 1 nodes to produce attention weights that indicate saliency of the segments. The number of segments in each video is fixed to T (32 in HMDB51 and

Table 1. Comparisons to the state-of-the-art ZSAR methods on the HMDB51 and the UCF101 datasets. Average accuracy are reported here. Visual Repr. and Semantic Repr. denote the visual features and semantic representations. D represents deep visual features and L represents low-level features. WV means the word vector of class names.

Method	Visual Repr.	Semantic Repr.	HMDB51 (26/25)	UCF101 (51/50)	UCF101 (81/20)
ESZSL [17]	D	WV	18.5 ± 2.0	15.0 ± 1.3	–
SJE [1]	L	WV	–	12.0 ± 1.2	–
ZSECOC [16]	L	WV	16.5 ± 3.9	13.7 ± 0.5	–
BiDiLEL [24]	D	WV	18.6 ± 0.7	18.9 ± 0.4	38.3 ± 1.2
GMM [14]	D	WV	19.3 ± 2.1	17.3 ± 1.1	–
TARN [2]	D	WV	19.5 ± 4.2	19.0 ± 2.3	36.0 ± 5.3
Ours	D	WV	$\mathbf{21.6 \pm 1.5}$	$\mathbf{20.5 \pm 1.2}$	$\mathbf{39.0 \pm 3.4}$

57 in UCF101). To obtain semantic embedding, two FC layers of size 4096 and 512 are applied on 300-dimensional word vectors generated by the pretrained skip-gram model [13] trained on the Google News dataset.

Visual features are extracted from the C3D network pretrained on the Sports-1M dataset [10] with 4096 dimension. The bidirectional GRU layers of hidden size 256 output 512-dimensional visual embedding. The constant α in saliency detection block is set to 0.5. The matching module performs segment-level matching between visual and semantic embedding. The aggregation module has two FC layers of size 256 and 1, outputting final classification results. We use Adam optimizer with learning rate 0.001 and gradient clipping 0.5 to train our network in an end-to-end manner. The batch sizes of 8 and 16 are used for HMDB51 and UCF101, respectively. All experiments are conducted on an Nvidia Tesla M40 card with the PyTorch. The network is trained for 3000 episodes and evaluated every 30 training episodes.

4.3   Results

Our experiments have two main goals: compare our method to previous work and investigate the performance of our method with dynamic attention sampling and saliency detection separately used. The first experiment is necessary to validate that DVAMN can outperform other approaches with similar inputs. The latter one will allow us to understand under what condition DVAMN can be particularly beneficial.

Comparison to the State of the Art. In order to have a common setting across the majority of works, we compare with works which use spatial and temporal video features as video representation and word vectors as semantic representation. Specifically, we do not compare with works that: (1) discover

Table 2. Accuracy of DVAMN model at different settings on the HMDB51 and the UCF101 datasets for zero-shot action recognition. The baseline is experimented under the condition that neither dynamics attention sampling module nor saliency detection block is used. We refer DS to dynamics attention sampling module and SD to saliency detection block in the table.

Method	HMDB51 (26/25)	UCF101 (51/50)	UCF101 (81/20)
BASELINE	19.5 ± 4.2	19.0 ± 2.3	36.0 ± 5.3
DVAMN(with DS, w/o SD)	19.8 ± 2.6	19.1 ± 2.2	33.7 ± 3.2
DVAMN(w/o DS, with SD)	20.4 ± 2.1	19.7 ± 2.4	35.4 ± 2.9
DVAMN(with DS, with SD)	$\mathbf{21.6 \pm 1.5}$	$\mathbf{20.5 \pm 1.2}$	$\mathbf{39.0 \pm 3.4}$

objects as visual features [8,9]; (2) require access to the data of unseen classes during training (transductive setting) [26,28]; (3) use auxiliary data to augment datasets [27,29]. This allows us to have a clear comparison to the literature. Table 1 summarizes the comparison results over the splits setting of HMDB51 (26/25), UCF101 (51/50) and UCF101 (81/20). As Table 1 shows, our architecture achieves the best results over the HMDB51 with about 2.1%, the UCF101 51/50 and 81/20 splits with almost 1.5% and 0.7%. Moreover, we apply different settings on DVAMN to discuss the necessity of dual visual attention mechanism.

Ablation Study. In Table 2, we show the results of our architecture on zero-shot action recognition using different settings. Neither dynamics attention sampling module nor saliency detection block is used in the baseline. In the second and third settings, either the dynamics attention sampling module or saliency detection block is chosen in the model. The second and the third row in Table 2 shows comparative improvement in accuracy with 0.3% and 0.9%, respectively. Furthermore, Table 2 demonstrates that the best performance is achieved with 2.1% improvement when utilizing both of them, namely dual visual attention mechanism. The results sufficiently show that dual visual attention mechanism can effectively establish the visual space related to semantic space and align video features with semantic features more accurately.

5   Conclusion

In this paper, we proposed a novel network named dual visual attention matching network (DVAMN) for zero-shot action recognition (ZSAR) which focuses on the effective depiction of videos. Our approach encouraged the set of frames with high content variance to describe the dynamics of an action, which alleviates redundancy in video frames and makes visual features sparse. Furthermore, we incorporated saliency detection block to detect saliency among segments, establishing a more focused visual space. Dual visual attention mechanism increases the robustness of the mapping function between visual space and semantic space. By extensive experiments on the HMDB51 dataset and the UCF101 dataset, we showed that our framework competitively improves the accuracy of zero-shot action recognition. In future work, we will focus on improving the efficiency and the robustness of DVAMN and consider efficient mechanisms for selecting informative segments to establish sparse and focused visual space for relational learning in zero-shot action recognition.

Acknowledgements. The work was supported by Shenzhen Science and Technology Foundation (JCYJ20170816093943197), the Science and Technology Program of Guangzhou, China (202002030263) and the Guangdong Basic and Applied Basic Research Foundation (2020A1515110997).

References

1. Akata, Z., Reed, S., Walter, D., Lee, H., Schiele, B.: Evaluation of output embeddings for fine-grained image classification. In: Proceedings of the IEEE Conference on Computer Vision and Pattern Recognition, pp. 2927–2936 (2015)
2. Bishay, M., Zoumpourlis, G., Patras, I.: Tarn: temporal attentive relation network for few-shot and zero-shot action recognition. arXiv preprint arXiv:1907.09021 (2019)
3. Carreira, J., Zisserman, A.: Quo vadis, action recognition? A new model and the kinetics dataset. In: Proceedings of the IEEE Conference on Computer Vision and Pattern Recognition, pp. 6299–6308 (2017)
4. Cho, K., et al.: Learning phrase representations using RNN encoder-decoder for statistical machine translation. arXiv preprint arXiv:1406.1078 (2014)
5. Dalal, N., Triggs, B.: Histograms of oriented gradients for human detection. In: 2005 IEEE Computer Society Conference on Computer Vision and Pattern Recognition (CVPR 2005), vol. 1, pp. 886–893. IEEE (2005)
6. Demirel, B., Gokberk Cinbis, R., Ikizler-Cinbis, N.: Attributes2classname: a discriminative model for attribute-based unsupervised zero-shot learning. In: Proceedings of the IEEE International Conference on Computer Vision, pp. 1232–1241 (2017)
7. Fu, Y., Hospedales, T.M., Xiang, T., Gong, S.: Learning multimodal latent attributes. IEEE Trans. Pattern Anal. Mach. Intell. **36**(2), 303–316 (2013)
8. Gao, J., Zhang, T., Xu, C.: I know the relationships: zero-shot action recognition via two-stream graph convolutional networks and knowledge graphs. In: Proceedings of the AAAI Conference on Artificial Intelligence, vol. 33, pp. 8303–8311 (2019)
9. Jain, M., Van Gemert, J.C., Mensink, T., Snoek, C.G.M.: Objects2action: classifying and localizing actions without any video example. In: Proceedings of the IEEE International Conference on Computer Vision, pp. 4588–4596 (2015)
10. Karpathy, A., Toderici, G., Shetty, S., Leung, T., Sukthankar, R., Fei-Fei, L.: Large-scale video classification with convolutional neural networks. In: Proceedings of the IEEE Conference on Computer Vision and Pattern Recognition, pp. 1725–1732 (2014)
11. Kuehne, H., Jhuang, H., Garrote, E., Poggio, T., Serre, T.: HMDB: a large video database for human motion recognition. In: 2011 International Conference on Computer Vision, pp. 2556–2563. IEEE (2011)

12. Lampert, C.H., Nickisch, H., Harmeling, S.: Learning to detect unseen object classes by between-class attribute transfer. In: 2009 IEEE Conference on Computer Vision and Pattern Recognition, pp. 951–958. IEEE (2009)

13. Mikolov, T., Sutskever, I., Chen, K., Corrado, G., Dean, J.: Distributed representations of words and phrases and their compositionality. arXiv preprint arXiv:1310.4546 (2013)

14. Mishra, A., Verma, V.K., Reddy, M.S.K., Arulkumar, S., Rai, P., Mittal, A.: A generative approach to zero-shot and few-shot action recognition. In: 2018 IEEE Winter Conference on Applications of Computer Vision (WACV), pp. 372–380. IEEE (2018)

15. Nguyen, P., Han, B., Liu, T., Prasad, G.: Weakly supervised action localization by sparse temporal pooling network. In: Proceedings of the IEEE Conference on Computer Vision and Pattern Recognition, pp. 6752–6761 (2018)

16. Qin, J., et al.: Zero-shot action recognition with error-correcting output codes. In: Proceedings of the IEEE Conference on Computer Vision and Pattern Recognition, pp. 2833–2842 (2017)

17. Romera-Paredes, B., Torr, P.: An embarrassingly simple approach to zero-shot learning. In: International Conference on Machine Learning, pp. 2152–2161 (2015)

18. Simonyan, K., Zisserman, A.: Two-stream convolutional networks for action recognition in videos. In: Advances in Neural Information Processing Systems, pp. 568–576 (2014)

19. Soomro, K., Zamir, A.R., Shah, M.: UCF101: a dataset of 101 human actions classes from videos in the wild. arXiv preprint arXiv:1212.0402 (2012)

20. Sung, F., Yang, Y., Zhang, L., Xiang, T., Torr, P.H., Hospedales, T.M.: Learning to compare: relation network for few-shot learning. In: Proceedings of the IEEE Conference on Computer Vision and Pattern Recognition, pp. 1199–1208 (2018)

21. Tran, D., Bourdev, L., Fergus, R., Torresani, L., Paluri, M.: Learning spatiotemporal features with 3D convolutional networks. In: Proceedings of the IEEE International Conference on Computer Vision, pp. 4489–4497 (2015)

22. Wang, L., Xiong, Y., Lin, D., Van Gool, L.: Untrimmednets for weakly supervised action recognition and detection. In: Proceedings of the IEEE Conference on Computer Vision and Pattern Recognition, pp. 4325–4334 (2017)

23. Wang, L., et al.: Temporal segment networks for action recognition in videos. IEEE Trans. Pattern Anal. Mach. Intell. **41**(11), 2740–2755 (2018)

24. Wang, Q., Chen, K.: Zero-shot visual recognition via bidirectional latent embedding. Int. J. Comput. Vision **124**(3), 356–383 (2017)

25. Wang, S., Jiang, J.: A compare-aggregate model for matching text sequences. arXiv preprint arXiv:1611.01747 (2016)

26. Xu, X., Hospedales, T., Gong, S.: Transductive zero-shot action recognition by word-vector embedding. Int. J. Comput. Vision **123**(3), 309–333 (2017)

27. Xu, X., Hospedales, T.M., Gong, S.: Multi-task zero-shot action recognition with prioritised data augmentation. In: Leibe, B., Matas, J., Sebe, N., Welling, M. (eds.) ECCV 2016. LNCS, vol. 9906, pp. 343–359. Springer, Cham (2016). https://doi.org/10.1007/978-3-319-46475-6_22

28. Xu, Y., Han, C., Qin, J., Xu, X., Han, G., He, S.: Transductive zero-shot action recognition via visually connected graph convolutional networks. IEEE Trans. Neural Netw. Learn. Syst. (2020)

29. Zhu, Y., Long, Y., Guan, Y., Newsam, S., Shao, L.: Towards universal representation for unseen action recognition. In: Proceedings of the IEEE Conference on Computer Vision and Pattern Recognition, pp. 9436–9445 (2018)

Dynamic Tuning and Weighting of Meta-learning for NMT Domain Adaptation

Ziyue Song, Zhiyuan Ma, Kaiyue Qi$^{(\boxtimes)}$, and Gongshen Liu$^{(\boxtimes)}$

School of Electronic Information and Electrical Engineering,
Shanghai Jiao Tong University, Shanghai 200240, China
{song_zy,presurpro,tommy-qi,lgshen}@sjtu.edu.cn

Abstract. Neural machine translation (NMT) systems fall short when training data is insufficient. For low-resource domain adaptation, meta-learning has proven to be an effective training scheme. It aims to find an optimal initialization that is easily adaptable to new domains. However, it is assumed that samples contribute equally in tasks and tasks contribute equally in the training task distribution, which deteriorates the performance of the meta-model. In the inner loop, we propose the dynamic tuning strategy to distinguish the tasks' adapting abilities and weight the loss according to the representativeness to discriminate tasks from the same domain. In the outer loop, to measure effects of each task on meta parameters, we calculate uncertainty-aware confidence and assign weights on meta-updating steps. Experiments show that the proposed approaches gain stable improvements in all domains (+1.35 BLEU points in maximum). We also analyze the ability of our strategies to alleviate domain imbalance in non-ideal settings.

Keywords: Meta-learning · Domain adaptation · NMT · Few-shot learning · Dynamic tuning · Dynamic weighting

1 Introduction

Neural Machine Translation (NMT) has achieved comparable results to human translation with abundant data [22]. However, general architectures perform poorly in low-resource scenarios. For new domains with a small amount of data, such as education and medicine, adapting from the general domain is a common solution. But it may lead to model over-fitting or catastrophic forgetting [21]. When there are multiple domains to be adapted simultaneously, solutions like data concatenation, domain control [13], weighted multi-model ensemble [19] and multi-stage fine-tuning [6] fail to capture domain-specific knowledge and ignore the domain divergence. Meta-learning, like Model-Agnostic Meta-Learning (MAML) [7], is a popular adaptation method. It's firstly used in image classification and now is also applied in Natural Language Generation (NLG) tasks, like NMT [15,20]. Instead of finding a local optimum, the meta-solution

© Springer Nature Switzerland AG 2021
I. Farkaš et al. (Eds.): ICANN 2021, LNCS 12895, pp. 576–587, 2021.
https://doi.org/10.1007/978-3-030-86383-8_46

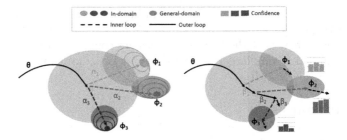

Fig. 1. Individualized meta-training steps in the inner loop (left) and the outer loop (right). **Left**: using dynamic learning rates rather than a fixed one helps adapt to new domains more efficiently. Updating directions of tasks also differ in the same domain. **Right**: the dynamic weighting helps discriminate meta-updating degrees.

leverages tasks to search an initialization for all domains. It is very efficient in few-shot settings where new domains can be quickly adapted. Even if more domains are added, the existing model can be reused for continual training.

However, vanilla MAML assumes that all tasks share the same adapting learning rate and contribute equally in the meta-updating process. It is too limited for real-world datasets because the size of the data set varies in different domains. Moreover, it is detrimental because various data distributions lead to considerable task divergence [4], which enlarges the bias in the final results. To avoid learning equally from meta-knowledge, meta-SGD [16] dynamically learns the learning rates of each layer at each step, at the cost of additional training parameters. Reweighted MAML [12] poses the weights by minimizing the nested losses. To overcome the challenge of class imbalance, [18] assigns weights to training instances according to the gradient directions. Bayesian TAML [14] leverages uncertainties to balance meta-learning. But none of those methods are applied to tackle NLG tasks.

In this paper, three strategies are proposed to improve MAML. Firstly, we dynamically tune the domain-specific learning rates in the inner loop to distinguish the adapting abilities among domains, as illustrated in the left side of Fig. 1. Secondly, we assign representativeness weights in the back-propagation process to distinguish individual sensitivity of tasks from the same domain. Thirdly, we propose the confidence weighting strategy in the outer loop, which utilizes model uncertainty to measure the confidence of the predictions and guides the meta-learner to choose degrees and directions of meta-updating automatically, as shown in the right side of Fig. 1. We conduct experiments on English-German parallel corpus in 10 domains and the proposed strategies achieve great improvements. Moreover, the experiments show that our weighting strategy can balance task distributions in non-ideal settings.

In summary, the contributions are threefold:

- To the best of our knowledge, this is the first work on modifying the intrinsic defects of unified-weighted methods in NLG tasks. We introduce three different dynamic tuning and weighting strategies to distinguish contributions of tasks to the meta parameters.

- We explore the ability of our strategies in imbalanced domains and make an empirical analysis.
- Experiments show that the three strategies are complementary to each other and the unified results largely improve the performance (up to +4.2%) compared to vanilla MAML.

2 Preliminaries

2.1 MAML

The goal of MAML [7] is to find an optimal initialization θ sensitive for new tasks. Instead of training from random initialization, a classical choice is to fine-tune the model pre-trained from the general domain. Each task T_n in batches sampled from $T = \{T_1, T_2, ..., T_N\}$, is composed of the support set $T_{n,support}$ and the query set $T_{n,query}$. We introduce an intermediate parameters ϕ_n for task T_n, which can be seen as a gradient update of θ,

$$\phi_n = \theta - \alpha \nabla_\theta \mathcal{L}_{T_{n,support}} (f_\theta), \tag{1}$$

where α denotes the inner learning rate, $\mathcal{L}$ denotes the loss function and f_θ denotes the model function. In the meta-optimization stage, the optimization goal is the minimal sum of losses in each model. Formally, it can be formulated as follows:

$$\theta = \theta - \beta \nabla_\theta \sum_{T_n \sim p(T)} \mathcal{L}_{T_{n,query}} (f_{\phi_n}), \tag{2}$$

where β is the meta learning rate. In test stage, we keep the value of θ unchanged.

2.2 Adaptive Module

Fine-tuning on additional adaptive layers [1,11] is cost-effective for domain adaptation in NMT. The vanilla Transformer only slightly outperforms the lightweight modules with high computation cost. The inputs of adapters are the outputs of each layer of Transformer, which are firstly layer-normalized to make it extensible for any kinds of base networks as

$$\tilde{z}_i^T = LayerNorm_T (z_i), \tag{3}$$

where z_i denotes the output of the i-th layer of Transformer with d dimension. Besides, each adapter module includes two projections, between which is a nonlinear activation function defined as

$$h_i^T = relu(W_{bd}^T \tilde{z}_i^T). \tag{4}$$

The down projection W_{bd} transfers d dimension into b dimension, while the up projection W_{db} converts the dimension into d, which can be seen as a bottleneck layer. Finally, the output is wrapped with a residual connection as follows,

$$x_i^T = W_{db}^T h_i^T + z_i. \tag{5}$$

Those self-contained adapters can be added straightly onto the primary structure.

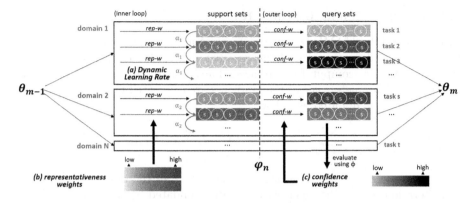

Fig. 2. Overview of three proposed strategies. Dynamic-tuning learning rates and representativeness weighting are applied in the inner loop, while confidence weighting is applied in the outer loop. They are independent of each other.

3 Methodology

3.1 Overview

Our work is based on meta-learning. This section introduces three strategies to improve the performance of MAML. As shown in Fig. 2, they are brought up from different perspectives and work on different stages of the meta-learning process. At each step, the meta-learning rate α is tuned dynamically according to the last training step, in order to achieve a more robust training for complex tasks. In the inner loop, the representativeness weighting is computed in advance. Accordingly, the loss function in (1) can be modified as

$$\mathcal{L}_{T_{support}} = \sum_{(\mathbf{x}_i, \mathbf{y}_i)} w_{rp,i} \log p\left(\mathbf{y}_i \mid \mathbf{x}_i\right), \tag{6}$$

where $(\mathbf{x}_i, \mathbf{y}_i)$ denotes a parallel sentence pair in the support set of task T, and $w_{rp,i}$ denotes the calculated representativeness weight. In the outer loop, each sentence pair's confidence weight is calculated online, using intermediate parameters ϕ. We implement it in loss function,

$$\mathcal{L}_{T_{query}} = \sum_{(\mathbf{x}_i, \mathbf{y}_i)} w_{cf,i} \sum_{j=1}^{J} w_{cf,i,j} \log P\left(\mathbf{y}_j \mid \mathbf{x}_i, \mathbf{y}_{<j}\right), \tag{7}$$

where $(\mathbf{x}_i, \mathbf{y}_i)$ stands for a parallel sentence pair in the query set of task T, J is the number of words in sentence $\mathbf{y}_j$ and $w_{cf,i}, w_{cf,i,j}$ indicate sentence-level and word-level confidence weights respectively. Those three methods are can be easily integrated because they are independent of each other.

3.2   Dynamic-Tuning Learning Rate

Inspired by alpha-MAML [2], we put forward a tuning strategy for the inner-loop learning rate, where only hyperparameter γ, initial learning rate α_0, and one extra gradient computation are needed additionally. In each iteration, after intermediate parameters ϕ and the gradients are calculated by (1), the dynamic learning rate α is updated as follows,

$$\alpha_{m+1} = \alpha_m + \gamma \sum_{T_n \sim p(T)} \nabla_{\phi_n} \mathcal{L}_{T_{n,query}} (f_{\phi_n}) \cdot \nabla_{\theta_{n-1}} \mathcal{L}_{T_{n,support}} (f_{\theta_{m-1}}), \qquad (8)$$

where m denotes the iteration number and n denotes the task number. In the next iteration $m + 1$, the inner learning rate is updated to α_{m+1}.

3.3   Representativeness Weighting

We calculate representativeness scores of sentences by measuring how well a sentence pair can represent its domain. A premise is that in each domain, tasks are in the same data distribution. To alleviate this hypothetical bias, sentences with higher scores are assigned with higher weights in (6). $w_{rp,i}$ is calculated by Wee's method [25] as

$$w_{rp,i} = \frac{sc_i - sc_{min}}{sc_{max} - sc_{min}}, \qquad (9)$$

where sc_{min}, sc_{max} denote the minimum and maximum scores among all sentences in the same domain respectively. The representativeness score sc_i is normalized to the range from 0 to 1.

Language Model Cross-entropy. Inspired by [23], we regard bilingual Language Model (LM) cross-entropy (ce) difference as the representativeness metric. Both source and target monolingual data are used for building language models respectively. It's formulated as

$$sc_i = H_G(\mathbf{x}_i) - H_{in}(\mathbf{x}_i) + H_G(\mathbf{y}_i) - H_{in}(\mathbf{y}_i). \qquad (10)$$

Note that $H_G(\mathbf{x}_i)$ represents the cross-entropy between source sentence x_i and the general-domain LM. Similarly, $H_{in}(\mathbf{y}_i)$ represents the cross-entropy between the target sentence y_i and the in-domain LM. Sentences with higher scores can better describe the in-domain data distribution. Each task is required to be tagged with a domain tag in advance.

Representation Distance Difference. We also use distance differences between sentence-to-in-domain representations and sentence-to-general-domain representations to score the representativeness. The Universal Sentence Encoder (USE) [5] is used to convert a variable-length sentence into an N-dimension sentence representation x_i, y_i. The average of sentence representations in one domain is noted as the domain representation d_{in}, d_G. Bilingual distance differences are used to cover discrepancy from the source language and the target language. Similar to (10), it is rewritten as

$$sc_i = D(d_G, \mathbf{x}_i) - D(d_{in}, \mathbf{x}_i) + D(d_G, \mathbf{y}_i) - D(d_{in}, \mathbf{y}_i). \tag{11}$$

Take $D(d_G, \mathbf{x}_i)$ as an example, which denotes the distance between the general domain and the source sentence. Then we normalize sc_i by (9).

Respectively, we adopt Euclidean Distance (*ed*) and Jensen-Shannon Distance (*js*) to calculate $D(d_G, \mathbf{x}_i)$. Precisely, *ed* is calculated from the domain representation and the sentence representation, while *js* is calculated from probability distributions converted from those two vectors.

3.4 Confidence Weighting

Model uncertainty is often used to decide whether the model can describe the data distribution [3]. Intuitively, the higher uncertainty the model has, the harder it is to learn. Following the work [24], we employ Monte-Carlo dropout sampling [9] to approximate Bayesian inference. We get confidence scores of sentences in the query set by passing them to the decoder for k times. Meta-updating is weighted proportionately in (7). Precisely, we evaluate confidence weighting at both sentence level and word level.

Sentence Level. We define $P(\mathbf{y}_i \mid \mathbf{x}_i, \phi_n)$ as the sentence translation probability, where ϕ_n signifies the model parameter updated in the inner loop. Further, we estimate the sentence-level confidence weighting $w_{cf,i}$ using the compound of expectation and variance [24] using K samples,

$$w_{cf,i}(\mathbf{x}_i, \mathbf{y}_i, \phi_n) = \left(1 - \frac{\mathrm{Var}\left\{P\left(\mathbf{y} \mid \mathbf{x}_i, \phi_n^{(k)}\right)\right\}_{k=1}^{K}}{\mathbb{E}\left\{P\left(\mathbf{y}_i \mid \mathbf{x}_i, \phi_n^{(k)}\right)\right\}_{k=1}^{K}}\right)^{\beta}, \tag{12}$$

where $\phi_n^{(k)}$ denotes the k-th pass through ϕ_n by random deactivation. β is a hyperparameter for distinguishing confidence values among different sentences. In other words, sentences with higher confidence are rewarded with larger weights.

Word Level. Likewise, the word-level confidence can be calculated as

$$w_{cf,i,j}(\mathbf{x}_i, \mathbf{y}_{i,j}, \phi_n) = \left(1 - \frac{\mathrm{Var}\left\{P\left(\mathbf{y}_{i,j} \mid \mathbf{x}_i, \mathbf{y}_{i,<j}, \phi_n^{(k)}\right)\right\}_{k=1}^{K}}{\mathbb{E}\left\{P\left(\mathbf{y}_{i,j} \mid \mathbf{x}_i, \mathbf{y}_{i,<j}, \phi_n^{(k)}\right)\right\}_{k=1}^{K}}\right)^{\beta}. \tag{13}$$

If weighted word losses are directly summed up to the sentence loss, the update will stall. Consequently, instead of using word-level confidence alone, we put it on top of sentence-level weighting. To maintain the same loss scale as the previous model, we normalize the word-level confidence using *softmax*. When both $w_{cf,i}$ and $w_{cf,i,j}$ equal to 1, Eq. (7) describes the loss modified by sentence-word-level confidence.

Time-Relative Function. Since the initial model is uncertain about predictions at the early phase of adaptation, confidence weighting may lead to a slow or even rambling update. Hence we do not distinguish hesitant samples and confident samples at first, but focus more on adaptation. With training proceeding, we pay more attention to uncertainty gradually. In each epoch t, the time function is defined as

$$\lambda(t) = \min\left(1, \sqrt{\frac{t - t_0}{t_1 - t_0}}\right). \tag{14}$$

In the beginning, all sentences contribute in the same loss scale, where the confidence weighting equals to 1 constantly. After t_0 epochs, the contribution of confidence weighting begins to increase according to the square-root growing function. We renew the confidence weighting by multiplying it by $\lambda(t)$. After t_1 epochs, the confidence weighting is fully applied.

4 Experiments

4.1 Datasets

Experiments are conducted on English-German parallel datasets. We use 6.4M sentence pairs to get a pre-trained model. The training data are sampled from CommonCrawl, Europarl, NewsCommentary, Wikititles and Rapid2019. We also collect 10 domains from Open Parallel Corpus[1] for meta-training, including ECB, EMEA, ELRC, JRC, KDE4, MultiUN (UN), QED, Tanzil, TED and Wikipedia. The average number of tokens per sentence in different domains varies from 9.3 to 38.3. To avoid imbalanced data distribution across domains, we measure the size of tasks by tokens. Each batch contains one task and each task contains a support set with 1K tokens and a query set with 4K tokens. We sample 100 tasks from each domain separately and get a training set with a total of 1K tasks. Both English and German corpora are tokenized by a joint SentencePiece[2] with a vocabulary of 40K subword units.

4.2 Implementation Details

We use the Transformer [22] as the pre-trained model and implement our methods based on THUMT[3]. Parameters follow the base setting, where embedding size = 512, feed-forward filter size = 2048, head number = 8, encoder-decoder number = 6, except for dropout = 0.1 and batch size = 4096. We adopt the same Adam optimizer and default hyperparameter settings as paper [22] (learning rate = 7e−4, warm-up steps = 4000). The hidden size is set to 32 in adaptive layers. MAML includes a fast adaptation stage (the inner loop) and a meta updating stage (the outer loop). In the first stage, we freeze the base transformer model while updating parameters in adaptive layers to avoid catastrophic forgetting [10]. In the outer loop, all parameters are updated to guarantee that meta-knowledge is fully learned. We implement first-order MAML, where the

---

[1] http://opus.nlpl.eu.
[2] https://github.com/google/sentencepiece.
[3] https://github.com/THUNLP-MT/THUMT.

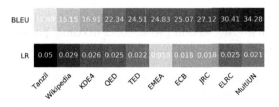

Fig. 3. Upper: color scales of the stable learning rates after dynamic-tuning in ten domains. **Bottom**: color scales of initial BLEU scores in ten domains. (Color figure online)

Table 1. Ablation results in methods and comparison in metrics.

Method	ECB	EMEA	ELRC	JRC	KDE4	UN	QED	Tanzil	TED	Wiki
Baseline										
MAML	33.81	44.27	43.96	38.75	21.40	45.86	28.44	18.27	29.41	15.97
1-dynamic tuning										
Shared α	**34.07**	44.59	43.75	38.94	21.59	45.84	**28.78**	18.43	29.90	15.87
Specific α	34.00	**44.62**	43.95	**39.07**	**21.61**	**46.32**	28.76	**18.65**	**30.13**	15.91
$\triangle$	+0.26	+0.35	–	+0.32	+0.21	+0.46	+0.34	+0.38	+0.72	–
2-representativeness weighting										
+ce	**33.87**	44.35	43.84	**39.13**	21.47	46.26	28.63	**18.67**	30.44	**16.15**
+ed	33.59	44.26	**44.04**	39.12	21.37	**46.38**	28.54	18.28	29.62	16.00
+js	33.61	**44.43**	43.85	39.05	**21.58**	46.30	**28.87**	18.58	29.94	15.98
$\triangle$	+0.06	+0.16	+0.10	+0.38	+0.18	+0.52	+0.43	+0.40	+1.03	+0.18
3-confidence weighting										
+sw	33.15	44.44	43.91	38.71	20.89	**49.04**	27.78	18.54	29.82	**16.09**
+s	**34.37**	**44.87**	**44.69**	**39.10**	21.19	46.42	**28.57**	**18.82**	30.10	15.93
$\triangle$	+0.56	+0.60	+0.73	+0.35	–	+3.18	+0.13	+0.55	+0.69	+0.12

inner-loop learning rate equals $1e-3$. Meta parameters are optimized in the outer loop, using the same optimizer as the pre-trained model, except for learning rate $= 5e-5$ and warm-up steps $= 3000$. For evaluation, We sample an adapting set with 1K tokens and a testing set with 3K sentences from each domain. We calculate BLEU [17] scores on testing sets to measure the effectiveness of meta-initialization.

4.3 Choices of Metrics

In this part, the experiments of three strategies are conducted respectively. Table 1 lists their performances after 10-epoch training. Results marked in bold are the best in the corresponding strategy relatively. Results that are no better than raw MAML are replaced with a dash.

Dynamic Tuning of Learning Rate. We adopt two tuning methods, including a sharing one among all domains (*shared* α) and a domain-specific one (*specific* α).

Both outperform raw MAML with a fixed learning rate. Especially, the latter one performs better on average and improve +0.72 maximal points in TED. Then we analyze relationships between learning rates and the difficulty of domain adaptation. Figure 3 illustrates two color scales of the stable learning rates after dynamic-tuning and initial BLEU scores in 10 domains. From light to dark, blue squares roughly represent the difficulty of adapting from easy to hard. Darker squares correspond to smaller learning rates. It's observed that as BLEU score increases, the corresponding learning rate decreases first and then rises. Empirically, for domains with high evaluation scores, higher learning rates allow larger gains from in-domain knowledge. Domains with low initial scores also require high learning rates. It is speculated that gains from new knowledge outweigh degradation from original knowledge. For other domains like EMEA, applying a conservative learning rate makes adaptation more robust.

Representativeness Weighting. For *ce* weighting, We reuse Kenneth's toolkit[4] to train LMs for each domain. We randomly sample 40K sentences as training subsets in the general domain and other ten domains separately. The same training dataset size and shared vocabulary ensure probability estimations given by LMs are not affected by extraneous factors. For *ed* and *js* metrics, we pre-train the model[5] on a general-domain dataset and get representation scores of sentences. As Table 1 shows, in most domains, representativeness weighting in three metrics improves the performances of raw MAML. Specifically, the improvement of scores in TED using *ce* metric achieves the best by +1.03 BLEU points. Among the three metrics, *ce* generally shows the best enhancement effect on average. Accordingly, we choose *ce* method as the best-performing representativeness weighting metric.

Confidence Weighting. We set forward propagation times to 10 in training time. In each iteration, we calculate the sentence-level (*conf-s*) and sentence-word-combined-level (*conf-sw*) confidence weighting online on query sets. We can see from Table 1 that the *conf-s* method leads to a maximal improvement on ELRC by +0.73 points. However, the performance in domains like KDE4 suffers degrading. Because meta-updating slows down for those with good initial scores, which helps produce general improvement among all tasks. In other words, KDE4 is over-learned in raw MAML. After confidence weighting and balancing, there is no more significant improvement. Unexpectedly, although *conf-sw* weighting brings significant improvement by +3.18 scores in MultiUN, it also leads to erratic fluctuation over domains. Therefore, we choose *conf-s* weighting as the outer-loop weighting strategy.

4.4   Overall Results

The highest scores are in bold in Table 2. **No fine-tuning** is evaluated directly on the pre-trained Transformer. Classical **fine-tuning** [8] mixes all support and query sets in ten domains to guarantee no additional data are used. Unsurprisingly, MAML outperforms the others with an average improvement of +2.94

---

[4] https://github.com/kpu/kenlm.
[5] https://tfhub.dev/google/universal-sentence-encoder-xling/en-de/1.

Table 2. Baseline and main results.

Method	BLEU	$\triangle_{Avg.}$	$\triangle_{Max.}$
Baseline			
No fine-tuning	23.21	–	–
Fine-tuning [8]	29.07	–	–
Mixed fine-tuning [6]	29.59	–	–
MAML [7]	32.01	–	–
Separate method (Ours)			
MAML+1 (specific α)	32.30	+0.29	+0.72
MAML+2 (rep-ce)	32.28	+0.27	+1.03
MAML+3 (conf-s)	32.41	+0.39	+0.73
Joint method (Ours)			
MAML+1,2	32.44	+0.43	+0.95
MAML+1,2,3	**32.71**	**+0.70**	**+1.35**

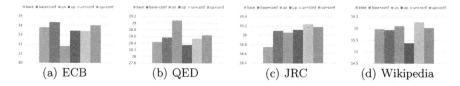

(a) ECB   (b) QED   (c) JRC   (d) Wikipedia

Fig. 4. BLEU scores in six different training settings. Blue columns represent balanced baseline settings (**base,base-conf**). Red columns reflect under-sampling and up-sampling settings (**un,up**). Green columns indicate the same settings in addition to the confidence weighting (**un-conf,up-conf**). (a) shows results in the domain with less training data, while (b) (c) (d) shows results in domains with rich training data. (Color figure online)

points compared to **mixed fine-tuning** [6]. As results in the last two column reveal, the three selected methods are all effective and efficient. The combination of 1 and 2 further improves MAML by up to +0.95 scores. After combining the confidence weighting, the joint method achieves a significant improvement among all domains by +1.35 points in maximum and +0.7 points on average. It demonstrates the three methods are complementary to each other.

4.5   Analysis of Domain Balancing

An imbalanced task distribution of training sets may lead to translation deteriorate. So we sample an equal number of training tasks in all domains to get through a balanced meta-updating process as the baseline (**base**). We conduct comparison experiments to prove the proposed strategy (**conf**) helps mitigate the degradation from multi-domain imbalance. To simulate uneven task distribution in real-world data, we keep an equal number of training tasks per domain except ECB, from which only half of the original number of tasks are sampled (**un**). We also up-sample ECB tasks to the same size as those in other domains (**up**). The final results are almost equal to the baseline. Besides, we analyze three other domains with abundant training data of the equal size.

As Fig. 4 shows, for ECB, **un**'s performance of raw MAML suffers great degradation. For other domains, under-sampling from ECB yields high performance on them. Empirically, the meta learner tends to draw less attention to domains with fewer shots and vice versa. Up-sampling in ECB helps alleviate the problem, as indicated in the dark red columns in Fig. 4(a) (b). However, Fig. 4(c) (d) indicate the manual up-sampling method is unstable to solve the domain imbalance problem. As shown in Fig. 4(b) (c) (d), when training with confidence weighting, the results in imbalanced and balanced data distributions are roughly the same. It is reasonable to conclude that our method can balance contributions across domains in a more intelligent and robust way than simple up-sampling. Besides, there is no significant effect on results whether up-sampling or not. It illustrates a dynamic training process, which guarantees balanced contributions from tasks. Thus, there is no excessive translation result fluctuation in specific domains.

5 Conclusion

In this work, we focus on the intrinsic defects of MAML and propose three novel improving strategies, including the dynamic tuning and the representativeness weighting in the inner loop, and the confidence weighting in the outer loop. We conduct experiments on ten domains and reveal our methods largely improve the few-shot adaptation results. They are easy to implement and combine. In particular, our strategies show the ability of alleviating imbalanced task distributions across varied domains automatically. This attribution offers the possibility of loosening the stringent prerequisites of an equal number of tasks in each domain and helps MAML to be further applied to a non-ideal environment.

Acknowledgement. This research work has been funded by the National Natural Science Foundation of China (Grant No. 61772337, U1736207).

References

1. Bapna, A., Firat, O.: Simple, scalable adaptation for neural machine translation. In: EMNLP/IJCNLP (1), pp. 1538–1548. Association for Computational Linguistics (2019)
2. Behl, H.S., Baydin, A.G., Torr, P.H.S.: Alpha MAML: adaptive model-agnostic meta-learning. CoRR abs/1905.07435 (2019)
3. Buntine, W.L., Weigend, A.S.: Bayesian back-propagation. Complex Syst. **5**(6), 603–643 (1991)
4. Cai, D., Sheth, R., Mackey, L., Fusi, N.: Weighted meta-learning. CoRR abs/2003.09465 (2020)

5. Cer, D., et al.: Universal sentence encoder for English. In: EMNLP (Demonstration), pp. 169–174. Association for Computational Linguistics (2018)
6. Chu, C., Dabre, R., Kurohashi, S.: An empirical comparison of simple domain adaptation methods for neural machine translation. CoRR abs/1701.03214 (2017)
7. Finn, C., Abbeel, P., Levine, S.: Model-agnostic meta-learning for fast adaptation of deep networks. In: ICML Proceedings of Machine Learning Research, vol. 70, pp. 1126–1135. PMLR (2017)
8. Freitag, M., Al-Onaizan, Y.: Fast domain adaptation for neural machine translation. CoRR abs/1612.06897 (2016)
9. Gal, Y., Ghahramani, Z.: Dropout as a Bayesian approximation: representing model uncertainty in deep learning. In: ICML JMLR Workshop and Conference Proceedings, vol. 48, pp. 1050–1059. JMLR.org (2016)
10. Gu, S., Feng, Y.: Investigating catastrophic forgetting during continual training for neural machine translation. In: COLING, pp. 4315–4326. International Committee on Computational Linguistics (2020)
11. Houlsby, N., et al.: Parameter-efficient transfer learning for NLP. In: ICML, Proceedings of Machine Learning Research, vol. 97, pp. 2790–2799. PMLR (2019)
12. Killamsetty, K., Li, C., Zhao, C., Iyer, R.K., Chen, F.: A reweighted meta learning framework for robust few shot learning. CoRR abs/2011.06782 (2020)
13. Kobus, C., Crego, J.M., Senellart, J.: Domain control for neural machine translation. In: RANLP, pp. 372–378. INCOMA Ltd. (2017)
14. Lee, H., et al.: Learning to balance: Bayesian meta-learning for imbalanced and out-of-distribution tasks. In: ICLR. OpenReview.net (2020)
15. Li, R., Wang, X., Yu, H.: MetaMT, a meta learning method leveraging multiple domain data for low resource machine translation. In: AAAI, pp. 8245–8252. AAAI Press (2020)
16. Li, Z., Zhou, F., Chen, F., Li, H.: Meta-SGD: learning to learn quickly for few shot learning. CoRR abs/1707.09835 (2017)
17. Papineni, K., Roukos, S., Ward, T., Zhu, W.: BLEU: a method for automatic evaluation of machine translation. In: ACL, pp. 311–318. ACL (2002)
18. Ren, M., Zeng, W., Yang, B., Urtasun, R.: Learning to reweight examples for robust deep learning. In: ICML, Proceedings of Machine Learning Research, vol. 80, pp. 4331–4340. PMLR (2018)
19. Sajjad, H., Durrani, N., Dalvi, F., Belinkov, Y., Vogel, S.: Neural machine translation training in a multi-domain scenario. CoRR abs/1708.08712 (2017)
20. Sharaf, A., Hassan, H., III, H.D.: Meta-learning for few-shot NMT adaptation. In: NGT@ACL, pp. 43–53. Association for Computational Linguistics (2020)
21. Thompson, B., Gwinnup, J., Khayrallah, H., Duh, K., Koehn, P.: Overcoming catastrophic forgetting during domain adaptation of neural machine translation. In: NAACL-HLT (1), pp. 2062–2068. Association for Computational Linguistics (2019)
22. Vaswani, A., et al.: Attention is all you need. In: NIPS, pp. 5998–6008 (2017)
23. Wang, R., Utiyama, M., Liu, L., Chen, K., Sumita, E.: Instance weighting for neural machine translation domain adaptation. In: EMNLP, pp. 1482–1488. Association for Computational Linguistics (2017)
24. Wang, S., Liu, Y., Wang, C., Luan, H., Sun, M.: Improving back-translation with uncertainty-based confidence estimation. In: EMNLP/IJCNLP (1), pp. 791–802. Association for Computational Linguistics (2019)
25. van der Wees, M., Bisazza, A., Monz, C.: Dynamic data selection for neural machine translation. In: EMNLP, pp. 1400–1410. Association for Computational Linguistics (2017)

Improving Transfer Learning in Unsupervised Language Adaptation

Gil Rocha$^{(\boxtimes)}$ and Henrique Lopes Cardoso

Laboratório de Inteligência Artificial e Ciência de Computadores (LIACC),
Faculdade de Engenharia, Universidade do Porto, Porto, Portugal
{gil.rocha,hlc}@fe.up.pt

Abstract. Unsupervised language adaptation aims to improve the cross-lingual ability of models that are fine-tuned on a specific task and source language, without requiring labeled data on the target language. On the other hand, recent multilingual language models (such as mBERT) have achieved new state-of-the-art results on a variety of tasks and languages, when employed in a direct transfer approach. In this work, we explore recently proposed unsupervised language adaptation methods – Adversarial Training and Encoder Alignment – to fine-tune language models on a specific task and language pairs, showing that the cross-lingual ability of the models can be further improved. We focus on two conceptually different tasks, Natural Language Inference and Sentiment Analysis, and analyze the performance of the explored models. In particular, Encoder Alignment is the best approach for most of the settings explored in this work, only underperforming in the presence of domain-shift between source and target languages.

Keywords: Deep neural networks · Natural Language Processing · Transfer learning · Unsupervised language adaptation

1 Introduction

Recent language models have obtained impressive results on a variety of NLP tasks, both on language-specific and cross-lingual settings [6]. Several recent studies have pointed out the remarkable ability of pre-trained multilingual language models on a variety of target languages, including less-resourced ones [9,21]. The usual approach consists on fine-tuning such language models for a specific task and language. For cross-lingual settings, direct transfer [11] (also known as zero-shot) is commonly used, relying on the resilience of language models in terms of their cross-lingual abilities, even after performing fine-tuning on a downstream task based on a source language.

In this paper, we investigate the potential of unsupervised language adaptation methods, namely Adversarial Training [3] and Encoder Alignment [4], to further improve the cross-lingual ability of language models. In unsupervised language adaptation, annotated resources on a source language S are available,

© Springer Nature Switzerland AG 2021
I. Farkaš et al. (Eds.): ICANN 2021, LNCS 12895, pp. 588–599, 2021.
https://doi.org/10.1007/978-3-030-86383-8_47

in the form $\langle X_S, Y_S \rangle$, where X_S stands for an input sequence in the source language and Y_S to the corresponding class label (annotation). For a target language T, however, no annotations are assumed to exist for training machine learning models with. The goal is to learn representations that are useful to perform a given task on S, while being also useful to perform the same task in the target language T (or even across multiple languages). We employ the aforementioned methods in two conceptually different tasks: Natural Language Inference (NLI) and Sentiment Classification. Furthermore, we explore the usage of in and out of domain data for adversarial training, and analyze the impact of this choice.

Our main contributions are: (a) we apply unsupervised language adaptation models on two NLP problems of different formulations – single sentence (Sentiment Classification) and sentence pair classification (NLI) – and show that these methods coupled with state-of-the-art multilingual language models can improve the cross-lingual ability of the models on different tasks; (b) we show how differences in the encodings proposed to tackle the NLI task impact the cross-lingual ability of the employed methods; (c) we analyze the impact of using different types of unlabeled data in adversarial training (in-domain vs out-domain data).

2   Related Work

NLI is one of the most widely used tasks to evaluate natural language understanding capabilities. Given two text fragments, "Text" (T) and "Hypothesis" (H), NLI aims at determining whether the meaning of H is *entailed, contradicted* or is *neutral* in relation with T [5]. To capture semantic-level relations between T and H, two generic formulations have been proposed. In the Cross-Encoder setting, each learning instance is represented in a single input sequence, where T and H are concatenated with a special delimiter token [6,16]. In the Siamese-Encoder setting, T and H are encoded separately. To capture the relations between T and H, sentence aggregation functions and attention mechanisms have been successfully applied [1,15]. Cross-lingual approaches have been used in NLI, namely by exploiting parallel corpora [12] and lexical resources [2]. These systems rely on multilingual resources and machine translation to explore projection [23] or direct transfer [11] approaches. Recently, a large-scale corpus for NLI covering 15 languages has been released, together with multilingual sentence encoder baselines [4].

Sentiment classification has also been addressed in cross-lingual settings, using similar techniques [8,18]. Some works explore parallel data resources to learn bilingual document representations [25] or to perform cross-lingual distillation [22]. Adversarial Deep Averaging Networks (ADAN) [3] have been proposed to avoid the need for parallel data, and employ adversarial training in an unsupervised language adaptation approach.

Following new trends in language modeling techniques, new state-of-art results have been obtained in several tasks, including NLI [6]. Multilingual language models are pre-trained in a similar way as their monolingual counterparts, without requiring specific cross-lingual objectives and alignment between languages. These models have shown impressive capabilities to generalize across a wide range of languages for a variety of downstream tasks [9, 21].

3   Methods

To address the task of unsupervised language adaptation, we explore: Adversarial Training and Encoder Alignment. In general, following an unsupervised language adaptation setting, in the training phase the model is fed with labeled data on the source language, but labeled data on target language is not available. To encourage the model to generalize well in cross-lingual settings, unlabeled data on the source and target languages are provided in the training phase.

3.1   Adversarial Training

The aim of adversarial training is to augment a neural network employed to address a task, with an adversarial component that constrains the model to satisfy a second objective. In a cross-lingual setting, we aim to make the model agnostic to the input language while learning to address a specific task.

As shown in Fig. 1, a neural network with adversarial training is composed of: a *Feature Extractor* $\mathcal{F}$ that maps the input sequence x to a feature space $\mathcal{F}(x)$, a *Task Classifier* $\mathcal{P}$ that given the feature representation $\mathcal{F}(x)$ predicts the labels for the task at hand, and a *Language Discriminator* $\mathcal{Q}$ that also receives $\mathcal{F}(x)$ as input and aims to discriminate the language of the input sequence. Given this setup, the two objectives employed in the neural network (*i.e.* the fine-tuning objective on a specific task applied to $\mathcal{F}$ and $\mathcal{P}$ and the adversarial objective applied to $\mathcal{F}$ and $\mathcal{Q}$) are meant to obtain an input representation in $\mathcal{F}$ that is suitable for the task at hand and at the same time transferable across languages.

Chen *et al.* [3] propose to minimize the Wasserstein distance $\mathcal{W}$ between the feature distributions of source and target instances according to the Kantorovich-Rubinstein duality [19], and show that this improves stability for hyperparameter selection. The adversarial component aims to maximize the following loss:

$$\mathcal{L}_{adv} \equiv \max_{\theta_q}(\mathbb{E}\left[\mathcal{Q}(\mathcal{F}(x_{src}))\right] - \mathbb{E}\left[\mathcal{Q}(\mathcal{F}(x_{tgt}))\right]) \tag{1}$$

where $\mathbb{E}$ corresponds to the average expected distribution.

For the task classifier component $\mathcal{P}$, we aim to minimize the negative log-likelihood of the target label for each source language example:

$$\mathcal{L}_{task} = -\sum_{i=0}^{N_{src}} y_i^{src} \cdot \log \hat{y}_i^{src}, \tag{2}$$

where y_i^{src} corresponds to the one-hot encoding of the class label for source input i and $\hat{y}_i^{src}$ to the softmax predictions of the model: $\hat{y}_i^{src} = \mathcal{P}(\mathcal{F}(x_i^{src}))$.

Finally, the goal of training the complete neural network is to minimize both the task classifier and adversarial component losses:

$$\mathcal{L}_{ADAN} = \mathcal{L}_{task} + \lambda(t)\,\mathcal{L}_{adv}, \tag{3}$$

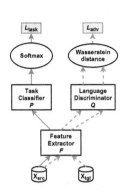

Fig. 1. ADAN architecture.

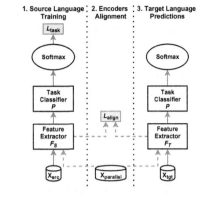

Fig. 2. EncAlign architecture.

where the hyper-parameter λ balances the importance of the adversarial component in the overall loss. Differently from Chen *et al.* [3], who use a constant value $\lambda(t) = 0.01$, we employ a λ schedule that increases with the number of epochs, which is used to make the adversarial component more important along time, while keeping a good performance on the task. Following Ganin and Lempitsky [7], λ starts at 0 and is gradually increased up to 1:

$$\lambda(t) = \frac{2}{1 + \exp(-10t)} - 1, \tag{4}$$

where t corresponds to the percentage of training completed.

3.2 Encoder Alignment (EncAlign)

The Encoder Alignment (EncAlign) method aims to align the encoder for the target language based on a fine-tuned encoder on the source language [4]. In this method, the target encoder learns to copy the source encoder representation based on parallel sentences, relying on the assumption that representations captured by the source encoder (which was previously fine-tuned for the task at hand) should be similar for the target encoder in high-dimensional space.

As shown in Fig. 2 the architecture is composed of a *feature extractor for the source language* $\mathcal{F}_S$ that maps input sequence x_{src} to a feature space $\mathcal{F}_S(x_{src})$, a *feature extractor for the target language* $\mathcal{F}_T$ that maps x_{tgt} to $\mathcal{F}_T(x_{tgt})$, and a *task classifier* $\mathcal{P}$ that given the feature representation $\mathcal{F}(x)$ predicts the labels for the task at hand.

EncAlign includes three steps: (a) fine-tuning using labeled data on the source language, (b) aligning source and target encoders with parallel sentences, and (c) inference on the target language. The first step corresponds to the conventional monolingual fine-tuning procedure: $\mathcal{F}_S$ and $\mathcal{P}$ are trained using labeled data in the source language. In the second step, the goal is to align a target encoder $\mathcal{F}_T$ based on the source encoder $\mathcal{F}_S$ learned in the previous step. Since we follow an unsupervised language adaptation setting, this procedure is performed with parallel sentences external to the task at hand (Sect. 4), as follows: given parallel sentences in the source and target languages, z_{src} and z_{tgt}, $\mathcal{F_T}$ learns to represent the input sequence z_{tgt} as close as possible in the feature space to the representation produced by $\mathcal{F_S}$ for the parallel sentence z_{src}. The alignment loss $\mathcal{L}_{align}$, proposed by Conneau et al. [4], is employed in this setting as follows:

$$\begin{aligned} \mathcal{L}_{align} = {} & dist(\mathcal{F_S}(z_{src}), \mathcal{F_T}(z_{tgt})) \\ & - \eta(dist(\mathcal{F_S}(z_{src}^{neg}), \mathcal{F_T}(z_{tgt})) + dist(\mathcal{F_S}(z_{src}), \mathcal{F_T}(z_{tgt}^{neg}))), \quad (5) \end{aligned}$$

where $(z_{src}^{neg}, z_{tgt}^{neg})$ correspond to negative examples (i.e. z_{src}^{neg} was randomly sampled from the parallel sentences dataset and does not correspond to a parallel sentence of z_{tgt}; similarly between z_{tgt}^{neg} and z_{src}), and η controls the weight of the negative sampling in the overall loss (following prior work [4], we use $\eta = 0.25$). As distance metric, we use the Euclidean distance $dist(x, y) = \|x - y\|_2$. When the alignment of the source and target encoder is completed, we can make predictions on the target language without further training (third step). The neural network employed to perform such predictions is composed of $\mathcal{F_T}$ and $\mathcal{P}$, resulting in: $\mathcal{P}(\mathcal{F_T}(x_{tgt}))$.

4   Corpora

NLI. The Cross-Lingual NLI corpus (XNLI) [4] is a large-scale corpus containing annotations for 15 languages. Each pair of sentences is annotated with one of three labels: Entailment, Contradiction or Neutral.

Sentiment Classification. We replicate the collection of datasets used by Chen et al. [3], covering English, Chinese and Arabic. For English, we use a balanced collection of $700k$ Yelp reviews [24] with ratings provided as labels (1–5), of which $650k$ reviews are used as the training set. To report in-language scores, we randomly split the original $50k$ reviews in the validation set into $25k$ for the test set and the remaining for the validation set (keeping label distributions).

The Chinese dataset is composed of $170k$ balanced hotel reviews [10], annotated with labels 1 to 5. A total of $10k$ reviews are used for validation, and the same amount as test set. As in-domain target language data used for Adversarial Training, we use the remaining $150k$ reviews (ignoring their labels).

For Arabic, we use the BBN Arabic Sentiment Analysis dataset [13], which contains 1200 sentences (600 validation + 600 test) from social media posts annotated with one of 3 possible labels ($-$, 0, $+$). We map the 1–5 labels from the English dataset into these labels: 4 and 5 are mapped to $+$, 1 and 2 to $-$, and 3 to 0. As unlabeled target language data, we use the validation set (without labels) during training (similar to Chen et al. [3]).

Parallel Sentence Resources. In all experiments reported in Sect. 5, the source language is English (*en*). As target languages, we report results on Chinese (*zh*), Arabic (*ar*), and German (*de*). Chinese and Arabic were chosen for being the available target languages on the Sentiment Classification corpora; for NLI, German was also chosen to study the effect of transfer learning across languages from the same family. To learn the alignment between English and target languages, as employed in the EncAlign method, we use public parallel sentence resources. We also use them as unlabeled out-of-domain large-scale data for the Adversarial Training method (even though it does not require parallel sentences). Similarly to Conneau *et al.* [4], we use the United Nations (UN) corpus [26] for Arabic and Chinese, and the Europarl corpus for German. For all experiments, we set the number of parallel sentences to 2 million.

5 Experiments

To evaluate the methods described in Sect. 3 for unsupervised cross-lingual settings, we have designed experiments on two conceptually different tasks: Cross-Lingual NLI (XNLI) and Sentiment Classification. The code for replicating our experiments is publicly available.[1]

5.1 Implementation Details

We use case-preserving WordPiece tokenization, as followed in the BERT implementation. No other pre-processing operations were involved.

For the Feature Extractor components $\mathcal{F}$, we employ the mBERT architecture [6] from HuggingFace library [20], using as fixed size sentence-level representation the 768 hidden units corresponding to the last Self-Attention layer for the "CLS" (classification) token. As suggested by Devlin *et al.* [6], this corresponds to the aggregate representation of the input sequence for classification tasks, which can be used even when the input sequence is composed of two different sentences. For the Task Classifier $\mathcal{P}$ and Language Discriminator $\mathcal{Q}$ components, we use a single fully-connected layer with dimension 128, followed by a Softmax layer for $\mathcal{P}$. For the source language experiments, we stop training once accuracy on the validation set does not improve for 3 epochs (early-stop criterion) or when 30 epochs are completed. For the remaining experiments (Adversarial and Encoder Alignment), we use an early-stop criteria of 1 epoch. We use a batch size of 16 for source language and ADAN training, and 64 for EncAlign.

In XNLI experiments, input sequences were trimmed to 128 tokens for the Cross-Encoder setting (similar to Devlin *et al.* [6]) and to 64 tokens in each sentence for the Siamese-Encoder setting. For the Sentiment Classification experiments, input sequences were trimmed to 100 tokens.

---

[1] https://github.com/GilRocha/icann2021-unsupervised-language-adaptation.

Table 1. XNLI accuracy scores for the Siamese-Encoder setting

Method	en	zh	ar	de
Direct Transfer	71.14	63.16	57.11	62.67
Adversarial (out-domain)		63.11	57.26	62.86
Adversarial (in-domain)		62.99	57.44	61.57
Encoder Alignment		**66.09**	**63.66**	**66.91**

Table 2. XNLI accuracy scores for the Cross-Encoder setting

Method	en	zh	ar	de
Direct Transfer [6]	81.48	63.80	62.10	**70.50**
Adversarial (out-domain)		68.15	62.65	68.92
Adversarial (in-domain)		68.09	**62.76**	68.47
Encoder Alignment		**68.50**	62.10	69.70

5.2 Analysis of Results for XNLI

Experimental results for the XNLI task are shown in Table 1 for the Siamese-Encoder setting, and in Table 2 for the Cross-Encoder setting. The EncAlign method is as described in Sect. 3.2. Adversarial (out-domain) and Adversarial (in-domain) correspond to the method presented in Sect. 3.1, the former using out-of-domain data in the language discrimination task, and the latter making use of in-domain data (automatic machine translations provided in the XNLI dataset). Direct Transfer [11] consists of employing a model fine-tuned on the source language to make predictions on the target language, without further training, and corresponds to current state-of-the-art approaches [6]. Since reporting single performance scores is insufficient to compare non-deterministic learning approaches [14], we report averages scores of 3 runs with different random seeds. The metric used is accuracy, given that all labels are equally represented.

Siamese-Encoder Setting. For Siamese-Encoding, Encoder Alignment obtains the best overall scores on all target languages. These scores surpass by a large margin the existing state-of-the-art (Direct Transfer).

We stress out that the Encoder Alignment approach requires a substantial amount of parallel sentences between the source and target languages, which might be difficult to obtain, with adequate quality, for some language pairs. However, taking into account the target languages explored in this work, we consider this a feasible approach even across different families of languages.

Adversarial Training present scores that are similar to Direct Transfer, meaning that the training procedure performed on top of the fine-tuned model on the source language did not contribute to obtain substantially better scores for the target languages. We noticed that the employed multilingual pre-trained language models yield sentence-level representations in different languages that are close to language-agnostic representations. To come to this conclusion, we followed the approach from Chen et al. [3] to analyze whether $\mathcal{F}$ layer can distinguish the input language. To this end, 500 sentences were randomly sampled from the validation set of each of the languages, and then we employed t-SNE with Principal Component Analysis (PCA) to reduce the representation into a two dimensional feature space. The Averaged Hausdorff Distance (AHD) [17] measures the distance between two sets of points, and the lower AHD observed values indicate that the $\mathcal{F}$ layer cannot distinguish the input language. Given the strong cross-lingual abilities of the pre-trained language models, we hypothesize that techniques such as adversarial training cannot provide additional cross-lingual ability because the auxiliary objective (to make $\mathcal{F}$ language agnostic) is already close to optimal from the outset.

Within Adversarial Training results, we observe that providing in-domain unlabeled data did not contribute for a better adaptation of the model on the target languages, yielding slightly worst results in most cases. As detailed in Sect. 3.1, Adversarial Training aims to make the model agnostic towards a specific characteristic of input sequences (in our case, their languages). Given that XNLI data contains learning instances from a variety of genres, we hypothesized that the fine-tuned model on the source language is already robust to the domain shift and, consequently, the domain of the unlabeled data provided during adversarial training is not a relevant factor. To validate this hypothesis, we designed a cross-domain experiment in which we employed instances from a single domain/genre (e.g. government) as the main task's (both train and test) data. Then, two experiments using Adversarial Training were performed, where we used as unlabeled data for the language discrimination auxiliary task (a) learning instances from the same genre (e.g. government) – Adversarial (in-domain), or (b) learning instances from another genre (e.g. telephone) – Adversarial (out-domain). The results obtained show that there is no significant difference in the overall scores, which refutes our initial hypothesis. While the genre of the data used for the language discrimination auxiliary task is not a relevant factor for language adaptation, it seems that this is not due to the cross-domain capabilities of the pre-trained model, but instead to some other phenomena.

Cross-Encoder Setting. For the source language (en), the Cross-Encoder setting works better than the Siamese-Encoder. BERT models benefit from jointly modeling T and H, being able to use attention and the special encoding of the input sequence (*i.e.* [CLS] T [SEP] H [SEP]) to capture the semantic relations between T and H for the NLI task.

For cross-lingual transfer, Direct Transfer scores are higher in the Cross-Encoder setting (+0.64% for zh, +4.99% for ar, and +7.83% for de), which we correlate with the higher scores obtained in the source language. On the Cross-Encoder setting, Direct Transfer is competitive compared to the remaining methods, with an exception for the zh in which all methods provide significant improvements. As previously detailed, compared to the Siamese-Encoder, Direct Transfer with Cross-Encoder was unable to boost the scores for zh as observed with the other languages, improvements that are only observed when employing the methods for unsupervised language adaptation. The Adversarial (in-domain) approach works on par with Adversarial (out-domain), for reasons we leave for future work to investigate, as detailed for the Siamese-Encoder setting. Based on a similar additional cross-genre experiment, we conclude that the differences in the observed scores for the cross-encoding setting are not significantly different. EncAlign obtains slightly better scores, within unsupervised language adaptation models, for *zh* and *de*, and lower for *ar*.

Siamese Vs Cross-Encoder. The Cross-Encoder provides the best scores, except for *ar* when using the EncAlign. In general, unsupervised language adaptation methods are able to improve the scores more consistently in the Siamese-Encoder setting. Remarkably, even with lower baseline scores obtained via Direct Transfer, after employing unsupervised language adaptation methods, the best scores obtained with the Siamese-Encoder (via EncAlign) are competitive to the best scores obtained with the Cross-Encoder (EncAlign for *zh* and *de*, and Adversarial (in-domain) for *ar*).

In sum, EncAlign is the unsupervised language adaptation method that provides the best overall results in most settings. Compared to the state-of-the-art results reported for this dataset (see Direct Transfer in Table 2), unsupervised language adaptation methods obtain better scores for *zh* and *ar* in both encodings and remain below for *de* with the Cross-Encoder, by a small margin (-0.8% in accuracy). These results support our hypothesis that unsupervised language adaptation methods are able to further improve the cross-lingual ability of language models. Given that the Cross-Encoder obtains better scores for the NLI task on the source language, improving the cross-lingual ability on this setting seems to be a promising direction.

5.3 Analysis of Results for Sentiment Classification

The results for Sentiment Classification are shown in Table 3. ADAN refers to Chen *et al.* [3], employing multilingual word embeddings and Deep Averaging Networks for the feature extractor component $\mathcal{F}$. The remaining architectures follow settings that are similar to the ones detailed in Sect. 5.2.

In the 5 labels setting, the dataset is balanced. In the 3 labels setting, classes are unbalanced in both target languages. We use accuracy as the evaluation metric, in order to allow comparing results with prior work, even though macro averaged metrics would be more suitable.

For Chinese, Adversarial Training (out-domain) and EncAlign methods obtained slightly worst scores compared to Direct Transfer. Adversarial (in-domain) yielded the best overall scores (substantially better than Direct Transfer).

Using EncAlign is not as promising, comparing with the NLI task, where both source and target languages share the domain (even if the XNLI dataset is composed of different domains, they overlap between languages). In the Sentiment Classification task, we perform the alignment of the target language feature extractor to the source language feature extractor (trained on Yelp reviews in English) and then ask the system to perform predictions on a different language and domain (Chinese hotel reviews). Such "hard alignment" between encoders is deemed to suffer from the domain shift in the target language.

As detailed in Sect. 3.1, Adversarial Training aims to make the model agnostic regarding a specific characteristic of input sequences (in our case, the language).

Table 3. Sentiment Classification accuracy scores

	5 labels		3 labels	
Method	*en*	*zh*	*en*	*ar*
ADAN [3]	–	42.49	–	54.54
Direct Transfer	62.29	39.26	79.71	52.83
Adversarial (out-domain)		37.88		49.78
Adversarial (in-domain)		**44.84**		53.05
Encoder Alignment		38.61		**55.72**

Intuitively, if the unlabeled data used to make the model agnostic to the input language comes from the same domain as the evaluation data, then the model will generalize well for the evaluation phase in the target language, as observed with the scores obtained for the Adversarial (in-domain) approach. However, if the data provided in the training and evaluation phases differ in other phenomena (such as domain, genre, or style), the model might not generalize well, as observed for the Adversarial (out-domain) approach. Finally, the scores obtained employing Adversarial (in-domain) surpass the scores reported using multilingual word embeddings (ADAN, in Table 3), indicating that approaches based on language models are able to improve the cross-lingual performance in this particular setup.

For Arabic, EncAlign obtains the best overall scores. As indicated above, the amount of labeled data for the *ar* language is very limited. Consequently, approaches that rely on in-domain resources lack a significant quantity of data to further improve the scores. Similar to *zh*, the result obtained for *ar* using Adversarial Training (out-domain) suffers from the additional domain shift between training and test data. In this case, Encoder Alignment is the only proposed method that improves the scores on the target language, reaching a new state-of-the-art in this setting.

6 Conclusions and Future Work

We have experimentally analyzed the efficacy of recently proposed unsupervised language adaptation methods (Adversarial Training and Encoder Alignment) on improving the cross-lingual ability of language models that are fine-tuned on a specific source language. We report our findings on two NLP tasks that require different problem formulations.

Our empirical evaluation allows us to conclude that the interplay of different factors decides the optimal configuration for a particular setup. Our main findings are the following: EncAlign is the most promising approach. We only found it to under-perform when there is a domain mismatch between the training set on the source language (for which we have labeled data) and the test set on the target language (in which we aim to evaluate the cross-lingual capabilities of the system). In that situation, Adversarial Training works better when making use of substantial amounts of in-domain unlabeled data.

Regarding encoding schemes employed for XNLI, the Cross-Encoder performs better for in-language and provides strong Direct Transfer baseline scores. For that reason, while unsupervised language adaptation techniques provide substantial improvements for the Siamese-Encoder over Direct Transfer, they do not show solid improvements with cross-encoding.

In future work, we aim to explore new approaches that can build on the strong baseline scores provided by the Cross-Encoder. We also envision to study the impact of these techniques on a wider spectrum of source and target languages. Finally, hyper-parameter tuning of different loss components is a challenging and computationally expensive task that we would like to explore in more detail.

Acknowledgment. Gil Rocha is supported by a PhD studentship (with reference SFRH/BD/140125/2018) from Fundação para a Ciência e a Tecnologia (FCT). This research is supported by LIACC (FCT/UID/CEC/0027/2020) and by project DARGMINTS (POCI/01/0145/FEDER/031460), funded by FCT.

References

1. Bowman, S.R., Angeli, G., Potts, C., Manning, C.D.: A large annotated corpus for learning natural language inference. In: Proceedings Conference Empirical Methods in Natural Language Processing, pp. 632–642. ACL, Lisbon, September 2015
2. Castillo, J.J.: A wordnet-based semantic approach to textual entailment and cross-lingual textual entailment. Int. J. Mach. Learn. Cybern. **2**(3), 177–189 (2011)
3. Chen, X., Sun, Y., Athiwaratkun, B., Cardie, C., Weinberger, K.Q.: Adversarial deep averaging networks for cross-lingual sentiment classification. TACL **6**, 557–570 (2018)
4. Conneau, A., et al.: XNLI: evaluating cross-lingual sentence representations. In: Proceedings of the EMNLP, pp. 2475–2485. ACL, Brussels (2018)
5. Dagan, I., Roth, D., Sammons, M., Zanzotto, F.M.: Recognizing Textual Entailment: Models and Applications. Synthesis Lectures on Human Language Technologies, Morgan & Claypool Publishers, San Rafael (2013)
6. Devlin, J., Chang, M.W., Lee, K., Toutanova, K.: BERT: Pre-training of deep bidirectional transformers for language understanding. In: Proceedings NAACL, pp. 4171–4186. ACL, Minneapolis June 2019
7. Ganin, Y., Lempitsky, V.: Unsupervised domain adaptation by backpropagation. In: Bach, F., Blei, D. (eds.) Proceedings International Conference on Machine Learning, vol. 37, pp. 1180–1189. PMLR, Lille (2015)
8. Kanclerz, K., Miłkowski, P., Kocoń, J.: Cross-lingual deep neural transfer learning in sentiment analysis. Procedia Comput. Sci. **176**, 128–137 (2020)
9. Karthikeyan, K., Wang, Z., Mayhew, S., Roth, D.: Cross-lingual ability of multilingual BERT: an empirical study. arXiv arXiv:1912.07840 (2020)
10. Lin, Y., Lei, H., Wu, J., Li, X.: An empirical study on sentiment classification of Chinese review using word embedding. In: Proceedings Pacific Asia Conference on Language, Information and Computation: Posters, pp. 258–266, Shanghai (2015)
11. McDonald, R., Petrov, S., Hall, K.: Multi-source transfer of delexicalized dependency parsers. In: Proceedings of the Conference on Empirical Methods Natural Language Processing, pp. 62–72. ACL, Edinburgh, July 2011

12. Mehdad, Y., Negri, M., Federico, M.: Using bilingual parallel corpora for cross-lingual textual entailment. In: Proceedings Association Computational Linguistics: Human Language Technologies, pp. 1336–1345. ACL, Portland (2011)
13. Mohammad, S.M., Salameh, M., Kiritchenko, S.: How translation alters sentiment. J. Artif. Int. Res. **55**(1), 95–130 (2016)
14. Reimers, N., Gurevych, I.: Reporting score distributions makes a difference: performance study of LSTM-networks for sequence tagging. In: EMNLP, pp. 338–348. ACL, Copenhagen (2017)
15. Reimers, N., Gurevych, I.: Sentence-BERT: sentence embeddings using Siamese BERT-networks. In: Proceedings EMNLP/IJCNLP, pp. 3982–3992. ACL, Hong Kong (2019)
16. Rocktäschel, T., Grefenstette, E., Hermann, K.M., Kociský, T., Blunsom, P.: Reasoning about entailment with neural attention. In: International Conference on Learning Representations, ICLR, San Juan, 2–4 May 2016
17. Shapiro, M., Blaschko, M.: On Hausdorff Distance Measures. Department of Computer Science, University of Massachusetts Amherst, Amherst (2004)
18. Thakkar, G., Preradovic, N.M., Tadic, M.: Multi-task learning for cross-lingual sentiment analysis. In: Proceedings of the 2nd International Workshop on Cross-lingual Event-centric Open Analytics, pp. 76–84 (2021)
19. Villani, C.: Optimal Transport: Old and New. Grundlehren der mathematischen Wissenschaften, Springer, Heidelberg (2008). https://doi.org/10.1007/978-3-540-71050-9
20. Wolf, T., et al.: Transformers: state-of-the-art natural language processing. In: Proceedings of the 2020 Conference on Empirical Methods in Natural Language Processing: System Demonstrations, pp. 38–45. ACL, Online October 2020
21. Wu, S., Dredze, M.: Beto, bentz, becas: The surprising cross-lingual effectiveness of BERT. In: Proceedings Empirical Methods Natural Language Processing and International Joint Conference Natural Language Processing, pp. 833–844. ACL, Hong Kong (2019)
22. Xu, R., Yang, Y.: Cross-lingual distillation for text classification. In: Proceedings of the 55th Annual Meeting of the Association for Computational Linguistics, vol. 1, pp. 1415–1425. ACL, Vancouver, July 2017
23. Yarowsky, D., Ngai, G., Wicentowski, R.: Inducing multilingual text analysis tools via robust projection across aligned corpora. In: Proceedings of the First International Conference on Human Language Technology Research (2001)
24. Zhang, X., Zhao, J., LeCun, Y.: Character-level convolutional networks for text classification. In: Proceedings International Conference on Neural Information Processing Systems. NIPS 2015, pp. 649–657. MIT Press, Cambridge (2015)
25. Zhou, X., Wan, X., Xiao, J.: Cross-lingual sentiment classification with bilingual document representation learning. In: Proceedings Annual Meeting Association Computational Linguistics, pp. 1403–1412. ACL, Berlin, August 2016
26. Ziemski, M., Junczys-Dowmunt, M., Pouliquen, B.: The united nations parallel corpus v1.0. In: Proceedings of the Tenth International Conference on Language Resources and Evaluation (LREC 2016), ELRA, Paris, May 2016

Sample-Label View Transfer Active Learning for Time Series Classification

Patrick Kinyua$^{(\boxtimes)}$ and Nicolas Jouandeau

Université Paris 8, PIF Department, Paris, France
patrick.gikunda@dkut.ac.ke, n@up8.edu

Abstract. In many real-world applications, Time Series Data are captured over the course of time and exhibit temporal dependencies that cause two or otherwise identical points of time to belong to different classes or exhibit different characteristic. Although time series classification has attracted increasing attention in recent years, it remains a challenging task considering the nature of data dimensionality, voluminousness and continuous updates. Most of existing Deep Learning methods often depend on hand-crafted feature extraction techniques, that are expensive for real-world time series data mining applications which in addition, require expert knowledge. In practice, training a quality classifier is highly dependent on large number of labeled samples which is mostly inadequate in real-world time series datasets. In this paper, we present a novel Deep Learning approach for time series classification problems, called Transfer Active Learning (TAL) which jointly evaluates informativeness and representativeness of a candidate sample-label pair. TAL learns to map each input into a latent space from both sample and sample-label views which is more effective. For similar tasks, TAL is able to reuse model skill with further reduction on feature extraction costs. Extensive experiments on both classification datasets and real-world prediction tasks demonstrate the efficiency of the proposed approach on exponential reduction of training cost.

Keywords: Transfer learning · Active learning · Time series classification

1 Introduction

The technological landscape is changing at high speed, with ability to capture, process and disseminate huge amount of time-based data faster than before. Increased demand for smart homes, self-drive cars, autonomous trading algorithms, smart transport network, smart farming, weather forecast etc., uses a datatype that measures how things change over time [1]. Its characterized by data points indexed in *time* order and defined as *Time Series Data* (TSD). The unique feature of TSD is that the new appended data does not overwrite previous

© Springer Nature Switzerland AG 2021
I. Farkaš et al. (Eds.): ICANN 2021, LNCS 12895, pp. 600–611, 2021.
https://doi.org/10.1007/978-3-030-86383-8_48

data entry in the database. Therefore, there is ability to track system behavior changes over time as database insert and not update. Collecting such event data can lead to large sized datasets over short period of time. On the other hand, managing or analyzing such huge amount of data for practical world application comes at great performance cost which requires high computation power and complex methods.

Deep Learning (DL) is a well known Machine Learning technique with promising results in many prediction tasks. In particular the Convolutional Neural Network (CNN) which sit at the core of most of the recent breakthroughs in computer vision and data mining related tasks [2]. This is attributed by their ability to learn hierarchies of abstract localized structured data [3]. Despite of the impressive performance in analyzing big data, deep learning methods requires vast amount of labeled data for training. The dependence on large training data necessitates to develop methods for cutting down the classifiers training cost. Research on techniques to reduce training costs include Active Learning (AL), Transfer Learning (TL) and other statistical methods. Despite successful use of DL for big data analysis, it has not widely been used on time series datasets. The reasons for this absence might be: a) it is only recently that DL was proven to work well for TSC and there is still much to be explored in building DL methods for mining TSD [4] and, b) there is a lack of big, general purpose TSC datasets like ImageNet or OpenImages.

TL is used to improve model performance by reusing skills acquired in other related tasks especially where the target task has inadequate training data e.g. a model trained to classify mangoes can be used to classify oranges. Currently TL is used in many real-world DL applications [5]. Training is first done on a ConvNet using a large dataset (e.g. ImageNet which contains 1.2M samples with 1000 classes), and then transferring the ConvNet either by fine-tuning or as fixed feature extractor on the target task. The performance of a classifier on the target task can greatly be improved by using its experience on similar tasks. The assumption is that the source task and the target task can share some hyper-parameters depending on their nature [6]. When the source and target tasks are unrelated, the knowledge transfer from source task may not be useful or even compromise the performance of a target task through a negative transfer.

In DL its considered that the model can maximize the learning performance by allowing it to choose the data points to learn from [7]. AL provides means for iterative selection of data points the model wants to learn from [8]. This means that for a classification task the model will not require all the data for training but instead pick the most effective data-points for training. This is done by evaluating either or both the informativeness or representativeness of a sample. Whereas random and uncertainty sampling are often used, they suffer especially when data has skewed categorical features which can result in selecting non-informative or redundant samples. Classical statistical theory techniques such as entropy and margin are used as an utility to measure sample informativeness, however, they often fails to capture the data distribution information [9]. In this paper, we propose a query selection strategy based on a combination of both uncertainty and information density. The querying strategy is then logically coupled with TL which we now refer as *Transfer Active Learning* (TAL).

The rest of the paper is organized as follows: Sect. 2 highlights related works on TL and AL; Sect. 3 presents our proposed Transfer Active Learning method; Sect. 4 describes the experimental setup and respective results are presented in Sect. 5; Sect. 6 presents our perspective and future interests.

2   Related Works

Time Series Classification (TSC) is a mapping task from the space of possible time based inputs to a probability distribution over the labels which can either be: a) an univariate time series $X = [x_1, x_2, ..., x_T]$ which is an ordered set of real values with the length of X being equal to the number of observable time-points T or; b) multiple time series $X = [X^1, X^2, ..., X^M]$ which consist of M observable per time-point with $X^i \in R^T$. We now define a time series dataset $D = (X_1, Y_1), (X_2, Y_2), ..., (X_N, Y_N)$ as a collection of pairs (X_i, Y_i) where X_i could either be univariate or multivariate with Y_i as its corresponding label. For a dataset containing K classes, the label vector Y_i is a vector of length K where each element $j \in [1, K]$ is equal to 1 if the class of X_i is j and 0 otherwise.

While the literature has sufficient works in which AL or TL is applied to TSD, a combination of the two remain comparatively unexplored. We attribute this to the fact that its only recent that TSC big datasets have began shaping up. In real-world TSC settings, the key motivation for use of AL is the dearth of training data. To account for lack of, or sparse training data, different techniques are used either together with user-intervention settings or a fully-model based. Considering the cost of training an effective model, methods such as random selection may result in models with unrepresentative truth of the actual data distribution. Therefore, it is particularly important to select samples that provide a distinct view of how the different classes are separated in the data with regions of greater uncertainty often sampled to define the decision boundaries [10].

Recurrent Neural Network(s) (RNN) are among the main methods used for time series forecasting. The recurrent based methods suffer the following limitations in the TSC tasks: a) they mainly predict an output for each time stamp in the time series [11]; b) when they train from long data series, they suffer vanishing gradient problem [12]; c) they have high computational requirement and hard to parallelize [13]. Successful application of DL in various domains motivates researchers to adopt different strategies to overcome RNN limitations in TSC tasks [4]. Majorly, these strategies are aimed at reducing the training data annotation cost using AL and transfer of model skill among tasks using TL.

Effectiveness AL strategies are based on sample selection techniques which include: a) uncertainty sampling with an objective to select an sample the classifier is most uncertain about as the most potential sample for labeling [14]; b) variance reduction aims at minimizing the model error rate by selecting samples with the minimum variance; c) expected gradient length that aims at querying sample that causes the maximal change to the current model [8]. The most common selection technique is the uncertainty sampling which considers the most uncertain sample for labeling [15]. In all cases the focus is to develop classification models with good generalization performance on unseen samples in the problem domain.

In representativeness-based approach the queried samples tend to provide a theoretical resemble of the overall distribution of the input space better. Unlike error-based methods which tend to improve the error behavior on the aggregate or uncertainty-based methods that query samples from most unknown regions, representativeness-based methods tend to avoid outlier-like samples by considering dense regions [15]. These methods combine both heterogeneity and representativeness behavior of the queried sample from the unlabeled set. Multiple approaches can be combined for query selection in active learning [16]. Such approaches exhibit: a) informative properties of the queried samples either close to the decision boundary of the learning model or far away from existing labeled samples in order to bring new knowledge about the feature space; b) representative behavior of the queried samples which should be less likely to be an outlier data and should be representative of the input space [16]. Natarajan and Laftchiev in their work combined TL and AL methods to predict personal thermal comfort. The method leverages domain knowledge from prior users and an AL strategy for new users that reduces the necessary size of the labeled dataset. When tested on real dataset from five users, their method achieves a 70% reduction in the required size of the labeled dataset as compared to the fully supervised learning approach [17]. The authors of [18] created an informativeness metric that considers the characteristics of time series data for defining instance uncertainty and utility. Their experiments on UCR archive presents a 50% reduction of training data. Using min-max approach the authors of [19] demonstrates viability of AL to reduce the annotation cost during training.

3 Transfer Active Learning

3.1 Transfer Learning

The TSC problem introduced in this paper is a semi-supervised classification problem. The two main strategies of deep TL include: a) using pre-trained models as feature extractors where objective is to leverage the pre-trained weight to extract features and only the final layer is replaced and; b) fine tuning pre-trained models by retraining some selected layers and freezing others. The question of weather to freeze pre-trained layers or use them as fixed feature extractor is determined by the size of available labeled set in the target settings that can be used for training. Under normal circumstance, when the labels in the target task are scarce, freezing is the best option to avoid overfitting. When the labels are sufficient then fine-tuning is a better choice. A DL model Θ with parameters and hyper-parameters can be represented as $Y \approx \Theta \vartheta(\theta|D)$ where θ are parameters and ϑ are hyper-parameters. D is training data and Y is class labels. The objective is to find estimate of parameters θ that optimizes some loss function L. The model performance based on loss function is dependent on ϑ, this implies that the parameters are also dependent on the hyper-parameters.

The parameters are learnt during training, but hyper-parameters are set of initial model variables set before start of training and they include: number and size of the ConvNet layers, learning rate, weight initialization, etc. A general TSC problem can be expressed with the following equation:

$$C_t = f(w \times X_{t-l/2:t+l/2} + b)|\forall t \in \{1, T\} \tag{1}$$

where C denotes the result of a convolution on a univariate time series X of length T with a filter w of length l, a bias parameter b and a final non-linear function f. To learn multiple discriminative features filters are applied on individual univariate time series. Using the same filter values w and b in ConvNets, we are able to find the result for all time stamps $t \in [1, T]$. This is possible by using weight sharing that enables the model to learn feature detectors that are invariant across the time array.

3.2   Sample Selection

We consider actively selecting samples in batches, where the selection must be constrained by some budget. Let x_i represents a sample and y_i where $y_i \in \{1, ...K\}$ represents the class label for x_i, class in D we use the class probability estimator to compute the estimate of a label. In order to avoid the problem of generalization of unseen samples and to learn an accurate model, we present a robust approach that uses both uncertainty and correlation utility. Since the boundary class regions are often those in which samples of multiple classes are present, they can be characterized by class label uncertainty or disagreements between different learners. However, this may not always be the case, because samples with greater uncertainty are not always representative of the data, and may lead to the selection of outlier data points. This situation is especially likely to occur in datasets that are very noisy. In order to address such issues, some models focus directly on the error itself, or try to find samples that are representative of the underlying data. Based on the application, analyst goal, and data distribution, different strategies have different tradeoffs and work differently.

Uncertainty Measure. In uncertainty sampling settings, the model attempts to select samples it is most uncertain about or what has not been seen so far. Application of the approach range from simple binary classification problem using Bayes classifier to multivariate classification using deep neural networks [10]. For the binary classification the naive Bayes classifier is normalized to ensure that the predicted probabilities sum up to 1. Therefore, the entropy objective function $En(X)$ for the binary class ($k = 2$) problem should be minimized and can be defined as follows:

$$En(\bar{X}) = \sum_{i=1}^{k} p_i - 0.5 \tag{2}$$

Great entropy value indicate greater uncertainty therefore the objective function should be maximized. In case of imbalanced data the uneven distributed classes are often associated with cost of misclassification denoted by w_i. Each probability p_i is replaced by a value proportional to $p_i \cdot w_i$, with the constant of the proportionality being determined by the probability values summing to 1. Given a label space Y the uncertainty measure f_u of independent and identically distributed sample considering the features and the label expressed as:

$$f_u(x) : \begin{cases} D^U \rightarrow R, & \text{(i) features view} \\ (D^U \times Y) \rightarrow R, & \text{(ii) features-label view} \end{cases} \quad (3)$$

to a real number space R. In 3: (i) the sample features are only considered in computing the sample uncertainty while in, (ii) a combination of sample label and features is considered in computing sample uncertainty. The three common metrics used to define uncertainty includes least confidence, sample margin and entropy. Least confidence consider the class label with the highest posterior probability with an objective function to decrease the error rate. Sample margin considers the first two most probable class labels using an objective function of decreasing the error rate. One major deficiency of both least confidence and sample margin is that they does not consider the output distributions for the remaining class labels in the set. Entropy considers class label over the whole output prediction distributions with an objective function to reduce the log-loss. Using entropy the uncertainty of an sample x_i in D^U can be defined as:

$$f_u(x_i) = \operatorname*{argmax}_i - \sum_i P(y_i|x_i) \log P(y_i|x_i) \quad (4)$$

where $P(y_i|x_i)$ denotes posterior probability of the sample x_i being a member of the ith class, which ranges over all possible labels y_i. For a binary classification task, the most potential samples are the ones with equal posterior probability with respect to all possible classes.

Correlation Measure. Majorly, AL query strategies use the uncertainty metric measure to evaluate the utility of a independent and identically distributed sample. However, when developing efficient AL methods, it is important to consider sample distribution information. The sample diversity information aids in querying most representative samples. This approach significantly improves the query performance while avoiding selecting outlier samples. Different algorithms for exploiting sample information exists. Majorly these algorithms are used in a multi-label learning tasks when an sample has more than one label. This setting is ideal for mining tasks on samples with complex structure. In this paper we focus on exploiting the pairwise similarities of samples, therefore the informativeness of a sample is weighed by average similarity to its neighbors. Let x_i and x_j be a pair of samples. To cope with the drawback of uncertainty based selection, we consider the diversity by evaluating the correlation of the samples.

Algorithm 1. Deep Transfer Active Learning (TAL)
_____

Input: labeled data D^L, available D^U **Parameter**: θ, loss L, budget m **Hyperparameters**: ϑ

Output: Θ

1: $D_{train}, D_{test} \leftarrow D^L$
2: $\Theta \leftarrow \vartheta\theta$ Initialize model
3: **while** $m \neq 0$ **do**
4:     **for** each x_i in D^U **do**
5:         $u \leftarrow f_u(x)$
6:         $c \leftarrow f_c(x)$
7:         Select sample $\hat{x} \leftarrow \underset{i}{\mathrm{argmax}}(u \cdot c)$
8:         Predict class $\hat{y} \leftarrow \Theta(\hat{x})$
9:         Update labeled set $D_{train} \leftarrow \hat{x}_t, \hat{y}_t$
10:        Compute query loss $L_{train} \leftarrow L(\hat{y}, y)$
11:     **end for**
12:     **for** $t = [1, T]$ **do**
13:         Get batch from D-train $x_t, y_t \leftarrow D_{train}$
14:         Get train loss on each batch $L \leftarrow L(\Theta(x_t), y_t)$
15:         $\theta \leftarrow:$ Update parameters
16:     **end for**
17:     Get batch from D-test $x, y \leftarrow D_{test}$
18:     Get test loss on test batch $L_{test} \leftarrow L(\Theta(x), y)$
19:     $\Theta \leftarrow:$ Update model
20:     $m \leftarrow m - 1$
21: **end while**
22: **return** Θ
_____

Given a label space Y the correlation measure f_c between a pair of samples x_i and x_j can be defined as:

$$f_c(x_i) = \frac{1}{D^U} \sum_{x_j \in D^U / x_i} f_c(x_i, x_j) \tag{5}$$

The value of $f_c(x_i)$ represents the sample density of x_i in the unlabeled set and $f_c(x_i, x_j)$ represents the mean of the correlation for all $j \neq i$. The larger the value, the more densely a sample is correlated with others. A low value of the correlation measure indicates an outlier sample which should not be considered for labeling.

Selecting the most informative and representative samples in a distribution is very critical for improving the generalization performance of the classifier. To do this we integrate correlation and uncertainty values together. The most effective sample to label by the current model can be expressed as a product of $f_u(x_i)$ and $f_c(x_i)$. As the uncertainty of a sample increases its potential for being selected for labeling increases. To do this we can rank the selected samples based on the utility value f_i: with the top ranked samples in each subset being the most effective samples to label. The query strategy discussed implemented in this paper is based on product utility $f_u(x_i) \cdot f_c(x_i)$. In Algorithm 1. learning starts from a small labeled set D^L with initialized parameters θ and hyperparameters ϑ for the model Θ and proceeds sequentially. For each iteration of AL, uncertainty $f_u(x)$ and correlation $f_i(x)$ for each candidate sample x_i are computed. Then using a heuristic combination, the most informative sample is selected for labeling. After that, the new labeled sample is directly added to the training set D^L and the model is updated. Classification on new data proceeds in batch mode while computing loss for every iteration. Specifically classification is now presented as a mapping function from the feature space F to the class label space Y which can be expressed as $P(x) : F \mapsto Y$. To improve the model performance, we use a reward utility function for a single task of labeling an sample x.

$$R(Y, x) = \sum_y p(Y = y|x) r(p, y|x) \tag{6}$$

where p represents the posterior probability of sample x belonging class y with $R()$ as a reward function. This formula accumulates the reward on each possible label y.

4   Experimental Setup

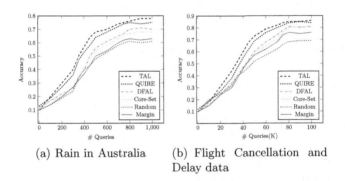

(a) Rain in Australia

(b) Flight Cancellation and Delay data

Fig. 1. Classification accuracy comparison on two real-world multivariate time-series datasets

To validate the effectiveness of the proposed approach, experiments were performed on both real-world application dataset for Rainfall in Austrialia & Flight Delay and Cancellation and 30 multivariate UEA Time Series classification[1] datasets. ResNet architecture was used to implement the experiments for two reasons: a) it has been adopted in other recent time series classifications [20]; b) it performs comparably well in a large number of cases [21]. This is because

---

[1] http://www.timeseriesclassification.com/.

skip-connections or residual connections are very efficiently with deeper networks by allowing gradient flow directly through the bottom layers. The residual connections allow skipping of multiple layers within deeper neural network [22]. In the experiment settings, the network main hyper-parameters are 4 residual modules, 8×32 kernels and 128 filters. For all convolutional and dense layers L2 regularization is used with 10^{-1} learning rate and categorical cross-entropy used as loss function. Accuracy is used as a performance metric by recording training loss and performance reporting on the testset. A sigmoid function was used as a decision boundary to return a probability value between 0 and 1. The boundary is set to $p \leq 0.5$ for no rain and $p \geq 0.7$ for rain. All experiments were implemented in python with scikit-learn, PyTorch and on NVidia K80 GPU. We further test our model on transfer skill using the 30 multivariate time-series datasets. For each of the 30 datasets, we randomly partition it into two subsets, 70% training samples and 30% testing samples. From the training set, 10% is sampled as labeled set and 100 samples are actively selected in every iteration. To ensure uniform fine-tuning, only the last layer parameters are adjsuted to match the target classes. In the end we ended up with total of 60 pre-trained models i.e. 30 pair for each of the UCR dataset with Rainfall dataset and for Flight dataset respectively.

The selected pre-trained models were then used to learn a more challenging real-world problem using a multivariate rainfall in Austrialia[2] and flight information[3] datasets. The objective is to predict whether or not it will rain tomorrow. The dataset contains daily weather observations from numerous Australian weather stations. We then empirically demonstrate that combining TL and Al greatly improves performance. Firstly, we test the proposed method on two real-world prediction datasets. On the Rainfall data we segregate it into categorical and numerical variables. The date variable is denoted by Date column with 1000 total data points. The 4 categorical variables include: Location, WindGustDir, WindDir9am and WindDir3pm. There are two binary categorical variables i.e. RainToday and RainTomorrow with RainTomorrow being the target variable. For categorical the date variable that has the highest cardinality of 3436 labels, we performed some feature engineering to deal with high cardinality problem. To do this we parse the date coded as object into datetime format. Then one hot encoding is performed on all variables while adding dummy variables on missing data. The other data pre-processing include removing of outliers and splitting the data at 80% for training and 20% for testing. To avoid overfitting, imputation was done over the training set and then propagated to the test set. For missing categorical values, assumed input was done with most frequent value.

A set of 100 and 10000 samples was randomly selected as initial annotated samples for rainfall and flight datasets respectively. On flight information dataset, similar to above data preparation was carried out. The predicted probabilities of k classes are *delay. divert* and *cancelation* which take the form $p_1, ...p_k$ based on current labeled examples. Since $k = 3$ the entropy is defined as follows:

---

[2] https://www.kaggle.com/jsphyg/weather-dataset-rattle-package.
[3] https://www.kaggle.com/usdot/flight-delays.

$$En(\bar{X}) = -\sum_{i=1}^{3} p_i.\log(p_i) \tag{7}$$

The data contain 100000 data samples spread over 9 categories. Each sample is linked to the cause which includes: cancellation reason, air-system delay, security delay, late-aircraft delay and weather delay.

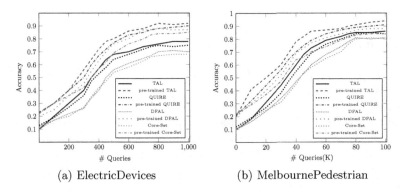

(a) ElectricDevices   (b) MelbournePedestrian

Fig. 2. Classification accuracy variation in percentage over the original accuracy and on pre-trained models. In (a) classification is done on Rain dataset using a model pre-trained on ElectricDevices dataset. Similarly, in (b) we compare the performance on Flight dataset using a model pretrained with MelbournePedestrian and (c) on FordB datasets

5   Results

We compare our proposed method with the following active learning methods in the experiments: a) Random selection; b) Margin based sampling; c) proposed method; d) QUIRE - a method inspired by the margin based active learning from the mini-max viewpoint with emphasize on selecting unlabeled instances that are both informative and representative [16]; e) DFAL method that selects unlabeled samples with the smallest perturbation. The distance between a sample and its smallest adversarial example better approximates the original distance to the decision boundary [23]; f) Core-Set for non-uncertainty based AL method [24]. Figure 1 shows classification accuracy of different AL methods with varied number of queries on each dataset. As expected the baseline methods are not as effective as hybrid methods. Both DFAL and Core-Set approach can outperform the Random and Margin methods but are worse than hybrid methods. Random and Margin approach have little improvement on both datasets. The performance of DFAL and Core-Set is impressive at the beginning, but loses the edge as the querying goes on. The proposed method performs 4% better than

QUIRE on both Rain and Flight datasets. We attribute the better performance
on the ability to jointly evaluate sample informativeness and representativeness
based on sample-label pair. In addition, the potential contribution of the current sample candidate is incorporated on the strategy for the subsequent query.
Figure 2 presents the comparison on skill transfer using three best performing datasets. Its evident that skill transfer significantly reduce the training cost
while achieving the same or better performance as compared to using un-trained
model.

6   Conclusion

In this paper, we propose a novel Active Learning approach that jointly evaluates
informativeness and representativeness for multivariate time series classification
problems. Observing the structured nature of time series data, we propose a
bi-objective method based on sample-label pair views which is considered more
effective in reducing annotation cost. Learning is enhanced by Transfer Learning to further lower the learning cost by reusing model skill among tasks. One
key phenomena observed with many UCR datasets is network overfitting which
we attribute to the small size of the datasets. Since deep networks are highly
dependent on large amount of training data, generating synthetic data can help
mitigate the data size challenge. Also its important to note that with emerging time series big data repositories, the challenge of data size is a hot topic
for many researchers. In the future, we will study other hybrid transfer active
learning approaches.

References

1. Gikunda, P.K., Jouandeau, N.: State-of-the-art convolutional neural networks for
   smart farms: a review. In: Arai, K., Bhatia, R., Kapoor, S. (eds.) CompCom 2019.
   AISC, vol. 997, pp. 763–775. Springer, Cham (2019). https://doi.org/10.1007/978-
   3-030-22871-2_53
2. Szegedy, C., et al.: The inception architecture for computer vision. In: Conference on Computer Vision and Pattern Recognition (CCVPR 2016), pp. 2818–2826.
   IEEE (2016)
3. Krizhevsky, A., Sutskever, I., Hinton, G.E.: ImageNet classification with deep convolutional neural networks. Commun. ACM **60**(6), 84–90 (2017)
4. Torres, J.F., et al.: A deep learning for time series forecasting: a survey. Big Data
   9(1), 3–21 (2021)
5. Gavves, E., et al.: Active transfer learning with zero-shot priors: reusing past
   datasets for future tasks. In: International Conference on Computer Vision (ICCV
   2015), pp. 2731–2739. IEEE (2015)
6. Raina, R., et al.: Self-taught learning: transfer learning from unlabeled data. In:
   International Conference on Machine Learning, pp. 759–766. ACM (2017)
7. Fu, Y., Zhu, X., Li, B.: A survey on instance selection for active learning. Knowl.
   Inf. Syst. **35**, 49–283 (2013). https://doi.org/10.1007/s10115-012-0507-8
8. Settles, B.: Active learning literature survey. Computer Sciences Technical Report
   1648, University of Wisconsin-Madison, Madison (2009)

9. Li, X., Guo, Y.: Adaptive active learning for image classification. In: Conference on Computer Vision and Pattern Recognition (CCVPR 2013), pp. 859–866. IEEE (2013)
10. Lewis, D.D., Catlett, J.: Heterogeneous uncertainty sampling for supervised learning. In: Machine Learning Proceedings, pp. 148–156 (1994)
11. Langkvist, M., Karlsson, L., Loutfi, A.: A review of unsupervised feature learning and deep learning for time-series modeling. Pattern Recogn. Lett. **42**, 11–24 (2014)
12. Pascanu, R., Mikolov, T., Bengio, Y.: Understanding the exploding gradient problem. CoRR, abs/1211.5063, vol. 417 (2012)
13. Pascanu, R., Mikolov, T., Bengio, Y.: On the difficulty of training recurrent neural networks. In: International Conference on Machine Learning, pp. 1310–1318. ACM (2013)
14. Lewis, D.D., Gale, W.A.: A sequential algorithm for training text classifiers. In: Croft, B.W., van Rijsbergen, C.J. (eds.) SIGIR 1994, pp. 3–12. Springer, London (1994). https://doi.org/10.1007/978-1-4471-2099-5_1
15. Settles, B., Craven, M.: An analysis of active learning strategies for sequence labeling tasks. In: Conference on Empirical Methods in Natural Language Processing, pp. 1070–1079 (2008)
16. Huang, S.J., Jin, R., Zhou, Z.H.: Active learning by querying informative and representative examples. IEEE Trans. Pattern Anal. Mach. Intell. **36**(10), 1936–1949 (2014)
17. Natarajan, A., Laftchiev, E.: A transfer active learning framework to predict thermal comfort. Int. J. Prognostics Health Manage. **10**(3), 1–13 (2019)
18. Peng, F., Luo, Q., Ni, L.M.: ACTS: an active learning method for time series classification. In: International Conference on Data Engineering (ICDE 2017), pp. 175–178 (2017)
19. Gikunda, P.K., Jouandeau, N.: Cost-based budget active learning for deep learning. In: 9th European Starting AI Researchers' Symposium co-located with 24th European Conference on Artificial Intelligence (ECAI 2020), vol. 2655 (2020)
20. Wang, Z., Yan, W., Oates, T.: May. time series classification from scratch with deep neural networks: a strong baseline. In: International Joint Conference on Neural Networks (IJCNN 2017), pp. 1578–1585 (2017)
21. Fawaz, H.I., et al.: Deep learning for time series classification: a review. Data Min. Knowl. Discov. **33**(4), 917–963 (2019)
22. He, K., Zhang, X., Ren, S., Sun, J.: Deep residual learning for image recognition. In: Conference on Computer Vision and Pattern Recognition (CVPR 2016), pp. 770–778 (2016)
23. Deng, Y., Chen, K., Shen, Y., Jin, H.: Adversarial active learning for sequences labeling and generation. In: International Joint Conference on Artificial Intelligence (IJCAI 2018), pp. 4012–4018 (2018). https://doi.org/10.24963/ijcai.2018/558
24. Sener, O., Savarese, S.: Active learning for convolutional neural networks: a coreset approach. In: 6th International Conference on Learning Representations (ICLR 2018)

Video Processing

Learning Traffic as Videos: A Spatio-Temporal VAE Approach for Traffic Data Imputation

Jiayuan Chen, Shuo Zhang, Xiaofei Chen, Qiao Jiang, Hejiao Huang$^{(\boxtimes)}$, and Chonglin Gu

The Department of Computer Science and Technology,
Harbin Institute of Technology (Shenzhen), Shenzhen, China
huanghejiao@hit.edu.cn

Abstract. In the real world, data missing is inevitable in traffic data collection due to detector failures or signal interference. However, missing traffic data imputation is non-trivial since traffic data usually contains both temporal and spatial characteristics with inherent complex relations. In each time interval, the traffic measurements collected in all spatial regions can be regarded as an image with more or fewer channels. Therefore, the traffic raster data over time can be learned as videos. In this paper, we propose a novel unsupervised generative neural network for traffic raster data imputation called *STVAE*, which works well robustly even under different missing rates. The core idea of our model is to discover more complex spatio-temporal representations inside the traffic data under the architecture of variational autoencoder (VAE) with Sylvester normalizing flows (SNFs). After transforming the traffic raster data into multi-channel videos, a Detection-and-Calibration Block (DCB), which extends 3D gated convolution and multi-attention mechanism, is proposed to sense, extract and calibrate more flexible and accurate spatio-temporal dependencies of the original data. The experiments are employed on three real-world traffic flow datasets and demonstrate that our network *STVAE* achieves the lowest imputation errors and outperforms state-of-the-art traffic data imputation models.

Keywords: Traffic data imputation · Unsupervised learning · Variational autoencoder · Normalizing flows · Attention mechanism · 3D gated convolution

1 Introduction

With the remarkable improvement of the capability of computation and storage, a vast amount of urban traffic data swarm into the Intelligent Transportation

This work is financially supported by National Key R&D Program of China under Grant No.2017YFB0803002 and National Natural Science Foundation of China under Grant No.61732022.

© Springer Nature Switzerland AG 2021
I. Farkaš et al. (Eds.): ICANN 2021, LNCS 12895, pp. 615–627, 2021.
https://doi.org/10.1007/978-3-030-86383-8_49

System (ITS) community. In order to enhance the efficiency of traffic management and reduce traffic congestion, many ITS applications (e.g., traffic flow forecasting [18] and traffic state prediction [3]) require the collected data should be complete without missing values. However, traffic data missing always happens when there are loop sensor failures or GPS signal interference, which may greatly affect the overall service of ITS. Therefore, restoring the missing values through traffic data imputation is of great significance. The biggest challenge is how to learn robust and accurate spatio-temporal correlations jointly in the context of incomplete spatio-temporal traffic data.

To address the above challenge for missing data imputation, many researchers have put forward the solutions including statistical methods, machine learning based approaches and deep learning based models. The statistical methods [8, 14] are widely used in traffic domain, but they only consider linear dependency of time, bringing about poor filling performance in practice. The machine learning based approaches such as PPCA [10], tensor-based methods [11] could capture nonlinear dynamic features in traffic data, but incapable of processing large-scale traffic datasets. In contrast, deep learning based methods, especially SDAE [5], GAN [3], VAE [1, 7, 12], are suitable to process mass data and can construct more nonlinear spatio-temporal patterns for incomplete traffic data. Nevertheless, both SDAE and GAN require complete data for training, which is difficult to meet in real data collection scenarios. In addition, few studies distinguish between the missing data and the observed data when extracting spatio-temporal correlations in traffic data.

In this paper, we study the problem of traffic data imputation through learning traffic data as videos. The urban transportation network is a map that can be divided into grids, each with multiple measurements (e.g., crowd inflow and outflow) collected at regular intervals. In each interval, the map can be regarded as an image snapshot with multi-channels, so that daily urban traffic data can be learned as a video. The merit of such representation can well preserve the temporal and spatial information of the raw traffic raster data to a great extent. Inspired by [1], we propose a novel generative neural network called *STVAE* for urban traffic data imputation with robust spatio-temporal representations modeled. It is designed based on variational autoencoder (VAE), which can learn the probabilistic densities of latent variables rather than deterministic hidden representations. This property enables VAEs to utilize incomplete datasets for model training in an unsupervised way. To further profile the distribution of traffic data, we adopt Sylvester normalizing flows (SNFs) [15] that maps an isotropic normal distribution to a more flexible one through a series of invertible transformations. In the encoder phase, we also design a module called Detection-and-Calibration Block (DCB). It is composed of a modified 3D gated convolution mechanism [2] to distinguish the missing part and observed part in incomplete traffic data, and an extended multi-attention mechanism to calibrate features in channel, spatial and temporal dimensions. Thus, our model could learn spatio-temporal features better, as our experimental results show.

The contributions of our work can be summarized as follows:

- We regard traffic raster map over time as videos from a distinctive perspective considering both spatial dependencies and temporal dependencies of traffic data simultaneously.
- We propose *STVAE*, a novel spatio-temporally aware generative neural network for missing traffic data imputation. To the best of our knowledge, this is the first work that applies VAE in conjunction with normalizing flows to handle traffic data imputation.
- We design a Detection-and-Calibration Block (DCB), which extends and integrates 3D gated mechanism and multi-attention mechanism, improving the accuracy of imputation.
- Extensive experiments show that our model exhibits better performance with high robustness in three real-world traffic datasets than state-of-the-art traffic imputation methods do.

The rest of this paper is organized as follows. Section 2 give some preliminaries. Section 3 presents the approach of *STVAE* in detail. Section 4 evaluates our model. Section 5 concludes the whole paper.

2   Preliminaries

In this section, we give the definition of raster data and state our problem.

2.1   Traffic Raster Data

Definition 1. *(spatio-temporal Traffic Raster Data* [6]*) A urban traffic network is divided by longitude and latitude into a raster map $\mathcal{G}$ which contains H rows and W columns. One grid in $\mathcal{G}$ represents one region. For each region (h, w), there are C different measurements at the same time. Thus, over the whole map $\mathcal{G}$, the observations in a fixed time span, which contains T time intervals, can be viewed as a tensor $x \in \mathbb{R}^{C \times T \times H \times W}$.*

In particular, we consider two measurements when studying traffic crowd flows: inflow and outflow [18]. Here we use $x[0, t, h, w]$ and $x[1, t, h, w]$ to denote the inflow and outflow of crowd in region (h, w) at time interval t, respectively.

2.2   Problem Statement

Under the MCAR assumption that data are missing completely at random [13], we consider that there are some i.i.d. incomplete traffic raster data making up a set $\mathbf{X} = \{x_i\}_{i=1}^n \in \mathcal{X}^n$ where $\mathcal{X} = \mathbb{R}^{C \times T \times H \times W}$. Correspondingly, we define another set $\mathbf{M} = \{m_i\}_{i=1}^n \in \mathcal{M}^n$, where $\mathcal{M} = \{0, 1\}^{C \times T \times H \times W}$ subject to $m_i[c, t, h, w] = 1$ if $x_i[c, t, h, w]$ is observed, otherwise $m_i[c, t, h, w] = 0$.

For tensor-based traffic data imputation, given an incomplete tensor x in the space $\mathcal{X}$, we divide it into the observed part x^{obs} and the missing part x^{mis}. The objective is to impute x^{mis} by utilizing x^{obs}.

3 Approach

In this section, we present the framework of our *STVAE* and introduce the major module (i.e., Detection-and-Calibration Block) in *STVAE* followed by the description of offline model training and online imputation.

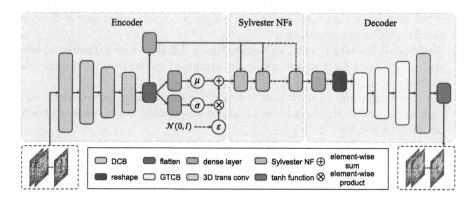

Fig. 1. The overall structure of *STVAE*. During training, the model learns the distribution of corresponding latent variable given each incomplete traffic sample. Based on the trained model, missing data can be imputed after passing through the model.

3.1 The Framework of *STVAE*

For traffic data imputation, we propose a spatial-temporal variational autoencoder (*STVAE*), containing three key points as follows: First, we apply VAE [9], which can learn the distribution of latent variables rather than deterministic hidden representations, to reconstruct data. Second, to enhance the scalability of approximate posterior in the context of amortized variational inference, we adopt Sylvester normalizing flows (SNFs) which transform a diagonal Gaussian distribution to a more flexible one through an iterative process. Third, it stacks Detection-and-Calibration Blocks (DCBs) in the encoder to sense the missing points, extract and calibrate spatio-temporal dependencies. *STVAE* is composed of three parts (Encoder, Sylvester NFs and Decoder), as shown in Fig. 1.

Encoder. At the outset, an input x is sent to the encoder and flows along four DCBs (*see Sect. 3.2*) in sequence, each is composed of 3D gated convolution and multi-attention mechanism, to generate the deterministic hidden representations denoted as h. Then, h is flattened and enters the fully-connected layers to parametrize mean μ and log deviation $\log \sigma^2$ for the initial latent variable z_0. Next, z_0 can be produced via the reparametrization trick [7] as follows:

$$z_0 = \mu + \sigma \otimes \varepsilon \tag{1}$$

where ε is sampled from $\mathcal{N}(0, I)$.

Sylvester NFs. From a probabilistic perspective, the output of the encoder z_0 obeys the posterior $q_0(z_0 \mid x) \sim \mathcal{N}(\mu, \sigma^2 I)$ which may limit the expression of the vanilla VAE [7]. To this end, Berg et al. [15] proposed Sylvester NFs which improve the flexibility of the proposal distribution $q_\varphi(z|x)$. We firstly draw z_0 from q_0, and then transform z_0 to z_K by a series of invertible mappings $f_K \circ \cdots \circ f_k \circ \cdots \circ f_1(z_0)$, where $f_k(z_{k-1}) = z_{k-1} + QRtanh(\widetilde{R}Q^T z_{k-1} + b)$. Parameters R, $\widetilde{R}$ are upper triangular matrix, Q is a column orthogonal matrix and b is a vector. All of them can be produced via a separate dense layer. The final log posterior $\log q_K(z_K|x)$ can be formulated as follows:

$$\log q_K(z_K|x) = \log q_0(z_0|x) - \sum_{k=1}^{K} \log \left| det\left(\frac{\partial f_k(z_{k-1})}{\partial z_{k-1}} \right) \right| \qquad (2)$$

where $det\left(\frac{\partial f_k(z_{k-1})}{\partial z_{k-1}}\right)$ is the Jacobian determinant of the transformation f_k. We make $q_\varphi(z \mid x) := q_K(z_K \mid x)$ and then specify z_K as the input of our decoder.

Decoder. The decoder utilizes one dense layer and reshape operation to transform z_K into a tensor, and then adopts 3D gated transposed convolution block (GTCB) for up-sampling. GTCB consists of 3D gated transposed convolution, batch normalization and $PReLU$ activation function. Eventually, the reconstructed sample $\hat{x}$ can be obtained through the last 3D transposed convolution and $tanh$ function.

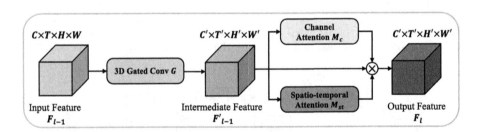

Fig. 2. The internal structure of the Detection-and-Calibration Block (DCB) where $\otimes$ denotes the element-wise product which obeys broadcasting rules.

3.2 Detection-and-Calibration Block

Detection-and-Calibration Block (DCB), as the main component of the encoder, contains two submodules, namely 3D gated convolution and multi-attention mechanism. Given a feature map $F_{l-1} \in \mathbb{R}^{C \times T \times H \times W}$ as the input of l-th DCB, by applying 3D gated convolution **G**, DCB initially detects which pixels in F_{l-1} contain invalid information (i.e., missing part previously) and pays more attention to valid ones to extract important features. Next, it further calibrates the feature maps by utilizing multi-attention mechanism that combines channel attention $\mathbf{M}_c$ with spatio-temporal attention $\mathbf{M}_{st}$. Therefore, we named it Detection-and-Calibration Block as shown in Fig. 2. The overall procedure of DCB is formulated as follows:

$$F'_{l-1} = \mathbf{G}(F_{l-1}) \tag{3}$$

$$F_l = \mathbf{M}_c(F'_{l-1}) \otimes \mathbf{M}_{st}(F'_{l-1}) \otimes F'_{l-1} \tag{4}$$

where $F'_{l-1} \in \mathbb{R}^{C' \times T' \times H' \times W'}$ is an intermediate feature map, $F_l \in \mathbb{R}^{C' \times T' \times H' \times W'}$ is the output of l-th DCB and $\otimes$ denotes element-wise product. Under the hood, we describe 3D gated convolution and multi-attention in the following.

3D Gated Conv. 3D gated convolution was proposed in [2] for video inpainting problem. We extend it to traffic data imputation so as to model spatio-temporal patterns more accurately. As we can see, there are two branches in Fig. 3. One is made up of 3D convolution and batch normalization followed by a *PReLU* activation function, which is used for nonlinear spatio-temporal modeling. The other is to allocation soft weights to the former by a sigmoid function, which behaves as a valve. Finally, they are merged by element-wise product.

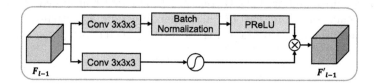

Fig. 3. The overview of 3D gated convolution. Note that we can replace convolution with transposed convolution and it becomes the module (GTCB) we use in decoder.

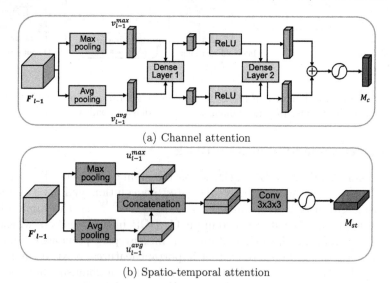

(a) Channel attention

(b) Spatio-temporal attention

Fig. 4. The overview of multi-attention mechanism consisting of channel attention and spatio-temporal attention.

Multi-attention Mechanism. Channel attention and spatial attention in convolutional block attention module (CBAM) [17] are widely applied for image classification. We extend CBAM from spatial field to spatio-temporal field, as shown in Fig. 4. In our multi-attention mechanism, channel attention is used to explore the correlation among channels such as traffic crowd inflows and outflows, while spatio-temporal attention is applied to enhance or suppress different spatio-temporal features. In a nutshell, multi-attention devotes to calibrating features computed by 3D gated convolution.

For channel attention, max-pooling and average-pool operations are performed on feature map F'_{l-1} to generate two channel-wise descriptors denoted as $v^{max}_{l-1} \in \mathbb{R}^{C \times 1 \times 1 \times 1}, v^{avg}_{l-1} \in \mathbb{R}^{C \times 1 \times 1 \times 1}$, both of which are sent to the same dense layer separately. The two dense layers are used for channel fusion by reducing the dimension of descriptors to C/r and then restore it back, r is the reduction ratio. Then we sum both descriptors up and feed it into the sigmoid to generate a channel attention map $\mathbf{M_c} \in \mathbb{R}^{C \times 1 \times 1 \times 1}$.

For spatio-temporal attention, max-pooling and average-pooling operations are still performed on feature map F'_{l-1} to generate two 3D maps denoted as $u^{max}_{l-1} \in \mathbb{R}^{1 \times T \times H \times W}, u^{avg}_{l-1} \in \mathbb{R}^{1 \times T \times H \times W}$. After that, they are concatenated and fed into 3D convolution layer followed by a sigmoid to generate a spatio-temporal attention map $\mathbf{M}_{st} \in \mathbb{R}^{1 \times T \times H \times W}$.

In the end, a new feature map F_l is gained by element-wise product of $\mathbf{M_c}$, $\mathbf{M}_{st}$ and F'_{l-1}, as shown above in Eq. 4.

3.3 Offline Model Training and Online Imputation

Offline Model Training. Since we can not obtain the real values of missing data, the object of *STVAE* is to minimize the negative log likelihood of the observed data. For the i-th sample x_i, the loss function can be written as follows:

$$
\begin{aligned}
\mathcal{F}(\theta, \varphi) &= -\log p_\theta(x^{obs}_i) \\
&= -\mathbb{E}_{q_0(z_0|x^{obs}_i)}[\log p_\theta(x^{obs}_i \mid z_K)] - \sum_{k=1}^{K} \log \left| det\left(\frac{\partial f_k(z_{k-1})}{\partial z_{k-1}} \right) \right| \\
&\quad + D_{KL}[q_0(z_0 \mid x^{obs}_i) \parallel p_\theta(z_K)]
\end{aligned}
\tag{5}
$$

where parameters (φ, θ) are parameterized by the encoder and the decoder respectively. In fomula (5), the first item is the reconstruction error of the observed part x^{obs}_i that can be calculated as $m_i \otimes \parallel x_i - \hat{x}_i \parallel_F$, the second item is from SNFs and the last item is the Kullback–Leibler distance between two densities. Thus, we can train *STVAE* offline by optimizing the Eq. 5.

Online Imputation. Once so-obtained, we can fill in an incomplete traffic raster sample x online like in Fig. 5. First of all, the missing values x^{mis} are initialized by zeroes and this operation hardly affects the filling effect because of the nonlinear transformation of neural network [9]. Then, the distribution of the latent variable is obtained through the encoder and SNFs. According to the conditional distribution $p_\theta(x^{mis} \mid x^{obs})$, we can impute x^{mis} iteratively by sampling from a Markov chain [12].

4 Evaluation

In this section, we introduce datasets and give model settings followed by evaluation metrics used in our experiments. Next, we compare *STVAE* with other baseline methods. Furthermore, ablation experiments are conducted to evaluate the effectiveness of each module in *STVAE*.

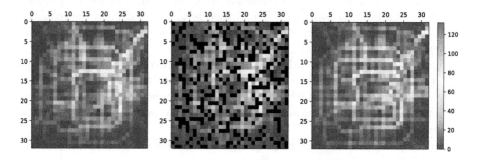

Fig. 5. An example of imputation effectiveness of our model on TaxiBJ described in Table 1. Green means unblocked, red means congested. **Left:** Truth-ground data matrix. **Mid:** Incomplete data with missing rate of 30%. **Right:** Reconstructed data. (Color figure online)

4.1 Datasets and Experiment Settings

Datasets. We conduct our experiments on three real-world datasets about traffic flows including TaxiBJ [18], TaxiNYC [4], BikeNYC [18], as shown in Table 1.

Table 1. Dataset description

Dataset	TaxiBJ [18]	TaxiNYC [4]	BikeNYC [18]
Data type	Taxi GPS	Taxi GPS	Bike rent
Location	Beijing	New York	New York
Time range	7/1/2013–10/30/2013 3/1/2014–6/30/2014 3/1/2015–6/30/2015 11/1/2015–4/10/2016	1/1/2010–12/31/2010 1/1/2012–12/31/2013	4/1/2014–9/30/2014
Time interval	30 min	1 h	1 h
Raster map size	(32, 32)	(10, 20)	(16, 8)
Number of time intervals	22,459	26,304	4392

All the datasets above are processed into raster data, as defined in Sect. 2.1, and missing values are manually randomly generated at different missing rates. Then, we treat the raster data of each day as a sample. Before the output of the *STVAE*, we apply *tanh* whose range is between -1 and 1 as our final activation function, so here we use the Min-Max normalization to scale the data to the interval $[-1, 1]$ correspondingly for data reconstruction. For each dataset, we split it into training and test sets at the ratio of $9 : 1$. On the basis of this, we choose 90% of the training data for model training, and the rest is used for validation. In order to avoid overfitting, we save the model with best performance on the validation set.

Model Configurations. We set the hyper-parameters of *STVAE* empirically as follows. The kernel size of each 3D convolution module is set to 3 and the stride of 3D convolution is 2. The length of Sylvester NFs is 5. The dimension of each latent variable z_k is fixed to 128. The number of out channels of each DCB is 16, 32, 64, 128 in ascending order. The reduction ratio r in channel attention mechanism is set to 2. We apply Adam optimizer with an initial learning rate of 10^{-3} to perform gradient descent. Furthermore, the total numbers of epochs for TaxiBJ, TaxiNYC and BikeNYC are 500, 200, 2000 separately.

Evaluation Metrics. To evaluate our approach and baseline methods quantitatively, we measure the fitness of truth-ground data and imputed data by refined normalized mean square error (NMSE) and refined root mean square error (RMSE) defined as

$$\text{NMSE} = \mathbb{E}\left\{ \frac{\|(1 - m) \otimes (x - \hat{x})\|_F}{\|(1 - m) \otimes x\|_F} \right\} \tag{6}$$

$$\text{RMSE} = \sqrt{\mathbb{E}\{ \|(1 - m) \otimes (x - \hat{x})\|_F \}} \tag{7}$$

where m is the mask tensor of observations defined in Sect. 2.2, $\otimes$ denotes the Hadamard product, $\|\cdot\|_F$ denotes the Frobenius norm.

4.2   Comparison and Analysis of Results

In this section, we compare our *STVAE* with five baseline methods: PPCA [10], HaLRTC [11], SDAE [5], MLP-VAE [1] and CombCN [16].

Figure 6 shows the comparisons of the above five baseline methods with our approach in NMSE and RMSE on three datasets with different missing rate. One important observation is that *STVAE* achieves a more competitive imputation performance than all baseline models including the state-of-the-art VAE-based model in the field of traffic data imputation (i.e., MLP-VAE), especially when missing rate is high. Besides excellent performance, our *STVAE* shows highly strong robustness even under different missing rate.

PPCA [10] is a Gaussian latent variable model where the relationship between latent and observed variables are linear. Nonlinearity can improve imputation performance because it can discover more complex spatio-temporal features. As a nonlinear model, *STVAE* performs much better than PPCA on all datasets.

HaLRTC [11] is a tensor-based method. It performs well at low missing rates on TaxiNYC and BikeNYC, as it considers the spatio-temporal features. Nevertheless, it shows overfitting when missing rate reaches 90%. In contrast, the objective function of *STVAE* (see Eq. 5) adds regularization terms to prevent overfitting, leading to good performance even under high missing rate.

Stacking denoising autoencoder (SDAE) [5] is a stacking model based on encoder-decoder reconstruction, which is not sensitive to the proportion of missing values in the data. Hence it is not surprising that the RMSE of SDAE increases more slowly than *STVAE* when the missing rate changes from 10% to 90%. However, SDAE requires the completeness of data for model training. In contrast, *STVAE* is an unsupervised network that can be trained by using incomplete traffic data.

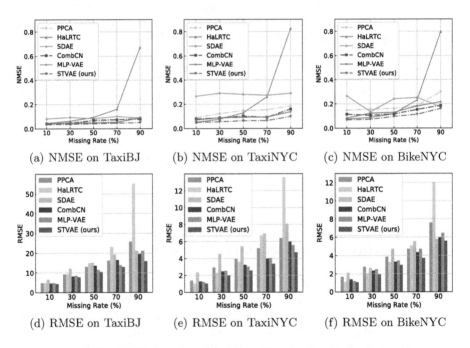

Fig. 6. Comparisons on performance of baselines and our approach.

CombCN [16] is a autoencoder-based network for video inpainting. It applies a 3D convolution subnetwork for temporal consistency and a 2D completion subnetwork for spatial modeling. In contrast, our network uses 3D gated convolution to focus on valid features. Meanwhile, our VAE-based method shows bigger generative ability against autoencoder-based ones. As a result, *STVAE* performs better than CombCN.

The VAE model proposed in [1] (MLP-VAE for short) is a deep Bayesian model that could learn stochastic mappings between observed data and latent variables. Thus, MLP-VAE discovers more flexible representations than other baselines. The overall performance of MLP-VAE on three datasets is second only to *STVAE*, as shown in Fig. 6. However, the Gaussian posterior impels the model underfitting. These results confirm our judgement that we should utilize Sylvester NFs to approximate the non-Gaussian posterior distribution. Moreover, Detection-and-Calibration Block is also more effective than multilayer perceptrons in sensing and extracting useful spatio-temporal information.

4.3   Ablation Study

In this section, we experimentally investigate the effects of two major modules in *STVAE*: Detection-and-Calibration Block (DCB) and Sylvester NFs (SNFs). We reconfigure *STVAE* to create three variants described as follows. 1) Pure-3DVAE: we remove both modules and only use vanilla 3D convolution in VAE. 2) *STVAE* w/o SNFs: SNFs is removed while DCB is preserved. 3) *STVAE* w/o DCB: DCB is replaced by vanilla 3D convolution layer while SNFs is preserved. Table 2 presents the NMSE and RMSE of *STVAE* and the three variants on three datasets with different missing rate. We observe that *STVAE* surpasses the other three variants on both evaluation metrics, which reflects the powerful modeling ability of DCB on incomplete spatio-temporal data and the effectiveness of SNFs in optimizing the posterior distribution of latent variables.

Table 2. Comparison of the *STVAE* and its variants on test sets. Each experiment was repeated 20 times to calculate its mean and deviation.

Missing rate	Models	TaxiBJ		TaxiNYC		BikeNYC	
		NMSE	RMSE	NMSE	RMSE	NMSE	RMSE
10%	Pure-3DVAE	0.0358 (0.0002)	4.34 (0.01)	0.0708 (0.0022)	1.21 (0.01)	0.0812 (0.0041)	1.17 (0.03)
	STVAE w/o SNFs	0.0344 (0.0002)	4.26 (0.01)	0.0571 (0.0010)	1.10 (0.01)	0.0748 (0.0030)	1.12 (0.02)
	STVAE w/o DCB	0.0347 (0.0003)	4.28 (0.02)	0.0654 (0.0005)	1.18 (0.01)	0.0706 (0.0010)	1.08 (0.01)
	STVAE	**0.0342 (0.0002)**	**4.24 (0.01)**	**0.0507 (0.0004)**	**1.04 (0.01)**	**0.0688 (0.0012)**	**1.07 (0.01)**
30%	Pure-3DVAE	0.0406 (0.0003)	8.01 (0.03)	0.0731 (0.0007)	2.25 (0.01)	0.0984 (0.0044)	2.24 (0.05)
	STVAE w/o SNFs	0.0395 (0.0003)	7.90 (0.04)	0.0617 (0.0010)	2.09 (0.02)	0.0743 (0.0042)	1.94 (0.06)
	STVAE w/o DCB	0.0398 (0.0003)	7.94 (0.03)	0.0704 (0.0004)	2.23 (0.01)	0.0784 (0.0038)	1.99 (0.05)
	STVAE	**0.0370 (0.0003)**	**7.66 (0.04)**	**0.0552 (0.0004)**	**1.97 (0.01)**	**0.0729 (0.0028)**	**1.92 (0.04)**
50%	Pure-3DVAE	0.0454 (0.0007)	11.12 (0.08)	0.0777 (0.0021)	2.86 (0.04)	0.1184 (0.0107)	3.24 (0.15)
	STVAE w/o SNFs	0.0432 (0.0003)	10.83 (0.04)	0.0678 (0.0014)	2.67 (0.03)	0.1031 (0.0054)	3.07 (0.08)
	STVAE w/o DCB	0.0410 (0.0004)	10.64 (0.05)	0.0779 (0.0009)	2.86 (0.02)	0.0972 (0.0026)	2.98 (0.04)
	STVAE	**0.0408 (0.0003)**	**10.53 (0.04)**	**0.0629 (0.0008)**	**2.57 (0.02)**	**0.0961 (0.0035)**	**2.96 (0.05)**
70%	Pure-3DVAE	0.0497 (0.0008)	13.57 (0.10)	0.0848 (0.0023)	3.86 (0.05)	0.1371 (0.0133)	4.09 (0.20)
	STVAE w/o SNFs	0.0474 (0.0008)	13.25 (0.12)	0.0751 (0.0018)	3.63 (0.04)	0.1204 (0.0094)	3.83 (0.15)
	STVAE w/o DCB	0.0477 (0.0006)	13.29 (0.08)	0.0802 (0.0011)	3.75 (0.02)	0.1179 (0.0053)	3.78 (0.09)
	STVAE	**0.0454 (0.0004)**	**12.96 (0.05)**	**0.0642 (0.0008)**	**3.36 (0.02)**	**0.1133 (0.0064)**	**3.71 (0.10)**
90%	Pure-3DVAE	0.0710 (0.0014)	18.85 (0.18)	0.1110 (0.0041)	5.00 (0.09)	0.1971 (0.0153)	6.16 (0.24)
	STVAE w/o SNFs	0.0505 (0.0007)	15.93 (0.11)	0.1026 (0.0039)	4.81 (0.09)	0.1710 (0.0183)	5.73 (0.31)
	STVAE w/o DCB	0.0536 (0.0009)	16.39 (0.13)	0.0989 (0.0023)	4.72 (0.05)	0.1732 (0.0072)	5.81 (0.12)
	STVAE	**0.0497 (0.0004)**	**15.80 (0.06)**	**0.0979 (0.0016)**	**4.70 (0.04)**	**0.1614 (0.0060)**	**5.58 (0.11)**

5   Conclusion

In this paper, we addressed the challenge of traffic data imputation task. We proposed an unsupervised generative neural network architecture $STVAE$, which can compute more complicated posterior distribution than vanilla VAE by leveraging Sylvester NFs. We designed a DCB module in encoder phase by extending both 3D gated convolution and multi-attention mechanism to detect more valuable pixels in feature maps, and extract and fine-tune more complex spatio-temporal patterns for high-dimensional traffic data. The results indicate that $STVAE$ is superior to the existing work in traffic data imputation.

References

1. Boquet, G., Morell, A., Serrano, J., Vicario, J.L.: A variational autoencoder solution for road traffic forecasting systems: missing data imputation, dimension reduction, model selection and anomaly detection. Transp. Res. Part C: Emerg. Technol. **115**, 102622 (2020)
2. Chang, Y.L., Liu, Z.Y., Lee, K.Y., Hsu, W.: Free-form video inpainting with 3D gated convolution and temporal PatchGAN. In: Proceedings of the IEEE International Conference on Computer Vision, pp. 9066–9075 (2019)
3. Chen, Y., Lv, Y., Wang, F.Y.: Traffic flow imputation using parallel data and generative adversarial networks. IEEE Trans. Intell. Transp. Syst. **21**(4), 1624–1630 (2019)
4. Donovan, B., Work, D.: New York City Taxi Trip Data (2010–2013) (2016). https://doi.org/10.13012/J8PN93H8
5. Duan, Y., Lv, Y., Liu, Y.L., Wang, F.Y.: An efficient realization of deep learning for traffic data imputation. Transp. Res. Part C: Emerg. Technol. **72**, 168–181 (2016)
6. Guo, S., Lin, Y., Li, S., Chen, Z., Wan, H.: Deep spatial-temporal 3D convolutional neural networks for traffic data forecasting. IEEE Trans. Intell. Transp. Syst. **20**(10), 3913–3926 (2019)
7. Kingma, D.P., Welling, M.: Auto-encoding variational bayes. arXiv preprint arXiv:1312.6114 (2013)
8. Li, Y., Li, Z., Li, L.: Missing traffic data: comparison of imputation methods. IET Intel. Transport Syst. **8**(1), 51–57 (2014)
9. Nazábal, A., Olmos, P.M., Ghahramani, Z., Valera, I.: Handling incomplete heterogeneous data using VAEs. Pattern Recogn. 107501 (2020)
10. Qu, L., Li, L., Zhang, Y., Hu, J.: PPCA-based missing data imputation for traffic flow volume: a systematical approach. IEEE Trans. Intell. Transp. Syst. **10**(3), 512–522 (2009)
11. Ran, B., Tan, H., Wu, Y., Jin, P.J.: Tensor based missing traffic data completion with spatial-temporal correlation. Physica A **446**, 54–63 (2016)
12. Rezende, D.J., Mohamed, S., Wierstra, D.: Stochastic backpropagation and approximate inference in deep generative models. In: International Conference on Machine Learning, pp. 1278–1286 (2014)
13. Rubin, D.B.: Inference and missing data. Biometrika **63**(3), 581–592 (1976)
14. Tak, S., Woo, S., Yeo, H.: Data-driven imputation method for traffic data in sectional units of road links. IEEE Trans. Intell. Transp. Syst. **17**(6), 1762–1771 (2016)

15. Van Den Berg, R., Hasenclever, L., Tomczak, J.M., Welling, M.: Sylvester normalizing flows for variational inference. In: 34th Conference on Uncertainty in Artificial Intelligence, pp. 393–402 (2018)
16. Wang, C., Huang, H., Han, X., Wang, J.: Video inpainting by jointly learning temporal structure and spatial details. In: Proceedings of the AAAI Conference on Artificial Intelligence, vol. 33, pp. 5232–5239 (2019)
17. Woo, S., Park, J., Lee, J.Y., So Kweon, I.: CBAM: convolutional block attention module. In: Proceedings of the European Conference on Computer Vision, pp. 3–19 (2018)
18. Zhang, J., Zheng, Y., Qi, D.: Deep spatio-temporal residual networks for citywide crowd flows prediction. In: Proceedings of the 31th AAAI Conference on Artificial Intelligence, pp. 1655–1661 (2017)

Traffic Camera Calibration via Vehicle Vanishing Point Detection

Viktor Kocur[✉] and Milan Ftáčnik

Faculty of Mathematics, Physics and Informatics of Comenius University
in Bratislava, Mlynská Dolina, 841 01 Bratislava, Slovakia
viktor.kocur@fmph.uniba.sk

Abstract. In this paper we propose a traffic surveillance camera calibration method based on detection of pairs of vanishing points associated with vehicles in the traffic surveillance footage. To detect the vanishing points we propose a CNN which outputs heatmaps in which the positions of vanishing points are represented using the diamond space parametrization which enables us to detect vanishing points from the whole infinite projective space. From the detected pairs of vanishing points for multiple vehicles in a scene we establish the scene geometry by estimating the focal length of the camera and the orientation of the road plane. We show that our method achieves competitive results on the BrnoCarPark dataset while having fewer requirements than the current state of the art approach.

Keywords: Camera calibration · Traffic surveillance · Deep learning

1 Introduction

Automatic traffic surveillance aims to provide information about the surveilled vehicles such as their speed, type and dimensions and as such is an important aspect of intelligent transportation system design. To perform these tasks an automatic traffic surveillance system requires an accurate calibration of the recording equipment.

Standard procedures of camera calibration require a calibration pattern or measurement of distances on the road plane which requires human intervention. Ideally the camera calibration should be carried out automatically. Traffic cameras can be automatically calibrated when observing straight road segments [9, 15]. The limitation to straight road segments makes the methods not applicable in various important traffic scenarios such as curved roads, intersections, roundabouts and parking lots. Methods not requiring straight road segments also exist, but they come with specific drawbacks such as requiring parallel curves to be visible on the road plane [6] or vehicles of specific make and model to be present in the scene [1, 2, 4].

In this paper we propose a camera calibration method which overcomes these limitations.[1] Our method works by estimating pairs of orthogonal vanishing

---

[1] We make the code available online: https://github.com/kocurvik/deep_vp/.

© Springer Nature Switzerland AG 2021
I. Farkaš et al. (Eds.): ICANN 2021, LNCS 12895, pp. 628–639, 2021.
https://doi.org/10.1007/978-3-030-86383-8_50

points for individual vehicles present in the scene. For this purpose we use a deep convolutional neural network based on the stacked hourglass architecture [13]. The outputs of the network are interpreted as heatmaps of the positions of vanishing points represented in the diamond space parametrization [7]. The diamond space parametrization is used as it provides a convenient way of representing the whole infinite projective space in a bounded diamond shaped space. To minimize the error caused by parametrizing an infinite space by a heatmap with finite resolution we opt to output multiple heatmaps for each of the vanishing points with each heatmap representing the original projective space at different scales. During inference we select the heatmap with the lowest error caused by the sampling of the grid around the maximum of the peak of the heatmap. Finally, the obtained pairs of vanishing points for vehicles across multiple frames are used to determine the horizon line of the road plane and the focal length of the camera in a simple manner inspired by the Theil-Sen estimator [18].

We evaluate our method on two different datasets of traffic surveillance videos of straight roads [16] and parking lots [2]. Our experiments show that our method is not competitive against more specialized methods which utilize the assumption of a straight road segment. However when considering the parking lot scenario where these methods are not applicable and our method achieves accuracy close to the existing state of the art approach [2] while not being limited by requiring vehicles of specific make and model to be present in the scene.

We also experimentally show that approaching this problem as a direct regression task of finding the two vanishing points leads to poor accuracy or even an inability for the network to converge depending on the used loss function. Additionally we also verify the impact of using multiple heatmaps instead of one for each vanishing point as well as the effect of training data augmentation.

2   Related Work

2.1   Traffic Camera Calibration

In the context of traffic surveillance the knowledge of the scene geometry can be used to infer the real-world positions of the surveilled vehicles. In order to obtain the geometry of the scene it is necessary to perform camera calibration. A review of available methods has been presented by Sochor et al. [16]. The review found that most published methods are not automatic and require human input such as drawing a calibration pattern on the road, using positions of line markings on the road or some other measured distances related to the scene.

An automated method based on the detection of two orthogonal vanishing points has been proposed by Dubská et al. [9]. The first vanishing point corresponds to the movement of the vehicles. Relevant keypoints are detected and tracked using the KLT tracker. The resulting lines connecting the tracked points between frames are accumulated using the diamond space accumulator [7]. The second vanishing point is accumulated from the edges of moving vehicles which are not aligned with the vehicle motion. The two orthogonal vanishing points are sufficient to compute the orientation of the road plane relative to the camera and the focal length. In order to enable the measurement of real-world distances in the road plane a scale parameter is determined by constructing 3D bounding boxes around detected vehicles and recording their mean dimensions which are compared to statistical data based on typical composition of traffic in the country. This method has been further improved [15] by fitting a 3D model of a known common vehicle to its detection in the footage. The detection of the second vanishing point is also improved by using edgelets instead of edges. These methods are fully automatic and provide accurate camera calibration, however they only work when the scene contains a straight segment of a road thus preventing their use in many important traffic surveillance scenarios such as intersections, roundabouts, curved roads or parking lots.

Corral-Soto and Elder [6] detect pairs of curves which are parallel in the scene and use the properties of parallel curves under perspective transformation to iteratively pair points on one curve with their counterparts on the other curve followed by the optimization of the camera parameters. This approach works for both straight as well as curved roads, but requires the presence of clear parallel curves in the scene which may not be present in many traffic surveillance scenarios.

More recent methods rely on detecting keypoints of specific vehicle models. The relative 3D positions of these keypoints are known for each vehicle and therefore they can be used to obtain the extrinsic camera parameters by using a standard PnP solver. Bhardwaj et al. [4] use extensive filtering and averaging to obtain the final extrinsic parameters. Bartl et al. [1] use iterative calculation of the focal length to minimize the weighted reprojection error of the keypoints followed by a least squares estimation of the ground plane. In a very similar approach [2] the authors employ differential evolution to obtain all of the camera parameters in a single optimization scheme. These methods require that few specific models of vehicles are present in the scene which may prevent their use in regions with different vehicle model composition or in areas with low traffic density. Only the last two methods do not require the focal length to be known therefore we use them as benchmarks for evaluation of our method.

2.2   Vanishing Point Detection

Since vanishing points provide important information about scene geometry a range of methods of detecting the vanishing points have been proposed. Methods usually rely on various estimation algorithms such as the Hough transform [12], EM [14], RANSAC [3] or deep neural networks [5].

In our approach we use a deep neural network to detect the vanishing points parametrized in the diamond space which was introduced in [7] where it was used in a modified scheme of the Cascaded Hough Transform [19] to obtain three orthogonal vanishing points of a Manhattan world. The diamond space parametrization is based on the PClines line parameterization [8] and it is defined as a mapping of the whole real projective plane to a finite diamond shaped space. We discuss the properties of this mapping in greater detail in Sect. 4.

3 Traffic Camera Geometry

Under perspective projection the images of parallel lines meet in the vanishing point corresponding to the direction of the lines. The knowledge of vanishing points can be utilized to constrain the camera calibration parameters. Each pair of vanishing points corresponding to two orthogonal directions places a single constraint on the focal length and the coordinates of the principal point. Therefore knowledge of three orthogonal vanishing points is sufficient to obtain the intrinsic camera parameters when no distortion and zero skew is assumed. If the position of the principal point is known or can be reliably assumed then only one such pair is sufficient. Additionally, the rotation of the camera with respect to the directions represented by the vanishing points can also be obtained and used to project image points onto the plane parallel with these directions.

Given the principal point $\mathbf{p} = (p_x, p_y)$ and two orthogonal vanishing points $\mathbf{u} = (u_x, u_y)$ and $\mathbf{v} = (v_x, v_y)$ expressed in the pixel coordinates and assuming zero skew and no distortion it is possible to calculate the focal length:

$$f = \sqrt{-(\mathbf{u} - \mathbf{p}) \cdot (\mathbf{v} - \mathbf{p})} = \sqrt{-(u_x - p_x)(u_y - p_y) - (v_x - p_x)(v_y - p_y)}. \quad (1)$$

Note that for some configurations this leads to imaginary focal lengths. In the calibration pipeline we simply discard any such configurations.

The two vanishing points also define the horizon line $\mathbf{h} = (h_a, h_b, h_c)$ in the image. The horizon line corresponds to a set of planes which are parallel with the directions corresponding to the two vanishing points. The shared normal of these planes can then be calculated using the intrinsic matrix K:

$$\mathbf{n} = K^T \mathbf{h} = \begin{pmatrix} f & 0 & 0 \\ 0 & f & 0 \\ p_x & p_y & 1 \end{pmatrix} \begin{pmatrix} h_a \\ h_b \\ h_c \end{pmatrix}. \quad (2)$$

It is possible to use the normal $\mathbf{n}$ to project any point $\mathbf{q} = (q_x, q_y)$ in the image to a point $\mathbf{Q}$ lying in a plane defined by the normal [15]:

$$\mathbf{Q} = -\frac{\delta}{\hat{\mathbf{q}} \cdot \mathbf{n}} \hat{\mathbf{q}}, \quad (3)$$

where $\hat{\mathbf{q}}$ denotes the vector $(q_x - p_x, q_y - p_y, f)$. Note that due to the ambiguity arising from the properties of single-view perspective geometry only the normal of the plane is established. In order for the projected points to have accurate metric properties it is necessary to include a scaling parameter δ which needs to be determined for the plane by using a real world measurement.

Since we want to detect two vanishing points for every vehicle we establish the directions to which these two would correspond to. In further text we will denote the vanishing point corresponding to the direction in which the vehicle is facing as the *first vanishing point*. The *second vanishing point* corresponds to the direction parallel with the road plane and orthogonal to the direction in which the vehicle is facing.

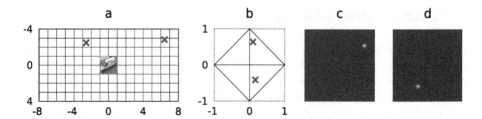

Fig. 1. a) The first (red cross) and the second (green cross) vanishing point in an image coordinate system where the dimensions of the bounding box of a vehicle span from -1 to 1 for both axes. b) The same vanishing points represented in the diamond space. c) The heatmap generated for the first vanishing point. d) The heatmap generated for the second vanishing point. (Color figure online)

4   Diamond Space

To represent the vanishing points we use the diamond space parametrization [7]. This parametrization can be thought of as a mapping of the infinite projective space to a bounded diamond-shaped space. The relation (4) shows the mapping of a point expressed in homogeneous coordinates from the original projective space to the diamond space. The relation (5) shows the opposite mapping.

$$(x, y, w)_o \rightarrow (-w, -x, \text{sgn}(xy)x + y + \text{sgn}(y)w)_d \tag{4}$$

$$(x, y, w)_d \rightarrow (y, \text{sgn}(x)x + \text{sgn}(y)y - w, x)_o \tag{5}$$

In our method we represent the diamond space with a heatmap of finite resolution. For use in the heatmap we also rotate the diamond space by $45°$ clockwise for a more efficient representation. The ground truth vanishing points are represented as 2D Gaussians with standard deviation of 1 pixel centered on the pixel closest to the actual position of the vanishing point in the diamond space. An example of the vanishing points in the original image space, the diamond space and the heatmaps is shown in Fig. 1.

Representing the diamond space with a heatmap of finite resolution brings about an issue of insufficient representation for some positions of vanishing points. In general the points which are farther from the origin in the original image space will have a greater systemic error arising from insufficient heatmap representation. To reduce the error the original image space can be scaled so that the vanishing points are closer to the origin. However just choosing only one scale may lead to problems since then the objects which were close to the origin before the scaling would occupy smaller space in the image space and thus also be represented with greater error in the heatmap. Therefore we choose to use multiple scales to represent each vanishing point. During inference we select the value from the heatmap with the smallest expected error.

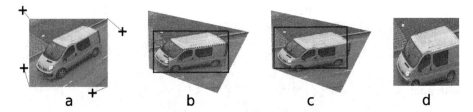

Fig. 2. Training data augmentation. a) The original image from the BoxCars116k dataset [17]. Each corner (black cross) is randomly perturbed to generate a perspective transformation. b) The image after perspective transformation. The transformation is also applied to the ground truth 3D bounding box (dotted red). 2D bounding box (solid blue) is constructed based on the 3D bounding box. c) The 2D bounding box is randomly perturbed. d) The image is cropped to only contain the contents of the perturbed 2D bounding box and rescaled to 128 × 128 pixels. (Color figure online)

5   Vanishing Point Detection Network

To detect the vanishing points we use the stacked hourglass network [13] with two hourglass modules. We make a slight modification in the initial convolutional layer of the network by not using stride. As input we use cropped images containing only one vehicle with the resolution 128 × 128 pixels. The network outputs 4 heatmaps with the resolution 64 × 64 representing the diamond space at four different scales (0.03, 0.1, 0.3, 1.0) for each of the two vanishing points.

For training we use the data from the BoxCars116k dataset [17]. The dataset contains images of vehicles which were cropped from traffic surveillance footage along with bounding box and vanishing point annotations. We split the dataset into 92 932 training images, 11 798 validation images and 11 556 testing images.

We train the network with L2 loss over the heatmaps with batch size of 32 using the Adam optimizer. We train with the learning rate at 0.001 for 60 epochs and 0.0001 for 15 epochs.

5.1   Augmentation

The BoxCars116k dataset contains only images of vehicles captured by 137 distinct traffic cameras. Therefore the distribution of vanishing point positions in the training data is limited. To remedy this problem we apply a random perspective transformation on the training images and vanishing points.

The perspective transform is constructed by warping the position of the corners of the original image by a random offset sampled from normal distribution with a standard deviation of 12.5 pixels. The transformation is applied on the image, the two vanishing points and the ground truth 3D bounding box. The transformed 3D bounding box is used to calculate the new 2D bounding box of the vehicle. Since vehicles are not planar objects this approach is not strictly geometrically correct, but it nevertheless results in better performance.

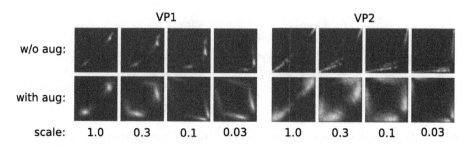

Fig. 3. The cummulative heatmaps for the first (left) and the second (right) vanishing point at the four scales used by the network averaged over 10 000 training samples without augmentation (top) and with augmentation (bottom). After applying augmentation the training data represent a much broader distribution of vanishing point positions.

During inference the network will work in a pipeline after detecting vehicles with an independent object detection network which may output badly aligned bounding boxes. We therefore also augment the 2D bounding box by shifting its corners with values sampled from uniform distribution spanning -5 to 5 pixels. We also horizontally flip the images with a 50 % probability. The whole process of augmentation of a single training sample is visualised in Fig. 2.

Applying the transformation significantly expands the range of vanishing point positions in the training data. This fact is visualised in Fig. 3.

6   Camera Calibration Pipeline

6.1   Vanishing Point Detection

To calibrate a traffic camera we receive its traffic surveillance footage and detect vehicles in the individual frames using the CenterNet object detector [20] pre-trained on the MS COCO dataset [11]. We only retain at most ten bounding boxes with greatest confidence. In scenes with vehicles which are static over multiple frames we discard the static vehicles after few initial detections. With the remaining bounding boxes we crop the detected vehicles from the frames and resize them to 128×128 pixels. These vehicle images are put into the vanishing point detection network to obtain the 4 heatmaps for each vanishing point. For each scale we find the row i^s and column j^s with the maximum heatmap value. We also find all of the positions in the heatmap for which the value is close to the maximum:

$$P^s = \left\{ (i,j) \mid H^s_{i,j} \geq 0.8 H^s_{i^s,j^s} \right\}, \tag{6}$$

where $H^s_{i,j}$ is the value in the i-th row and j-th column in the heatmap for scale s. We then calculate an approximate measure of accuracy of the vanishing point for each scale:

$$D^s = \frac{1}{|P^s|} \sum_{(i,j) \in P^s} \frac{||\mathbf{v}^s_{i,j} - \mathbf{v}^s_{i^s,j^s}||}{||\mathbf{v}^s_{i^s,j^s}||}, \tag{7}$$

where $|P^s|$ is the number of elements in the set, $\mathbf{v}_{i,j}^s$ is the vanishing point corresponding to the i-th row and j-th column in the heatmap for scale s in the image coordinate system in which the vehicle bounding box is centered at the origin and spans -1 to 1 in both dimensions (see Fig. 1-a). We then select the vanishing point $\mathbf{v}_{i^s,j^s}^s$ with the scale which minimizes D^s. The measure D^s incorporates both the inaccuracies arising from the finite resolution of the heatmaps as well as the uncertainty caused by unclear peaks in the heatmaps. Finally we convert the vanishing point position from the coordinate system tied to the bounding box to the image coordinate system of the whole frame.

6.2   Calculating the Scene Geometry

Given the pairs of vanishing points we first calculate the focal lengths for each pair using (1). We discard all pairs which result in an imaginary focal length. We determine the focal length to be the median of the remaining focal lengths. Assuming zero skew, no distortion and the principal point of the camera to be in the centre of the image we can now construct the intrinsic camera matrix K.

We determine the horizon using an approach inspired by the Theil-Sen estimator [18]. For each pair of vanishing points we determine the line defined by them and express it in the form of $y = mx + k$. We take the median of all the values of m and denote it $\hat{m}$. For each vanishing point $\mathbf{v} = (v_x, v_y)$ we calculate $q = v_y - v_x\hat{m}$. Again we calculate the median of the values of q and denote it as $\hat{q}$. The final horizon can be expressed with the equation $y = \hat{m}x + \hat{q}$ which is equivalent to the line $\mathbf{h} = (\hat{m}, -1, \hat{q})$. We can then use (2) to determine the normal of the road plane.

7   Evaluation

We evaluate our method on two traffic surveillance datasets BrnoCompSpeed [16] and BrnoCarPark [2]. Both datasets contain information about measured distances between pairs of points on the road plane along with their pixel positions in the images. We use this information to evaluate the accuracy of our methods. Since our method does not produce a scale we use two distinct evaluation metrics which are scale independent. The first metric is calculated by comparing the relative differences of ratios of two different measurements defined as

$$r_{i,j} = \frac{\left| \frac{\hat{d}_i}{\hat{d}_j} - \frac{d_i}{d_j} \right|}{\frac{\hat{d}_i}{\hat{d}_j}}, \tag{8}$$

where $\hat{d}_i$ is the i-th ground truth distance measurement and the d_i is the i-th measurement based on the projection (3) for the given method. To obtain

Table 1. The mean calibration errors on the BrnoCompSpeed dataset [16].

Method	MRDE	OSMAPE
DubskaCalib [9]	8.54	5.67
SochorCalib [15]	3.83	2.57
DeepVPCalib (ours)	14.95	13.74

Table 2. The mean calibration errors on the BrnoCarPark dataset [2].

Method	MRDE	OSMAPE
LandmarkCalib [2]	8.17	5.29
PlaneCalib [1]	21.99	13.17
DeepVPCalib (ours)	8.66	6.83

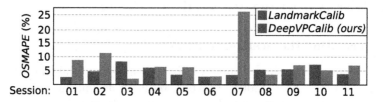

Fig. 4. The *OSMAPE* errors for *LandmarkCalib* [2] and our method *DeepVPCalib* for individual sessions in the BrnoCarPark dataset.

the calibration error for a single camera we compute the mean over all of the possible pairs of ground truth measurements. We denote this error as *MRDE*. As the second metric we use the mean absolute percentage error:

$$E_{MAPE} = \frac{100}{n} \sum_{i=1}^{n} \frac{|d_i - \hat{d}_i|}{\hat{d}_i} \qquad (9)$$

with the scale parameter δ determined using linear programming such that the value of (9) is minimal. We denote this error as *OSMAPE*.

7.1 BrnoCompSpeed

The BrnoCompSpeed dataset [16] contains 21 hour-long videos of a freely flowing traffic on straight road segments captured from overpasses. For evaluation we use the split C of the dataset which contains 9 videos. For our method we only use every tenth frame of the video and we only process the first 1500 frames.

We compare our method denoted as *DeepVPCalib* with the fully automatic methods from [9] which we denote as *DubskaCalib* and [15] which we denote as *SochorCalib*. The results are available in the Table 1. Compared with the state of the art our method underperforms. However we note that the methods *DubskaCalib* and *SochorCalib* both only work if the observed vehicles move along a straight road segment and our method does not have this limitation.

7.2 BrnoCarPark

The BrnoCarPark dataset [2] contains footage from 11 different cameras observing parking lots. The footage is available both in video form and as frames which were directly used for camera calibration by the method which was published alongside the dataset. To perform a better comparison with this method which we denote as *LandmarkCalib* we also apply our method denoted as *DeepVP-Calib* on the selected frames and not the whole video. The number of frames per session ranges from 152 to 45 282, but we opt to use at most the first 5000 frames. In the comparison we also include the results from [1] which we denote as *PlaneCalib*. The results for the methods are available in the Table 2. We also include the *OSMAPE* errors for the individual sessions from the BrnoCarPark dataset for *LandmarkCalib* and our method in Fig. 4.

The calibration error for our method is comparable to the error of the state of the art method *LandmarkCalib* while not requiring any vehicles of specific models to be present in the surveilled scene. Our method is thus more easily usable in wider range of geographical regions where the composition of vehicle models might be such that any traffic surveillance footage might contain only a very limited number of the required vehicles to be visible.

7.3 Ablation Studies

Table 3. Results of the various ablation experiments discussed in Subsect. 7.3.

#	Type	Backbone	Scales	Loss	Augmentation Perspective	BBox shift	Flip	BrnoCompSpeed *MRDE*	*OSMAPE*	BrnoCarPark *MRDE*	*OSMAPE*
1	Heatmap	Hourglass	0.03, 0.1, 0.3, 1.0	L2	12.5 px	5 px	Yes	**14.94**	13.74	**8.66**	**6.83**
2	Heatmap	Hourglass	0.03, 0.1, 0.3, 1.0	L2	5 px	5 px	Yes	18.29	16.09	12.41	9.15
3	Heatmap	Hourglass	0.03, 0.1, 0.3, 1.0	L2	n/a	n/a	No	19.17	**10.49**	19.67	12.85
4	Heatmap	Hourglass	0.03	L2	12.5 px	5 px	Yes	17.62	18.00	20.43	15.84
5	Heatmap	Hourglass	0.1	L2	12.5 px	5 px	Yes	23.89	16.60	20.91	17.61
6	Regression	ResNet50	1.0	nL1	12.5 px	5 px	Yes	19.48	18.85	20.45	14.54
7	Regression	ResNet50	1.0	nL1	5 px	5 px	Yes	15.54	15.00	19.64	14.88
8	Regression	ResNet50	1.0	nL2	12.5 px	5 px	Yes	20.61	21.26	21.27	16.71
9	Regression	ResNet50	1.0	nL2	5 px	5 px	Yes	15.31	20.99	21.60	15.58
10	Regression	Hourglass	1.0	nL1	12.5 px	5 px	Yes	18.35	17.67	22.62	16.60

To verify that the various components of our method are necessary to obtain the final results we trained several models with some of the components removed. The results on the split C of the BrnoCompSpeed dataset [16] and the BrnoCarPark dataset [2] for all of these models are presented in Table 3. The model for the main method presented in this article is labeled as model 1.

The results for models 1–3 show that augmentation reduces the calibration error especially on the BrnoCarPark dataset. To verify that the use of four heatmaps at different scales is beneficial we trained models 4 and 5 with only one heatmap for each vanishing point. As expected using just a single scale leads to increased error.

We also tested an approach based on direct regression of the positions of vanishing points. We ran multiple experiments training networks with L1 or L2 loss directly applied to the outputs of the network, but we were unable to get the networks to converge to any meaningful results even with extensive grid search for optimal hyperparameters. We therefore opt use normalized losses. The normalized L2 loss has the form of:

$$loss_{nL2}(\mathbf{v}, \hat{\mathbf{v}}) = \frac{||\mathbf{v} - \hat{\mathbf{v}}||^2}{||\hat{\mathbf{v}}||^2}, \tag{10}$$

where $\mathbf{v}$ is the output vanishing point and $\hat{\mathbf{v}}$ is ground truth vanishing point. The normalized L1 loss is analogous to (10). For models 6–9 we used the standard ResNet50 backbone [10] trained for 60 epochs with learning rate of 0.001, 20 epochs with learning rate of 0.0001 and 20 epochs with the learning rate of 0.00001. For model 10 we used the stacked hourglass network with 2 hourglass modules outputting a 64 channel heatmap followed by global pooling and several fully connected layers and the same training schedule as the other heatmap models. The regression models achieve significantly worse results than the main heatmap model. This indicates that using the diamond space heatmap representation is beneficial for detecting the vanishing points of vehicles for camera calibration.

8   Conclusion

In this paper we have presented a semi-automatic method for camera calibration based on a deep neural network detecting vanishing points of vehicles traffic surveillance footage. Our method achieves competitive results on the BrnoCarPark dataset [2] while not requiring a presence of specific vehicle models in the surveilled scene. Compared to the current state of the art our method can be applied in a wider range of use-cases without the need for further training.

In the current form our method does not provide the scaling parameter which would enable metric measurements in the road plane. A single known distance in the road plane can be used to obtain this parameter. In future we aim to make this process automatic. Our method already detects the relative orientations of vehicles via vanishing points it may be possible to use existing approaches based on 3D bounding box statistics [9] or scaling of 3D CAD models of vehicles [15].

Acknowledgments. The authors would like to thank Zuzana Kukelova for her valuable comments. The authors gratefully acknowledge the support of NVIDIA Corporation with the donation of GPUs. This work was funded by grant no. UK/214/2021 provided by the Comenius University in Bratislava.

References

1. Bartl, V., Juránek, R., Špaňhel, J., Herout, A.: PlaneCalib: automatic camera calibration by multiple observations of rigid objects on plane. In: 2020 Digital Image Computing: Techniques and Applications (DICTA), pp. 1–8. IEEE (2020)
2. Bartl, V., Špaňhel, J., Dobeš, P., Juránek, R., Herout, A.: Automatic camera calibration by landmarks on rigid objects. Mach. Vis. Appl. **32**(1), 1–13 (2021)
3. Bazin, J.C., Pollefeys, M.: 3-line RANSAC for orthogonal vanishing point detection. In: 2012 IEEE/RSJ International Conference on Intelligent Robots and Systems, pp. 4282–4287. IEEE (2012)

4. Bhardwaj, R., Tummala, G.K., Ramalingam, G., Ramjee, R., Sinha, P.: AutoCalib: automatic traffic camera calibration at scale. ACM Trans. Sens. Netw. (TOSN) **14**(3–4), 1–27 (2018)

5. Chang, C.K., Zhao, J., Itti, L.: DeepVP: deep learning for vanishing point detection on 1 million street view images. In: 2018 IEEE International Conference on Robotics and Automation (ICRA), pp. 4496–4503. IEEE (2018)

6. Corral-Soto, E.R., Elder, J.H.: Automatic single-view calibration and rectification from parallel planar curves. In: Fleet, D., Pajdla, T., Schiele, B., Tuytelaars, T. (eds.) ECCV 2014. LNCS, vol. 8692, pp. 813–827. Springer, Cham (2014). https://doi.org/10.1007/978-3-319-10593-2_53

7. Dubská, M., Herout, A.: Real projective plane mapping for detection of orthogonal vanishing points. In: Proceedings of the British Machine Vision Conference 2013, pp. 90.1–90.10. British Machine Vision Association (2013)

8. Dubská, M., Herout, A., Havel, J.: PClines-line detection using parallel coordinates. In: CVPR 2011, pp. 1489–1494. IEEE (2011)

9. Dubská, M., Herout, A., Sochor, J.: Automatic camera calibration for traffic understanding. In: Proceedings of the British Machine Vision Conference, vol. 4, p. 8. BMVA Press (2014)

10. He, K., Zhang, X., Ren, S., Sun, J.: Deep residual learning for image recognition. In: Proceedings of the IEEE Conference on Computer Vision and Pattern Recognition, pp. 770–778 (2016)

11. Lin, T.-Y., et al.: Microsoft COCO: common objects in context. In: Fleet, D., Pajdla, T., Schiele, B., Tuytelaars, T. (eds.) ECCV 2014. LNCS, vol. 8693, pp. 740–755. Springer, Cham (2014). https://doi.org/10.1007/978-3-319-10602-1_48

12. Matessi, A., Lombardi, L.: Vanishing point detection in the Hough transform space. In: Amestoy, P., et al. (eds.) Euro-Par 1999. LNCS, vol. 1685, pp. 987–994. Springer, Heidelberg (1999). https://doi.org/10.1007/3-540-48311-X_137

13. Newell, A., Yang, K., Deng, J.: Stacked hourglass networks for human pose estimation. In: Leibe, B., Matas, J., Sebe, N., Welling, M. (eds.) ECCV 2016. LNCS, vol. 9912, pp. 483–499. Springer, Cham (2016). https://doi.org/10.1007/978-3-319-46484-8_29

14. Nieto, M., Salgado, L.: Simultaneous estimation of vanishing points and their converging lines using the EM algorithm. Pattern Recogn. Lett. **32**(14), 1691–1700 (2011)

15. Sochor, J., Juránek, R., Herout, A.: Traffic surveillance camera calibration by 3D model bounding box alignment for accurate vehicle speed measurement. Comput. Vis. Image Underst. **161**, 87–98 (2017)

16. Sochor, J., et al.: Comprehensive data set for automatic single camera visual speed measurement. IEEE Trans. Intell. Transp. Syst. **20**(5), 1633–1643 (2018)

17. Sochor, J., Špaňhel, J., Herout, A.: Boxcars: improving fine-grained recognition of vehicles using 3-D bounding boxes in traffic surveillance. IEEE Trans. Intell. Transp. Syst. **20**(1), 97–108 (2018)

18. Theil, H.: A rank-invariant method of linear and polynomial regression analysis. Indag. Math. **12**(85), 173 (1950)

19. Tuytelaars, T., Proesmans, M., Van Gool, L.: The cascaded Hough transform. In: Proceedings of International Conference on Image Processing, vol. 2, pp. 736–739. IEEE (1997)

20. Zhou, X., Wang, D., Krähenbühl, P.: Objects as points. arXiv preprint arXiv:1904.07850 (2019)

Efficient Spatio-Temporal Network with Gated Fusion for Video Super-Resolution

Changyu Li[1,2], Dongyang Zhang[2,3], Ning Xie[2,3], and Jie Shao[2,3(✉)]

[1] Guizhou Provincial Key Laboratory of Public Big Data, Guizhou University,
Guiyang 550025, China
[2] Center for Future Media, University of Electronic Science and Technology of China,
Chengdu 611731, China
{changyulve,dyzhang}@std.uestc.edu.cn, shaojie@uestc.edu.cn
[3] Sichuan Artificial Intelligence Research Institute, Yibin 644000, China

Abstract. Video super-resolution (VSR) has drawn much attention in research community recently. As the typical three-dimensional (3D) signals, how to exploit spatio-temporal features in videos effectively and efficiently is critical for VSR. Most methods explore spatio-temporal feature through optical flow estimation and motion compensation. Although promising, these methods suffer from the trade-off between model performance and complexity. In this paper, we present a novel efficient spatio-temporal network (denoted as "ESTN") for VSR, which is designed to separately encode spatial and temporal video frames through two parallel streams. In particular, several 2D and 3D convolutions are utilized to encode on the central frame and consecutive frames for feature extraction. Besides, for the better ability of adaptive alignment at the feature level, instead of direct addition, we propose to learn a gate module consisting of deformable convolution to fuse the spatial and temporal features from the two streams. Such design enables efficient spatio-temporal exploration and maintains a lightweight model. Experiments demonstrate that the proposed ESTN achieves competitive or even better performance than its competitors with similar parameters.

Keywords: Video super-resolution · Deep learning · Spatio-temporal features

1 Introduction

Recently, super-resolution (SR) has obtained a lot of attention both in the research and industrial communities, due to the vast demand for high-quality images and videos. The target of SR is to generate the corresponding high-resolution (HR) images or videos from its low-resolution (LR) counterparts. However, SR is a well-known inverse problem since each LR input may have multiple HR solutions.

© Springer Nature Switzerland AG 2021
I. Farkaš et al. (Eds.): ICANN 2021, LNCS 12895, pp. 640–651, 2021.
https://doi.org/10.1007/978-3-030-86383-8_51

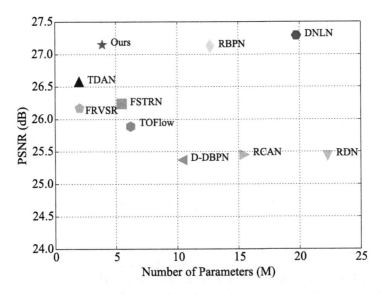

Fig. 1. Trade-off between parameter and accuracy of various methods. The average PSNR results of different methods are tested on the Vid4 dataset with ×4 factor.

As a kind of classical two-dimensional (2D) signals, with the booming of deep learning, single image super-resolution (SISR) has been extensively studied by existing methods [2–4]. As a pioneer, Dong et al. [3] firstly proposed an end-to-end trainable architecture, named SRCNN, to learn the nonlinear mapping between LR and HR images. In pursuit of better performance, some existing networks concentrated on wider or deeper network designs. Inspired by the idea of back-projection, D-DBPN [5] was proposed to calculate reconstruction error between HR and LR features through multiple up- and down-sampling layers. Besides, benefiting from both the residual learning and dense learning, RDN [22] can restore high-quality HR images. Borrowing the idea from the squeeze-and-excitation (SE) block [7], Zhang et al. [21] integrated the attention mechanism into SR network and proposed a very deep network named RCAN. In an unsupervised manner, Guo et al. [4] proposed a dual regression strategy to reduce the possible mapping space by introducing additional constraints on LR data. Although promising, the methods mentioned above suffer from enormous parameters, hindering their applications to real world problems, and how to design a lightweight network becomes a current rising trend [8,20].

Compared with the great progress made in SISR, video super-resolution (VSR) is still an open issue. Although vanilla SISR methods can be directly applied to video frames by treating them as single images, the final results are often beyond our expectation [17]. As shown in Fig. 1, the leading SISR methods, such as D-DBPN [5], RDN [22] and RCAN [21], fail to obtain pleasing results in the VSR task and cost huge parameters. This is because videos are typical 3D signals, and apart from the spatial relation, abundant detailed information is

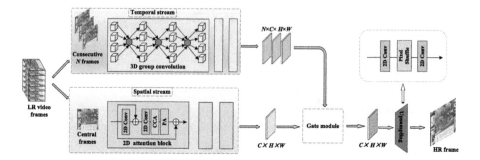

Fig. 2. Architecture of our proposed efficient spatio-temporal network (ESTN). PReLU is used as the activation function in our network. For simplicity, we omit PReLU here.

available by exploring intrinsic temporal information among multiple LR video frames. To overcome the limitation of SISR, optical flow estimation and motion compensation are used for temporal modeling [1,6,19]. However, due to the existence of motion blur and occlusion, it is still a challenging problem for reliable optical flow estimation. Besides, the optical flow algorithm is notorious for huge computational overhead, e.g., RBPN [6] is a huge model with unbearable computing costs. Moreover, instead of optical flow, Sajjadi et al. [14] proposed a fully end-to-end trainable network, named FRVSR. By dividing each 3D filter into the product of two 3D filters, FSTRN [11] was proposed to expedite the running time for VSR.

Although the above models obtain fascinating results, their model sizes and computation complexities could be unfavorable. Thus, it is imperative to develop certain models to simultaneously maintain high accuracy and computational efficiency. To this end, in this paper we build an efficient model to achieve the trade-off between the effectiveness and efficiency for VSR task. As shown in Fig. 2, the proposed architecture is a unified network with two branches. In particular, one branch (spatial stream) extracts the spatial features of the key frame while the other branch (temporal stream) models the temporal features of the consecutive frames. In contrast to the traditional methods [11,14], the performance of VSR can be enhanced by explicitly leveraging both spatial and temporal information. Meanwhile, a gate module based on deformable convolution is proposed to conduct motion alignment at feature level, which integrates the two streams into a compact feature for the final HR reconstruction. Overall, the main contributions of our work can be summarized as the following aspects:

– We propose a novel efficient spatio-temporal network (ESTN) as shown in Fig. 2, which decouples the joint spatio-temporal feature extraction in two separate streams for better network regularization.
– We propose a novel gate module to adaptively fuse the features from the two streams into an informative feature representation.

– The proposed ESTN is not only efficient, achieving competitive or even better performance than its competitors, but also entails small number of parameters.

2   Related Work

2.1   Single Image Super-Resolution

The target of single image super-resolution (SISR) is to restore an HR image from its LR observation. With the development of deep learning, the CNN based methods have achieved dramatic advantages over conventional methods based on feature engineering for SISR. As the seminal work, Dong et al. [3] firstly applied CNN to SISR and proposed a 3-layer end-to-end convolutional neural network named SRCNN. Since then, extensive learning-based SISR methods have emerged. To further increase CNN depth, Zhang et al. [22] proposed a very deep network named RDN. Meanwhile, the residual learning and dense learning were exploited to ease training difficulty. Then, Zhang et al. [21] proposed RCAN with the help of channel attention, and in particular, residual-in-residual (RIR) structure in RCAN showed great capacity in high frequency detail recovery. Guo et al. [4] proposed a dual regression strategy to reduce the possible mapping space by introducing additional constraints on LR data, which can be extended to unsupervised manner.

2.2   Video Super-Resolution

For the VSR task, how to model temporal information is very important. Some classical methods exploited optical flow to estimate motion between consecutive frames, such as RBPN [6]. However, such methods suffer from huge computation and inaccurate optical flow estimation. Besides, some methods adopted 2D CNN with early fusion to exploit temporal information. TDAN [15] utilized several 2D convolutional layers to aggregate multi-frames feature. DNLN [17] utilized non-local operation to capture long-range dependencies among frames, and then aggregated the correlations maps with several 2D convolution. Moreover, 3D CNN with slow fusion was also widely used in VSR methods. For example, FSTRN [11] utilized stacked 3D filters to extract both spatial and temporal information, which move both in temporal and spatial axis directions concurrently. In contrast to the above methods, the proposed ESTN explicitly exploits the spatial and temporal information through two branches, leading to a complementary effect to each other for enhanced results.

3   Proposed Method

As shown in Fig. 2, the proposed ESTN is composed of three components: temporal stream, spatial stream and gate module. The spatial stream focuses on static appearance features from a still image, while temporal part captures the motion information across the consecutive frames, providing the additional clue for image restoration. Moreover, the gate module integrates the two streams into one compact representation for final HR restoration. In the following, we give the details of the three components.

3.1 Temporal Stream

Due to the existence of additional clue (time), how to exploit the motion flow among consecutive frames is the key issue for the VSR task [1]. One common solution is to apply extra 2D CNNs to learn motion features based on pre-extracted optical flow. Such design is simple and intuitive, and most methods [1,6] adopt it for temporal modeling. However, extracting optical flow is notorious for its time and storage cost, especially in the case of HR videos. Besides, optical flow estimation is often inaccurate. The follow-up solution is 3D based CNNs, such as C3D [16], which is capable of directly learning spatio-temporal features from N consecutive frames. However, it still suffers from low computational efficiency due to the high-dimensional 3D kernels [12].

Overall, the temporal stream can be formulated as:

$$Stream_{temp} = NET_{temp}(I^{LR}_{[t-N,t+N]}), \tag{1}$$

where $I^{LR}_{t-N,t+N}$ is the consecutive $2N+1$ LR input frames in the form of $\{I^{LR}_{t-N}, \ldots, I^{LR}_{t-1}, I^{LR}_{t}, I^{LR}_{t+1}, \ldots, I^{LR}_{t+N}\}$, NET_{temp} denotes the temporal stream network which is a typical modification based on ResNeXt [18] and $Stream_{temp} \in R^{(2N+1) \times H \times W \times C}$ is the output feature. Instead of optical flow extraction, we employ the 3D convolution for temporal modeling. To avoid the heavy computational cost, we take the advantage of 3D version ResNeXt [18], which adopts the idea of grouped convolution [10]. ResNeXt [18] shows that, in addition to the dimensions of depth and width, cardinality denoted as the size of the set of nonlinear transformations is also an essential factor to improve the performance and it is more effective to increase the model capacity by increasing cardinality than going deeper or wider. Thus, we replace the 2D convolutions with 3D convolutions in ResNeXt and the cardinality is set to 16. In this way, we can minimize computational complexity and model size brought by 3D convolution and maximize the model performance.

3.2 Spatial Stream

The target of spatial stream is to capture the global statistics from a single video frame, which is strongly associated with high frequency detail restoration. Actually, spatial stream is similar to SISR to some extent, which can be formulated as:

$$Stream_{spat} = NET_{spat}(I^{LR}_{t}), \tag{2}$$

where I^{LR}_{t} is the central frame in $I^{LR}_{t-N,t+N}$, and the output $Stream_{spat} \in R^{C \times H \times W}$ is the 2D feature maps. As shown in Fig. 2, the core of spatial stream is the 2D attention block incorporating attention mechanism in terms of channel-wise and pixel-wise. Initially, the channel attention (CA) is based on SE block [7] and most SR methods [21] based on CA only utilize global average to capture the global information, which shows less informative for areas with intense contrast, such as edges. Thus, we propose the contrast-aware channel attention (CCA). Concretely, the global average pooling is replaced by the summation of standard deviation and mean which evaluate the contrast degree of a feature map. Given $X \in R^{H \times W \times C}$ as the input feature, H and W indicate spatial resolution, C is the channel number and x_c is the c-th element of X. Therefore, the contrast information can be obtained by:

$$
\begin{aligned}
w_c &= CCA(x_c) \\
&= \sqrt{\frac{1}{WH} \sum_{(i,j)\in x_c} (x_c^{(i,j)} - \frac{1}{WH} \sum_{(i,j)\in x_c} x_c^{(i,j)})^2} \\
&\quad + \frac{1}{WH} \sum_{(i,j)\in x_c} x_c^{(i,j)}.
\end{aligned}
\tag{3}
$$

Similarly, w_c is the c-th element of attention weight W, which is used to recalibrate X through channel-wise multiplication.

Besides CCA, we also consider the pixel attention (PA) in the proposed 2D attention block. Referring to Fig. 2, F (the output feature of CCA) is directly used as input feature for PA, which can be formulated as:

$$
P = \sigma(Conv(\delta(Conv(F)))), \tag{4}
$$

where σ and δ stand for the sigmoid and PRelu activation function, and the two convolution layers change the shape from $H \times W \times C$ to $H \times W \times 1$. Finally, to obtain the output, element-wise multiplication is applied for input F and weight map P. By means of CCA and PA together, the flexibility of the model is increased to deal with all sorts of information.

3.3 Gate Module

In order to integrate the temporal and spatial streams, a novel gate module is proposed. As shown in Fig. 3, each clip from temporal stream has the same resolution as spatial stream, and both of the two streams are used to perform frame alignment through the DC module. Particularly, the DC module is based on deformable convolutions [23]. In contrast to normal convolution, deformable convolution is operated on irregular positions with dynamic weights to achieve adaptive sampling. Moreover, to increase the size of receptive field, the hierarchical feature fusion block (HFFB) [9] is used in the DC module. HFFB adopts the spatial pyramid structure with dilated convolutions to effectively enlarge receptive field with low computational cost. Next, offsets are generated, which are fed to deformable convolution for deformable kernel. Benefiting from the

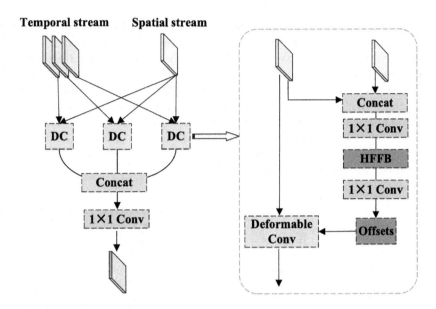

Fig. 3. Scheme of the gate module. The output feature is used for HR image restoration by sub-pixel convolution.

large receptive field and deformable kernel, our gate module can address complicated and large motions between frames well. Through N DC modules, the neighboring temporal feature maps can be well aligned at feature level. Then, concatenation operation followed by 1×1 convolution is used to generate fused feature maps. Overall, the function of the gate module can be formulated as:

$$Stream_{fusion} = NET_{gate}(Stream_{spat}, Stream_{temp}), \qquad (5)$$

where $Stream_{fusion} \in R^{C \times H \times W}$ is used for final HR frame restoration.

4   Experiment

4.1   Implementation Details

In all our experiments, we focus on $4 \times$ SR factor. Vimeo-90k [19] is used as the training dataset, which includes various and complex real-world scenarios. In particular, Vimeo-90k [19] is a relatively large dataset, containing 64612 training clips, and each clip is composed of consecutive seven frames, with a fixed resolution of 448×256. With regard to evaluation, we compare our model with other state-of-the-art methods on the Vid4 dataset [13] which is composed of four video clips: *city, walk, calendar* and *foliage*. Following most VSR methods, we utilize peak signal-to-noise ratio (PSNR) and structural similarity (SSIM) computed on the brightness Y channel to evaluate the model performance. Moreover, the network takes seven consecutive frames as input and we adopt $\mathcal{L}_1$ as the loss function, with the help of Adam optimization for model parameter update. Initially, the learning rate is set to 0.0001 for all layers. Besides, we implement our ESTN with the PyTorch framework and all experiments are conducted on one NVIDIA RTX GPU.

Table 1. Ablation study of each proposed component. "Block" denotes the 2D attention block in spatial stream, "Temp" denotes the temporal stream and "Gate" stands for the proposed gate module as shown in Fig. 3. The results are tested on *Foliage* clips on the Vid4 dataset. For fairness, all the comparison methods are designed to have nearly the same numbers of parameters.

Model	1	2	3	4
Block		√	√	√
Temp			√	√
Gate				√
PSNR	24.28	25.25	26.13	26.22
#Para	2686803	2700228	3491204	3840552

4.2 Model Analysis

To demonstrate the superiority of the two-stream architecture, we conduct an ablation study by considering the different components of our proposed ESTN. Referring to Table 1, as each component mounts to the model gradually, the performance gets better and better, manifesting the effect of each component to boost the performance. Model-1 is a baseline, which is only built upon the channel attention block introduced in [21]. Comparing Model-1 and Model-2, the effect of the proposed 2D attention block is evident. Based on Model-2, Model-3 is equipped with one additional key component (temporal stream). However, Model-3 shows significant performance improvement, which demonstrates the importance of temporal modeling for VSR. Beyond that, we also make an visual comparison in terms of the temporal profile which shows the temporal consistency between consecutive frames. Following [14], the temporal profile is generated by taking the same horizontal row of pixels from consecutive frames and stacking them vertically to form a new image. As shown in Fig. 4, we can find that the method of bicubic interpolated frames demonstrates the worst results because of poor image details. Although, the model without temporal stream produces a clearer structure than interpolation-based method, it still suffers from over-smooth, meaning that image SR methods will lead to severe temporal discontinuity for video SR. As to our two-stream method, it produces the best temporal profile with more sharp structure, showing its strong capacity to capture long-range dependencies. Finally, comparing Model-3 with full Model-4, instead of gate module, the two-stream fusion can be achieved by simple element-wise addition in Model-3, and we can find that the gate module indeed contributes to boost the PSRN improvement and can deal well with complicated and large motions between frames due to deformable convolutions and large receptive field.

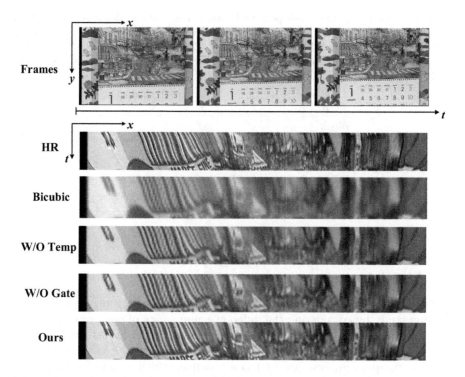

Fig. 4. Temporal profile comparison of different methods on the *Calendar* clip from Vid4. Zooming in for better visual quality.

4.3 Comparisons with State-of-the-Art Methods

In this section, both quantitatively and qualitatively, we compare our ESTN with other state-of-the-art methods, among which, DBPN [5], RDN [22] and RCAN [21] are leading SISR methods, while FSTRN [11], TDAN [15] and FRVSR [14] are VSR methods. As shown in Table 2, measuring on PSNR and SSIM indices, our method surpasses all the other methods by a large margin on the Vid4 benchmark. Generally speaking, it is noteworthy that SISR methods are barely satisfactory than VSR methods, which further indicates the importance of temporal modeling for video based tasks. We compare our model with other methods in terms of running time to verify the high efficiency of our work in Table 3. For the sake of fairness, all the models are tested on one NVIDIA RTX GPU. Table 3 shows that our model represents a great surpass on running time in comparison with other methods. Beyond that, the qualitative results are also shown in

Fig. 5. Due to the contrast-aware channel attention and pixel attention in spatial stream, which can capture the contrast degree of a feature map and impose the attention on each image pixel, our method can deal with the intense changed area, such as structure, texture, and edge. Referring to Fig. 5, we can find that the proposed ESTN is competent to recover the most pleasing SR frames with sharper and clearer edges. Besides, the trade-off between PSNR and parameter amount is better visualized in Fig. 1, indicating that ESTN is lightweight as performing favorably against the existing methods with less computational cost.

Table 2. PSNR and SSIM results evaluated on the Vid4 dataset with scale factor ×4.

Methods	City	Walk	Calendar	Foliage	Average
	PSNR/SSIM	PSNR/SSIM	PSNR/SSIM	PSNR/SSIM	PSNR/SSIM
Bicubic	25.13/0.601	26.06/0.798	20.54/0.571	23.50/0.566	23.81/0.634
DBPN [5]	25.80/0.682	28.64/0.872	22.29/0.715	24.73/0.661	25.37/0.732
RDN [22]	26.03/0.694	28.61/0.872	22.32/0.721	24.76/0.663	25.43/0.736
RCAN [21]	26.06/0.694	28.64/0.873	22.33/0.723	24.77/0.664	25.45/0.738
FSTRN [11]	26.31/0.698	28.69/0.891	22.27/0.718	25.02/0.675	25.57/0.745
TDAN [15]	26.99/0.757	29.50/0.890	22.98/0.756	25.51/0.717	26.24/0.780
FRVSR [14]	27.65/0.805	29.70/0.899	23.44/0.784	25.97/0.753	26.69/0.810
Ours	**27.75/0.805**	**30.71/0.911**	**23.96/0.807**	**26.22/0.756**	**27.16/0.819**

Table 3. Running time comparison with ×4 scale. All the methods are evaluated on the same mechine.

Model	D-DBPN	RDN	RBPN	FSTRN	TDAN	RCAN	Ours
Time (ms)	221	166	443	144	162	330	133

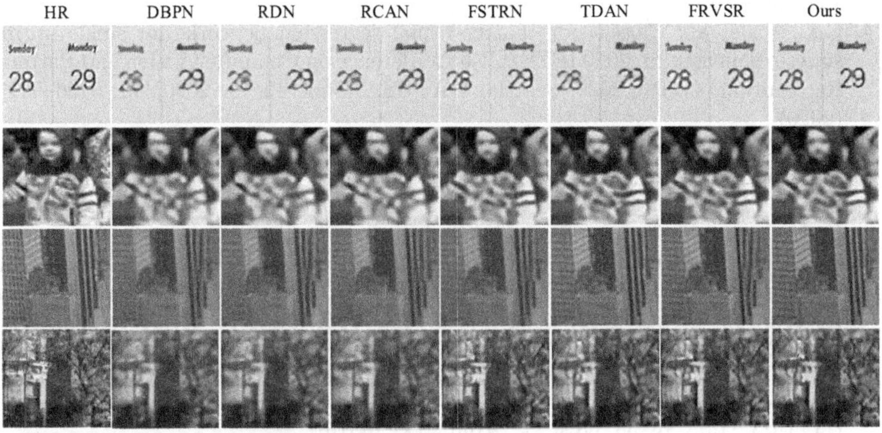

Fig. 5. Visual results of different video SR methods on the Vid4 dataset, for 4 × upscaling. Zooming in for better details.

5   Conclusion

This paper proposes a novel efficient spatio-temporal network (ESTN) for video SR, which typically contains two branches, spatial stream and temporal stream used for incorporating the appearance and motion information from consecutive frames. Specifically, the channel attention combined with pixel attention is integrated into the novel 2D attention block in spatial stream. 3D group convolution is used in the temporal stream to reduce the parameters. To fuse the two streams, we also propose the gate module based on deformable convolution for adaptive alignment at the feature level. The experiments reveal the superiority of the proposed method in terms of effectiveness and efficiency.

Acknowledgments. This work is supported by Major Scientific and Technological Special Project of Guizhou Province (No. 20183002) and Sichuan Science and Technology Program (No. 2021JDRC0071 and No. 2019YFG0535).

References

1. Caballero, J., et al.: Real-time video super-resolution with spatio-temporal networks and motion compensation. In: 2017 IEEE Conference on Computer Vision and Pattern Recognition, CVPR 2017, Honolulu, HI, USA, July 21–26, 2017, pp. 2848–2857 (2017)
2. Chen, Y., Chen, Y., Xue, J., Yang, W., Liao, Q.: Lightweight single image super-resolution through efficient second-order attention spindle network. In: IEEE International Conference on Multimedia and Expo, ICME 2020, London, UK, July 6–10, 2020, pp. 1–6 (2020)
3. Dong, C., Loy, C.C., He, K., Tang, X.: Learning a deep convolutional network for image super-resolution. In: Computer Vision - ECCV 2014 - 13th European Conference, Zurich, Switzerland, September 6-12, 2014, Proceedings, Part IV, pp. 184–199 (2014)
4. Guo, Y., et al.: Closed-loop matters: dual regression networks for single image super-resolution. In: 2020 IEEE/CVF Conference on Computer Vision and Pattern Recognition, CVPR 2020, Seattle, WA, USA, June 13–19, 2020, pp. 5406–5415 (2020)
5. Haris, M., Shakhnarovich, G., Ukita, N.: Deep back-projection networks for super-resolution. In: 2018 IEEE Conference on Computer Vision and Pattern Recognition, CVPR 2018, Salt Lake City, UT, USA, June 18–22, 2018, pp. 1664–1673 (2018)
6. Haris, M., Shakhnarovich, G., Ukita, N.: Recurrent back-projection network for video super-resolution. In: IEEE Conference on Computer Vision and Pattern Recognition, CVPR 2019, Long Beach, CA, USA, June 16–20, 2019, pp. 3897–3906 (2019)
7. Hu, J., Shen, L., Sun, G.: Squeeze-and-excitation networks. In: 2018 IEEE Conference on Computer Vision and Pattern Recognition, CVPR 2018, Salt Lake City, UT, USA, June 18–22, 2018, pp. 7132–7141 (2018)
8. Hui, Z., Gao, X., Yang, Y., Wang, X.: Lightweight image super-resolution with information multi-distillation network. In: Proceedings of the 27th ACM International Conference on Multimedia, MM 2019, Nice, France, October 21–25, 2019, pp. 2024–2032 (2019)

9. Hui, Z., Li, J., Gao, X., Wang, X.: Progressive perception-oriented network for single image super-resolution. Inf. Sci. **546**, 769–786 (2021)
10. Krizhevsky, A., Sutskever, I., Hinton, G.E.: Imagenet classification with deep convolutional neural networks. Commun. ACM **60**(6), 84–90 (2017)
11. Li, S., He, F., Du, B., Zhang, L., Xu, Y., Tao, D.: Fast spatio-temporal residual network for video super-resolution. In: 2019 IEEE Conference on Computer Vision and Pattern Recognition, CVPR 2019, Long Beach, CA, USA, June 16–20, 2019, pp. 10522–10531 (2019)
12. Li, X., et al.: STH: spatio-temporal hybrid convolution for efficient action recognition. CoRR abs/2003.08042 (2020)
13. Liu, C., Sun, D.: On bayesian adaptive video super resolution. IEEE Trans. Pattern Anal. Mach. Intell. **36**(2), 346–360 (2014)
14. Sajjadi, M.S.M., Vemulapalli, R., Brown, M.: Frame-recurrent video super-resolution. In: 2018 IEEE Conference on Computer Vision and Pattern Recognition, CVPR 2018, Salt Lake City, UT, USA, June 18–22, 2018, pp. 6626–6634 (2018)
15. Tian, Y., Zhang, Y., Fu, Y., Xu, C.: TDAN: temporally-deformable alignment network for video super-resolution. In: 2020 IEEE/CVF Conference on Computer Vision and Pattern Recognition, CVPR 2020, Seattle, WA, USA, June 13–19, 2020, pp. 3357–3366 (2020)
16. Tran, D., Bourdev, L.D., Fergus, R., Torresani, L., Paluri, M.: Learning spatiotemporal features with 3D convolutional networks. In: 2015 IEEE International Conference on Computer Vision, ICCV 2015, Santiago, Chile, December 7–13, 2015, pp. 4489–4497 (2015)
17. Wang, H., Su, D., Liu, C., Jin, L., Sun, X., Peng, X.: Deformable non-local network for video super-resolution. IEEE Access **7**, 177734–177744 (2019)
18. Xie, S., Girshick, R.B., Dollár, P., Tu, Z., He, K.: Aggregated residual transformations for deep neural networks. In: 2017 IEEE Conference on Computer Vision and Pattern Recognition, CVPR 2017, Honolulu, HI, USA, July 21–26, 2017, pp. 5987–5995 (2017)
19. Xue, T., Chen, B., Wu, J., Wei, D., Freeman, W.T.: Video enhancement with task-oriented flow. Int. J. Comput. Vis. **127**(8), 1106–1125 (2019)
20. Zhang, H., Jin, Z., Tan, X., Li, X.: Towards lighter and faster: learning wavelets progressively for image super-resolution. In: MM '20: The 28th ACM International Conference on Multimedia, Virtual Event / Seattle, WA, USA, October 12–16, 2020, pp. 2113–2121 (2020)
21. Zhang, Y., Tian, Y., Kong, Y., Zhong, B., Fu, Y.: Residual dense network for image super-resolution. In: 2018 IEEE Conference on Computer Vision and Pattern Recognition, CVPR 2018, Salt Lake City, UT, USA, June 18-22, 2018, pp. 2472–2481 (2018)
22. Zhang, Y., Tian, Y., Kong, Y., Zhong, B., Fu, Y.: Residual dense network for image super-resolution. In: 2018 IEEE Conference on Computer Vision and Pattern Recognition, CVPR 2018, Salt Lake City, UT, USA, June 18–22, 2018, pp. 2472–2481 (2018)
23. Zhu, X., Hu, H., Lin, S., Dai, J.: Deformable convnets V2: more deformable, better results. In: IEEE Conference on Computer Vision and Pattern Recognition, CVPR 2019, Long Beach, CA, USA, June 16–20, 2019, pp. 9308–9316 (2019)

Adaptive Correlation Filters Feature Fusion Learning for Visual Tracking

Hongtao Yu and Pengfei Zhu$^{(\boxtimes)}$

College of Intelligence and Computing, Tianjin University, Tianjin 300350, China
{yuhongtao,zhupengfei}@tju.edu.cn

Abstract. Tracking algorithms based on discriminative correlation filters (DCFs) usually employ fixed weights to integrate feature response maps from multiple templates. However, they fail to exploit the complementarity of multi-feature. These features are against tracking challenges, e.g., deformation, illumination variation, and occlusion. In this work, we propose a novel adaptive feature fusion learning DCFs-based tracker (AFLCF). Specifically, AFLCF can learn the optimal fusion weights for handcrafted and deep feature responses online. The fused response map owns the complementary advantages of multiple features, obtaining a robust object representation. Furthermore, the adaptive temporal smoothing penalty adapts to the tracking scenarios with motion variation, avoiding model corruption and ensuring reliable model updates. Extensive experiments on five challenging visual tracking benchmarks demonstrate the superiority of AFLCF over other state-of-the-art methods. For example, AFLCF achieves a gain of 1.9% and 4.4% AUC score on LaSOT compared to ECO and STRCF, respectively.

Keywords: Visual tracking · Discriminative correlation filters · Feature fusion · Adaptive temporal smoothing

1 Introduction

Recently, discriminative correlation filters (DCFs)-based trackers attract increasing interest because of their high accuracy and efficiency [1,2]. Multi-feature fusion and model update via temporal constraints all contribute remarkable improvement of accuracy and robustness to DCFs-based trackers. Although they achieve great progress, existing DCFs-based trackers still have two limitations, i.e., the non-robust feature representation and poor adaptability to various scenarios.

This work was supported by the National Key Research and Development Program of China under Grants 2019YFB2101904, the National Natural Science Foundation of China under Grants 61732011 and 61876127, the Natural Science Foundation of Tianjin under Grant 17JCZDJC30800, and the Applied Basic Research Program of Qinghai under Grant 2019-ZJ-7017.

I. Farkaš et al. (Eds.): ICANN 2021, LNCS 12895, pp. 652–664, 2021.
https://doi.org/10.1007/978-3-030-86383-8_52

Abundant feature responses provide robust object representations for visual tracking [2,10,29]. Feature response is the correlation of target feature maps and discriminative filters. The quality of the response affects the accuracy of target localization significantly. The usual approach for feature responses fusing is to add multiple response maps together via equal weights [6,7,9–11,19]. This rough combination of multi-feature responses improves the quality of target representation to some extent, but destroys the original advantages owned by each response map. For example, the responses of handcrafted features (such as the histogram of oriented gradient, HOG) contain sharp peaks for accurate localization. But the localization has similar distractors [27]. In contrast, the responses represented by convolution neural networks (CNNs) have high-confidence semantics, while they work for coarse localization [10,27] merely. The naive addition of the two reduces the discrimination and accurate localization expressed in the original responses. Staple [29] combines HOG and color histogram by the artificial weights, which makes full use of the discrimination of the HOG response and the robustness of the color histogram to target deformation. The artificial weights [29] are selected by numerous comparative experiments on the validation set and are hard to generalize to other datasets with different video attributes. MCCT [26] and UPDT [27] employ self-evaluation of the responses to resolve the fusion weights for shallow and deep feature responses. Evaluation-based methods avoid the extra cost of fine-tuning and are superior to fixed fusion weights. Yet, the evaluation of response fusion largely depends on the current tracking quality, leading to a risk of tracking drift. Besides, temporal constraints maintain reliable model updates, which in turn brings about robust feature response. The temporal smoothing [6,7,19,20] can overcome model corruption caused by unreliable object representation. Specifically, the temporal smoothing penalty influences the direction and speed of model updates to resist occlusion [6] or appearance variation [7,20] in a short time. STRCF [6] introduces the temporal regularization to DCFs-based tracker, alleviating model corruption caused by occlusion. Considering the interference of background clutter, ARCF [7] imposes consistency constraint on the response maps between adjacent frames during filters learning. So the filter coefficients can update according to object information rather than the background. LADCF [20] enforces temporal consistency via restricting the filter model to update around its historical value. Then, the obtained object representation is robust to temporal appearance variations. However, the fixed temporal smoothing penalty [6,7,20] ignores the variability of tracking cases and only allows for few scenarios. A video sequence usually contains various scene variations, such as illumination variations, occlusion, and others. We conclude two challenges: how to form robust target responses and maintain gentle model updates.

To this end, we propose the adaptive feature response fusion learning (AFL) for the DCFs-based tracking model. We fuse the handcrafted and deep responses via the learned weight. The weight is solved by adaptive supervised learning instead of artificial settings or self-evaluation. The optimal fusion weight fully exploits the complementary advantages of multi-responses and gains a robust target representation. Inspired by the peak to sidelobe ratio (PSR) [1], we design the adaptive temporal smoothing (ATS) penalty to ensure gentle model updates in motion variation. Our AFLCF performs favorably against the state-of-the-art trackers on five tracking benchmarks including OTB-2015 [12], VisDrone2019 [13], Nfs [14], DTB70 [15], and LaSOT [16].

The **contributions** of our work are summarized as follows: (1) Our AFLCF learns a robust fusion weight for the ensemble of multiple feature responses, which is supervised by the Gaussian-shaped target response. The fused response map obtains the complementary characteristics of semantic information and spatial localization represented by deep and handcrafted feature responses, respectively. (2) We propose the ATS penalty that updates according to changing tracking cases, thus ensuring reliable model updates. (3) Our tracker achieves remarkable performance on five different visual object tracking benchmarks.

2   Proposed Method

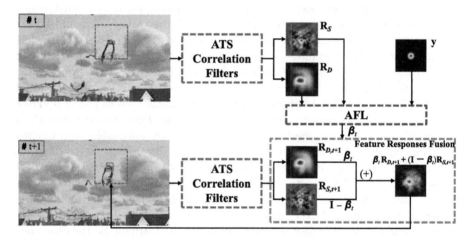

Fig. 1. A flowchart of the proposed AFLCF tracker. AFLCF consists of adaptive feature response fusion learning (AFL) and adaptive temporal smoothing (ATS) modules.

The overall flowchart of our AFLCF is illustrated in Fig. 1. We utilize DCFs with ATS to update the model and solve discriminative filters. The AFL module learns the optimal fusion weight for the ensemble of HOG (R_S) and CNNs (R_D) feature response maps in the current frame. Then, the solved weight β_t is utilized for multi-feature response ensemble in the next frame. STRCF [6] is our DCFs-based baseline model. The objective function of STRCF is shown in Eq. (1),

$$E(\mathbf{w}) = \frac{1}{2} \left\| \sum_{k=1}^{K} \mathbf{x}^k * \mathbf{w}^k - \mathbf{y} \right\|^2 + \frac{1}{2} \sum_{k=1}^{K} \left\| \mathbf{M} \odot \mathbf{w}^k \right\|^2 + \frac{\lambda}{2} \left\| \mathbf{w} - \mathbf{w}_{t-1} \right\|^2, \quad (1)$$

where $\mathbf{x} = \left[\mathbf{x}^1, \ldots, \mathbf{x}^K\right]$ denotes the input feature map consisting of K channels. $\mathbf{w}$ is the correlation filter to be learned. $*$ denotes the convolution operator. $\mathbf{y}$ is Gaussian-shaped ground truth. The second term is the spatial regularization quoted from SRDCF [3]. $\odot$ stands for element-wise product. $\|\mathbf{w}_t - \mathbf{w}_{t-1}\|^2$ is the temporal smoothing constraint utilized to prevent the model from corrupting, and λ is its penalty factor.

2.1   Adaptive Feature Response Fusion Learning

The robust object representation with explicit semantics and spatial details enables the tracker to defeat various object variations [10,11,27,29]. Our tracker introduces HOG and CNNs for feature responses integration. CNNs have more channels than HOG. Hence, the desired correlation filters should be solved by two independent DCFs training processes. The training samples are $\mathbf{x}_D = \left[\mathbf{x}_D^1, \ldots, \mathbf{x}_D^{K_1}\right]$ and $\mathbf{x}_S = \left[\mathbf{x}_S^1, \ldots, \mathbf{x}_S^{K_2}\right]$. The desired filters are denoted as $\mathbf{w}_D$ and $\mathbf{w}_S$ corresponding to $\mathbf{x}_D$ and $\mathbf{x}_S$. The feature response maps are written as the convolution of samples $\mathbf{x}$ and filters $\mathbf{w}$. As shown in Eq. (2), R_D and R_S denote the CNNs and HOG feature response maps, respectively. In order to facilitate the fusion of response maps, R_D and R_S are reshaped into the same size.

$$\mathrm{R}_D = \sum_{k=1}^{K_1} \mathbf{x}_D^k * \mathbf{w}_D^k; \mathrm{R}_S = \sum_{k=1}^{K_2} \mathbf{x}_S^k * \mathbf{w}_S^k. \tag{2}$$

Innovatively, our AFLCF learns an optimal fusion weight for HOG and CNNs feature response maps by ridge regression with the supervision of Gaussian-shaped ground truth. The objective function is formulated as

$$E(\boldsymbol{\beta}_t) = \frac{1}{2} \|\mathbf{y} - [\boldsymbol{\beta}_t \odot \mathrm{R}_D + (\mathrm{I} - \boldsymbol{\beta}_t) \odot \mathrm{R}_S]\|^2 + \frac{\mu}{2} \|\boldsymbol{\beta}_t - \tilde{\boldsymbol{\beta}}\|^2, \tag{3}$$

where $\boldsymbol{\beta}_t$ denotes the optimal feature response fusion matrix. $\tilde{\boldsymbol{\beta}}$ is a reference of $\boldsymbol{\beta}_t$, which provides a priori information for $\boldsymbol{\beta}_t$ and avoids model degradation. We set $\tilde{\boldsymbol{\beta}} = \mathrm{I}/2$, and I is the identity matrix. μ is a hyperparameter. The second term is a regularization used to prevent the model from over-fitting.

For computing efficiency, Eq. (2) can be transformed into the Fourier domain using Parseval's theorem. Then it is written as

$$\widehat{\mathrm{R}}_D = \sum_{k=1}^{K_1} \widehat{\mathbf{x}}_D^k \odot \widehat{\mathbf{w}}_D^k; \widehat{\mathrm{R}}_S = \sum_{k=1}^{K_2} \widehat{\mathbf{x}}_S^k \odot \widehat{\mathbf{w}}_S^k. \tag{4}$$

Correspondingly, Eq. (3) can be written as

$$E(\boldsymbol{\beta}_t) = \frac{1}{2} \left\|\widehat{\mathbf{y}} - \left[\boldsymbol{\beta}_t \odot \widehat{\mathrm{R}}_D + (\mathrm{I} - \boldsymbol{\beta}_t) \odot \widehat{\mathrm{R}}_S\right]\right\|^2 + \frac{\mu}{2} \left\|\boldsymbol{\beta}_t - \tilde{\boldsymbol{\beta}}\right\|^2. \tag{5}$$

Finally, the closed-form solution of β_t in the objective function (5) can be solved and written as

$$\beta_t = \left[\mu I + \left(\text{diag}\left(\widehat{R}_D - \widehat{R}_S\right)\right)^T \text{diag}\left(\widehat{R}_D - \widehat{R}_S\right)\right]^{-1} Q$$

$$= \frac{Q}{\mu I + \left(\widehat{R}_D - \widehat{R}_S\right)^T \odot \left(\widehat{R}_D - \widehat{R}_S\right)}, \tag{6}$$

where $Q = \left[\left(\widehat{R}_D - \widehat{R}_S\right)^T \odot \left(\widehat{y} - \widehat{R}_S - \mu\tilde{\beta}\right)\right]$. The fraction denotes element-wise division.

2.2   Adaptive Temporal Smoothing

DCFs-based trackers utilize temporal smoothing penalty to enforce the similarity of current filters with the previous to prevent the model from corruption [6,20]. Intuitively, as the target meets violent appearance variations and occlusion, the desired filter coefficients should not be overly similar to the previous. In that way, the strength of temporal smoothing penalty should be decreased appropriately. Otherwise, the opposite is true.

To verify the hypothesis, we conduct the comparisons of different strengths of temporal smoothing penalty. As shown in Fig. 2, adjacent frames of 42 and 43 undergo significant illumination variation and occlusion. Then, the weak penalty ($\lambda = 1$) achieves a better IOU (intersection over union) score than the larger ($\lambda = 10, 20, 30$). Even worse, larger penalties cause tracking drift in the subsequent frames. The larger penalties force adjacent filter coefficients to be too similar to adjust to complex and changing scenes. A similar case appears in frames 78 and 79. On the contrary, frames 55 and 56 contain slight illumination variation. The parameter $\lambda = 30$ obtains better IOU than small parameter settings. Therefore, the optimal scheme of λ should be capable of adapting to the changing situations. Based on the above analysis and experimental verification, we formulate the ATS penalty inspired by PSR [1]. PSR is a measure of the quality of feature response.

$$\lambda = \frac{\gamma}{e^{|R_{\max,t-1} - R_{\max}| + \|R_{t-1} - R\|_2} - 1}, \tag{7}$$

where $R = \beta_t \odot R_D + (I - \beta_t) \odot R_S$ denotes the feature response integrated by adaptive fusion weight β_t, and β_t is learned by the AFL module. R_{t-1} represents the response fusion of the previous frame. γ is a hyperparameter. $R_{\max,t-1}$ and $R_{\max}$ denote the maximum confidence of previous and current frame.

The first term $|R_{\max,t-1} - R_{\max}|$ in function $e^{(\cdot)}$ represents the gap of maximum confidence between adjacent frames. $\|R_{t-1} - R\|_2$ denotes the difference of overall response quality between adjacent frames. In fact, maximum confidence and overall response quality can measure the accuracy and robustness of a tracker respectively [1,27]. The maximum confidence decides the target localization, while overall response map with target semantics affects the discrimination

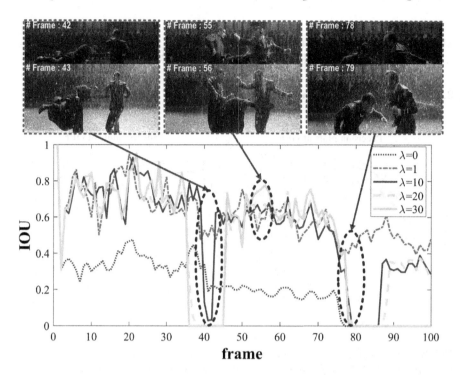

Fig. 2. The sequence (*"Matrix"*) with scenario and target variations is used to evaluate the IOU (intersection over union) score of different penalty factors in each frame.

between foreground and background. Therefore, the performance variation of the tracker in adjacent frames can be measured by $e^{|R_{\max,t-1}-R_{\max}|+\|R_{t-1}-R\|_2}$. It also indirectly reflects the degree of scenario variations in the current few frames. Therefore, we summarize that $e^{|R_{\max,t-1}-R_{\max}|+\|R_{t-1}-R\|_2}$ will become larger as the degree of tracking scene variation increases, while λ gets smaller. Otherwise, it would be the other way. The ATS penalty realizes adapting to changing tracking scenarios and maintaining reliable model updates.

3 Experiments

To verify the effectiveness of our AFLCF, we conduct comparisons with other state-of-the-art trackers on five different types of visual object tracking benchmarks, i.e., OTB-2015 [12], VisDrone2019 [13], Nfs [14], DTB70 [15], and LaSOT [16]. We take one pass evaluation (OPE) to test the performance of trackers based on two criteria: distance precision and overlap success. Red, green and blue fonts indicate the top-3 results, respectively.

Implementation details: we take HOG as the handcrafted feature and Conv4 from VGG-M as a deep representation. We use 7×1.01 scales for scale estimation. ADMM iteration is set to 2. The hyperparameter μ in Eq. (3) is set to 0.01. γ

in Eq. (7) is set to 17. Our AFLCF is implemented on MATLAB2017a with MatConvNet toolbox. All the experiments run on a PC equipped with an Intel i7-8700 CPU, 16 GB RAM, and a single NVIDIA GTX 1050Ti GPU.

3.1   Quantitative Evaluation

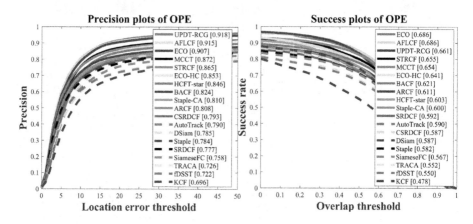

Fig. 3. Performance evaluation on OTB-2015 Benchmark. The legend shows the distance precision at 20 pixels and the area-under-curve (AUC) of each tracker.

OTB-2015 [12] is one of the most popular visual object tracking benchmarks. It contains 100 fully annotated sequences and 11 video attributes. We compare AFLCF against 18 state-of-the-art trackers on the dataset, as shown in Fig. 3. Table 1 reports the top-5 trackers by AUC. UPDT-RCG [18] improves UPDT [27] by replacing cosine window with Gaussian-shaped mask, and utilizes ResNet-50 [28] for feature extraction. It achieves the best precision and surpasses our tracker by 0.3% with a speed of 0.7 FPS (frame per second). Our AFLCF is 0.8% ahead of the third-best ECO [9]. Owning to the AFL module, our tracker equipped with VGG-M obtains the best AUC. Our AFL is superior to evaluation-based methods [18,26]. AFLCF achieves 2.5% and 4.2% AUC gain than UPDT-RCG [18] and MCCT [26], and equals ECO [9]. Figure 4 exhibits the visual comparison of AFLCF and related methods. AFLCF achieves outstanding tracking performance.

   VisDrone2019 [13] is a drone-based tracking dataset with 35 short-term and 25 long-term video sequences. We compare our AFLCF with 17 state-of-the-art trackers. Table 2 reports that AFLCF obtains almost the best performance with the precision of 0.553 and AUC score of 0.407. Specifically, our AFLCF achieves the best precision. There is only a little gap of 0.2% AUC score between AFLCF and ECO [9]. AFLCF surpasses MCCT [26], ASRCF [5], and DeepSTRCF [6] by 1.6%, 2.9%, and 2.3% in terms of AUC score respectively. AFL module contributes to the performance boost as it integrates more robust target representation. Besides, AFLCF shows obvious superiority over long-term tracker, i.e.,

AFLCF UPDT-RCG [18] MCCT [26] AutoTrack [19] STRCF [6] ECO-HC [9]

Fig. 4. Visualization evaluation of AFLCF and five related trackers on the *Soccer, Skiing, Bird1*, and *Biker* sequences. Our method successfully tracks the targets undergoing occlusion, deformation, illumination variation, appearance change, and other challenges in these video sequences. Moreover, our AFLCF achieves the best performance compared to five state-of-the-art trackers [6,9,18,19,26].

PTAV [17] and LCT [30]. ATS module ensures reliable updates of the filters to avoid filters from corruption caused by scenario variations in long-term tracking. AutoTrack [19] with automatic spatio-temporal regularization achieves excellent performance in drone-based tracking methods. However, our AFLCF greatly surpasses AutoTrack [19] both in accuracy and AUC score.

Table 1. Performance comparisons of top-5 trackers on OTB-2015. * represents running on GPU.

Method	ECO [9]	UPDT-RCG [18]	STRCF [6]	MCCT [26]	**AFLCF**
AUC	0.686	0.661	0.655	0.644	0.686
Precision	0.907	0.918	0.865	0.872	0.915
FPS	10.2*	0.7*	22.5	5.2*	5.3*

Nfs [14] is the first higher frame rate visual object tracking benchmark that contains 100 videos, 380K frames in total. The adjacent frames change smoothly in the higher frame rate video. Usually, the trackers with temporal smoothing constraints [6,7,19] perform well. To verify the effectiveness of our ATS

Table 2. Performance comparisons on VisDrone2019 Benchmark.

Method	AUC	Precision	Method	AUC	Precision
SRDCF [3]	0.345	0.479	TRACA [23]	0.326	0.446
LCT [30]	0.263	0.420	STRCF [6]	0.369	0.517
fDSST [8]	0.312	0.406	DeepSTRCF [6]	0.384	0.514
Staple [29]	0.347	0.488	CSRDCF [21]	0.336	0.464
PTAV [17]	0.327	0.425	HCFT-star [11]	0.304	0.437
ECO [9]	0.409	0.550	ARCF [7]	0.367	0.513
BACF [4]	0.358	0.506	ASRCF [5]	0.378	0.542
Staple-CA [22]	0.351	0.489	AutoTrack [19]	0.350	0.488
MCCT [26]	**0.391**	**0.545**	**AFLCF**	0.407	0.553

Table 3. Performance evaluation on Nfs Benchmark.

Method	AUC	Precision	Method	AUC	Precision
SRDCF [3]	0.413	0.489	TRACA [23]	0.362	0.431
KCF [2]	0.292	0.391	MCCT [26]	0.459	0.532
fDSST [8]	0.324	0.382	STRCF [6]	**0.508**	**0.638**
SiameseFC [25]	0.496	0.603	ARCF [7]	0.446	0.510
Staple [29]	0.422	0.496	ASRCF [5]	0.483	0.577
ECO-HC [9]	0.468	0.582	HCFT-star [11]	0.376	0.430
ECO [9]	0.480	0.581	AutoTrack [19]	0.441	0.511
Staple-CA [22]	0.396	0.474	UPDT-RCG [18]	0.519	0.668
BACF [4]	0.433	0.503	**AFLCF**	0.530	0.676

mechanism, we compare our AFLCF against 17 state-of-the-art trackers. Table 3 exhibits that AFLCF obtains the best performance of both precision and AUC score among all related methods. AFLCF obtains 2.2%, 8.4%, 1.1%, and 5% AUC score gain than STRCF [6], ARCF [7], UPDT-RCG [18], and ECO [9], respectively. The ATS method of AFLCF shows a significant advantage over fixed temporal penalty [6,7] because our mechanism performs self-regulation according to scenario variations and has better scene adaptability.

DTB70[15] is a tracking benchmark of high diversity, containing 70 sequences captured by drone cameras. We compare AFLCF with 17 state-of-the-art trackers. In Table 4, we summarize the precision and AUC score of the 18 trackers. AFLCF shows apparent superiority over other trackers. With the aid of AFL and ATS modules, AFLCF outperforms ECO [9], MCCT [26], and ASRCF [5] by the precision of 3.8%, 2.1%, and 6.3%, respectively. Even compared to ARCF [7] and AutoTrack [19] designed for drone-based tracking, AFLCF shows obvious superiority owing to reliable model updates achieved by the ATS module.

LaSOT [16] is a large-scale visual tracking benchmark with 1,400 sequences, 3.5M manually annotated frames in total. To verify the robustness of our AFLCF on the large dataset, we compared it with 35 default trackers reported in [16]. Table 5 reports the best five DCFs-based trackers. ATS promotes accurate model updates in large-scale video tracking. Therefore, AFLCF achieves the best performance and precedes the state-of-the-art ECO [9].

Table 4. Performance comparisons on DTB70 Benchmark.

Method	AUC	Precision	Method	AUC	Precision
SRDCF [3]	0.367	0.522	MCCT [26]	**0.497**	0.739
KCF [2]	0.280	0.467	CSRDCF [21]	0.437	0.643
Staple [29]	0.338	0.488	STRCF [6]	0.437	0.649
fDSST [8]	0.342	0.508	TRACA [23]	0.360	0.523
BACF [4]	0.401	0.593	ARCF [7]	0.469	0.685
ECO-HC [9]	0.455	0.668	HCFT-star [11]	0.415	0.616
DSiam [24]	0.483	0.718	ASRCF [5]	0.469	0.697
ECO [9]	0.502	**0.722**	AutoTrack [19]	0.479	0.715
Staple-CA [22]	0.354	0.512	**AFLCF**	0.506	0.760

Table 5. Accuracy comparisons of top-5 DCFs-based trackers on LaSOT [16].

Method	ECO [9]	STRCF [6]	ECO-HC [9]	TRACA [23]	**AFLCF**
AUC	0.340	0.315	0.311	0.285	0.359
Precision	0.298	0.292	0.272	0.237	0.322

3.2 Attribute-Based Evaluation

Figure 5 reports the overlap success plots of six video attributes. In extreme appearance variation (illumination variation, deformation, and occlusion), AFLCF achieves the obvious gain of 4.1%, 4.5%, and 2.4% than MCCT [26]. AFL module contributes to more robust object representation than evaluation-based and artificial setting-based methods (MCCT [26], Staple-CA [22], and STRCF [6]). In the tracking scenarios with apparent variation, i.e., background clutter, fast motion, illumination variation, and camera motion, AFLCF improves STRCF [6] by 9.4%, 7.8%, 3.6%, and 3.1%, respectively. This is mainly owed to the ATS penalty that ensures gentle and reliable model updates in tracking variation. Besides, the adaptive temporal smoothing utilized in AFLCF obtains more advanced performance than automatic temporal regularization in Auto-Track [19].

3.3   Ablation Study

To verify the effectiveness of the adaptive response fusion learning (AFL) and adaptive temporal smoothing (ATS) for our tracker, we conduct the ablation study on the OTB-2015 [12] dataset. The Baseline stands for the method that does not apply adaptive response fusion learning and adaptive temporal smoothing modules. Baseline+AFL contains Baseline and AFL module. Baseline+ATS is the combination of the Baseline and ATS module. AFLCF consists of Baseline, ATS, and AFL. Table 6 reports the comparison results. Compared with the Baseline, ATS and AFL modules achieve obvious improvement of both precision and AUC. Besides, AFLCF improves the baseline methods by 7% and 3.6% in terms of precision and AUC, respectively.

Table 6. Ablation Study on the OTB-2015 Benchmark [12].

Method	Baseline	Baseline+AFL	Baseline+ATS	**AFLCF**
AUC	0.650	0.672	0.681	**0.686**
Precision	0.845	0.890	0.897	**0.915**

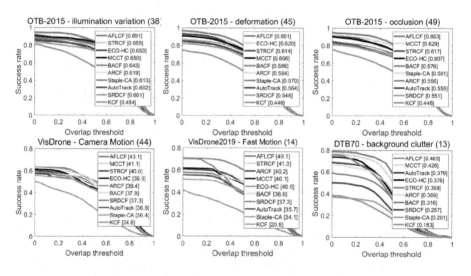

Fig. 5. Attribute-based comparison on six tracking scenarios, i.e., illumination variation, deformation, occlusion, camera motion, fast motion, and background clutter. The legend shows the AUC (area-under-curve) score of each tracker.

4   Conclusions

To our knowledge, our work is the first attempt to fuse feature response maps from multiple templates with the supervision of Gaussian-shaped target response. Compared with the previous methods that use artificial weights to fuse the multi-responses, our AFLCF integrates a more robust target representation via the learned weights. AFLCF obtains the complementary advantages of multi-feature responses. We design the adaptive temporal smoothing penalty to avoid model corruption. It maintains reliable model updates according to the variations of target and tracking scene. Extensive experiments on five different visual tracking benchmarks (including long-term, short-term, drones, and higher frame rate tracking) report that our AFLCF achieves advanced performance compared to state-of-the-art methods. The outstanding performance also verifies the generalization of AFLCF for different tracking scenarios.

References

1. Bolme, D.S., Beveridge, J.R., Draper, B.A., Lui, Y.M.: Visual object tracking using adaptive correlation filters. In: IEEE Conference on Computer Vision and Pattern Recognition, pp. 2544–2550 (2010). https://doi.org/10.1109/CVPR.2010.5539960

2. Henriques, J.F., Caseiro, R., Martins, P., Batista, J.: High-speed tracking with kernelized correlation filters. IEEE Trans. Pattern Anal. Mach. Intell. **37**(3), 583–596 (2015). https://doi.org/10.1109/TPAMI.2014.2345390

3. Danelljan, M., Häger, G., Khan, F.S., Felsberg, M.: Learning spatially regularized correlation filters for visual tracking. In: IEEE International Conference on Computer Vision, pp. 4310–4318 (2015). https://doi.org/10.1109/ICCV.2015.490

4. Galoogahi, H.K., Fagg, A., Lucey, S.: Learning background-aware correlation filters for visual tracking. In: IEEE International Conference on Computer Vision, pp. 1144–1152 (2017). https://doi.org/10.1109/ICCV.2017.129

5. Dai, K., Wang, D., Lu, H., Sun, C., Li, J.: Visual tracking via adaptive spatially-regularized correlation filters. In: CVPR, pp. 4670–4679 (2019). https://doi.org/10.1109/CVPR.2019.00480

6. Li, F., Tian, C., Zuo, W., Zhang, L., Yang, M.: Learning spatial-temporal regularized correlation filters for visual tracking. In: CVPR, pp. 4904–4913 (2018). https://doi.org/10.1109/CVPR.2018.00515

7. Huang, Z., Fu, C., Li, Y., Lin, F., Lu, P.: Learning aberrance repressed correlation filters for real-time UAV tracking. In: ICCV, pp. 2891–2900 (2019). https://doi.org/10.1109/ICCV.2019.00298

8. Danelljan, M., Häger, G., Khan, F.S., Felsberg, M.: Discriminative scale space tracking. IEEE Trans. Pattern Anal. Mach. Intell. **39**(8), 1561–1575 (2016). https://doi.org/10.1109/TPAMI.2016.2609928

9. Danelljan, M., Bhat, G., Khan, F.S., Felsberg, M.: ECO: efficient convolution operators for tracking. In: CVPR, pp. 6931–6939 (2017). https://doi.org/10.1109/CVPR.2017.733

10. Ma, C., Huang, J., Yang, X., Yang, M.: Hierarchical convolutional features for visual tracking. In: ICCV, pp. 3074–3082 (2015). https://doi.org/10.1109/ICCV.2015.352

11. Ma, C., Huang, J., Yang, X., Yang, M.: Robust visual tracking via hierarchical convolutional features. IEEE Trans. Pattern Anal. Mach. Intell. **41**(11), 2709–2723 (2019). https://doi.org/10.1109/TPAMI.2018.2865311

12. Wu, Y., Lim, J., Yang, M.: Object tracking benchmark. IEEE Trans. Pattern Anal. Mach. Intell. **37**(9), 1834–1848 (2015). https://doi.org/10.1109/TPAMI.2014.2388226

13. Zhu, P., Wen, L., Du, D., Bian, X., Hu, Q., Ling H.: Vision Meets Drones: Past, Present and Future. arXiv preprint arXiv:1804.07437 (2020)
14. Galoogahi, H.K., Fagg, A., Huang, C., Ramanan, D., Lucey, S.: Need for speed: a benchmark for higher frame rate object tracking. In: ICCV, pp. 1134–1143 (2017). https://doi.org/10.1109/ICCV.2017.128
15. Li, S., Yeung, D.: Visual object tracking for unmanned aerial vehicles: a benchmark and new motion models. In: AAAI, pp. 4140–4146 (2017)
16. Fan, H., et al.: LaSOT: a high-quality benchmark for large-scale single object tracking. In: CVPR, pp. 5374–5383 (2019). https://doi.org/10.1109/CVPR.2019.00552
17. Fan, H., Ling, H.: Parallel tracking and verifying: a framework for real-time and high accuracy visual tracking. In: ICCV, pp. 5487–549 (2017). https://doi.org/10.1109/ICCV.2017.585
18. Li, F., Wu, X., Zuo, W., Zhang, D., Zhang, L.: Remove cosine window from correlation filter-based visual trackers: when and how. IEEE Trans. Image Process. **29**, 7045–7060 (2020). https://doi.org/10.1109/TIP.2020.2997521
19. Li, Y., Fu, C., Ding, F., Huang, Z., Lu, G.: AutoTrack: towards high-performance visual tracking for UAV with automatic spatio-temporal regularization. In: CVPR, pp. 11920–11929 (2020). https://doi.org/10.1109/CVPR42600.2020.01194
20. Xu, T., Feng, Z., Wu, X., Kittler, J.: Learning adaptive discriminative correlation filters via temporal consistency preserving spatial feature selection for robust visual object tracking. IEEE Trans. Image Process. **28**, 7949–7959 (2019). https://doi.org/10.1109/TIP.2019.2919201
21. Lukežič, A., Vojíř, T., Čehovin Zajc, L., Matas, J., Kristan, M.: Discriminative correlation filter tracker with channel and spatial reliability. Int. J. Comput. Vis. **126**(7), 671–688 (2018). https://doi.org/10.1007/s11263-017-1061-3
22. Mueller, M., Smith, N., Ghanem, B.: Context-aware correlation filter tracking. In: CVPR, pp. 1387–1395 (2017). https://doi.org/10.1109/CVPR.2017.152
23. Choi, J., et al.: Context-aware deep feature compression for high-speed visual tracking. In: CVPR, pp. 479–488 (2018). https://doi.org/10.1109/CVPR.2018.00057
24. Guo, Q., Feng, W., Zhou, C., Huang, R., Wan, L., Wang, S.: Learning dynamic Siamese network for visual object tracking. In: ICCV, pp. 1781–1789 (2017). https://doi.org/10.1109/ICCV.2017.196
25. Bertinetto, L., Valmadre, J., Henriques, J.F., Vedaldi, A., Torr, P.H.S.: Fully-convolutional Siamese networks for object tracking. In: Hua, G., Jégou, H. (eds.) ECCV 2016. LNCS, vol. 9914, pp. 850–865. Springer, Cham (2016). https://doi.org/10.1007/978-3-319-48881-3_56
26. Wang, N., Zhou, W., Tian, Q., Hong, R., Wang, M., Li, H.: Multi-cue correlation filters for robust visual tracking. In: CVPR, pp. 4844–4853 (2018). https://doi.org/10.1109/CVPR.2018.00509
27. Bhat, G., Johnander, J., Danelljan, M., Khan, F.S., Felsberg, M.: Unveiling the power of deep tracking. In: Ferrari, V., Hebert, M., Sminchisescu, C., Weiss, Y. (eds.) ECCV 2018. LNCS, vol. 11206, pp. 493–509. Springer, Cham (2018). https://doi.org/10.1007/978-3-030-01216-8_30
28. He, K., Zhang, X., Ren, S., Sun, J.: Deep residual learning for image recognition. In: CVPR 2016, pp. 770–778 (2016). https://doi.org/10.1109/CVPR.2016.90
29. Bertinetto, L., Valmadre, J., Golodetz, S., Miksik, O., Torr, P.H.S.: Complementary learners for real-time tracking. In: CVPR, pp. 1401–1409 (2016). https://doi.org/10.1109/CVPR.2016.156
30. Ma, C., Yang, X., Zhang, C., Yang, M.: Long-term correlation tracking. In: CVPR, pp. 5388–5396 (2015). https://doi.org/10.1109/CVPR.2015.7299177

Dense Video Captioning for Incomplete Videos

Xuan Dang, Guolong Wang, Kun Xiong, and Zheng Qin[✉]

Tsinghua University, Beijing, China
{dx18,wanggl16,xk18,qingzh}@mails.tsinghua.edu.cn

Abstract. Incomplete video or partially-missing video situations are rarely considered in video captioning research. Previous approaches are mainly trained and evaluated on complete video clip datasets where all the events involved are thoroughly observed. In this work, we formulate the issue of video content description for partially-missing videos. To tackle this challenge, we propose a Visual-Semantic Embedding with Context (VSEC) module to capture the missing visual content by jointly embedding the constructed contextual visual representation and corresponding textual annotation. We further employ a transformer-based captioning network to generate complete and coherent descriptions for the incomplete video. To validate the effectiveness of our method, we construct a new dataset based on ActivityNet Caption to imitate incomplete video situations in reality, named as ActivityNet Caption-P. We train and test our method both on ActivityNet Caption-P and achieve outstanding performances in most metrics.

Keywords: Video caption · Dense · Incomplete

1 Introduction

With the recent advances in deep learning technology, describing multi-events happened in a relatively long-range video has become possible, named as dense video captioning task [13]. Though existing methods [4,6,13,17,23,28,31] have achieved acceptable performance in datasets such as ActivityNet Captions [13] and VideoStory [6], they all generate sentences when complete videos are provided. But on real-world conditions, we encounter situations when the video at hand is incomplete caused by partial occlusion, background clutter, cameras dithering, data corruption during transfer, and etc. Here the word *incomplete* essentially means that part of the video content is missing. In this work, we tackle the problem of multi-event description generation for partially-missing videos.

Inspired by a frequent occurrence in our daily lives that we can always correctly get the whole message delivered by a video even though we are interrupted during watching, we aim to design a model with a similar ability to get complete descriptions from partially-missing videos. The model has the following rationales: contents in a video are interrelated at two granularities: 1) the events that

© Springer Nature Switzerland AG 2021
I. Farkaš et al. (Eds.): ICANN 2021, LNCS 12895, pp. 665–676, 2021.
https://doi.org/10.1007/978-3-030-86383-8_53

happened in a video are related, they usually are parts of the whole story in a video; 2) activities or changes in an event are closely related, and one activity always followed with another in some set pattern. Therefore, we can always speculate the content of a missing video segment from its surrounding parts.

Specifically, we propose a VSEC (Visual-Semantic Embedding with Context) module which can extract expressive feature representations from incomplete videos. We firstly divide a long video to N segments and construct the visual feature of a segment with its surrounding segments instead of the segment itself. In this way, the visual representation of each segment is based on its contextual video contents, as if we cannot see the target segment. And partially missing situation can be extended to be pervasive within the scope of the entire video through such an operation. Secondly, we relate each video segment with a corresponding annotated sentence to construct segment(incomplete)-sentence(complete) pairs. Then, the VSEC module is used to learn a joint embedding of the incomplete visual and complete semantic representations to compensate for the information loss in visual representations. We then design an end-to-end framework combining the VSEC module with a transformer-based captioning module to produce descriptions for the entire video.

In order to imitate the incomplete video situations in reality, we construct a new dataset by artificially cutting off part of each video at random points of time in the ActivityNet Caption dataset, named as ActivityNet Caption-P. We train and verify our method both on ActivityNet Caption-P.

The main contributions of this paper are three-folded:

(1) We formulate the problem of multi-event description generation for incomplete videos.
(2) We propose a VSEC module which utilizes the incomplete visual information and semantic implications to derive a relatively complete and coherent representation for the incomplete video.
(3) The proposed model outperforms the comparisons in terms of description generation for partially-missing videos, especially in the completeness and coherence of the video content description.

2 Related Work

2.1 Dense Video Captioning

Over the past years, the task of video captioning has been expanded from "one sentence description for short videos" to "multi-sentence description for long videos". Dense video captioning task was first introduced in [13] as a two-stage pipeline: temporal proposal generation for event detection and event description generation for each event. Based on such a framework, following works made efforts mainly in three aspects: 1) Reference [17,31] further explored interactions between the two sub-tasks to improve the overall performances of both; 2) Reference [28] focused on generating more expressive feature representations; 3) Reference [16,20,23,29] tried to make the generated sentences seem more like a unified whole paragraph than independent sentences which are prone to be redundant or incoherent. There are also works exploit the additional audio modality to improve the captioning ability [8,9,25], which are out of our discussion in this paper.

Compared with the prior works, our method is different mainly in two aspects: 1) our method is specifically designed for incomplete video situations which are rarely considered in past researches; 2) Though some works [4,28] have exploited the role of context information in video captioning tasks, they typically use context information as supplementary. Different from them, we boldly abandon the original visual observation and purely use the context information to construct the visual representation. By such a cut-out and fill-in operation in visual feature construction, we want to achieve an imitation of human ability to supplement incomplete visual contents.

2.2 Partially Observed Video Analysis

Though incomplete videos are rarely considered in video captioning tasks, there are researches using partially-observed videos in other action analysis related tasks, such as action prediction [2,3,11,12,14,15] and partial action recognition [24,30]. Action prediction aims to predict the next action with the future video unobserved, and partial action recognition is to classify the action with partially observed video. The above mentioned tasks are different from ours mainly in two aspects: 1) They have much more strict restrictions than our task. To be specific, the former restricts the missing part of the video to be future actions, and the latter restricts the task to be an action classification problem. While our method is designed for incomplete videos could have missing part at any time and has a more complicated task to describe the whole events with sentences. 2)They mainly aim at solving the problem of activity recognition/classification while we look at a bigger picture for long video content understanding.

3   Proposed Method

Our model aims to provide relatively complete and coherent captions for incomplete long videos. In order to realize such an ability, we combine our proposed VSEC module with a Transformer-based captioning module, named as VSEC-T. In the VSEC module, we firstly design a novel context-based visual feature construction which utilizes context video contents to represent the target video segment. The context-based design means to align with the partially-missing assumption that the target video segment is missing and explore the hidden information lies in its surrounding segments. Then we relate each segment with a corresponding annotated sentence as a visual-semantic pair and apply a visual-semantic embedding network to leverage the textual information to jointly capture the assumed missing visual content. In the captioning module, we employ an encoder-decoder network with transformer units to generate captions for the given video. The overall framework of our model is illustrated as in Fig. 1.

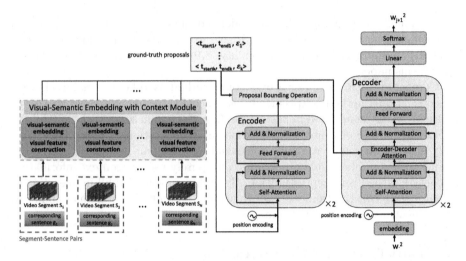

Fig. 1. The overall structure of our proposed VSEC-T model. Firstly, we relate each video segment with a corresponding descriptive sentence to construct a segment-sentence pair. Then we construct the video segment features with context video contents and use the VSEC module to learn more expressive features for incomplete video contents. Finally, we employ an encoder-decoder network with transformer units to generate captions for the given video.

3.1   Visual-Semantic Embedding with Context Module

Firstly, we divide each video into N segments with equal length of time, denoted as $S_i (i = 1, 2, ..., N)$, and produce a visual feature for each segment by operating mean-pooling on all frame-level features in it following [21], denoted as $\mathbf{s}_i \in \mathbb{R}^{D_v}$, where D_v is the feature dimension same as that of the frame-level features. Then, we construct a new feature for S_i with its surrounding segments instead of the segment itself. The number of neighboring segments, denoted as N_c, will be set according to the experimental results. Suppose we use the neighboring one segment at each side of S_i, i.e., $N_c = 1$, we perform mean-pooling over $\mathbf{s}_{i-1}$ and $\mathbf{s}_{i+1}$ to combine the two features into one, denoted as $\bar{\mathbf{s}}_i \in \mathbb{R}^{D_v}$. Through such a cut-out and fill-in operation, we expand the partially-missing condition to all segments. And because N (set to between 10 and 50) is much bigger than the average number of events in a video, which is 3.65, we can approximately assume that our feature construction made almost all the events have a part missing, and the percentage of the missing part in an event is different because of the different length of the events in a video, which is just in line with our incomplete assumption in real scenarios.

Let a video V contain a set of events $\xi = \{e_1, ..., e_K\}$ with corresponding descriptions $G = \{g_1, ..., g_K\}$ and g_k is a sentence with N_k words. We denote the textual feature of the n-th word in g_k as $\mathbf{w}_k^n \in \mathbb{R}^{D_w}$. Then we employ a transformer encoder unit to encode the words in a sentence and use the hidden state of the first word as the representation for the whole sentence, denoted as $\mathbf{g}_k \in \mathbb{R}^{D_w}$. We relate each segment with a sentence which describes the event happened in this segment as a visual-semantic pair. Partially appearance of an event also counts here. If there are more than one event appear in one segment, we choose the longer one appeared in this segment, and if the entire segment is covered by more than one event, we choose the event with a closer center point of time to this segment. Given the constructed visual representation with incomplete visual content and the corresponding textual representation for the related complete event, we want to leverage the complete information in $\mathbf{g}_k$ to make $\bar{\mathbf{s}}_i$ to express more. Inspired by the joint embedding approach which has been successfully applied in relative tasks [18,19,21], we learn a joint embedding space for the visual and textual representations where they will be close if and only if they are semantically relative. We use two transformation matrix $\mathbf{T}_s \in \mathbb{R}^{D_e \times D_v}$ and $\mathbf{T}_g \in \mathbb{R}^{D_e \times D_w}$ to derive the projection for them on the joint embedding space, respectively. Then the embedded features of the visual and textual representations, denoted as $\bar{\mathbf{s}}_{ie} \in \mathbb{R}^{D_e}$ and $\mathbf{g}_{ke} \in \mathbb{R}^{D_e}$ respectively can be derived by

$$\bar{\mathbf{s}}_{ie} = \mathbf{T}_s \bar{\mathbf{s}}_i \quad and \quad \mathbf{g}_{ke} = \mathbf{T}_g \mathbf{g}_k \tag{1}$$

Though there are multiple ways to measure the relevance between the visual and textual content, we use the simple but effective way which computes the distance between their projection on the common space [21]. Thus, the relevant loss between a video segment and its corresponding descriptive sentence now can be defined as

$$\begin{aligned} \mathcal{L}_r(S_i, g_k) &= \| \bar{\mathbf{s}}_{ie} - \mathbf{g}_{ke} \|^2 \\ &= \| \mathbf{T}_s \bar{\mathbf{s}}_i - \mathbf{T}_g \mathbf{g}_k \|^2 \end{aligned} \tag{2}$$

The total relevant loss between the entire video and all corresponding sentences can be defined as following

$$\mathcal{L}_r(S, G) = \sum_{i=1}^{N} \mathcal{L}_r(S_i, g_k) \tag{3}$$

By minimizing the relevance loss, the embedded visual representation is expected to express more information than just using given video content. An illustration of our VSEC module workflow for segment S_i is shown in Fig. 2

We concatenate the learned embedding of each segment to represent the whole video, which can be represented as $[\bar{\mathbf{s}}_{1e}, \bar{\mathbf{s}}_{2e}, ..., \bar{\mathbf{s}}_{Ne}]$. Then we use the following encoder-decoder captioning module with transformer unit to generate descriptions for the video.

3.2   Encoder-Decoder Captioning Module with Transformer Unit

Transformer is an attention-based architecture designed to improve the performance of neural machine translation applications [26]. The attention mechanism used in transformer made it possible to learn an attention value between any two of the input value, thereby overcoming the memory loss problem in traditional RNN-based methods. The foundation of the transformer architecture is

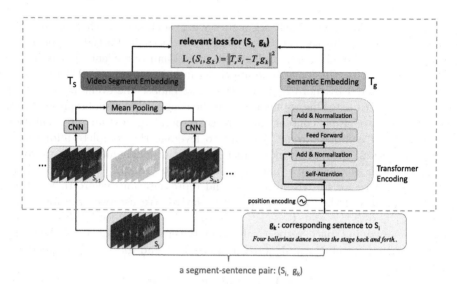

Fig. 2. An illustration of our Visual-Semantic Embedding with Context (VSEC) module. We relate each video segment with a corresponding descriptive sentence to construct a segment-sentence pair. In the video processing streamline, the visual features of each segment is firstly extracted by CNN networks, then we use the features of the surrounding segments, S_{i-1} and S_{i+1}, instead of the segment itself to represent S_i by a mean pooling operation. In the sentence processing streamline, we use a transformer encoder unit to extract the textual feature of the corresponding sentence. Then we learn a joint embedding space for the visual and textual representation pairs by the transformation matrix T_v and T_s to generate a more expressive feature for segment S_i.

the scaled dot-product attention. Given query matrix Q, key matrix K and value matrix V, the attention value is computed as:

$$A(Q, K, V) = softmax\left(\frac{QK^T}{\sqrt{d_k}}\right)V \qquad (4)$$

The multi-head attention is obtained by combining H paralleled scaled dot-product attention. Each single scaled dot-product attention layer is called a "head". The multi-head attention output can be derived by

$$MultiHead(Q, K, V) = Concat(head_1, ..., head_h)W^O \qquad (5)$$

$$head_i = A(QW_i^Q, KW_i^K, VW_i^V) \qquad (6)$$

where W_i^Q, W_i^K, W_i^V and W^O are the projection matrices. When the query, key and value matrix are all the same, it is called self-attention and when the query matrix is different from the key and value matrix, it is called cross-attention [26].

In order to capture as much contextual information as possible from the visual features of given videos, we use the transformer unit to design our encoder-decoder based captioning model. The basic construction of the encoder and decoder module are the same with the modules used in [31]. While in [31], they use the visual features of the video frames as input to the encoder, we use the output of our VSEC module instead. As the feature dimension of each video segment has been compressed to be D_e, and the time dimension of an entire video has been compressed to be the number of segments in a video (i.e., N), our model can work in a more effective way. Positional encoding is first added to our VSEC module output $\bar{s}_{ie} \in \mathbb{R}^{D_e} (i = 1, ..., N)$ to form the encoder input, denoted as $F^0 = \{f_1^0, ..., f_N^0\}$. In each layer, the self-attention module takes the input from the previous layer and apply multi-head attention and then a feed forward layer. Residual connection and layer-normalization are applied for each layer (more details of the transformer-based encoder can be find in [31]).

With the output of the encoder, we want to generate descriptions for each event respectively. And because the input features dimension of the encoder have been compressed to be N at the dimension of time, we cannot use the ground-truth time stamps to restrict the decoder input to the features in the same event. We design a proposal bounding network to separate features of different events. We choose the encoded features of the most relevant segments with the given event proposal as the decoder input. When the time overlap between a segment and an event is over half of the length of a segment, we choose it as a relevant segment to the given event. Otherwise, we discard the feature of this segment for caption generation of the event.

Then, with the chosen features relevant to an event, we use the decoder to generate corresponding sentences. The decoder also follows the transformer architecture(see more in [31]), as shown in Fig. 1.

4   Experiment

4.1   Dataset

ActivityNet Captions. ActivityNet Captions [13] is the most commonly used dataset in dense video captioning tasks. It contains 20k videos with an average number of 3.65 events contained in each video. The train/val/test splits for ActivityNet Captions are 0.5:0.25:0.25. There are 10,009 videos in train set and 4,917 videos in Validation set.

ActivityNet Captions – P. In order to simulate the incomplete video situation in real scenarios, we construct a new dataset based on ActivityNet Captions, named as ActivityNet Captions-P. ActivityNet Captions-P is composed of partially-missing video by artificially cutting off part of each video in the original ActivityNet Captions, where we only use the videos in train and validation set. For a video with length T, we set the truncation window length (l_{win}) to be the same as the length of the longest event in this video, and the position of the truncation window at five random time between $\frac{1}{2}l_{win}$ and $(T - \frac{1}{2}l_{win})$, respectively. As a result, the total number of the videos and annotations in ActivityNet Captions-P are five times as much as the number in ActivityNet Captions. Statistically, there are only 389 events remain complete after our truncation operation, and 189,848 events are partially or totally cut off. Most of the missing parts are cross-event.

4.2   Data Preprocessing

We follow the same feature extraction strategy for video frames as in [31]. We extract the 1-D appearance and optical flow features at 2FPS to represent the videos. Specifically, we use the output of the "Flatten-673" layer in ResNet-200 [7] as the appearance features; we extract optical flows from 5 continuous frames, and take the feature vectors of the "global-pool" layer in BN-Inception [10] as the optical flow features. Both networks are pretrained on the ActivityNet dataset [1] for the task of action recognition. Then the appearance and optical flow features are concatenated and further encoded with a linear layer.

4.3   Implementation Details

For our VSEC module, we set the dimension of the visual-semantic embedding space to 1024. We set the number of the divided segments in a video and the number of neighboring segments at each side of the target segment to be parameters fixed by the experimental results. For the captioning module, we follow the settings in [31]. Specifically, the transformer model dimension is set to 1024, the hidden size of feed-forward layer to 2048, the number of attention heads to 8, and the number of transformer layers to 2.

4.4   Experimental Results

Baseline. In order to evaluate the performance of our model, we use the Masked-Transformer model as our baseline model. Masked-Transformer was proposed in [31] for dense video caption and has been widely used as a baseline model in related researches. Considering our VSEC-T model mainly aims at compensating the missing video content to generate more complete descriptions and do not generate event proposals, we use the ground-truth proposals instead of the generated proposals in Masked-Transformer to fairly compare their performances. We run both models on ActivityNet Captions-P dataset.

Quantitative Evaluation. We use the most commonly-used evaluation metrics: BLEU@{1,2,3,4} [22], METEOR [5], and CIDEr [27] to compare the captioning performance of our model and the baseline model. The results of our model with different settings of N and N_c and the baseline model are shown in Table 1. We can see that our model outperforms the baseline in most metrics.

Qualitative Examples. We find our model can generate more complete, relevant and coherent descriptions for incomplete videos than baseline model. Through human evaluation, we find that when $N = 20$, $N_c = 1$, the generated descriptions of our model are the best. In the example shown with the corresponding video in Fig. 3, we can see that the sentences from the baseline model

Table 1. Video captioning results including Bleu@{1,2,3,4}, CIDEr and METOR for our model and baseline model on the constructed ActivityNet Captions-P dataset. We report performances obtained from ground-truth proposals for both. For our VSEC-T model, we report performances with different settings of N as 10, 15, 20, 30, 40 and 50 and N_c as 1, 2, 3 and 4, respectively. Top scores are highlighted.

Method		Bleu@1	Bleu@2	Bleu@3	Bleu@4	CIDEr	METOR
Masked transformer		19.35	8.24	3.12	1.31	37.53	8.95
VSEC-T	$N = 10, N_c = 1$	19.63	8.17	3.43	1.77	38.56	9.14
	$N = 10, N_c = 2$	19.24	7.74	3.18	1.57	36.64	9.05
	$N = 10, N_c = 3$	19.93	8.11	3.27	1.56	37.69	9.19
	$N = 15, N_c = 1$	20.75	8.90	3.97	**2.02**	42.72	9.33
	$N = 15, N_c = 2$	20.44	8.75	3.76	1.73	41.60	9.17
	$N = 15, N_c = 3$	18.79	7.90	3.22	1.49	36.42	8.86
	$N = 20, N_c = 1$	21.23	**9.50**	**4.01**	1.85	42.62	**9.57**
	$N = 20, N_c = 2$	20.54	9.02	3.82	1.87	43.07	9.39
	$N = 20, N_c = 3$	20.49	9.02	3.84	1.87	43.02	9.27
	$N = 30, N_c = 1$	**20.94**	9.33	3.94	1.93	**44.19**	9.32
	$N = 30, N_c = 2$	20.23	8.90	3.65	1.73	41.56	9.01
	$N = 30, N_c = 3$	19.56	8.34	3.36	1.66	40.38	8.81
	$N = 30, N_c = 4$	20.56	9.13	3.86	1.85	42.40	9.25
	$N = 40, N_c = 1$	19.66	8.59	3.51	1.72	40.80	8.83
	$N = 40, N_c = 2$	19.80	8.71	3.76	1.80	42.09	8.84
	$N = 40, N_c = 3$	20.06	8.75	3.63	1.79	41.29	8.91
	$N = 40, N_c = 4$	19.96	8.83	3.74	1.85	41.87	8.89
	$N = 50, N_c = 1$	20.24	9.04	3.92	1.76	35.94	9.31
	$N = 50, N_c = 2$	19.37	8.32	3.38	1.68	39.75	8.65
	$N = 50, N_c = 3$	20.12	8.99	3.83	1.83	42.00	9.07
	$N = 50, N_c = 4$	19.46	8.63	3.70	1.82	40.85	8.68

can not only correctly describe the event with missing video content, but also generate unrelated and incoherent sentences for events which are fully-observed. In contrast with the severe degradation of the baseline model's performance with partially-missing videos, our proposed method can generate relatively precise and coherent sentences for events whether with complete video content or not.

4.5   Ablation Analysis

We compare the performance of our model with different settings of N and N_c, which are the number of the divided segments in a video and the number of neighboring segments at each side of the target segment used to construct the

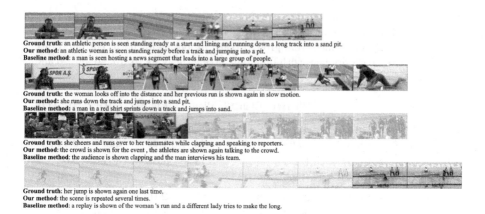

Ground truth: an athletic person is seen standing ready at a start and lining and running down a long track into a sand pit.
Our method: an athletic woman is seen standing ready before a track and jumping into a pit.
Baseline method: a man is seen hosting a news segment that leads into a large group of people.

Ground truth: the woman looks off into the distance and her previous run is shown again in slow motion.
Our method: she runs down the track and jumps into a sand pit.
Baseline method: a man in a red shirt sprints down a track and jumps into sand.

Ground truth: she cheers and runs over to her teammates while clapping and speaking to reporters.
Our method: the crowd is shown for the event , the athletes are shown again talking to the crowd.
Baseline method: the audience is shown clapping and the man interviews his team.

Ground truth: her jump is shown again one last time.
Our method: the scene is repeated several times.
Baseline method: a replay is shown of the woman 's run and a different lady tries to make the long.

Fig. 3. An example of a partially-missing video and its corresponding descriptions from ground-truth annotations, generated by our method and baseline method. We use sampled frames to illustrate the whole video where the faded ones are the missing part. In the above video, there are four sub-events happened and two of them are partially-missing. The words in red are those not relevant to the video content. Our method can generate relatively complete and precise descriptions for the events whether it is partially-missing or not, while there are much more errors in the baseline model results. (Color figure online)

visual feature of a segment, respectively. The results are shown in Table 1 and the line chart of the three evaluation criteria with different settings of N and N_c are shown in Fig. 4. We find that though the best settings of the above two parameters is different based on different evaluation criteria, our model performs best when N is not too small or too big and the two hyper-parameters, N and N_c, are weakly sensitive to each other.

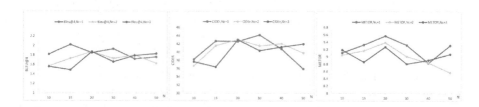

Fig. 4. The line chart of the results of our model. We compare Bleu@4, CIDEr, and METOR with different settings of N and N_c.

5   Conclusion

In this paper, we formulate the issue of dense video captioning for incomplete videos. To tackle the challenge of incomplete video content description, we propose a novel visual-semantic embedding with context module to construct visual representation by using the surrounding video segments instead of the segment itself, and leverage the complete textual information in corresponding annotated sentences for the event to produce a better representation carrying more information which can compensate the missing video content. We further apply an encoder-decoder captioning module with transformer units to produce descriptions for the whole video. We construct a new dataset with incomplete videos based on ActivityNet Captions to train and evaluate our proposed model, and achieve outstanding performances in most metrics compared with the baseline model.

References

1. Caba Heilbron, F., Escorcia, V., Ghanem, B., Carlos Niebles, J.: Activitynet: a large-scale video benchmark for human activity understanding. In: Proceedings of the IEEE Conference on Computer Vision and Pattern Recognition, pp. 961–970 (2015)
2. Cai, Y., Li, H., Hu, J.F., Zheng, W.S.: Action knowledge transfer for action prediction with partial videos. In: Proceedings of the AAAI Conference on Artificial Intelligence, vol. 33, pp. 8118–8125 (2019)
3. Cao, Y., et al.: Recognize human activities from partially observed videos. In: Proceedings of the IEEE Conference on Computer Vision and Pattern Recognition, pp. 2658–2665 (2013)
4. Chen, S., et al.: Activitynet 2019 task 3: exploring contexts for dense captioning events in videos. arXiv preprint arXiv:1907.05092 (2019)
5. Denkowski, M., Lavie, A.: Meteor universal: language specific translation evaluation for any target language. In: Proceedings of the Ninth Workshop on Statistical Machine Translation, pp. 376–380 (2014)
6. Gella, S., Lewis, M., Rohrbach, M.: A dataset for telling the stories of social media videos. In: Proceedings of the 2018 Conference on Empirical Methods in Natural Language Processing, pp. 968–974 (2018)
7. He, K., Zhang, X., Ren, S., Sun, J.: Deep residual learning for image recognition. In: Proceedings of the IEEE Conference on Computer Vision and Pattern Recognition, pp. 770–778 (2016)
8. Iashin, V., Rahtu, E.: A better use of audio-visual cues: dense video captioning with bi-modal transformer. arXiv preprint arXiv:2005.08271 (2020)
9. Iashin, V., Rahtu, E.: Multi-modal dense video captioning. In: Proceedings of the IEEE/CVF Conference on Computer Vision and Pattern Recognition Workshops, pp. 958–959 (2020)
10. Ioffe, S., Szegedy, C.: Batch normalization: accelerating deep network training by reducing internal covariate shift. arXiv preprint arXiv:1502.03167 (2015)
11. Kong, Y., Gao, S.: Action prediction from videos via memorizing hard-to-predict samples. In: AAAI (2018)
12. Kong, Y., Tao, Z., Fu, Y.: Deep sequential context networks for action prediction. In: Proceedings of the IEEE Conference on Computer Vision and Pattern Recognition, pp. 1473–1481 (2017)
13. Krishna, R., Hata, K., Ren, F., Fei-Fei, L., Niebles, J.C.: Dense-captioning events in videos. In: ICCV, pp. 706–715 (2017)
14. Lai, S., Zheng, W.S., Hu, J.F., Zhang, J.: Global-local temporal saliency action prediction. IEEE Trans. Image Process. **27**(5), 2272–2285 (2017)

15. Lan, T., Chen, T.-C., Savarese, S.: A hierarchical representation for future action prediction. In: Fleet, D., Pajdla, T., Schiele, B., Tuytelaars, T. (eds.) ECCV 2014. LNCS, vol. 8691, pp. 689–704. Springer, Cham (2014). https://doi.org/10.1007/978-3-319-10578-9_45

16. Lei, J., Wang, L., Shen, Y., Yu, D., Berg, T.L., Bansal, M.: Mart: memory-augmented recurrent transformer for coherent video paragraph captioning. arXiv preprint arXiv:2005.05402 (2020)

17. Li, Y., Yao, T., Pan, Y., Chao, H., Mei, T.: Jointly localizing and describing events for dense video captioning. In: Proceedings of the IEEE Conference on Computer Vision and Pattern Recognition, pp. 7492–7500 (2018)

18. Miech, A., Laptev, I., Sivic, J.: Learning a text-video embedding from incomplete and heterogeneous data. arXiv preprint arXiv:1804.02516 (2018)

19. Miech, A., Zhukov, D., Alayrac, J.B., Tapaswi, M., Laptev, I., Sivic, J.: Howto100m: learning a text-video embedding by watching hundred million narrated video clips. In: Proceedings of the IEEE International Conference on Computer Vision, pp. 2630–2640 (2019)

20. Mun, J., Yang, L., Ren, Z., Xu, N., Han, B.: Streamlined dense video captioning. In: Proceedings of the IEEE Conference on Computer Vision and Pattern Recognition, pp. 6588–6597 (2019)

21. Pan, Y., Mei, T., Yao, T., Li, H., Rui, Y.: Jointly modeling embedding and translation to bridge video and language. In: Proceedings of the IEEE Conference on Computer Vision and Pattern Recognition, pp. 4594–4602 (2016)

22. Papineni, K., Roukos, S., Ward, T., Zhu, W.J.: Bleu: a method for automatic evaluation of machine translation. In: Proceedings of the 40th Annual Meeting of the Association for Computational Linguistics, pp. 311–318 (2002)

23. Park, J.S., Rohrbach, M., Darrell, T., Rohrbach, A.: Adversarial inference for multi-sentence video description. In: Proceedings of the IEEE Conference on Computer Vision and Pattern Recognition, pp. 6598–6608 (2019)

24. Qin, J., et al.: Binary coding for partial action analysis with limited observation ratios. In: Proceedings of the IEEE Conference on Computer Vision and Pattern Recognition, pp. 146–155 (2017)

25. Rahman, T., Xu, B., Sigal, L.: Watch, listen and tell: multi-modal weakly supervised dense event captioning. In: Proceedings of the IEEE International Conference on Computer Vision, pp. 8908–8917 (2019)

26. Vaswani, A., et al.: Attention is all you need. Adv. Neural. Inf. Process. Syst. **30**, 5998–6008 (2017)

27. Vedantam, R., Zitnick, C.L., Parikh, D.: Cider: consensus-based image description evaluation. In: 2015 IEEE Conference on Computer Vision and Pattern Recognition (CVPR), pp. 4566–4575 (2015)

28. Wang, J., Jiang, W., Ma, L., Liu, W., Xu, Y.: Bidirectional attentive fusion with context gating for dense video captioning. In: Proceedings of the IEEE Conference on Computer Vision and Pattern Recognition, pp. 7190–7198 (2018)

29. Xiong, Y., Dai, B., Lin, D.: Move forward and tell: a progressive generator of video descriptions. In: Proceedings of the European Conference on Computer Vision (ECCV), pp. 468–483 (2018)

30. Xu, K., Qin, Z., Wang, G.: Recognize human activities from multi-part missing videos. In: 2016 IEEE International Conference on Multimedia and Expo (ICME), pp. 1–6. IEEE (2016)

31. Zhou, L., Zhou, Y., Corso, J.J., Socher, R., Xiong, C.: End-to-end dense video captioning with masked transformer. In: Proceedings of the IEEE Conference on Computer Vision and Pattern Recognition, pp. 8739–8748 (2018)

Modeling Context-Guided Visual and Linguistic Semantic Feature for Video Captioning

Zhixin Sun[1], Xian Zhong[1,2(✉)], Shuqin Chen[1], Wenxuan Liu[1], Duxiu Feng[3], and Lin Li[1,2]

[1] School of Computer Science and Technology, Wuhan University of Technology, Wuhan 430070, China
{sunzx_jdi,zhongx,csqcwx0801,lwxfight,cathylilin}@whut.edu.cn
[2] Hubei Key Laboratory of Transportation Internet of Things, Wuhan University of Technology, Wuhan 430070, China
[3] ZhongQianLiYuan Engineering Consulting Co., Ltd., Wuhan 430071, China

Abstract. It has received increasing attention to exploiting temporal visual features and corresponding descriptions in video captioning. Most of the existing models generate the captioning words merely depend on video temporal structure, ignoring fine-grained complete scene information. And the traditional long-short term memory (LSTM) in recent models is used as decoder to generate sentences, the last generated states in previous hidden ones are used to do work directly to predict the current word. This may lead to the predicted word highly related to the last generated hidden state other than the overall context information. To model the temporal aspects of activities typically shown in the video and better capture long-range context information, we propose a novel video captioning framework via context-guided semantic features model (CSF). Specifically, to maximum information flow, several previous and future information are aggregated to guide the current token by the semantic loss in the encoding and decoding phase respectively. The visual and linguistic information are corrected by fusing the surrounding information. Extensive experiments conducted on **MSVD** and **MSR-VTT** video captioning datasets demonstrate the effectiveness of our method compared with state-of-the-art approaches.

Keywords: Video captioning · Visual semantic feature · Linguistic semantic feature · Semantic loss · Attention mechanism

1 Introduction

With the development of deep neural networks in computer vision, audio, and natural language processing, video captioning has drawn increasing attention in

This work was supported in part by the Fundamental Research Funds for the Central Universities of China under Grant 191010001 and in part by the Hubei Key Laboratory of Transportation Internet of Things under Grant 2020III026GX.

© Springer Nature Switzerland AG 2021
I. Farkaš et al. (Eds.): ICANN 2021, LNCS 12895, pp. 677–689, 2021.
https://doi.org/10.1007/978-3-030-86383-8_54

recent years. Video captioning aims to generate a descriptive sentence for a given video, which can be applied to areas such as video retrieval, video recommendation, and autopilot assistance. Recent visual captioning approaches [1–7] follow the encoder-decoder architecture. The encoder is used to extract visual features, and the decoder dynamically selects the visual features to generate sentences.

Recent researches on video captioning mainly focus on sequence learning-based methods [8–15] with the convolutional neural network (CNN) as the encoder and the long-short term memory (LSTM) as the decoder. Despite the significant improvements have achieved by previous methods, current approaches still exist some limitations, more concretely: *1) As for the visual part, the visual features obtained through the attention mechanism are simply used as the input of LSTM without further exploring the interaction between these features, such as* [5,6,8,9]. Spatio-temporally attention mechanism [5,6] and graph convolution network (GCN) [8,16] are modeled to fully exploit the temporal-spatial information and characterize the video feature. In detail, Zhang *et al.* [8] proposed an object-relational graph-based encoder to extract the relation information between the objects based on GCN. Wang *et al.* [9] utilized a fusion network to fuse the features of different modalities. Chen *et al.* [10] designed a gated attention recurrent unit to pick up the main video content features guided by the motion features. Tan *et al.* [17] applied a visual reasoning module network, which divided the visual features into three categories and selected the most relevant features dynamically. However, all of these methods are designed to extract visual features before attention mechanisms, redundant visual features make it difficult to explore the relationship between visual features. *2) In terms of language, due to the unidirectional way of generating sentences, it is still difficult to merge future context information across time.* To deal with long-range dependencies between words, Zhu *et al.* [18] proposed a model named DenseLSTM to sufficiently utilize previous generated hidden state to solve the problem that long-range information is easy to be forgotten, which make the predicted word more related to partial information other than the overall scene information without taking advantage of the future predicted information. Wei *et al.* [19] proposed a reconstruction network reproduce the video features by capitalizing on the hidden state to promote the lexical feature fully express the visual content. The decoder serves to construct visual and linguistic temporal dependencies to generate sentences, how to take all visual and linguistic contextual information is still to be explored.

To address the aforementioned problems, we propose context-guided visual and linguistic semantic features model (CSF) for video captioning. Initially, appearance features, motion features, and fine-grained object features are extracted by pre-trained networks and selected by attention mechanism dynamically. Then, we employ an LSTM to model the temporal scene information supervised by the visual semantic information, which is extracted in the manner of sliding window. We use one-dimensional convolution over current visual feature and several neighboring ones to extract the visual semantic information for the current token. The semantic loss is built between the current hidden state and the visual semantic information to encourage LSTM to learn comprehensive contextual information during training. At last, in the similar manner, several previous, current, and future hidden states are directly used to predict the current word by the linguistic semantic loss in training phase. More abundant information is used and the predicted word is more related to the context. The model can memorize the long-range relation between sequence features with a larger receptive field.

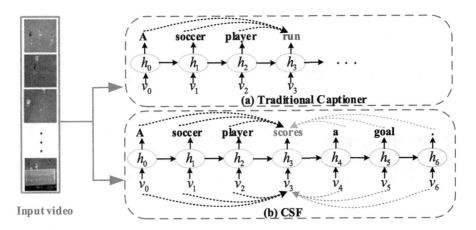

Fig. 1. The illustration of the proposed CSF.

To generate a sentence with accurate and rich semantics matching to the video, the contextual visual and lexical information is essential for each word. As illustrated in Fig. 1, the traditional captioner predicts the current word based on the previous sentence "A soccer player" during training, which will easily generate incorrect word (*e.g.* "runs") when encountering similar situations in the inference stage. Therefore, the future predictions "goal" and the visual feature corresponding to word "soccer" and "goal" may contain more critical information for current word, and should be considered to improve the scene understanding ability of captioning model.

Our main contributions can be summarized threefold:

- In the encoding stage, we construct dependencies between visual features after attention mechanism under the guidance of a sequence of the temporal scene information. And to the best of our knowledge, we are the first to further integrate the visual features obtained through attention mechanisms.
- In the decoding stage, by directly incorporating previous and future temporal scene information, more complete information is utilized to predict each word, which makes the generated sentences more accurate. The linguistic semantic loss is used to guide it to memorize more long-range dependence information.
- Extensive experiments on two public video captioning benchmark datasets demonstrate the effectiveness and superiority in the proposed approach.

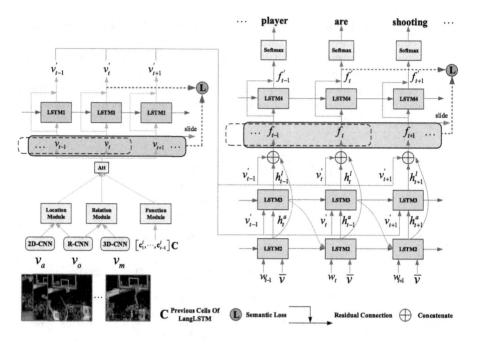

Fig. 2. An overview of our framework of CSF.

2   Methodology

An overview of our proposed video captioning model is shown in Fig. 2. Appearance, motion, and fine-grained objects features are extracted by diverse pretrained networks. A visual semantic LSTM (LSTM1) is used to capture the previous, current, and future predicted visual feature extracted by the attention mechanism supervised by the visual semantic loss, which work as a late fusion of visual features. Besides, we employ a two-layer LSTM structure [20], including an attention LSTM (LSTM2) and a language LSTM (LSTM3), to model the temporal dependence of attention mechanism and words. And then we model visual semantic expression and linguistic semantic expression to acquire temporal visual information and long-range lexical information by a linguistic semantic LSTM (LSTM4).

2.1   Encoder

In the encoder stage, for each video, we sample n frames evenly, and extract appearance features v_a by 2D-CNN, M object features v_o by R-CNN, and motion features v_m by 3D-CNN for each frame. Inspired by [17], we divide the visual features into three categories by location module, relation module, and function module, and the attention mechanism is used to select this visual features, we denote it as $V = \{v_0, v_1, \ldots, v_t, \ldots, v_m\}$, where m is the length of sentence.

To model the dependencies of visual features extracted by the attention mechanism, LSTM1 is used to correct current visual features through the residual connection, formally:

$$v'_t = v_t + h^v_t \qquad (1)$$

$$h^v_t, c^v_t = \textbf{LSTM1}(v_t, h^v_{t-1}) \qquad (2)$$

where v'_t is the revised visual feature that incorporates contextual information, h^v_t and c^v_t are the hidden state and cell state of LSTM1 at step t, v_t denotes the visual feature after attention mechanism.

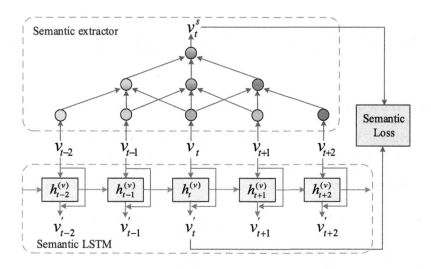

Fig. 3. The visual semantic loss at step t.

Semantic loss is used to ensure the bidirectional message passing of contextual semantic information to alleviate the unidirectional defect of generating sentences. We use one-dimensional convolution with a receptive field of 3 (previous, current, and future one) to extract contextual semantic information. Meanwhile, we expand the receptive field by stacking multiple layers of convolution operation:

$$v^s_t = \text{Conv}^{t+k}_{t=t-k}(v_t) \qquad (3)$$

where Conv refers to multi-layer convolution operation, $k + 1 + k$ stands for its receptive field, and v^s_t denotes the context information extracted by convolution operation at step t. A semantic loss is used between v^s_t and v'_t to ensure that the LSTM1 fuse both previous and future information as shown in Fig. 3, formally:

$$\mathcal{L}_{vs} = \frac{1}{m} \sum_{t=1}^{m} \psi(v^s_t, v'_t) \qquad (4)$$

where m denotes the number of words in the ground-truth caption, the semantic loss is measured by $\psi(\cdot)$, which is simply chosen as the Euclidean distance.

2.2   Decoder

In the decoder stage, as proposed in [20], we use two-layer LSTM (denotes as LSTM2 and LSTM3) to model the time dependence of attention mechanism and words:

$$h_t^a, c_t^a = \mathbf{LSTM2}([\overline{v}, w_{t-1}, h_{t-1}^l], h_{t-1}^a) \qquad (5)$$

$$h_t^l, c_t^l = \mathbf{LSTM3}([v_t', h_t^a], h_{t-1}^l) \qquad (6)$$

where $\overline{v}$ is the global visual feature obtained by averaging different frames, w_{t-1} is the last generated word embedding, v_t' is the revised visual semantic features in (1), h_t^a and h_t^l are the hidden state of LSTM2 and LSTM3 respectively. To obtain the semantic feature expression of the generated words, we fuse the hidden states in the two-layer LSTM and the corrected visual features v_t' by cascading:

$$f_t = W_1(v_t' \oplus h_t^a \oplus h_t^l) \qquad (7)$$

where W_1 is trainable parameters, f_t refers to the semantic feature corresponding to the word at step t.

Similar to visual feature expression, we combine several previous and future semantic features to guide the generation of current token which is similar to Fig. 3. LSTM4 combined with residual connection is used and the revised semantic features denotes as f_t'. In detail, the multi-layer one-dimensional convolution is used to extract context semantic features of lexical, and we denote it as f_t^s. The linguistic semantic loss is built between f_t' and f_t^s and denoted as $\mathcal{L}_{ls} = \frac{1}{m} \sum_{t=1}^m \psi(f_t^s, f_t')$. We apply LSTM4 model in a sliding window manner to slide over a video to exploit the long-range context information.

We get the probability of generating words through the corrected features of lexical:

$$p_t = \mathrm{softmax}(W_2(f_t')) \qquad (8)$$

where the W_2 is trainable parameter.

2.3   Training

We train our network model with maximum likelihood estimation. Given a video and the corresponding ground-truth caption $W = [w_1, w_2, \ldots, w_t]$, the object is to minimize the negative log-likelihood to produce the correct description sentence:

$$\mathcal{L}_{ce} = -\sum_{t=1}^T \log P(w_t \mid v, w_0, w_1, \ldots, w_{t-1}) \qquad (9)$$

In addition, to balance the unidirectionality of generated sentences and the bidirectionality of contextual semantic information, a visual semantic loss and a linguistic semantic loss mentioned above are used. We also use part-of-speech (POS) tag labels as supervisory information to guide the attention mechanism to choose the correct module information, which is denoted as $\mathcal{L}_{pos}$, please refer to [17] for more mathematical details. Therefore, the overall loss function of our model is given by:

$$\mathcal{L} = \mathcal{L}_{ce} + \lambda_{pos}\mathcal{L}_{pos} + \lambda_{vs}\mathcal{L}_{vs} + \lambda_{ls}\mathcal{L}_{ls} \tag{10}$$

where λ_{pos}, λ_{vs}, and λ_{ls} are three trade-off weights.

Table 1. The comparison results on **MSVD**. **Bold** and blue numbers are the best and second-best results respectively. "†" indicates that the method is reproduced by us.

Method	Venue	B-4	M	R	C
DenseLSTM [18]	MM '19	50.4	32.9	69.9	72.6
GRU-EVE [6]	CVPR '19	47.9	35.0	71.5	78.1
POS + CG [9]	ICCV '19	52.5	34.1	71.3	88.7
OA-BTG [1]	CVPR '19	56.9	36.2	–	90.6
MARN [12]	CVPR '19	48.6	35.1	71.9	92.2
Two-stream [13]	TPAMI '20	54.3	33.5	–	72.8
SAAT [14]	CVPR '20	46.5	33.5	69.4	81.0
BiLSTM-CG [11]	NPL '20	53.3	35.2	71.6	84.1
STGCN [16]	CVPR '20	52.2	36.9	**73.9**	93.0
ORG-TRL [8]	CVPR '20	54.3	36.4	**73.9**	95.2
CR + RR [15]	AAAI '20	**57.0**	36.8	–	96.8
RMN [17] †	IJCAI '20	54.1	36.6	73.5	93.2
CSF (ours)		56.3	**37.3**	**73.9**	**100.2**

3   Experimental Results

3.1   Datasets and Evaluation Metrics

MSVD dataset consists of 1,970 short clips selected from Youtube, each video has roughly 41 English descriptions. To be consistent with previous works, we split the dataset into 3 subsets, 1,200 clips for training, 100 clips for validation, and the remaining 670 clips for testing.

MSR-VTT dataset contains 10,000 video clips, and each video corresponds to 20 English sentence descriptions. Following the existing work, we take 6,513 clips for training, 497 clips for validation, and 2,990 clips for testing.

For a fair evaluation, we adopt four popular metrics for the captioning task, including BLEU-4 (B-4), METEOR (M), ROUGE (R), and CIDEr (C). All metrics represent the score between the predicted sentence and the ground-truth sentence, The higher the indicator score is, the better the generated sentences will be. CIDEr is proposed for captioning tasks specifically and is considered

Table 2. The comparison results on **MSR-VTT**. **Bold** and blue numbers are the best and second-best results respectively. "†" indicates that the method is reproduced by us.

Method	Venue	B-4	M	R	C
DenseLSTM [18]	MM '19	38.1	26.6	–	42.8
OA-BTG [1]	CVPR '19	41.4	28.2	–	46.9
MARN [12]	CVPR '19	40.4	28.1	60.7	47.1
GRU-EVE [6]	CVPR '19	38.3	28.4	60.7	48.1
POS + CG [9]	ICCV '19	**42.0**	28.2	**61.6**	48.7
Two-stream [13]	TPAMI '20	39.7	27.0	–	42.1
VideoTRM [2]	MM '20	38.8	27.0	–	44.7
BiLSTM-CG [11]	NPL '20	39.1	27.7	59.9	46.4
STGCN [16]	CVPR '20	40.5	28.3	60.9	47.1
RMN [17] †	IJCAI '20	39.2	27.8	59.9	46.7
CSF (ours)		41.7	**28.5**	61.2	**48.8**

more consistent with human judgment. BLEU-4 mainly focuses on the fraction of n-grams between the ground-truth and the generated sentences, while METEOR and CIDEr consider the synonyms for n-grams and key information of the sentences, respectively.

3.2 Implementation Details

Feature Extraction. In our experiments, Inception-ResNet-V2 (IRV2) [21] and I3D [22] are used to extract appearance features and motion features respectively, then we equally space 26 features for each video. The IRV2 is trained on ILSVRC-2012-CLS image classification dataset and the I3D is trained on **Kinetics** action classification dataset. We adopt Faster-RCNN [23] to extract 36 region features for each sampled frame. And we truncate the captions with more than 26 words and zero pad the captions with less than 26 words.

Training Details. Our model is optimized by Adam optimizer, the initial learning rate is set to 0.0001 and it is divided by 3 every 5 epochs. The coefficient λ_{pos}, λ_{vs}, and λ_{ls} in loss functions are 0.1, 1.0, and 1.0, respectively. The size of the receptive field is set to 5. And the hidden size of the LSTM is 1,024 for the **MSVD** dataset and 1,300 for the **MSR-VTT** dataset. During testing, we use a beam search with size 2 for the final caption generation.

3.3 Comparison with the State-of-the-Arts

To verify the validity of our approach, we make comparisons with the state-of-the-arts for video captioning on both **MSVD** and **MSR-VTT**. As shown in Tables 1 and 2, on **MSVD** dataset, our model achieves the best performance

Table 3. Sensitivity experiment of receptive field on **MSVD**.

Size	MSVD			
	B-4	M	R	C
3	53.8	36.2	72.6	92.8
5	**55.1**	37.1	74.1	**99.4**
7	54.8	**37.8**	74.3	99.1
9	54.0	37.6	**74.4**	98.8

Table 4. The ablation results on **MSVD** and **MSR-VTT**.

Method	MSVD		MSR-VTT	
	B-4	C	B-4	C
Baseline	54.1	93.2	39.2	46.7
w VS	54.9	98.3	41.3	48.2
w LS	55.1	99.4	40.6	47.6
w VS + LS	**56.3**	**100.2**	**41.7**	**48.8**

as far as we know for METEOR, ROUGE, and CIDEr. It is worth noting that our model has a substantial improvement for the CIDEr metric, and we can conclude that the sentences generated by our model can more fully represent the main content of the ground-truth and has the highest semantic similarity with it. The main reason can be summarized as that our model achieves better semantic feature representation by fusing bidirectional contextual information. On **MSR-VTT** dataset, our proposed method outperforms compared methods on METEOR and CIDEr, which once again confirms that our approach can extract overall context information. For the BLEU-4 metric, our model is slightly lower than CR + RR [15] and POS + CG [9] on **MSR-VTT**, which can be attributed the prior knowledge is used for objects relation reasoning in [15], and in the POS + CG [9], the two-stage training method is used to obtain the pos feature to guide sentence generation.

3.4 Study on Trade-Off Parameter

Trade-Off Parameter of Receptive Field. We design one-dimensional convolution with a receptive field of n ($n = 3, 5, 7, 9$) to extract temporal information. And then we conduct experiment on linguistic semantic LSTM (LSTM4) on **MSVD** dataset and the result is shown in Table 3. As the receptive field increasing, the indicator show a trend from rise to decline. We analysis that when the receptive field is too small, sufficient contextual semantic information cannot be extracted. And in oversized receptive field, the zero pad added on the border becomes noise may mislead the network. We set the receptive field to 5.

Trade-off Parameter λ_{vs} and λ_{ls}. With different λ_{vs} and λ_{ls} values in (10), the obtained values of BLEU-4 and CIDEr metric are given in Fig. 4. First, it can be concluded again that adding the visual semantic loss ($\lambda_{vs} > 0$) and linguistic semantic loss ($\lambda_{ls} > 0$) do improve the performance of video captioning. Second, The values of parameter λ_{vs} and λ_{ls} can improve model performance in a large range especially on the CIDEr metric, but there is still a trade-off between the cross-entropy loss and the semantic loss. Thus, a lot of experiments are needed to balance the contributions of the two loss to get the best results. Taking all the metrics together, we empirically set λ_{vs} to 5.0 for the VS module and set

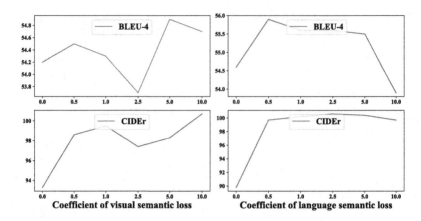

Fig. 4. Effects of the trade-off parameter λ_{vs} and λ_{ls} in terms of BLEU-4 and CIDEr on **MSVD**.

λ_{ls} to 1.0 for the LS module mentioned above. For the combination of the two modules, we set the value of both λ_{vs} and λ_{ls} to 1.0.

3.5 Ablation Study

To show the effectiveness of our proposed method, we separate our model for ablation study on **MSVD** and **MSR-VTT** with BLEU-4 and CIDEr as representative metrics, as shown in Table 4. There are several different settings: 1) **baseline**: The baseline module means the reasoning module is applied to extract the visual feature and the two-layer LSTM is used to generate the sentences. It follows the cross-entropy criterion and no semantic LSTM is used. 2) **VS** module refers to the visual semantic LSTM (LSTM1) and the visual semantic loss are applied to realize the semantic expression of visual features. 3) **LS** module refers the linguistic semantic LSTM (LSTM4) and the linguistic semantic loss are applied to realize the semantic expression of lexical features. 4) **VS + LS** refers both the LSTM1 and LSTM4 combined with semantic loss are used.

It can be viewed from Table 4, both the semantic expression of visual features and the semantic expression of linguistic features can make a great improvement on both BLEU-4 and CIDEr metrics than the baseline, and the combination of the two modules achieves the best results. In detail, the VS module improves the BLEU-4 and CIDEr scores by 0.8%, 5.1% and 2.1%, 1.5% on **MSVD** and **MSR-VTT** datasets, respectively. The LS module improves these metrics by 1.0%, 6.2% and 1.4%, 0.9% on two datasets, respectively. The combination of the two modules improves the BLEU-4 and CIDEr scores by 2.2% and 7.0% on **MSVD** dataset, and 2.5% and 2.1% on **MSR-VTT** dataset. All these results indicate the effectiveness of our proposed captioning model of both VS and LS and the combination of the two modules.

3.6   Qualitative Results

We compare the difference among baseline, ground-truth, and our model in Fig. 5. Orange represents the words with rich semantics. As expected, our model generates sentences with richer semantics. Specifically, in the first example, our model can integrate contextual semantic information to generate the word "interview" instead of "speak" with more accurate semantics. The second example can also illustrate that, thanks to the fusion of contextual visual and linguistic information, our model can generate semantically rich words such as "basketball player" and "slam dunk".

GT: a woman is being interviewed for tv.
Baseline: a woman is speaking in a room.
Ours: a woman is interviewed.

GT: a celtics basketball player makes a slam dunk.
Baseline: a man is playing a basketball game.
Ours: a basketball player makes a slam dunk.

Fig. 5. The generated captioning of ground-truth, baseline, and our method on **MSR-VTT** dataset.

4   Conclusion

In this paper, we propose a novel video captioning model that combines visual and linguistic semantic features. A visual semantic LSTM is used to construct the relation information between visual features obtained through the attention mechanism. Correspondingly, a linguistic LSTM is used to explore the connection between context words. Semantic loss is designed to ensure that visual semantic LSTM and linguistic semantic LSTM merge bidirectional contextual semantic information. Both of the two strategies can exactly improve the performance of caption generation for videos.

References

1. Zhang, J., Peng, Y.: Object-aware aggregation with bidirectional temporal graph for video captioning. In: Proceedings of the IEEE/CVF Conference on Computer Vision and Pattern Recognition (CVPR), pp. 8327–8336 (2019)
2. Chen, J., Chao, H.: VideoTRM: pre-training for video captioning challenge 2020. In: Proceedings of the 28th ACM International Conference on Multimedia (MM), pp. 4605–4609 (2020)
3. Karayil, T., Irfan, A., Raue, F., Hees, J., Dengel, A.: Conditional GANs for image captioning with sentiments. In: Tetko, I.V., Kůrková, V., Karpov, P., Theis, F. (eds.) ICANN 2019. LNCS, vol. 11730, pp. 300–312. Springer, Cham (2019). https://doi.org/10.1007/978-3-030-30490-4_25

4. Fan, Y., Xu, J., Sun, Y., Wang, Y.: A novel image captioning method based on generative adversarial networks. In: Tetko, I.V., Kůrková, V., Karpov, P., Theis, F. (eds.) ICANN 2019. LNCS, vol. 11730, pp. 281–292. Springer, Cham (2019). https://doi.org/10.1007/978-3-030-30490-4_23

5. Chen, J., Pan, Y., Li, Y., Yao, T., Chao, H., Mei, T.: Temporal deformable convolutional encoder-decoder networks for video captioning. In: Proceedings of Conference on Artificial Intelligence (AAAI), pp. 8167–8174 (2019)

6. Aafaq, N., Akhtar, N., Liu, W., Gilani, S.Z., Mian, A.: Spatio-temporal dynamics and semantic attribute enriched visual encoding for video captioning. In: Proceedings of IEEE/CVF Conference on Computer Vision and Pattern Recognition (CVPR), pp. 12487–12496 (2019)

7. Xu, J., Yao, T., Zhang, Y., Mei, T.: Learning multimodal attention LSTM networks for video captioning. In: Proceedings of ACM International Conference on Multimedia (MM), pp. 537–545 (2017)

8. Zhang, Z., et al.: Object relational graph with teacher-recommended learning for video captioning. In: Proceedings of the IEEE/CVF Conference on Computer Vision and Pattern Recognition (CVPR), pp. 13278–13288 (2020)

9. Wang, B., Ma, L., Zhang, W., Jiang, W., Wang, J., Liu, W.: Controllable video captioning with POS sequence guidance based on gated fusion network. In: Proceedings of the IEEE/CVF International Conference on Computer Vision (ICCV), pp. 2641–2650 (2019)

10. Chen, S., Jiang, Y.G.: Motion guided spatial attention for video captioning. In: Proceedings of the AAAI Conference on Artificial Intelligence (AAAI), pp. 8191–8198 (2019)

11. Chen, S., Zhong, X., Li, L., Liu, W., Gu, C., Zhong, L.: Adaptively converting auxiliary attributes and textual embedding for video captioning based on BiLSTM. Neural Process. Lett. **52**, 2353–2369 (2020)

12. Pei, W., Zhang, J., Wang, X., Ke, L., Shen, X., Tai, Y.W.: Memory-attended recurrent network for video captioning. In: Proceedings of the IEEE/CVF Conference on Computer Vision and Pattern Recognition (CVPR), pp. 8347–8356 (2019)

13. Gao, L., Li, X., Song, J., Shen, H.T.: Hierarchical LSTMs with adaptive attention for visual captioning. IEEE Trans. Pattern Anal. Mach. Intell. **42**, 1112–1131 (2019)

14. Zheng, Q., Wang, C., Tao, D.: Syntax-aware action targeting for video captioning. In: Proceedings of the IEEE/CVF Conference on Computer Vision and Pattern Recognition (CVPR), pp. 13096–13105 (2020)

15. Hou, J., Wu, X., Zhang, X., Qi, Y., Jia, Y., Luo, J.: Joint commonsense and relation reasoning for image and video captioning. In: Proceedings of the AAAI Conference on Artificial Intelligence (AAAI), vol. 34, no. 07, pp. 10973–10980 (2020)

16. Pan, B., et al.: Spatio-temporal graph for video captioning with knowledge distillation. In: Proceedings of the IEEE/CVF Conference on Computer Vision and Pattern Recognition (CVPR), pp. 10870–10879 (2020)

17. Tan, G., Liu, D., Wang, M., Zha, Z.J.: Learning to discretely compose reasoning module networks for video captioning. In: Proceedings of International Joint Conference on Artificial Intelligence (IJCAI) (2020)

18. Zhu, Y., Jiang, S.: Attention-based densely connected LSTM for video captioning. In: Proceedings of the 27th ACM International Conference on Multimedia (MM), pp. 802–810 (2019)

19. Zhang, W., Wang, B., Ma, L., Liu, W.: Reconstruct and represent video contents for captioning via reinforcement learning. IEEE Trans. Pattern Anal. Mach. Intell. **42**(12), 3088–3101 (2019)

20. Anderson, P., et al.: Bottom-up and top-down attention for image captioning and visual question answering. In: Proceedings of the IEEE Conference on Computer Vision and Pattern Recognition (CVPR), pp. 6077–6086 (2018)
21. Szegedy, C., Ioffe, S., Vanhoucke, V., Alemi, A.: Inception-v4, inception-ResNet and the impact of residual connections on learning. In: Proceedings of Conference on Artificial Intelligence (AAAI), vol. 31, no. 1 (2017)
22. Carreira, J., Zisserman, A.: Quo vadis, action recognition? A new model and the kinetics dataset. In: Proceedings of the IEEE Conference on Computer Vision and Pattern Recognition (CVPR), pp. 6299–6308 (2017)
23. Ren, S., He, K., Girshick, R., Sun, J.: Faster R-CNN: towards real-time object detection with region proposal networks. IEEE Trans. Pattern Anal. Mach. Intell. **39**(6), 1137–1149 (2016)

Correction to: Short Text Clustering with a Deep Multi-embedded Self-supervised Model

Kai Zhang, Zheng Lian, Jiangmeng Li, Haichang Li, and Xiaohui Hu

Correction to:
Chapter "Short Text Clustering with a Deep Multi-embedded Self-supervised Model" in: I. Farkaš et al. (Eds.): *Artificial Neural Networks and Machine Learning – ICANN 2021,* **LNCS 12895, https://doi.org/10.1007/978-3-030-86383-8_12**

Due to an oversight, the second affiliation of three co-authors was omitted in the originally published version. The revised version has the correct affiliations of all co-authors.

The updated version of this chapter can be found at
https://doi.org/10.1007/978-3-030-86383-8_12

© Springer Nature Switzerland AG 2022
I. Farkaš et al. (Eds.): ICANN 2021, LNCS 12895, p. C1, 2022.
https://doi.org/10.1007/978-3-030-86383-8_55

Author Index

Printed in the United States
by Baker & Taylor Publisher Services